PHYSICS

BJU PRESS

Greenville, South Carolina

I Sing the Mighty Power of God

I sing the mighty pow'r of God,
    that made the mountains rise;
That spread the flowing seas abroad,
    and built the lofty skies.
I sing the wisdom that ordained
    the sun to rule the day;
The moon shines full at His command,
    and all the stars obey.

I sing the goodness of the Lord,
    that filled the earth with food;
He formed the creatures with His Word,
    and then pronounced them good.
Lord, how Thy wonders are displayed,
    where'er I turn my eye:
If I survey the ground I tread,
    or gaze upon the sky!

There's not a plant or flow'r below,
    but makes Thy glories known;
And clouds arise, and tempests blow,
    by order from Thy throne;
While all that borrows life from Thee
    is ever in Thy care,
And ev'rywhere that man can be,
    Thou, God, art present there.

                  Isaac Watts

PHYSICS

Second Edition

R. Terrance Egolf

Linda Shumate

PHYSICS
Second Edition

R. Terrance Egolf, CDR, USN (Retired)
Linda Shumate

Contributing Author
Franklin Hall, M.A.

First Edition Authors
Rosemary A. Lasell
Paul Wilt

Consultants
Richard A. Seeley, M.S.
Fredrick J. Hall, M.Div

Editor
Michael Santopietro

Project Manager
David Harris

Compositor/Design
PreMedia ONE
Elly Kalagayan
Dan VanLeeuwen

Photo Acquisition
Brenda Hansen
Joyce Landis
Susan Perry

Illustration
PreMedia ONE
John Bjerk
Aaron Dickey

Produced in cooperation with the Bob Jones University Division of Natural Science of the College of Arts and Science and Bob Jones Academy.

Acknowledgment
Isaac Newton quotes on pages 124, 126, and 128. From Isaac Newton, Florian Cajori ed., *Mathematical Principles of Natural Philosophy and His System of the World*. Edited/translated by Andrew Motte, by permission of Regents of the University of California. © Regents of the University of California, 1934, 1962.

CONTENTS

Introduction . . . . **viii**

UNIT 1: A Framework

CHAPTER 1 Physics: A Biblical Framework **1**

1A: Science and Pseudoscience 1   Facet: Scientific Models 13

1B: A Brief History of Science 6   1D: What is Physics? 15

1C: Scientific Methodology 11   Facet: Limitations of Science 16

CHAPTER 2 Measurement **23**

2A: The Dimensions of Physics 23   2C: Truth in Measurements and

Facet: Just a Second 26   Calculations 35

2B: Principles of Measurement 30   2D: Problem Solving 40

Facet: Vernier Scales 34

UNIT 2: Classical Mechanics

CHAPTER 3 Motion in One Dimension **49**

3A: Describing Motion 49   3B: The Equations of Motion 61

Facet: Hypersonic Jets 56   Biography: Galileo 67

CHAPTER 4 Vectors and Scalars **73**

4A: Properties of Vectors and Scalars . . . 73   4C: Operations with Vectors:

4B: Operations with Vectors: Analytical Techniques 80

     Geometric Techniques 76

CHAPTER 5 Motion in a Plane **93**

5A: Kinematics of   Facet: Shot Put Release Angles 107

     Two-Dimensional Motion 93   Biography: Sir Isaac Newton 110

5B: Projections 98

CHAPTER 6 Dynamics **113**

6A: The History of Dynamics 113   Facet: Flight and Newton's Third Law . 130

6B: Forces 117   Biography: Aristotle 135

6C: Newton's Laws of Motion 124

CHAPTER 7 Circular Motion **137**

7A: Circular Motion 138   7C: Universal Gravitation 151

7B: Dynamics of Circular Motion 146   Facet: Roller Coasters 160

CHAPTER 8 Applying Newton's Laws **163**

8A: Simplifying Problems 163   8D: More Applications 179

8B: Transmitting Mechanical Forces . . . 166   Facet: Artificial Gravity 188

8C: Friction 174

CHAPTER 9 Work and Energy **191**

9A: Work 192   9C: Total Mechanical Energy 207

9B: Energy 198   Facet: Hydroelectric Dams 212

CHAPTER 10 Conservation of Energy **215**

10A: Total Mechanical Energy 215

Facet: Conserving Energy 221

10B: Simple Machines 222

CHAPTER 11 Momentum **233**

11A: Introducing Momentum 233

11B: Collisions 239

Facet: Gravity Assists 247

11C: Center of Mass and Angular
Momentum 249

CHAPTER 12 Periodic Motion **257**

12A: Simple Harmonic Motion 258

12B: Periodic Motion and
the Pendulum 262

Facet: Foucault Pendulum 266

12C: Oscillations in the Real World . . . 269

12D: Waves 272

Facet: Earthquakes 276

UNIT 3: Thermodynamics and Matter

CHAPTER 13 Properties of Matter **285**

13A: Theories of Matter 286

13B: States of Matter 290

Facet: Ice Skating 294

Facet: Relative Humidity 298

CHAPTER 14 Expansion and Temperature **303**

14A: Thermal Properties 303

Facet: Ball and Ring 309

14B: Measuring Temperature 309

14C: Gas Laws 314

Biography: Count Rumford 324

Biography: James Joule 325

CHAPTER 15 Thermal Energy and Heat **327**

15A: Theories of Heat 327

15B: Thermal Energy and Matter 330

15C: Mechanisms for Heat Transfer . . . 339

Biography: Nicolas Sadi Carnot 344

CHAPTER 16 Thermodynamic Laws **347**

16A: The Zeroth and First Laws 347

16B: The Second and Third Laws 355

16C: Entropy and Its Consequences . . . 361

Facet: Growth and the Second Law . . . 364

CHAPTER 17 Fluid Mechanics **369**

17A: Hydrostatics: Fluids at Rest 369

Facet: Maple Syrup Hydrometers 381

17B: Hydrodynamics: Fluids in Motion 382

Facet: Aerodynamically Designed Cars 389

UNIT 4: Electromagnetics

CHAPTER 18 Electric Charge **397**

18A: Electrification 397

Biography: William Gilbert 399

18B: Detecting Electric Charge 402

Facet: Static Electricity 402

Biography: Michael Faraday 410

CHAPTER 19 Electric Fields **413**

19A: Modeling the Electric Field 413

19B: Capacitors 419

Facet: Kinds of Capacitors 425

CHAPTER 20 Electrodynamics **429**

20A: Current, Voltage, and Resistance . . 429

20B: Electrical Circuits 436

20C: Electrical Safety 445

Facet: Fuel Cells 452

CHAPTER 21 Magnetism **455**

21A: Describing Magnetism 455   21B: Electromagnetism and Charges . . . 464

Facet: Paleomagnetism and the   21C: Electromagnetism and Conductors 469
　　　 Age of the Earth 462

CHAPTER 22 Electromagnetism **477**

22A: Currents and Magnetic Fields 477   Facet: Diesel Locomotives 498

22B: Alternating Current 484   Biography: James Clerk Maxwell 499

22C: AC Circuit Characteristics 490

UNIT 5: Geometric Optics and Light

CHAPTER 23 Light and Reflection **503**

23A: Light and the Electromagnetic   Biography: Christiaan Huygens 511
　　　 Spectrum 503   23C: Reflection and Mirrors 513

23B: Sources and Propagation of Light . . 507

CHAPTER 24 Refraction **527**

24A: Theory of Refraction 527   Facet: Fresnel Lenses 545

24B: Application of Refraction—Lenses 534

CHAPTER 25 Wave Optics **547**

25A: Wave Interference 547   25B: Diffraction 558

Biography: Thomas Young 550   25C: Polarization of Light 562

Facet: Lasers 556   Facet: Optical Testing 565

CHAPTER 26 Using Light **569**

26A: Intensity and Color 569   Facet: Microscopy 578

26B: Optical Instruments 576

UNIT 6: Modern Physics

CHAPTER 27 Relativity **587**

27A: Galilean Relativity 587   27C: General Relativity 602

Biography: Albert Einstein 592   Facet: Gravitational Red Shift 604

27B: Special Relativity 593

CHAPTER 28 Quantum Physics **609**

28A: Quantum Theory 609   28C: Modern Atomic Models 619

Biography: Max Planck 611   Facet: Electron Microscope 620

28B: Quantum Mechanics and the Atom 614   Biography: Niels Bohr 627

CHAPTER 29 Nuclear Physics **629**

29A: Radiation and Radioactivity 629   29C: Nuclear Reactions 642

29B: Radioactive Decay 636   29D: Elementary Particles 646

Facet: Radioactive Halos 639   Facet: Cosmic Rays 653

Appendixes **654**

Glossary **668**

Index **683**

Photo Credits **689**

Introduction

You are about to embark upon an academic expedition of discovery unlike any you may have experienced in the past. While you are learning the principles of physics, the foundational laws of God's creation will be unveiled to you. Every student who is a child of God needs to recognize the hand of the Creator in His works, whether in the complexity of the structure of the atom or in the predictability of the motions of celestial objects. Understanding physics will allow you to appreciate the arc of a soccer ball in the air, the pitch of a violin string as it is plucked, the inner workings of a DVD player, the array of colors in a rainbow, and many more ordinary and not-so-ordinary events in your physical world.

Physics is the study of the interactions of energy and matter. In a high school course such as this, the topics discussed represent a survey of only the most essential material. Advanced theoretical physics involves the use of high-level mathematics, which you will not see unless you choose to pursue a college science or technological course of study. In this course, you will need only your algebraic skills to understand the fundamental principles of physics. This simplicity is a testimony to the basic orderliness of creation that allows us to model complex ideas in simple mathematical formulas.

What Is Christian About Physics?

One question that the title *Physics for Christian Schools* could raise is this: "What is *Christian* about physics?" Some people have developed the idea that higher mathematics and science have little to do with the Bible or the Christian life. They think that because physics deals with scientific facts, or because it is not pervaded with evolutionary ideas, there is no need to study it from a Christian perspective. This kind of thinking ignores a number of important facts to the Christian: First, *all* secular science is pervaded by mechanistic, naturalistic, and evolutionistic philosophy. Learning that the laws of mechanics as they pertain to a baseball in flight are just the natural consequences of the way matter came together denies the wisdom and power of our Creator God. These naturalistic laws are extended to demonstrate how the universe came into being without the involvement of the same God. Second, physics as taught in the schools of the world contradicts the explicit teachings of God's Word pertaining to Creation, the Flood, and the processes that shaped the world we see today. Trying to believe both secular physics and the Bible leaves you in a state of confusion that will weaken your faith in God's Word.

We have listed a number of reasons that studying physics from a *Christian* perspective will strengthen your faith and prepare you for a life of service for Christ:

1. Your knowledge of and faith in God and His Word can be increased by a detailed study of creation. When God wished to show Job His wisdom, power, and greatness, He gave Job a tour of the universe that He had created (Job 38–41). After seeing the intricacies, the grandeur, and the splendor of the universe, Job declared, "I know that thou canst do every thing, and that no thought can be withholden from thee. . . . I have heard of thee by the hearing of the ear: but now mine eye seeth thee" (Job 42:2, 5). Like Job, you can learn new things about God through a study of His creation.

2. Christians can be more effective for the Lord when they combine strong academics with genuine faith in God. Daniel, Moses, Paul, and Luke are examples of men who used their good education to serve the Lord. No matter what career the Lord has planned for you, you should prepare as well as you can. Physics is a necessary part of the preparation for many vocations.

3. Physics offers unique opportunities for the development of Christian character in your life. Discipline, diligence, accuracy, organization, inquisitiveness, and thoroughness are just a few of the personal qualities that you can develop from your study of physics.

4. Christian students must be aware of scientific evidences for Creation. Often physics is involved in the debate on the origin of the universe (cosmology). You should be informed about the laws of thermodynamics, the motions of objects, the decay of atoms, and the consequences of the theories of relativity. While impressive scientific conversation will not win unsaved men to Christ, it is important that you handle and speak of scientific facts correctly. Christians should be able to show that their faith is sound, reasonable, and superior to any alternative.

5. Spiritual truths must serve as moral guidelines when physics is applied in our society. Today's physicists can do many things. They can develop high-speed communications systems, build immense telescopes to study the stars, and harness the potential of nuclear energy. Yet sometimes the things scientists *can* do are not the things they *should* do. Wisdom, discernment, and sound values—things not learned from academic textbooks or through scientific methodology—must guide these decisions. The world needs dedicated Christians who are qualified to speak out about these important topics.

6. Christians have been instructed to subdue and care for God's creation (Gen. 1:28). A knowledge of physics will help you know how to use, yet not abuse, what God has provided. Even if you are not a scientist, you are a member of a society that is grappling with the issues of unrestrained urban growth, ever more limited energy resources, light and noise pollution, the consequences of icecap melting, and atmospheric ozone depletion. Being a good steward means becoming informed of the issues and taking an active part in the use of God's creation.

Your Approach to this Course

Physics has the reputation of being a difficult course, and it can be if you are not willing to devote the effort to learn and practice the new concepts. You must always recall that you should not be studying physics for your own personal advancement but for God's ultimate glory. "Whether therefore ye eat, or drink, or whatsoever ye do, do all to the glory of God" (I Cor. 10:31). Keeping this in mind will provide you the necessary motivation.

We hope that your study of physics will bring you an increased understanding of the world around you, a solid preparation for future studies, and a heightened interest in science. More importantly, may your studies this year help you to honor Christ and to do His will.

A Framework

The science called physics began in the dim ages before the time of Christ, when natural philosophers tried to make sense of the world around them. Studying the history of science in general and physics in particular will reveal that real strides in scientific investigation did not occur until phenomena and processes were observed critically and described mathematically.

Every course of study must have a beginning. Just as a day begins with a sunrise and a building must start with a foundation, you must establish a firm basis for the subject of physics. Unit 1 will provide you with a biblical philosophical framework that is necessary to appreciate the beauty, complexity, and orderliness of the physical universe. The methods of real science will be discussed, as well as its limitations. The wide-open fields of study and employment available to the student of physics will also be presented. The mathematics needed in this course are not exotic, but you may encounter unfamiliar applications of math you may have already studied, so you will study the math concepts that are essential to introductory physics.

1A	Science and Pseudoscience	1
1B	A Brief History of Science	6
1C	Scientific Methodology	11
1D	What is Physics?	15
Facet:	Scientific Models	13
Facet:	Limitations of Science	16

One scientist's idea of the appearance of the earth during the first day of Creation

In the beginning God created the heaven and the earth.
And the earth was without form, and void; and darkness was upon the face of the deep.
And the Spirit of God moved upon the face of the waters.
And God said, Let there be light: and there was light.
And God saw the light, that it was good: and God divided the light from the darkness.
And God called the light Day, and the darkness he called Night.
And the evening and the morning were the first day.

Genesis 1:1–5

Physics: A Biblical Framework

> *Blessed is the man that walketh not in the counsel of the ungodly, nor standeth in the way of sinners, nor sitteth in the seat of the scornful. But his delight is in the law of the Lord; and in his law doth he meditate day and night.* Psalm 1:1–2

The influence of science and its applications is pervasive in today's society and economy. Our modern civilization depends on electricity, medical technology, agricultural hybridization, food distribution systems, and computer networks. For better or for worse, we are living in an increasingly complex technological world that depends on the fruits of scientific research.

Before we plunge into this introductory physics course, we should consider a few questions. What is true science? How can we recognize false science? How should a Christian view science and its discoveries? What is physics?

Is it important to be able to tell the difference between true science and false science? How can a Christian avoid walking in the counsel of the ungodly if he does not quickly recognize the difference?

1A SCIENCE AND PSEUDOSCIENCE

1.1 False Science

In I Timothy 6:20, the apostle Paul admonishes the young pastor Timothy with the words "keep that which is committed to thy trust, avoiding profane and vain babblings, and oppositions of science falsely so called." The word "science" is better translated as "knowledge," specifically the esoteric knowledge that became the religion of the Gnostics. Paul was warning Timothy to guard against teachings that opposed the Christian faith and scriptural wisdom but were being taught under the guise of knowledge. This is still an appropriate warning today, since there is much false science being passed off as real science.

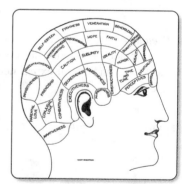

1-1 A phrenologist's model (chart)

Let's look at some examples.

- During the early years of the nineteenth century, German physician Franz Joseph Gall established the field of *phrenology*, which claimed that a person's character and intellectual abilities depend on the shape of his cranium (the part of the skull containing the brain).

- In 1989, a pair of physicists announced that they had observed nuclear fusion (the process believed to power stars) occurring *at room*

Franz Joseph Gall (1758–1828) was a German physiologist and neuroanatomist.

1-2 A machine that automatically measured a "patient's" skull and produced a phrenology report

A **phenomenon** is anything that the mind can perceive, whether it is proven to be real or imaginary. In physics, it means anything perceptible by the senses or by instruments.

A **prejudice,** as used in this textbook, is any personal bias formed without examining the facts of the case or with no knowledge of the issue.

Pseudoscience is any work that masquerades as science but fails to meet the standards of experimental science.

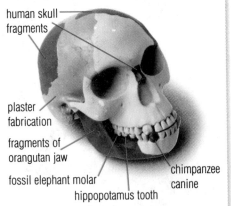

human skull fragments

plaster fabrication

fragments of orangutan jaw

fossil elephant molar

hippopotamus tooth

chimpanzee canine

1-3 After 40 years on display in the British Museum of Natural History, the fossil of Piltdown Man was revealed to be a fake.

temperature in their lab. Such a discovery could herald the advent of cheap, inexhaustible energy.

- In 1999, an author published a theory that humans are the evolutionary end products of undetectable virus-like information replicators called "memes" and that our perception of existence is an illusion.

- In 2001, an experimenter patented plans for a power generator that produced more energy than it uses—another seemingly endless energy source.

These claims are just a sampling of the hundreds of similarly stunning "discoveries" reported in recent years. You may find them hard to fathom, yet many of the people involved earnestly believed that they were reporting scientific findings. Each one of the discoveries above was shown by competent scientific review to be unreproducible or an outright fraud. If you have a good understanding of what true science is, you will probably not be deceived by popular reporting of bad science and will dismiss certain "discoveries" as being misinterpreted, false, or even worse, purposely faked.

1.2 Types of False Science

False science can take many forms. In one form, people offer no authority other than themselves or like-minded individuals for their so-called scientific claims. The **phenomenon** or device that they describe does not obey basic laws of science, cannot be reproduced by other scientists, or is eventually shown to be an outright falsification. This form of false science is often called "crackpot" science. People who engage in "crackpot" science are deceivers interested only in notoriety and making money or are themselves deluded by extreme **prejudice.** Physical healing by crystals, aliens among us, and machines that miraculously diagnose illnesses over great distances are examples of crackpot science.

Another form of false science is properly called "pseudoscience." Pseudoscience is much more difficult for the average person to recognize. Some practitioners of pseudoscience are intelligent crackpots who surround their ideas with scientific trappings to make them seem publicly respectable and at the same time incomprehensible to all but knowledgeable scientists. Phrenology is just one of many examples of medical pseudoscience (also called "quackery"). Pseudoscience includes the work of people who are well meaning but do not follow good scientific methodology. Usually their discoveries contradict some well-established scientific theory or prediction that makes their claim suspect to begin with. The cold fusion experiment falls into this category. It is difficult to tell if these researchers purposely falsified their data or if they were just mistaken. Proper scientific methodology eventually reveals the error in either case.

You have to know what true science is before you can reject something as false science. One scientist has hypothesized that the matter in the universe began as an unimaginably large sphere of water (more than two light years in diameter!). The scientist is also a Creationist who holds to a literal interpretation of Genesis chapters 1 and 2. His theory goes on to explain how this created matter became distributed throughout the universe. Many Creationists consider this theory to be exceptionally effective because it explains numerous astronomical observations and is completely consistent with Scripture, even though it leaves some questions unanswered. Evolutionists consider such an idea to be preposterous and place *any* attempt to scientifically support supernatural Creation firmly in the category of pseudoscience. Many Christians, on the other hand, consider the big bang theory and evolutionistic fossil dating based on the geologic column to be pseudoscience. The way you define science ultimately depends on your religious worldview.

The previous statement may surprise you. The following discussion will give you a better appreciation for how insubstantial the term "science" really is.

1.3 What Is Science?

Science can refer to any body of knowledge gained from observing the natural world (for example, the science of entomology) as well as the activities involved in acquiring scientific knowledge. Within the last century, "science" has gained an institutional meaning similar to the word "government." For example, some people think that "science" will find a cure for AIDS some day.

Science involves more than just the observation of nature. Philosophers claim that they can discern the purpose of life by *observing* the relationships of living things in nature. A painter intently *observes* the texture of the bark on a beech tree so that she can accurately convey the sense of "beech tree-ness" in a painting. No one would mistake either of these activities as natural science.

Since so many fields of interest are included in scientific endeavor, a concise definition of "science" is impossible. Many people claim that true science includes a number of basic characteristics, some of which are listed in Table 1-1. Can you think of any others?

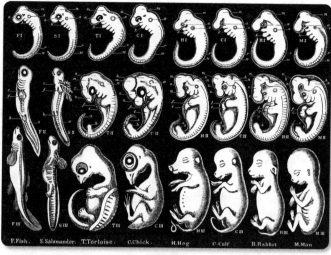

1-4 Haeckel's diagram of evolutionary embryonic development has been thoroughly discredited.

TABLE 1-1
Characteristics of Science
• New theories are tested against previously acquired facts.
• Theories are falsifiable—no principle is safe from being proven false by new information.
• That which is studied must be observable, either directly or indirectly, and can be measured or described.
• Concepts can be expressed as mathematical relationships (formulas or equations).
• A phenomenon can be properly studied only if it can be repeatedly observed in some way.
• The facts are analyzed and theories are made objectively, that is, without personal prejudice.
• Reports are not published until they have been reviewed by other workers trained and familiar with the research.

1.4 Objectivity

One of the characteristics listed in Table 1-1 is objectivity. Most people understand scientific **objectivity** to mean that scientists are not influenced by *a priori* knowledge or cultural and personal prejudices. This view is carefully cultivated in the minds of the public by writers of science articles in newspapers and popular magazines and in the minds of students through secular science textbooks. There is also a general impression that scientists and the scientific establishment are somehow immune to pettiness, feelings, and peer pressure. Because all men have a sin nature, this impression is inaccurate. Scientists are human, and their pursuit of scientific knowledge is subject to human **biases** and all of the weaknesses of human character.

A priori knowledge is information that is known before a problem is investigated.

A personal **bias** is an inclination or "leaning" based on personal experience or other influences in one's life. Biases prevent decisions from being completely impartial.

A better description of objectivity states that an objective scientist will fairly consider everything that bears on a particular question, will not reject legitimate data, interpretations, or criticism out of hand, and will not twist data to promote his own agenda. A redeemed Christian scientist should be motivated to seek the truth in order to return glory to God. For him, purposefully avoiding objectivity would be dishonest and sinful.

As you can see, no one can be truly neutral when it comes to interpreting data. A scientist's bias cannot be separated from the scientific process, nor should it be. His biases determine the kind of science that he will produce.

A person's biases arise from his underlying philosophical view of the world and the way it works. One's fundamental worldview is also called his **paradigm.** A person may have numerous biases for each facet of his life, but he will probably have only one underlying paradigm that shapes his biases. His paradigm is important because it influences not only his formulation of scientific theories but also his view of just about every other human activity that involves rules, laws, and standards.

A person's fundamental paradigm is not built on physical facts but on what he thinks is eternal and self-existent (not dependent on anything else for its existence). In other words, one's worldview depends on what he believes is God, has god-like properties, or is divine. A non-Christian may believe in the self-existence of inert matter or some immaterial energy. Note that unless a reliable authority (such as the Bible) is used to establish your worldview, you have only your own wisdom to depend on. The natural heart and mind are not reliable sources of decisions and information. "The heart is deceitful above all things, and desperately wicked: who can know it?" (Jer. 17:9). The concepts of eternality and self-existence are religious in nature—not scientific.

Proverbs 1:7 says "The fear of the Lord is the beginning of knowledge." The word *fear* means more than mere fright or reverential respect for God. Rather, it suggests complete submission to Him. The word *beginning* implies "first-ness," or primacy, in the sense of the head of a person or the foundation of a building. Therefore, one must be redeemed to even begin to have real knowledge. A belief in God is fundamental to an appreciation of the orderliness of the universe. Why can scientific rules often be expressed as precise mathematical formulas? When we try to predict new findings, we are assuming that the rules of nature don't change. God caused the universe. This cause-and-effect pattern is an underlying principle of all science. When a scientist sees an effect, he can be assured that there is a cause because the universe is inherently predictable. While an unbelieving scientist can observe the world's orderliness, he can't really know why it is so. A Christian knows it is so because God is unchanging and the laws of His creation don't change unless He intervenes through miracles.

1.5 Comparing Worldviews

In modern science, there are two prominent paradigms that underlie the formation of theories and interpretation of data—the materialistic worldview and the biblical worldview. Many authors in both camps have written volumes arguing the strengths of their worldview. We will briefly compare the views of both paradigms on several issues.

Origins—The materialist begins with the presupposition that everything now in existence arose from inert matter through naturalistic evolutionary processes. He completely rejects the idea of intelligent design and denies that a supernatural creator was involved with the formation of the physical universe. If he sees design in nature, then he says that the naturalistic evolutionary process has the inherent ability to create order out of disorder. The naturalist-materialist sees evolutionary processes at work everywhere, but he is incapable of demonstrating the natural mechanisms that produce more complex systems from simpler arrangements of parts.

A redeemed person with a biblical scientific worldview affirms that there is a Creator-God and that the Bible is His inerrant and providentially preserved message to His fallen creatures. The biblical Creationist believes that the universe was spoken into existence by God relatively recently during six literal days as described in Genesis chapters 1 and 2. He understands that the whole universe gives evidence of intelligent design. He also knows that the Bible does not speak on every scientific subject, but anything that is mentioned bearing on natural phenomena is scientifically correct.

Mankind and morality—The materialist views mankind as an accidental end product of biological evolution—a natural creature that is not morally responsible to anyone higher than himself. He believes that those "higher" attributes of humanness such as language, love, intellect, philosophy, and religion came into being because they somehow gave mankind greater advantage in the battle for the survival of the fittest. This view of man as nothing more than a highly evolved animal has led to perversions of law and justice as well as horrendous human atrocities during the world wars in the last century and in regional conflicts that continue to this day.

A biblicist believes that man was originally created perfect in God's image. Man fell in sin and now bears a marred image of God and suffers the curse of death. He believes that he is morally responsible to God because he was created by Him and that answers to questions of morality must be based on the character of God. Humans are unique creations in God's image and have individual value to God that should preclude inhumane treatment, involuntary experimentation, sterilization, euthanasia, and abortions.

In addition, the biblical framework of science establishes that the original face of the earth was completely obliterated in the Noachian Flood and that the vast majority of landforms and oceans now visible were a direct consequence of that flood. A Creationist sees complete harmony between his explanations of nature and the scientific descriptions in the Bible.

You can see that a scientist's worldview, and thus his science, are strongly affected by his religious presuppositions. A regenerated Christian scientist has great advantages over an unsaved scientist when conducting scientific investigations. The Holy Spirit can direct him in the wisdom of God as he does his research. This direction does not come in the form of audible voices but may occur through inclinations and hunches, governed by a desire to see God's truth vindicated. Throughout history many scientific discoveries have been attributed by Christian scientists to divine guidance.

1.6 True Science?

You should have concluded by now that it is not possible to objectively or operationally define "science." True science is defined by the ultimate purpose for which the knowledge is gained and the use to which it is put, in addition to having the attributes of science such as those listed in Table 1-1. Recognizing that the God of the Bible is the Creator of the universe and all that is in it, we are to use our God-given faculties to subdue creation for His glory according to His commandments. "Thou art worthy, O Lord, to receive glory and honour and power: for thou hast created all things, and for thy pleasure they are and were created" (Rev. 4:11).

Much valuable science is being accomplished by unbelievers today because they use the biblically scientific principles of cause-and-effect that God designed

1-5 Survivors of attempted genocide in World War II.

How do we know God's original creation of man was perfect? "So God created man in his own image. . . . And God saw every thing that he had made, and, behold, it was very good" (Gen. 1:27, 31).

into nature, even as they deny that He exists. We can only guess what might be learned if they were seeking His glory rather than man's. "Because that, when they knew God, they glorified him not as God, neither were thankful; but became vain in their imaginations, and their foolish heart was darkened. Professing themselves to be wise, they became fools" (Rom. 1:21–22).

1A   Section Review Questions

1. Discuss two properties of crackpot science.

2. Why is *pseudoscience* so difficult to define?

3. Referring to Table 1-1, for each of the properties of science listed, give an area of study that is not normally considered science for which the property is also true. The text gives several examples.

4. What *a priori* knowledge does a Creationist use when he writes a rebuttal to an evolutionist's claim that a certain dinosaur was an ancestor to birds? Is his argument scientific using this approach? Is his argument reasonable?

5. List five of your biases. Which of your biases depend on your biblical worldview? Be prepared to discuss in class the basis for each of your biases.

6. Research and write or give a brief report about a crackpot discovery announced within the past year.

7. Explain how scientific reports about present-day observable phenomena can sometimes include unscientific statements.

8. Why is it important for Creationists to evaluate current scientific data and its interpretation within the framework of Scripture rather than skewing what the Bible says to conform to popular scientific thought?

1B   A BRIEF HISTORY OF SCIENCE

1.7   From Genesis to the Greeks

According to Genesis chapter 1, God created Adam on the sixth day of Creation week and gave him dominion over all of the earth. The *dominion mandate,* or the commandment to occupy and make use of the earth, included the need to study creation in order to learn how to best utilize the earth's resources for the benefit of mankind. This was necessary because, though Adam was created as an unfallen, intelligent creature, he did not have all knowledge.

There remains no contemporary record of the advance of man's knowledge of the natural world during the period from Creation until well after the Flood, because God destroyed nearly all evidence of civilization in the Flood. The Bible, which was written after the Flood, is the only existing document that contains any scientific information from before the Flood. We can be assured of the accuracy of this information because its source is God Himself. However, nothing in Scripture mentions *how* scientific knowledge increased or anything about ancient scientific methodology. The earliest documentation of scientific discoveries comes from secular histories during the centuries prior to the first advent of Christ.

Before the sixteenth century A.D., the study of nature was more philosophy than actual science. Beginning with the Greeks in the sixth century B.C., abstract reasoning was the dominant method for interpreting reality and arriving at truth. The philosophies of Plato and Aristotle considered the material world to be imperfect and constantly changing. To actually observe natural phenomena to determine the laws of nature through experimentation was generally believed to be infinitely inferior to the purity of rational thought or reasoning. However, even at this early date a few Greek experiments were documented.

1.8 Deductive Reasoning

The form of reasoning perfected by the Greek philosopher Aristotle is called **deductive reasoning.** Deductive reasoning, as Aristotle understood it, begins with a statement called a *premise* (which may have been developed *inductively*—see below). Through a succession of logical statements, one arrives at a conclusion that must be true (assuming that all of the preceding statements are true and the form of the argument is valid). Because many premises of Aristotle and his adherents down through the centuries were based on prejudices, opinions, or what they believed to be *a priori* truths that were actually wrong, Aristotelian philosophical science was full of erroneous **deductions.**

There was great resistance to the pagan Greek philosophy of nature by both Jews and the early Christians. However, during the Middle Ages, the influential Dominican friar Thomas Aquinas effectively argued that, since both reason and truth come from God, Aristotle's reasonings were not incompatible with Scripture. This view, failing to account for human limitations and depravity, set the Catholic Church on a misguided course that the scientific revolution of the following centuries would later expose.

Church fathers applied Aristotelian deductive reasoning to biblical passages in order to discern scientific truths in Scripture. The result was a rigid, literal interpretation of biblical passages that were obviously describing the *appearance* of natural phenomena. For example, in Numbers 2:3 the Sun is said to be rising. The Catholic Church considered that verse to mean that the Sun was physically moving around the earth, not that it just appeared to be moving because of the earth's rotation. Church scholars deduced that the earth must be the center of the universe, as ancient geocentric theory had always held. *Geocentrism* thus became a fundamental scientific doctrine of the Catholic Church.

During the Middle Ages, nearly all academic thought was subject to the Catholic Church because of the church's pervasive influence in secular as well as religious matters. Any scientific idea that was contrary to Aristotle's teachings was considered heretical, and, as a result, scientific work in Western cultures was ignored or suppressed for nearly two thousand years after Aristotle.

1.9 Inductive Reasoning

Philosophizing about nature began to give way to modern-style science in the late 1500s through the efforts of men who recognized that their observations of common events, such as falling objects and the swinging of lamps in cathedrals, did not agree with Aristotelian reasoning. In 1590, the Italian scientist Galileo Galilei published his *De motu* (*On Motion*) which refuted Aristotle's physics. A few decades later the English philosopher Francis Bacon attacked Aristotelian logic in his *Novum Organum* (*The New Organon*), published in 1620. He proposed that true knowledge could be acquired only by collecting specific information about a problem, then drawing conclusions from the data.

1-8 Galileo Galilei (1564–1642) was a prominent Italian inventor, mathematician, and scientist. See more biographical information on page 67.

Aristotle (384–22 B.C.) was a Greek philosopher. See more biographical information on page 135.

Deductive reasoning begins with a series of logical statements and ends with a true conclusion, assuming the supporting statements are true and the form of the argument is valid.

1-6 Thomas Aquinas (1225–74) was an Italian philosopher and theologian.

Geocentrism is the belief that the earth is at the center of the visible universe, around which all heavenly bodies revolve.

1-7 An early chart based on geocentrism

Francis Bacon (1561–1626) was an English philosopher, courtier, and member of Parliament. He established the philosophical basis for the advancement of science during the Renaissance.

Inductive reasoning begins with a collection of observations from which a general conclusion is obtained.

The *heliocentric theory* states that the Sun is the center around which all visible heavenly objects, other than the stars, revolve.

1-9 James Ussher (1580–1655), in addition to being Archbishop of Armagh, Church of Ireland, was a world traveler and theologian.

This thinking was the basis for what came to be known as **inductive reasoning** or **induction.**

Induction shows what is *most probable,* or *likely* true, while deduction leads to what must be true. To illustrate this process, assume that, after observing many instances of water boiling at 100 °C, you propose a rule that water always boils at 100 °C. This rule was developed inductively and would be valid as long as water continued to boil at 100 °C. But if someone observed water boiling at a different temperature, because of changes in atmospheric pressure or the presence of dissolved substances in the water, the rule would no longer agree with observations. Therein lies the strength of the inductive method: as observations reveal flaws in a man's statement of a natural rule, the rule can be changed or replaced to explain the observations.

Once a general physical principle was inductively developed, one could use deduction to predict unknown consequences of the principle, then perform further observations to see if the predictions were true. This inductive-deductive method was originally attributed to Aristotle, but he never developed definite rules for induction. To make the observation process more efficient, the pioneering scientists of Galileo's day contrived carefully designed tests of their ideas—what we would today call **experiments.** Initially, the reaction of the Catholic Church to observations that contradicted its ancient and large body of Aristotelian dogma was outright rejection and persecution of the scientists. After Galileo published his *Dialogue Concerning the Two Chief World Systems—Ptolemaic and Copernican* in 1632, he was called before the Inquisition, declared a heretic, and forced to publicly recant his belief in the *heliocentric* theory of Copernicus.

1.10 Renaissance—Changing Views

In spite of opposition, the tide had turned regarding man's view of the world (although the Catholic Church continued to officially deny the Copernican theory until 1922), and scientific thinking would never again be the same. In the ensuing centuries, man's understanding of the natural world grew slowly initially, but then more rapidly as discovery built upon discovery, involving scientists from many countries. It is believed that the Protestant Reformation in Europe was the impetus for many scientific advances and inventions toward the end of the Renaissance.

It was during the Renaissance that Archbishop James Ussher used contemporary astronomical observations and biblical genealogies to calculate that Creation occurred in 4004 B.C. (making the world about 5600 years old at the time). His date of Creation came close to being accepted as religious dogma, and his chronology of Bible history has been printed frequently in Bibles up through the present time. The expanding body of scientific knowledge was accompanied by new inventions (such as the telescope, microscope, regulated clock, and many others), changing social, political, and economical structures in Europe and the Americas, and the waxing and waning of various philosophical worldviews. During this time, many great Christian men were instrumental in establishing the key physical and biological principles upon which modern science grew.

Most men and women of science prior to the nineteenth century would have readily agreed that nature bore the imprint of intelligent design, a conclusion that would imply there had to be a Designer (the **teleological** evidence that God exists). In the writings of those scientists are many statements describing the awe and privilege they experienced as they revealed the formerly hidden mysteries of God's creation. In fact, Johannes Kepler went so far as to say that his work as a scientist was "merely thinking God's thoughts after Him." Many scientists believed studying nature through science was one of the best ways to glorify God.

1.11 Evolutionary Challenges

The 1700s and 1800s saw a significant change in the philosophical basis for scientific activities, particularly for those areas called the physical sciences. The success of that giant of science, Sir Isaac Newton, in using simple principles of cause-and-effect mechanics to explain many physical phenomena encouraged people to view other sciences in the same way. A mechanical approach to chemistry and biology was attempted in the eighteenth century and was accompanied by the rise of new social philosophies that emphasized the preeminence of reasoning. Scientists began to reason that the continuity of life "forms" or "types" from the lowliest bacterium to humans indicated that there might be some relationship between similar forms. The stage was set for the acceptance of many theories that directly conflicted with biblical Creation and the age of the earth established by Ussher.

1-10 Isaac Newton (1643–1727) was an English physicist, mathematician, and inventor. See more biographical information on page 110.

1-11 Sir Charles Lyell (1797–1875) was a British geologist.

Geology was the first branch of science to come under attack. In 1779, Compte de Buffon claimed that 75,000 years had passed since Creation. In 1785 the proposal by James Hutton that all natural causes were unchanged throughout history established the basis for the theory of *uniformitarianism*. In 1842, Charles Lyell claimed that the earth was at least hundreds of millions of years old. Although many godly scientists countered these and other theories with solid biblical and scientific evidence, the unbiblical naturalistic views gained wide acceptance, particularly in European spheres of influence.

The work that caused the most significant change in the viewpoint of science in the nineteenth century was Charles Darwin's *On the Origin of Species* (1859), which claimed to establish the process, although not the mechanism, for biological evolution. During the following century and a half, the evolutionary paradigm has become the basis for interpreting all physical, biological, and human systems.

Many factors have contributed to the wide acceptance of evolutionary theory. The rapid introduction and improvement of many new technologies tended to make man believe he was less dependent on God's providence. The rise of secular humanism, which replaced God with man as the ultimate object of worship, adopted naturalistic biological and cosmological evolution as essential scientific principles. These factors coincided with the loss of power and influence of the mainline Christian denominations. The authority of the Bible was systematically destroyed in the minds of millions of people through the influence of higher criticism (critical analysis of the origin of the Scriptures that attacked their inspiration and undermined the key biblical doctrines), which itself was linked to humanistic thought. Because evolutionary theory assumed that natural processes operated completely free of metaphysical or supernatural forces, it rapidly found acceptance in nearly all scientific circles as the only scientific explanation for the origin of all things.

Georges Louis Leclerc Comte de Buffon (1707–88) was a French mathematician and scientist.

Uniformitarianism, the concept that all processes have continued in the past as they appear today, is the basis for evolutionistic theory. It explicitly denies miracles in the past because miracles do not appear to be occurring today.

James Hutton (1726–97) was a Scottish geologist, chemist, and naturalist.

1.12 Science Today—A Philosophical View

Today, many (one could say most) scientists believe that *scientific naturalism* (also called *naturalism, naturalistic materialism,* or *scientific materialism*) is the foundational view of science. Many secular science writers emphatically state that supernatural origins and metaphysical processes are not part of the "real" world and therefore are not scientific. Since neither human senses nor instruments can detect

1-12 Charles Darwin (1809–82) was a British naturalist who is called the Father of Evolution.

the spiritual realm and the existence of God is not verifiable through naturalistic observations, many scientists claim either that God does not exist or that He operates completely outside of the physical world. Because of the effectiveness of Creationist efforts to counter evolutionistic bias in public education and the media, some evolutionists have modified their position. They now say that belief in both evolutionary science and God is possible but attempt to show that God is not needed for evolution to occur, claiming He is irrelevant to all things in the natural world. They prefer to place God and spiritual matters in a separate compartment from science and the "real world." Once again, this perspective is from a *philosophical* and *religious* position, not a scientific one.

Atheistic scientists, sociologists, and government officials have been so successful promoting their worldview that many believe that science and religious thought have nothing in common. Additionally, it is believed that since science deals with the "real world," man can depend on science to provide answers to real-world questions, while religion has little relevance to physical-world matters. Through your study of physics and other sciences from a Christian perspective, you will realize how wrong this belief is.

You may think that science is corrupt and not worthy of serious study by Christians today. Nothing could be further from the truth! While the philosophical basis for much of today's science may be unbiblical, the facts of God's creation that are being discovered each day are still facts waiting to be interpreted by God's people. No matter who discovers it, or under what circumstances, all truth is God's truth because He is Truth (Deut. 32:4; John 3:33). Throughout history, God has given many men and women of all nations great powers of perception and intelligent, inquisitive minds to discern the truths of His creation. These individuals, both believers and nonbelievers, are fulfilling the dominion mandate of Genesis 1:28. There are more opportunities to study God's world through science today for His glory than at any time in history. Christians must not abandon this spiritual battlefield (and mission field) to the atheistic naturalists.

The late Stephen Jay Gould, an influential twentieth century evolutionist, proposed the idea that religion and science are not incompatible realms of human endeavor in his "nonoverlapping magisteria approach" (NOMA).

1B   Section Review Questions

1. Whose philosophy of the natural world governed scientific thought for over two thousand years?

2. According to the laws of logic, deductive reasoning will always result in a true conclusion. What conditions must exist for this to occur?

3. Give one example of an erroneous scientific conclusion that was a result of using a biblical passage written with the language of appearance as a premise for Aristotelian deduction.

4. Discuss the differences between deduction and induction.

5. What great seventeenth-century discovery in physics set the stage for the rise of evolutionary theory in the eighteenth and nineteenth centuries?

●6. Consider the following series of statements:

(Premise)	a. The belief that there is only one reality is an illusion.
(Implying)	b. There are multiple realities.
(Implying)	c. Death is merely the transition from one reality to another.
(Implying)	d. Death is an illusion.
(Concluding that)	e. Murder is not morally wrong.

What kind of logical reasoning is this? Is the conclusion in this case true? Explain, giving an authoritative reason for your answer.

●7. Every time a physics student visited her grandparents in Maine for the past five years, it rained at least once during her stay. She is going to visit them again next weekend and she expects it will probably rain. What logical process did she use to come to this conclusion? Does this process guarantee that her conclusion is correct? Explain your answer.

●8. Explain what is meant when we say that Darwin described the process *but not the mechanism* of naturalistic evolution. You may have to do some research on the faults of Darwinism in order to answer this question.

1C SCIENTIFIC METHODOLOGY

1.13 The Scientific Method

You probably have heard of the scientific method. What is the scientific method? First, the scientific method is not a series of specific steps that must be followed every time to arrive at scientific truth. Every different kind of scientific question requires a different method to find an answer. A better phrase to refer to the way scientific research is performed could be **scientific methodology.**

Scientific methodology helps scientists describe nature with as much objectivity as possible. The goal is to produce a workable explanation of how a natural process occurs. *Workable* implies that the explanation actually accounts for the data. It does not imply that the explanation is the absolutely "true" description of nature. Nor does it imply that the explanation will endure: new facts sometimes contradict the explanation, forcing scientists to seek a better one.

Scientific methodology is the general approach to conducting experiments and making observations. It includes forming an objective, meticulously recording data, and describing how the observations are to be carried out. Scientific methodology combines inductive and deductive logic, imagination, communication, and creativity. Scientists use inductive reasoning when, after many observations, they propose an explanation. The following discussion highlights the major elements of one kind of scientific method using a historical illustration.

1.14 Observations

Observations are seen, heard, smelled, felt, or tasted. They involve any or all of the five senses. You see the blue-white light of the star Sirius (Alpha Canis Majoris) in the constellation Canis Major; you hear the characteristic call of a hermit thrush (*Catharus guttatus*); you smell the pungent skunk cabbage (*Symplocarpus foetidus*); you feel the bristles on the underside of an earthworm (*Lumbricus terrestris*); and you taste the bitterness of the limonoid present in orange juice.

In the eighteenth century, observations about heat were explained by the theory that heat was a weightless, invisible fluid called *caloric*. Whenever a hot slab of iron was placed in contact with a cold slab, the hot iron cooled and the cold iron heated up. Scientists said that the caloric fluid flowed from the hot iron into the cold iron. As the hot iron lost caloric, it cooled; meanwhile, the cold iron heated up because it gained additional caloric.

Late in the 1700s, a physicist discovered a fact that the caloric theory could not explain. Count Rumford, Bavaria's minister of war, was watching drills bore holes in cannon barrels at a cannon foundry when he noticed that the drills produced incredible heat—enough to boil water. He estimated that the amount of heat

Scientific **observations** are the raw data, measurements, and descriptions of phenomena that are used to support or refute scientific hypotheses, theories, and laws.

1-13 You observe Sirius with your sense of sight.

1-14 A cannon boring device similar to that observed by Count Rumford

Sir Benjamin Thompson (1753–1814), Count Rumford, was an Anglo-American physicist. See more biographical information on page 333.

produced by the drilling operation was more than enough to melt down the entire cannon if it could be applied to the metal again! In other words, more heat was produced while drilling out the gun bore than could have initially been contained in the metal. Rumford's observation was an overwhelming contradiction of the caloric theory. This type of disagreement between an established theory and an observation is one kind of **scientific problem.** Scientific problems provide infinite opportunities for further research. An answer to one scientific problem usually introduces problems in several other areas. Identifying a scientific problem is the starting point for developing a methodology to solve it.

1.15 Hypotheses

A **hypothesis** is the scientist's best explanation that accounts for all known observations of a natural phenomenon.

In response to a scientific problem a scientist states a **hypothesis** (plural: hypotheses)—his best guess at an explanation that accounts for the observation. The hypothesis provides the direction for finding the solution. Formation of a hypothesis is a process influenced by the scientist's underlying bias. A born-again Christian has a different worldview than an unsaved scientific naturalist, so it is not surprising that Creationists and evolutionists often form different hypotheses to explain the same phenomena and data.

What are the qualities of a good hypothesis? First, the hypothesis must be reasonable. It would have been silly for Rumford to hypothesize that the heat released by the drill was so great because physical laws are different in Bavaria. Rumford's own hypothesis, that heat is not a fluid but instead involves the motion of particles, is reasonable.

As we learn more about microbiology, even the so-called simple forms of life are being found to be astonishingly complex. And we frequently find "irreducible complexity," i.e., six, eight, ten or more symbiotic functions—relationships that cannot exist without the simultaneous existence of the others. Gradual evolution, allegedly small changes over a period of decades or centuries, cannot explain such relationships.

Second, the hypothesis must be testable. Scientists test hypotheses by making predictions from them that additional observations and experiments can prove or disprove. A hypothesis that cannot be tested is not scientific, though it may be true. No hypothesis of the origin of life is scientific, because none can be tested. Creation *is* more reasonable, however, because the probability of even the simplest forms of life coming spontaneously into existence is exceedingly small.

To account for the greater heat released by a dull drill than by a sharp one, it was theorized that dull drills stimulate metals to produce more caloric. Although this statement explains the observation, nothing measurable can be predicted, and thus, it cannot be tested. Rumford's hypothesis, however, makes specific, measurable predictions. The amount of heat produced in cannon boring should be proportional to the work done in turning the drill. Rumford's experiments in the cannon foundry showed this hypothesis to be true.

Third, the hypothesis should not contradict well-established principles. This condition may be violated under one of two conditions. First, the new hypothesis must explain observations better than the old theory; Rumford's hypothesis, that heat is the transfer of energy by particle motion, was a definite improvement over the caloric theory. Second, the new hypothesis is acceptable if it restricts an old theory to special cases. Einstein's theory of special relativity, for example, restricted the applicability of Newton's laws of motion to objects traveling at speeds much less than the speed of light.

Fourth, the hypothesis must explain all current observations and predict new ones. This quality constitutes the goal of science: to relate known data to newly discovered facts and to point the way to future research. The "particle motion" hypothesis of heat explains all the known facts about heat. It also predicts that heat involves a form of energy that is interchangeable with other forms of energy. The same amount of electric energy, for example, should always produce the same amount of thermal energy. The prominent nineteenth-century inventor James Joule later confirmed this prediction after a series of experiments. Nevertheless, even a hypothesis that makes true predictions is not infallible or unfalsifiable.

Fifth, the hypothesis should be as simple as possible. This is not a hard-and-fast rule. Our present knowledge of God's creation indicates that the significant processes in the universe can be expressed by simple rules. It is true that some complex processes cannot be reduced to simple rules because God is not restricted by this principle. But when two hypotheses are equally good, scientists usually choose the simpler one, consistent with their worldview and biases.

James Prescott Joule (1818–89) was an English physicist and inventor. See more biographical information on page 332.

FACETS of PHYSICS

Scientific Models

Many phenomena cannot be observed directly. They are too small (atoms), too large (the galaxy), occur too rapidly (explosions), too slowly (plate movements in the earth's crust), are too complex (weather systems), or are physically unobservable (the interior of the earth). To study and visualize these kinds of things, scientists often propose a **model** as a hypothesis. For example, accurate computer models of the atmosphere have dramatically improved weather forecasting. A model might be any combination of words, equations, pictures, or computer simulations. It is an educated guess about an unobservable phenomenon. Astronomers have created various models to account for the structure of galaxies that involve dark matter—material that cannot be seen but whose presence is suspected due to its gravitational effects. As with a hypothesis, scientists make predictions about

A computer model of the outer core of the earth (blue wireframe) containing the magnetodynamic inner core (yellow). This model was developed by measuring the speed of earthquake waves through the core.

what would be seen in nature if the model were correct. These predictions are then compared to what is actually observed. Anyone proposing hypotheses of the early Earth must resort to models of the earth's atmosphere, crust, and oceans.

1-15 Hypotheses may be verified, revised, or discarded through experimentation.

1.16 Testing the Hypothesis

After forming a hypothesis, a scientist tests its validity with an experiment or some other method to make observations (e.g., performing surveys, reviewing existing data, or conducting exploration). In order to show cause-and-effect relationships clearly, he attempts to restrict and control the number of variable factors affecting his observations. He cannot ignore observations that may contradict the hypothesis. If the hypothesis makes incorrect predictions, it must be revised or discarded. If it fails miserably, the scientist may need to reevaluate his original assumptions. He may even arrive at completely unexpected conclusions that suggest new approaches for his research.

Discarding a pet hypothesis can be painful and humiliating, and a researcher, because of his fallen nature, is often tempted to alter the facts to fit his hypothesis. To discourage this dishonesty, other scientists review the researcher's work before it is published. The reviewers usually find any mistakes in the experimental design or interpretation that must be corrected. This peer review process maintains high credibility for scientific work all over the world for the vast majority of research. However, this system of checking objectivity *can* break down when the reviewers have the same prejudices as the researcher.

The path that a hypothesis takes to becoming a scientific law can vary greatly depending on the circumstances. Usually it is a step-wise process. After the hypothesis has been tested and found not to fail, it gains stature as a **theory** in the scientific community when it is published. The theory becomes a tool for solving other scientific problems. Yet a theory is no more infallible than a hypothesis—it simply has withstood testing. If a well-established theory makes many correct predictions over time, it comes to be known as a **law**. Laws are the most reliable ideas in science. However, laws are still falsifiable. New observations can prove any scientific concept wrong, no matter how widely it is accepted or how long it has been established. But observations can never prove a scientific concept right beyond any need for further testing! Scientists can say only that a theory explains the observations that have been made so far.

How long should it take for a hypothesis to gain the stature of a theory and then ascend to the role of a scientific law? That would depend on the availability of research funds, the number of qualified scientists studying the problems, and the difficulty of generating valid test conditions. You might find it interesting to watch the scientific community for a few years to answer this question for yourself.

Theories are hypotheses that have successfully predicted new observations. Scientific **laws** have been tested extensively and provide general rules for a large number of phenomena.

1C   Section Review Questions

1. Explain why the scientific methodology used to solve various scientific problems does not involve the same steps in every case.

2. What is the importance of the hypothesis to the scientific method of solving problems?

3. List the five characteristics of a good hypothesis.

4. Which of the following scientific questions would require a model to investigate?

   a. the lift characteristics of a new aircraft wing design

   b. the interior structure of the Sun

   c. the structural response of a building design to an earthquake

   d. a new design for a computer microprocessor chip

5. Do all theories become laws? Explain.

◉6. Explain why the results of scientific research are *workable* rather than the *truth*.

7. After doing some additional research, list three phenomena (measurable by instruments) that our bodies are not designed to detect in any form. Do not include phenomena that are similar to ones that we can sense but are outside of the normal range of our senses (e.g., infrared radiation).

8. Some evolutionists consider the evolutionary worldview to be so essential to understanding science and everything else in the physical world that they believe it should be called the *fact* of evolution. Explain why this idea is not scientific.

1D WHAT IS PHYSICS?

1.17 Revealing God's Order

You are about to embark on an adventure. The study of physics reveals the wonderful orderliness of God's creation—so orderly that it can be comprehended in terms of relatively simple principles (mathematical formulas). The realm of physics encompasses everything from the tiny elementary particles that combine to form the nuclei of atoms to the immense clusters of thousands of galaxies containing billions of stars in the expanse of the universe. Physics is important because through it mankind learns how creation actually works. It satisfies our God-given curiosity about nature. Seeing that God does "great things and unsearchable; marvelous things without number" (Job 5:9), men have dedicated their lives to unraveling the rich mysteries of creation.

As we mentioned at the beginning of this chapter, physics is one of the foundational sciences. As with any science, it includes aspects of mathematics, reasoning, philosophy, and effective communication. The word *physics* comes from the Latin word *physica,* which is itself derived from the Greek *fusika* meaning the study or science of natural things. Until the mid-1800s, physics was called *natural philosophy*, but this phrase fell out of favor with the desire to separate the natural sciences from purely philosophical disciplines.

Physics can be subdivided into two general groupings of topics based on the time in history when they emerged as distinct disciplines. **Classical physics** was developed during the period from 1600 to 1900, and **modern physics** encompasses all discoveries from 1900 to the present. Since this division is somewhat arbitrary, there is some overlap of the two periods between 1890 and 1920.

Limitations of Science

During each phase of every scientific methodology, scientists must make personal choices. Science, therefore, is only as trustworthy as the men who participate in it. Using a scientific methodology does not guarantee that an explanation of nature is objective. Nor can all the questions that face mankind be answered by science. Science has four major limitations.

First, science is limited to phenomena that can be observed. **Instruments** increase man's ability to observe by extending the range of his natural senses, and they can give him sensory abilities he was not created with. Instruments can measure only what is either presently occurring or what are present effects caused by past events (e.g., radiocarbon dating a human fossil or observing supernovas of distant stars). No instrument can ever directly observe past events. Neither do instruments interpret the data that they measure. The same measurements are recorded whether an instrument is used by a Creationist or an evolutionist.

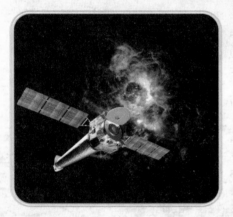

Instruments that detect phenomena outside the range of human senses: left, the Chandra x-ray observatory; above, the SNO neutrino detector

Third, science has difficulty defining and explaining the causes of natural phenomena. Many basic but abstract natural concepts and processes can be defined only through operational definitions. *Operational definitions* prescribe specific measurable tests that must be passed in order for a definition to apply. For example, the operational definition of "matter" is anything that has measurable mass and volume, with the methods of measurements specified. The problem with such definitions is that they often do not provide all the possible meanings of a term or they involve words that are circularly related to the term being defined (e.g., mass and matter).

Second, science changes because man's knowledge is incomplete. What we know today is not as complete and correct as what we will know tomorrow. Science differs from nearly every other human endeavor because it aggressively pursues self-correction. In this physics course you will learn of several well-established theories that had to be replaced with more usable theories as new observations showed that the earlier theories were inadequate. The only sure truths are found in God's Word, which is settled forever in heaven (Ps. 119:89). Scientific "truths" established by human reasoning and observation are tentative and are abandoned if someone proves them wrong. But the Bible, written through the inspiration of an omniscient God, can never be proved wrong.

Earth and the Moon form a gravitational system.

Neither can scientific study explain *why* natural phenomena occur. For instance, scientists can describe the characteristics of gravity and how masses interact, but it is unlikely that they will ever be able to explain what motivates masses to attract each other or why they are not repelled instead. The Bible reveals that God directs all things to please Him: "Thou art worthy, O Lord, to receive glory and honour and power: for thou hast created all things, and for thy pleasure they are and were created" (Rev. 4:11).

Fourth, questions about man's purpose and condition are beyond the realm of true science. This limitation deserves special emphasis. For example, science cannot solve

- questions of morality, because God gives principles of conduct by direct revelation—"for we walk by faith, not by sight" (II Cor. 5:7);
- questions of human sociology, because man's nature is very complex and marred by sin (Jer. 17:9);
- questions of origins (as mentioned earlier), because no one was present at Creation to record the events—"Through faith we understand that the worlds were framed by the word of God, so that things which are seen were not made of things which do appear" (Heb. 11:3).

The evolution-Creation debate illustrates a stormy controversy that cannot be solved through science. The origin of life cannot be repeated or observed, and neither belief is scientifically falsifiable. Further, evolution and Creation have applications to morality and sociology, so conflict between the competing systems is to be expected. Followers of both systems are heavily biased by faith in their respective beliefs, and this bias influences how they interpret their observations of natural phenomena. It must be understood that both paradigms are based on faith, not on science, but the objects of faith are different. Evolutionists trust in an "evolutionary principle" said to be present in all matter, but Creationists trust in God and in His Word.

Science cannot solve questions of human sociology.

1.18 Classical Physics

Classical physics began with the overturning of Aristotelian mechanical concepts in the seventeenth century. Sir Isaac Newton, building on the discoveries of Kepler and Galileo before him, described how objects moved *(kinematics)* and developed laws that explained the causes of motion *(dynamics)*. Newton's combination of kinematics and dynamics formed a body of the most comprehensive scientific laws in history, which changed mankind's view of nature forever. Kinematics and dynamics together are called **mechanics.** Mechanical theory includes the description of motion, how forces affect motion, and motion in lines, arcs, and circles. Closely associated with the state of motion are the concepts of mechanical energy and the momenta of objects. Another related branch of mechanics that deals with the distribution of forces within unmoving and rigid bodies is called *statics.* Newton incorporated many static principles into his development of dynamics.

The effects of electrical charges and magnetism have been known since the time of the ancient Greeks, and men investigated electricity and magnetism as separate phenomena throughout the eighteenth and nineteenth centuries. Benjamin Franklin assigned the names of positive and negative charges. Charles Coulomb discovered the rule governing the forces between charges. Christian Oersted showed the relationship between electrical current and magnetism, and Michael Faraday demonstrated the opposite phenomenon, where alternating magnetic fields can produce electrical currents. In 1864, James Clerk Maxwell discovered that electricity and magnetism were two aspects of the same phenomenon, called **electromagnetism.** From this body of foundational theory came the prediction of

Mechanics is the study of both how and why motion occurs.

Charles Coulomb (1736–1806) was a French mathematician and theoretical physicist.

Michael Faraday (1791–1867) was a British physicist. See more biographical information on page 420.

Electromagnetism is the study of electricity and magnetism.

1-16 One of Watt's practical steam engines

Thermodynamics is the study of the relationship and the conversion of heat to other forms of energy.

Sadi Carnot (1796–1832) was a French mathematician and mechanical engineer. See more biographical information on page 354.

Albert Einstein (1879–1955) was a German-American theoretical physicist. See more biographical information on page 596.

Quantum mechanics is the study of matter and radiation on an atomic level.

electromagnetic waves, which were eventually discovered and found to compose visible light.

One significant development of the Industrial Revolution was the invention of the steam engine. The study of **thermodynamics** came into existence as scientists attempted to understand this new power source. James Watt invented the first practical steam engine. Sadi Carnot developed an understanding of engine efficiency and promoted the caloric theory of heat. James Prescott Joule studied the transmission of mechanical energy and heat transfer. Others developed the concept of entropy and the statistical kinetic theory of particle motion. These efforts eventually led to the establishment of the laws of thermodynamics that became the third body of fundamental physical principles.

1.19 Modern Physics

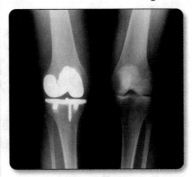

1-17 Scientific discoveries lead to practical technological applications, such as medical x-rays.

The age of modern physics dawned late in the nineteenth century at a point when some physicists believed that nothing important was left to be discovered. The overarching principles of Newtonian mechanics, electromagnetism, and thermodynamics were believed to be all there was to know about the physical world, with only some details remaining to be worked out. A series of discoveries around the turn of the century made scientists realize that the universe was much more complex than they had thought. Penetrating x-rays and radioactivity showed new aspects of both electromagnetic radiation and matter that were not previously anticipated. The discovery of the electron demonstrated that the atom could be broken down into smaller parts. The atomic nucleus and the atom's electron structure were discovered a few years later. Following these discoveries was the realization that energy comes in packets called *quanta*. The discovery of the photoelectric effect led to the idea that electromagnetic energy is quantized in photons.

In the midst of these dizzying discoveries, Albert Einstein concluded that the centuries-old, rock-firm Newtonian mechanics was only a limited case of a more general concept of motion in the universe. All of these new discoveries were collected together within a broad new area of physics called **quantum mechanics** that has provided fertile ground for research for many decades. Physicists in the latter part of the twentieth century devoted significant effort to finding a "law of everything" that relates the four fundamental forces of gravity, electromagnetism, and the strong and weak nuclear forces in a Grand Unification Theory. This goal has proved to be tantalizingly elusive.

You should not get the impression from this discussion that there is an "old" physics and a "modern" physics, or that classical physics is out of date. Scientists and engineers still use classical physics daily. The physics used to send space probes to the planets and to design airbags for cars is classical physics.

1.20 Areas of Physics

You have been introduced to the development of the science of physics, but what exactly do physicists do today? A *very* long list could be made of occupations that involve physics and physicists. Instead, we will list some of the more prominent fields in which physicists work. You will note that physics overlaps most of the main branches of science in some way.

Astrophysics—combines astronomy and physics. Astrophysicists are interested in the movements and structures of celestial bodies, from meteoroids to galaxies. They are responsible for calculating the trajectories of natural and man-made space objects.

Geophysics—combines geology and physics. Geophysicists study the interior and crustal structure and dynamics of the earth, including the oceans and atmosphere. They are interested in both the cause of geomagnetism and its effects as it extends into the near-Earth regions of space. Some geophysicists become involved in the study of the structure of the other planets and moons in our solar system.

Biophysics—combines biology and physics. Biophysicists are interested in the mechanics of living organisms, especially man. They design noninvasive detectors, such as MRI and CAT scanners, to image the interior of the body. Biophysicists are also involved in developing radiation protection for nuclear workers as well as in using radiation therapy to treat diseases.

Atmospheric physics—meteorology. Meteorologists are working to develop accurate models of the atmosphere in order to improve weather forecasting and to make long-term predictions regarding global climatology. Meteorologists are also working to understand the formation and action of destructive weather such as tornadoes and hurricanes.

Nuclear and High Energy Physics (HEP)—investigates the ultimate structure of matter. Using particle accelerators, these physicists study the nature of the nuclei of atoms and the numerous elementary particles that make up the nucleus. Nuclear physicists develop more efficient nuclear fuel arrays for power and propulsion reactors. One of the main objectives of high-energy particle physics is to develop fusion as a plentiful and economical commercial energy source. Another objective is to identify the relationships of the fundamental forces of nature in an attempt to develop a Grand Unification Theory.

Applied Physics—develops new technologies. Physicists work in numerous capacities as they develop new products and applications for physical phenomena. Many are, or work closely with, mechanical, electrical, and civil engineers. A sampling of such activities includes electro-optical communications research, developing new composite materials, building more crash-proof cars, designing faster microprocessors, improving satellite communications, and designing aircraft, space vehicles, ships, and submarines. Physicists are involved in the design of devices from pacemaker batteries to nuclear reactors and from hearing aids to space surveillance radar systems.

Students who graduate with a college degree in physics can seek employment in almost any sector of the economy and work in countries all over the world. A quick search of the Internet will reveal a huge selection of job and educational opportunities in physics and related fields. However, no matter how important finding a job may be, you should remember that the *most* important reason for learning about the physical universe is to return glory to God. The ministry of a preacher

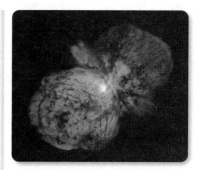

1-18 Nebulous bubbles of dust and gases surround the remnant star of Eta Carinae after it went supernova in 1843.

1-19 Geophysicists have discovered the complex nature of the magnetic field that protects the earth from the solar wind.

1-20 An MRI machine

1-21 A computer-generated model of a hurricane system

1-22 A commercial nuclear power plant

1-23 An artist's concept of the U.S. Navy's USS *Virginia*-class nuclear submarine

or missionary could be greatly enhanced by a sound knowledge of physics. Focus on the eternal purpose of your studies, and God will meet your temporal needs. "But seek ye first the kingdom of God, and his righteousness; and all these things shall be added unto you" (Matt. 6:33).

1D Section Review Questions

1. What differentiates classical physics from modern physics?
2. Name the three major theoretical areas of classical physics.
3. What is the ultimate law that physicists are striving to discover in the twenty-first century?
4. What branch of physics produces the practical basis for devices and products that are useful to the average citizen?

In Terms of Physics

phenomenon	1.2		observation	1.14
prejudice	1.2		scientific problem	1.14
objectivity	1.4		hypothesis	1.15
bias	1.4		theory	1.16
paradigm	1.4		law	1.16
deductive reasoning	1.8		classical physics	1.17
deduction	1.8		modern physics	1.17
inductive reasoning	1.9		instrument	1.17
induction	1.9		mechanics	1.18
experiment	1.9		electromagnetism	1.18
teleological	1.10		thermodynamics	1.18
scientific methodology	1.13		quantum mechanics	1.19

Review Questions

1. Discuss the characteristics of "crackpot" science.

2. Why is it difficult to tell pseudoscience from true science?

3. Does an explicit Christian faith in God disqualify a scientist's activity from being true science? Explain your answer.

4. Describe one activity or characteristic that you usually associate with scientific inquiry.

⊛5. What is the main difference between science conducted by a Christian versus that conducted by a non-Christian?

⊛6. Why did the Roman Catholic Church adopt the paganistic teachings of Aristotle as the basis for scientific doctrine during the Middle Ages?

7. What is the difference between inductive and deductive reasoning?

8. Under what conditions may a controversial scientific hypothesis legitimately contradict established scientific theories or laws?

⊛9. Is the field of science a legitimate vocation for a Christian? Explain.

⊛10. Why are scientific problems valuable?

⊛11. Discuss the scientific methodology used to illustrate Count Rumford's historic investigation of heat. Also list or describe the steps in another version of the scientific method that you find in some other science textbook, encyclopedia, or on the Internet. Explain why it is possible to have more than one version of the scientific method.

12. Suppose you are studying what happens when a rolling ball collides with an identical stationary ball. Which of the following observations should probably be made if you are interested in developing a hypothesis that will predict the motions of the balls after the collision?
   a. the weight of the balls
   b. the material of which the balls are made
   c. the color of the balls
   d. the direction of each ball after the collision
   e. the time of day that the balls collided
   f. the initial speed of the balls
   g. the amount of light in the room

13. Discuss at least three ways scientists may perform scientific observations.

True or False (14–18)

14. The development of a hypothesis is completely objective, controlled by the data at hand.

15. Scientists can prove that a theory is correct.

16. A theory may be workable without being absolutely true.

17. A scientific law may be discarded if contradicting data arise.

18. Every scientific hypothesis is affected by the religious worldview of the scientist.

19. List three limitations of science.

20. Which of these statements is within the boundaries of *experimental* science?
   a. Dinosaurs were extinct long before man appeared.
   b. Most fossils in the world today were formed during the Flood.
   c. It is possible to change some characteristics of a species by careful breeding.
   d. Nature wants to become better.
   e. The United States should develop a laser-based defense system against intercontinental ballistic missiles.

⊛⊛21. Write a brief report discussing how failure of a Christian to accept the biblical presentation of Creation in six literal days and a world-covering flood as historical facts can lead to loss of faith in other key biblical and salvation doctrines.

⊛⊛22. Discuss the various ways of defining a concept. If available, refer to BETTER THINKING AND REASONING (BJU Press, 1995), Chapters 1 and 2. Is it possible to develop a reversible definition for "science"? Explain.

⊛⊛23. List some of the ministries and careers you might be involved in fifteen years from now. How could the information and life perspectives that you glean from this course contribute to the fields on your list?

2A	The Dimensions of Physics	23
2B	Principles of Measurement	30
2C	Truth in Measurements and Calculations	35
2D	Problem Solving	40
Facet:	Just a Second	26
Facet:	Vernier Scales	34

New Symbols and Abbreviations

SI	2.2	mole (mol)	2.7
meter (m)	2.5	significant digit (SD)	2.13
second (s)	2.6	radian (rad)	2.19
kilogram (kg)	2.7		

Astronomers can measure the distance between Earth and the Moon to a few centimeters—a relative error of a billionth of a percent.

Measurement

We expect people in certain occupations (for example, engineers, doctors, pastors, teachers, and scientists) to adhere to the highest standards of truth and accuracy as they practice their professions. However, God holds each one of us responsible to be truthful in all of our dealings. It is one of Satan's lies that a desired result is so important that it justifies a little deception.

2A THE DIMENSIONS OF PHYSICS

2.1 Introduction

Four centuries before Christ, the Greeks reasoned from their presuppositions regarding the nature of the world to unravel its mysteries. Their commonsense solutions were accepted for nearly two thousand years. But when scientists during the Renaissance stopped trusting traditional "science" and began testing their observations with new instruments and improved mathematical systems, they discovered that God's universe was far more orderly—often in unexpected ways—than they had thought.

The essence of physics is to measure the observable world and describe the principles that underlie everything in creation. Physical theoretical principles are usually expressed or defined by mathematical *formulas*. As instruments and measurements improve, scientists must constantly refine their theories to fit the data. To ensure that he has the best possible data, a physicist must know what to measure, how to measure it, how to record his measurements accurately, and how to use the measurements to produce mathematically meaningful results.

2-1 Ancient measuring instruments included balance scales, knotted cords, and hourglasses.

2.2 The SI

Early in history, each nation maintained its own system of currency, weights, and measures. With the rapid growth of scientific fields during the eighteenth and nineteenth centuries, nations recognized the need for worldwide standardized units.

A *formula* is an equation that orders the relationship between given physical quantities.

France took the standardization initiative in 1799 by establishing a decimal metric system consisting of the kilogram and the meter. The metric concept gradually gained wide acceptance among scientists.

Throughout the nineteenth and twentieth centuries, researchers sought to identify new standard units of measure and to more accurately define the existing ones. In 1960 the current metric system was renamed the *Système International d'Unités,* from which we get the abbreviation **SI.** It consists of seven base units and many other units derived from the base units. As new fields of science are established, new derived units continue to be added to the SI.

2.3   Meanings of *Dimension*

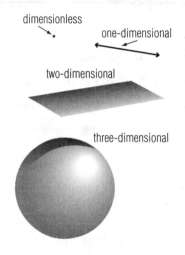

2-2 A point has no dimensions. A line exists in one dimension, a plane in two, and a sphere in three dimensions.

The word **dimension** has two meanings when used in physics. The first refers to the number of spatial coordinates required to describe an object. A point is dimensionless, a line occupies one dimension, and a plane exists in two dimensions. A volume of space or a physical object can exist only in three dimensions. The image on a TV screen is two-dimensional, but a horse (*Equus caballus*) is three-dimensional.

The second meaning of *dimension* refers to a *measurable physical quantity.* Examples include length, time, mass, resistivity, hardness, and momentum. Each dimension is described by a specific type of unit or combination of units. For example, length can be associated only with units of length (meter, foot, etc.), and density is limited to mass per unit volume (kg/m^3 or lbm/ft^3).

2.4   Fundamental Dimensions

Christians should approach the study of any science with the presupposition that God's creation will reflect the characteristics of its Creator. One basic quality of God is that He is three Persons in One—a unity of three, none of which can exist separately from the other two.

In a similar way, the physical universe consists of three dimensions—**space, time,** and **matter**—which are intricately and seamlessly bound together. One cannot exist if either of the others is missing.

2.5   Space and Length

Space is the apparently limitless three-dimensional volume within which all matter exists and all physical phenomena occur. Matter must exist and be present in order for us to perceive the size of space. If you looked into a space that contained no matter, you would see nothing. You would not know if you were looking into an atom-sized or a galaxy-sized space! The physical separation of entities in space (distance) or the size of an object (which itself occupies space) are measured in units of length.

The metric unit of length, the **meter (m),** was originally defined as one-ten-millionth (10^{-7}) of the distance between the earth's north geographic pole and the equator, as measured along the meridian (line of longitude) passing through Paris. Because such a distance could not be reproduced in a laboratory, much less

2-3 The vast universe contains billions of galaxies such as Galaxy NGC 1512.

measured accurately, scientists fabricated a corrosion-proof platinum-iridium bar bearing two scribe marks one meter apart. This *standard meter* was retained in the International Bureau of Weights and Measures at Sèvres, France, under stringent environmental conditions that prevented expansion and contraction due to temperature changes.

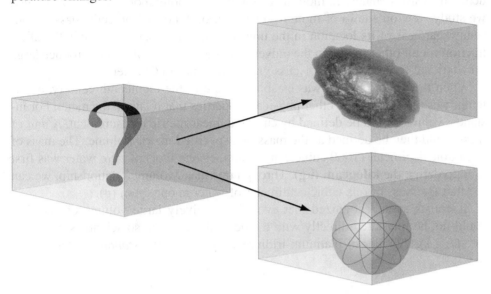

2-4 How large is a given space? The presence of matter fixes the size of space.

Even well-protected physical objects can be affected by environmental variables, so throughout the twentieth century scientists worked to abandon all references to dimensional standards based on man-made artifacts. At the same time, such objects became less useful as increased accuracy and precision were required. Scientists preferred instead to use extremely accurate and stable natural phenomena that could be measured in any properly equipped laboratory. In 1983, the meter was given its present definition in the SI as the distance traversed by light in a vacuum in *exactly* $\frac{1}{299\,792\,458}$ second.

2.6 Time

Have you ever tried to define the concept of time? Think about it. Everyone *knows* what time is, but try to define it! One attempt at a definition could be, *"Time is a nonphysical continuum that orders the sequence of events and phenomena."* (A *continuum* is any expansive entity without gaps or breaks.) A moment of time originates in the indeterminable future, is experienced in the instantaneous present, and immediately moves into the past where its significance can be understood only by one's reflecting upon past events. Notice the seamless, three-in-one character of time—again revealing the triune nature of its Creator. Time is best defined by specifying the method of measurement that clearly differentiates it from all other physical phenomena.

The SI unit of time is the **second (s).** The second was originally defined as $\frac{1}{86\,400}$ of a mean solar day (24 h + 60 min/h + 60 s/min = 86 400 s). Unfortunately, the rotational period of the earth is not stable enough to define the second to the precision required for modern scientific purposes. In the mid-1950s, scientists invented very accurate timekeeping devices that used the resonant vibrations of a pure element to provide a time standard. A second is now defined as exactly 9 192 631 770 cycles of oscillating gaseous cesium-133 atoms. "Atomic clocks" are accurate to within three microseconds (μs) per year.

Platinum-iridium alloy is a precious metal that is extremely corrosion resistant.

The moral standards of the world are constantly changing. Only the Bible provides us with an unchanging moral standard, because it is based on the unchanging holiness of God.

The SI unit of length is the **meter (m).** A meter is the distance traversed by light in a specified time interval.

2-5 A platinum-iridium alloy meter standard

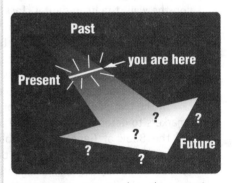

2-6 Our position in time is always the present. Our progress in time is always forward. Only in our memory can we experience the past.

The SI unit of time is the **second (s).** A second is the time interval of a specified number of vibrations of gaseous cesium-133 atoms resonating within special instruments.

Measurement **25**

2.7 Matter and Mass

As stated earlier, space would be dimensionless without the presence of matter. Matter is often described as "that which occupies space and has mass." We know what space is, but what about mass? The *mass* of an object is a measure of its tendency to resist a change in motion. You will learn more about this attribute when we study Newton's laws of motion in Chapter 5. A normal object's mass is constant regardless of its location in the universe, and that mass has gravitational attraction for all other matter in the universe. Large amounts of matter produce large gravitational forces. We will discuss gravity in detail in Chapter 7.

When the metric system was established late in the eighteenth century, an attempt was made to define other physical properties that relate to the meter. For instance, volume could be defined by cubing a linear metric measurement. A unit of mass could then be defined as the mass of a specified metric volume. The mass of one cubic decimeter (1000 cubic centimeters, or one liter) of pure water was first used to define the **kilogram (kg).** Through this mass-volume relationship, we can see that the mass of one cubic centimeter of water is one gram (g).

Scientists soon recognized that even the relatively large volume of one liter could not be measured directly with the desired precision, so scientists manufactured a 1 kg cylinder of platinum-iridium alloy (called the *standard kilogram*) to

2-7 A modern cesium-beam time standard maintained by the National Institute of Science and Technology

FACETS of PHYSICS

Just a Second

Stop for a moment (a second?) and look around you. How many displays of time do you see? Your watch, the clock on the wall, the little digital clock down in the corner of your computer desktop, your PDA. More than at any other time in history, we order our lives today by the clock. Busy personal schedules, manufacturing assembly lines, power generation, international financial transactions, GPS navigation, and communications all require knowing the correct time.

In the dim beginnings of history, man resorted to celestial markers to determine the passage of time. In the first chapter of the Book of Genesis, the Bible records that a day was defined as one complete cycle of the Sun across the sky. Days added to weeks, months, and years. Determining an accurate annual calendar was a long and difficult process that culminated in the Gregorian calendar, which was established in 1582 but not fully adopted by some countries, for political reasons, until the twentieth century.

Accurately measuring smaller increments of time proved to be even more challenging. Early scientists, and particularly astronomers, recognized the need to define smaller standard time units—first the hour, then the minute and second. They started with various devices such as calibrated candles, water clocks, and hour glasses. Eventually large, delicate, stationary clock-like instruments located in observatories were constructed. Public time was provided by ponderous clock tower mechanisms, many that were correct only twice a day. Some clocks had only the hour hand. There was little pressing need for accurate portable timepieces until the latter part of the sixteenth century. The necessity for accurately knowing the time came from what may seem like an unexpected quarter—ship navigation.

Far-flung merchant enterprises during the Age of Discovery required accurate ocean navigation to ensure the safe arrival of passengers and cargo in the shortest possible time. Knowing the ship's latitude (its position north and south on the globe) was easily determined from observing the Sun's angle above the horizon. Determining a vessel's longitude (its position east or west) was dependent on knowing the time at a reference

replace the water standard. Its dimensions and the alloy mix of 90% platinum and 10% iridium are reproducible to a much higher level of precision than was possible with the water standard. The standard kilogram is retained at the International Bureau of Weights and Measures in Sèvres, France. It is the *only* SI standard that is still based on a manufactured object. Even this may eventually change. Scientists are currently investigating ways to define the kilogram in terms of the **mole (mol),** which is the SI unit for Avogadro's number of particles (N_A).

The seven fundamental units of the SI are listed in Table 2-1. Note that there is a difference between the name of a physical dimension and its unit of measure. There is also a difference between the unit symbol and the mathematical variable symbol used to represent the dimension in physics formulas. The meter, second, and kilogram (and combinations of these units) will be the only units of concern

2-8 Even an asteroid can produce enough gravity to have its own moon. Here, the 1.5 km diameter rock called Dactyl orbits the asteroid 243 Ida.

The standard **kilogram (kg)** is the mass of a cylinder manufactured from a precious alloy. It is the only SI standard based on a manufactured object.

Lorenzo Romano Amedeo Carlo Avogadro (1776–1856) was an Italian lawyer, scientist, and mathematician.

$$1 \text{ mole (mol)} = N_A = 6.022 \times 10^{23} \text{ particles}$$

point, such as the port of departure, and subtracting that time from local time obtained from the position of the Sun. The poor accuracy of shipboard clocks made longitude determination very inaccurate. The result was position errors of hundreds of miles, and many ships were lost due to navigational errors.

After decades of failures, the first successful marine *chronometer* was built by an English carpenter and self-taught clockmaker, John Harrison, in 1736. With further refinements, the problem of finding longitude at sea was solved, and the manufacturing of precision timepieces and small portable watches became an established technology.

All clocks contain a component that subdivides time into regular intervals. Early clocks relied upon weight-driven pendulums which were subject to inaccuracies due to friction and pendulum errors. Later clocks and chronometers used spring-driven rotational escapements and free pendulums.

The advent of electronics early in the twentieth century was the motivation for the discovery of a more accurate and precise timekeeper. A clock containing a quartz crystal oscillator was invented at the Bell Laboratories in 1927.

Throughout this developmental period, however, the second continued to be defined by scientists as $\frac{1}{86\,400}$ of a mean solar day. Although the invention of the quartz clock proved that the earth's rotation was irregular, quartz crystals could not be manufactured with sufficient uniformity to become a replacement time reference.

The development of the atomic clock in the 1950s provided the stable and reliable time standard scientists required. Since 1967, the second has been defined as the time for gaseous cesium-133 atoms to resonate exactly 9 192 631 770 times

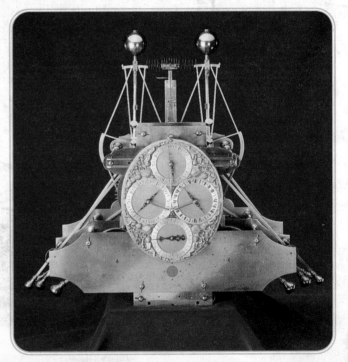

The first marine chronometer

inside a specially designed instrument. These extremely accurate and precise atomic clocks, which are installed in laboratories across the globe, provide the reference time signals for the modern technologies of today.

Excuse me, do you have the time?

2-9 One of the duplicate kilogram standards maintained at national bureaus of weights and measures around the world

for the chapters dealing with motion and mechanics in Unit 2. The other SI units will be introduced later as needed.

TABLE 2-1			
Dimension	Fundamental SI Unit	Unit Symbol	Variable Symbol
length	meter	m	d, r, s, l
time	second	s	t
mass	kilogram	kg	m
thermodynamic temperature	kelvin	K	T
amount of substance	mole	mol	n
electric current	ampere	A	I
luminous intensity	candela	cd	I_L

Units indicate kinds of dimensions in data, and *variable symbols* represent dimensional quantities in formulas.

SI derived units are obtained from combinations of SI base units or other approved or derived units.

TABLE 2-2		
Prefix	Symbol	Factor
exa-	E	$\times 10^{18}$
peta-	P	$\times 10^{15}$
tera-	T	$\times 10^{12}$
giga-	G	$\times 10^{9}$
mega-	M	$\times 10^{6}$
kilo-	k	$\times 10^{3}$
hecto-	h	$\times 10^{2}$
deka-	da	$\times 10^{1}$
(base)	—	$\times 10^{0}$
deci-	d	$\times 10^{-1}$
centi-	c	$\times 10^{-2}$
milli-	m	$\times 10^{-3}$
micro-	μ	$\times 10^{-6}$
nano-	n	$\times 10^{-9}$
pico-	p	$\times 10^{-12}$
femto-	f	$\times 10^{-15}$
atto-	a	$\times 10^{-18}$

The SI unit prefixes are derived from Greek, Latin, and Norwegian roots descriptive of the corresponding powers of ten.

2.8 SI Derived Units

Scientists deal with many more dimensional quantities than those listed in Table 2-1. For example, there is no fundamental unit for area. However, area can be represented by the square of length, l^2, which has the SI unit m². Similarly, speed is represented by the dimension of distance per unit time (d/t) with units of m/s. Such dimensions and their associated units are derived from the SI fundamental units and are called **SI derived units.** Some derived units have been given special names and symbols. Examples include the SI unit for force, the *newton* (N) (1 N = 1 kg·m/s²), and the SI unit for electrical charge, the *coulomb* (C) (1 C = 1 A·s).

Some SI derived units are obtained from other derived units. For example, the SI unit for work and energy, the *joule* (J), is defined as

$$1 \text{ J} = 1 \text{ N·m}.$$

2.9 Unit Conversions and Analysis

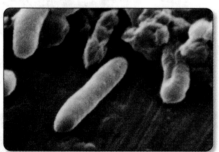

2-10 The size of a bacterium is measured in micrometers (μm).

Is the size of a virus more appropriately reported as 0.000 000 5 m or 500 nm? The basic SI units and their derived units often are not conveniently sized for use in every conceivable measurement. The founders of the metric system recognized that a method was needed to provide larger and smaller increments of the base units that would be convenient for any measurements that might be required.

Prefixes were developed that, when combined with a given *base unit,* indicate the size of the new unit. Table 2-2 lists the named metric system prefixes, the prefix symbols, and the factor by which the base unit is multiplied. Note that, for consistency, the gram is the *base* unit of mass. However, the SI *fundamental* mass unit is the kilogram. The prefixes that we will use most frequently are *kilo-, centi-,* and *milli-.*

It is often necessary in science to perform calculations with similar dimensions that have different-sized units or non-SI units. All similar dimensions should be converted to the same units before you calculate. You accomplish *unit conversion* by multiplying the quantity by a **conversion factor,** a fraction that contains both the new and the original units. In order to preserve the size of the original measurement, the conversion factor must be *equivalent to one*—its numerator and denominator are equal-sized quantities. To find the correct ratio, we start with an equation relating the two units.

For example,

$$1 \text{ m} = 100 \text{ cm}.$$

The factor required to convert *from* meters *to* centimeters is found by dividing both sides by 1 m.

$$\frac{\cancel{1 \text{ m}}}{\cancel{1 \text{ m}}} = \frac{100 \text{ cm}}{1 \text{ m}}$$

$$1 = \frac{100 \text{ cm}}{1 \text{ m}}$$

Note that the unit we are converting to is on top. If we needed to convert from centimeters to meters, the conversion factor would be $^{1\,m}/_{100\,cm}$. Both ratios are equal to 1 because the numerator and denominator are equivalent.

EXAMPLE 2-1

Finding a Conversion Factor

Convert 22.4 L to mL.

Solution:

$$1 \text{ L} = 1000 \text{ mL} \Rightarrow \frac{1 \cancel{\text{L}}}{1 \cancel{\text{L}}} = \frac{1000 \text{ mL}}{1 \text{ L}} \begin{array}{l} \leftarrow \text{required units} \\ \leftarrow \text{given units} \end{array}$$

$$22.4 \text{ L} = \frac{22.4 \cancel{\text{L}}}{1} \times \frac{1000 \text{ mL}}{1 \cancel{\text{L}}} = 22\,400 \text{ mL}$$

When several simultaneous unit conversions are required, such as when converting mi/h to m/s, we may multiply the appropriate conversion factors together in any order, taking advantage of the commutative property of multiplication. We use a mathematical aid called the *unit analysis bridge* to help with conversion factor unit cancellation and to organize the solutions to physics problems.

Look at Figure 2-11. The horizontal line corresponds to the fraction bar (representing the *division operation*) in any fractional numbers involved in a calculation. The vertical lines correspond to *multiplication signs* between factors in the calculation. Identical units that appear on opposite sides of the horizontal line may be canceled as usual. To see how this process works, study the following example.

After cancellation, the result is obtained by multiplying and dividing the remaining numbers and units. The key advantage of the bridge method is the convenient arrangement of factors, which eases the cancellation of like units. When solving physics problems using the bridge utility, we can often check the setup of the solution by analyzing the units remaining after cancellation to verify that the resulting dimensional units are the ones required for the solution. This process is called *dimensional analysis,* or more commonly, **unit analysis.** Valid physics equations must have the same dimensions on both sides of the equal sign. Dimensional inequality is a guarantee that the solution is incorrect. Of course, the equation can have equivalent dimensions and still have an incorrect arithmetic solution. Dimensional analysis is only a check on the setup of the solution.

Problem-Solving Strategy 2.1

In calculations, the prefix of an SI dimensional unit may be replaced by the exponential multiplier of the base unit.

The product of any number and 1 is the number.

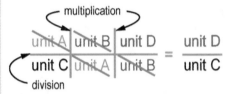

2-11 The unit analysis bridge

Problem-Solving Strategy 2.2

In a conversion factor, the unit you are converting *to* is in the same position relative to the fraction bar as the unit you are converting *from* in the given quantity.

Problem-Solving Strategy 2.3

The first factor in the unit analysis bridge should be the given data. The remaining data and conversion factors may be inserted in any order to cancel the given units and substitute the required units in their place.

$$volume = volume$$

$$V_{sphere} \text{ (in m}^3) = \tfrac{4}{3}\pi r^3 \text{ (in m}^3)$$

2-12 A physics equation must have the same dimensions on both sides of the equation.

EXAMPLE 2-2

Using the Bridge: Unit Analysis

Convert 120. km/h to meters per second.

Solution:

Standard notation:

$$120. \; \frac{km}{h} \times \frac{1000 \text{ m}}{km} \times \frac{1 \text{ h}}{60 \text{ min}} \times \frac{1 \text{ min}}{60 \text{ s}} = 0.033\overline{3} \; \frac{m}{s}$$

Bridge utility:

$$\boxed{\begin{array}{c|c|c|c} 120. \text{ km} & 1000\,\text{m} & 1 \text{ h} & 1 \text{ min} \\ \hline h & km & 60 \text{ min} & 60\,\text{s} \end{array}} = \boxed{0.033\overline{3}\,\frac{\text{m}}{\text{s}}}$$

given data   required units

2A Section Review Questions

1. Why did scientists seek to develop a standard metric system of measurement in the eighteenth and nineteenth centuries?

2. What are three dimensional quantities required to perceive the universe?

3. What is the current definition of the meter?

4. List the seven fundamental dimensional units of the SI, their unit symbols, and their variable symbols. Explain the difference between the unit symbol and the corresponding variable symbol.

5. What is an SI derived unit? Give an example.

⊛6. Discuss three disadvantages of using a manufactured object as the standard meter bar.

⊛7. Write a definition of time that includes a way of measuring it.

⊛8. Discuss the two meanings of the term "dimension" in physics.

⊛9. Convert the following quantities to the indicated units using the unit analysis bridge utility.

    a. 6.5 ft to m   **c.** 325 m/s to km/h

    b. 3.25 km to m   **d.** 1040 kg/m^3 to g/cm^3

⊛10. Convert the speed of light to astronomical units (ua) per year.

2B PRINCIPLES OF MEASUREMENT

2.10 Instruments

God designed the human senses to provide all essential information for daily life. Early in history, many linear units of measure were based on the dimensions of the human body (e.g., "pace," "cubit," "span," and "hand"). As society and technology advanced, more accurate methods for measurement were required. Using their God-endowed intellectual capabilities, men invented the necessary tools to extend

2-13 It is likely that Noah had to use a variety of measuring instruments for the construction of the ark.

their senses and refine their perception. These tools, collectively called *instruments,* permitted more exacting measurements.

Instruments are invaluable to modern scientific research because they permit the accurate and precise measurement of quantities that unaided human perception can only estimate or that are completely outside the range of human senses. For example, lengths and distances can be measured with rulers as well as with lasers, radar, and sonar. Astronomical distances can be measured from stellar red shifts obtained with telescopic spectrometers, and atomic distances can be measured using instruments such as tunneling

2-14 Lasers are finding practical uses as range finders and level references at building construction sites.

electron microscopes and x-ray diffractometers. Besides measuring length, time, and mass, instruments have been developed to measure force, speed, temperature, light intensity, magnetism, electric fields, frequency, and a host of other physical properties.

All modern instruments have at least one limitation—they are all man-made. No matter how accurately an instrument can measure and how skillful the user, the tiny variations in dimensions and quality during its construction and the limitations of human perception guarantee that there will always be some error in the measurement.

2.11 Accuracy

How can you be sure when you measure the length of a pencil with a ruler that the observed length is the actual length? The question is answered by a qualitative term called **accuracy.** Whenever we make a measurement, we compare the object being measured to the graduated scale of an instrument. The instrument is accurate if it produces measurements very close to the actual value.

The accuracy of an instrument depends on the quality of its original design and construction *and* how well it is maintained. Most scientific instruments can be adjusted so that the error of their measurements is kept within an acceptable limit. Periodic *calibration checks* are performed that compare an instrument's output to a highly accurate *prime standard* to verify that it is accurate at one or more points on its scale. Zeroing a laboratory balance is a simple instrument calibration.

Obviously, the accuracy of a measurement also reflects the skill of the operator. The contribution of human error to a measurement can be reduced by averaging the results of several measurements, taken either by the same or different persons.

To make a *quantitative* assessment of the accuracy of measured data, we must compare the measurement to the *accepted value* or *actual value* and determine the **error.** Error is expressed as the simple difference of the observed and accepted values. The sign of the result indicates whether the observed value is larger (+) or smaller (−) than the accepted value. The *magnitude* of the error, or *absolute error,* is the absolute value of the difference.

2-15 A seismograph recording an earthquake

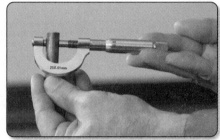

2-16 A micrometer caliper permits high-precision measurements of small lengths.

Small permissible variations in the accuracy of manufactured parts are called *manufacturing tolerances.*

2-17 A student performs a zero calibration of a mechanical laboratory balance.

Accuracy is a *qualitative* evaluation of how close a measurement is to the actual value.

A *calibration check* of an instrument compares an indicated measurement with the corresponding accepted value.

The *actual value* is a theoretical concept indicating the result of measuring with a perfect instrument. An *accepted value* is the agreed-upon value established for a particular physical quantity by an authoritative source.

Error = observed − accepted
Absolute error =
|observed − accepted|

Knowing the error of measured data is not always as useful as knowing how large the error is compared to the accepted value. In this case, we want to find the relative or **percent error**, which gives the measurement error as a percentage of the accepted value.

$$\text{percent error} = \frac{\text{observed value} - \text{accepted value}}{\text{accepted value}} \times 100\% \qquad (2.1)$$

You can also find the *absolute percent error* by using the absolute error in the numerator of Equation 2.1.

EXAMPLE 2-3

To Measure is to Err: Percent Error

A student measured the density of a copper cylinder to be 8.83 g/cm³. The accepted value for the density of copper metal is 8.94 g/cm³. (a) What is the error of the student's measurement? (b) What is the percent error?

Solution:

a. Error = measured − accepted = 8.83 g/cm³ − 8.94 g/cm³
   Error = −0.11 g/cm³

b. $\text{percent error} = \dfrac{\text{measured} - \text{accepted}}{\text{accepted}} \times 100\%$

$\text{percent error} = \dfrac{8.83 \text{ g/cm}^3 - 8.94 \text{ g/cm}^3}{8.94 \text{ g/cm}^3} \times 100\%$

$\text{percent error} = -1.23\%$

The negative error and percent error indicate that the measured density was less than the accepted density.

2.12 Precision

Precision is a *qualitative* evaluation of how exactly a measurement can be made.

Modern handheld calculators can display calculation results to 11 or more digits. Is it appropriate to report your homework problem results to this many figures? The answer is obviously no, but why not? The reason rests on the **precision** of the measured data given in the problem.

Precision is a *qualitative* term that describes the exactness of a number or measured data. Some physical quantities are known exactly because of the way in which they are defined. The speed of light and the length of a meter are two such quantities. Other numbers can be counted. When we count the number of people in the classroom, we know the number *exactly*. There are other mathematical quantities that are not exact, but we may express them to any degree of exactness we desire by specifying the number of decimal places to include. These are called irrational numbers. Their values are exact only when we write them using special symbols reserved for irrational numbers, such as the square root bracket ($\sqrt{}$), the Greek letter π (ratio of a circle's circumference to its diameter, *C/d*), and the letter *e* (base of natural logarithms). When an irrational number is written as a decimal, the number of decimal places indicates the precision of the irrational number. *Irrational numbers have potentially unlimited precision.*

Measured data does not have unlimited precision. When we make a measurement with a mechanical instrument, we read the major and minor scale division marks of the instrument. The digits in the measurement based on the scale marks

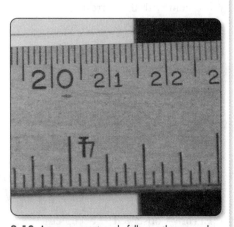

2-18 A measurement rarely falls exactly on a scale marking. The position of the measurement on the scale must be estimated.

are known for certain. But measurements almost never fall exactly on an instrument scale mark, so we must estimate the fraction of the distance between the marks. **When you use a mechanical *metric* instrument (one with scale subdivisions based on tenths), measurements should be estimated to the nearest $\frac{1}{10}$ of the smallest *decimal* increment.** For most measurements made in a high school laboratory, this estimated digit is the last (right-most) digit in the measurement. With finer graduations, more certain digits are possible, and the measured data potentially has greater precision (see Figure 2-19). The maximum precision of an instrument is limited by its finest scale subdivisions.

The last digit that has any significance in a measurement is *estimated*. (See the discussion of significant digits in the next section.) The human act of measurement introduces some uncertainty into every measurement. We will assume for this course that the **uncertainty** of any measurement is plus-or-minus half of the place value of the estimated digit. For a meter stick whose smallest subdivision is the millimeter (0.001 m), the uncertainty is ± 0.5 mm or ± 0.0005 m. The uncertainty of most measurements you will make will be much smaller than this. In actual research, uncertainty is determined by a statistical process.

The **relative uncertainty** compares the size of the uncertainty to the size of the measurement. For two measurements taken with the same instrument, a smaller measurement has a larger relative uncertainty than a larger one because the uncertainty is the same for both. The larger the relative uncertainty, the greater the error in calculated results. One way to alleviate this problem with small measurements is to use an instrument equipped with a **vernier scale,** which refines the precision of the installed instrument scale.

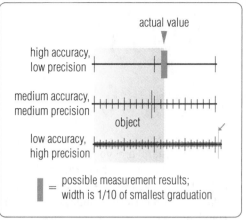

2-19 The finer the scale markings, the more precise the measurement. However, accuracy depends on instrument calibration as well as on the fineness of scale graduation.

A very precise measurement may still be inaccurate if it is far from the actual or accepted value.

Uncertainty in measurements is related to the precision of the instrument.

Relative uncertainty of a measurement increases as the size of the measurement decreases for a given instrument.

A **vernier scale** is an auxiliary scale on an instrument that increases its precision by one or more decimal places.

2B Section Review Questions

1. Discuss why all measured data contains some error.

2. Describe one method for quantifying accuracy.

3. To what decimal place will you report measurements taken with the following instruments?

   a. a graduated cylinder with 1 mL subdivisions

   b. a meter stick with 1 mm subdivisions

   c. a laboratory balance with 0.1 g subdivisions

   d. a thermometer with 2 °C subdivisions

4. Which volume is reported with greater precision, 87.0 mL or 87.04 mL? Why?

⊛5. Discuss some possible reasons that God did not create man with extended senses such as ultrasonic hearing or infrared vision.

⊛6. You and your lab partner determined the density of a sample of pure mercury to be 13.52 g/cm^3. The accepted value for the density of mercury is 13.60 g/cm^3. Determine (a) the error, (b) the absolute error, and (c) the percent error of your result.

⊛7. You obtain two measurements with a meter stick marked in millimeters: 35.94 cm and 2.47 cm.

   a. Which measurement will have the highest relative uncertainty?

   b. Calculate both relative uncertainties.

Fine-Tuning Measurements—
Vernier Scales

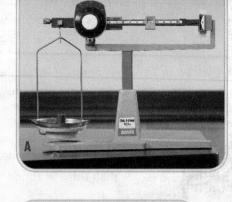

A

ow sharp are your eyes? Could you measure the size of a grain of sand (dimensions on the order of 0.1 mm) if you had a standard meter stick? Probably, if you have average eyesight. Now, do you think you could measure the size of a large biological cell (around 0.001 mm) using the same instrument? You probably could not even see the cell. However, machinists who build precision mechanical equipment must often make measurements that are precise to hundredths or even thousandths of a millimeter.

Scientists also need to estimate measurements to much finer tolerances than is possible using simple visual estimation. Instruments that measure to such extreme precision use a specially calibrated auxiliary scale called a **vernier scale.** The term *vernier* (vur′nē-êr) is used in honor of the seventeenth-century French engineer, architect, and surveyor Pierre Vernier, who developed an instrument to more accurately measure angles for land surveys. Modern vernier scales are based on his invention. They are commonly used in micrometers, calipers, analytical balances, fine focusers for optical instruments, electrical potentiometers, and other precision measuring devices.

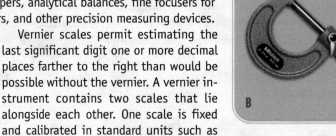

B

Vernier scales permit estimating the last significant digit one or more decimal places farther to the right than would be possible without the vernier. A vernier instrument contains two scales that lie alongside each other. One scale is fixed and calibrated in standard units such as millimeters or hundredths of an inch. We will call this the *standard scale.* The other scale that actually indicates the measurement, called the *vernier,* slides alongside the standard scale. It is subdivided in a similar fashion as the standard scale with one difference—10 subdivisions on the vernier scale exactly spans one less than an integer multiple of 10 subdivisions on the standard scale. For example, 10 vernier subdivisions could span either 9 standard subdivisions or 49 standard subdivisions.

The number of vernier subdivisions determines the precision to which the instrument can be read. If there are 10 vernier subdivisions, then the instrument can produce readings to $\frac{1}{10}$ (0.1) of the standard scale's smallest subdivision. If there are 50 vernier subdivisions, then the instrument can be read to $\frac{1}{50}$ (0.02) unit. Notice that with a vernier scale, it is possible to accurately estimate the last significant digit to a greater precision than just $\frac{1}{10}$ of the smallest standard metric scale calibration.

Figures C and D show how the vernier scale functions. The vernier scale scribe mark along the scale that lines up exactly with a standard scale subdivision mark is the vernier decimal reading. This decimal value is added to the value of the *smaller standard scale* subdivision near the index line (zero mark) of the vernier scale.

Vernier scales are just one of the tools scientists use to improve the precision with which they observe and measure the physical world. However, you should remember that any instrument is subject to errors built into it as well as to damage and wear. Man's observations in this universe will always be limited by the finite precision and accuracy of his instruments.

C The reading is 10.1 mm.

D The reading is 10.7 mm.

2C TRUTH IN MEASUREMENTS AND CALCULATIONS

2.13 Introducing Significant Digits

A scientist's integrity and professional reputation depend on the accuracy and lack of ambiguity in his observations and reports. If a scientist's observations and data cannot be reproduced by others using the same instruments and techniques, he loses the regard of his peers (along with the potential for future research funding). Scientists are careful to report their data only to the precision warranted by the instruments or methods used.

Precision is indicated by the kind and placement of digits reported in the data. These digits are called the significant figures, or **significant digits (SDs).** Only the digits that are certain (based on the instrument scale calibrations) and one estimated digit (in tenths of the smallest subdivision) may be reported as significant digits. Consequently, the last (right-most) significant digit is the estimated digit when recording measured data. Zeros are added as placeholders before or after the significant digits, as appropriate, to locate the decimal point.

What is the temperature indicated by the thermometer in the photo? You should report 98.1 °C.

2.14 Rules for Determining Significant Digits in Data

How can we tell which digits are significant when we read someone else's data? This is important to know, especially if a mathematical operation is to be performed using the data. Rules have been developed which make this process fairly easy. Each rule below is followed by one or more examples.

Rule 1—SDs apply only to measured data.

The following numbers are not *measured* data. They are considered to be *exact,* and therefore SDs are not applicable:

- Counted or pure numbers

$$2 \quad \text{one dozen} \quad \pi \quad \sqrt{2}$$

- Ratios that are *exactly* 1 by definition

$$60 \text{ s/1 min}$$
$$100 \text{ cm/1 m}$$

Rule 2—All nonzero digits in measured data are significant.

324.4 m	(4 SDs)
62 °C	(2 SDs)

Rule 3—All zeros between nonzero digits in measured data are significant.

100.3 mm	(4 SDs)
7.025 μm	(4 SDs)

Rule 4—For measured data *containing a decimal point:*

- **All zeros to the right of the last nonzero digit (trailing zeros) *are* significant.**

30.0 s	(3 SDs)
6.500 L	(4 SDs)

To be valid, a scientific experiment must be repeatable. This means that it yields the same results, within an acceptable range, each time it is repeated. Repeatability implies reliability. As a general rule, simple measuring procedures are more repeatable than complex ones.

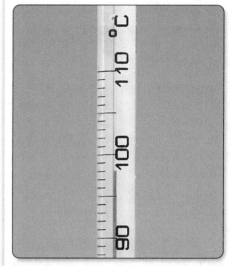

2-20 Read thermometers to ¹⁄₁₀ of the smallest decimal graduation.

Significant digits (SDs) in a measurement consist of all certain digits and the one estimated digit.

Problem-Solving Strategy 2.5

SD rules regarding zeros vary depending on whether the decimal point is written or is assumed due to placeholding zeros.

The position of the estimated digit in measured data containing trailing zeros is uncertain unless it has a decimal point.

Measurement **35**

- **All zeros to the left of the first nonzero digit (leading zeros) *are not* significant.** They are merely placeholders for the decimal point.

$$0.040 \text{ m} \qquad \text{(Only last two digits are SDs.)}$$

Rule 5—For measured data *lacking* a decimal point:

- **All trailing zeros *are not* significant.**

$$1500 \text{ m} \qquad \text{(2 SDs)}$$

We cannot tell if this measurement was estimated to the nearest 100 m, 10 m, or 1 m. Therefore, we can be assured only that at least the two nonzero digits are SDs. (SD Rule 2)

- **Scientific notation (see Appendix G) shows *only* significant digits in the decimal part of the expression.**

$$1.50 \times 10^3 \text{ m} \qquad \text{(3 SDs)}$$

This notation indicates that the measurement was estimated to the nearest 10 m.

- **In this textbook, a decimal point following the last zero indicates that the zero in the ones place *is* significant.**

$$1500. \text{ m} \qquad \text{(4 SDs)}$$

This measurement was estimated to the nearest 1 m, so all four digits are SDs.

Mathematical Operations with Measured Data

2.15 Calculator Precision

Why do we need to know about significant digits? Let's calculate the density of a stone with a mass of 5.85 g and a volume of 4.3 cm³.

$$\text{density} = \frac{\text{mass}}{\text{volume}} = \frac{5.85 \text{ g}}{4.3 \text{ cm}^3} \doteq 1.360\,465\,116 \text{ g/cm}^3$$

The result as displayed indicates that the density calculation is precise to at least a billionth (10^{-9}) g/cm³. However, the volume was known only to the nearest tenth (10^{-1}) cm³ and the mass only to the nearest hundredth (10^{-2}) g. If we compare the relative uncertainties of the mass, the volume, and the calculated density, we find that the density displayed above has a far smaller relative uncertainty than the measured data. A scientifically valid calculation should have a relative uncertainty that is about the same as the uncertainties of the contributing factors.

Scientists often perform calculations with measured data to derive other data, so they have established some additional rules in order to make sure that their results do not have more (or less) precision than the original data. The rules for mathematical operations that follow are convenient for high school work, but they do not give the scientifically acceptable precision in every case. Scientists doing research must determine the precision for each calculation based on approved statistical methods for a given kind of calculation and the kind of measured data used.

2.16 Math Rule 1—Adding and Subtracting Measured Data

A. Measured data must be the same kind of data and have the same units before they can be added or subtracted. (You cannot add apples and oranges!)

2-21 This calculator can display 11 digits in its answer.

In this textbook, the symbol "$\doteq$" means that a truncated or rounded calculator result follows.

What if a town had a sign on its outskirts that read as follows:

Population	3,501
Altitude	2,955
Tumbleweeds	4,687
Total	11,143

Would there be any practical use for this total?

B. The sum or difference of measured data cannot have greater precision than the *least* precise measurement. Round results to the place value of the estimated digit in the least precise quantity.

EXAMPLE 2-4

Adding SDs: Length

Add 1.54 m and 60 cm.

Solution:
First convert to common units.

$$\frac{1.54 \ \text{m}}{} \left| \frac{100 \ \text{cm}}{1 \ \text{m}} \right. = 154 \ \text{cm}$$

Add the lengths in a column.

$$
\begin{array}{r}
15\underline{4} \ \text{cm} \\
+ \ 6\underline{0} \ \text{cm} \\
\hline
21\underline{4} \ \text{cm} \approx 2\underline{1}0 \ \text{cm}
\end{array}
$$

Estimated digits are underlined. The sum must be rounded to the estimated place of the *least* precise number—to the nearest 10 cm. Review rounding rules in Appendix G.

> **Problem-Solving Strategy 2.7**
>
> When adding or subtracting numbers in a column, draw a vertical line to the right of the estimated digit in the *least* precise data. In the answer, round to the digit's place to the *left* of this line.

> In this textbook, the symbol "$\approx$" means that a calculated result has been rounded to the appropriate number of significant digits.

2.17 Math Rule 2—Multiplying and Dividing Measured Data

A. The product or quotient of measured data cannot have more SDs than the measurement with the fewest SDs. Round results as necessary. (The density that we calculated on the opposite page should be rounded to 2 SDs, or $\approx 1.4 \ \text{g/cm}^3$, since the volume, $4.3 \ \text{cm}^3$, has only 2 SDs.)

EXAMPLE 2-5

Multiplication of SDs: Area

Find the area of a rectangular solar cell array that is 10.3 m long and 3.7 m wide.

Solution:

$$A = l \times w = lw = (10.3 \ \text{m})(3.7 \ \text{m})$$

$$A = 3\underline{8}.11 \ \text{m}^2$$

Since 3.7 m has only 2 SDs, only 2 SDs are allowed in the result.

$$A = 38 \ \text{m}^2$$

Study the explanation provided in the margin for this rule.

$$
\begin{array}{r}
10.\textcircled{3}\,\text{m} \\
\times \ 3.\textcircled{7}\,\text{m} \\
\hline
\textcircled{7\ 2\ 1} \\
3\ 0\,\textcircled{9} \\
\hline
3\,\textcircled{8}.1\,\textcircled{1}\,\text{m}^2 \approx 38 \ \text{m}^2
\end{array}
$$

Estimated digits or the results of operations with estimated digits are circled. As with the original measured data, the final answer can have only one estimated digit, so it is rounded as shown, which gives 2 SDs.

B. The product or quotient of measured data and a pure number should not have more or less precision than the original measurement.

Examples: $7 \times 1.75 \ \text{cm} = 12.25 \ \text{cm}$ (*not* 12.3 cm)

  $1.35 \ \text{cm} \div 7 \approx 0.19 \ \text{cm}$ (*not* 0.193 cm)

2.18 Math Rule 3—Compound Calculations

You should know how to efficiently use a modern scientific calculator. Multiple math operations can be completed simultaneously using nested parentheses and other functions. *Compound calculations* involve two or more simultaneous math operations.

A. If the operations in a compound calculation are all of the same kind (either addition/subtraction **or** multiplication/division), complete the operations simultaneously using standard order of operations before rounding to the correct significant digits (Math Rule 1). This practice eliminates excessive rounding errors (see Example 2-6).

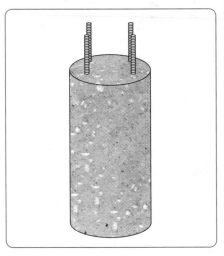

2-22 Concrete street lamp base

Note that the last two factors in the "bridge" convert the given units of volume and density to the required units of mass.

EXAMPLE 2-6

Compound Calculations: Base-ic Math

The base for a street lamp is a concrete cylinder 0.85 m in diameter and 1.5 m tall (see Figure 2-22). Calculate the mass of the lamp base in kilograms if the density of concrete is 2.83 g/cm³.

Solution:

The mass of an object may be found from the product of its density (ρ) and its volume (V).

Given: $d = 0.85$ m; $r = d/2$; $h = 1.5$ m; $\rho = 2.83$ g/cm³

Formulas: $V = \pi r^2 h$; $m = \rho V = \rho(\pi r^2 h)$

	ρ	pi	r^2	h	m³ → cm³	g → kg	
$m = $	$\dfrac{2.83 \text{ g}}{\text{cm}^3}$	π	$\dfrac{0.85 \text{ m} \times 0.85 \text{ m}}{2 \times 2}$	1.5 m	$\dfrac{10^6 \text{ cm}^3}{1 \text{ m}^3}$	$\dfrac{1 \text{ kg}}{10^3 \text{ g}}$	$\doteq 2408.82$ kg

Since this is a compound multiplication/division calculation, the result cannot have more SDs than the measured data with the fewest SDs (2). Therefore,

$$m \approx 2400 \text{ kg.} \qquad \text{(2 SDs allowed)}$$

B. If a solution to a problem requires the combination of both addition/subtraction and multiplication/division operations, rounding the intermediate solutions may introduce excess rounding errors into the final answer.

(1) For intermediate calculations, you should underline the estimated digit in the result and retain *at least* one extra digit beyond the estimated digit. *Do not round to the extra digit.*

(2) Round the final calculation to the correct significant digits according to the applicable math rules, taking into account the underlined estimated digits in the intermediate answers. Carefully examine Example 2-7 to see how this works.

Special subscripts are often used to clearly identify variables and expressions in equations.

EXAMPLE 2-7

Compound Calculations: Sum of Areas

Three students are assigned the task of calculating the total floor area of the school's science lab. The first student finds that the area of the main lab floor (A_m) is 9.3 m by 7.6 m. Meanwhile, the second student measures the floor area of the chemical storage room (A_s) to be 3.35 m by 1.67 m. The third student determines that the closet floor area (A_c) is 93.5 cm by 127.6 cm.

What is the total floor area in square meters?

Solution:

Given: l_m = 9.3 m, w_m = 7.6 m, l_s = 3.35 m, w_s = 1.67 m, l_c = 127.6 cm, w_c = 93.5 cm

The total area (A_{total}) is equal to the sum of all three areas, or $A_{total} = A_m + A_s + A_c$.

The area of any rectangular region is the product of length and width.

$$A_{total} = (lw)_m + (lw)_s + (lw)_c$$

Convert the closet measurements to meters.

$$l_c = \frac{127.6 \text{ cm}}{} \left| \frac{1 \text{ m}}{100 \text{ cm}} \right. = 1.276 \text{ m}$$

$$w_c = \frac{93.5 \text{ cm}}{} \left| \frac{1 \text{ m}}{100 \text{ cm}} \right. = 0.935 \text{ m}$$

$A_{total} = (9.3 \text{ m} \times 7.6 \text{ m}) + (3.35 \text{ m} \times 1.67 \text{ m}) + (1.776 \text{ m} \times 0.935 \text{ m})$

$A_{total} = 70.68 \text{ m}^2 + 5.5945 \text{ m}^2 + 1.66056 \text{ m}^2$

(Underlined digits show last SD according to Math Rule 2.)

$$
\begin{array}{rl}
A_{total} = & 70.6 \quad \text{m}^2 \\
& 5.594 \quad \text{m}^2 \\
& 1.660 \quad \text{m}^2 \\
\hline
& 77.854 \quad \text{m}^2 \approx 78 \text{ m}^2 \qquad \text{(precise to 1 m}^2\text{, 2 SDs allowed)}
\end{array}
$$

2.19 Math Rule 4—Angles and Trigonometry

A. Angles in the SI are measured in derived units called radians. A **radian (rad)** is the plane angle that subtends a circular arc equal in length to the radius of the circle (see Figure 2-23). There are 2π radians in 360 degrees. Standard degree units for angles (°) are not part of the SI, nor are they metric. In this textbook, angles may be expressed in either degrees or radians depending on the application. Angles measured with a protractor should be reported to the nearest 0.1 degree.

B. The process for converting between degrees and radians is discussed in Appendix G but is summarized here.

degrees to radians:

$$\theta_{radians} = \theta_{degrees} \times \frac{\pi}{180°}$$

radians to degrees:

$$\theta_{degrees} = \theta_{radians} \times \frac{180°}{\pi}$$

These conversion factors are definitions, so the new angles retain the number of SDs of the original measurement.

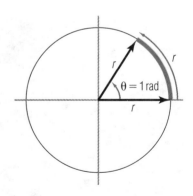

2-23 Definition of a radian

Sine, cosine, and tangent ratios are nonlinear functions of angles, so simple significant digit rules are not possible. Trigonometric ratios expressed to four decimal places will generally permit determining angles to within 0.1° precision.

C. Report angles resulting from trigonometric calculations to the lowest precision of any angles given in the problem. For example, if an angle given in the problem is stated in whole degrees, then any angles resulting from calculations should be expressed to the nearest whole degree.

D. For simplicity, assume that trigonometric ratios for angles given in this textbook are pure numbers, so SD restrictions do not apply. Consequently, the SDs of trigonometric problem solutions will depend on the precision of the nonangular data given in the problem rather than the precision of the trigonometric ratios.

2C Section Review Questions

1. In each of the following quantities, which place contains the estimated digit? How many significant digits are in each?

   a. 84 g   **b.** 100 mL   **c.** 3   **d.** 0.0020 m

2. Express the following measurements in scientific notation (see Appendix G if necessary). Be sure to include the units in your answers.

   a. 803 000 m   **b.** 30.75 cm   **c.** 0.15 mL   **d.** 0.05030 s

3. Which of the following measurements is the most precise?

   a. 7.40×10^3 cm   **b.** 7400 cm   **c.** 7400. cm

4. Using a calculator, find the decimal value of the following trigonometric expressions. Record your answers according to the rules in this section.

   a. sin 53°   **b.** tan 28°   **c.** cos $\pi/3$ **d.** sin π

5. Convert the following angles to the other angular form.

   a. 37°   **b.** 125°   **c.** $2\pi/3$ **d.** $\pi/8$

6. Using a flexible tape measure, you measure the circumference of the shaft of a machine bolt to be 5.98 cm. Calculate the bolt's diameter, adhering to the significant digit rules given in this section.

7. Report the answers for the following calculations using the correct number of SDs. Show your work, including the answers prior to rounding.

   a. 74.63 cm + 12.0 cm   **c.** 110 m × 5 m

   b. 8.6 L − 667 mL   **d.** 10.7 s ÷ 3

8. Find the unknown side and angles for the right triangle in the diagram. Report your answers with the correct SDs and include the appropriate units.

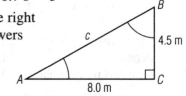

2D   PROBLEM SOLVING

2.20 The Need for Problem Solving

To inquisitive scientists, questions about the structure or behavior of the universe are problems to be solved, and every problem has its own unique solution. You will be presented with a variety of problems throughout this course—not to make solving them more difficult, but to develop skills that will help you solve many problems that you will face. Solutions to problems may require using one or more of the techniques that will be discussed in this section.

The correct solution to a physics problem—or any problem, for that matter—is usually obtained by following an orderly process that leads eventually to the answer. This is not to say that you cannot occasionally stumble onto the answer, but it is not likely, and much effort can be wasted along the way. The four general steps that follow outline a process that can be applied to finding a solution to any physics problem.

2.21 Read the Exercise to Understand the Problem

This is the most important step, though students frequently rush through it. Read the problem statement completely through once to gain an overview of the problem and determine what you are being asked to find.

Read the problem a second time slowly and carefully to note what information has been given and to consider the principles relating to the problem. Identify the known and unknown variables and write down the data stated in the exercise. Also, watch for information that has nothing to do with solving the problem.

Make a simple sketch to help organize the information. It does not have to be an artistic Rembrandt, but neither should it be an abstract Picasso! The essential objects, surfaces, paths, vectors, values, and directions should be indicated in their correct relationships. Show the unknown using the correct variable symbol.

2.22 Determine the Method of Solution

Problem solutions may require physical constants, data from tables, conversion factors, or other information that is not expressed in the exercise.

Write down the formulas needed to solve the problem that relate the variables you identified in the first step. Sometimes you will have to perform one or more intermediate calculations to provide the missing information, so more than one formula may be required.

Consider the best tool for finding a solution. Some methods that will yield an answer include the use of the following:

algebra	analogy
estimation	reasoning
tables	guess and check
graphs	

In this course, you will usually have to show a mathematical or graphical solution. Even with this restriction, there may be several possible approaches.

2.23 Substitute and Solve

Solve the formula for the unknown variable to isolate it on one side of the equation; then substitute the known values.

Write down each mathematical step on your paper before performing the calculations. Do not get into the habit of completing a solution on your calculator and writing down just the answer. Writing out the calculation first allows you to perform unit analysis, and it guides your mathematical operations. Later, if your answer is wrong, you can quickly locate your mistake. This step also permits your teacher to identify conceptual problems when grading your paper.

2-25 Doing everything in an orderly fashion is a biblical principle that can be practiced every time you do homework (I Cor. 14:40).

We often try to solve life's problems by addressing the superficial symptoms without trying to understand the underlying spiritual problem. Don't settle for merely changing your behavior without letting God change your heart.

Order of magnitude indicates
the nearest power of 10 that approximates a given quantity. 85 000
($\sim 10^5$) is four orders of magnitude
larger than 43 ($\sim 10^1$).

TABLE 2-3

Orders of Magnitude

Object	Size
Quasar, distance (10^9 ly)	$\sim 10^{24}$ m
Galaxy, diameter (10^5 ly)	$\sim 10^{21}$ m
Nearest star, distance (~ 4 ly)	$\sim 10^{17}$ m
Solar system, diameter	$\sim 10^{13}$ m
Earth's orbit, diameter	$\sim 10^{11}$ m
Moon's orbit, diameter	$\sim 10^9$ m
Earth's diameter	$\sim 10^7$ m
Large city	$\sim 10^5$ m
City park	$\sim 10^2$ m
Human	$\sim 10^0$ m
Insect	$\sim 10^{-2}$ m
Cell	$\sim 10^{-5}$ m
DNA molecule	$\sim 10^{-7}$ m
Inorganic molecule	$\sim 10^{-9}$ m
Atom	$\sim 10^{-10}$ m
Atomic nucleus	$\sim 10^{-15}$ m

Work to make your solutions neat and orderly. The written work that you turn
in for grading should be a methodical, step-wise process to the solution.

2.24 Check Your Answer for Reasonableness

a. This step is the one most often missed by students. Reread the problem. Have
you answered the right question? Does the result have the expected **order of
magnitude?** Does the answer have the correct dimensional units? Performing these checks will provide additional assurance that the answer is correct
(or at least reasonable).

b. Mentally make an estimation of the magnitude of the answer. You can usually
estimate a quantity to at least some power of 10 that is within a factor of 2
or 3 of the correct calculator result. This kind of estimation is often called
an *order of magnitude estimation.*

c. Be sure to simplify units, especially if there are derived units that stand for
more complex combinations of SI units. For example, the newton (N) can replace kg·m/s^2.

d. Express results in the correct number of significant digits. For mathematical
solutions, you should carefully follow the rules discussed in Sections 2B and
2C for maintaining the correct number of SDs in your solution.

EXAMPLE 2-8

Estimation

You calculate that a car traveling at a legal U.S. highway speed is moving 2.5 m/s. You should question this result, since a person can easily
run at that rate. To check your solution by estimation, you figure that,
since there are about 2 km/mi, a car moving at 55 mi/h is traveling
about 110 km/h, or about 10^5 m/h. There are 3600 seconds per hour,
which is about 4×10^3 s/h. Mentally divide the distance by the time to
find the speed. First subtract the exponents of the powers of 10.

$$\frac{1 \times 10^{\overset{2}{\cancel{5}}}\ \text{m}}{\cancel{h}} \left| \frac{\cancel{h}}{4 \times 10^{\cancel{3}}\ \text{s}} \right. = \frac{100\ \text{m}}{4\ \text{s}}$$

The result is about 25 m/s. Since your estimate is ten times larger than
the calculated result, you need to recheck your original work.

2D Section Review Questions

1. When beginning a physics exercise, what should you consider doing in
order to organize the information and visualize the circumstances?

2. In order to find the mass of a cubic meter of sea water, what information would you need to look up?

3. In most cases, which of the solution methods discussed in Section 2D
would be the least likely to produce an accurate answer?

4. How many orders of magnitude larger is 7250 m than 83 m?

5. Discuss several checks that you can perform to give greater assurance
that your solution is correct.

Problem Statement for Questions 6–10: Firewood is purchased in units called cords. A cord of wood is a stack 4.00 ft high, 4.00 ft wide, and 8.00 ft long. The density of sugar maple wood is about 6.8×10^2 kg/m^3. What is the volume of a half-cord of sugar maple wood in cubic meters?

6. What quantity are you asked to find?

7. Make a sketch of the problem showing the required information.

8. Discuss any additional information that you must provide.

9. State any information that may be extraneous.

⦿10. Solve the problem.

Chapter Review

In Terms of Physics

SI (International System of Units)	2.2	unit analysis	2.9
dimension	2.3	accuracy	2.11
space	2.4	error	2.11
time	2.4	percent error	2.11
matter	2.4	precision	2.12
meter (m)	2.5	uncertainty	2.12
second (s)	2.6	relative uncertainty	2.12
kilogram (kg)	2.7	vernier scale	2.12
mole (mol)	2.7	significant digit (SD)	2.13
derived unit	2.8	radian (rad)	2.19
conversion factor	2.9	order of magnitude	2.24

Problem-Solving Strategies

2.1 (page 29) In calculations, the prefix of an SI dimensional unit may be replaced by the exponential multiplier of the base unit.

2.2 (page 29) In a conversion factor, the unit you are converting *to* is in the same position relative to the fraction bar as the unit you are converting *from* in the given quantity.

2.3 (page 29) The first factor in the unit analysis bridge should be the given data. The remaining data and conversion factors may be inserted in any order so as to cancel the given units and substitute the required units in their place.

2.4 (page 30) The division operation is the inverse of multiplication. To indicate division by a quantity using the bridge utility, invert the quantity *and its unit;* then write it as one of the factors in the bridge.

2.5 (page 35) SD rules regarding zeros vary depending on whether the decimal point is written or is assumed due to place-holding zeros.

2.6 (page 36) Use scientific notation when trailing zeros make the precision of measured data ambiguous.

2.7 (page 37) When adding or subtracting numbers in a column, draw a vertical line to the right of the estimated digit in the *least* precise data. In the answer, round to the digit's place to the *left* of this line.

2.8 (page 38) For problems requiring more than one calculation, do not round the intermediate answers, but note the last SD place for each answer as if you were. Round only the final answer, taking into account the SD placement in each of the intermediate answers.

2.9 (page 40) Consider trigonometric ratios of angles to be pure numbers for the purpose of determining the SDs of problems involving angles.

2.10 (page 41) Read the problem carefully and make a sketch that accurately relates all known information. Note information that is extraneous.

2.11 (page 41) Select the appropriate method to solve the problem. Usually, you will be required to show a mathematical or graphical solution.

2.12 (page 41) Solve the applicable formula(s) for the unknown quantity and substitute the known information. Show as much work as necessary so that another person can follow your reasoning.

2.13 (page 42) Check the answer for reasonableness.

1. Name the three essential dimensions of the universe.

2. Name two SI base units that have exact definitions.

3. In typeset text, what is the difference between the abbreviation for meters and the variable symbol for mass used in a formula?

4. Write the conversion factor needed to convert 10.052 kg to grams.

5. What factor should normally be the first term in the bridge utility when performing unit analysis?

6. List three instruments each for measuring time and distance, in increasing order of precision.

7. If the error for a classroom mass balance is –0.75 g, is the mass of a sample indicated by the balance more or less than the actual mass?

8. Which is more precise, $\sqrt{2}$ or 1.414 213 562? Explain your answer.

9. A Celsius thermometer is calibrated in 1° increments. To what decimal place should you report a temperature obtained with this thermometer?

10. State the number of significant digits in each of the following quantities:
    a. 103 kg   c. 1.607×10^{-2} A   e. 0.0034 °C   g. 30 μm
    b. 1500. mL   d. 2000 km   f. 10 eggs   h. tan(70°)

11. Round the following quantities to 3 significant digits:
    a. 23 432 cm   b. 0.000 329 5 m   c. 10.009 N   d. 7659 kg

12. True or false: The calculated density of an object having a mass of 23.80 g and a volume of 24.0 cm^3 is 0.992 g/cm^3. Justify your conclusion.

13. Give several examples of the kinds of additional information that may be required to solve physics exercises.

14. Discuss some reasons why you should write down all mathematical steps in a solution to a physics exercise.

15. By how many orders of magnitude do the following quantities differ?
    a. 4500 ly vs. 8.6×10^6 ly   c. 160 kg vs. 2.0 g
    b. 8.32×10^6 L vs. 0.023 L   d. 22 m vs. 12 755 km

16. Estimate the following:
    a. The length of your shoe in centimeters.
    b. The number of times your heart beats in one minute.
    c. The area of your desk in square centimeters.
    d. The distance from your school to your house in meters.

⊙17. Convert the following measured data to the units indicated. Use scientific notation where appropriate.
    a. 2.9 cm to m   c. 235 mL to L
    b. 35.70 Hz to MHz   d. 2325 ft to km

⊙18. Convert the density 4230 mg/mm^3 to kg/m^3. Use the bridge utility for unit conversion.

⊙19. The accepted value for the mass of one electron is 9.11×10^{-31} kg. Two experimenters separately measured the mass of the electron three times each. The first experimenter obtained the following masses: 7.31×10^{-29} kg, 7.29×10^{-29} kg, and 7.32×10^{-29} kg. The other experimenter obtained 8.5×10^{-31} kg, 9.6×10^{-31} kg, and 9.1×10^{-31} kg. Which experimenter's results were more accurate? Which were more precise?

⊙20. Perform the required operations on the following quantities. Express your answers using scientific notation, the correct number of SDs, and the correct units.
    a. $(6.234 \text{ m})(4.91 \times 10^{-5} \text{ N})$
    b. $(1.17 \times 10^3 \text{ mL}) + (2.315 \times 10^2 \text{ mL})$
    c. $(32 \text{ m})(9.45 \times 10^6 \text{ s}^{-1})$
    d. $(3.14 \times 10^{23} \text{ ly}) - (7.18 \times 10^{20} \text{ ly})$
    e. $(2.32 \times 10^{-4} \text{ m})(4.331 \times 10^{-3} \text{ m})$
    f. $(7.913 \times 10^{-4} \text{ kg}) \div (9.1 \times 10^{-7} \text{ m})$
    g. $(7.5812 \text{ g}) - (2.1 \times 10^{-1} \text{ g})$
    h. $(2.32 \times 10^{-4} \text{ m}) \div (4.331 \times 10^{-3} \text{ m})$
    i. $(3.0 \times 10^8 \text{ m/s}) \div (3.661 \times 10^{10} \text{ s}^{-1})$
    j. $(9.81 \text{ m/s}^2)(7.3 \times 10^1 \text{ kg})$

⊙21. Find the long side (hypotenuse) and the area of a flower bed in the shape of a right triangle that has legs measuring 12.5 m and 10. m.

⊙22. A board for a shelf is 2.31 m long and 33.4 cm wide.
    a. What is the area of the shelf in square meters?
    b. What is the perimeter of the shelf in centimeters?
    c. Which will reduce the area of the shelf more: (1) cutting 3.4 cm from its width, or (2) cutting 0.31 m from its length?

⊙23. A rectangular swimming pool is 10.0 m wide and 15.0 m long. The water depth varies over the length of the pool in three 5.0 m zones. The zone under the diving board is 3.0 m deep. The shallow zone at the other end is 1.0 m deep. The zone in the middle has a sloping bottom changing from 3.0 m to 1.0 m deep.
    a. What is the volume of the pool in cubic meters?
    b. What is the volume of the pool in liters?
    c. The density of water at pool temperature is 1.0 g/cm^3. What is the mass, in kilograms, of the entire volume of water in the pool?
    d. If a swimmer is allotted 5.0 m^2 of surface swimming space, how many swimmers may be in the pool at one time?

⊙⊙24. Describe the current dimensional standard for each of the seven fundamental SI units. Use the National Institute of Standards and Technology (NIST) website or another authoritative source.

⊙⊙25. Evaluate the statement, "Numerical scientific data may contain only significant digits."

⊙⊙26. Write an essay discussing the history of the metric system. Include in your discussion the development of the definitions for the seven fundamental SI units.

⊙⊙27. Estimate the number of German shepherds in your state. List any assumptions you make and the sources of information that you used to make the assumptions.

Classical Mechanics

Classical mechanics had its beginnings with the work of Galileo in the sixteenth century. The height of the classical mechanics revolution came in the following century when Sir Isaac Newton discovered his three laws of motion. One of his greatest accomplishments was deducing the law of universal gravitation by analyzing the motion of the Moon around the Earth. For nearly 300 years, Newton's laws were the basis for understanding mechanical phenomena.

In Unit 2, you will learn how straight-line, two-dimensional, circular, and periodic motions are measured. You will delve into the nature of forces and how they can change motion and do work. You will also learn of the great mechanical conservation laws and the way various forms of energy are interchanged in order to conserve the total energy in the universe.

3A	Describing Motion	49
3B	The Equations of Motion	61
Facet:	Hypersonic Jets	56
Biography:	Galileo	67

New Symbols and Abbreviations

distance (d, s)	3.3	acceleration (a)	3.7
displacement (d)	3.3	average acceleration ($\bar{a}$)	3.7
average speed ($\bar{v}$)	3.4	difference vector (Δv)	3.7
velocity (v)	3.6	gravitational acceleration (g)	3.14
average velocity ($\bar{v}$)	3.6		

A U.S. Navy F/A-18C Hornet fighter-bomber lands aboard an aircraft carrier. The pilot experiences significant negative acceleration in one dimension as the plane rapidly comes to a stop.

Motion in One Dimension

> *Beloved, now are we the sons of God, and it doth not yet appear what we shall be: but we know that, when he shall appear, we shall be like him; for we shall see him as he is. And every man that hath this hope in him purifieth himself, even as he is pure.* 1 John 3:2–3

Did you know that the life of a Christian is a good example of one-dimensional motion? If you have believed on Jesus Christ and received Him as your Savior, the goal of your daily life should be Christlikeness—thinking and acting to emulate God's character according to His will for you. Christians strive to move forward in their walk with Christ, but sometimes it seems as if no progress is being made, and at times, Christians can even be guilty of backsliding. However, it is our responsibility to continue moving forward on the path of sanctification. The verses above give us the promise that eventually, by God's grace, this process will be completed when we enter His presence.

3A · DESCRIBING MOTION

3.1 Introduction

Electrons moving through a tungsten light bulb filament generate light photons. These photons reflect off the letters on this page as your eyes scan across them and strike sensitive cells in your eyes. The retinas in your eyes generate charges that rush along the optic nerves to relay images to your brain. Meanwhile, the sunlight streaming in through the window gradually changes direction throughout the day as the earth spins around on its axis. The great wind and ocean current systems of the earth spiral in response to the planet's rotation. Each night, the Moon waxes or wanes during its monthly orbit of the earth. The giant gaseous planets of our solar system revolve ponderously in their decades- or even centuries-long orbits about the Sun. Distant galaxies retreat from us at hundreds and thousands of kilometers per second.

The universe is full of motion. The study of motion is called **mechanics.** Mechanics includes all kinds of motion, from simple one-dimensional motion along a straight line to complex three-dimensional space flight. This chapter investigates motion, speed and velocity, and acceleration.

3-1 Motion is everywhere in the universe.

3.2 What Is Motion?

Motion is the change of position with time.

What is motion? Men have spent their lives searching for the answer. Simply put, motion is a change of position. Scientists have divided mechanics into two branches: **kinematics,** to describe *how* objects move, and **dynamics,** to explain the *causes* of motion. Early Greek philosophers began their study of motion with the wrong question. They sought the causes of motion without understanding how things move in the first place. As a result, many of their ideas seem ridiculous. *How* motion occurs must be known before its *causes* can be understood.

Kinematics is the description of *how* things move. **Dynamics** is the study of the *causes* of motion.

3.3 Mathematical Representations of Motion

If you were dropped into the middle of the ocean and you swam for an hour, how could you prove that you had changed position? Despite all your hard work, you might have swum in a circle. The only way to find out is to compare your first and last positions to a reference point, like an island. If your distance or direction from the island changed, then you could prove that you had moved.

To represent motion along a straight line, you identify the object's position on a number line. The number line is a one-dimensional coordinate system. Each position is compared to a reference point called the *origin*. Positions to the *left* of the origin are assigned *negative* values, and positions to the *right* are assigned *positive* values. For example, an object 4 m to the left of the origin has a position of -4 m; an object 9 m to the right of the origin has a position of 9 m. The axis of the number line may be oriented in any direction, including vertically. *Up* is usually considered *positive* and *down* is usually considered *negative*.

A variable with a subscript of 0 indicates an initial or earlier condition with respect to time. The symbol without a subscript indicates a later value relative to the initial value.

An object has moved if at one time its position is x_0 and at another time its position is x. Notice that each position is associated with a *time:* position x_0 occurs at time t_0 and position x occurs at time t. The object's position at each time can be represented by an *ordered pair,* written (t_0, x_0) or (t, x).

Suppose you swim twenty-one lengths in a 50-meter pool. Because each length is 50 m, the distance you swim is 21×50 m $= 1050$ m. But notice that after swimming 1050 m, you end up only 50 m from the starting point (at the opposite end of the pool). The distance, or length, of the path (s) often differs from the net change in position. This change in position between two distinct points is called **displacement (d).** In this example, $s = 1050$ m but $\mathbf{d} = 50$ m forward from the starting point.

The symbol s is the path distance traveled during a time interval. The path may be straight, curved, circular, or segmented.

When a quantity contains only one piece of information, such as a distance of 1050 m or a temperature of -50 °C, it is a *scalar* quantity. A scalar value can be positive, negative, or zero. Scalars are symbolized by italicized letters, such as T for absolute temperature. When a quantity can be completely described only by stating both a *magnitude* and a *linear direction* (like 50 m forward, which is two pieces of information), the quantity is called a *vector*. Vectors are represented by bold roman characters, such as **d** for displacement. We will discuss vectors and their properties in more detail in Chapter 4. With motion along a straight line, the direction of a vector quantity is indicated by a positive or negative sign before the numerical magnitude of the vector. Forward motion is usually positive, and backward motion is usually negative.

Scalar quantities contain one piece of information—a numerical value. **Vector** quantities contain two pieces of information—a numerical value and a direction. **Magnitude** is the absolute value of a numerical value and is always positive.

In *one-dimensional motion,* the scalar symbol with the subscript of the coordinate axis (e.g., v_x) is identical to the more general vector notation for the same quantity (e.g., **v**).

Mathematical operations with vectors require special rules that you will learn in Chapter 4. To simplify vector operations in one-dimensional motion, we use a special notation that consists of the *scalar* symbol standing for the quantity with a subscript indicating the coordinate axis it is parallel to. For example, a displacement in the x-direction is symbolized by d_x. This notation represents *only* one-dimensional vectors, whereas the more general bold vector symbol can represent vectors in one, two, or three dimensions. The symbols d_x and **d** are equivalent if **d** is parallel to the x-axis.

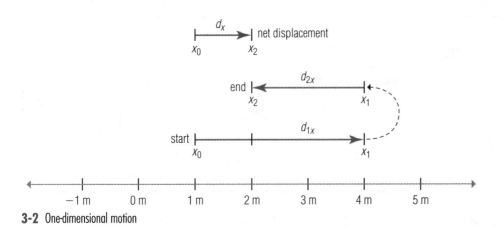

3-2 One-dimensional motion

In Figure 3-2 an object begins at position $+1$ m at time t_0. At time t_1, it reaches $+4$ m to the right of its original position. Then it moves left until it reaches $+2$ m at t_2. The object's displacement (d_x) for the interval t_0 to t_2 is its *net* change in position; that is,

$$d_x = d_{1x} + d_{2x}$$
$$d_x = (x_1 - x_0) + (x_2 - x_1) = x_2 - x_0$$
$$d_x = (+3 \text{ m}) + (-2 \text{ m})$$
$$d_x = +1 \text{ m, or 1 m to the right of the original position.}$$

The object has actually traveled over a path from x_0 to x_1 and back to x_2. The path distance is equal to the sum of the individual displacement magnitudes. Displacement magnitude is the absolute value of the difference of positions. This is true for linear motion in any number of dimensions.

$$s = |d_x| = |x - x_0|$$

The path distance for the object in Figure 3-2 is determined as follows:

$$s = |d_{1x}| + |d_{2x}|$$
$$s = |(+4 \text{ m}) - (+1 \text{ m})| + |(+2 \text{ m}) - (+4 \text{ m})|$$
$$s = |+3 \text{ m}| + |-2 \text{ m}| = 3 \text{ m} + 2 \text{ m}$$
$$s = +5 \text{ m}$$

The object's *displacement* is $+1$ m, but it traveled a total *distance* of 5 m.

People tend to think of their lives as linear paths. They get comfortable in their routines and are caught by surprise when things happen that disrupt their lives. God cautions us to be always vigilant and to prepare for the unexpected.

"If thou hast run with the footmen, and they have wearied thee, then how canst thou contend with horses? And if in the land of peace, wherein thou trustedest, they have wearied thee, then how wilt thou do in the swelling of Jordan?" (Jer. 12:5).

EXAMPLE 3-1

Picnic at the Lake: Displacement

A family leaves its campsite and travels by car to the picnic area on the north side of the lake. A sign at the campsite says that the picnic area is 2 km north by boat, but the children notice that the car's odometer advances by 10 km during the trip.

a. What is the car's displacement from the campsite when the family reaches the picnic area?

b. How far has the car traveled?

c. When the family returns to the campsite by the same road, what will be the car's displacement relative to its original starting point?

d. How far will the car have traveled?

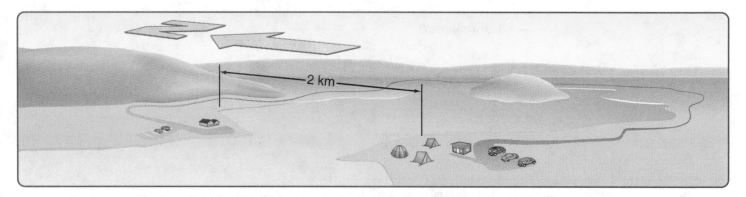

3-3 The distance across the lake is not the same as the distance traveled by car around the lake.

Solution:

Position an imaginary x-coordinate axis so that it passes through the initial and final positions, with north positive.

a. The car's displacement is the straight-line distance and direction from its starting point to its final position. The car is 2 km north of its starting point.

b. The car has traveled 10 km, as its odometer shows.

c. When the car returns to the campsite, it will be at its starting position, so its displacement is zero. For this answer, it is unnecessary to specify a direction since a zero displacement is the same in any direction.

d. The car has traveled a total of 10 km + 10 km, or 20 km.

3-4 Map of the road between Gumine and Lalibu

EXAMPLE 3-2

Visiting Lalibu: Displacement in PNG

A map of Papua New Guinea (PNG) shows the two towns of Lalibu and Gumine separated by the mountains of the Western Highlands. Lalibu is 100 km due west of Gumine. A missionary visiting Lalibu from Gumine, traveling the only main road between the two towns, discovers that his odometer reading increases by 190 km. How far did he travel? What is his displacement for the trip?

Solution:

Draw the x-axis as an east-west line between Lalibu and Gumine, with its origin at Gumine. East is positive. The missionary has traveled 190 km. However, he is now only 100 km from his starting point. His displacement is 100 km west ($d_x = -100$ km).

3.4   Rate of Motion—Speed

You probably saw ordered pairs like (t, x) in algebra. In that class you plotted each pair as a point on a graph. You can also plot the (t, x) ordered pair on a graph called a **position-time graph.** This type of graph shows much useful information in a simple form.

For example, Figure 3-5a is the position-time graph of a train on an east-west track between Ames and Marshalltown, Iowa. The graph shows several (t, x) ordered pairs. From looking at this graph, you might think that the train jumped from point to point (city to city). Obviously this is impossible. The points are only

Problem-Solving Strategy 3.1

When graphing time-dependent data, always plot time on the horizontal axis and plot the dependent variable on the vertical axis.

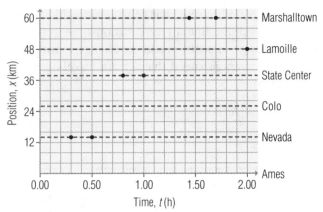

3-5a Plot of specific train positions and their times

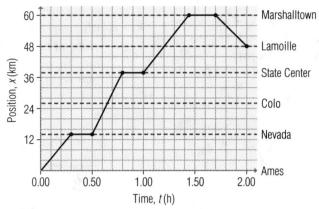

3-5b Continous plot of train position with time

the recorded positions of the train at certain times. If the train moved smoothly from point to point, you can draw lines connecting the points, as Figure 3-5b shows.

From Figure 3-5c you can find the displacement over a given time period. For example, suppose you want to know the displacement of the train from 0.60 h to 0.80 h. The graph shows that at $t_1 = 0.60$ h, the train's position is $x_1 = 22$ km east of its departure station. At $t_2 = 0.80$ h, its position is $x_2 = 38$ km east (State Center). The train's displacement during that period is the change in x during the interval. That is,

$$d_x = x_2 - x_1$$
$$d_x = (+38 \text{ km}) - (+22 \text{ km})$$
$$d_x = +16 \text{ km}.$$

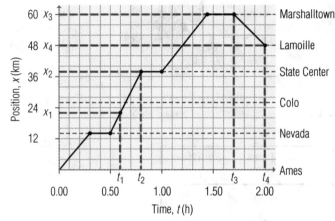

3-5c Time and position coordinate pairs for determining displacement

At 0.80 h, the train is 16 km farther east than it was at 0.60 h.

What if the displacement is negative? For instance, what is the displacement of the train from $t_3 = 1.70$ h to $t_4 = 2.00$ h? At t_3, $x_3 = 60$ km east (Marshalltown); at t_4, $x_4 = 48$ km east (Lamoille). The displacement is as follows:

$$d_x = x_4 - x_3$$
$$d_x = (+48 \text{ km}) - (+60 \text{ km})$$
$$d_x = -12 \text{ km}$$

In this answer, the negative displacement means that the train has moved backward (west) toward the origin.

A position-time graph also helps to determine the speed at which an object moves. The *average speed* ($\overline{v}$) of a train is the ratio of the distance traveled $|x - x_0|$ over an interval of time ($t - t_0$). The absolute value sign is required because distance is always positive. If the train *does not change its direction*, its average speed is

$$\overline{v} = \frac{|x - x_0|}{t - t_0}.$$

The symbol Δ (Greek letter delta) conventionally means "change in," so Δx means "the change in position x" or $x - x_0$. On a position-time graph, Δx is the difference in height between some position (x) and the initial position (x_0). The "change in time" (Δt, or $t - t_0$), which is always positive (unless you are using a time machine), is the horizontal distance between the points on the timeline marking the beginning and the end of the time interval. In any linear graph, the change in

God often describes our spiritual lives in dynamic terms—walking, running, seeking, and so on. Even the wicked move—but in the wrong direction.

"The wicked, through the pride of his countenance, will not seek after God: God is not in all his thoughts" (Ps. 10:4).

Problem-Solving Strategy 3.2
Whenever you must find the change of a quantity, always subtract the initial value from the final value.

The Greek letter "Δ" stands for the change in a quantity over a specified time interval. For example,

$$\Delta x = x - x_0.$$

Motion in One Dimension **53**

If the object changes direction one or more times during the time interval, then the average speed is found by dividing the total path distance by the elapsed time.

$$\bar{v} = \frac{s}{\Delta t},$$

where $s = |x_1 - x_0| + |x_2 - x_1| + \ldots + |x_n - x_{n-1}|$.

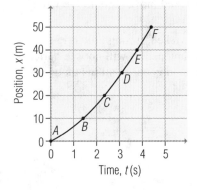

3-6a This skier must maximize her downhill velocity in order to win.

Flag	Time (s)	Position (m)
Start	0.00	0.0
1	1.41	10.0
2	2.35	20.0
3	3.11	30.0
4	3.76	40.0
5	4.34	50.0

A *tangent line* is a line that intersects a curve at a single point without crossing the curve.

The limb of a planet is the visible edge of the planetary disc seen against the background of space.

height between two points (the *rise*) divided by the change in horizontal distance (the *run*) between the points is the *slope* of the line. In the position-time graph, the slope of the graph is the ratio of the change in position and the change in time. The magnitude of the slope represents the average speed that a train travels between two points. The formula for average speed is

$$\bar{v} = \frac{|\Delta x|}{\Delta t}. \qquad (3.1)$$

EXAMPLE 3-3

Downhill Skiing: Instantaneous Velocity

A physics student, who also is a competition skier in Vermont, skis directly down a straight, 50.0 m long hillside that has a flag every 10.0 m. Each time she passes a flag, her coach records the time on a stopwatch. The table in the margin gives the time the skier passed each flag, and the position-time graph plots her motion. Her initial position is at the top of the hill, and the initial time is at the start of the descent. She passes the Start flag moving at 5.0 m/s. Find the skier's average speed (a) for the entire trip; (b) between flags 1 and 4; and (c) between flags 2 and 3.

Solution:

$$\bar{v} = \frac{|\Delta x|}{\Delta t}$$

a. $\bar{v}_{run} = \dfrac{|50.0\ m - 0.0\ m|}{4.34\ s - 0.00\ s} = \dfrac{50.0\ m}{4.34\ s} \doteq 11.5\underline{2} \approx 11.5\ m/s$

b. $\bar{v}_{1-4} = \dfrac{|40.0\ m - 10.0\ m|}{3.76\ s - 1.41\ s} = \dfrac{30.0\ m}{2.35\ s} \doteq 12.7\underline{6} \approx 12.8\ m/s$

c. $\bar{v}_{2-3} = \dfrac{|30.0\ m - 20.0\ m|}{3.11\ s - 2.35\ s} = \dfrac{10.0\ m}{0.76\ s} \doteq 13.1\underline{5} \approx 13.2\ m/s$

3.5 Instantaneous Speed

As Example 3-3 shows, the average speed changes depending on the interval chosen. The skier's speed increases as her position changes from the top of the hill to the bottom. Suppose you want to know her speed at any one moment (her *instantaneous speed*). For example, what is her speed at the moment she passes the third flag (3.11 s, 30.0 m)? The slope of the position-time curve at that point is her speed. But how do you find a slope at one point on a nonlinear curve?

One method to estimate the slope is to construct a tangent line segment at a single point on the curve and measure the tangent's slope. If you could view the point on the graphed curve with a zoom-microscope, you would notice that as you increased the magnification power, the length of the curve segment visible in the eyepiece becomes shorter and the curve appears to become straighter. At maximum magnification, the curve is indistinguishable from the straight-line tangent at that point.

To help you understand this concept, consider the curve of the earth. Viewing it from the Moon, you see that the limb of the earth is clearly circular; but as you approach the earth, the horizon becomes more sweeping and the curvature less distinct. By the time you touch down on land, the horizon appears to be flat. This small segment of the earth's 40 000 km circumference that you can see is essentially parallel to the tangent line at the point upon which you stand.

In a similar way, you can estimate the skier's instantaneous speed at the third flag by "zooming in" on the curve around the flag. A very rough estimate for her speed at this point can be made by finding the skier's average speed for the entire run. (This calculation was made in Example 3-3.) As depicted on the position-time graph (Figure 3-7a), your estimate would be the line connecting the top and the bottom points of the ski run. But the slope of this line is only roughly parallel to the tangent at the third flag (Point *D*).

To improve your estimate, try constructing a line between the endpoints of a shorter curve segment, for example, from Point *B* to Point *E*. This line is more nearly parallel to the curve at the third flag; so its slope provides a better estimate for the instantaneous speed at Point *D*. However, you can improve the estimate even more. Figure 3-7c shows a line through Points *C* and *E* that is nearly parallel to the curve at Point *D*. Thus its slope of 13.2 m/s is a good estimate of the slope of the tangent (and therefore the instantaneous speed) at the third flag. In summary, you estimate the instantaneous speed at a given point by calculating the average speed of a small interval containing that point—the smaller the interval, the better the estimate. This is the graphical basis of differential calculus.

The difference between average and instantaneous speed is important to automobile drivers, since the speed limit applies to instantaneous speed, not average speed. To maintain a particular speed (usually the speed limit) on long trips, drivers often use cruise control. Although this device can maintain a certain average speed, it cannot guarantee that the automobile's instantaneous speed will always equal the speed for which it is set. Suppose, for instance, that a driver is traveling along a hilly road and the speed limit is 90 km/h (about 55 mph). The cruise control is set at that speed when the car begins to climb a hill. As the car climbs, it requires more fuel to maintain a speed of 90 km/h. When it reaches the top of the hill, the need for fuel abruptly decreases. If the cruise control cannot react quickly enough to reduce the fuel supply, the car speeds up and breaks the speed limit even though the cruise control was set at 90 km/h (Figure 3-8).

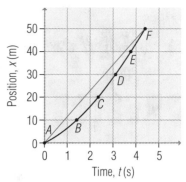

3-7a First estimate

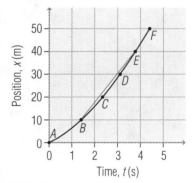

3-7b A better estimate

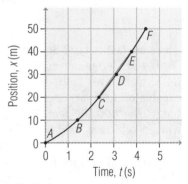

3-7c The best estimate

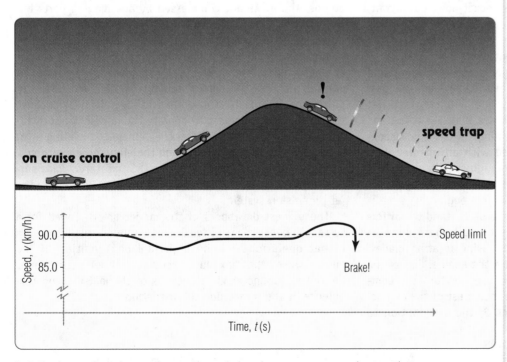

3-8 The diagram shows the car and terrain. The graph shows how instantaneous speed varies with time.

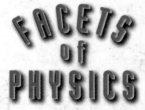

Hypersonic Jets
SCRAM!!!

How fast is fast? Scientists around the world are trying to redefine our definition of a quick trip. The fastest jet airplane ever built, the SR-71 *Blackbird* spy plane, was capable of 3.5 times the speed of sound (Mach 3.5) at altitudes over 100 000 feet. The speed record for any type of winged vehicle is held by the X-15 rocket-powered aircraft, at Mach 6.7. Rockets that launch space probes are capable of much higher speeds, but they cannot operate in the atmosphere at those speeds without being destroyed.

Rockets carry both fuel (usually liquid hydrogen) and oxidizer (usually liquid oxygen), which allows them to operate independently of the atmosphere. In contrast, jet engines rely on atmospheric oxygen to burn their fuel and are therefore classified as "air breathing." NASA is now seeking to develop a vehicle equipped with an air-breathing engine capable of operating at very high speeds in the thin upper reaches of the earth's atmosphere. The key may be the **scramjet** *(supersonic combustion ramjet)* engine.

A little background is necessary to understand the differences among these propulsion systems. Modern conventional *jet* engines compress the air using rotating blades, creating high fuel-air pressure inside the engine. Part of the energy of the combustion gases is used to turn the compressor. The exhaust gases provide the thrust of the engine. As inlet air speed approaches Mach 3, the contribution to thrust practically disappears and actually reduces the engine's efficiency. At speeds above Mach 3, the compressor is irrelevant.

Scientists began researching *ramjet* technology during World War II for use in high-speed missiles. In a ramjet engine, the air is not compressed by any moving parts but simply by the speed of the vehicle forcing (ramming) the air into the engine intake, where the internal geometry creates compression. Because it has no moving parts, a ramjet engine is lighter and mechanically simpler than a normal jet engine. However, a ramjet vehicle must have a booster propulsion system to accelerate it above Mach 3, because the compression required to operate the ramjet is not available below this speed. Though the vehicle is traveling at or above Mach 3, the airflow through the engine remains below the speed of sound (subsonic). The U.S. military's *Talos* surface-to-air missile deployed from 1959 to 1979 used a ramjet propulsion system.

The biggest drawbacks of the ramjet are the need for a system to boost the vehicle to a high enough speed to begin ramjet operation and the high fuel consumption of the engine. Also, the maximum speed for ramjet operation is limited to around Mach 5 because of thermal-induced turbulence at the intake due to air friction.

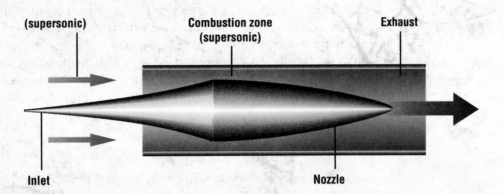

(supersonic)   Combustion zone (supersonic)   Exhaust

Inlet   Nozzle

The scramjet is a recent design that takes advantage of high-temperature materials and computer-aided engineering in order to operate at speeds up to Mach 10 (ten times the speed of sound). It is called a supersonic ramjet because, in contrast to the older ramjets, the air continues flowing through the engine at hypersonic speeds (much greater than the speed of sound). Scramjets have some of the same drawbacks as ramjets, but if the program is successful, the payoffs could be tremendous.

Scramjets could replace current rocket-propulsion launch vehicles for small space payloads (such as communications satellites). Rocket propulsion is extremely expensive and inefficient, requiring a large, multistage rocket. Even the space shuttle, NASA's premier reusable launch vehicle, must jettison its huge fuel tank, which then falls back to Earth and burns up.

Because it is theoretically more efficient and does not need to carry its oxidizer, a scramjet launcher achieving the same thrust would be lighter than a conventional launch vehicle. This would allow the payloads to increase while the launch-vehicle costs decrease.

Scramjets are also being considered for high-speed intercontinental commercial and passenger aircraft. Many improvements in metallurgy, fuels, and manufacturing processes must occur before this will become a reality.

The potentially great speed of a scramjet vehicle obviously has military applications. One application could be in land- or sea-based missiles used as quick interceptors of an enemy's ballistic missile.

3.6   Velocity

You learned earlier that a distance such as "15 km" is a scalar because a single piece of information completely describes it. But a displacement such as "3 km west" is a vector because it requires two pieces of information for its description. Like distance, a given speed such as "30 km/h" is a scalar. When a direction is designated as well (for example, "10 m/s due west"), the quantity is a vector called **velocity (v).** To calculate average speed, you used the total distance traveled. To find the *average velocity* ($\overline{\mathbf{v}}$), you must find the ratio of the displacement (**d**) to the time interval (Δt) during which the displacement occurred:

Average velocity is the displacement divided by the time interval:
$$\overline{\mathbf{v}} = \frac{\mathbf{d}}{\Delta t}$$

$$\overline{\mathbf{v}} = \frac{\mathbf{d}}{\Delta t} \qquad (3.2)$$

In one-dimensional motion, velocity is parallel to a coordinate axis (just like displacement), so it can be denoted with a scalar symbol and a subscript. For example, the velocity vector parallel to the y-axis is denoted v_y.

What is the difference between $|\Delta x|$ in the average speed equation and **d** in the average vector velocity equation? The distance $|\Delta x|$ (or s) is always positive. Displacement (either **d** or d_x) may be positive or negative in one-dimensional motion and represents only the direct distance between two positions. The *magnitudes* of s and **d** will be equal if the object travels in only one direction. Thus a car traveling east on a straight road might have an average speed of 50 km/h and also a velocity of 50 km/h due east. However, if the moving object changes direction during the time interval, its speed no longer equals the magnitude of the average velocity over the time interval. Consider, for instance, the earlier example of a campsite and a picnic area across the lake. Suppose the family car travels along the 10 km road at a speed of 50 km/h and took 12 min, or $\frac{1}{5}$ h. Because the car's ultimate (net) change in position was only 2 km and not 10 km, the average velocity vector is 10 km/h due north.

Average velocity vector = 2 km north $\div \frac{1}{5}$ h = 10 km/h north.

A position-time graph shows the change in an object's velocity. For example, turn to Figure 3-5b on page 53 and find the points on the graph where the slope of the line changes dramatically. The value of the slope (including the sign) between each pair of these points equals the average velocity for that interval. You can calculate velocities for the train by finding the slope of each of these four intervals:

$$0 \text{ h to } 0.30 \text{ h: } \overline{v}_x = \frac{d_x}{\Delta t} = \frac{14 \text{ km} - 0 \text{ km}}{0.30 \text{ h} - 0 \text{ h}}$$
$$\overline{v}_x \doteq +46.6 \text{ km/h}$$

$$0.50 \text{ h to } 0.80 \text{ h: } \overline{v}_x = \frac{d_x}{\Delta t} = \frac{38 \text{ km} - 14 \text{ km}}{0.80 \text{ h} - 0.50 \text{ h}}$$
$$\overline{v}_x = +80. \text{ km/h}$$

$$1.00 \text{ h to } 1.44 \text{ h: } \overline{v}_x = \frac{d_x}{\Delta t} = \frac{60. \text{ km} - 38 \text{ km}}{1.44 \text{ h} - 1.00 \text{ h}}$$
$$\overline{v}_x = +50. \text{ km/h}$$

$$1.70 \text{ h to } 2.00 \text{ h: } \overline{v}_x = \frac{d_x}{\Delta t} = \frac{48 \text{ km} - 60 \text{ km}}{2.00 \text{ h} - 1.70 \text{ h}}$$
$$\overline{v}_x = -40. \text{ km/h}$$

3-9 Trains often go straight for miles on tracks in the Midwest United States.

The average velocity for the last interval is negative because the train reverses direction. During the intervals $0.30-0.50$ h, $0.80-1.00$ h, and $1.44-1.70$ h, the train has stopped at a station, and the initial and final positions are unchanged, so the train's speed is zero.

Figure 3-10 shows this data plotted on a **velocity-time graph.** On such a graph, lines above the horizontal zero velocity line show positive velocity (moving in the positive direction from the origin). Curves below this line are negative velocities (moving in the negative direction).

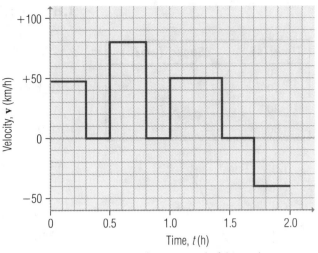

3-10 A velocity-time graph of the train between Ames and Lamoille stations, Iowa

3.7 Acceleration—Changing Velocity

The last problem was easy because it assumed that the train's velocity was uniform during each interval and that the velocity jumped instantly from one interval to the next (for instance, at 0.50 h the train instantly gained 80 km/h!). But you know from driving in a car that velocity changes frequently and smoothly—not in instantaneous jumps. This change in velocity with time is called **acceleration (a).** *Average acceleration* (**ā**) is the change in velocity (Δ**v**) divided by the time interval (Δt) of the velocity change.

The quantity Δ**v** is a **difference vector.** You will learn more about difference vectors in Chapter 4. When an acceleration occurs, the initial and final velocities can have different magnitudes *and* different directions. However, in one-dimensional motion, the initial and final velocities must be parallel to the line of motion (but may be in opposite directions). Therefore, Δ**v** can be expressed in scalar-subscript notation as

$$\Delta v_x = v_x - v_{0x}.$$

This statement means that the change in velocity in the *x*-direction is the difference of the initial and final velocities in the *x*-direction. The velocity difference points in the direction of the algebraic difference of the two velocities.

Consequently, acceleration is just the ratio of the velocity difference vector and the time required to change velocities:

$$\overline{\mathbf{a}} = \frac{\Delta \mathbf{v}}{\Delta t} \tag{3.3}$$

Acceleration is a vector of magnitude $\Delta v / \Delta t$ pointing in the same direction as Δ**v.** In one-dimensional motion, $\overline{a}_x$ may be positive, negative, or zero. A *positive* acceleration shows either an increase of speed in the positive direction or a decrease of speed in the negative direction. A *negative* acceleration shows a decrease of speed in the positive direction or an increase of speed in the negative direction. Zero acceleration means that velocity is constant.

You can find acceleration from a velocity-time graph. An object's average acceleration for an interval is the slope of the line joining the endpoints of the interval on the graph. Figure 3-11 is a hypothetical velocity-time graph of a car on a straight street. Note that speed varies nonuniformly with time. The average acceleration between times t_0 and t_1 is the slope of the dashed line. The *instantaneous acceleration* at a point is the slope of the tangent to the velocity-time curve at that point. Average acceleration does not have to equal the instantaneous acceleration from moment to moment.

Do not confuse a velocity-time curve with a position-time curve. The slope of a *position-time* curve is the *velocity;* the slope of a *velocity-time* curve is the *acceleration.*

Recall that scalar-subscript notation is reserved *only* for vectors that lie parallel to the coordinate axis along which position is measured.

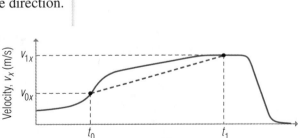

3-11 Speed normally varies continuously with time.

3.8 Uniformly Accelerated Motion

Time (s)	Speed (km/h)
0	5
10	42
20	79
30	116

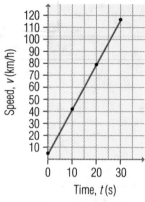

3-12 Velocity-time graph of the coasting car

EXAMPLE 3-4

Out of Gas: Uniform Acceleration

A man driving a car on a straight road runs out of fuel at the top of a long, smooth hill. There is a gas station in the town in the valley at the bottom of the hill. The driver does not use his brakes while coasting in order to get as close as possible to the town. Assume that the car coasts freely down the hill to the level valley below. The driver notes his speedometer reading every 10 s. His readings are tabulated in the table, and Figure 3-12 shows a velocity-time graph constructed from the driver's data. What is the car's average acceleration for the trip down the hill?

Solution:

Assume that the car's coordinate axis is parallel to the slope of the hill and positive is downhill. The origin is placed at the top of the hill. Since the car's speed is in one direction, its speed can be plotted directly on a velocity-time graph. You can find the average acceleration by calculating the slope of the line joining the point of initial speed (0 s, 5 km/h) and the final speed (30 s, 116 km/h).

$$\bar{a}_x = \frac{\Delta v_x}{\Delta t} = \frac{(v - v_0)_x}{\Delta t}$$

$$\bar{a}_x = \frac{116 \text{ km/h} - 5 \text{ km/h}}{30.\text{ s} - 0 \text{ s}} = \frac{111 \text{ km/h}}{30.\text{ s}} = \frac{3.7 \text{ km}}{\text{h·s}}$$

Notice that the units of hours and seconds are mixed in the answer. Convert the acceleration to the proper SI units (m/s^2) as follows:

$$\bar{a}_x = \frac{3.7 \text{ km}}{\text{h·s}} \left| \frac{1 \text{ h}}{3600 \text{ s}} \right| \frac{1000 \text{ m}}{1 \text{ km}} \doteq 1.02 \text{ m/s}^2$$

$$\bar{a}_x \approx 1.0 \text{ m/s}^2 \text{ downhill} \qquad \text{(2 SDs allowed)}$$

Uniformly accelerated motion exists when the acceleration of an object is constant.

In this example, the velocity-time curve is a straight line, so its slope is constant. The car's acceleration must also be constant. Therefore, the average acceleration ($\bar{a}_x$) is the same as the instantaneous acceleration (a_x). This kind of motion is called **uniformly accelerated motion.** We will study the significance of uniformly accelerated motion in the next section.

3A Section Review Questions

1. Mathematical descriptions of change of position, rate of motion, and acceleration of objects fall within what area of physics?

2. Under what one condition may an object's displacement and the distance it has traveled between two positions be equal?

3. What is the difference between average speed and instantaneous speed?

4. State the kinematic quantity that is equivalent to the slope of
   a. a position-time graph.
   b. a velocity-time graph.

5. What kind of motion does a negative acceleration describe?

6. Sketch and label the following graphs:
   a. Velocity-time graph of a delivery truck traveling at a uniform 50 km/h.
   b. Position-time graph of a fire ant (*Solenopsis invicta*) crawling the length of a 150 mm pencil during a 3-minute interval at a uniform rate.
   c. Velocity-time graph of a freight yard switch engine initially moving at 8 km/h. During a 20 s interval the engine slows uniformly to a stop and then rests for 1 minute. During the next 20 s interval it uniformly speeds back up in the opposite direction to 8 km/h.

7. Sketch a velocity-time graph showing a continuously varying speed during a time interval. Prove using the sketch that, for at least one point during the interval, the instantaneous velocity equals the average velocity for the interval.

8. Assume that a yo-yo drops 1.0 m vertically from its owner's hand at a uniform speed of 0.80 m/s along its string. At the end of the string, it changes direction in 0.10 s and rises toward the owner's hand at a uniform speed of 0.60 m/s. Compute the average acceleration of the yo-yo at the bottom end of the string as it changes direction.

3B THE EQUATIONS OF MOTION

Many problems in physics involve approximately constant acceleration, represented as a straight portion of a velocity-time graph. The acceleration at any point along the straight-line segment is simply the slope of the line. Since this approximation of motion is common, we will develop equations to analyze it.

3.9 The First Equation of Motion

The acceleration formula Equation 3.3 can be rewritten as

$$\overline{a}_x = \frac{v_x - v_{0x}}{\Delta t}.$$

What if you know the initial velocity and that the acceleration is constant but you want to find the final velocity? Solving for v_x,

$$\overline{a}_x \Delta t = v_x - v_{0x}, \text{ or}$$

$$v_x = v_{0x} + \overline{a}_x \Delta t. \tag{3.4}$$

This is the **first equation of motion.** This equation is also true for two or three dimensions using standard vector notation. However, for simplicity, we will retain the scalar-subscript form of vector notation for the equations of motion because we will always analyze motion with respect to only one coordinate axis at a time.

In algebra, you learned the working equation for a straight line:

$$y = mx + b,$$

where m is the line's slope and b is the point where the line crosses the y-axis (y-intercept). Similarly, the equation

$$v_x = v_{0x} + \overline{a}_x \Delta t,$$

when rearranged as

$$v_x = + \overline{a}_x \Delta t + v_{0x},$$

The x-subscripts indicate that these vector quantities are all parallel to the same coordinate or position axis.

Problem-Solving Strategy 3.3
The first equation of motion is
$$v_x = v_{0x} + \overline{a}_x \Delta t.$$

is the equation of a straight line on a velocity-time graph. Its slope is $\bar{a}_x$ and it crosses the velocity axis at v_{0x}. If you know the slope of a line and one point on the line, you can construct a linear graph. Therefore, if you know an object's constant acceleration and its velocity at a given time, you can construct a velocity-time graph.

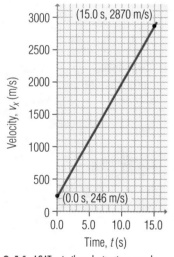

3-13 Launching an ASAT missile from an Air Force F-15 Eagle.

3-14 ASAT missile velocity-time graph

EXAMPLE 3-5

Antisatellite Missile: Graphing Velocity Versus Time

During the 1980s the United States developed a special missile system to destroy enemy spy satellites in low Earth orbits. The weapon was called the air-launched antisatellite missile, or air-launched ASAT missile. It was fired from an Air Force F-15 Eagle aircraft at a precise angle, speed, and altitude toward a region in the sky ahead of the target satellite. The missile destroyed the satellite by ramming into it at high speed. No explosives were used.

The missile launches itself when the aircraft is 11.6 km high at an angle of 65.0° to the ground (see Figure 3-13). The initial speed of the missile is 246 m/s when its engine fires. The average acceleration of the missile during the 15.0 s powered portion of its flight is 175 m/s².

Draw a velocity-time graph for the missile's flight while its engine is burning. Assume the missile flies a straight path for the entire 15.0 s interval.

Solution:

Choose the x-coordinate axis parallel to the missile's path, increasing upward. The origin is at the point of launch. Find the coordinates of two points on the line.

Given: $v_{0x} = +246$ m/s, $\bar{a}_x = +175$ m/s²
Substitute the known information into the first equation of motion.

$$v_x = v_{0x} + \bar{a}_x \Delta t$$

$$v_x = (+246 \text{ m/s}) + (+175 \text{ m/s}^2)\Delta t$$

Find two points on the line:
1. For $t_0 = 0.0$ s, $v_{0x} = +246$ m/s; one point is (0.0 s, 246 m/s).
2. For $t = 15.0$ s, $\Delta t = 15.0$ s − 0.0 s = 15.0 s.

$$v_x = (+246 \text{ m/s}) + (+175 \text{ m/s}^2)(15.0 \text{ s})$$

$$v_x \doteq +2871 \text{ m/s}$$

$$v_x \approx 2870 \text{ m/s} \qquad \text{(3 SDs allowed)}$$

The second point is (15.0 s, 2870 m/s).
Figure 3-14 is a velocity-time graph of the line through the two points.

3.10 Determining Displacement Algebraically

In Example 3-5, it is possible to calculate how far the missile flies during the powered portion of its flight from the information given. The displacement of the missile is the distance it flies from the time the missile is released until its engine cuts out. Knowing that

$$\bar{v}_x = d_x / \Delta t,$$

you can conclude that

$$d_x = \bar{v}_x \Delta t. \qquad (3.5)$$

You already know the change in time (Δt), and the average velocity during an interval is

$$\bar{v}_x = \frac{v_{0x} + v_x}{2}. \qquad (3.6)$$

Substitute Equation 3.6 into Equation 3.5.

$$d_x = \left(\frac{v_{0x} + v_x}{2}\right)\Delta t$$

$$d_x = \tfrac{1}{2}(v_{0x} + v_x)\Delta t \qquad (3.7)$$

From Example 3-5, you know that the initial velocity of the missile is $v_{0x} = +246$ m/s, its final velocity is $v_x = +2870$ m/s, and the powered flight time interval is $\Delta t = 15.0$ s. Substituting this information into Equation 3.7, you find that:

$$d_x = \tfrac{1}{2}[(+246 \text{ m/s}) + (+2870 \text{ m/s})](15.0 \text{ s}) \doteq +23\,\underline{3}70 \text{ m}$$

$$d_x \approx +23.4 \text{ km} \qquad \text{(3 SDs allowed)}$$

Therefore, the missile flies about 23.4 km powered by its rocket engine.

3.11 Determining Displacement Graphically

The area under the curve of a velocity-time graph is numerically equal to the displacement of the moving object. Figure 3-15 is the velocity-time graph of the ASAT missile. Because it is a straight line, the shaded area can be divided into a triangle and a rectangle. The area of the rectangle is base × width. For the rectangular area in the graph, the base = Δt, and the width = $v_{0x} - 0$ m/s, as shown. For any triangle, area is $\tfrac{1}{2}$(base × height). Again, the base of the triangular area = Δt, and its height = $v_x - v_{0x}$.

The total area of the region under the graph is the sum of the rectangular and triangular areas:

$$A_{\text{total}} = A_{\square} + A_{\triangle}$$

$$A_{\text{total}} = v_{0x}\Delta t + \tfrac{1}{2}(v_x - v_{0x})\Delta t \qquad (3.8)$$

Factoring and combining terms gives

$$A_{\text{total}} = \tfrac{1}{2}(v_{0x} + v_x)\Delta t.$$

Notice the similarity between this and Equation 3.7, which we have already used to calculate an answer of 23.4 km.

$$d_x = \tfrac{1}{2}(v_{0x} + v_x)\Delta t$$

Both methods yield the same answer.

3.12 The Second Equation of Motion

Sometimes it is not an object's displacement but its position that you want to find. For one-dimensional motion, the position is related to displacement by the equation

$$d_x = x - x_0,$$

Problem-Solving Strategy 3.4

The average velocity during an interval is one-half the sum of the initial and final velocities.

$$\bar{v}_x = \frac{v_{0x} + v_x}{2}$$

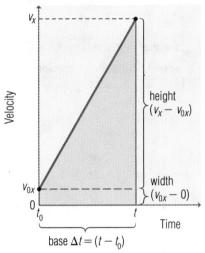

3-15 Using area on a velocity-time graph to estimate displacement

The area of the region under a graph is often related in some way to the product of the variables that produce the graph. This is the basis for integral calculus.

or rearranged,

$$x = x_0 + d_x.$$

You have just seen in Equation 3.8 that

$$d_x = v_{0x}\Delta t + \frac{1}{2}(v_x - v_{0x})\Delta t.$$

Recall that

$$\overline{a}_x = \frac{v_x - v_{0x}}{\Delta t}, \text{ or}$$

$$\overline{a}_x\Delta t = v_x - v_{0x}.$$

Substituting $\overline{a}_x\Delta t$ for $v_x - v_{0x}$ in Equation 3.8 gives

$$d_x = v_{0x}\Delta t + \frac{1}{2}(\overline{a}_x\Delta t)\Delta t, \text{ or}$$

$$d_x = v_{0x}\Delta t + \frac{1}{2}\overline{a}_x(\Delta t)^2. \tag{3.9}$$

Problem-Solving Strategy 3.5

The second equation of motion is

$$d_x = v_{0x}\Delta t + \frac{1}{2}\overline{a}_x(\Delta t)^2.$$

Substituting the definition of displacement for d_x in Equation 3.9 and solving for x, we obtain

$$x = x_0 + v_{0x}\Delta t + \frac{1}{2}\overline{a}_x(\Delta t)^2. \tag{3.10}$$

Equations 3.9 and 3.10 are two forms of the **second equation of motion.** Observe that this formula is a quadratic equation in the form $y = c + bx + ax^2$, where the time interval (Δt) is the independent variable and the position (x) is the dependent variable. The average acceleration ($\overline{a}_x$), initial velocity (v_{0x}), and initial position (x_0) are constants. A plot of time-position ordered pairs using this equation will produce a parabolic curve.

3.13 The Third Equation of Motion

A third equation is useful when the time interval is not known or is irrelevant. Since

$$\overline{a}_x = \frac{v_x - v_{0x}}{\Delta t}, \text{ then}$$

$$\Delta t = \frac{v_x - v_{0x}}{\overline{a}_x}. \tag{3.11}$$

Also, you know from Equation 3.7 that

$$d_x = \frac{1}{2}(v_{0x} + v_x)\Delta t.$$

Now substitute Equation 3.11 for Δt above, and rearrange the terms to get

$$d_x = \frac{v_x + v_{0x}}{2} \cdot \frac{v_x - v_{0x}}{\overline{a}_x},$$

and simplify:

Problem-Solving Strategy 3.6

The third equation of motion is

$$d_x = \frac{v_x^2 - v_{0x}^2}{2\overline{a}_x}.$$

$$d_x = \frac{v_x^2 - v_{0x}^2}{2\overline{a}_x} \tag{3.12}$$

$$x = x_0 + \frac{v_x^2 - v_{0x}^2}{2\overline{a}_x} \tag{3.13}$$

Equations 3.12 and 3.13 are two forms of the **third equation of motion.** You need to learn only the displacement form of the second and third equations of motion since the position form can be easily derived from the definition of displacement.

The three equations of motion are helpful in solving most problems in *straight-line, constant acceleration* motion.

	Second Equation	**Third Equation**
Displacement form	$d_x = v_{0x}\Delta t + \frac{1}{2}\,\bar{a}_x\,(\Delta t)^2$	$d_x = \dfrac{v_x^2 - v_{0x}^2}{2\bar{a}}$
Position form	$x = x_0 + v_{0x}\Delta t + \frac{1}{2}\,\bar{a}_x\,(\Delta t)^2$	$x = x_0 + \dfrac{v_x^2 - v_{0x}^2}{2\bar{a}_x}$

Problem-Solving Strategy 3.7

Use the *displacement form* of the equations of motion when only displacement is involved in a problem. Use the *position form* of the equations if position information is involved.

EXAMPLE 3-6

Stopping the Train: Negative Acceleration

A train is traveling at 26.8 m/s. If it brakes at a constant acceleration of -0.500 m/s², how far will the train continue to roll from the time it starts to brake until it comes to a complete stop?

Solution:

Assume the x-coordinate axis is oriented parallel to the train's motion and the train is moving in the positive direction. You want to find the *displacement* (d_x) for the unknown time interval (Δt) in which the train's velocity decreases from 26.8 m/s to 0.0 m/s.

Given: $\bar{a}_x = -0.500$ m/s², $v_{0x} = 26.8$ m/s, and $v_x = 0.0$ m/s.

3-16 Long freight trains have a large mass and are hard to stop.

Easier Solution:
Substitute the known values into the third equation of motion:

$$d_x = \frac{v_x^2 - v_{0x}^2}{2\bar{a}_x}$$

$$d_x = \frac{(0\ \text{m/s})^2 - (26.8\ \text{m/s})^2}{2(-0.500\ \text{m/s}^2)}$$

$$d_x = \frac{-718.2\ \text{m}^2/\text{s}^2}{-1.000\ \text{m/s}^2} \doteq 718.2\ \text{m}$$

$$d_x \approx 718\ \text{m} \quad \text{(3 SDs allowed)}$$

More Difficult Solution:
First, solve for the time interval while the train was braking, using the first equation of motion:

$$v_x = v_{0x} + \bar{a}_x\Delta t$$

$$0\ \text{m/s} = 26.8\ \text{m/s} + (-0.500\ \text{m/s}^2)\Delta t$$

$$\Delta t = \frac{0\ \text{m/s} - 26.8\ \text{m/s}}{-0.500\ \text{m/s}^2}$$

$$\Delta t = 53.6\ \text{s}$$

Now substitute Δt into the second equation of motion:

$$d_x = v_{0x}\Delta t + \frac{1}{2}\bar{a}_x(\Delta t)^2$$

$$d_x = (26.8\ \text{m/s})(53.6\ \text{s}) + \frac{1}{2}(-0.500\ \text{m/s}^2)(53.6\ \text{s})^2$$

$$d_x = 1436\ \text{m} - 718.2\ \text{m}$$

$$d_x \doteq 718\ \text{m}$$

$$d_x \approx 720\ \text{m} \quad \text{(2 SDs allowed)}$$

Both methods of solving this problem give essentially the same answer, but the first method is much more direct. It is almost always quicker to use one equation rather than two. Also, using two equations rather than one can introduce excessive rounding errors. This illustrates one way that a systematic approach to problem solving can save you time and improve precision.

3.14 Free Fall

One common example of uniformly accelerated motion is **free fall,** the fall of an object under the influence of gravity alone with negligible air resistance. The magnitude of the acceleration due to gravity $|g|$ is almost constant near the earth's surface. We will assume that it is exactly 9.81 m/s². By convention, for problems involving gravitational acceleration, *up* is considered positive. Therefore, the gravitational acceleration vector at the earth's surface is $\mathbf{g} = g_y = -9.81$ m/s². The subscript notation assumes that positive y is up.

The equations of motion can be adapted to problems involving free fall. The following equations are given only to show you that gravitational acceleration can be substituted directly for average acceleration in the equations of motion. Therefore, memorizing another set of equations is not necessary.

First equation: $\qquad\qquad\qquad v_y = v_{0y} + g_y\Delta t$

Second equation: $\qquad\qquad d_y = v_{0y}\Delta t + \tfrac{1}{2}g_y(\Delta t)^2$

Third equation: $\qquad\qquad d_y = \dfrac{v_y^2 - v_{0\,y}^2}{2g_y}$

3-17 Skydivers experience free fall for the initial portion of their descent.

Free fall is the vertical motion of an object under the influence of gravity alone.

Problem-Solving Strategy 3.8

Use the symbol g_y to represent the vector $\mathbf{g}$ when all motion is parallel to the y-axis. Indicate the magnitude of $\mathbf{g}$ using $|g|$, $|g_y|$, or $|\mathbf{g}|$.

3-18 Dropping stones into a canyon is *not* recommended.

EXAMPLE 3-7

Dropping a Stone: Free Fall

A physics student drops a stone from a scenic overlook on the rim of a canyon into the canyon below. (*Note: Dropped* means that the stone had zero initial velocity.) (a) Ignoring air resistance, what is its velocity after 4.0 s? (b) How far did the stone fall in that time?

Solution:
Given: $v_{0y} = 0$ m/s; $\Delta t = 4.0$ s; $a_y = g_y = -9.81$ m/s²

Find v_y when $t = 4.0$ s and d_y for the interval from $t = 0$ s to $t = 4.0$ s.

Assume that the y-coordinate system is oriented vertically, positive upward, and the origin is at the point where the stone is released.

To find v_y, use the first equation of motion.

$$v_y = v_{0y} + g_y\Delta t$$
$$v_y = (0 \text{ m/s}) + (-9.81 \text{ m/s}^2)(4.0 \text{ s})$$
$$v_y \doteq -39.2 \text{ m/s}$$
$$v_y \approx -39 \text{ m/s}$$

The negative sign indicates that the stone is falling downward.

To find the vertical displacement, use the second equation of motion.

$$d_y = v_{0y}\Delta t + \tfrac{1}{2}g_y(\Delta t)^2$$
$$d_y = (0 \text{ m/s})(4.0 \text{ s}) + \tfrac{1}{2}(-9.81 \text{ m/s}^2)(4.0 \text{ s})^2$$
$$d_y \doteq 0 \text{ m} - 78.4 \text{ m}$$
$$d_y \approx -78 \text{ m}$$

The negative sign of the displacement indicates that the stone moved in the downward (negative y) direction.

Galileo
(1564–1642)

Near the end of the Middle Ages, Western scholars rediscovered Aristotle's works. Soon Aristotle gained such prestige that few scholars even thought of contradicting his teachings. After the noted theologian Thomas Aquinas wrote a book reconciling Aristotle's philosophical and scientific ideas with church tradition, even the Roman Catholic church hierarchy was reluctant to believe Aristotle to be wrong.

This was the intellectual climate into which Galileo Galilei, an Italian, was born. Galileo became a student of mathematics. He was curious, and he had the unusual idea (for his time) that the way to find out about the world was to observe it. He experimented to find out if heavy objects really do fall faster than lighter ones and concluded that they do, but only because of differences in air resistance. Upon studying astronomy, Galileo became convinced that Copernicus was right in his theory that the earth revolves around the Sun. Since Aristotle's idea that the earth is motionless had been venerated for so long, Galileo's belief made him unpopular with the intellectual leaders of his day. These leaders, especially the older ones, had trouble believing that the philosophy they had followed for so long might be flawed. They persuaded some Catholic priests to declare to their parishioners that Galileo's beliefs about the earth contradicted the Bible and church tradition. Galileo was then denounced before the Inquisition, and Copernicus's book was banned until it could be "corrected."

Although Copernicus was held to be in error, his theory was not declared heresy, and it was still legal to discuss his theory. Galileo wanted to present Copernicus's idea to the public, so he obtained permission from the pope to write a book presenting both views of the earth. The pope agreed, on the condition that Galileo include a statement that, however convincing Copernicus's theory was, it was contrary to divine revelation and therefore wrong.

Galileo's book presented Copernicus's theory to be so convincing that, if the Church had not spoken, any thinking person would have believed it. It was written as a dialogue, with the Catholic Church's official position defended by a character named Simplicio. Even the pope's statement, almost in a direct quotation, was attributed to Simplicio. The Catholic Church, angered by this insult, insisted that Galileo again face the Inquisition. This time Galileo was convicted as a heretic and put under house arrest for the rest of his life, and his book was banned. It is interesting that the pope never declared Copernicus's theory heretical even after Galileo's conviction. However, the Catholic Church hierarchy officially denied the Copernican theory until 1922, and Galileo himself was not exonerated until 1999.

Galileo published one more book after his trial, an account of his experiments with motion. His experiments and deductions form the basis for modern kinematics.

3B Section Review Questions

1. Write the first equation of motion. Under what condition(s) is this equation valid?

2. What quantities must be known in order to find the displacement of an object using the second equation of motion?

3. Which equation of motion should you use if you do not know the elapsed time of an object's motion?

4. Discuss the differences, if any, between the one-dimensional quantities $\mathbf{d}$, $|\mathbf{d}|$, d, and d_x.

● 5. You throw a ball for your dog Procyon to fetch. "Pro" runs from right next to you to the ball lying 22 m away and brings it directly back to you. The round trip takes 11 s. You pat his head and give him a treat.
   a. What was the net change in Procyon's position?
   b. What was the total distance covered by the dog?
   c. What was the dog's average speed?
   d. What was the dog's average velocity for the entire interval?

● 6. The eccentric orbit of Pluto carries it as far as 7.304×10^9 km from the Sun every 248 Earth years. At this distance, how long, in hours, will it take a photon of light traveling at 3.00×10^8 m/s to travel between the Sun and Pluto?

● 7. A triple-A dragster can accelerate from zero to 440. km/h in 5.0 s. The adjacent table records the vehicle's speed logged by a recording device every second during the race heat.
   a. Draw a velocity-time graph for the dragster using the data in the table.
   b. Was the acceleration uniform for the entire heat? Explain your answer.
   c. Calculate the average acceleration for the 5.0 s interval.

● 8. Refer to Example 3-5. After the engine of the air-launched ASAT missile shuts down, assume that the missile coasts along a straight path. What is the maximum distance from the satellite at which the engine could cut out and still ensure that the missile is traveling at least 100. m/s upon collision with the target? Assume that it experiences a uniform acceleration of -8.89 m/s^2 parallel to its path after the engine quits. Give your answer in both kilometers and statute miles.

Dragster Velocity	
Time (s)	Speed (km/h)
0.0	0
1.0	20.0
2.0	60.0
3.0	120.0
4.0	240.0
5.0	440.0

Chapter Review

In Terms of Physics

mechanics	3.1	velocity-time graph	3.6
kinematics	3.2	acceleration (**a**)	3.7
dynamics	3.2	difference vector	3.7
displacement (**d**)	3.3	uniformly accelerated motion	3.8
position-time graph	3.4	free fall	3.14
velocity (**v**)	3.6		

Problem-Solving Strategies

3.1 (page 52) When graphing time-dependent data, always plot time on the horizontal axis and plot the dependent variable on the vertical axis.

3.2 (page 53) Whenever you must find the change of a quantity, always subtract the initial value from the final value.

3.3 (page 61) The first equation of motion is
$$v_x = v_{0x} + \bar{a}\Delta t.$$

3.4 (page 63) The average velocity during an interval is one-half the sum of the initial and final velocities.
$$\bar{v}_x = \frac{v_{0x} + v_x}{2}$$

3.5 (page 64) The second equation of motion is
$$d_x = v_{0x}\Delta t + \tfrac{1}{2}\,\bar{a}_x(\Delta t)^2.$$

3.6 (page 64) The third equation of motion is
$$d_x = \frac{v_x^2 - v_{0x}^2}{2\bar{a}_x}.$$

3.7 (page 65) Use the *displacement form* of the equations of motion when only displacement is involved in a problem. Use the *position form* of the equations if position information is involved.

3.8 (page 66) Use the symbol g_y to represent the vector **g** when all motion is parallel to the y-axis. Indicate the magnitude of **g** using $|g|$, $|g_y|$, or $|\mathbf{g}|$.

Review Questions

1. What does the study of kinematics describe?

2. What is the net change of position called?

3. How would you determine the average speed of a trip to Grandma's house?

4. How does increasing the number of data points on a position-time graph improve your ability to estimate instantaneous velocity at any point on the curve?

5. Describe the motion of an object if its velocity-time graph is horizontal.

6. What kind of information is determined by the slope of a velocity-time graph?

7. What is the general shape of a graph produced by
   a. the first equation of motion?
   b. the second equation of motion?

8. What is the area under a velocity-time graph, bounded by a specific time interval, numerically equal to?

9. a. Identify one common form of uniform acceleration.
   b. Give an example of an object that would exhibit this kind of acceleration.

10. What vector has a magnitude equal to the instantaneous speed?

11. Which of the following quantities are *always* positive?
    a. v c. $|\vec{v}|$ e. x g. $|\vec{g}|$
    b. d_x d. $\bar{v}_y$ f. g_y h. a

12. State which of the following are vector quantities. Which are scalars?
    a. distance traveled   e. acceleration
    b. displacement   f. change in velocity
    c. average velocity   g. change in position
    d. instantaneous speed   h. change in time

True or False (13–19)

13. Only magnitudes that are positive are scalar quantities.

14. Average speed is never negative.

15. The slope of a line between two points on a velocity-time graph is the average velocity for that time interval.

16. Average speed is the total distance traveled divided by the time interval for the trip.

17. You must know the time interval (Δt) in order to use the third equation of motion.

18. The scalar g and the vector $\mathbf{g}$ are the same thing.

19. The average speed and the magnitude of the average velocity vector are equal only when total path distance and the magnitude of displacement are the same.

⊚20. A railroad freight yard switch engine on a straight track travels 1000. m to the end of the track, then back 600. m, then forward 300. m.
   a. What is the switch engine's final displacement from its starting position?
   b. How far did it travel?

⊚21. A sprinter in training runs along a straight 50.0 m track. At every 10.0 m is an observer with a stopwatch. The sprinter begins to run at the sound of the starting gun. The observers start their stopwatches at that same instant. Each observer stops his stopwatch as the runner passes him. After one trial

run, the observers' stopwatches have the readings listed in the following table.
   a. Construct a position-time graph for this sprint.
   Calculate the sprinter's average speed for
   b. the entire run.
   c. the first 30.0 m.
   d. the last 20.0 m.

Observer	Time (s)
start	0.00
1	1.10
2	2.20
3	3.30
4	4.10
5	4.90

The following graph represents the motion of a trolley along a straight track. Use the information presented here to answer questions 22–27.

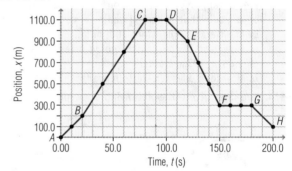

22. During which interval did the trolley move the farthest?
    a. A to B c. E to F
    b. B to C d. F to G

23. During which interval was the trolley's average speed the greatest?
    a. A to B c. E to F
    b. B to C d. F to G

⊚24. Calculate the trolley's average speed for the interval in the answer to question 23.

⊚25. What is the trolley's speed at (90 s, 1100 m)?

⊚26. During which interval(s) is the trolley's speed 15.0 m/s?

⊚27. Using the position-time graph above, sketch a velocity-time graph for the trolley. Does the trolley undergo constant acceleration?

⊚28. An auto manufacturer boasts that its newest sports car model "goes from 0 to 60 (mph) in 5.00 seconds!"
    a. What is the car's acceleration (in m/s^2)?
    b. How far does it travel as it accelerates from 0.0 mph to 60.0 mph (in meters)?

⊚29. You are stuck on a two-lane road behind a truck traveling 64.4 km/h (40 mph). You pull into the oncoming lane to pass and immediately begin accelerating at 2.50 m/s^2. After passing the truck, 6.10 s have elapsed. You stop accelerating and pull back into your lane. How far down the road did you travel in the oncoming lane (in meters) during the time interval?

Motion in One Dimension   **69**

Use the velocity-time graph below to answer questions 30–32.

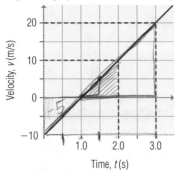

Time, *t* (s)

●30. What happens to the object's direction of motion at 1.00 s?

●31. Calculate
   a. the average velocity from 0.50 s to 1.50 s.
   b. the average acceleration from 0.50 s to 1.50 s.

●32. What is the displacement
   a. between 0.00 s and 1.00 s?
   b. between 1.00 s and 2.00 s?
   c. between 0.00 s and 3.00 s?

●33. A science fiction author describes a space vehicle launching system consisting of a long linear track built at an angle into a mountain range on the Moon. The space vehicle is mounted →not moving on a cradle that is accelerated by a magnetic field generated by the track at a constant 100. m/s².
   a. How fast is the vehicle going after it has traveled the 2.88×10^4 m track?
   b. How long does it take to reach this speed?

●●34. You are given the following four position-time graphs:

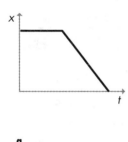

a

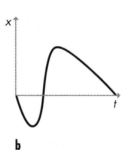

b

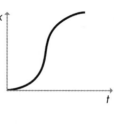

c

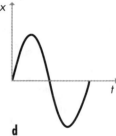

d

Indicate which of the velocity-time graphs below corresponds best with each of the position-time graphs by pairing the appropriate letters.

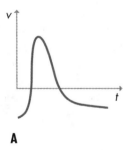

A

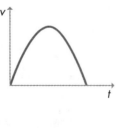

B

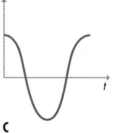

C

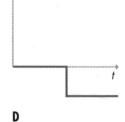

D

●●35. Your average running speed is 8.0 m/s and your best friend's is 7.0 m/s. You both are 65 m from the front door of the school when the class bell rings, at which time you both sprint to get to class. How far are you in front of your friend when you reach the door of the school?

4A	Properties of Vectors and Scalars	73
4B	Operations with Vectors: Geometric Techniques	76
4C	Operations with Vectors: Analytical Techniques	80

New Symbols and Abbreviations

vector angle (θ)	4.2
reference angle (α)	4.8
position vector (r)	4.8

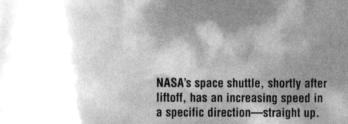

NASA's space shuttle, shortly after liftoff, has an increasing speed in a specific direction—straight up.

Vectors and Scalars

> *[Paul was a] Hebrew of the Hebrews; as touching the law, a Pharisee; Concerning zeal, persecuting the church; touching the righteousness which is in the law, blameless. But what things were gain to me, those I counted loss for Christ. Yea doubtless, and I count all things but loss for the excellency of the knowledge of Christ Jesus my Lord: for whom I have suffered the loss of all things, and do count them but dung, that I may win Christ. Philippians 3:5–8*

Paul's life contains a striking contrast. Before his conversion he was Saul, zealous in persecuting God's people; yet we know him as Paul, the great apostle and missionary, leading many to personal faith in Christ. He was always zealous, but on the road to Damascus God dramatically changed Paul's direction. The world may teach that sincerity and personal integrity are the supreme personal values. However, sincerely and faithfully pursuing selfish or sinful goals cannot please God. If Christ has saved you, your ultimate fulfillment is found in submitting your life to God and pursuing His course for you as zealously as you may have pursued your own desires in the past. In other words, it is not just the magnitude of your zeal but its direction that reveals whether your life is being lived well.

4A PROPERTIES OF VECTORS AND SCALARS

4.1 Defining Vectors and Scalars

A manager gives a customer directions to her store: "Three blocks north and four blocks west of City Hall." The customer's trip can be represented by two arrows on a map, each having a specific length and direction.

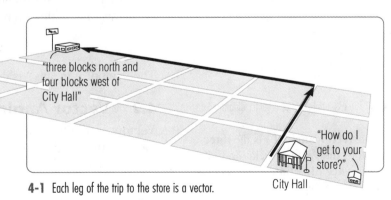

4-1 Each leg of the trip to the store is a vector.

4-2 The velocity vector describes both speed and direction.

An action done without a guiding purpose is like a scalar—it has no direction. The same action done with a specific purpose is like a vector. Our actions are similar to positive vectors and negative vectors. First Corinthians 10:31 says, "Whether therefore ye eat, or drink, or whatsoever ye do, do all to the glory of God." In other words, "Let all your action-vectors glorify God."

The *magnitude* of a quantity is the absolute value of its numerical value. As such, magnitudes are always positive. It is important to remember that while all magnitudes are scalars, not all scalars are magnitudes.

The magnitude of **A** is written *A* or |**A**|. The magnitude of a vector is also called its *modulus*. The magnitude of a vector is graphically indicated by the relative length of the vector arrow.

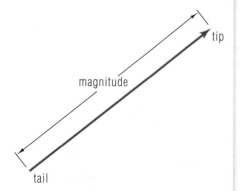

4-4 The essential parts of a vector diagram

The **vector angle (θ)** is the direction of the vector in relation to its **reference direction,** the direction of zero angle.

An artillery officer knows that the muzzle velocity of a fired cannon projectile is 930 m/s. He must raise the barrel of his cannon to an elevation of 33° in order to hit a target. The initial speed of the projectile can be represented by a long arrow labeled 930 m/s at an angle of 33° above a horizontal line.

An engineer for a tire manufacturing company is designing a new, low-profile radial tire. In order to understand how the tire will respond to hard braking, he sketches a diagram of the various forces at work on the tire. He draws arrows representing the weight of the vehicle, the angular momentum of the wheel, braking torque, and the tire's friction with the road. Large quantities are represented by long arrows and small ones by short arrows. The diagram helps the engineer understand the dynamics needed to create a computer model of the tire design.

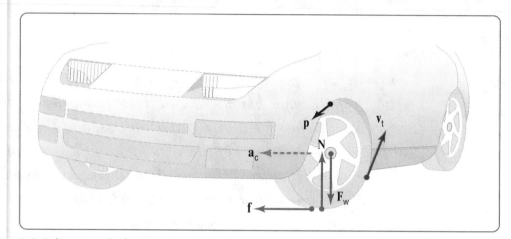

4-3 Each quantity in the diagram is represented by a vector.

In each case above, **vectors** are used to study quantities that have two pieces of information—a specific numerical value and a direction. Physicists use vectors to study dimensions such as displacement, velocity, force, acceleration, and magnetic fields. Vectors are drawn as arrows on *vector diagrams.*

Physicists also work with scalars. A **scalar** is a quantity that can be described completely by only a single numerical piece of information. Time, mass, volume, and temperature are all scalars because, unlike vectors, they have no direction associated with them. Some scalars are only positive, such as mass and time. Others, such as temperature, may be positive or negative, depending on the instant of measurement.

A vector symbol is printed with boldface roman type, **A.** Since writing boldface letters is difficult, you may use a regular letter with an arrow over it, $\vec{A}$. Upright lines around the vector, |**A**|, stand for its absolute value and specify its magnitude. As a shorthand notation for the magnitude, either the italicized letter or the absolute value sign will be used; that is,

$$A \equiv |\mathbf{A}| \text{ (handwritten, } A \text{ or } |\vec{A}|\text{)}.$$

The symbol ≡ means "is defined as."

4.2 Vector Angles

Vectors in diagrams provide important information. The tail of the arrow can indicate the starting point of motion or the point at which a force is physically applied. The length of the arrow is proportional to the magnitude of the vector quantity. Vector diagrams also include directional information indicated by the orientation of the arrow. The **vector angle,** labeled with the Greek letter θ, is given a numerical value in degrees or radians measured from a fixed direction. This **reference direction** is usually determined by convention or is selected for convenience in working the problem. This is especially important in two- and three-dimensional vectors. For one-dimensional vectors, direction can be indicated by a plus or minus sign, as stated in Chapter 3.

There are some *conventions* (generally agreed-upon rules) for establishing reference directions for angular measurement. This will be especially applicable to two-dimensional motion, which you will study in the next chapter. In the Cartesian (two-dimensional) plane, the reference direction is the positive *x*-axis. Positive angles are measured counterclockwise and negative angles are measured clockwise around the origin. There is no limit to the size of geometric angles. An angle can be measured in either degrees or radians.

Map directions are always referenced to geographic north, which by convention is located at the *top* of the map or diagram. Angles referenced to true north are indicated by a capital "T" in place of the degree symbol. Map directions are usually measured clockwise from north. For example, 315T is northwest. When using north as the reference direction, angles are always positive and are reported only in degrees from 0° to 360°. This textbook will always give map directions using three digits (which is standard practice for navigational purposes). In this system, north is 000T, northeast is 045T, south is 180T, and so on. On the other hand, if a problem gives an angle with the degree symbol (e.g., 120°), then you know that a Cartesian reference direction is being used.

4.3 Transporting Vectors

Drawing accurate vector diagrams requires a protractor and a ruler. For most physics problems, it is sufficient to sketch the approximate magnitude and orientation of each vector. Unless specifically requested, sketches don't have to be made exactly to scale.

Vectors are equal if they have both the same magnitude and the same direction. Examine vectors **A** and **B** in Figure 4-7. Even though they are at different locations in the diagram, they have identical magnitudes and they point in the same direction, so **A** = **B**. Remember that vectors are models of the physical quantities that they represent. If necessary for the sake of a problem solution, it is perfectly permissible to transport a vector from one location to another in a diagram as long as you maintain its original length and orientation (see Figure 4-8).

A marine biologist is periodically monitoring the location of a submerged humpback whale (*Megaptera novaeangliae*) using a sonic transponder. He plots its

Cartesian angles

$\theta = 100°$

$\theta = -30°$

Map angle

250T

r

4-5 The method for measuring vector angles depends on the kind of angle.

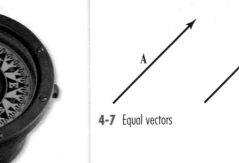

4-6 A compass rose shows the cardinal and intercardinal directions, as well as degrees.

Map direction angles referenced to true north are followed by the letter "T." Some maps include magnetic north as a north reference. Angles in this system are followed by the letter "M."

Modern directional compass scales are subdivided into 360°. The four principal compass directions, or *cardinal directions,* are represented by angles that are 90° apart: north is 000T, east is 090T, south is 180T, and west is 270T. *Intercardinal* directions (NE, SE, SW, and NW) are 45° from their associated cardinal directions.

Another way of reporting map directions is based on the acute angle north or south of either east or west. For example, southwest is called 45° south of west and northeast is 45° north of east.

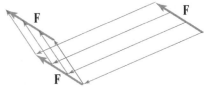

A   B

4-7 Equal vectors

F   F

F

4-8 Vector transport

Problem-Solving Strategy 4.1

Vectors in a diagram can be transported from one point to another as long as their lengths and orientations remain unchanged.

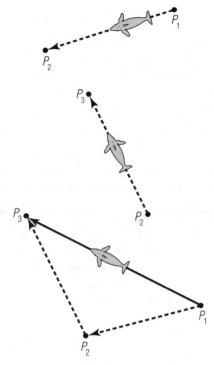

4-9 Positions and displacement of a whale

If you had a map of the lands around the Aegean Sea, how many vectors could you construct from the following passage? "And when they had seen the brethren, they comforted them, and departed [from Philippi]. Now when they had passed through Amphipolis and Apollonia, they came to Thessolonica, where was a synagogue of the Jews" (Acts 16:40; 17:1).

position at three different times on a nautical chart. The whale's position with time is shown in Figure 4-9. The distance the whale moves between the first and second positions is represented by the length of the line segment between P_1 and P_2, written P_1P_2. The whale's *displacement* is indicated by an arrow drawn from P_1 to P_2 (labeled $\overrightarrow{P_1P_2}$). Displacement is a vector quantity because it specifies both distance and direction. Distance is a scalar because only the magnitude of the displacement, a single piece of information, is given.

If the whale moves on to P_3, the total distance that it traveled through the water from P_1 is the sum of the distances P_1P_2 and P_2P_3. Since the total distance is length alone, it is still just a scalar quantity. The whale's total displacement, however, is the vector from P_1 to P_3. The length of this vector is clearly different from the distance that the whale traveled. Vectors cannot be added by simple arithmetic. Fortunately, vector addition is not difficult to understand.

4A Section Review Questions

1. What is the key difference between scalar and vector quantities?
2. State whether the following quantities are scalars, vectors, or neither.
   a. force
   b. one-way sign
   c. pressure
   d. π (3.14 . . .)
   e. distance
   f. displacement
   g. velocity
   h. speed
   i. mass
   j. temperature
   k. Wednesday
   l. density
3. When dealing with vectors, what is the difference between **B** and *B?* Write another symbol that is equivalent to *B*.
4. Compare and contrast the properties of angles measured in a Cartesian coordinate system to those measured on a map.
 ⊛5. Why can you transport a vector from one place to another in a vector diagram?

4B OPERATIONS WITH VECTORS: GEOMETRIC TECHNIQUES

When more than one quantity is involved in a problem, vectors often need to be added, subtracted, multiplied, and even divided. They may be multiplied and divided by scalars as well. Mathematicians have developed methods to simplify these operations and yield useful results. The geometric techniques of vector math can be done quickly with only a ruler and a pencil.

4.4 Adding Equal Vectors by Scalar Multiplication

The simplest case of vector addition is adding a vector equal to itself. Graphically, this is accomplished by placing the tail of the second vector at the tip of the first. Their sum is the vector drawn from the tail of the first vector to the tip of the second (see Figure 4-10). The vector sum, called the **resultant,** is twice as long as the original vector and oriented in the same direction.

$$\mathbf{V} + \mathbf{V} = 2\mathbf{V}$$

The bold plus sign indicates vector addition. Note that this vector sum is also the product of a scalar number and the original vector. When a vector is multiplied by a positive scalar number, its length (magnitude) changes but its direction stays the same. The magnitude of vector $2\mathbf{V}$ is $2V$.

In Figure 4-11, if $\mathbf{V}$ is a vector 6 units long, $2\mathbf{V}$ would be a vector 12 units long pointing in the same direction as $\mathbf{V}$. If the scalar multiplier is negative, the resulting vector points in the opposite direction from the original. Thus, $-\mathbf{V}$ (that is, $-1 \times \mathbf{V}$) is a vector 6 units long pointing in the opposite direction from $\mathbf{V}$. Scalar coefficients may also be fractions, so $-\frac{1}{2}\mathbf{V}$ (or $-\frac{\mathbf{V}}{2}$) is a vector 3 units long oriented opposite to $\mathbf{V}$.

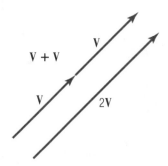

4-10 Adding a vector to itself—simple scalar multiplication

> The sum or difference of two or more vectors is called the **resultant.**

4.5 Adding Unequal Vectors Graphically

How are two different vectors added—for example, $\mathbf{A} + \mathbf{B}$? If they have different magnitudes but the same direction, then their resultant ($\mathbf{R}$) is a vector whose magnitude is $R = A + B$, and it has the same direction as the original vectors. Again, the resultant vector extends from the tail of the first vector to the tip of the second. Observe that there is no difference in the resultant $\mathbf{R}$ between the operations $\mathbf{A} + \mathbf{B}$ and $\mathbf{B} + \mathbf{A}$ (see Figure 4-12). In other words, vector addition is *commutative.*

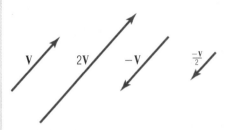

4-11 Scalar multiples of vector $\mathbf{V}$

In most physics problems, however, vectors are usually unequal—they don't have the same magnitude and rarely point in the same direction. Let's look at a two-dimensional example. A student pilot flies 30 km north and then 40 km west on her solo flight, shown in Figure 4-13. The two flight legs are displacements that can be represented by vectors. To find the plane's net displacement from the airfield at the end of the second leg, you must sum the two leg vectors. Using the graphical method, each leg is drawn to scale on a sheet of paper or a map. Since vectors can be transported without changing their mag-

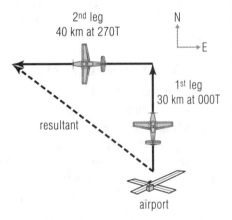

2nd leg
40 km at 270T

1st leg
30 km at 000T

resultant

airport

4-13 When adding vectors, the resultant stretches from the tail of one vector to the tip of the other.

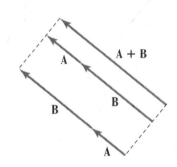

4-12 Commutative character of vector addition

nitude or direction, you can place the vectors head to tail in either order. As before, the sum of the two vectors is the resultant vector that stretches from the tail of one vector to the tip of the other.

The resultant is identical to the path the aircraft could have taken to fly directly from the airport to its position at the end of the second leg. You can find the magnitude of the resultant displacement vector graphically by measuring its length with a ruler and applying the map's scale factor. The airplane is found to be 50 km from the airfield. Notice that this is *not* the arithmetic sum of 30 km and 40 km. When the directions of the vectors are not identical, the magnitude of the resultant is not equal to the sum of the magnitudes of the two legs,

$$R \neq A + B.$$

The direction of the resultant is also different from either of the original vectors. You can measure the compass direction (vector angle) from the airport using

> The resultant of two vectors can often be represented as the diagonal of a parallelogram with the sides formed by the two vectors placed tail to tip.

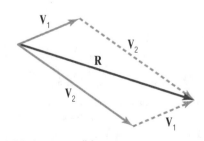

4-14 A vector parallelogram

a protractor. Measured clockwise from north using navigational angles, the aircraft bears 307T from the airport.

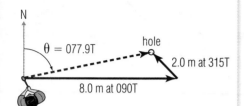

N

$\theta = 077.9T$

hole

2.0 m at 315T

8.0 m at 090T

4-15 What would have been the right first putt?

EXAMPLE 4-1

Putting Around with Golf

A golfer is attempting to sink a putt. The ball rolls 8.0 m to the east and misses the hole. He putts again 2.0 m to the northwest and the ball finally drops into the cup. How far and in what direction should he have putted the ball in order to sink the shot on the first attempt?

Solution:

Refer to Figure 4-15. Draw vectors to scale that represent each putt and place them so that the tail of the second touches the head of the first (northwest is halfway between west and north, or 315T). Then draw the resultant vector from the tail of the first vector to the head of the second. According to the diagram scale, the resultant vector has a length of about 6.9 m. The angle as measured from north (the top of the diagram) using a protractor is 077.9T.

Note that when the resultant must be measured from a diagram, each vector must be drawn accurately to scale.

Three or more vectors may be added in a similar way. Place the tail of each vector at the tip of the previous one. Their resultant is the vector drawn from the tail of the first to the tip of the last. Because vector addition is commutative, the order of the vectors does not matter.

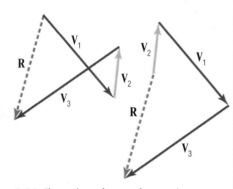

4-16 The resultant of a sum of vectors does not depend on the order of addition.

> A vector's opposite is a vector with the same length pointing in the opposite direction.

4.6 Vector Subtraction

In the expression $\mathbf{A} - \mathbf{B}$, the bold negative sign indicates vector subtraction. Recall from algebra that subtraction is equivalent to adding the opposite of a number—subtracting 3 is the same as adding -3. The vector expressions $\mathbf{A} - \mathbf{B}$ and $\mathbf{A} + (-\mathbf{B})$ are equivalent. Remember that a vector's opposite has the same length but points in the opposite direction. Once you have found the opposite vector, vector addition is carried out to find the resultant. Note that, as with arithmetic subtraction, vector subtraction is *not* commutative; that is, $\mathbf{A} - \mathbf{B}$ is not the same as $\mathbf{B} - \mathbf{A}$ (see Figure 4-17).

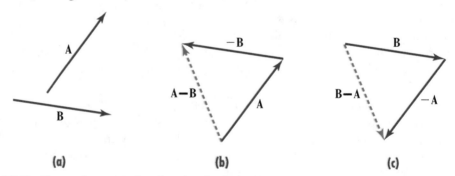

4-17 The difference of two vectors depends on the order of subtraction.

A second method exists for easily finding the graphical difference of two vectors. Place the two vectors tail to tail, then draw the resultant vector from the tip of the second vector in the difference equation to the tip of the first.

A **difference vector** is the resultant of a vector subtraction operation (see Example 4-2). This method is useful for determining the *change* in a vector quantity.

> **Problem-Solving Strategy 4.2**
>
> Quickly find the difference of two vectors ($\mathbf{A} - \mathbf{B}$) by placing them tail-to-tail and drawing the **difference vector** from the tip of the second ($\mathbf{B}$) to the tip of the first ($\mathbf{A}$).

We saw a difference vector in Chapter 3 when we determined average acceleration, where

$$\bar{\mathbf{a}} = \frac{\Delta \mathbf{v}}{\Delta t}.$$

In this case, $\Delta \mathbf{v}$ is the difference vector of the velocity measured at two different times.

EXAMPLE 4-2

Vector Subtraction
Given the vectors **C** and **D**, find **C−D**.

Solution—Two Methods:
Adding the opposite of the second to the first:

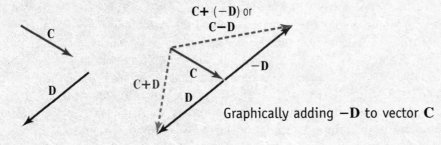

Graphically adding −**D** to vector **C**

Direct vector subtraction:

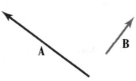

Note that the vector **C−D** is the same using either method.

4B Section Review Questions

1. What is the sum of two or more vectors called?

2. Compare the vector 4**B** with the vector **B**.

3. Compare the vector **A** and the vector $-\frac{\mathbf{A}}{3}$.

Copy vectors **A** and **B** onto your paper. Refer to them for Questions 4–6.

⊚4. Graphically find the resultant of the following operations.

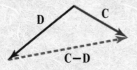

   a. **A+B**

   b. **B+A**

   c. **A−B**

   d. 2**B+A**

⊚5. Describe the difference between the vector **A−B** and the vector **B−A.**

⊚6. Determine the vector 2**B−2A** using the second method discussed in this section for finding the difference of two vectors. Describe how the vector 2**B−2A** compares to the vector **B−A.** Show how this illustrates the distributive property of equality.

⊚7. You are given two vectors, **F** and **G,** so that $F = 1$ unit and $G = 2$ units. Sketch the vector sum **F+G** so that $|\mathbf{F+G}| = 1$.

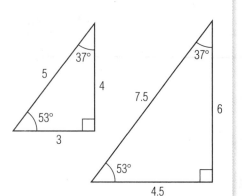

4-18 Similar triangles

The shorter sides of a right triangle are called the *legs* of the triangle. The *hypotenuse* is always opposite the right angle.

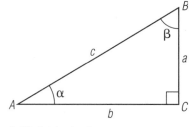

4-19 A right triangle

The angle measure, such as α, β, or θ, is called an *argument* of the trigonometric ratios sin, cos, and tan, and must be included. You should not say, "The sine is 0.866," but rather, for an angle α, "The sine of α is 0.866."

The *adjacent side* to an acute angle in a right triangle is the shorter of the two sides that form the angle. The longer side is the hypotenuse.

OPERATIONS WITH VECTORS: ANALYTICAL TECHNIQUES

Graphical vector addition using sketches provides quick, convenient results but lacks precision and accuracy. Even carefully drawn vector diagrams are subject to drawing errors. For physics, mathematical techniques involving trigonometry allow you to perform vector addition with much greater accuracy.

4.7 Solving Triangles

In solving two-dimensional vectors, you will have to solve for unknown quantities in vector triangles using the mathematical tools taught in algebra and geometry. To help you find the measure of angles and lengths of sides in triangles, you must know the rudiments of trigonometry. We will review some of the more important trigonometric concepts here.

One useful principle of trigonometry is that two triangles are **similar triangles** when the three angles of one triangle equal the corresponding three angles of the other triangle. The angles, not the sides, determine similarity. Compare the triangles in Figure 4-18, which have different side lengths but equal corresponding angles. Upon closer inspection you should see that the lengths of the sides are indeed related: the ratios between corresponding sides of triangles A and B are all the same. In fact, all similar triangles have proportional sides. Physicists and mathematicians use this important relationship to solve vector triangles.

By far the most common triangle found in physics problems is the **right triangle,** which has one 90° angle. Right triangles possess several properties that make it easy to determine unknown sides and angles. Study Figure 4-19 carefully before you proceed further. The two acute angles, which are less than 90°, are labeled A and B. The measure of angle A is α and the measure of angle B is β. Side a is opposite acute angle A, while side b is opposite acute angle B. The hypotenuse, the line segment opposite the right angle C, is labeled c.

Because the sum of the interior angles of a triangle is always 180°, you can determine an unknown acute angle in a right triangle if you know the measure of the other acute angle. Similarly, if you know two of the sides of a right triangle, you can find the third according to the Pythagorean theorem.

$$a^2 + b^2 = c^2 \qquad (4.1)$$

A significant benefit of using a right triangle is that one of the angles is *always* known—the right angle. As long as either two sides or an acute angle and a side are known, all the angles and all the sides in a right triangle can be determined mathematically. This can be accomplished by taking advantage of the proportionality of similar triangles. The values of ratios of corresponding sides in similar triangles are always the same. The ratios of the hypotenuse or sides that are either adjacent or opposite to a specified angle are given special names.

The **sine** of an acute angle is the ratio of the opposite side's length to the length of the hypotenuse. From Figure 4-19, you can identify the sines of angles α and β, which are written "sin α" and "sin β."

$$\sin\alpha = \frac{\text{opposite}}{\text{hypotenuse}} = \frac{a}{c}$$

$$\sin\beta = \frac{\text{opposite}}{\text{hypotenuse}} = \frac{b}{c}$$

Similarly, the **cosine** of an angle is the ratio of the length of the angle's adjacent side to the length of the hypotenuse. The cosine of α is written "cos α."

$$\cos\alpha = \frac{\text{adjacent}}{\text{hypotenuse}} = \frac{b}{c}$$

$$\cos\beta = \frac{\text{adjacent}}{\text{hypotenuse}} = \frac{a}{c}$$

The **tangent** of an acute angle is the ratio of the length of the side opposite the angle to the length of its adjacent side. The tangent of α is written "tan α."

$$\tan\alpha = \frac{\text{opposite}}{\text{adjacent}} = \frac{a}{b}$$

$$\tan\beta = \frac{\text{opposite}}{\text{adjacent}} = \frac{b}{a}$$

The values of sine, cosine, and tangent for any angle are easily calculated using modern calculators.

When the names of trigonometric ratios are spoken, never use their abbreviated forms but rather their complete names—sine, cosine, and tangent.

EXAMPLE 4-3

Finding an Unknown in a Right Triangle

Study Figure 4-20. You are given one side, a, and the measure of one acute angle, β, in a right triangle. Solve for side b and the measure of acute angle A, α.

Solution:
1. You can find side b, using the tangent for angle β.

$$\tan\beta = \frac{\text{opposite}}{\text{adjacent}}$$

$$\tan 17° = \frac{b}{a} = \frac{b}{8.2}$$

$$b = 8.2 \tan 17°$$

$$b \doteq 2.5$$

Whenever possible, perform all math operations involving trigonometric ratios simultaneously.

2. You can find the value of α, knowing that the angles of any triangle add up to 180°.

$$\alpha + \beta + 90° = 180°, \text{ so}$$

$$\alpha = 180° - 90° - 17°$$

$$\alpha = 73°$$

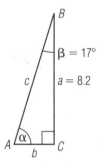

4-20 If you know one side and one acute angle in a right triangle, you can find the other unknowns.

Given only the sides of a right triangle, how can you find a measure of the two unknown acute angles? You could draw a scale model and measure the angles with a protractor, but this technique is not precise. Instead, it is better to use one of your calculator's inverse trigonometric operators (usually shown as a secondary function to the corresponding trig ratio keys—$\sin^{-1}$, $\cos^{-1}$, and $\tan^{-1}$). The argument for these operators is the ratio of the appropriate sides, and the output is the corresponding angle.

Given the right triangle in Figure 4-21, suppose you want to find the measure of the angle whose opposite side is 7 units and whose adjacent side is 8 units.

$$\tan\alpha = \frac{\text{opposite}}{\text{adjacent}} = \frac{7}{8} = 0.8750$$

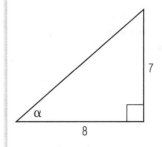

4-21 Finding α from its tangent ratio

Using the inverse tangent function on your calculator gives you α directly:

$$\alpha = \tan^{-1}\left(\frac{7}{8}\right)$$

$$\alpha \doteq 41.2°$$

The inverse tangent equation above may be read "Alpha is the angle whose tangent is 0.8750." The exponent -1 in an inverse trigonometric operator does not mean "reciprocal," so do not invert the expression.

4.8 Vector Components

Before we look at the analytical methods for adding and subtracting vectors, we first need to develop the concept of **component vectors.** Every vector lying in a plane is the resultant of the sum of two special vectors called component vectors or **components.** To understand the relationship of components to the original vector, imagine two flashlights shining on a vector as in Figure 4-22. Note that the light beams are parallel to the vertical and horizontal axes of the vector's coordinate system. The shadows of the vector cast by the flashlights are projected onto each axis. These shadows represent the component vectors. In fact, components are often called *projections* of the vector onto the coordinate axes. Components are parallel to the coordinate axes and are therefore perpendicular to each other.

Adding vectors analytically involves combining their x-components and their y-components to obtain the components of the resultant vector. Quickly finding the components of a vector is therefore very important. This is where trigonometry comes into the picture. When a vector and its components are sketched together in a triangle, they form a *vector triangle*. The vector triangle for any given vector is a right triangle because the component vectors that form the legs (shorter sides) of the triangle are perpendicular to each other. As long as the vector's length and angular direction are known, you can find its components using trigonometry.

Vector components are labeled with the original vector's symbol and include a subscript of the axis to which the component is parallel. For example, the x-component of the displacement vector $\mathbf{d}$ is $\mathbf{d}_x$.

The angle used when determining a vector's components is usually the acute angle between the vector and its horizontal coordinate axis (that is parallel with the horizontal component). We will call this the **reference angle (α).** The reference angle may or may not be the same as the vector angle (θ) depending on the reference direction used for the problem. Regardless of the orientation of the original vector, the reference angle is always a positive acute angle within the vector triangle. When the reference angles for several vectors are required in a problem, you should label each angle with its vector name as a subscript (e.g., α_A, α_B).

The scalar values of the component vectors are called their **scalar components.** While the magnitude of a component vector is always positive, the scalar value may be positive, negative, or zero, depending on the orientation of the original vector. The sign indicates the direction of the component relative to its coordinate axis. Positive scalar components point up and to the right. Negative

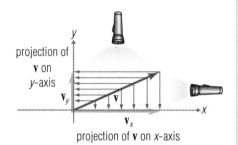

4-22 The projection of components onto the coordinate axes

TABLE 4-1		
Relationship Between θ and α		
Quadrant	θ	α
I	$\theta = \alpha$	$\alpha = \theta$
II	$180° - \alpha$	$180° - \theta$
III	$180° + \alpha$	$\theta - 180°$
IV	$360° - \alpha$	$360° - \theta$
NE	$90° - \alpha$	$90° - \theta$
SE	$90° + \alpha$	$\theta - 90°$
SW	$270° - \alpha$	$270° - \theta$
NW	$270° + \alpha$	$\theta - 270°$

scalar components point down and to the left. The one-dimensional vectors discussed in Chapter 3 had only one component—the vector itself.

When the component vectors of a given vector have been identified, we say that the vector has been *resolved into its components*. Completing the following steps will help you resolve vectors:

- *Select a convenient coordinate system.* Usually the coordinate axes will be vertical and horizontal, but other orientations are possible depending on the situation (see Figure 4-23).
- *Note the kind of directional system used,* Cartesian or map.
- *Determine the reference angle (α) for the vector.* Identify this angle and determine its magnitude from the geometry of the problem.
- *Find the magnitude of the horizontal scalar component:* $|V_x| = V \cos \alpha$, *where V is the magnitude of the vector* **V.**
- *Find the magnitude of the vertical scalar component:* $|V_y| = V \sin \alpha$.
- *Assign the correct signs to the scalar components based on their orientations to the coordinate axes.* The positive reference directions in the Cartesian system are the positive *x*- and *y*-axes. In the map system, positive is north and east. Always refer to the problem sketch when assigning scalar component signs.

Problem-Solving Strategy 4.3

Make it a habit to quickly identify the reference angles for all vectors in a problem. Resolving vector components then becomes a straightforward process.

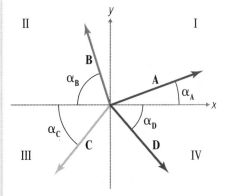

4-23 Reference angles of vectors in each quadrant

Problem-Solving Strategy 4.4

When resolving a vector, always assign the signs of its scalar components based on the original vector's orientation in the problem sketch. Always start the solution to a problem with an accurate sketch.

EXAMPLE 4-4

Resolving Vector Components

Vector **L** points at an angle of 37° above the *x*-axis of a graph and has a length of 15 units. Find the lengths of its components.

Solution:

Sketch vector **L** at the approximate vector angle given in the problem (see Figure 4-24). The vector forms the hypotenuse of a vector right triangle. The two unknown component vectors are the legs of the triangle.

The reference angle α is the acute angle between the horizontal side and the hypotenuse. Label the resulting right triangle. Once this is done, find the lengths of the two sides.

$$\sin \alpha = \frac{\text{opposite}}{\text{hypotenuse}}$$

Substituting the values for α and the hypotenuse gives

$$\sin 37° = \frac{|L_y|}{15}.$$

Solve for $|L_y|$.

$$|L_y| = (15 \text{ units}) \sin 37°$$

$$|L_y| \doteq 9 \text{ units}$$

Similarly,

$$\cos \alpha = \frac{\text{adjacent}}{\text{hypotenuse}} \quad \text{or} \quad \cos 37° = \frac{|L_x|}{15}.$$

Solve for $|L_x|$.

$$|L_x| = (15 \text{ units}) \cos 37°$$

$$|L_x| \doteq 12 \text{ units}$$

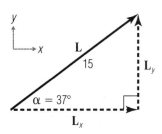

4-24 Find the vector components.

The original vector **L** can be thought of as the vector sum of vector **L**$_x$ and **L**$_y$.

$$\mathbf{L} = \mathbf{L}_x + \mathbf{L}_y, \text{ where}$$

L$_x$ = 12 units long pointing in the positive *x*-direction, and
L$_y$ = 9 units long pointing in the positive *y*-direction.

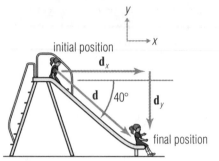

4-25 What are the components of the girl's displacement?

EXAMPLE 4-5

Resolving Displacement on a Slide

A little girl slides down a playground slide 4.0 m long. The slide forms a 40° angle with the horizontal. Determine the vertical and horizontal components of the girl's displacement from the top of the slide to the bottom.

Solution: Refer to Figure 4-25.

The girl's displacement, **d,** is the hypotenuse of the vector right triangle formed by the vertical and horizontal components, **d**$_y$ and **d**$_x$ respectively. From the problem diagram, we can see that $d = 4.0$ m and $\alpha = 40.°$.

$$|d_y| = d \sin \alpha = (4.0 \text{ m})\sin 40.° \doteq 2.\underline{5}7 \text{ m downward}$$

$$d_y \approx -2.6 \text{ m (downward is negative } y)$$

$$|d_x| = d \cos \alpha = (4.0 \text{ m})\cos 40.° \doteq 3.\underline{0}6 \text{ m to the right}$$

$$d_x \approx +3.1 \text{ m (right is positive } x)$$

The girl's displacement components are 2.6 m downward and 3.1 m to the right.

Vectors described using a map system of directions are resolved in the same way as vectors in a Cartesian coordinate system. The cardinal map directions must first be redefined as Cartesian axes. The east-west line becomes the *x*-axis and the north-south line the *y*-axis. Resolve the vectors as usual. Components may be labeled **V**$_{N-S}$ and **V**$_{E-W}$ to indicate that they are map vector components.

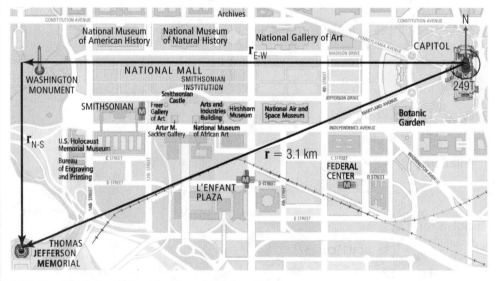

4-26 Map of the National Mall and its environs, Washington, D.C.

EXAMPLE 4-6

Where Is the Jefferson Memorial?

The Thomas Jefferson Memorial lies on a compass bearing 249T from the United States Capitol Building in Washington, D.C. Its distance from the Capitol is 3.1 km. Determine the components of the Jefferson Memorial's position vector (**r**) relative to the Capitol Building.

Solution: Refer to Figure 4-26.

The monument's position vector angle θ is 249° from north, or 249T. In order to find the vector components, the reference angle α must be determined.

Recall that α is the acute angle between the vector and the horizontal axis. For map problems, the horizontal axis is the east-west direction. Therefore, the reference angle in this problem is measured from the west direction to the position vector: $\alpha = 270T - 249T = 21°$. (Note that the difference of two directions is an *angle* and not a direction.)

Now the east-west and north-south components may be easily found.

$$|r_{N\text{-}S}| = r \sin \alpha = (3.1 \text{ km})\sin 21° \doteq 1.\underline{1}1 \text{ km}$$

$$r_{N\text{-}S} \approx -1.1 \text{ km (the north-south component points south)}$$

$$\mathbf{r}_{N\text{-}S} = 1.1 \text{ km south}$$

$$|r_{E\text{-}W}| = r \cos \alpha = (3.1 \text{ km})\cos 21° \doteq 2.\underline{8}9 \text{ km}$$

$$r_{E\text{-}W} \approx -2.9 \text{ km (the east-west component points west)}$$

$$\mathbf{r}_{E\text{-}W} = 2.9 \text{ km west}$$

A **position vector** (**r**) is a vector in two- or three-dimensional space that locates the position of a system. The vector's tail is placed at the origin of the coordinate system and its head is at the position of the system.

Component vectors may be summed to find the original vectors. This may seem to be an obvious statement, but you will need this skill to determine resultant vectors. First, let us look at a graphical example.

EXAMPLE 4-7

Obtaining Vectors from Their Components

Sketch the vectors that have the following components:

a. $K_x = 2$, $K_y = 5$ b. $P_x = 3$, $P_y = -1$

Solution:

a.

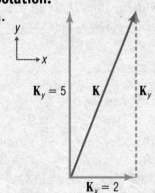

b.

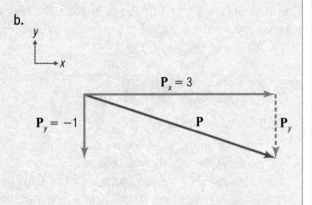

Three-dimensional vectors can be resolved into three components. In this case, the components are parallel to the x-, y-, and z-axes. By convention, the x- and y-axes lie in the horizontal plane and the z-axis is vertical. The following example shows how we obtain a three-dimensional vector from its components.

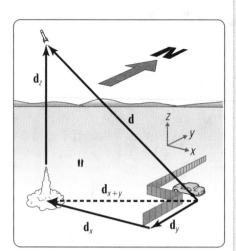

4-27 Displacement in three dimensions

A scalar component is to a component vector what the brand name "Boeing 747" is to the airplane itself. A vector is a physical phenomenon, and its vector symbol represents that physical entity. You cannot use a vector symbol in a strictly arithmetic operation any more than you can use the physical airplane in a sentence about the plane. Arithmetic and trigonometric operations work only with pure numbers. That is why you must convert components to equivalent scalar values before using them in nonvector equations.

EXAMPLE 4-8

Rocket Launch

A physics student carries a model rocket from his car 200. m south along a sidewalk and 250. m west into a field to the launch site. He then launches the rocket to an altitude of 450. m. Find the rocket's distance from the car when it reaches its highest altitude.

Solution:

If the car is considered the origin in a three-dimensional coordinate system, the x-axis marks the east-west direction, the y-axis marks the north-south direction, and the z-axis marks the vertical direction.

The final displacement is the vector sum of these three components:

$$\mathbf{d} = \mathbf{d}_x + \mathbf{d}_y + \mathbf{d}_z$$

The three scalar components are $d_x = -200.$ m, $d_y = -250.$ m, and $d_z = +450.$ m.

Examine Figure 4-27. The displacement of the rocket at its launch point on the ground is the vector sum of $\mathbf{d}_x$ and $\mathbf{d}_y$. The magnitude of this horizontal displacement is found using the Pythagorean theorem.

$$|\mathbf{d}_x + \mathbf{d}_y| = |\mathbf{d}_{x+y}| = \sqrt{d_x^2 + d_y^2} = \sqrt{(-200.\ \text{m})^2 + (-250.\ \text{m})^2}$$

$$|\mathbf{d}_{x+y}| \doteq 320.1\ \text{m}$$

Using this result as the base for the vertical displacement vector triangle, you can use the Pythagorean theorem again to find the magnitude of total displacement of the rocket, which is the distance you are looking for.

$$|\mathbf{d}_x + \mathbf{d}_y + \mathbf{d}_z| = |\mathbf{d}| = \sqrt{d_{x+y}^2 + d_z^2} = \sqrt{(320.1\ \text{m})^2 + (450.\ \text{m})^2}$$

$$|\mathbf{d}| \doteq 552.2\ \text{m}$$

$$d \approx 552\ \text{m}$$

4.9 Operations with Components

Components enable you to add or subtract vectors with great precision. Instead of using a protractor and a ruler, you use a calculator. Let's look at an example that allows us to put this tool to use.

EXAMPLE 4-9

Still Putting Around

Our impressive golfer attempts to sink a putt on the next hole. The ball rolls 4.0 m toward 070.0T and misses the hole (again). His second putt rolls 1.5 m toward 320.0T and drops into the cup. In what direction and how far should he have putted the ball to make the shot on the first attempt?

Solution: Refer to Figure 4-28.

Sketch the putt displacement vectors and label their magnitudes and vector angles. Develop the habit of sketching coordinate axes at the *tail* of each vector. Since the directions of the putts referred to compass directions, the axes should be north-south and east-west. Draw in the legs of the vector triangles, using each vector as a hypotenuse of its respective triangle. The legs must be parallel to the coordinate axes. Label the legs of the triangles to identify the components. The reference angles of the two displacement vectors are as follows:

$$\alpha_A = 90.0T - 70.0T = 20.0°$$

$$\alpha_B = 320.0T - 270T = 50.0°$$

The scalar components of **A** are A_{N-S} and A_{E-W}:

$$\left|A_{N-S}\right| = A \sin \alpha_A = (4.0 \text{ m})\sin 20.0° \doteq 1.\underline{3}6 \text{ m}$$

$$A_{N-S} = +1.\underline{3}6 \text{ m or } 1.\underline{3}6 \text{ m north}$$

$$\left|A_{E-W}\right| = A \cos \alpha_A = (4.0 \text{ m})\cos 20.0° \doteq 3.\underline{7}5 \text{ m}$$

$$A_{E-W} = +3.75 \text{ m or } 3.\underline{7}5 \text{ m east}$$

The scalar components of **B** are B_{N-S} and B_{E-W}:

$$\left|B_{N-S}\right| = B \sin \alpha_B = (1.5 \text{ m})\sin 50.0° \doteq 1.\underline{1}4 \text{ m}$$

$$B_{N-S} = +1.\underline{1}4 \text{ m or } 1.\underline{1}4 \text{ m north}$$

$$\left|B_{E-W}\right| = B \cos \alpha_B = (1.5 \text{ m})\cos 50.0° \doteq 0.\underline{9}6 \text{ m}$$

$$B_{E-W} = -0.\underline{9}6 \text{ m or } 0.\underline{9}6 \text{ m west}$$

The resultant **R** = **A** + **B**. But **R** also has components, so **R** = **R**$_{N-S}$ + **R**$_{E-W}$ (resultant vector components).

$$\mathbf{R}_{N-S} = \mathbf{A}_{N-S} + \mathbf{B}_{N-S} \text{ and } \mathbf{R}_{E-W} = \mathbf{A}_{E-W} + \mathbf{B}_{E-W}$$

$$R_{N-S} = 1.\underline{3}6 \text{ m} + 1.\underline{1}4 \text{ m} = +2.\underline{5}0 \text{ m (or } 2.\underline{5}0 \text{ m north)}$$

$$R_{E-W} = 3.\underline{7}5 \text{ m} + (-0.\underline{9}6 \text{ m}) = 2.\underline{7}9 \text{ m (or } 2.\underline{7}9 \text{ m east)}$$

From this information, the resultant that represents the putt that the golfer *wanted* to make can be found (see Figure 4-29).

The magnitude of **R** is the distance of the putt. Use the resultant scalar components in the Pythagorean theorem:

$$R = \sqrt{R_{N-S}^2 + R_{E-W}^2} = \sqrt{(2.\underline{5}0 \text{ m})^2 + (2.\underline{7}9 \text{ m})^2}$$

$$R \doteq 3.\underline{7}4 \text{ m}$$

$$R \approx 3.7 \text{ m}$$

Calculate the reference angle of the putt:

$$\alpha_R = \tan^{-1}\left(\frac{\text{opposite}}{\text{adjacent}}\right)$$

$$\alpha_R = \tan^{-1}\left(\frac{\left|R_{N-S}\right|}{\left|R_{E-W}\right|}\right) = \tan^{-1}\left(\frac{\left|+2.\underline{5}0 \text{ m}\right|}{\left|+2.\underline{7}9 \text{ m}\right|}\right)$$

$$\alpha_R \doteq 41.\underline{8}6°$$

$$\alpha_R \approx 41.9°$$

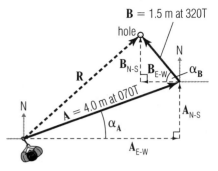

4-28 Find the resultant analytically.

The negative sign is used because the component B_{E-W} points west.

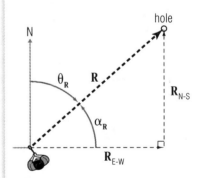

4-29 Adding resultant components

Compute the reference angle to the same precision as the precision of any angles given in the problem.

The original putt directions were map directions given with respect to north. The direction of **R** is 41.9° north of east. To find the compass direction, subtract 41.9° from 090.0T:

$$\theta_R = 048.1T$$

The required putt should have been 3.7 m in the direction of 048.1T.

Example 4-9 covers every step required for analytically adding two vectors. Notice that you must convert vectors to scalar components in order to perform algebraic or trigonometric operations on them. At each step your solutions should show the calculations and the values of the variables.

Because vectors can be used for displacements, velocities, accelerations, or forces, they can have a variety of different units. When you add or subtract two vectors, make sure that they have the same units.

4C Section Review Questions

1. What property of similar triangles allows the development of trigonometric ratios for any acute angle in a right triangle?

2. What two geometric principles permit you to find an unknown angle or an unknown side in a right triangle?

3. Name (give the symbols for) the two-dimensional Cartesian component vectors of vector **F.**

●4. Solve the right triangles for the indicated sides or angles.

a.

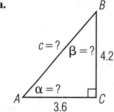

c.

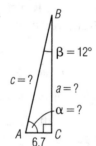

b.

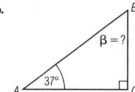

d.

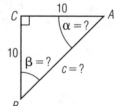

●5. Determine the Cartesian vector components of a vector, **r**, that is 3.4 units long at an angle of 132°.

●6. Using a ruler and a protractor, *graphically* add a vector that is 3.0 cm long at an angle of 230° to another vector that is 4.5 cm long at an angle of 0°.

●7. *Analytically* find $F_1 + F_2$ if F_1 is 8.50 units at an angle of 30.0° and F_2 is 10.25 units at an angle of 60.0°.

●8. Find the difference vector $F_1 - F_2$ in Question 7 using analytical techniques.

In Terms of Physics

vector	4.1
scalar	4.1
vector angle (θ)	4.2
reference direction	4.2
resultant	4.3
difference vector	4.6
similar triangles	4.7
right triangle	4.7

sine θ	4.7
cosine θ	4.7
tangent θ	4.7
component vectors	4.8
components	4.8
reference angle (α)	4.8
scalar components	4.8

Problem-Solving Strategies

4.1 (page 75) Vectors in a diagram can be transported from one point to another as long as their lengths and orientations remain unchanged.

4.2 (page 78) Quickly find the difference of two vectors ($\mathbf{A} - \mathbf{B}$) by placing them tail-to-tail and drawing the **difference vector** from the tip of the second ($\mathbf{B}$) to the tip of the first ($\mathbf{A}$).

4.3 (page 83) Make it a habit to quickly identify the reference angles for all vectors in a problem. Resolving vector components then becomes a straightforward process.

4.4 (page 83) When resolving a vector, always assign the signs of its scalar components based on the original vector's orientation in the problem sketch. Always start the solution to a problem with an accurate sketch.

Review Questions

1. What is the difference between a vector and a scalar?

2. State three physical quantities that can be represented by vectors. State three quantities that are scalars.

3. How will you show the difference between a vector symbol and its magnitude in your handwritten work? Give an example of each.

4. Write the angular compass directions for the four intercardinal points on a compass.

5. A model rocket is launched with an initial acceleration of 110 m/s² at an angle of 90.0° to the ground. Sketch a diagram of the rocket's initial acceleration vector (**a**) and appropriately label the sketch with all known information.

6. Draw vector **D**, shown in the diagram on your paper, and then draw the following scalar multiples of the vector.

   a. 4**D**   **b.** ⅓**D**   **c.** −3**D**   **d.** −½**D**

7. Consider two parallel vectors **M** and **N** that have magnitudes of 7 units and 5 units respectively. Show the directions that they must point relative to each other so that

   a. $|\mathbf{M} + \mathbf{N}| = 12$ units.

   b. $|\mathbf{M} + \mathbf{N}| = 2$ units.

   c. $|\mathbf{M} - \mathbf{N}| = 2$ units.

   d. $|\mathbf{M} - \mathbf{N}| = 12$ units.

8. Draw three vectors that, when added together graphically, give a zero resultant vector.

9. Given the components $\mathbf{T}_x$ and $\mathbf{T}_y$, write the formula you would use to find the magnitude of **T**.

10. Sketch the horizontal and vertical components of the following vectors and name them appropriately.

    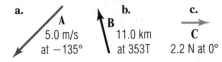

    a. A 5.0 m/s at −135°   **b.** B 11.0 km at 353T   **c.** C 2.2 N at 0°

11. Write the names of the vertical and horizontal scalar components of the following vector quantities in symbols and indicate their signs:

    a. **v** = 1300 m/s at an angle of −25°

    b. **d** = 4.0 m bearing 250T

    c. **F** = 13 N at an angle of 78°

    d. **a** = 20 m/s² in a direction of 330T

12. Discuss the meanings of the symbols α and θ for specific angles in vector analysis. Under what conditions may they refer to the same angle?

13. When adding two vectors analytically, why must you use the same coordinate system for finding the components of both vectors?

14. Discuss the difference between a scalar component and its magnitude.

15. A physics student is observing her horse (*Equus caballus*) in a rectangular field. Using herself as the origin, she notes the horse's initial and final position vectors ($\mathbf{r}_i$ and $\mathbf{r}_f$, respectively) over a 5-minute period. How could she indicate symbolically the total change of the horse's position during the time interval?

16. Carefully draw the vector diagrams for each of the following situations and graphically determine the final displacement vector.
  a. A car drives 8.0 km west, then turns southwest for 3.0 km, and then drives south for an additional 4.0 km.
  b. An airplane flies 100.0 km north, then turns east and flies 100.0 km, then turns southwest and flies 141.4 km.

17. If the scalar components of **A** are A_x and A_y, write the formula for determining the magnitude of **A**.

18. Write the trigonometric formula for finding the reference angle of **A** in Question 17.

19. If $|D_x| = 3.5$ cm and $|D_y| = 4.7$ cm, calculate D. Show your work.

20. What are the x- and y-scalar components of the force vector **F**, which is 35.0 N oriented at 216.0°? Show your work.

21. Each of the following vectors has been written in terms of the designated scalar components. Sketch each resultant vector, give its magnitude, and calculate the reference angle, α, and the vector angle, θ. An example problem is worked out to show what is expected in your written work.

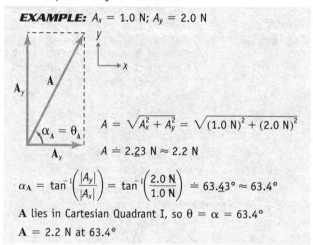

EXAMPLE: $A_x = 1.0$ N; $A_y = 2.0$ N

$$A = \sqrt{A_x^2 + A_y^2} = \sqrt{(1.0\text{ N})^2 + (2.0\text{ N})^2}$$

$$A \doteq 2.2\underline{3}\text{ N} \approx 2.2\text{ N}$$

$$\alpha_A = \tan^{-1}\left(\frac{|A_y|}{|A_x|}\right) = \tan^{-1}\left(\frac{2.0\text{ N}}{1.0\text{ N}}\right) \doteq 63.\underline{4}3° \approx 63.4°$$

A lies in Cartesian Quadrant I, so $\theta = \alpha = 63.4°$

A = 2.2 N at 63.4°

  a. $A_x = 3.0$ m; $A_y = -3.0$ m
  b. $B_x = -2.00$ N; $B_y = 8.00$ N
  c. $C_{E-W} = -12.0$ m/s; $C_{N-S} = -5.0$ m/s
  d. $D_x = 0.0$ m/s²; $D_y = 6.0$ m/s²

22. Resolve each vector into its component vectors.

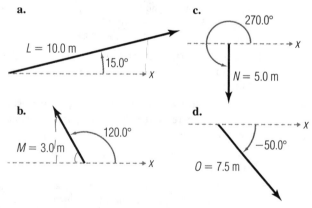

a.
$L = 10.0$ m
15.0°

c.
270.0°
$N = 5.0$ m

b.
120.0°
$M = 3.0$ m

d.
−50.0°
$O = 7.5$ m

23. Use the components of **L**, **M**, **N**, and **O** in the previous problem to determine the following vectors:
  a. **L** + **M**   **d.** **L** − **N**
  b. **M** + **N**   **e.** **M** − **O**
  c. **N** + **O**   **f.** **L** + **M** + **N** + **O**

24. A midfielder on the soccer team kicks the ball 20.0 m NW to his striker, who immediately kicks it 10.0 m NE into the goal. Describe the compass direction and distance that the midfielder would have had to kick the ball in order to place it in exactly the same point in the goal. Show your work.

25. Are the following properties true for vector addition? Support your answers with illustrative sketches. For each true relationship, give the name of the property of equality represented by the operation.
  a. **A** + **B** = **B** + **A**
  b. **A** − **B** = **B** − **A**
  c. (**A** + **B**) + **C** = **A** + (**B** + **C**)
  d. $n(\mathbf{A} + \mathbf{B}) = n\mathbf{A} + n\mathbf{B}$

26. A common housefly (*Musca domestica*) in a top corner of a 7.0 m × 10.0 m × 3.5 m room walks to the lowest opposite corner of the room.
  a. What will its net displacement be after its epic trip?
  b. If the fly walks only along the edges of the room, what is the shortest distance it must walk?
  c. If it is not restricted to the edges, what is the distance of the most direct route the fly must walk to reach its destination?
  d. Name the surfaces that the fly crosses in part *c*.

5A	The Kinematics of Two-Dimensional Motion	93
5B	Projections	98
Facet:	Shot Put Release Angles	107
Biography:	Sir Isaac Newton	110

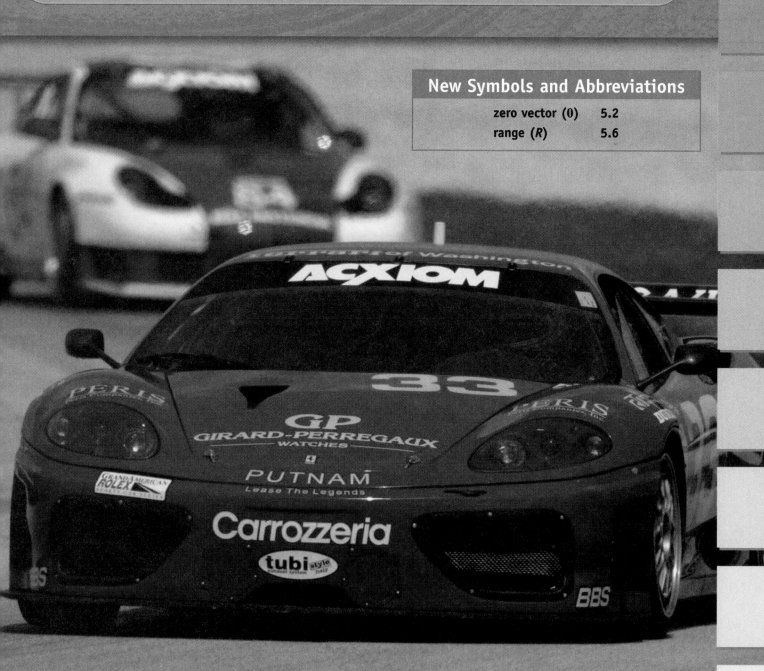

New Symbols and Abbreviations

| zero vector (0) | 5.2 |
| range (R) | 5.6 |

Formula race cars move in two dimensions on a flat track.

Motion in a Plane

And the hand of the Lord was there upon me; and he said unto me, Arise, go forth into the plain, and I will there talk with thee. Then I arose, and went forth into the plain: and, behold, the glory of the Lord stood there, as the glory which I saw by the river of Chebar: and I fell on my face. Ezekiel 3:22, 23

The prophet Ezekiel was commanded by God to go into the flatlands that lie between the Tigris and Euphrates Rivers in present-day Iraq. The two-dimensional motion of his journey is an example of what this chapter is about. The Scripture passage also reminds us of Ezekiel's obedience to God's commandments, and that God rewarded Ezekiel with the opportunity to directly experience His glory here on Earth. For us, God's power and glory are continually revealed through His creation and the ordinances that He established, making the study of physics and every other natural science possible.

5A KINEMATICS OF TWO-DIMENSIONAL MOTION

5.1 Position Vectors in Two Dimensions

Analyzing position and motion in two dimensions is only slightly more complicated than analyzing one-dimensional motion. Positions, displacements, velocities, and accelerations in two dimensions are all vector quantities that can be resolved into their components. You then can analyze the components using the equations of motion derived in Chapter 3.

In straight-line motion, you located an object by its position on a one-dimensional number line using coordinates. In two-dimensional motion, position is determined using the rectangular Cartesian coordinate system. Conventionally, the x-axis is horizontal and the y-axis is vertical, but any orientation and labeling is permissible as long as positions, directions, and angles are measured consistently within a given problem.

Position in two dimensions is determined by the position vector ($\mathbf{r}$). The tail of the position vector is at the origin of the coordinate system, and the head is at the

Problem-Solving Strategy 5.1

The orientation of coordinate axes depends on the kind of motion being analyzed. When solving two-dimensional problems, it is usually best to align one of the axes parallel to the straight-line motion or to a force causing a change in motion.

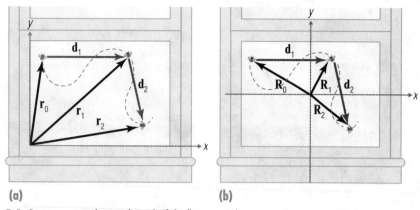

(a)   (b)

5-1 Position vectors showing the path of the fly on a window

object's location. Figure 5-1 (a) shows the path of a fly on a window pane. At time t_0, the fly is at the tip of $\mathbf{r}_0$; at time t_1, the fly is at the tip of $\mathbf{r}_1$; at time t_2, the fly is at the tip of $\mathbf{r}_2$. The origin is at the lower left corner of the window. Figure 5-1 (b) shows the same path of the fly, but now the origin is at the center of the window. Although the positions of the fly are identical, the position vectors are different when measured from the different reference points. A **frame of reference** is simply the coordinate system within which motion is measured or observed. There is an infinite number of reference frames and, as far as we can tell, there is no fixed "center of the universe" frame of reference from which all motion may be measured.

> A **frame of reference** or **reference frame** is a coordinate system within which motion is measured. A frame of reference may be stationary or, if it is moving, it can either have constant velocity or be accelerating.

5.2   Displacement in Two Dimensions

The displacement (**d**) of an object is its change in position represented by the difference vector $\Delta\mathbf{r}$. The displacement for an interval of time is

$$\mathbf{d} = \Delta\mathbf{r} = \mathbf{r} - \mathbf{r}_0,$$

where **r** is the position vector at the end of an interval and $\mathbf{r}_0$ is the initial position vector. For instance, the fly's displacement ($\mathbf{d}_1$) for the interval t_0 to t_1 in Figure 5-1 is

$$\mathbf{d}_1 = \mathbf{r}_1 - \mathbf{r}_0$$

for Figure 5-1 (a), or

$$\mathbf{d}_1 = \mathbf{R}_1 - \mathbf{R}_0$$

for Figure 5-1 (b).

The displacement is the same, no matter which reference frame is used.

> Two-dimensional displacement:
> $$\mathbf{d} = \mathbf{r} - \mathbf{r}_0$$

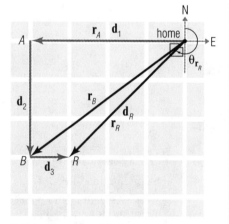

5-2 Position vectors of the jogger

> The **zero vector (0)** is a vector with a magnitude of zero. It is required when determining displacements from the origin.

EXAMPLE 5-1

Jogging Around

A beginning jogger starts from home and jogs four blocks west to corner A, turns and jogs three blocks south to corner B, and then turns east and jogs for another block before stopping to rest at point R. Using the jogger's home as the origin of the coordinate system, draw the jogger's complete path and the position vectors to the corners A and B, and position R.

   a. What is the jogger's displacement from corner A to corner B?

   b. What is his displacement from home to the resting spot?

   c. What is the total distance that he jogged?

Use the unit "block" for all answers.

Solution: Refer to Figure 5-2.

   a. $\mathbf{d}_2 = \mathbf{r}_B - \mathbf{r}_A$

     $\mathbf{d}_2 = 3$ blocks south

   b. $\mathbf{d}_R = \mathbf{r}_R - \mathbf{0} = \mathbf{r}_R$ (The **zero vector (0)** is any vector with zero magnitude. The position vector of the origin is **0**.)

Determine the scalar components of $\mathbf{r}_R$:

$r_{R_{E-W}} = -3$ blocks (west)

$r_{R_{N-S}} = -3$ blocks (south)

$$r_R = \sqrt{(-3 \text{ blocks})^2 + (-3 \text{ blocks})^2} = \sqrt{18 \text{ blocks}^2}$$

$$r_R = 3\sqrt{2} \text{ blocks} \doteq 4.2 \text{ blocks}$$

$$\alpha = \tan^{-1}\left(\frac{|r_{R_{N-S}}|}{|r_{R_{E-W}}|}\right) = \tan^{-1}\left(\frac{|-3 \text{ blocks}|}{|-3 \text{ blocks}|}\right) = 45°$$

Since $\mathbf{r}_R$ is in the southwest quadrant, $\theta = 270T - \alpha$.

$$\theta = 270T - 45° = 225T \text{ (southwest)}$$

The rest position is about 4.2 blocks southwest.

c. Total distance jogged, d_{total}:

$$d_{total} = d_1 + d_2 + d_3$$

$$d_{total} = (4 \text{ blocks} + 3 \text{ blocks} + 1 \text{ block})$$

$$d_{total} = 8 \text{ blocks}$$

5.3 Velocity and Speed in Two Dimensions

Average vector velocity and average speed between two positions in the Cartesian plane are determined just as they are in one dimension. Use vector notation when describing average vector velocity in two dimensions.

Average Velocity: $\quad \overline{\mathbf{v}} = \dfrac{\Delta \mathbf{r}}{\Delta t} = \dfrac{\mathbf{d}}{\Delta t}$

Average Speed: $\quad \overline{v} = \dfrac{s}{\Delta t}$

The *instantaneous velocity vector*, which is the velocity of an object at any given moment, always points in the direction that the object is moving at that point in time. Figure 5-3 shows the instantaneous velocity of a car traveling around a curve. At time t_0 the vector points east, but at time t_1 the vector points north. In two- or three-dimensional motion, direction is indicated by the vector angles.

The instantaneous speed is equal to the magnitude of the instantaneous velocity, $v = |\mathbf{v}|$. For instance, in Figure 5-3, if $|\mathbf{v}_0|$ equals $|\mathbf{v}_1|$, and $|\mathbf{v}|$ remains constant through the curve, the speed of the car is constant. Depending on the path an object travels, the average speed ($\overline{v}$) and the magnitude of the average velocity $|\overline{\mathbf{v}}|$ may be quite different. These quantities are equal only if $s = |\mathbf{d}|$.

EXAMPLE 5-2

Swimming Across the River

A boy tries to swim due north across a river that flows due east at a speed of 4.0 km/h (v_r). He swims at a speed of 3.0 km/h (v_b), heading directly across the river. His father is standing and watching from the north bank. What is the boy's velocity with respect to his father in km/h?

Solution:
Refer to Figure 5-4a. Orient the rectangular coordinate system east-west and north-south.

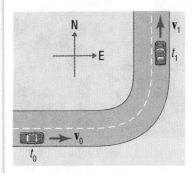

5-3 A car rounding a curve has a different instantaneous velocity at each point.

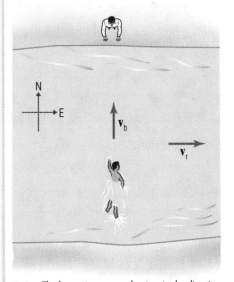

5-4a The boy swims across the river in the direction of his father.

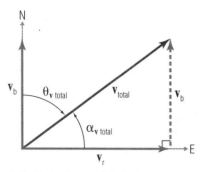

N

5-4b Vector diagram of the boy's component and resultant velocities

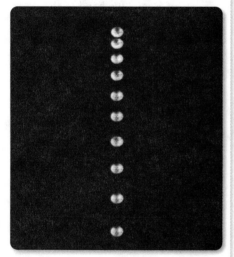

5-5 Acceleration in a straight line. The velocity of the golf ball changes in magnitude but not in direction.

Average Acceleration Vector:

$$\bar{\mathbf{a}} = \frac{\mathbf{v} - \mathbf{v}_0}{\Delta t}$$

Problem-Solving Strategy 5.2

While positive and negative signs are sufficient to indicate the direction of one-dimensional motion, vector angles must be used to indicate the directions of two-dimensional displacements, velocities, and accelerations. Signs *do* indicate the directions of the scalar components of two-dimensional motion, however, because the components are one-dimensional vectors.

The boy's total velocity with respect to his father is the vector sum of his swimming velocity and the river's velocity:

$$\mathbf{v}_{total} = \mathbf{v}_b + \mathbf{v}_r$$

A vector diagram of the situation is shown in Figure 5-4b. Since $\mathbf{v}_b$ and $\mathbf{v}_r$ are parallel to the N-S and E-W axes respectively, they are the vector components of the boy's total velocity. Therefore, you can use scalar components to calculate the magnitude and direction of the total velocity vector.

Total speed:

$$v_{total} = \sqrt{(v_{E-W})^2 + (v_{N-S})^2} = \sqrt{v_r^2 + v_b^2}$$

$$v_{total} = \sqrt{(+4.0 \text{ km/h})^2 + (+3.0 \text{ km/h})^2}$$

$$v_{total} = 5.0 \text{ km/h}$$

Boy's direction:

$$\alpha_{v_{total}} = \tan^{-1}\left(\frac{|v_b|}{|v_r|}\right) = \tan^{-1}\left(\frac{|+3.0 \text{ km/h}|}{|+4.0 \text{ km/h}|}\right)$$

$$\alpha_{v_{total}} \doteq 36.\underline{86}° \approx 36.9°$$

The vector $\mathbf{v}_{total}$ is in the northeast quadrant, so $\theta = 090T - \alpha$.

$$\theta_{v_{total}} = 090T - 36.9° = 53.1T$$

The father sees his son moving at 5.0 km/h in the direction 053.1T.

5.4 Acceleration in Two Dimensions

A vector can change in any of three ways: magnitude only, direction only, or both magnitude and direction. Most real motion involves the third kind of change. Recall from Chapter 3 that acceleration is the rate of change of the velocity vector ($\Delta \mathbf{v}$). The **average acceleration** vector ($\bar{\mathbf{a}}$) is equal to the velocity difference divided by the time interval of the velocity change,

$$\bar{\mathbf{a}} = \frac{\mathbf{v} - \mathbf{v}_0}{\Delta t} = \frac{\Delta \mathbf{v}}{\Delta t}. \tag{5.1}$$

The direction of the average acceleration is always the same direction as the velocity difference vector, regardless of the direction of motion. For example, as a car slows to a stop at a traffic light, the magnitude of its final velocity is less than the magnitude of its initial velocity, so $\Delta \mathbf{v}$ points in the opposite direction from the car's motion. Therefore, as the car slows, $\bar{\mathbf{a}}$ points in the opposite direction from $\mathbf{v}$. When a moving object changes *direction,* its acceleration vector is always at an angle to its path. As with velocity, acceleration direction is indicated by a vector angle.

The *instantaneous acceleration* of an object is its acceleration at a particular moment. Its vector also points in the direction of the instantaneous velocity difference vector. In two-dimensional motion, determining the instantaneous change in velocity can be fairly complicated without the use of calculus. We will limit the discussion of acceleration to problems where either direction or speed vary, but not both. As with one-dimensional acceleration, if the acceleration is uniform, then the average and instantaneous accelerations are the same.

EXAMPLE 5-3

Taking the Curve

A car heading due east at 90.0 km/h enters a curve in the road. The curve ends with the car heading due north at 90.0 km/h. Determine the car's average acceleration vector ($\bar{\mathbf{a}}$) if the turn took 5.0 s.

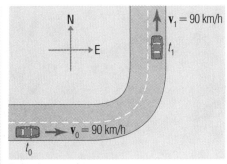

5-6a A car accelerates on a curve

Solution:

A diagram of the car on the curve is given in Figure 5-6a. Note the orientation of the initial and final velocity vectors.

You know that $\mathbf{v}_0 = 90.0$ km/h east, $\mathbf{v}_1 = 90.0$ km/h north, and $\Delta t = 5.0$ s.

Determine the velocity difference vector
$\Delta \mathbf{v} = \mathbf{v}_1 - \mathbf{v}_0 = \mathbf{v}_1 + (-\mathbf{v}_0)$
(see Figure 5-6b).

The scalar components of the velocity difference vector are as follows:

$\Delta v_{\text{N-S}} = v_{1_{\text{N-S}}} - v_{0_{\text{N-S}}}$

$\Delta v_{\text{N-S}} = (+90.0 \text{ km/h}) - (0.0 \text{ km/h})$

$\Delta v_{\text{N-S}} = +90.0$ km/h

$\Delta v_{\text{E-W}} = v_{1_{\text{E-W}}} - v_{0_{\text{E-W}}}$

$\Delta v_{\text{E-W}} = (0.0 \text{ km/h}) - (+90.0 \text{ km/h})$

$\Delta v_{\text{E-W}} = -90.0$ km/h

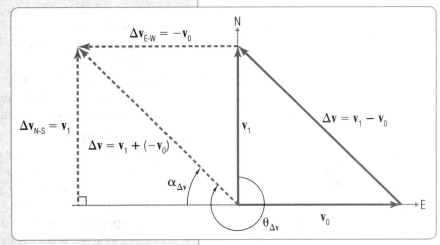

5-6b Vector diagram for the car on a curve

$$\Delta v = \sqrt{(\Delta v_{\text{E-W}})^2 + (\Delta v_{\text{N-S}})^2}$$
$$\Delta v = \sqrt{(-90.0 \text{ km/h})^2 + (+90.0 \text{ km/h})^2}$$
$$\Delta v \doteq 127.\underline{27} \text{ km/h}$$

The magnitude of average acceleration is determined by substituting the known information into the magnitude form of Equation 5.1:

$$\bar{a} = \frac{\Delta v}{\Delta t} = \frac{127.27 \text{ km/h}}{5.0 \text{ s}} \left| \frac{1000 \text{ m}}{1 \text{ km}} \right| \frac{1 \text{ h}}{3600 \text{ s}}$$

$\bar{a} \doteq 7.\underline{0}7 \text{ m/s}^2$

$\bar{a} \approx 7.1 \text{ m/s}^2$

Determine the orientation of $\bar{\mathbf{a}}$:

$$\alpha = \tan^{-1}\left(\frac{|\Delta v_{\text{N-S}}|}{|\Delta v_{\text{E-W}}|}\right) = \tan^{-1}\left(\frac{|+90.0 \text{ km/h}|}{|-90.0 \text{ km/h}|}\right) = 45.0°$$

$\bar{\mathbf{a}}$ is in the northwest quadrant, so $\theta = 270.0\text{T} + \alpha$

$\theta_{\bar{a}} = 270.0\text{T} + 45.0° = 315.0\text{T}$ (northwest)

The average acceleration of the car during the 5.0 s interval is about 7.1 m/s² toward the northwest.

5A   Section Review Questions

1. In physics, two-dimensional kinematic position is most efficiently represented by what quantity?

2. Describe how displacement is calculated in two-dimensional motion. Include the formula you would use in your answer.

3. Discuss the difference between the quantities represented by the symbols $\mathbf{v}$, $\overline{\mathbf{v}}$, $|\overline{\mathbf{v}}|$, and $\overline{v}$.

4. How is the *direction* of the average acceleration vector, $\overline{\mathbf{a}}$, determined if you are given the initial and final velocities?

◉5. A visitor to Washington, D.C. walks directly from the White House along Pennsylvania Avenue to Constitution Avenue. The distance is 2.00 km in the direction 110.T. He turns left on Constitution Avenue and walks due east for an additional 0.87 km. When he stops, the dome of the U.S. Capitol Building is 0.32 km due south of his position. He walks to the Capitol Building and enters the Rotunda under the dome. What is his displacement from the White House?

◉6. Famous quarterback Fleet O. Foote is attempting a quarterback draw by running up the middle toward his goal line at 15 ft/s (4.6 m/s). A linebacker hits him squarely, and their combined final velocity is 12 ft/s (3.7 m/s) at an angle of +120.° from the quarterback's original direction. The time it took to change direction at impact was 0.20 s. Choose an appropriate horizontal coordinate system and calculate the average acceleration vector of the quarterback during the tackle.

◉7. An athlete sprints a circular 400.0 m track in 50.0 s. Assume he runs at a constant speed.
   a. What is his speed over the entire course?
   b. What is the magnitude of his instantaneous velocity at any point along the circuit?
   c. If he starts facing west and runs counterclockwise, what is his velocity halfway around the circle?
   d. What is the average acceleration for one lap of the track?

5-7 The lava bombs of a volcano are ballistic projectiles.

> A **projectile** is any flying object that is given an initial velocity and then continues its flight affected only by external forces.

> The path of a projectile is its **trajectory.** The path of the unpowered portion of a projectile's flight is called its *ballistic* trajectory.

5B   PROJECTIONS

5.5   Projectiles

Flip a coin into the air—heads or tails? Fire a fast tennis ball deep into your opponent's left corner with your racquet—30, love! A competition high diver performs a perfect one-and-a-half gainer before completing a clean entry into the water. Mt. Aetna on the island of Sicily erupts, tossing fiery bombs of molten rock into high arcs. Each of these situations describes a kind of projectile.

A **projectile** is any flying object that is given an initial velocity and then allowed to fall under the influence of external forces, such as gravity. In other words, a projectile can be propelled for only the first part of its flight, if at all. Projectiles include those objects usually associated with weapons, such as bullets, bombs, and missiles, as well as any other object that falls or is launched through air or space.

A projectile's path is called its **trajectory.** The unpowered portion of a projectile's flight is called a ballistic trajectory. A ballistic projectile is assumed to have only gravitational force acting on it. For simplicity's sake, we will disregard the effects of air resistance unless otherwise noted. This is a reasonable assumption for relatively slow-moving, small objects.

5.6 Horizontal Projections

In Chapter 3 you studied the free fall of a dropped projectile. The path of motion was linear, and there was only one component of motion, parallel to the *y*-axis. A one-dimensional coordinate system was adequate for this kind of motion. However, if you kick a pebble off the top of a stairway, for example, you have another dimension to consider. As the pebble leaves the toe of your shoe, it is moving *horizontally*. In the total absence of gravity (and ignoring air resistance), the pebble would continue at the same velocity indefinitely. However, because of the presence of gravity, the pebble also begins to fall *vertically*.

A **horizontal projection** is motion in which an object is initially propelled horizontally and then allowed to fall in a ballistic trajectory. The kinematics of the vertical and horizontal components of motion are completely separate, but they occur simultaneously to give a smooth, continuous path. Figure 5-8 compares the position of a dropped object with a second object projected horizontally at the same instant. You can see that at each time, the vertical positions of both objects in the stroboscopic photo are identical. This shows that the vertical component of motion does not depend on the horizontal component. The total velocity of a projectile at any time after launch is the vector sum of the horizontal and vertical velocity components.

The horizontal and vertical components of projectile kinematics (displacement, velocity, and acceleration) are treated as two separate one-dimensional problems to which we can apply the equations of motion learned in Chapter 3. Two quantities often of interest in horizontal projections are the time of flight, Δt, and the magnitude of the horizontal displacement, $|\mathbf{d}_x|$, called the **range (R).**

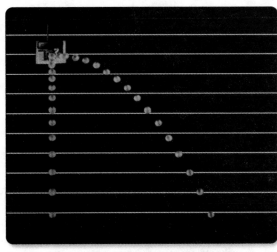

5-8 An experiment with golf balls demonstrates that horizontal motion and vertical motion are independent of each other.

> **Problem-Solving Strategy 5.3**
> Each component of motion can be analyzed completely independently from the other components using the equations of motion.

A **horizontal projection** is motion in which an object is initially propelled horizontally and then allowed to fall ballistically.

The **range (R)** of a projectile is the magnitude of the horizontal displacement of a projection.

The horizontal velocity of a projectile is assumed to be constant.

5.7 Projectile Motion: Horizontal Components

A projectile ideally is assumed to experience no horizontal acceleration. (This is not really a valid assumption because air resistance tends to slow projectiles significantly. However, for small, slow-moving projectiles, it is a good approximation.) Zero acceleration in the horizontal direction means that the horizontal velocity is constant. The first equation of motion,

$$v_x = v_{0x} + \overline{a}_x \Delta t,$$

becomes the horizontal scalar component equation

$$v_x = v_{0x},$$

where v_x and v_{0x} are the *x*-components of the projectile's velocity. Since the velocity's *x*-component is confirmed to be constant, it will simply be called v_x.

The second equation of motion,

$$x = x_0 + v_{0x}\Delta t + \tfrac{1}{2}\overline{a}_x(\Delta t)^2,$$

becomes the horizontal scalar component equation

$$x = x_0 + v_x\Delta t,$$

where x and x_0 are the x-components of the projectile's position vectors **r** and $\mathbf{r}_0$, respectively. This equation is often rearranged into its displacement form,

$$d_x = x - x_0 = v_x \Delta t. \tag{5.2}$$

The third equation of motion is meaningless when the acceleration is zero, since you would be dividing by zero.

5.8 Projectile Motion: Vertical Components

You learned in Chapter 4 that all objects experience a uniform downward acceleration due to gravity near the earth's surface. The scalar value of this acceleration is

$$g = -9.81 \text{ m/s}^2.$$

The initial vertical velocity of a horizontal projection is zero. The final vertical velocity of the projectile is due solely to the amount of time it has to fall. Choosing the positive direction to be upward and using the y-scalar components of vectors, the equations of motion take the forms listed below.

First Equation: $\qquad\qquad\qquad v_y = g_y \Delta t \qquad\qquad\qquad (5.3)$

Second Equation: $\qquad\qquad d_y = v_{0y}\Delta t + \tfrac{1}{2}g_y(\Delta t)^2 \qquad\qquad (5.4)$

Third Equation: $\qquad\qquad\qquad d_y = \dfrac{v_y^2}{2g_y} \qquad\qquad\qquad (5.5)$

Problem-Solving Strategy 5.4

Do not learn a second set of motion equations for projectiles and free-fall problems. Use the equations in the forms presented in Chapter 3. Solve the equation for the unknown and then substitute the known scalar values (including their signs).

Problem-Solving Strategy 5.5

Remember when working with gravitational acceleration to substitute the *negative* scalar value of g for a in the equations of motion.

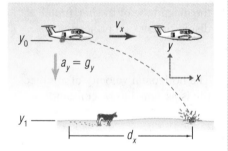

5-9 Aircraft dropping a bale of hay—disregard air resistance

EXAMPLE 5-4

Baling Out: A Horizontal Projection

A rancher in an airplane flying level at an altitude of 100. m and a speed of 190. km/h drops a bale of hay to hungry cattle stranded by a sudden snowstorm out on the range. If air resistance is negligible, how far in meters does the bale travel horizontally before it strikes the ground?

Solution:
Choose a Cartesian coordinate system with y positive upward and x positive in the direction of the plane's movement. Assume that the origin is at the position of the plane when the bale is released.

The horizontal distance that the bale traveled is the range, R, or the magnitude of the bale's horizontal displacement, $|d_x|$. You know the following information: $y_0 = 0$ m; $y_1 = -100.$ m (the ground); $v_{0y} = 0$ m/s; $v_{0x} = v_{1x} = v_x = +190.$ km/h; $g_y = -9.81$ m/s^2; $g_x = 0$ m/s.

From Equation 5.2, you know that

$$d_x = x - x_0 = v_x \Delta t. \tag{1}$$

You know v_x, but what is Δt? The time that the bale takes to fall vertically to the ground is the time that it is able to travel horizontally, so its free-fall time must be calculated. Use the second equation of motion for this:

$$y_1 = y_0 + v_{0y}\Delta t + \tfrac{1}{2}g_y(\Delta t)^2 \tag{2}$$

The first two terms on the right side are zero, so Equation 2 reduces to

$$y_1 = \tfrac{1}{2}g_y(\Delta t)^2.$$

Solve for Δt:

$$\Delta t = \sqrt{\frac{2y_1}{g_y}} \qquad (3)$$

Substitute the known values into Equation 3 to find the free-fall time:

$$\Delta t = \sqrt{\frac{2(-100.\text{ m})}{-9.81 \text{ m/s}^2}}$$

$$\Delta t \doteq 4.5\underline{1}5 \text{ s}$$

Substitute this time into Equation 1 to solve for range:

$$R = |d_x| = |v_x \Delta t| = |(190.\text{ km/h})(4.5\underline{1}5 \text{ s})|$$

$$R = \frac{190.\text{ km}}{\text{h}} \left| \frac{4.5\underline{1}5 \text{ s}}{} \right| \frac{1 \text{ h}}{3600 \text{ s}} \left| \frac{1000 \text{ m}}{1 \text{ km}} \right.$$

$$R \doteq 23\underline{8}.2 \text{ m}$$

$$R \approx 238 \text{ m}$$

The bale travels horizontally about 238 m from the time it is dropped to the time it hits the ground.

Problem-Solving Strategy 5.6
You must convert all quantities having the same dimension to the same unit to take advantage of cancellation in the unit-analysis phase of problem solving.

5.9 Frames of Reference in Projections

To a person flying in the plane that dropped the hay, it would look as if the hay were dropping straight down. A person on the ground, however, would say that the hay dropped in an arc. Their perceptions are different because they are comparing the motion to two different references. The person on the plane uses the plane as a reference with which to compare the movement of the hay bale. The hay is moving horizontally with the same speed as the plane; so it has zero horizontal velocity with respect to the plane. Since it does not move horizontally with respect to the plane, it appears to fall straight down due to the vertical component of its motion.

The person on the ground uses the ground as a reference with which to compare the bale's motion. The hay is moving horizontally with respect to the ground at a constant speed of 190. km/h. Therefore, during its fall it moves a considerable distance horizontally with respect to the ground. This is an example, as we discussed earlier, of the same motion being observed in different reference frames.

5-10 A B-1B bomber dropping free-fall bombs—projectiles

EXAMPLE 5-5

Fallen Rock Zone: A Horizontal Projection

A rock rolls down a talus slope and bounces horizontally off a ledge at the top of a vertical cliff. Its horizontal velocity as it becomes airborne is 10.0 m/s. It strikes the ground 30.0 m from the base of the cliff. (a) How high is the cliff? (b) What is the rock's speed just before it hits the ground?

Solution:

As usual, choose the coordinate system so that y is positive upward and x is positive away from the cliff. In this problem, let the origin be at ground level directly below the ledge (see Figure 5-11 on the next page).

Talus is the broken rock that accumulates at the bottom of cliffs. It is the result of mechanical weathering of a cliff.

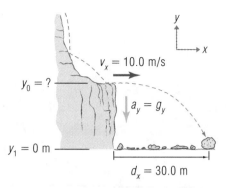

$v_x = 10.0$ m/s

$y_0 = ?$

$a_y = g_y$

$y_1 = 0$ m

$d_x = 30.0$ m

5-11 A rock falls from a cliff in a horizontal projection.

You know the following information: $d_x = +30.0$ m; $y_1 = 0$ m; $v_x = +10.0$ m/s; $g_y = -9.81$ m/s^2; and $g_x = 0$ m/s.

a. The vertical distance of fall is a function of the time of fall. The horizontal distance covered is a function of the same time interval and the horizontal velocity. Therefore, you can rearrange Equation 5.2 to find the time of fall.

$$\Delta t = \frac{d_x}{v_x}$$

$$\Delta t = \frac{30.0 \text{ m}}{10.0 \text{ m/s}}$$

$$\Delta t = 3.00 \text{ s}$$

Substitute the time of fall into the second equation of motion in the vertical (Equation 5.4) to find the initial vertical position, y_0.

$$y_1 = y_0 + v_{0y}\Delta t + \tfrac{1}{2}g_y(\Delta t)^2$$

Both y_1 and v_{0y} are zero, so the equation reduces as follows:

$$y_0 = -\tfrac{1}{2}g_y(\Delta t)^2$$

$$y_0 = -\tfrac{1}{2}(-9.81 \text{ m/s}^2)(3.00 \text{ s})^2$$

$$y_0 \doteq +44.\underline{1}4 \text{ m}$$

$$y_0 \approx +44.1 \text{ m}$$

b. In order to find the total speed of the rock, you must find its horizontal and vertical velocity components and sum them to find the resultant velocity at the instant it strikes the ground.

You already know the rock's horizontal velocity component, v_x. The vertical component can be calculated from the first equation of motion in the vertical (Equation 5.3) and the time of fall, Δt.

$$v_{1y} = v_{0y} + g_y\Delta t$$

$$v_{1y} = 0 \text{ m/s} + (-9.81 \text{ m/s}^2)(3.00 \text{ s})$$

$$v_{1y} = -29.\underline{4}3 \text{ m/s}$$

$$v_{1y} \approx -29.4 \text{ m/s}$$

To determine the speed, the magnitude of the final velocity resultant, use the Pythagorean theorem.

$$v_1 = \sqrt{v_{1x}^2 + v_{1y}^2} = \sqrt{(+10.0 \text{ m/s})^2 + (-29.\underline{4}3 \text{ m/s})^2}$$

$$v_1 \doteq 31.\underline{0}8 \text{ m/s}$$

$$v_1 \approx 31.1 \text{ m/s}$$

The rock's speed just before it hits the ground is about 31.1 m/s.

5.10 Projection at an Angle

While some projectiles do begin with only horizontal velocity, most projectiles are launched at an angle. The horizontal and vertical accelerations are assumed to be the same as for horizontal projections, but the initial vertical velocity, $\mathbf{v}_{0y}$, is no longer zero. This difference makes analyzing the motion slightly more complicated.

For example, javelin throwers launch their javelins at an angle with the horizontal. Their objective is to throw the javelin as far, horizontally, as possible. In other words, they try to maximize the range of their throws.

Launching at an angle introduces an additional step—you must calculate the vector components of the initial velocity of the projectile in both the *x* and *y* directions. Figure 5-12 shows these vector components for a velocity vector making an angle θ with the positive *x*-axis. As the figure shows, the horizontal initial velocity component is

$$v_{0x} = v_0 \cos \theta,$$

while the vertical initial velocity component is

$$v_{0y} = v_0 \sin \theta.$$

You must remember to substitute these expressions for the initial velocity terms in the equations of motion for the corresponding components. Once again, do not memorize a separate set of motion equations for projections at angles.

5.11 Graphing Projectile Motion

A projectile's vertical and horizontal displacement components can be easily calculated from the second or third equations of motion once its initial velocity components are identified. If displacement is measured from the launch point, the projectile's vertical and horizontal *position coordinates* can be determined from the position forms of the equations of motion. The trajectory of the projectile may be plotted on a graph from ordered pairs of vertical and horizontal position coordinates for various times after launch. Time-ordered coordinates are *ordered triples* (e.g., Δt, *x*, *y*).

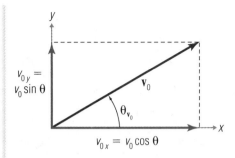

5-12 Resolving a velocity vector that is at an angle to the horizontal

"David . . . took thence a stone, and slang it, and smote the Philistine in his forehead . . . ; and he fell upon his face to the earth" (I Sam. 17:49). This stone is one of the most famous projectiles in history.

Problem-Solving Strategy 5.7

When plotting the trajectory of a projectile, set up a table with the variables time, *x*-position, and *y*-position. Both *x*- and *y*-coordinates are time-dependent variables in a two-dimensional position plot.

EXAMPLE 5-6

Flight of an Arrow: Graphing Motion

A young archer shoots an arrow at an angle of 45.0° from the horizontal with an initial speed of 41.6 m/s. (a) Ignoring air resistance, plot the vertical velocity of the arrow versus time every second for 6 s. (b) Then plot the height of the arrow versus range every second for 6 s.

Solution:

Choose a coordinate system with *y* positive upward and *x* positive to the right. Place the origin of the coordinate system at the point that the arrow leaves the bow string (see Figure 5-13).

a. The vertical velocity component is calculated from the first equation of motion.

$$v_y = v_{0y} + g_y \Delta t$$

$$v_y = v_0 \sin \theta + g_y \Delta t$$

$$v_y = (41.6 \text{ m/s}) \sin 45.0° + (-9.81 \text{ m/s}^2) \Delta t \quad (1)$$

The vertical velocity component is calculated for each time interval from 0 to 6 seconds. The results are shown in Table 5-1 in the margin.

5-13 An archer shoots an arrow for maximum range.

Arrow Flight Data			
Time (s)	v_y (m/s)	x (m)	y (m)
0	+29.4	0.0	0.0
1	+16.6	29.4	24.5
2	+9.8	58.8	39.2
3	0.0	88.2	44.1
4	−9.8	117.7	39.2
5	−19.6	147.1	24.5
6	−29.4	176.5	0.0

TABLE 5-1 Arrow flight data

Motion in a Plane   **103**

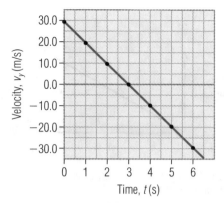

5-14 Arrow's vertical velocity versus time

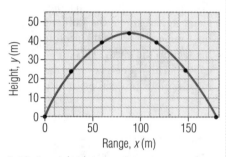

5-15 Arrow's height versus range

5-16 The distance of a pass depends on time of flight and horizontal velocity.

b. The position of the arrow is plotted using an ordered pair of x- and y-coordinates for each second of the flight. The x-coordinate is obtained from the second equation of motion in the horizontal. The y-coordinate is obtained from the second equation of motion in the vertical.

The x-coordinate:

$$x = x_0 + v_{0x}\Delta t + \tfrac{1}{2}a_x(\Delta t)^2$$
$$x = x_0 + (v_0 \cos \theta)\Delta t + \tfrac{1}{2}g_x(\Delta t)^2$$
$$x = 0 \text{ m} + (41.6 \text{ m/s})\cos 45.0°\Delta t + \tfrac{1}{2}(0 \text{ m/s}^2)(\Delta t)^2$$
$$x = (41.6 \text{ m/s})\cos 45.0°\Delta t \qquad (2)$$

The y-coordinate:

$$y = y_0 + v_{0y}\Delta t + \tfrac{1}{2}g_y(\Delta t)^2$$
$$y = y_0 + (v_0 \sin \theta)\Delta t + \tfrac{1}{2}g_y(\Delta t)^2$$
$$y = 0 \text{ m} + (41.6 \text{ m/s})\sin 45.0°\Delta t + \tfrac{1}{2}(-9.81 \text{ m/s}^2)(\Delta t)^2$$
$$y = (41.6 \text{ m/s})\sin 45.0°\Delta t + \tfrac{1}{2}(-9.81 \text{ m/s}^2)(\Delta t)^2 \qquad (3)$$

The values of each coordinate are shown in Table 5-1. A smooth plot of these points produces the trajectory shown in Figure 5-15.

Notice in Example 5-6 that the arrow's vertical velocity is zero at 3 s, and the arrow reaches its maximum height at 3 s. In other words, the arrow stops going up and starts going down, which means its vertical motion must be stopped for an instant. In that same instant, the arrow reaches the peak of its flight. These facts apply to the flight of any projectile: at the peak of its trajectory, the vertical velocity is zero.

Ignoring air resistance and wind, the time it takes a projectile to go from a given height to its peak is the same time that it takes the projectile to fall from its peak to its original height. For example, 1 s before the peak of its flight, the arrow is at 39.2 m. Exactly 1 s after the peak, it has fallen back to a height of 39.2 m.

Another helpful fact about projectile flight is that the vertical speed is the same at corresponding heights. When the arrow rose to a height of 24.5 m, its vertical velocity was 19.6 m/s upward. When the arrow had fallen back to the height of 24.5 m, its vertical velocity was 19.6 m/s downward. The vertical speed was the same at the corresponding height.

What kind of curve is the graph of the ballistic trajectory in Figure 5-15? You can identify the category of the curve by the form of Equations 2 and 3 in the previous example. The independent variable, Δt, is squared in the second equation of motion. When the highest power of the independent variable is two, then the equation is called a *quadratic function* and its graph is a *parabola*.

5.12 Range Determination

In any projection, the range is determined by the time the projectile takes to fall to the ground (its time of flight) and its initial horizontal velocity. The time of flight is a function of the projectile's initial vertical velocity and its initial height above the ground. These factors are important regardless of whether you are throwing a winning pass in a football game or firing a cannon shell at a distant target.

EXAMPLE 5-7

Range of a Bullet: Projectile Motion

A target match rifle is clamped at a 30.0° angle to the horizontal so that the muzzle is 1.00 m above the ground. At t_0, when the bullet exits the barrel, it has a velocity of 370. m/s. (a) How long does it take the bullet to hit the ground? (b) What is the bullet's range?

Solution:

We select a Cartesian coordinate system oriented vertically and horizontally with x positive away from the rifle and y positive upward. Place the origin at ground level directly below the muzzle of the rifle.

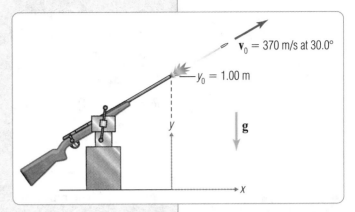

5-17 A bullet is fired from a rifle to determine its range at a specified angle.

a. As soon as the bullet exits the barrel, gravity accelerates the bullet downward. The time of flight is determined by how long the bullet takes to travel the vertical distance to the peak of the trajectory and then fall to the ground.

You know the following information: $v_0 = 370.$ m/s at 30.0°; $y_0 = +1.00$ m; $y_1 = 0.00$ m; and $g_y = -9.81$ m/s^2.

You could work this problem in three steps by first calculating the time to the peak of the trajectory, then finding the height of the peak, and finally finding the time for the bullet to fall from the peak to the ground. This process requires using the first, third, and second equations of motion, in that order.

A more direct and elegant solution in one step involves calculating the elapsed time of the entire flight by solving the second equation of motion using the quadratic formula. Let's see how this works:

The standard form of a quadratic function is

$$f(x) = ax^2 + bx + c.$$

Since the final position of the bullet (y_1) is 0 m, you can rewrite the second equation of motion in the vertical in the form of a quadratic function and set it equal to zero:

$$y_1 = y_0 + v_{0y}\Delta t + \tfrac{1}{2}g_y(\Delta t)^2 = 0 \text{ m}$$

$$\tfrac{1}{2}g_y(\Delta t)^2 + v_{0y}\Delta t + y_0 = 0 \text{ m} \tag{1}$$

Compared to the standard form, $\Delta t = x$, $\tfrac{1}{2}g_y = a$, $v_{0y} = b$, and $y_0 = c$.

The quadratic formula has the form

$$x = \frac{-b \pm \sqrt{b^2 - 4ac}}{2a}.$$

Solve Equation (1) for the unknown Δt using the quadratic formula:

$$\Delta t = \frac{-v_{0y} \pm \sqrt{v_{0y}^2 - 4(\tfrac{1}{2}g_y y_0)}}{2(\tfrac{1}{2}g_y)} = \frac{-v_{0y} \pm \sqrt{v_{0y}^2 - 2g_y y_0}}{g_y}$$

$$\Delta t = \frac{-v_0 \sin\theta \pm \sqrt{(v_0 \sin\theta)^2 - 2g_y y_0}}{g_y}$$

$$\Delta t = \frac{-(370. \text{ m/s})\sin 30.0° \pm \sqrt{[(370. \text{ m/s})\sin 30.0°]^2 - 2(-9.81 \text{ m/s}^2)(+1.00 \text{ m})}}{-9.81 \text{ m/s}^2}$$

Problem-Solving Strategy 5.8

For any quadratic equation, $0 = ax^2 + bx + c$, you can solve for x using the formula

$$x = \frac{-b \pm \sqrt{b^2 - 4ac}}{2a}.$$

$\Delta t \doteq 37.\underline{7}2$ s or -0.005 s

Time can only be positive, so the time of flight for the bullet is about 37.7 s.

b. The range (R) of the bullet is simply the product of the bullet's horizontal velocity and the time of flight:

$R = v_x \Delta t$

$R = (v_0 \cos \theta) \, \Delta t = (370. \text{ m/s})(\cos 30.0°)(37.\underline{7}2 \text{ s})$

$R \doteq 12\,\underline{0}86$ m

$R \approx 12\,100$ m

Problem-Solving Strategy 5.9

Use a quadratic solution of the second equation of motion to quickly obtain total time of flight for a projectile.

"Jonathan cried after the lad, and said, Is not the arrow beyond thee?" (I Sam. 20:37). Jonathan used range as a code to convey a message to David when he was in danger because of Saul.

The range of the bullet calculated in Example 5-7 (more than 7.5 miles) is quite long compared to the typical range of a rifle bullet. This is because air resistance was ignored for this problem. Research has shown that a hunting rifle bullet can lose as much as 20% of its muzzle velocity in the first 100 m of flight, so the assumption is not realistic. Ignoring air resistance helps clarify the basic motion concepts without unnecessarily complicating the calculations.

The maximum range that a projectile can attain for a given initial speed depends on its launch angle. You could determine this angle experimentally, or you could solve a series of problems with trial angles as in Example 5-7. Geometrically, the maximum theoretical range for a fixed launch speed is attained with a launch angle of 45°, assuming no air resistance. Real projectiles traveling at high speeds never reach their maximum theoretical range because of drag, but this angle still yields the maximum attainable range in the absence of wind or changes in air density. Range can also be increased by elevating the launch point. This is one reason armies try to capture and hold hills in battles.

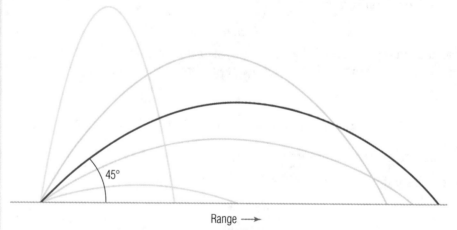

45°

Range ⟶

5-18 Range varies with launch angle for a fixed initial speed. Maximum range is obtained at an angle of 45°.

5B Section Review Questions

1. Is a remote-controlled gas-powered model airplane a projectile? Explain your answer.

2. What kinematic quantities are assumed to be constant when analyzing horizontal projections?

3. At the same instant that a pebble is dropped from a tree house platform, another is kicked horizontally off the platform.
   a. Which pebble has the higher total velocity?
   b. Which one will hit the ground first?

4. Given a fixed initial projectile speed, v_0, what factors can be changed in order to maximize projectile range?

⊙5. A marble resting near the edge of a 0.90 m high table is given an initial horizontal speed of 1.24 m/s. What will be its horizontal range from the table's edge when it strikes the floor?

⊙6. A rancher-pilot is dropping hay bales from an airplane to feed his cattle. The plane is flying level at an altitude of 60.0 m and a speed of 185 km/h. If a hay bale is dropped under these conditions, how far will it travel horizontally, neglecting air resistance?

⊙7. The American bullfrog (*Rana catesbeiana*) can jump a distance of nearly 15 times its length! If a bullfrog starts on a horizontal log and leaps with a velocity of 4.40 m/s at an angle of 37.0° to the horizontal, what distance can it cover?

⊙8. During the 1998 Olympics, an all-time record shot put was made— 22.47 m. If the athlete released the shot at a height of 2.25 m with a velocity of 14.3 m/s at an angle of 35.0° to the ground, what was its vector velocity when it impacted the ground at the end of its flight?

FACETS of PHYSICS

Shot Put Release Angles

You have learned in this chapter that the range of a horizontal projection is maximized when the release angle is 45°. There are circumstances, however, in which this is not necessarily true.

The problem is biomechanics—how the muscles coordinate to produce the force that propels the shot on its trajectory. The classic relationship of range to projection angle assumes that the initial velocity is constant. However, in the shot put, the initial velocity is affected by other variables. The human body is structured to create more force in the horizontal direction than in the vertical direction. Even highly trained athletes can bench-press more weight than they can lift in a shoulder press. A larger release angle measured to the horizontal causes a greater percentage of the total force to come from the comparatively weaker vertically directed muscles. Thus, a larger release angle decreases the total force that the athlete can exert on the shot, resulting in a lower initial velocity and a shorter range of the shot.

The best release angle is different for each athlete, varying with his size, strength, and throwing technique, and may be anywhere from 26° to 38°.

In Terms of Physics

frame of reference	5.1	trajectory	5.5
zero vector (**0**)	5.2	ballistic	5.5
average acceleration ($\bar{\mathbf{a}}$)	5.4	horizontal projection	5.6
projectile	5.5	range (*R*)	5.6

Problem-Solving Strategies

5.1 (page 93) The orientation of coordinate axes depends on the kind of motion being analyzed. When solving two-dimensional problems, it is usually best to align one of the axes parallel to the straight-line motion or to a force causing a change in motion.

5.2 (page 96) While positive and negative signs are sufficient to indicate the direction of one-dimensional motion, vector angles must be used to indicate the directions of two-dimensional displacements, velocities, and accelerations. Signs *do* indicate the directions of the scalar components of two-dimensional motion, however, because the components are one-dimensional vectors.

5.3 (page 99) Each component of motion can be analyzed completely independently from the other components using the equations of motion.

5.4 (page 100) Do not learn a second set of motion equations for projectiles and free-fall problems. Use the equations in the forms presented in Chapter 3. Solve the equation for the unknown and then substitute the known scalar values (including their signs).

5.5 (page 100) Remember when working with gravitational acceleration to substitute the *negative* scalar value of *g* for *a* in the equations of motion.

5.6 (page 101) You must convert all quantities having the same dimension to the same unit to take advantage of cancellation in the unit-analysis phase of problem solving.

5.7 (page 103) When plotting the trajectory of a projectile, set up a table with the variables time, *x*-position, and *y*-position. Both *x*- and *y*-coordinates are time-dependent variables in a two-dimensional position plot.

5.8 (page 105) For any quadratic equation, $0 = ax^2 + bx + c$, you can solve for *x* using the formula

$$x = \frac{-b \pm \sqrt{b^2 - 4ac}}{2a}.$$

5.9 (page 106) Use a quadratic solution of the second equation of motion to quickly obtain total time of flight for a projectile.

Review Questions

1. An air traffic controller reports to the pilot of an inbound passenger aircraft the range and direction from his control tower to the plane. What kinematic quantity does this information represent?

2. A displacement in two-dimensional motion is the vector difference of what two quantities?

3. The direction of the average velocity vector ($\bar{\mathbf{v}}$) is the same as the direction of what related vector?

4. At what point does a rocket begin its ballistic trajectory?

5. What initial velocity component is always zero for a horizontal projection?

6. What simplifying assumption is made regarding the horizontal velocity of a projectile? Why is this not an accurate assumption?

7. **a.** What is the independent variable in a trajectory position graph?

   b. How are the *x*- and *y*-coordinates related to the independent variable?

8. What is the most direct method for determining the time of flight of a projectile, given its initial velocity and launch height?

9. At what angle to the ground should a batter hit the ball in order to maximize the range of the hit?

10. What kind of trajectory do droplets of water from a garden hose follow?

11. What is the magnitude of a projectile's vertical velocity at the peak of its trajectory?
    a. It equals the magnitude of the horizontal velocity.
    b. It is zero.
    c. It equals 9.81 m/s.
    d. None of the above.

12. A bullet is fired from a rifle and hits a target some distance away. Ignoring air resistance, compare the speed of the bullet upon impact with its original muzzle velocity
    a. when the target is the same height as the muzzle.
    b. when the target is 2 m higher than the muzzle.

13. Explain why a war plane dropping a bomb from a low altitude needs to turn away from its original path as soon as it releases the bomb.

True or False (14–23)

14. An acceleration is always a change in speed.

15. A position vector is constructed from an object's initial position to its final position.

16. Range is a vector.

17. In two-dimensional motion, the acceleration vector often points in a direction other than parallel to the direction of instantaneous velocity.

18. The horizontal and vertical components of an object's motion can be analyzed independently of each other.

19. A passenger on the Amtrak train *California Zephyr* and the driver of a car waiting for the train to pass through a crossing observe each other in the same reference frame.

20. The horizontal velocity of a projectile under the influence of only gravity is assumed to be constant.

21. The vertical speed of an object projected at a positive angle first increases and then decreases during its flight.

22. A spinning Frisbee is a projectile.

23. The graphs of quadratic functions are parabolas.

⊚24. A motorized float in a Thanksgiving Day parade is traveling north at a speed of 16.0 km/h. Part of a "Pilgrim" woman's role in the thematic vignette is to walk across the float to call the men to dinner. She walks at a speed of 3.22 km/h east as the float passes the judge's booth, which is on the west side of the street. What is the woman's velocity with respect to the judge?

⊚25. A car going 110. km/h southwest reaches a curve in the road. Assume that the curve is an arc of a circle. After rounding the curve, the car's velocity is 110. km/h west. The car takes 2.50 s to round the curve.
   a. What is the car's acceleration?
   b. What is the length of the car's path on the curve?
   c. If the car instead maintained a steady speed of 100. km/h, how long would it take to round the curve?

⊚26. An aviation crew is flying an airplane to deliver supplies to a remote village. They intend to drop-deliver a cargo container attached to a parachute. The aircraft is flying at 165 km/h due east as the air crew drops the 100. kg crate. The parachute fails to open and the crate reaches the ground 4.5 s after it was released.
   a. What is the crate's vertical speed just before it hits the ground?
   b. How high was the plane when the container was released?
   c. The plane was 0.5 km west of a lake when it released the crate. Did the crate land in the lake?

⊚27. An astronaut on the Moon ($g_M = -1.63$ m/s^2) kicks a stone horizontally off a 45.0 m cliff with a speed of 2.50 m/s.
   a. How far from the base of the cliff does the stone land?
   b. How far from the base of the cliff would the stone have landed on Earth ($g_E = -9.81$ m/s^2, assuming no air resistance)?

⊚28. A skydiver drops from an airplane that is 2000. m above the ground. An automatic timer opens his parachute exactly 10.0 s after he leaves the plane.
   a. How far from the ground is he when his parachute is fully open, if it takes 1.0 s to open? (Assume no change in velocity until the chute is fully open.)
   b. In the next 10.0 s, the skydiver's speed decreases uniformly to 10.0 m/s. How far does he fall during this time? *(Hint: $d = \bar{v}\Delta t$)*
   c. After slowing, the skydiver's speed is constant until he lands on the ground. How long does the entire jump take?

⊚29. A rifle's muzzle velocity is 400. m/s. The marksman's shoulder is 1.70 m high and the rifle is 1.50 m long.
   a. What is the rifle's maximum possible range, neglecting air resistance?
   b. Suppose the marksman fires at the broad side of a barn (height of the roof = 15.0 m) holding the rifle at a 20.0° angle above the horizontal. The ridge of the barn's roof is 100. m away from the end of the muzzle. Does the bullet strike the barn, or does it go over the barn?
   c. What is the maximum initial vertical velocity component the bullet can have if it is to just graze the peak of the barn's roof at the midpoint of its trajectory?
   d. At what angle would the marksman have to fire in order to have the initial vertical velocity component found in part *c?*

⊚30. A professional baseball player throws a ball as hard as he can at an angle of 30.0° above the horizontal. At the same instant, his 10-year-old son throws an identical ball directly upward at exactly half the speed of his father's pitch. The boy is standing on a mound so that both balls are released at the same height. Which ball hits the ground first?

⊚31. Two archers are trying to see who can obtain a greater range. The first archer shoots his arrow at a velocity of 65.0 m/s, 10.0° from the horizontal. The arrow is released 1.40 m above the ground. The other archer shoots his arrow from the same height at a velocity of 48.0 m/s, 45.0° from the horizontal. Who will win the contest?

⊚32. A diver leaves the diving board, 10.0 m above the water, at a velocity of 12.0 m/s, 60.0° above the horizontal. If he takes 0.75 s to complete a somersault, how many full somersaults can the diver complete before hitting the water?

⊚33. Draw a circle of 5.00 cm radius at the top of a piece of graph paper. The graphic scale of the drawing is 1 cm = 1 m. Assume a marble revolves at a uniform rate around the center of the circle once every 16.0 s. Directly below the circle, construct a velocity-time graph for the *x*-component of the marble's velocity vector. Choose at least 8 evenly spaced points around the circle to help define the velocity-time curve. Label the graph axes $\bar{v}_x$ and *t*. Below this graph, construct a velocity-time graph for the *y*-component of the marble's velocity vector using the same time-coordinate axis. Label the vertical axis $\bar{v}_y$.
   a. Calculate the average scale speed of the marble.
   b. What is the instantaneous speed of the marble at any time?
   c. What is the average velocity for one complete revolution?
   d. What kind of graphs do the two velocity components describe?
   e. Discuss how the two graphs are related to each other; that is, When is one maximum compared to the other? When is one zero compared to the other? What is one curve doing when the other is decreasing?, and so forth.

Sir Isaac Newton
(1642–1727)

Undoubtedly, the most important man in the history of physics is Sir Isaac Newton. Besides devising the law of universal gravitation and laying the foundation for modern dynamics, Newton experimented extensively with optics and invented the most profound mathematical system yet devised—calculus—now used to describe the basic laws of science and engineering.

Three months before Newton was born, his father died, and the sickly child himself was not expected to live past his first day. He survived, however (85 years), and as he grew up, he became so fascinated with books and experiments that he often forgot his chores. He later studied mathematics at Cambridge University, one of the two best universities in England. After graduating in 1665, Newton returned home briefly because a plague had closed the university. In the leisure of the family farm, he began to formulate his theory of gravity and improve his theory of light and color.

Soon after he returned to Cambridge, he invented the first reflecting telescope, which brought him to the attention of the leading scientists of his time. In 1669, when he was only 26, Newton became a mathematics professor at Cambridge, a position that he held for much of his life. Three years later, England's most important scientific association, the Royal Society, heard of Newton's telescope and elected him a member. At that time, Newton published a paper about his theory of light and color. This paper drew Newton into a heated controversy, which deeply offended him and made him reluctant to publish more of his theories.

During the following years, Newton secretly matured his theory of gravity using differential calculus. In the early 1680s, a student named Edmund Halley asked him to analyze the problem of the orbits of comets. When Halley learned that Newton had already solved the problem and formulated a comprehensive theory of gravity, he insisted that Newton publish the work with Halley's own money. So in 1687 Newton's greatest work, *Philosophiae Naturalis Principia Mathematica* (*Mathematical Principles of Natural Philosophy*, called the *Principia* for short) was published.

Here Newton set forth the principles of dynamics, the theory of universal gravitation, and many applications of these principles. He defined for the first time such concepts as force and mass. This work is the foundation of all modern dynamics. For ordinary objects, Newton's laws have not been modified in the 300 years since the *Principia* was published.

The publication of the *Principia* established Newton as the foremost scientist in England. When Newton represented Cambridge at the famous convention that deposed the Catholic king James II, Newton came to the attention of politicians as well. As a result he was elected to Parliament in 1689 and served two years. Later Newton was employed at the mint to help redesign England's coins. In 1699 he became Master of the Mint and eventually gave up his professorship at Cambridge to move to London. He gained such a reputation as a counterfeit expert that his name became a byword among counterfeiters, who saw many of their fellows hanged on Newton's account.

Honors continued to come his way. In 1703 he was elected president of the Royal Society, and in 1705 Queen Anne knighted him, the first scientist so honored.

Although Newton's last years brought him fame, prestige, and time for experimenting and writing, they also brought controversies. The worst one persisted throughout the latter 25 years of Newton's life and divided mathematicians in England from those on the Continent. The problem arose because Newton, who had originated differential calculus early in the 1670s, waited until 1704 to publish it. Meanwhile, a German named Gottfried Wilhelm von Leibniz had developed a similar calculus independently and published it in 1684. Not surprisingly, each man accused the other of plagiarism. The ensuing quarrel did not slacken until both men were dead.

Like many scientists of his day, Newton had a wide range of interests and a critical distrust of unsound hypotheses. This attitude enabled him to contribute much to several areas of physics. Newton formulated modern dynamics, gave the first statement of the modern theory of color, and contributed to the study of light, astronomy, and metallurgy. He was both a good experimenter—one who observes—and a good theorist—one who explains. All great scientists after him, including Einstein and Bohr, gave tribute to his inspiring example.

6A	The History of Dynamics	113
6B	Forces	117
6C	Newton's Laws of Motion	124
Facet:	Flight and Newton's Third Law	130
Biography:	Aristotle	135

New Symbols and Abbreviations

summation (Σ)	6.6	pound-force (lbf)	6.16
weight (F_w)	6.8	pound-mass (lbm)	6.16
force of X on Y ($F_{X \to Y}$)	6.15		

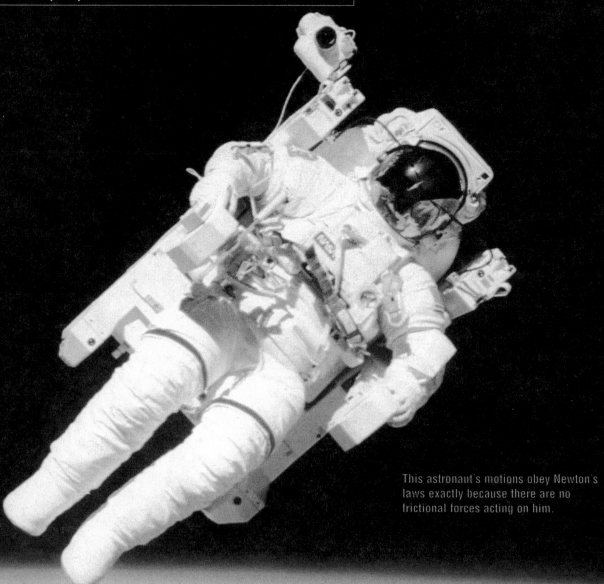

This astronaut's motions obey Newton's laws exactly because there are no frictional forces acting on him.

Dynamics 6

The question "Why?" is one of the most important in all of human inquiry. It demands an answer showing cause and effect. There are hundreds of instances in Scripture where the Lord asks "Why?" In Mark 11:31 the intelligent but perplexed Jewish religious leaders were trying to decide how to answer this anticipated question from Jesus. Why, indeed, did these men reject the many evidences that John the Baptist's words were the very words of life? What was the cause of their rejection? Have you rejected God's words? And even if you are a Christian, you must be ready to give an answer if asked why you believe in Jesus Christ. How did you come to accept God's gift of life?

"And why call ye me Lord, Lord, and do not the things which I say?" (Luke 6:46).

6.1 Introduction

What causes motion? This question captivated men for many years. Over two thousand years ago, the Greek philosopher Aristotle addressed this question as part of a plan to organize all the knowledge of his day into a coherent whole. Unfortunately, he attacked the question from the wrong perspective. Instead of first describing *how* objects move, he tried to explain *why* they move. This unfortunate choice resulted in incorrect ideas that hindered philosophers and scientists for hundreds of years afterward. In this chapter we show how those ideas were corrected and provide a strong foundation for dynamics.

6A THE HISTORY OF DYNAMICS

6.2 Early Greek Theories

The philosophers of Aristotle's day believed that objects could exhibit two types of motion: natural and unnatural. *Natural motion* was thought to be caused by some internal quality of an object that made it seek a certain "preferred" position without any application of force. For example, heated air rises, so its natural motion would be rising. Rocks fall; therefore, their natural motion would be falling. In this scheme of the universe, natural motions were thought to allow objects to reach their natural place, or place of rest. For heated air, the natural

Greek natural philosophy attributed organic "inclination," or purposeful activity, to inanimate matter. In other words, the belief was that matter "knows" the kind of motion it should have. Any motion contrary to its natural inclination was considered *unnatural* motion.

place was the heavens; for rocks, the natural place was the center of the earth. All motion other than natural motion was called *unnatural motion* and was thought to require applied force to be sustained.

Natural motions were divided into two subcategories: *terrestrial motions* (motions near Earth) and *celestial motions* (the motions of planets). This division of motions arose from the pagan Greeks' dualistic view of nature. The dualistic universe is divided into the celestial and the mundane; the divine and the mortal; the perfect and the imperfect. Celestial motions were thought to be purely circular. They represented the untouchable perfection of the celestial realm that was ruled by mathematics and pure reason. Terrestrial motion included all other forms of natural and unnatural motion. These two types of motion were considered to be completely separate from and unrelated to each other. This division helped to explain how heavenly bodies "automatically" moved in perfect circular paths.

According to Aristotle, an object's "heaviness" determined how "vigorously" it sought its natural place. A heavy object would move more vigorously toward the surface of the earth than a light object. Conversely, an object's vigorousness would measure its "heaviness." Were these conclusions supported by experience? It seemed so at the time.

Remember that Greek philosophy placed deductive reasoning on an infinitely higher plane than experimentation. Even if experimental data contradicted a belief derived from reasoning, one could say that the experiment itself was faulty, since it was subject to the presupposed corruption of the material world.

6.3 Galileo's Investigations

Galileo chose the proper perspective from which to study the problem of motion. His first task was to collect facts and establish an accurate *description of motion.* This description is known as kinematics. From this foundation, he inductively developed workable theories of dynamics (no scientist uses a strictly deductive approach in his work).

Suppose a feather, a tennis ball, a baseball, and a bowling ball were dropped from the top of a 30 m tower. How long would each object fall? Obviously, the bowling ball would hit the ground first, but why? If Aristotle's concept is correct, it is because heavier objects seek their natural positions near the center of the earth more vigorously than do lighter objects. Extending this logic, the speed of each object should be determined by its mass. The baseball, which has more than twice the mass of the tennis ball, should fall at least twice as fast. The bowling ball, with about fifty times more mass, should fall fifty times as fast.

Table 6-1 records the actual time required for each object to fall. Notice that the heavier objects fall at nearly the same rate. After a certain density is reached, the "heaviness" of the object has little effect. In this respect, Aristotle's theory of natural motions was shown to be wrong.

If a wind should blow during the experiment, the feather would be blown off course, and its falling time would be greatly increased. The falling times of the other objects would also be affected, but not as much as the feather's. Apparently, the presence of air has a great effect on the rate at which objects, especially light objects, fall.

Astronauts have performed experiments to test this conclusion. Colonel David Scott dropped a hammer and a feather in the vacuum at the Moon's surface. Without the presence of air, both objects fell at the same rate and hit the ground at the same time. Even though Galileo could not perform this experiment, he reached the same conclusion by considering the effects of air resistance on falling objects.

Aristotle believed that unnatural motions could only be maintained by the continuous application of force. Suppose you slide a large box of books from the

TABLE 6-1		
Object	Mass (g)	Time (s)
Feather	14	15
Tennis Ball	57	3.0
Baseball	142	2.8
Bowling Ball	6804	2.6

6-1 The famous experiment on the Moon proving Galileo's conjecture that all objects would fall at the same rate in the absence of air

foot of your bed into your closet. From experience you know that when you stop pushing, the box stops almost immediately. This observation seems to show that Aristotle was correct.

Do other observations confirm this conclusion? Consider the flight of an arrow. The arrow begins to move when a bowstring applies a substantial amount of force. Yet the arrow still moves after it leaves the bow. According to Aristotle, some force must continue to push the arrow or it will stop and fall to the ground. Where can the force come from? Aristotle said that the air separates in front of the arrow, allowing the arrow to pass through easily. In the wake of the arrow, he said, the air snaps closed and provides the force to keep the arrow moving.

To disprove Aristotle's explanation, you would need to perform this experiment in a vacuum. If air provides the force to sustain the arrow's motion, the arrow should immediately cease moving if released in a vacuum. Instead, it continues on almost the same path as before. Since Aristotle was committed to his theory, he had to endow the air with special, unobservable properties to account for the arrow's motion. This kind of hypothesis, based on conjecture rather than observation, is called an *ad hoc* hypothesis. Aristotle's explanation was a direct consequence of the Greek philosophers' erroneous belief that inert matter had attributes of conscious organisms.

6-2 Disproving Aristotle's theory of the cause of an arrow's flight

The belief that inert matter exhibits purpose or intent is an example of an *ad hoc* hypothesis.

6.4 The Principle of Inertia

Galileo studied unnatural motion with an experiment similar to the following. Suppose that an unpolished metal cube rests on a wooden plane. If you give the cube a push, it soon stops moving. If the cube is polished and the plane is oiled, polished metal, the cube travels a greater distance. Theoretically, if the cube and the plane could be made totally frictionless, the cube could slide forever without slowing down.

Now, suppose you arrange a frictionless, flexible, U-shaped strip of material as in Figure 6-3. This strip acts as a track for a frictionless ball. If you release the ball at the top of the track at the left end, the ball accelerates down the incline, rolls along the bottom of the curve, and then encounters a rising incline. The ball slows down as it rolls up the incline and stops when it reaches its original height (the reason this occurs will be considered in Chapter 9).

What happens if the slope of the right end of the track is decreased and the experiment is performed again? The ball again decelerates up the right-hand incline, but its deceleration is less than before. It continues moving until it reaches its original height, although it has to move farther horizontally to reach that height. If the right-hand end of the track were bent to be completely horizontal, the ball would not experience any deceleration. It would naturally continue at a constant velocity for an indefinite amount of time and distance.

From such concepts, Galileo inductively asserted that the "natural" state of motion of an object could include "moving" as well as "resting." Galileo's **principle of inertia**

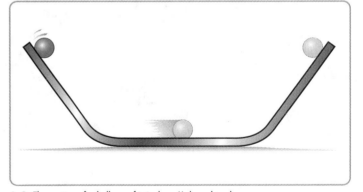

6-3 The motion of a ball on a frictionless, U-shaped track

states that an object will continue in its original state of motion unless some outside agent acts on it. A moving object does *not* require a continuous push to maintain a constant velocity. In fact, just the opposite is true! A push causes a *change* in an object's state of motion. This conclusion was truly amazing considering the limitations of the experimental apparatus of the time. The concept of inertia has served as the cornerstone of dynamics ever since Galileo.

Galileo's **principle of inertia** states that an object at rest tends to stay at rest, and an object in motion will continue in its current state of motion unless an outside agent acts on it.

Perhaps there is no better example of spiritual inertia than Pharaoh at the time of the Israelites' exodus (Exod. 5–14). God had to exert tremendous pressure on Pharaoh and his people through the plagues before Pharaoh would submit.

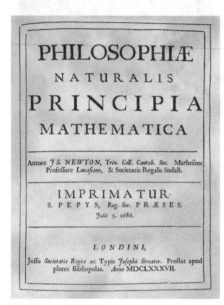

PHILOSOPHIÆ
NATURALIS
PRINCIPIA
MATHEMATICA

Autore JS. NEWTON, Trin. Coll. Cantab. Soc. Matheseos
Professore Lucasiano, & Societatis Regalis Sodali.

IMPRIMATUR.
S. PEPYS, Reg. Soc. PRÆSES.
Julii 5. 1686.

LONDINI,
Jussu Societatis Regiæ ac Typin Josephi Streater. Prostat apud
plures Bibliopolas. Anno MDCLXXXVII.

6-5 Newton's *Principia*

6.5   Sir Isaac Newton

Newton carried forward the conceptual investigations of Galileo. By the time he was able to devote his considerable powers of deduction to the problem, much information about the movement of planets had been documented by the great astronomer Tycho Brahe, and Johannes Kepler had formulated his three laws of celestial mechanics (see Chapter 7). Newton gave credit to these and other scientists that had gone before him for preparing the way for his discoveries.

6-4 Tycho Brahe (1546–1601) was a Danish astronomer who revolutionized astronomical observations. His discovery that comets were higher than the Moon and that planets moved in noncircular orbits did much to complete the destruction of the dualistic belief in the perfection of the celestial spheres.

Newton's work grew from his study of gravitational attraction between astronomical bodies using differential calculus, which he had developed for the purpose. Eventually, he published his findings, which included his three laws of motion that defined the concepts of force, mass, and momentum, in his famous work *Principia* (short for *Philosophiae Naturalis Principia Mathematica*, or *Mathematical Principles of Natural Philosophy*). Newtonian dynamics have been validated repeatedly for ordinary objects over the three centuries since *Principia*. Only within the last century have scientists discovered that Newtonian physics does not hold true for particles that are extremely tiny (subatomic sized) and extremely fast (moving at a significant fraction of the speed of light).

6A   Section Review Questions

1. According to Aristotle, how much faster should a 10 kg rock drop than a 1 kg rock? What was his explanation for the fact that rocks drop and smoke rises?

2. Which scientist first stated the principle of inertia?

3. When modern geologists find that the age obtained for a rock sample using radiological dating techniques does not agree with other supposedly credible dating methods, they hypothesize reasons why the date is not in agreement with the expected age. What kind of explanations are these?

4. What did Galileo conclude was the "natural" state of motion for an object?

5. State the conditions under which Newton's laws do not apply.

⊚⊙6. Research the quotation attributed to Sir Isaac Newton in which he admits that his discoveries would not have been possible without the contributions of the great scientists who proceeded him.

⊚⊙7. Locate a copy of *Principia* in a library or bookstore. What interesting facts are associated with the original publication of this book?

6.6   Summing Forces

Everyone intuitively understands that a push or a pull must be exerted in order for an object to be moved. "Pushes and pulls" belong to the class of physical phenomena called **forces.** All matter is held together (sustained) by forces; matter exerts forces and has forces imposed on it. You may associate the word "force" with human activities and machines. Immense man-made forces are required to lift the space shuttle into orbit or to nudge an aircraft carrier against a pier. However, these forces pale in comparison to the inconceivably powerful natural forces unleashed by a hurricane or the eruption of a solar flare (see Figure 6-6). You also consciously and unconsciously employ small forces in your daily life routines when you press the keys on a computer keyboard, breathe, swallow, walk, sit, pull a sweater on, or write.

Forces have magnitudes and direction. They can easily be represented by vectors. Forces, for example, obey vector addition rules. Multiple forces can also act on an object simultaneously. In order to describe and predict how a force will affect an object's motion, we must know *where* the force is being applied. We will address that question later, but for now we will assume that the object on which forces are acting is a particle—that it has no physical size and all forces act on the same point. In later chapters dealing with circular motion and torques, we will consider what happens when forces do not act on the same point on an object.

Even though forces can be pushes or pulls, we will diagram forces as only pulls on the point of application of the force. This convention will help you avoid confusion when solving problems. The SI derived unit of force is the newton (N).

The effects of simultaneous forces acting on an object are of primary interest in physics. Therefore, we will assign the vector sum of simultaneous forces the symbol

$$\Sigma \mathbf{F} \equiv \mathbf{F}_1 + \mathbf{F}_2 + \ldots + \mathbf{F}_n.$$

The Greek capital letter sigma (Σ) is the mathematical summation symbol that stands for the sum of all terms represented by the variable following it.

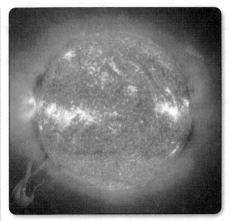
6-6   A solar flare contains unimaginable forces.

A **force** is an agent that causes or opposes changes in motion of matter, or tends to do so.

The SI unit of force is the newton (N). $1 \text{ N} = 1 \frac{\text{kg} \cdot \text{m}}{\text{s}^2}$

The symbol $\Sigma \mathbf{F}$ represents the sum of all simultaneous forces acting on an object. It is equivalent to the expression "net force."

6.7   Balanced and Unbalanced Force Sums

Forces do not always cause a change in motion. In Figure 6-7 (a), two people are pushing on a refrigerator equally hard but in opposite directions. As you can expect, no motion occurs. If the sum of all forces acting on an object is zero, then we say that the forces are balanced. **Balanced forces** cancel each other. The following mathematical expression is true for balanced forces:

$$\Sigma \mathbf{F} = 0 \Leftrightarrow \Sigma \mathbf{F}_x = 0, \text{ and } \Sigma \mathbf{F}_y = 0$$

In other words, when the force sum on an object is zero, the sums of the *x*- and *y*-components are also zero.

In figure b, one man has a cramp in one arm, so he cannot push as hard as the other

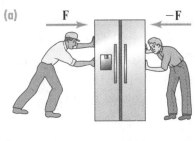

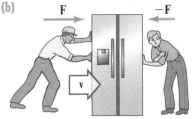

6-7   Only *unbalanced* forces cause motion.

When the sum of all simultaneous forces acting on an object equals zero, the forces are **balanced.**

The symbol $\Leftrightarrow$ in mathematical statements means "if and only if."

man. The refrigerator moves toward the man who pushes less. **Unbalanced forces** change an object's state of motion.

A pair of balanced forces on an object is represented by two vectors with the same magnitude but opposite directions. When these vectors are summed, they add up to a zero resultant vector. If two forces act at less than 180° to each other, you can use either the geometric-addition method or the analytical technique of vector resolution to determine the vector sum. Three or more forces acting on an object may sum to zero, but it is more likely that there will be a net resultant. The presence of unbalanced forces means that the force vectors do not cancel. Example 6-1 shows how unbalanced forces are used to move a ship in a desired direction.

EXAMPLE 6-1

Canal Mules: Summing Forces

Figure 6-8 shows a large ship entering the Gatún lock at the north end of the Panama Canal. The figure also shows two powerful "mule" engines running on railway tracks on both sides of the lock. They are attached to the ship with steel cables that allow them to pull the ship slowly into the lock basin as they move along the tracks.

If each mule engine exerts a pull of 1.50×10^5 N at an angle of 55.0° on either side to the front end (bow) of the ship, what is the magnitude and direction of the sum of the towing forces on the ship? Assume that the forces act at the same point on the centerline of the ship.

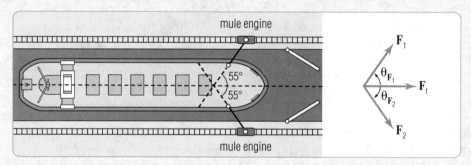

6-8 Vector diagram of mule engines towing a ship into a Panama Canal lock

Solution:

Choose a coordinate axis parallel to the length of the ship.

Given: $\mathbf{F}_1 = 1.50 \times 10^5$ N at $\theta_{\mathbf{F}_1} = +55.0°$;

$\mathbf{F}_2 = 1.50 \times 10^5$ N at $\theta_{\mathbf{F}_2} = -55.0°$

Find the total towing force ($\mathbf{F}_{total}$).

You know that the total towing force is the resultant of the vector sum of the two towing forces:

$$\mathbf{F}_{total} = \mathbf{F}_1 + \mathbf{F}_2$$

Every vector can be resolved into the sum of its components:

$$\mathbf{F}_{total} = \mathbf{F}_{total\,x} + \mathbf{F}_{total\,y}$$

The resultant components are the vector sums of the corresponding component vectors:

$$\mathbf{F}_{total} = \mathbf{F}_{1\,x} + \mathbf{F}_{2\,x}$$
$$\mathbf{F}_{total} = \mathbf{F}_{1\,y} + \mathbf{F}_{2\,y}$$

These equations are one-dimensional vector equations, so they may be expressed as scalar components in order to complete the algebra operations.

1. Identify reference angles: $\alpha_{F_1} = \theta_{F_1}$; $\alpha_{F_2} = -\theta_{F_2}$

2. Determine scalar components:

$F_{1x} = F_1 \cos \alpha_{F_1} = (1.50 \times 10^5 \text{ N})\cos 55.0° \doteq +0.86\underline{0}3 \times 10^5 \text{ N (at 0°)}$

$F_{1y} = F_1 \sin \alpha_{F_1} = (1.50 \times 10^5 \text{ N})\sin 55.0° \doteq +1.2\underline{2}8 \times 10^5 \text{ N (at 90°)}$

$F_{2x} = F_2 \cos \alpha_{F_2} = (1.50 \times 10^5 \text{ N})\cos 55.0° \doteq +0.86\underline{0}3 \times 10^5 \text{ N (at 0°)}$

$F_{2y} = F_2 \sin \alpha_{F_2} = (1.50 \times 10^5 \text{ N})\sin 55.0° \doteq -1.2\underline{2}8 \times 10^5 \text{ N (at 270°)}$

(signs of results are based on the geometry of the problem)

3. Find the towing force sum components:

$F_{\text{total } x} = 2(+0.86\underline{0}3 \times 10^5 \text{ N}) \approx +1.72\underline{0}6 \times 10^5 \text{ N (at 0°)};$

$F_{\text{total } y} = (+1.2\underline{2}8 \times 10^5 \text{ N}) + (-1.2\underline{2}8 \times 10^5 \text{ N}) = 0 \text{ N}$

4. The magnitude of the towing force is found by using the Pythagorean theorem:

$|\mathbf{F}_{\text{total}}| = F_{\text{total}} = \sqrt{F_{\text{total } x}^2 + F_{\text{total } y}^2} = \sqrt{(+1.72\underline{0}6 \times 10^5 \text{ N})^2 + (0 \text{ N})^2}$

$F_{\text{total}} \doteq 1.72\underline{0}6 \times 10^5 \text{ N}$

$F_{\text{total}} \approx 1.721 \times 10^5 \text{ N}$

The direction of $\mathbf{F}_{\text{total}}$ is the same as its x-component: 0°.

$\mathbf{F}_{\text{total}} = 1.721 \times 10^5 \text{ N}$ directly in front of the ship.

6.8 The Equilibrant Force

When a force sum is zero, then any one of the contributing forces exactly balances all of the others. A force that balances one or more other *concurrent* forces is called an **equilibrant force** (see Figure 6-9). The equilibrant of a set of unbalanced forces is a vector having the same magnitude as the vector sum of the other unbalanced forces but pointing in the opposite direction. In equation form, this means

$$\mathbf{F}_{\text{equil.}} = -\Sigma \mathbf{F}_{\text{other}},$$

where $\mathbf{F}_{\text{equil}}$ is the equilibrant force and $\mathbf{F}_{\text{other}}$ are all the other unbalanced forces acting simultaneously on the object. You can use the equilibrant concept to determine an unknown force. If the sum of all forces on an object is zero, then any unknown force must be the equilibrant of all of the known forces.

The **weight** of an object ($\mathbf{F}_w$) is the force of gravity acting on it. This is a common force that must often be considered in mechanics problems. Example 6-2 demonstrates how weight can be an equilibrant force.

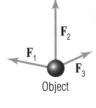

Object

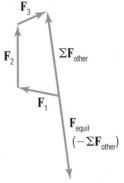

6-9 Resultant and equilibrant vectors

The term "concurrent" means "acting at a single point."

The **equilibrant** of a set of unbalanced forces is a vector having the same magnitude as the unbalanced force sum but pointing in the opposite direction.

The people in Nazareth, where Jesus grew up, failed to respect Him as people from other regions did. "And he could there do no mighty work, save that he laid his hands upon a few sick folk, and healed them" (Mark 6:5). Is disbelief the "equilibrant force" to miracles? No. Jesus can, but won't, do mighty works for those who do not believe in Him. He chooses to do mighty works based on His omniscience and omnipotence and not on what we (often mistakenly) think is best.

The **weight** ($\mathbf{F}_w$) of an object is the gravitational force on the object's mass.

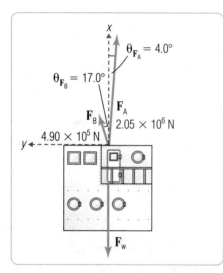

6-10 Force diagram for Example 6-2

EXAMPLE 6-2

Unknown Weight: Finding an Equilibrant

A shipyard has finished constructing a modular aluminum-and-steel superstructure for a new merchant ship and is preparing to install it on the hull. Naval architects need to have a close approximation of the weight of the structure in order to calculate its effect on the buoyancy of the hull after it is installed.

In order to weigh the superstructure, two shipyard cranes lift and suspend it as the architects note the angle and lifting force exerted by the crane cables on the superstructure. The forces and angular information are provided in Figure 6-10. Determine the total lifting force. The equilibrant of the lifting force is the unknown weight of the superstructure, $\mathbf{F}_w$.

Solution:

Choose a Cartesian coordinate system with axes horizontal and vertical. There are several ways to orient the coordinate system, but the least confusing is as shown in the figure. *Cable angles are measured from the vertical direction.*

From the figure, you know that $\mathbf{F}_A = 2.05 \times 10^6$ N at $-4.0°$ and $\mathbf{F}_B = 4.90 \times 10^5$ N at $+17.0°$.

Strategy: $\mathbf{F}_w$ is the equilibrant of the total lifting force:

$$\mathbf{F}_w = -\mathbf{F}_{lift}$$

The total lifting force is the vector sum of the two crane forces:

$$\mathbf{F}_{lift} = \mathbf{F}_A + \mathbf{F}_B$$

1. Determine the reference angles: The angles between the cables and the vertical are measured at the attachment point on the superstructure.

$\alpha_{\mathbf{F}_A} = 4.0°; \; \alpha_{\mathbf{F}_B} = 17.0°$

2. Determine the force sum components.

$|F_{Ax}| = (2.50 \times 10^6 \text{ N})\cos 4.0° \doteq 2.0\underline{4}5 \times 10^6 \text{ N (up)}$

$|F_{Ay}| = (2.50 \times 10^6 \text{ N})\sin 4.0° \doteq 0.14\underline{3}0 \times 10^6 \text{ N} = 1.4\underline{3}0 \times 10^5 \text{ N (right)}$

$|F_{Bx}| = (4.90 \times 10^5 \text{ N})\cos 17.0° \doteq 4.6\underline{8}5 \times 10^5 \text{ N} = 0.46\underline{8}5 \times 10^6 \text{ N (up)}$

$|F_{By}| = (4.90 \times 10^5 \text{ N})\sin 17.0° \doteq 1.4\underline{3}2 \times 10^5 \text{ N (left)}$

3. Determine the lifting force resultant components:

$F_{lift \, x} = F_{A \, x} + F_{B \, x} = (+2.0\underline{4}5 \times 10^6 \text{ N}) + (+4.6\underline{8}5 \times 10^6 \text{ N})$

$F_{lift \, x} \doteq 6.7\underline{3}0 \times 10^6 \text{ N} \approx 6.73 \times 10^6 \text{ N}$

$F_{lift \, y} = F_{A \, y} + F_{B \, y} = (+1.4\underline{3}0 \times 10^5 \text{ N}) + (-1.4\underline{3}2 \times 10^5 \text{ N})$

$F_{lift \, y} \doteq -0.0\underline{0}2 \times 10^5 \text{ N} \approx 0 \text{ N}$

4. The lifting force is $\mathbf{F}_{lift} = 6.73 \times 10^6$ N straight up since the horizontal component is essentially zero.

5. The superstructure's weight is the equilibrant to the sum of the lifting forces:

$$\mathbf{F}_w = -\mathbf{F}_{lift}$$
$$\mathbf{F}_w = -6.73 \times 10^6 \text{ N (straight down)}$$

6.9 Types of Forces

Gravitational Force

All forces can be classified as either fundamental forces or mechanical forces. The **fundamental forces** are inherently part of the atomic building blocks of matter. Scientists have identified four fundamental forces in the universe, from which all other forces are derived. Of the four, two were discovered as recently as the twentieth century.

The first to be recognized and mathematically described (by Sir Isaac Newton during the seventeenth century) was the **gravitational force.** This long-range force is proportional to the mass of objects and exerts its influence over theoretically infinite distances. It is the dominant force that organizes the structure of astronomical bodies, controls the motions of planets around the Sun, and governs the celestial processes of the universe. All objects exert gravitational force on all other objects. You will study the effects of gravitational force in association with nearly all mechanics topics in Units 2 and 3, as well as its importance to modern physics in Unit 6.

6-11 Both Earth and the Moon attract each other through the gravitational force.

Moon's gravity

Earth's gravity

Electromagnetic Force

The existence of magnetic and electric forces had been known for millennia and studied to some extent in the seventeenth century, but these forces were not recognized as being two components of the same fundamental force until the mid-nineteenth century when James Clerk Maxwell formulated the unifying theory of electromagnetism. The **electromagnetic force,** as it is now called, is also a long-range force. You have seen long-range electrostatic discharges in the form of lightning, and the *Voyager* space probes have detected the Sun's magnetic field billions of miles from its source. The electromagnetic forces also act at very short distances. They are responsible for chemical bonding and other interactions of atoms and molecules. The electromagnetic force governs the fine structure of matter because it determines many of the physical and all of the chemical properties of substances. You will study electric fields in Chapter 19 and magnetism in Chapter 21. Combined electromagnetic theory will be studied in Chapter 22 and in Unit 6.

6-12 A lightning strike is a large concentration of electromagnetic forces.

Strong and Weak Nuclear Forces

Early in the twentieth century, two short-range fundamental forces were discovered during the development of the quantum model of the atom. These forces act only over very short distances, less than a few femtometers (10^{-15} m), which is comparable to the diameter of protons and neutrons in the nuclei of atoms, and thus they are called nuclear forces. The **strong nuclear force** binds together the particles forming the nucleus while the **weak nuclear force** holds certain elementary nuclear particles together and is involved in several modes of nuclear decay. You may study these two nuclear forces in advanced physics courses.

The four fundamental forces are
- Gravitational force
- Electromagnetic force
- Strong nuclear force
- Weak nuclear force

6.10 Classification of Forces

Prior to the last century, forces seemed to be easily classifiable into those that could act through space without physical contact and those that could be transmitted only through physical contact between objects.

6-13 Action-at-a-distance forces are thought to emanate from certain objects and matter in general.

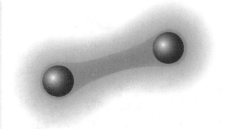

6-14 The field theory states that noncontact forces arise from the kind of space surrounding certain objects and matter.

Noncontact Forces

In one model, gravitational and electromagnetic forces are called **action-at-a-distance** forces to emphasize that no physical contact is necessary for transmission of the force. This model assumes that each kind of force is a property of the associated matter, which can somehow be exerted at a distance without requiring physical contact.

Another approach to explaining **noncontact forces** is the **field theory.** This theory proposes that the gravitational influence of matter and the attraction or repulsion of magnetic or charged particles are properties of the space surrounding the matter (the field). The matter exerts forces on other massive, magnetic, or charged particles when they are placed in their respective fields.

The development of quantum mechanics after World War II produced yet another model that tries to account for all of the fundamental forces. It suggests that **virtual particles** are exchanged between matter with like characteristics. The exchange of the virtual (nonmaterial) carrier particles supposedly creates the force. Virtual particles have been identified for the electromagnetic force (photon) and nuclear forces (gluon), but no particle has yet been identified that carries the gravitational force (graviton?). You will study some aspects of these three force models in later chapters.

6-15 Particle exchange theories suggest that noncontact forces exist because matter exchanges nonmaterial particles associated with each kind of force. Here, photons are exchanged.

The **noncontact forces** are gravity and the electromagnetic forces. Gravity displays only attraction, while electromagnetic forces are both attractive and repulsive. No physical contact is required to transmit or exert these forces.

Contact forces are derived from the interaction of electromagnetic forces among particles of matter. Contact forces require physical contact of objects in order to be transmitted or exerted.

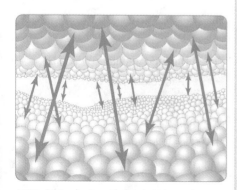

6-16 Friction begins with the electromagnetic attraction and irregularities of particles at the surfaces of objects.

Contact Forces

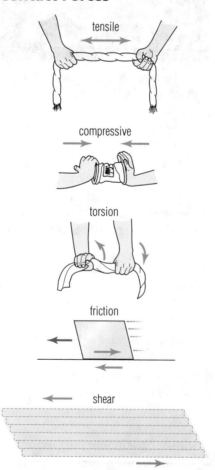

tensile

compressive

torsion

friction

shear

6-17 Mechanical or contact forces

As mentioned earlier, the electromagnetic force is responsible for both the structure of the atom and the chemical bonding between atoms. The electromagnetic forces in the electron orbital structures of atoms can either attract or repel other atoms. Most atoms have some tendency to attract and hold on to nearby atoms, but no two atoms can occupy the same space at the same time. When aggregates of atoms (any kind of matter) are forced together or pulled apart, the particles exert forces between each other. This gives rise to the concept of **contact** or **mechanical forces** (forces that are transmitted only by physical contact between objects).

Contact forces include **tension** or **tensile forces** (forces that tend to pull things apart), **compressive forces** (forces that tend to push things together or to crush), **torsion** (twisting forces), **friction** (forces that oppose motion between two objects in contact), and **shear forces** (forces that tend to cause layers within matter to slide past one another). Mechanical forces and the gravitational force will be the forces of primary concern in Units 2 and 3.

According to Scripture, the advent of forces in creation must have been during the very first instants of time (Gen. 1:1). All four fundamental forces appeared as matter was brought into existence—gravity from the matter itself; electromagnetic forces in the chemical bonds between the atoms of the matter; and the nuclear forces that held the atomic nuclei together. The material world continues to exist as a direct result of God's providence in sustaining these forces. "For by him [Jesus] were all things created, that are in heaven, and that are in earth. . . . and by him all things consist" (Col. 1:16–17).

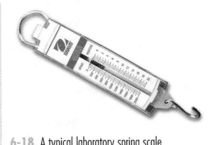

6-18 A typical laboratory spring scale

6.11 Measuring Forces

Forces may be measured using a variety of instruments. A typical force-measuring instrument in a high-school laboratory is the *spring scale,* shown in the photograph. The displacement of the spring, indicated by the change in position of the pointer from zero, is directly proportional to the force applied to the spring, so the scale is calibrated in appropriate increments to indicate the applied force.

Materials engineers use a type of force-sensing detector called a *load cell* to measure large forces applied to material samples. It consists of a crystalline substance, such as pure quartz, that produces an electrical signal when the crystal is squeezed by the applied force.

Forces may also be inferred using instruments such as *pressure gauges, ballistic pendulums,* and *accelerometers.* These and other instruments will be discussed later in this course.

In the classroom, a **force table,** shown in Figure 6-20, demonstrates the addition of force vectors. When the attachment ring is stationary directly above the center of the table, the applied forces are balanced. With this apparatus, the equilibrant force can be determined directly by finding the weight of the suspended mass generating the force.

6-19 A pressure gauge indicates the force per unit area within a fluid.

6-20 A force table

6B   Section Review Questions

1. Give a definition of "force" that considers its potential to change motion.

2. Name and briefly describe five mechanical forces.

3. Write the symbol that represents the resultant of simultaneous concurrent forces.

4. **a.** What is the force called that exactly balances a nonzero force sum acting on an object?

   b. What is its magnitude and direction compared to the original force sum?

5. List the four fundamental forces and briefly describe each.

◉6. Three concurrent forces act simultaneously on a point: $\mathbf{F}_1 = 15.0$ N at 25.0°, $\mathbf{F}_2 = 22.0$ N at 73.0°, and $\mathbf{F}_3 = 8.0$ N at 265.0°. Determine the force sum, $\Sigma\mathbf{F}$.

◉7. Your dog Rover suddenly exerts a force of 200 N on his leash. What is the equilibrant of this force vector?

◉8. A force board is set up in a physics laboratory and three spring scales are hooked to the central ring and the notched perimeter of the table. The first scale reads 12.5 N at an angle of 0.0° and the second scale reads 17.3 N at an angle of −34.0°. If the three forces are balanced, what are the reading and angle of the third spring scale?

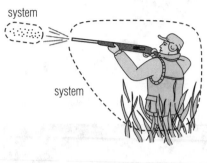

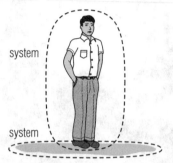

6-21 Several systems. The "system" is everything inside the boundary.

The word "system" has another meaning as well. In the phrase "system of forces" it refers to the set of related forces acting on an object. There can also be a system of forces acting on a system.

Problem-Solving Strategy 6.3

After sketching a physics problem, determine the system in the problem, draw a circle around it, and identify all forces acting *on* the object(s) inside the circle.

6.12 Systems and Newton's Laws

Dynamics is the study of the effect of forces on the motion of objects. The central principles of dynamics have been summarized in a series of statements known as **Newton's laws of motion.** The term *laws* does not mean that these principles hold true in every case or that, like God's laws, they are immutable. Scientific laws are man's attempt to organize his observations of the universe in a sensible and predictable way. Consequently, they are imperfect and subject to change. A law is a general principle that usually has been experimentally validated and that the majority of scientists accept as true for nearly all practical situations; any exceptions are usually well understood or else have not yet been discovered.

Before discussing Newton's laws, you must understand what they apply to. Physicists study the mechanics of **systems.** You may think of a system as a complex, interconnected collection of parts such as an electrical system, a plumbing system, or even the postal system. In physics, a system may have one or more parts. We will usually study systems with only one or two objects to minimize the complexity of the problem. The "system" concept changes the way we define the word "object." An object may be a single, discrete particle, but it could also be a rock, a quantity of gas molecules, a flock of geese, a space shuttle, a planet, or even you.

A physical system is contained within an imaginary boundary that separates it from its surroundings. Everything applicable to the problem and affecting the system exists within the system boundary. Everything outside the boundary may be ignored if it doesn't somehow act on the system. Developing the skill to correctly identify "the system" in a problem will vastly simplify the solutions to physics exercises.

6.13 Newton's First Law of Motion

Newton's first law generalizes Galileo's principle of inertia. Newton wrote in *Principia, "Every body continues in its state of rest, or of uniform motion in a right line, unless it is compelled to change that state by forces impressed upon it."* A more modern version of this statement is: A system at rest will remain at rest, and a moving system will move continuously with a constant velocity (constant speed and direction) unless acted on by outside unbalanced forces.

> The first law establishes two conditions:
>
> - If all external forces on a system are balanced, then its velocity remains constant (that is, no acceleration occurs).
>
> - If all forces acting on a system are *not* balanced, then a nonzero resultant force exists and the velocity changes (an acceleration occurs).

Mathematically, the first law can be expressed as

$$\Sigma \mathbf{F} = 0 \Leftrightarrow \mathbf{a} = 0.$$

This statement means that the sum of all forces on a system is zero if and only if the acceleration of the system is zero.

An equivalent statement of the first law is

$$\Sigma \mathbf{F} \neq 0 \Leftrightarrow \mathbf{a} \neq 0.$$

The first law defines the condition of a zero force sum on a system. However, zero total force does *not* mean that there are no forces on a system; only that their vector sum, or resultant, is zero.

Everyday experience sometimes seems to contradict the first law of motion. For example, if you do not continue to pedal a bike, it seems to slow down and stop "naturally." What causes this? From the first law you can predict that there is some unbalanced force that slows the object—friction.

Even when an unbalanced force is imposed on the system, the system tends to resist a change in motion. This tendency was identified by Galileo as inertia. The concept of mass was developed by scientists to quantify inertia.

When the vector sum of all forces on a system is zero, it is said to be in **mechanical equilibrium.** A system in equilibrium has a constant velocity. A system at rest (in a given reference frame) is in equilibrium with a constant velocity equal to zero. We will refer to mechanical equilibrium frequently in our study of mechanics.

Newton's first law predicts that without unbalanced forces, moving objects will travel in straight lines. When the young shepherd David defeated the Philistine hero Goliath, he propelled the pebble with a sling similar to the one shown in Figure 6-22. The principles of operation of a sling are simple, but putting them into practice is not. When the sling is whirled in a circle, the sling applies an unbalanced force to the stone that causes it to move in a circular path. When one of the thongs is released, the unbalanced force is removed from the pebble. It begins a straight-line trajectory in the direction of its instantaneous velocity at the moment of release.

6.14 Newton's Second Law of Motion

The second law is the most general of the three laws because it gives an operational working definition of the concept of force. It tells us how to measure a force, assuming that we can measure the mass and acceleration of a system. Before we state this law, let us consider an experiment that illustrates the relationship between mass, acceleration, and unbalanced forces.

Case 1: If you pull a 1 kg mass across a frictionless surface with *constant* force as measured with a spring scale, you observe a change in velocity (an acceleration) of the mass.

6-23a Case 1—Force causes a mass to accelerate.

Case 2: If you pull the same mass with a larger constant force you note a larger acceleration. If you repeat with many different forces, you begin to recognize a pattern.

6-23b Case 2—Acceleration is proportional to the force.

6-22 The velocity of a pebble is continually changing until it is released.

Case 3: If you pull the same mass with two identical forces acting in *opposite* directions, you observe that no acceleration occurs, since $\Sigma \mathbf{F} = 0$.

6-23c Case 3—Zero net force causes zero acceleration.

Case 4: If you pull the same mass with two identical forces in the *same* direction, you observe that the acceleration is twice as great as with one force.

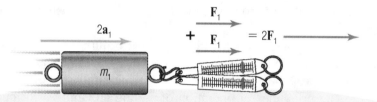

6-23d Case 4—Acceleration is *directly* proportional to force.

Case 5: If you pull a 2 kg mass with the same force as in Case 1, you note that the acceleration is half as great.

6-23e Case 5—Acceleration is inversely proportional to mass.

In the *Principia*, Newton wrote, *"The change in motion is proportional to the motive force impressed; and is made in the direction of the right line in which that force is impressed."*

In other words:

> The acceleration of a system is directly proportional to the sum of the forces (resultant force) acting on the system and is in the same direction as the resultant.

In a mathematical form, the second law is written

$$\mathbf{a} = \frac{\Sigma \mathbf{F}}{m}.$$ (6.1)

From Equation 6.1 you can see that the acceleration of a system is inversely proportional to the mass of the system. In other words, if you double the mass of a dump truck by adding rocks to the dump box, a given force applied by its engine will produce an acceleration only half as great as for the empty truck.

The second law states that acceleration is the response of the mass of the object to an unbalanced force. The force is the cause, and acceleration is the response of the mass. The SI unit of force, the newton (N), is formally defined as follows:

6-24 The forces needed to accelerate a Navy aircraft to takeoff speed are provided by a steam catapult.

1 N is the *total force* that, when applied to exactly 1 kg, produces an acceleration of exactly 1 m/s². In SI base units,

$$1 \text{ N} = \frac{1 \text{ kg·m}}{\text{s}^2} = 1 \text{ kg·} \frac{1 \text{ m}}{\text{s}^2}.$$

The second law equation is usually written

$$\Sigma \mathbf{F} = m\mathbf{a}. \tag{6.2}$$

Newton's second law establishes that a system's acceleration is proportional to the net force imposed upon it and is inversely proportional to its mass:

$$\Sigma \mathbf{F} = m\mathbf{a}$$

EXAMPLE 6-3

Acceleration from Force: The Second Law

A 4400 kg truck accelerates straight ahead at 2.0 m/s². What is the net force on the truck?

Solution:

Choose a coordinate system parallel to the acceleration of the truck, which is parallel to the ground. Since there is no acceleration perpendicular to the ground, we can ignore vertical forces for now.

Applicable formula: $\Sigma \mathbf{F} = m\mathbf{a}$

$$\Sigma F_x = (4400 \text{ kg})(+2.0 \text{ m/s}^2) = +8800 \text{ N}$$

The force sum is 8800 N straight ahead.

6-25 An accelerating truck

EXAMPLE 6-4

Great Shot: Quantifying Inertia—Mass

An athlete in the shot put event exerts a force with a magnitude of 200. N on the shot. If the magnitude of the shot's acceleration is 20. m/s² in the direction of its path, what is the mass of the shot? Assume that the athlete accelerates the shot in a straight line.

Solution:

Choose a coordinate axis parallel to the motion of the shot. Only motion along this axis will be considered.

$$F_x = ma_x$$

$$m = \frac{F_x}{a_x} = \frac{+200. \text{ N}}{+20. \text{ m/s}^2} = \frac{+200. \text{ kg·m/s}^2}{+20. \text{ m/s}^2}$$

$$m = 10. \text{ kg}$$

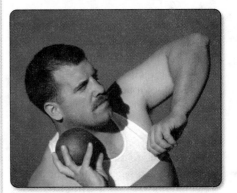

6-26 A shot-putter

The previous two examples were one-dimensional situations. However, the second law applies in two and three dimensions as well. The second law formula is a vector equation. Consequently, both sides can be resolved into corresponding component equations:

$$\Sigma F_x = ma_x$$

$$\Sigma F_y = ma_y$$

$$\Sigma F_z = ma_z \text{ (for three-dimensional situations)}$$

Problem-Solving Strategy 6.4

To determine the net force vector acting on a system, multiply each component of acceleration by the mass to find the net force components. Sum the components to find the resultant force.

Dynamics **127**

6.15 Newton's Third Law of Motion

Newton recognized the need for one additional postulate to completely state the fundamental laws of dynamics. The existence of the third law can be demonstrated using two spring scales, as described below.

Each scale is a separate system. Hook the spring scales together and pull them in opposite directions. Both record the same force when the springs are slightly stretched and held stationary. The force read on the right-hand scale is being exerted by the left-hand scale and vice versa. This shows that there are two forces involved in each spring scale system: the force applied *by the scale* on something outside its system, and the force acting *on the scale* from outside the system.

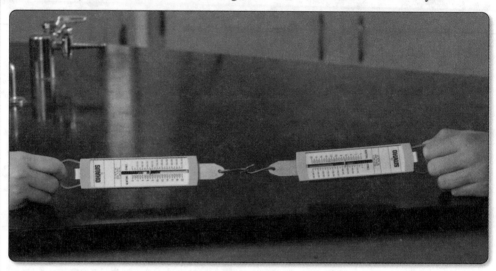

6-27 An action-reaction force pair

Newton's third law describes this balancing process. In his own words, *"To every action there is always opposed an equal reaction: or, the mutual actions of two bodies upon each other are always equal, and directed to contrary parts."* While the English translation of Newton's *Principia* used the words "action" and "reaction," these terms are not completely descriptive. Action-reaction force pairs can exist even if there is no visible action involved. To express this concept in more familiar terms:

> If system X exerts a force on system Y, then Y exerts a force of the same magnitude on X but in the opposite direction.

We denote the force of object X acting on Y by $\mathbf{F}_{X \rightarrow Y}$, where the arrow means "acts on." The other force in the action-reaction force pair is $\mathbf{F}_{Y \rightarrow X}$. In equation form, the third law can be expressed as

$$\mathbf{F}_{X \rightarrow Y} = -\mathbf{F}_{Y \rightarrow X}.$$

Forces have four properties as a consequence of the third law that you must keep in mind when solving physics problems:

- *All* forces occur in pairs.
- Each force in an action-reaction pair has the same magnitude.
- Each force acts in the opposite direction in line with the other force of the pair.
- Each force acts on a *different* system.

Newton's third law reveals that every force is always accompanied by a second force acting in the opposite direction. It has been said, "You cannot touch without being touched."

"My people are destroyed for lack of knowledge: because thou hast rejected knowledge, I will also reject thee" (Hos. 4:6). The knowledge that Israel rejected was the knowledge of God. Reject God and He will reject you.

Students who don't understand the fourth property often have difficulties solving physics problems. According to the third law of motion, *every* force in nature is one of a pair of forces. Each force of an **action-reaction force pair** acts on a different system. When analyzing acceleration using the second law equation, you must first isolate the system of interest. Identify the action-reaction pairs, and include in your diagram only those forces that are acting *on* the system, not the forces that the system exerts on something else.

EXAMPLE 6-5

Cargo in Space: The Third Law

A robotic arm attached to the International Space Station is used for construction of the space station and the positioning of modular cargo containers in various external storage locations. While unloading a space shuttle, the arm exerts a force of 50. N on a 250. kg cargo container.
a. What force does the cargo container exert on the robotic arm?
b. What is the magnitude of the acceleration of the cargo container?
c. If the arm sustains this force for 5.0 seconds in a straight path, how fast will the container be moving?

Solution: See Figure 6-28b.
a. The force the manipulator arm exerts on the container is

$$F_{a \to c} = 50. \text{ N}.$$

Therefore, the reaction force component of the container on the arm that is in line with the arm's force is

$$F_{a \to c} = 50. \text{ N in the opposite direction.}$$

b. The system of interest is the cargo container. The only force acting on the container is the force exerted by the manipulator arm. (For now, we can ignore the force exerted by Earth's gravity, since it affects everything at the space station equally.)

$$F_{a \to c} = m_c a_c$$

$$a_c = \frac{F_{a \to c}}{m_c} = \frac{50. \text{ N}}{250. \text{ kg}} = \frac{50. \text{ kg} \cdot \text{m/s}^2}{250. \text{ kg}}$$

$$a_c = 0.20 \text{ m/s}^2$$

c. To find the container's speed after 5.0 seconds, use the first equation of motion,

$$v = v_0 + a\Delta t.$$

Assume the container started at rest (in the space station's frame of reference), so that $v_0 = 0$ m/s.
Substitute: $v = 0$ m/s $+ (0.20 \text{ m/s}^2)(5.0 \text{ s})$
$v = 1.0$ m/s

Considering the large mass of the cargo container, the operator of the robotic arm must allow enough time to slow the container at its destination to avoid a damaging collision!

6-28a The Canadarm 2 mechanical arm serving at the International Space Station

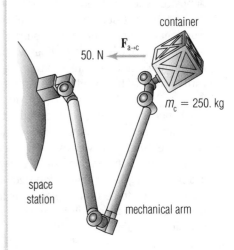

6-28b A manipulator arm moving a freight container

FACETS of PHYSICS

Flight and Newton's Third Law

At 412 775 kg (910 000 lb), the Boeing 747-400ER jumbo jet is the largest commercial airliner in the world. In this chapter you learned that for every action there is an equal and opposite reaction (Newton's third law). If the reaction you desire is to get such a large plane into the air, the engines must generate an enormous amount of thrust.

While the jet is idling in line, waiting its turn to take off, the engine's turbines are rotating, generating an amount of thrust that is balanced by the force of the brakes on the wheels. If you have ever flown on a commercial jet, you remember the feeling as the jet rounds the end of the runway. You can hear the engines revving, and then you feel the explosive burst of power as the brakes are released. What is happening? As the engine exerts a great force on the exhaust gas by blowing it out the back nozzle, the exhaust gases exert a tremendous force on the plane, propelling it forward. When it is traveling fast enough to generate sufficient lift, the plane rises into the air. The 747-400 typically takes off at 290 km/h (180 mi/h), a speed attained by four engines that can generate up to 269 677 N (62 100 lb) of thrust each.

Thrust can be generated in two basic ways: taking a large amount of air and increasing its speed a little, or taking a smaller amount of air and increasing its speed greatly. The modern Turbofan gas turbine engine combines both methods and is highly efficient.

The 747-400 lands at a speed of 260 km/h (160 mi/h), but how does the plane stop once it's on the ground? There are two systems deployed to assist the brakes in slowing the plane. First, spoilers raised on the top surface and flaps lowered on the trailing edge of the wing provide wind resistance.

Second, thrust reversers enable the plane's engines to help slow the plane. After the plane touches down, you hear what may sound like the engines revving in reverse. Remember, however, that jet engines can't run backward. Instead, the thrust reversers mechanically deploy flaps into the exhaust to deflect the exhaust gases toward the front of the plane. This creates a strong net-braking force on the plane. The wing flaps, the brakes, and the thrust reversers all help slow plane after it lands.

Applied physics, along with other disciplines, has been very important in the proliferation and development of aviation. To show you how far we have come, the 45 m economy section of a 747-400 is longer than the Wright brothers' first flight at Kitty Hawk, North Carolina!

6.16 Weight and Mass

Weight

The matter of the earth mysteriously produces a measurable attraction force on objects near the earth. This force keeps your feet planted firmly on the ground. Even more mysteriously, every object produces a gravitational force on every other object in the universe. Gravitational attraction between objects is related to the distance between them as well as their masses, so small and distant objects exert far less force than massive and near objects. We will investigate gravitational theory in greater depth in Chapter 7.

As we have discussed, the force of planetary gravitational attraction on an "average-sized" object is called the weight of the object, $\mathbf{F}_w$. Weight is directly proportional to the object's mass. In equation form,

$$F_w \propto m, \text{ or}$$

$$F_w = km,$$

where k is a proportionality constant with units of N/kg. If we analyze the base units associated with k, (kg·m/s^2)/kg, we find that they simplify to m/s^2, which is the unit of acceleration. Substituting the symbol a for k into the weight equation gives $F_w = am$ or, after transposing the right side,

$$F_w = ma, \tag{6.3}$$

which has a form similar to Newton's second law. However, rather than describing the response of a mass to a force sum, this equation assigns a value to a particular force—the weight of the object. In Chapter 3 you learned that the acceleration due to gravity for freely falling objects is $\mathbf{g} = 9.81$ m/s^2 straight down. The local acceleration due to gravity is the response of the mass of an object to the gravitational force acting on it (its weight). It is common practice to substitute $\mathbf{g}$ for acceleration in the vector form of Equation 6.3:

$$\mathbf{F}_w = m\mathbf{g} \tag{6.4}$$

Thus, a mass of 1 kg has a weight of 9.81 N and a mass of 10 kg has a weight of 98.1 N. Equation 6.4 is true no matter what the condition or orientation of the object. You can calculate the weight of an object using Equation 6.4 even if it is *not* falling. The weight of a rock of mass m resting on the ground is equal to $m|g|$. An object moving at "normal" speeds has the same weight as it does when stationary. (This is *not* the case for objects moving at a significant fraction of the speed of light—see Chapter 27.) For objects within a few thousand meters of the earth's surface, the proportionality between weight and mass is nearly constant over the entire planet.

Equation 6.4 is a vector equation. The magnitude of an object's weight vector is equal to $|mg|$ and the weight vector points in the same direction as the gravity vector, which is normally considered to be straight down (toward the center of the earth). The scalar component form of Equation 6.4 is

$$F_{w\,y} = mg_y.$$

There are other units for weight besides the newton. The most familiar unit in the United States is the pound, which is equivalent to the English Engineering System (EES) unit of force, the **pound-force (lbf).** One pound-force is equivalent to the weight, measured at the earth's surface, of a mass equal to one EES **pound-mass (lbm).** None of the measurement units based on the EES commonly used in the United States are approved by NIST for use with SI units in scientific research. We will normally avoid using EES units in this textbook.

> **Problem-Solving Strategy 6.5**
>
> Calculate the weight of an object by using the formula $\mathbf{F}_w = m\mathbf{g}$. The direction of the weight vector is always in the same direction as $\mathbf{g}$.

6-29 Sport utility vehicles are heavier than standard passenger cars.

EXAMPLE 6-6

A Massive Problem: Weight and the Second Law

What is the mass of an 8000. N SUV?

Solution:

$$F_w = |mg|$$

$$m = \frac{F_w}{|g|} = \frac{8000.\ N}{9.81\ m/s^2}$$

$$m \doteq 815.4\ kg$$

$$m \approx 815\ kg$$

Mass is always a positive scalar quantity.

EXAMPLE 6-7

Weighing in on Mars: A different g

If Mars has a gravitational acceleration of 3.69 m/s², how much would the SUV in Example 6-6 weigh there?

Solution:

$$F_w = |mg_{Mars}| = |(815.4\ kg)(-3.69\ m/s^2)|$$

$$F_w \doteq 3008\ N$$

$$F_w \approx 3010\ N$$

Mass

The method for measuring mass should be familiar to you from other science courses. A typical mechanical high-school balance is a seesaw-like device that compares the force of gravity on an unknown mass to the force on known masses. A mass balance would not work in orbit where there is no apparent gravity. Depending on the type of balance, the known masses may be individual calibrated masses or, more commonly, sliders on a calibrated arm. When the known masses are adjusted in order to align the balance at the zero index mark, the sum of the known masses must equal the unknown mass.

Other kinds of instruments can measure mass. Electronic scales measure the weight of an object and digitally display the corresponding mass based on Equation 6.3.

Trace amounts of unknown substances in a sample of matter can be identified from their characteristic molecular masses using a mass spectrometer. This instrument accelerates the molecules using a known force in a vacuum chamber to determine their masses.

Some of the force-measuring instruments mentioned in Section 6B can be configured to measure mass as well. Several mass-measuring instruments are pictured in Figure 6-30.

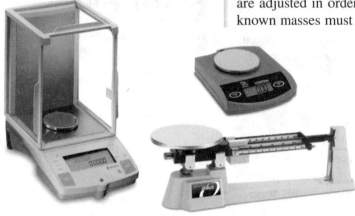
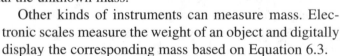

6-30 Various mass-measuring instruments

6C Section Review Questions

1. Batter up! With his feet firmly planted on the ground, the batter twists his body during the swing and exerts a force on the bat as it strikes the ball, sending the ball deep into the outfield. Sketch a diagram that defines the "system" for the purpose of describing the ball's motion at the instant it is hit. Include in your sketch any vector quantities that you think are acting *on* the system.

2. If an airliner is maintaining constant altitude, constant speed, and constant direction, what can you conclude about the forces acting on it?

3. Does the force you use to roll a bowling ball down the lane depend on the ball's acceleration, or does the ball's acceleration depend on the force that you exert? Explain your answer.

4. While standing, a physics student exerts a force of 625 N straight down on the ground. What other force must exist according to Newton's third law of motion? Describe the magnitude and the orientation of the force and identify the system upon which it acts.

⊛5. Which of the following will have the highest initial acceleration: a 63.3 kg human sprinter who can exert a traction force of 576 N, a 768 kg "funny car" dragster that exerts 19 700 N of traction, or a 16 600 kg Navy F/A-18 fighter-bomber catapulted with a force of 512 000 N? Show your work.

⊛6. Compute your weight in newtons. (1 lbm = 0.454 kg)

⊛7. The gravitational acceleration at the surface of the Moon is 1.62 m/s². If you can barely lift 195 lbf on Earth, what would be the maximum weight that you could lift on the Moon? Give your answer in newtons.

⊛8. A 0.170 kg puck rests on frictionless ice. Two hockey players strike the puck simultaneously. One force is 350. N exerted at 020.0T, and the other force is 600. N exerted at 065.0T. What is the magnitude and direction of the puck's acceleration at the instant that it is struck?

Chapter Review

In Terms of Physics

principle of inertia	6.4	noncontact force	6.10
forces	6.6	field theory	6.10
balanced forces	6.7	virtual particles	6.10
unbalanced forces	6.7	contact forces	6.10
equilibrant force	6.8	force board	6.11
weight ($\mathbf{F_w}$)	6.8	Newton's laws of motion	6.12
fundamental force	6.9	system	6.12
gravitational force	6.9	mechanical equilibrium	6.13
electromagnetic force	6.9	action-reaction force pair	6.15
strong nuclear force	6.9	pound-force (lbf)	6.16
weak nuclear force	6.9	pound-mass (lbm)	6.16

Problem-Solving Strategies

6.1 (page 117) Always sketch forces as *pulls* in a force vector diagram.

6.2 (page 119) If one of a resultant's components is zero, then the resultant has the same magnitude and direction as the nonzero component.

6.3 (page 124) After sketching a mechanics problem, determine the system in the problem, draw a circle around it, and identify all forces acting *on* the object(s) inside the circle.

6.4 (page 127) To determine the net force vector acting on a system, multiply each component of acceleration by the mass to find the net force components. Sum the components to find the resultant force.

6.5 (page 131) Calculate the weight of an object by using the formula $\mathbf{F}_w = m\mathbf{g}$. The direction of the weight vector is always in the same direction as $\mathbf{g}$.

6.6 (page 132) When calculating the weight of an object, it is permissible to report only its magnitude unless you need its vector form for solving other portions of the problem.

Review Questions

1. What common phenomenon likely accounts for the Greek natural philosophers' observation that objects naturally tend to slow down and that heavier objects fall faster than lighter objects?

2. What did Galileo conclude about the natural tendency of any moving object?

3. Which of the following statements would Newton disagree with? Why?
   a. An object's motion changes in response to an unbalanced force.
   b. For every action, there is an equal and opposite reaction.
   c. A force is needed to maintain constant velocity.
   d. The acceleration of an object is proportional to the net force acting on it and is inversely proportional to its mass.

4. Galileo referred to an "agent" that causes a change in the state of motion of an object. What is this agent?

5. At what point in the Creation event did the fundamental forces likely appear?

6. What force causes the transmission of light?

7. Which model of noncontact force suggests that the space around an object is involved in conveying the force to another object?

8. The sum of all forces on an object is called
   a. the weight vector.   c. the equilibrant force.
   b. the resultant force.   d. the reaction force.

9. What is the minimum number of forces that must act on a system in order to have
   a. an unbalanced force on the system?
   b. a balanced force on the system?

10. What do you know about a system of unknown forces if it has a nonzero equilibrant?

11. The diagram below shows a raccoon (*Procyon lotor*) resting on a log.
    a. If you need to find the forces acting on the raccoon, what is the system?
    b. What is the system if you must find the forces on the log?
    c. Identify the system if you must determine the force exerted by both the raccoon and the log together.

12. A hockey puck slides across the ice with a constant velocity. Does this mean that there is no force acting on the puck? Explain your answer.

13. According to Newton's second law, doubling a net force on a system of constant mass will have what result?

14. What characteristic of the acceleration vector (**a**) of a system is identical to the force sum vector ($\Sigma \mathbf{F}$) acting on the system?

15. A 22.0 N book rests on a table. According to Newton's third law, what force does the table exert on the book?

16. A hot-air balloon is floating at a constant altitude of 3000 m.
    a. Against what is the total weight of the balloon, basket, passengers, and hot air acting, according to Newton's third law of motion?
    b. Describe the reaction force to this combined weight.

17. What is the unit of weight in the SI? in the EES?

18. Do all mass-measuring devices involve a balance of some sort? Explain your answer.

True or False (19–28)

19. Sir Isaac Newton originated the concept of inertia.

20. Forces are really only pulls applied by one system on another.

21. All forces on an object cause a change in the object's motion.

22. Every fundamental force is at work in a flower vase sitting on a shelf.

23. It is possible for a scientific law to *not* hold true in every case where it is tested.

24. All three laws of motion apply in every instance where a force is exerted on a system.

25. A heavier object has a larger gravitational force acting on it than a lighter object does. Therefore, according to Newton's second law, it will experience a greater acceleration toward the ground than the lighter.

26. The "action" force never acts on the same system as the "reaction" force.

27. The value of *g* on the Moon is smaller than on Earth because objects have smaller masses on the Moon than on Earth.

28. Astronauts in the International Space Station can use a standard laboratory balance when finding the mass of an object.

⊚29. Three cattlemen are handling an unruly calf (*Bos taurus*). All three have ropes around the calf's neck. One man is directly in front of the calf, exerting a force of 120. N. A second man is at a right angle to the calf on its right, pulling with a force of 110. N. The third man is on the calf's left and ahead of it at a 30.0° angle from the rope of the first man, pulling with a force of 116 N. If the calf does not move, what forces is it exerting?

⊚30. A 6.67×10^{-2} kg leopard frog (*Rana pipiens*) jumps into the air with an acceleration of 0.300 m/s² straight up. What is the net force on the frog as it jumps?

⊚31. Three men are trying to push a 3.08×10^3 kg car up an icy hill. The owner pushes with a force of 430. N straight up the hill. Each of the other men pushes on a rear corner of the car with a force of 708 N at an angle of 45.0° from the direction the driver is pushing. The other forces on the car can be represented by a force of 800. N straight downhill. What is the magnitude of the acceleration of the car and in which direction?

32. A 25.0 kg bicycle is traveling at a speed of 10.0 m/s.
  a. If the bike comes to rest in 2.0 s, what is the magnitude of its acceleration?
  b. What net force is needed to provide this acceleration?
  c. How far will the bike travel in the 2.0 s?

33. The earth's mass is $m_E = 5.97 \times 10^{24}$ kg. An object of mass m_o falls from a height of 100. m, where it was at rest.
  a. How long does the object fall?
  b. Write the symbol for the force that the earth exerts on the object.
  c. Write the symbol for the force that the object exerts on the earth.
  d. What is the formula for determining the acceleration of the earth toward the object? Use the symbols from parts *b* and *c*.
  e. What is the value of m_o if the earth moves 1.00×10^{-15} cm toward the object during the object's fall toward the earth?

34. The earth exerts a force of 1.00 N on an object in free fall. What is the object's mass?

35. What is the weight of a 5 g mass in newtons? (Assume that the mass is at rest at the surface of the earth.)

36. A mass m_1 has an acceleration a_1 when it is acted upon by a force of magnitude F. If a second mass, $m_2 = 3m_1$, is acted upon by the same force F, what is its acceleration compared to a_1?

37. A mass of 2.50 kg hangs from a string.
  a. What force does the mass exert on the string?
  b. What force does the string exert on the mass?

38. A hunting bow imparts an acceleration of 5300. m/s² to a 29.3 g arrow at an angle of 15.0° to the horizontal. What is the horizontal component of the force exerted by the bow on the arrow?

39. A raindrop's terminal velocity (the velocity at which the falling drop stops accelerating due to air resistance) is $v = -1.0$ m/s. If the force of air resistance can be modeled by the equation $\mathbf{F}_r = -b\mathbf{v}$, and $b = 4.9 \times 10^{-3}$ kg/s, what is the raindrop's mass?

40. You are test driving a brand new 1350 kg BMW sports car. You find that the car accelerates from 0 to 96.0 km/h in 5.8 s along a level stretch of interstate (speed limit 60 mph). Assuming that the acceleration is constant, solve the following problems.
  a. Calculate the magnitude of the net force that produces the acceleration.
  b. Calculate the distance the car covers during the period of acceleration.
  c. If the road is inclined at an angle of 7.0° to the horizontal but the car can still attain 96 km/h in 5.8 s, is the magnitude of the net force the same as in part *a*? How does it compare? Explain.

41. A 1.50×10^3 kg car is pulling a 600. kg trailer. Leaving a traffic stop, the car accelerates at 0.200 m/s².
  a. What is the magnitude of the total force on the trailer?
  b. What is the magnitude of the total force on the car?
  c. What is the magnitude of the total force on the system consisting of both car and trailer?

42. A soccer ball of mass 0.380 kg bounces off the head of the sweeper. The incoming speed is 14.0 m/s and the outgoing speed is 17.0 m/s in the opposite direction. The ball is in contact with the player's head for 0.080 s.
  a. Determine the magnitude of the ball's average acceleration during the moment of the play.
  b. What is the magnitude of the force provided by the player's head to cause this acceleration?

43. Research and write a brief report about the discovery of calculus. Include a discussion of the contention between Newton and Leibniz.

Aristotle

The ancient Greeks spent much time trying to explain their surroundings, but their philosophical approach led to some unusual conclusions. For example, at one time there were two schools of thought about motion. One school said that all things are in continual motion; the other said that motion is impossible. However, some philosophers disagreed with both schools. Among them was Aristotle (384–322 B.C.).

Aristotle used his reason to explain motion. He began with what he thought was obvious—heavy objects fall, and the heavier they are, the faster they fall. He reasoned from this fact that every object moves to its natural place and stops. The heavier an object is, the more vigorously it seeks its resting place. Therefore, a falling object's speed is proportional to its weight. Since "everybody knew" that heavy objects fall faster than light objects, Aristotle thought it unnecessary to experiment to prove his theory.

Aristotle applied his reason to a wide variety of other subjects. In logic, he invented the deductive syllogism, which we still use today to test hypotheses and theories. His writings have had a profound influence on Western thought in fields as diverse as psychology, ethics, speech, poetry, logic, physics, metaphysics, astronomy, meteorology, and biology. Usually he used his "common-sense" philosophy, but he was not completely opposed to experimentation for his biological treatises. In light of his philosophical approach and broad range of topics, it is not surprising that a few of his ideas were incorrect. What is surprising is the insight and wisdom he possessed in spite of the limitations of his day.

7A	Circular Motion	138
7B	Dynamics of Circular Motion	146
7C	Universal Gravitation	151
Facet: Roller Coasters		160

New Symbols and Abbreviations

angular speed (ω)	7.4	torque (τ)	7.12
tangential velocity (v_t)	7.4	moment arm (L)	7.12
centripetal acceleration (a_c)	7.5	semi-major axis (R)	7.14
angular velocity (ω)	7.6	period (T)	7.14
angular acceleration (α)	7.8	Keplerian constant (K)	7.14
tangential acceleration (a_t)	7.9	astronomical unit (ua)	7.14
centripetal force (F_c)	7.11	universal gravitational constant (G)	7.15

The rings of Saturn are composed of innumerable particles of dust and ice moving silently in great, nearly circular paths around the planet.

Circular Motion

7

> *Canst thou bind the sweet influences of Pleiades, or loose the bands of Orion? Canst thou bring forth Mazzaroth in his season? or canst thou guide Arcturus with his sons?* Job 38:31–32

This wonderful passage in the ancient Book of Job draws attention to the fact of God's immense power over the structure and motions of the universe. Anyone can follow the circular path of the stars, such as Arcturus, and constellations, such as Orion, through the heavens each night. People from the dawn of history until well into the sixteenth century believed that the heavens revolved daily around the earth. Even during these early times the heavens were an impressive testimony to God's creative and sustaining power. Once it was realized how large the heavens truly were, that fact gave even greater glory to God. The verse also includes some subtle references to other celestial circular motions. The "Mazzaroth," or zodiacal signs, revolve with the seasons, reflecting the earth's stately annual motion around the Sun. And the "sweet influences of the Pleiades" refers to the gravitational grouping of this asterism as it moves through space. In the face of such majestic creative acts, man is insignificant and helpless apart from his Creator. The circular motions of the heavens are just one more testimony that our God is the one true God.

7-1 The Pleiades is a group of about a dozen bright stars and at least 500 dim stars in the constellation Taurus, bound together gravitationally.

7.1 Introduction

In the preceding chapters we examined the properties of linear motion and simple two-dimensional projections that occur when a system is uniformly accelerated. Many other forms of motion exist in the "real world." One kind of motion that is extremely common and very important today is *rotational* or *circular* motion. Life in the modern world would be inconceivable without circular motion. We would not have wheels, printing press rollers, computer hard drives, compact disk players, steam turbines, electric motors, gas engines, or ferris wheels, to name just a few things.

Certain natural systems approximate circular motion. Ancient Greeks believed that the celestial motions of the heavenly bodies followed "perfect" circles. Kepler showed that the celestial motions were elliptical instead. However, most planetary orbital motion is nearly circular to a first approximation. Planets also spin on their axes so that a point on a planet's surface describes a circular path. Humans have

7-2 The Galilean moons of Jupiter clearly demonstrate the mechanics of celestial circular motion.

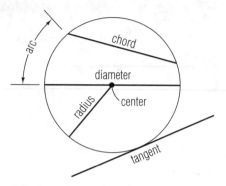

7-3 Geometric parts of a circle

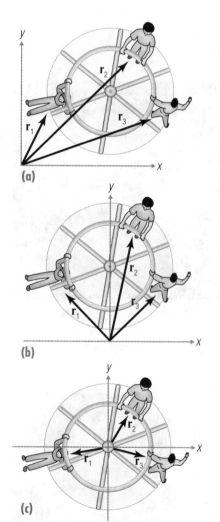

(a)

(b)

(c)

7-4 Three possible coordinate systems for describing circular motion. Figure (c) shows the system that is simplest to use.

mimicked celestial motions with artificial satellites such as the International Space Station and the Iridium communications satellites.

We will begin our study of circular motion by describing how systems move in a circular path—the kinematics of circular motion. Then, as we did for linear motion, we will investigate some of the causes of circular motion, including gravity.

7A CIRCULAR MOTION

7.2 Defining Circular Motion

Circular motion occurs when an object moves in a two-dimensional circular path. There are two general categories of circular motion. If an object rotates about an axis that passes through the object itself, then we say the object is spinning or has **spin.** If the object circles an axis that does *not* pass through the object itself, then we say that it *revolves* about an axis in **orbital motion.** The earth experiences both spin and orbital motion. Orbital motion is the more general case, but the formulas that will be developed can be applied to points on spinning objects as well.

You may want to stop for a moment here to review some terms used for circles. A *chord* is any straight line that connects two points on a circle. A *tangent* is a line that touches the circle in only one point and does not lie in the interior of the circle. An *arc* is a curved line segment that is part of a circle (see Figure 7-3).

7.3 Establishing Position in Circular Motion

By definition, the positions of an object experiencing circular motion lie in a plane. What is the best method for describing the position in circular motion? Physicists use position vectors for noncircular two-dimensional motion. How do position vectors work for circular motion?

A child is playing on a merry-go-round on a playground. He sits still while the merry-go-round moves at a constant speed. Since the merry-go-round rotates, the child travels in a circle. How would you describe his motion?

You should begin by choosing a coordinate system to use as a position reference. Figure 7-4 shows three possible coordinate systems to use. Which is simplest? Figure 7-4 (a) and (b) (and any coordinate system in which the origin is not centered on the rotational axis of the circle) give position vectors of different lengths and angles for different points in the child's path. The coordinate system in Figure 7-4 (c), however, gives position vectors with the same length for every point of the child's path. This is the simplest coordinate system for analyzing circular motion.

The position vector of a rotating object at any time t_n is $\mathbf{r}_n$. The magnitude of $\mathbf{r}$ is just the radius of the circular path, r. The position of the radial vector is uniquely described by its angle, θ. Therefore, the position of an object in a circular path can be given by the ordered pair (r, θ). These are known as *polar coordinates.*

In Chapter 2 we introduced the SI unit for angular measurement—the *radian* (rad). Recall the definition of a radian:

> One radian is equal to the central angle of a circle that subtends an arc of the circle's circumference whose length is equal to the length of the radius of the circle.

There are exactly 2π rad in one complete circle. If necessary, review the discussion in Chapter 2 addressing degree-radian conversions and other properties of radian angle measurement.

The kinematics of circular motion are analogous to those of linear motion. Where motion in straight-line kinematics is measured in units of length, such as the meter, change of position in circular motion is measured in angular units. An angle θ is the angular position of the position vector from a reference axis or reference direction and is analogous to the linear position x on a number line. It can have either a positive or negative value.

7.4 Speed and Velocity in Circular Motion

The symbol ω represents the time-rate of change of angular position, or the **angular speed.** The angular speed can be described by the angular change divided by time, or $\Delta\theta/\Delta t$. This is a scalar quantity. In common usage, angular speed may be expressed as the number of rotations per unit time. Each rotation is called a *revolution*. Angular speed thus has units of *revolutions per unit time* (e.g., revolutions per minute [rpm], such as indicated by an engine tachometer). Rotational rate may have decimal values, such as 3.4 rpm. If angular speed is constant, then the rotating object is experiencing **uniform circular motion.**

In science, angular speed is expressed in *radians per second* because radians and seconds are the SI units for angle and time, respectively. When an object experiences uniform circular motion, its angular speed indicates the number of radians that a point in a rotating object passes through each second. Example 7-1 demonstrates a simple conversion from revolutions per minute to SI units.

EXAMPLE 7-1

Angular Speed Conversion
Convert 45 rpm to rad/s.

Solution:
Use the unit analysis bridge utility. The first term in the bridge is the given quantity.

$$\frac{45 \text{ rev}}{\text{min}} \left| \frac{1 \text{ min}}{60 \text{ s}} \right| \frac{2\pi \text{ rad}}{\text{rev}} \approx 4.7 \text{ rad/s or } 4.7 \text{ s}^{-1}$$

The radian is a dimensionless SI unit, so "rad" is often omitted from angular speed units. Therefore, "rad/s" may appear as just "s^{-1}."

The velocity vector of a particle in circular motion is tangent to the circular path of the particle. This velocity is called **tangential velocity (v_t).** The tangential velocity vector is perpendicular to the radius of the circular path and, thus, to the position vector of the rotating object.

The magnitude of tangential velocity is *tangential speed,* $v_t = |\mathbf{v_t}|$. Tangential speed is equal to the length of arc traveled along the circular path in one second, or

$$v_t = \frac{\widehat{l}}{\Delta t}.$$

Arc length ($\widehat{l}$) is equal to the radius of the circle times the change of angle, $r \times \Delta\theta$, where $\Delta\theta$ is in radians. Therefore, average tangential speed, $\bar{v}_t$, is determined by

$$\bar{v}_t = \frac{r\Delta\theta}{\Delta t}. \tag{7.1}$$

Tangential speed has linear motion units of m/s.

An angular position cannot be considered a position vector in the same way that a number line coordinate can represent a one-dimensional position vector. Although angles and angular changes have magnitude and direction, they cannot be represented by straight arrows and they do not obey all of the rules for vector operations. Consequently, sums and differences of angles are scalar and not vector quantities.

As with the symbol for any scalar quantity, the Greek letter omega (ω) is not bold when representing angular speed.

Angular speed is the time-rate of change of angular position:

$$\omega = \frac{\Delta\theta}{\Delta t} = \frac{\theta - \theta_0}{\Delta t}$$

It may be expressed in revolutions per unit time (rpm) or radians per second (s^{-1}).

God asked Job, "Have you ever controlled the rotational speed of the earth so as to control when the Sun comes up?" This could be a humbling question for any person. "Hast thou commanded the morning since thy days, and caused the dayspring to know his place?" (Job 38:12).

The tangential velocity of a rotating object is perpendicular to the object's position vector:

$$\mathbf{v_t} \perp \mathbf{r}$$

Problem-Solving Strategy 7.1

Tangential speed can be calculated by the formula $v_t = r\omega$ as well as by $v_t = r\Delta\theta/\Delta t$.

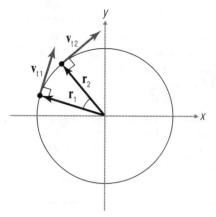

7-5a Velocity vectors for $\mathbf{r}_1$ and $\mathbf{r}_2$

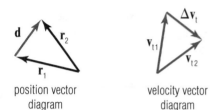

position vector
diagram

velocity vector
diagram

7-5b Adding vectors to form a triangle

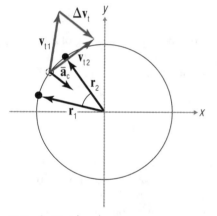

7-5c Centripetal acceleration

Centripetal is derived from the Latin words for "center-seeking."

7.5   Acceleration in Circular Motion

Referring back to the merry-go-round, in uniform circular motion the child's speed is constant, but his tangential velocity vector continually changes direction. Thus, the child experiences an acceleration. The question is, what is the magnitude and the direction of his acceleration? You can find his average acceleration for an interval just as you did in earlier chapters:

$$\overline{\mathbf{a}} = \frac{\Delta \mathbf{v}}{\Delta t} \tag{7.2}$$

Figure 7-5a shows the velocity vectors for positions $\mathbf{r}_1$ and $\mathbf{r}_2$. Figure 7-5b shows the displacement vector triangle and velocity difference vector triangle as well. Since $\mathbf{v}_{t\,1} \perp \mathbf{r}_1$, $\mathbf{v}_{t\,2} \perp \mathbf{r}_2$, $v_{t\,1} = v_{t\,2} = v_t$, and $r_1 = r_2 = r$, the two triangles in Figure 7-5b are similar. Therefore, the ratios of the corresponding sides are equal:

$$\frac{\Delta v_t}{d} = \frac{v_t}{r}$$

Multiplying both sides by d gives

$$\Delta v_t = \frac{v_t d}{r}.$$

Substituting this expression for Δv in Equation 7.2, you get

$$\overline{a} = \frac{v_t d / r}{\Delta t} = \frac{v_t d}{r \Delta t}, \text{ so}$$

$$\overline{a} = \frac{v_t}{r} \cdot \frac{d}{\Delta t}.$$

But, for very small time intervals, $d/\Delta t$ is just v_t. So

$$\overline{a} = \frac{v_t}{r} \cdot v_t, \text{ or}$$

$$\overline{a} = \frac{v_t^2}{r}. \tag{7.3}$$

Equation 7.3 is an expression for the *magnitude* of the acceleration experienced by the child rotating on the merry-go-round in uniform circular motion.

For very small time intervals (Δt), the tangential velocity at the beginning of the interval, $\mathbf{v}_{t\,1}$, is almost parallel to the velocity at the end of the interval, $\mathbf{v}_{t\,2}$. The difference vector $\Delta \mathbf{v}_t$ is essentially perpendicular to both $\mathbf{v}_{t\,1}$ and $\mathbf{v}_{t\,2}$ and points in the direction of $-\mathbf{r}$. Since $\mathbf{a} = \Delta \mathbf{v}/\Delta t$, $\mathbf{a}$ also points in the direction of $-\mathbf{r}$. That is, the child's instantaneous acceleration vector always points toward the center of the child's circular path. Such acceleration is called **centripetal acceleration ($\mathbf{a}_c$).**

During uniform circular motion, the magnitude of the child's centripetal acceleration is constant at

$$a_c = \frac{v_t^2}{r} \text{ m/s}^2.$$

In other words, the magnitude of centripetal acceleration is directly proportional to the square of the tangential speed and inversely proportional to the radius of the circular motion. The direction of $\mathbf{a}_c$ is opposite to the child's position vector direction, toward the center of the circle. This is true for all circular motion at constant radius and speed.

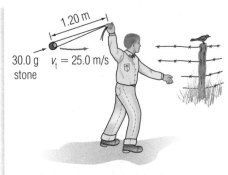

30.0 g $v_t = 25.0$ m/s
stone

7-6 Slinging stones in residential areas is not recommended!

EXAMPLE 7-2

Slings and Stones: Centripetal Acceleration

A boy in a corn field is getting ready to scare a crow with a 30.0 g stone thrown from a sling. As he swings the sling horizontally over his head in a circle, the stone has a tangential speed of 25.0 m/s. The sling is 1.20 m long from hand to stone. What is the magnitude of the centripetal acceleration on the stone?

Solution:

Note that Equation 7.3 does not involve the mass of the revolving object, so the mass of the stone is irrelevant.

Substitute the known values into Equation 7.3:

$$a_c = \frac{v_t^2}{r} = \frac{(25.0 \text{ m/s})^2}{1.20 \text{ m}}$$

$$a_c \doteq 520.8 \text{ m/s}^2$$

$$a_c \approx 521 \text{ m/s}^2$$

We can also use the angular speed to determine centripetal acceleration. First, we must express the tangential speed in terms of angular speed.

$$v_t = r\omega \qquad (7.4)$$

If we substitute Equation 7.4 for v_t into Equation 7.3 we obtain

Unit analysis of Equation 7.4:

$$\frac{m}{s} \Rightarrow m \cdot \frac{\langle rad \rangle}{s} = \frac{m}{s}$$

(Radians are dimensionless in terms of SI base units.)

$$a_c = \frac{(r\omega)^2}{r} = \frac{r^2\omega^2}{r}$$

$$a_c = r\omega^2. \qquad (7.5)$$

Equation 7.5 can be converted directly into a vector equation for centripetal acceleration:

$$\mathbf{a}_c = -\mathbf{r}\omega^2 \qquad (7.6)$$

Note that the negative sign indicates that the centripetal acceleration is oriented opposite to the direction of the position vector, **r,** which is consistent with the empirical observations on the merry-go-round.

Equation 7.6 is the vector equation for centripetal acceleration. Equation 7.3 is a non-vector equation that provides only the magnitude of the centripetal acceleration.

7.6 Angular Velocity

Angular speed is the scalar measure of the number of radians through which the position vector of an object moves per unit time. It may also be considered the magnitude of a related vector quantity called **angular velocity (ω).** A *uniform* angular velocity implies that the rate *and* direction of this aspect of circular motion is constant. How can a vector of circular motion be constant?

Here we introduce the **right-hand rule of circular motion.** Wrap the four fingers of your right hand around the axis in the direction of rotation of the object. Your extended thumb held at a right angle to the fingers will point in the direction of the angular velocity vector (see Figure 7-7). This is one of the few occasions where you will be concerned about the third dimension, defined by the z-axis, which is parallel to the axis of rotation. The angular velocity vector points in the

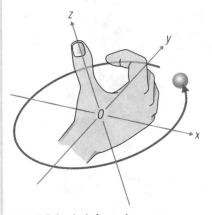

7-7 Right-hand rule for circular motion

Circular Motion   **141**

positive z-direction for counterclockwise rotation and in the negative z-direction for clockwise rotation. Similarly, positive scalar values of $\boldsymbol{\omega}$ indicate counterclockwise rotation, and negative values indicate clockwise rotation. The magnitude of $\boldsymbol{\omega}$ is indicated by arrow length, as is usual for vectors. The vector's orientation does not change unless the direction of circular motion changes.

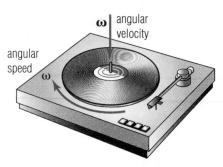

7-8 An LP turntable

Average angular velocity and instantaneous angular velocity are equal for uniform circular motion, unlike many other vector quantities of circular motion, such as tangential velocity and centripetal acceleration. Why is this so?

(a)

(b)

7-9 All man-made rotating systems must experience angular acceleration at some time.

EXAMPLE 7-3

An LP Record: Angular Velocity

In past years, before the advent of cassette tapes and audio CDs, long playing (LP) records were popular with music listeners. An LP rotates clockwise on a record player turntable at $33\frac{1}{3}$ rpm. Determine the magnitude and direction of the angular velocity vector of an LP record.

Solution:
The angular speed is $\omega = 33\frac{1}{3}$ rpm $= 33.\overline{3}$ rpm and rotation is clockwise. Convert rpm to radians per second as follows:

$$\omega = \frac{33.\overline{3}\ \text{rev}}{\text{min}} \left| \frac{1\ \text{min}}{60\ \text{s}} \right| \frac{2\pi\ \text{rad}}{1\ \text{rev}} \doteq 3.4\underline{9}0\ \text{rad/s}$$

$$\omega \approx 3.49\ \text{s}^{-1}$$

Applying the right-hand rule for circular motion to the record as it is revolving, the right thumb points downward. Therefore, angular velocity is

$$\boldsymbol{\omega} = -3.49\ \text{s}^{-1}.$$

7.7 Nonuniform Circular Motion

Uniform circular motion in rotating systems does not just pop into existence. When you start the engine of your car, its rotating parts start at rest, and at some later time they are rotating at a constant angular speed as the engine is idling. A ferris wheel stops in order for passengers to get on or off, but it takes some time for it to get up to speed during the ride and then to slow back down to a stop again. Every man-made mechanical device that rotates must change its speed of rotation at some time or another. In the same manner, the rotation of the earth itself is slowing at a rate of about 2 milliseconds per day over the course of every hundred years or so. These examples demonstrate that **nonuniform circular motion** is a natural and common consequence of any circular motion.

We will not present the derivation of kinematic equations for nonuniform circular motion due to the level of the mathematics required. However, we can discuss some of the properties from an intuitive perspective and analyze specific examples. At the end of this section we will present some summary equations of circular kinematics that will look surprisingly familiar.

7.8 Angular Acceleration

Let us consider how the motion of a compact disc changes in a CD player when you press the Play button. The angular velocity of any point on the CD increases from zero to some nonzero value. The change of angular velocity is called **angular** acceleration (α). Average angular acceleration ($\overline{\alpha}$) can be expressed in the same form as average linear acceleration,

$$\overline{\alpha} = \frac{\Delta\omega}{\Delta t} = \frac{\omega - \omega_0}{\Delta t}.$$ (7.7)

Units are in radians per second squared, or s^{-2}.

The magnitude of α is proportional to the difference of the angular velocities at the beginning and end of the time interval, Δt, and its direction is in the direction of the angular velocity difference vector, $\Delta\omega = \omega - \omega_0$. Since ω is oriented along the axis of rotation (z-axis), then α also coincides with the rotational axis. The sign of angular acceleration indicates the direction of change of angular velocity, just as for rectilinear motion.

Angular acceleration (α) is the time-rate of change of angular velocity. Unit analysis for Equation 7.7:

$$\frac{\langle rad\rangle}{s^2} \Rightarrow \frac{\langle rad\rangle/s}{s} = \frac{\langle rad\rangle}{s^2} = s^{-2}$$

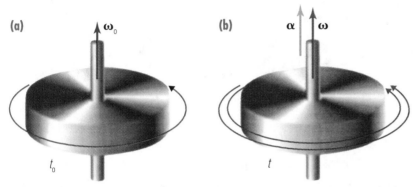

7-10 Angular acceleration is parallel to the rotational axis. It points in the same direction as the change of angular velocity.

7.9 Tangential Acceleration

We have already discussed how the direction of a revolving particle's tangential velocity changes in response to the centripetal acceleration of the particle in uniform circular motion. How do we describe the change of tangential speed as a revolving particle speeds up or slows down in nonuniform circular motion? The time-rate of change of the *magnitude* of tangential velocity is called **tangential acceleration (a_t).** The following equation defines tangential acceleration:

$$\overline{a}_t = \frac{\Delta v_t}{\Delta t}$$

If you haven't experienced it already, you will—it is the feeling that you are running in circles and accomplishing nothing. Take comfort in I Corinthians 15:58: "Therefore, my beloved brethren, be ye stedfast, unmoveable, always abounding in the work of the Lord, forasmuch as ye know that your labour is not in vain in the Lord."

Using Equation 7.4, we can replace Δv_t to obtain the following derivation:

$$\overline{a}_t = \frac{v_t - v_{t0}}{\Delta t} = \frac{\omega r - \omega_0 r}{\Delta t}$$

$$\overline{a}_t = \frac{(\omega - \omega_0)r}{\Delta t} = \frac{\Delta\omega r}{\Delta t} = \frac{\Delta\omega}{\Delta t}\cdot r$$

$$\overline{a}_t = \overline{\alpha}r$$ (7.8)

Calculus shows us that as the time interval approaches zero, Equation 7.8 becomes the instantaneous tangential acceleration

$$a_t = \alpha r.$$ (7.9)

The direction of instantaneous tangential acceleration is, by definition, tangent to the circular path at the particle's position. If tangential speed is increasing, then tangential acceleration is in the same direction as rotation. If tangential speed is decreasing, tangential acceleration points in the opposite direction of rotation.

EXAMPLE 7-4

The Frisbee Factor: Acceleration in Nonuniform Circular Motion

You fling a Frisbee from rest to 360. rpm in 0.10 s. The disk is 23.0 cm (0.23 m) in diameter. Calculate (a) the angular acceleration and (b) the tangential acceleration of a point on the edge of the Frisbee.

Solution:

First, convert angular speed to SI units:

$$\frac{360.\ \text{rev}}{\text{min}} \left| \frac{2\pi}{\text{rev}} \right| \frac{1\ \text{min}}{60\ \text{s}} \doteq 37.\underline{6}9\ \text{s}^{-1}$$

a. Since the Frisbee started from rest, $\Delta\omega = 37.\underline{6}9\ \text{s}^{-1}$ (the sign depends on whether you throw right or left handed).

Angular acceleration is calculated from Equation 7.7:

$$\overline{\alpha} = \frac{\Delta\omega}{\Delta t} = \frac{37.\underline{6}9\ \text{s}^{-1}}{0.10\ \text{s}}$$

$$\overline{\alpha} \doteq 37\underline{6}.9\ \text{s}^{-2}$$

$$\overline{\alpha} \approx 377\ \text{s}^{-2}$$

b. The radius of the Frisbee is (0.23 m)/2. Using Equation 7.8 and the result from part a, you get:

$$\overline{a}_t = \overline{\alpha}r = (37\underline{6}.9\ \text{s}^{-2})(0.23\ \text{m})/2$$

$$\overline{a}_t \doteq 4\underline{3}.3\ \text{m/s}^2$$

$$\overline{a}_t \approx 43\ \text{m/s}^2$$

7.10 Equations of Circular Motion

Three equations of circular motion can be developed that are directly analogous to their counterparts for linear motion. In these equations, angular velocity (ω) is substituted for linear velocity, and angular acceleration (α) is substituted for linear acceleration. Recall that the angle, θ, is analogous to the linear position coordinate, x. These equations are valid only for conditions of uniform acceleration.

The initial angular position θ_0 is conventionally considered to be 0 rad. The circular equations do not exist as vector equations because of the nonvector nature of angular measurement.

TABLE 7-1

Equations of Uniformly Accelerated Motion

Equation	Linear Motion	Circular Motion
First	$v = v_0 + \overline{a}\Delta t$	$\omega = \omega_0 + \overline{\alpha}\Delta t$
Second	$d = v_0\Delta t + \frac{1}{2}\overline{a}(\Delta t)^2$	$\Delta\theta = \omega\Delta t + \frac{1}{2}\overline{\alpha}(\Delta t)^2$
Third	$d = \dfrac{v^2 - v_0^2}{2\overline{a}}$	$\Delta\theta = \dfrac{\omega^2 - \omega_0^2}{2\overline{\alpha}}$

Motor Startup Time

An electric motor normally runs at 1800 rpm. Assuming a uniform angular acceleration of 64 s^{-2}, how many seconds will the motor require to go from stopped to full speed?

Solution:

Use the first equation of circular motion because it contains all of the known information and the unknown variable (Δt).

First, we must convert the common angular speed to SI units:

$$\omega = \frac{1800 \text{ rev}}{\text{min}} \left| \frac{1 \text{ min}}{60 \text{ s}} \right| \frac{2\pi}{1 \text{ rev}} \doteq 188 \text{ s}^{-1}$$

Rearrange the first equation of motion to solve for Δt:

$$\Delta t = \frac{\omega - \omega_0}{\alpha}$$

Substitute and solve for Δt.

$$\Delta t = \frac{188 \text{ s}^{-1} - 0 \text{ s}^{-1}}{64 \text{ s}^{-2}} = \frac{188/\text{s}}{64/\text{s}^2}$$

$$\Delta t \doteq 2.93 \text{ s}$$

$$\Delta t \approx 2.9 \text{ s}$$

7-11 Large AC motors may take several seconds to attain full speed when starting.

7A Section Review Questions

1. Describe the coordinate system that is usually chosen for analyzing circular motion and state at least one advantage for this choice.

2. What position variable in circular motion is analogous to the number-line position variable in rectilinear motion?

3. What is the difference between uniform circular motion and nonuniform circular motion? Give an example of each.

4. Why can circular equations of motion be written only in the scalar form?

⊛5. A light private aviation helicopter's main rotor blades rotate at approximately 450 rpm. What is the angular speed of the blades in the rotor in SI units?

⊛6. The main rotor blades of the helicopter in question 5 are 5.0 m long from the drive hub to tip. What is the tangential speed of the blades (a) at the drive hub, (b) 1.0 m from the hub, and (c) at the blade tip.

⊛7. The rotor blades of a helicopter on the ground with its engine idling rotate at 48 rpm. If it takes 12 s for the pilot to increase rotor speed to 460 rpm when taking off, what is the angular acceleration on the rotor blades?

⊛8. If it takes the earth 23 h 56 min 4 s to rotate once on its axis, how large an angle, in degrees and radians, will it rotate through in 1.00 hour?

⊛⊛9. What is the centripetal acceleration of a person standing on the equator of the earth?

7.11 Centripetal Force

We saw in Section 7A that any object in uniform orbital motion about an axis experiences centripetal acceleration ($\mathbf{a}_c$) toward the center of rotation. We also know that only when unbalanced forces act on a system will acceleration occur, and the direction of the acceleration is in the direction of the net force on the system ($\Sigma\mathbf{F}$). The unbalanced force sum that produces this acceleration is called **centripetal force ($\mathbf{F}_c$).** This is one of several types of center-seeking or *central forces.*

Let us first imagine that a puck-like object on a perfectly frictionless surface is connected by a string to a vertical rod rigidly imbedded in the surface so that the puck and string can freely rotate around the rod. The puck is set to rotating around the rod in orbital motion. We may assume that any forces acting parallel to the rod (the axis of rotation), and thus perpendicular to the surface of rotation, cancel out. Therefore, the net force acting on the rotating puck consists only of the force exerted by the string that causes centripetal acceleration.

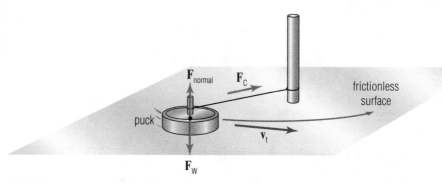

7-12 The string exerts a centripetal force on the puck.

Newton's second law is expressed in the equation

$$\Sigma\mathbf{F} = m\mathbf{a}.$$

From this, we can develop the second-law centripetal force equation by substituting the appropriate centripetal terms:

$$\mathbf{F}_c = m\mathbf{a}_c \qquad (7.10)$$

Equation 7.10 would be more useful to find the magnitude of the centripetal force on the puck if we could use easily measurable parameters of the system described above. We can substitute Equation 7.3 from the previous section for a_c, which gives us the magnitude for $\mathbf{F}_c$:

$$F_c = ma_c = m\left(\frac{v_t^2}{r}\right)$$

$$F_c = \frac{mv_t^2}{r} \qquad (7.11)$$

Equation 7.11 permits us to use easily measurable quantities to calculate the *magnitude* of the centripetal force that produces circular motion for a given object under a specific set of circumstances.

Centripetal force can be provided by a string, as we'll see in the following example. A planet may exert a centripetal force on a satellite through gravity (see Section 7C).

EXAMPLE 7-6

A Tethered Model Airplane: Centripetal Force

A model airplane enthusiast is flying his gas-powered plane in a circle tethered to a 10.0 m control line (see Figure 7-13a). The model has a mass of 0.90 kg and it takes 4.0 s to complete one circuit. (a) What is the magnitude of the force that the modeler must exert on the control line? (b) If he shortens the line to 5.0 m to land the airplane without reducing its flying speed, what force will he have to exert on the line?

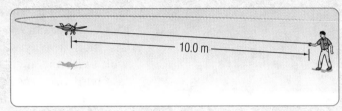

7-13a The modeler exerts a centripetal force through the control line to the model.

Solution:

a. You must find the tangential speed (v_t) before you can use Equation 7.11. You know that the distance of one circuit is the circumference of the circle $C = d = v_t\Delta t$. You can solve this equation for v_t after finding the circumference.

$$C = v_t\Delta t$$

$$v_t = \frac{C}{\Delta t} = \frac{2\pi r}{\Delta t} = \frac{2\pi(10.0 \text{ m})}{4.0 \text{ s}}$$

$$v_t \doteq 15.7 \text{ m/s}$$

Substitute tangential speed into Equation 7.11 to solve for centripetal force as follows:

$$F_c = \frac{mv_t^2}{r} = \frac{(0.90 \text{ kg})(15.7 \text{ m})^2}{10.0 \text{ m}}$$

$$F_c \doteq 22.1 \text{ N}$$

$$F_c \approx 22 \text{ N}$$

b. In this part, m and v_t are the same as above, but $r = 5.0$ m.

$$F_c = \frac{mv_t^2}{r} = \frac{(0.90 \text{ kg})(15.7 \text{ m})^2}{5.0 \text{ m}}$$

$$F_c \doteq 44.3 \text{ N}$$

$$F_c \approx 44 \text{ N}$$

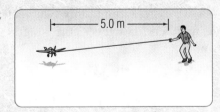

7-13b Landing the model plane

As expected, when the modeler shortened the line to half its length, he had to exert twice the force to maintain the same speed.

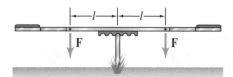

7-14 An empty seesaw balances at its center.

7.12 Torque

Accelerated *linear* motion results from a nonzero net force, but what causes accelerated *rotation?* To answer this question, let us perform some experiments with a playground seesaw.

How do you balance a seesaw? If it is empty, you can position the seesaw so that its center is the pivoting point. In this position the seesaw remains at rest. Since half of the seesaw is on either side of the pivot, each side of the board is attracted to the earth by an equal force. When the pivot is not at the center of the board, the longer side of the board is attracted to the earth with a greater force and the seesaw rotates.

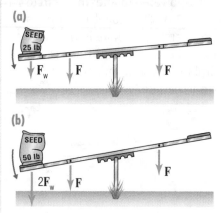

7-15 A greater weight causes greater angular acceleration.

Suppose you place a 25 lb bag of grass seed on one side of a balanced seesaw. What happens? The seesaw rotates, with the weighted side going down. If you place a 25 lb bag on the end of a balanced seesaw and a 50 lb bag on the end of another identical seesaw, what do you observe? The more heavily weighted seesaw rotates with twice the angular acceleration of the less burdened seesaw. Therefore, you can conclude that angular acceleration is proportional to force (**F**).

Suppose identical 25 lb bags of seed are placed on two identical balanced seesaws. One bag is placed at an end of one seesaw and the second bag is placed halfway between the pivot and an end of the other seesaw. You observe that the seesaw with the bag at the end rotates with twice the angular acceleration of the other seesaw. Therefore, angular acceleration is proportional to the force's position vector (**r**)—the distance from the pivot to the place that the force is applied.

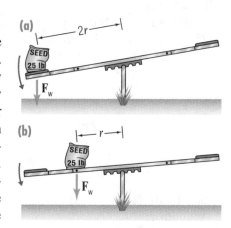

7-16 Angular acceleration increases with distance of the applied force from the pivot point.

> A **torque** exists when a force is exerted on a system at a distance from the system's point of rotation.

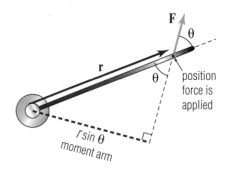

7-17 Force and its moment arm

> The **moment arm** is often called a **lever arm** for simple lever

You can see that angular acceleration is proportional to both the force and the force's position vector. Their product yields the vector quantity called **torque (τ).** It is not possible to multiply two vectors together algebraically since they are not true numbers. However, the magnitude of a torque can be calculated by the formula

$$\tau = rF \sin \theta, \qquad (7.12)$$

where r is the magnitude of the position vector pointing from the center of rotation to the point at which the force is applied, F is the magnitude of the applied force, and θ is the smallest angle ($<180°$) between vectors **r** and **F** when they are positioned tail to tail. The quantity $r \sin \theta$ is called the **moment arm (L)** of a torque. See Figure 7-18 to see the relationship of torque, moment arm, and force. The unit of torque is the m·N.

You can see from the figure that the magnitude of torque is the product of the component of the position vector that is perpendicular to the applied force (the moment arm) and the force. Maximum torque is obtained from a given force when it is perpendicular to its position vector. If the applied force is parallel to its position vector, the torque is zero.

Torque has the same role in rotational motion that force has in linear motion. That is, *angular acceleration is produced by unbalanced torques.* When the sum of all torques on a system is zero, it will rotate either at a constant angular speed or not at all. The condition of zero net torques is called **rotational equilibrium.** Mathematically,

$$\Sigma\tau = 0 \text{ m}\cdot\text{N}$$

is the condition for rotational equilibrium, just as Newton's first law established that

$$\Sigma\mathbf{F} = 0 \text{ N}$$

is the condition for translational equilibrium.

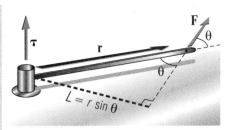

7-18 The geometry of a torque

EXAMPLE 7-7

Balancing Act: Summing Torques

Two boys on a playground want to balance a seesaw. One boy weighs 240 N, and the other boy weighs 260 N. The seesaw is 3.6 m long and of negligible mass. Assuming that the boys sit at the ends of the seesaw, how far from the middle must they place the pivot point in order to balance?

Solution:

Assume that the seesaw is horizontal when balanced, which means the boys' position vectors are horizontal. The boys' vertical weight vectors are then perpendicular to their position vectors. Therefore, the torque magnitude formula reduces to

$$\tau = rF \sin 90° = rF.$$

Let boy 1 be the lighter boy and boy 2 be the heavier boy.

$$F_1 = 240 \text{ N} \qquad r_1 = L_1$$
$$F_2 = 260 \text{ N} \qquad r_2 = L_2$$
$$L_1 + L_2 = l_s \text{ (length of seesaw)} \qquad (1)$$

Since the seesaw is in rotational equilibrium, the magnitudes of the torques must be equal:

$$\Sigma\tau = \tau_1 + \tau_2 = 0$$
$$\tau_1 = -\tau_2$$
$$|\tau_1| = |-\tau_2|$$
$$L_1F_1 = L_2F_2$$

This equation, called the **law of moments,** is true for any balanced seesaw. Rearranging,

$$\frac{F_1}{F_2} = \frac{L_2}{L_1}.$$

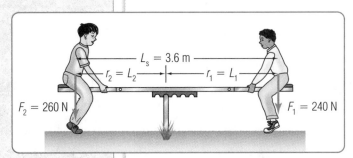

7-19 Balancing a seesaw when one child is heavier than the other

Multiplying both sides by L_1,

$$L_1 \cdot \frac{F_1}{F_2} = L_2. \tag{2}$$

Substituting Equation 2 for L_2 into Equation 1 gives

$$L_1 + \frac{L_1 F_1}{F_2} = l_s.$$

Factoring out L_1,

$$\left(1 + \frac{F_1}{F_2}\right)L_1 = l_s.$$

Now solve for L_1:

$$\left(1 + \frac{240 \text{ N}}{260 \text{ N}}\right)L_1 = 3.6 \text{ m}$$

$$L_1 = (3.6 \text{ m})\left(\frac{260 \text{ N}}{260 \text{ N} + 240 \text{ N}}\right)$$

$$L_1 \doteq 1.\underline{8}7 \text{ m} \approx 1.9 \text{ m}$$

The pivot point is about 1.9 m from the lighter boy. The midpoint of the seesaw is 3.6 m/2, or 1.8 m. Therefore, the pivot is about 0.1 m from the middle on the side of the heavier boy.

7B   Section Review Questions

1. What force causes centripetal acceleration? What is the name given to this class of forces?

2. What measurable properties of a revolving system determine the magnitude of centripetal force?

3. Under what conditions does rotational equilibrium exist?

4. What is a torque? Give an example.

⦿5. Calculate the centripetal force that acts on a 35.0 kg child riding a merry-go-round at a distance of 2.2 m from the spin axis. The merry-go-round is spinning at 15 rpm.

⦿6. A 73.0 kg father and his 45.0 kg son are sitting on opposite sides of a 4.00 m seesaw at the playground. The pivot point is fixed at the midpoint of the board. Where must the father position himself to horizontally balance the seesaw if his son sits at the end of the board?

⦿7. An auto mechanic is struggling to turn a bolt in the engine compartment of a car. He applies a force of 180. N on a wrench 18.0 cm from the bolt. Due to space restrictions, he must exert the force at an angle of 30.0° from the end of the wrench. How much torque is he actually applying to the bolt?

⦿8. What is the centripetal force acting on a 66.0 kg person standing on the equator of the earth?

UNIVERSAL GRAVITATION

Centripetal accelerations change the direction of the motions of objects from straight lines to curves. Different forces can cause centripetal acceleration, but perhaps the most important centripetal force is gravity. One of Newton's greatest achievements was to connect gravitation with other forces that cause circular motion. Newton's work was a breakthrough in a long line of studies of the motions of the planets.

7.13 Ancient Ideas

If you had not been taught that the earth orbits the Sun, it would probably seem obvious to you that the earth is the unmoving center of the universe. Except during an earthquake, you do not feel the earth move. Any clear day or night you can watch the Sun, Moon, and stars travel around the earth. This logical idea is called the **geocentric** (or earth-centered) theory.

From earliest history men have tried to explain the motions of the planets. Priests in Mesopotamia learned to predict the motions of the planets. They often used this knowledge to deceive their kings with astrology. The ancient Greeks, including Aristotle, were among the first to use the predictions to try to explain the planets' motions. The earth was, naturally, the center of Aristotle's universe.

One man who disagreed with the geocentric theory was Aristarchus of Samos, who lived in the third century before Christ. He thought that the Sun was the center of the universe and that all the planets, including the earth, revolved around it. His sun-centered, or **heliocentric,** theory seemed ridiculous to his contemporaries.

During Roman times a Greek scholar named Ptolemy modified Aristotle's theory to fit the actual motions of the planets. The planets do not travel in perfectly circular orbits around the earth, as the ancient astronomers could see even without telescopes. Among other complicated ideas, Ptolemy's theory included the idea that planets traveled in small circles called *epicycles* as they traveled around the earth in their principal *deferent* orbits. Concepts like this made Ptolemy's theory almost correspond to the observations of the planets' motions.

In the Middle Ages, Europeans lost most of the scientific knowledge that the Greeks and Romans had acquired. The crusades reintroduced to Europe the writings of Aristotle and Ptolemy that had been preserved by the Arab cultures. At first, Ptolemy's geocentric theory was accepted almost universally. Ptolemy's theory, however, did not quite agree with all astronomical observations. Astronomers tried to alter the theory to fit the observations. One astronomer,

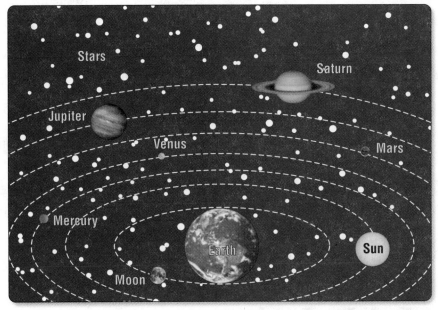

7-20 The geocentric cosmos

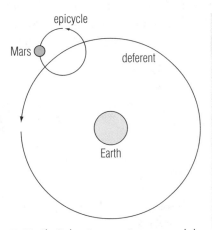

7-21 The Ptolemaic geocentric system needed epicycles to approximate the observed motions of the planets.

Aristarchus of Samos (ca. 310–230 B.C.) was a Greek mathematician and astronomer who was the first proponent of the heliocentric theory. He also described the daily axial rotation of the earth.

Claudius Ptolemaeus (second century A.D.) was an Alexandrian astronomer and geographer.

Ptolemy believed that he needed to add epicycles to the geocentric theory in order to explain why the planets farther from the Sun than Earth exhibited retrograde (backward) motion at certain points in their orbits.

7-22 The original cover of Copernicus's paper describing his heliocentric theory

Nicolaus Copernicus (1473–1543) was a Polish astronomer.

Tycho Brahe (1546–1601) was a Danish astronomer who built an elaborate observatory.

7-23 Johannes Kepler (1571–1630) was a German mathematician who postulated the three laws of celestial mechanics.

Nicolaus Copernicus, concluded that the geocentric idea itself was faulty. He found it simpler to explain the observations by saying that the planets orbit the Sun in circular orbits. Copernicus's heliocentric theory was not much more accurate than Ptolemy's geocentric theory, but it was much simpler.

The Danish astronomer Tycho Brahe disagreed with both Copernicus and Ptolemy. His own theory placed the earth, unmoving, at the center of the universe, the Sun orbiting the earth, and all the other planets orbiting the Sun. Tycho's greatest contribution to astronomy was his detailed, accurate observations of the heavens. Near the end of his life, Brahe hired Johannes Kepler, a mathematician, to interpret his observations.

7.14 Kepler's Laws

Kepler began the task of organizing Brahe's data with the idea that the data were more reliable than any theory. He tried several theories but rejected them all—one because one of its predictions differed from Brahe's data by an angle of eight-sixtieths of a degree (about $\frac{1}{4}$ the width of a full moon). Kepler devised theories of his own about the planet's motions, which also did not work. Finally, he came to the conclusion that a foundation of astronomical theory was wrong—the idea that the planets move at constant speeds. He united the idea of varying speed with the heliocentric theory to formulate **Kepler's laws** of planetary motion.

Kepler's second law states that the position vector of a planet travels through equal areas in equal times if the vector is drawn from the Sun. In Figure 7-24 (a), area A equals area B, and the planet takes the same time to travel from position 1 to position 2 as it takes to travel from position 3 to position 4. The planet travels farther from 1 to 2, so it must be traveling faster in that interval.

If the planet's speed varies, yet its position vector travels through equal areas in equal times, the planet's orbit cannot be circular, and it must be farther from the Sun when it is traveling more slowly. A planet in a circular orbit would always be the same distance from the Sun. *Kepler's first law* states that the orbit of a planet is an ellipse with the Sun at one focus, as in Figure 7-24 (b).

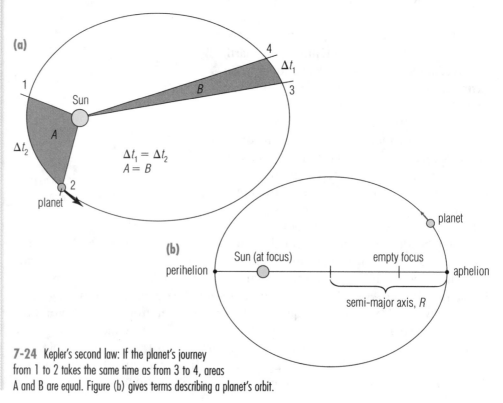

7-24 Kepler's second law: If the planet's journey from 1 to 2 takes the same time as from 3 to 4, areas A and B are equal. Figure (b) gives terms describing a planet's orbit.

Kepler's third law relates the size of each planet's orbit to the time it takes to complete one orbit. The size of the orbit is represented by its **semi-major axis (R)**, which is half of the distance between the planet's nearest position to the Sun (its *perihelion*) and the planet's farthest position from the Sun (its *aphelion*). The time the planet takes to complete an orbit is called its **period (T)**. Kepler found that, for every planet in the solar system, the ratio of the cube of the semi-major axis and the square of the period is the same. In symbols,

$$\frac{R^3}{T^2} = K, \tag{7.13}$$

where K is called the **Keplerian constant.** Each orbital system has its own K. For instance, the moons of Jupiter have one K, while the planets in the solar system have a different K. Within each system, however, K is the same.

We can determine the value of K for solar planets by substituting the known values of the earth's orbit into Equation 7.13. The semi-major axis must be expressed in terms of **astronomical units (ua).** One astronomical unit is the mean distance of Earth from the Sun. The semi-major axis of Earth's orbit can be assumed to be 1 ua, and its period is, of course, one year. The Keplerian constant for Earth (and any planet orbiting the Sun) is then just

$$K_E = \frac{R^3}{T^2} = \frac{(1\ \text{ua})^3}{(1\ \text{y})^2} = 1\ \text{ua}^3/\text{y}^2.$$

If we rearrange Equation 7.13, we can then develop a simple method for determining the orbital period for any of the other planets in the solar system if we know their semi-major axes in ua:

$$T^2 K_E = R^3$$
$$T^2 = R^3/K_E$$
$$T^2 = R^3(1\ \text{y}^2/\text{ua}^3) \tag{7.14}$$

This equation is good only for planets, comets, asteroids, etc., orbiting the Sun. It cannot be used for any planetary satellite or moon.

The **period (T)** of an object that repeats a specific motion, such as an orbit, is the length of time to complete one entire cycle. The units of T are time units. The SI unit for time is the second. In equations relating to planetary orbits, the period should be expressed in units of years.

Problem-Solving Strategy 7.3
Remember when using Kepler's constant $K = \text{y}^2/\text{ua}^3$ in his third law equation that R must be in terms of astronomical units and T must be in years.

EXAMPLE 7-8

Period of Saturn: Kepler's Third Law

The semi-major axis of Saturn's orbit is 1.427×10^9 km. What is its orbital period in years?

Solution:

We are given the semi-major axis of Saturn's orbit, $R = 1.427 \times 10^9$ km, and we are asked to find Saturn's orbital period in (earth) years. We must first convert Saturn's semi-major axis dimension into astronomical units (ua) in order to use Equation 7.14.

$$R_{\text{Saturn}} = \frac{1.427 \times 10^9\ \cancel{\text{km}}}{}\ \left|\ \frac{1\ \text{ua}}{1.496 \times 10^8\ \cancel{\text{km}}}\ \doteq 9.53\underline{8}7\ \text{ua}\right.$$

Substituting for R_{Saturn} in Equation 7.14:

$$T^2_{\text{Saturn}} = (9.53\underline{8}7\ \text{ua})^3\ \text{y}^2/\text{ua}^3$$

$$T^2_{\text{Saturn}} \doteq 867.\underline{8}9\ \text{y}^2\ (\text{ua}^3\ \text{cancels})$$

$$\sqrt{T^2_{\text{Saturn}}} = \sqrt{867.\underline{8}9\ \text{y}^2} \doteq 29.4\underline{5}9\ \text{y}$$

$$T_{\text{Saturn}} \approx 29.46\ \text{y}$$

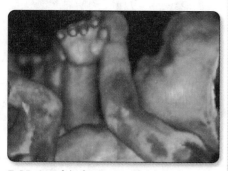

7-25 One of the few times in life when a human is almost completely without the sensation of gravity.

7-26 The orbital motion of the Moon was the basis for Newton's unraveling the secret of gravitation.

7.15 Newton

Since the time of Creation, people have gazed at the sky and tried to make sense of the motions of the wandering stars and planets, and how these relate to the great celestial lights, the Sun and the Moon. We in modern times have grown up with the notion that gravity controls the motions of celestial bodies, but this concept was completely unknown and unsuspected until the time of Sir Isaac Newton in the seventeenth century.

Humans are subject to the force of gravity from the time of conception (although, in the womb, buoyancy of amniotic fluid somewhat offsets the effects of gravity on the fetus). Whether we're conscious of it or not, the force of gravity is a part of our life on Earth until our last breaths are taken. Prior to Newton, it was impossible to imagine what things would be like without gravity. Even today, only astronauts in orbit and sky divers in free fall can experience the sensation of the absence of gravity.

Newton recognized that the force that causes an apple to fall to the ground was the same force that caused the Moon to fall toward the earth. Whether the apple landed on his head or even contributed to conceptualizing his theory is doubtful. Through experimental observations and inspired guesses, he was able to extend this idea to the rest of the solar system.

Newton had the benefit of being able to refer to the observations and conclusions of the great astronomers Tycho Brahe, Johannes Kepler, and Sir Edmund Halley. These men had compiled an immense amount of data that demonstrated conclusively that planets, moons, and comets moved in elliptical orbits. Newton, therefore, had a number of premises to work with as he began to consider the causes for orbital motion.

- He recognized that the Moon accelerates toward the earth because of gravitational force but misses it because of its sideways orbital motion. Therefore, the gravity that causes the acceleration exists not only on Earth but also at least as far as the Moon.

- His experiments revealed that the gravitational force between two objects depends on two physical properties—distance and mass.

Newton started with the formula for centripetal acceleration.

$$a_c = \frac{\overline{v}^2}{r}$$

Knowing the period, *T*, and the orbital radius, he calculated the centripetal acceleration caused by the mass of the earth both at the surface of the earth and at the Moon's center of mass. Newton's genius lay in his recognizing the relationship between these centripetal accelerations and the radial distances at which these values were calculated. He found that though the radius of the earth (the distance between the earth's center of mass and the surface) was $\frac{1}{60}$ of the distance between the centers of mass of the earth and the Moon, the centripetal acceleration at the Moon was $\frac{1}{3600}$ of that at the surface of the earth [$\frac{1}{3600} = (\frac{1}{60})^2$]. From this he concluded that the gravitational force between two objects is proportional to the inverse square of the distance between them.

$$F_g \propto \frac{1}{r^2}$$

Next, he realized that the gravitational force between two masses must be dependent on both masses symmetrically so that interchanging their positions would not change the result. Mathematically, the force could depend on either the sum ($M + m$) or the product (Mm) of the two masses, raised to some power *n*, so

7-27 Newton believed that the masses of both the earth and the Moon were involved in the gravitational force between them.

$$F_g = k(M + m)^n \quad \text{or} \quad F_g = k(Mm)^n,$$

where k is an unknown proportionality constant. Since $F_g = ma_g$,

$$ma_g = k(M + m)^n \quad \text{or} \quad ma_g = k(Mm)^n.$$

Dividing both sides by m,

$$a_g = \frac{k(M + m)^n}{m} \quad \text{or} \quad a_g = \frac{k(Mm)^n}{m}.$$

Galileo had shown that the acceleration due to gravity alone of mass M on a mass m was independent of the smaller mass, m. Everything falls toward M with the same acceleration, ignoring air resistance. In the first formula for a_g, there is no exponent n that eliminates dependence on m. In the second formula for a_g, only an exponent of $n = 1$ eliminates a dependence on m. Therefore, he concluded that

$$F_g \propto Mm.$$

When he combined gravity's two dependencies on mass and distance, the resulting proportionality became

$$F_g \propto \frac{Mm}{r^2}.$$

Newton's proportionality can be made an equation by simply including a proportionality constant, which conventionally is called G:

$$F_g = G\frac{Mm}{r^2} \tag{7.15}$$

The symbol G is called the **universal gravitational constant.** Do not confuse the constant G with the planetary acceleration due to gravity, g. The value of G was not known in Newton's time. It would not be determined until more than a century later. Equation 7.15 is called **Newton's law of universal gravitation.**

$$a_g = \frac{k(Mm)^1}{m} = kM$$

Newton's **law of universal gravitation** states that the gravitational force between two objects is directly proportional to their two masses and inversely proportional to the distance between their centers of mass squared. The proportionality constant in Newton's law of universal gravitation is G.

7.16 Properties of the Law of Universal Gravitation

The law of universal gravitation allows us to predict the gravitational force between two objects of given masses at a specified distance, but it does not indicate why gravitation exists, how it is propagated, or what gravitation is. In addition, the law is valid only for point-like masses. If the objects have appreciable size or are irregular in shape or composition, then the formula is only an approximation of the force between them unless they are very much farther apart than their physical sizes. Newton, using integral calculus that he discovered for the purpose, showed that finite spheres with spherical mass distribution (not necessarily *uniform* mass distribution) could be considered point masses. For example, the mass of an empty spherical steel water tank (the kind that stores your town's water, perhaps) could be considered to be concentrated at the center of the sphere for the purposes of determining the force of gravity on it.

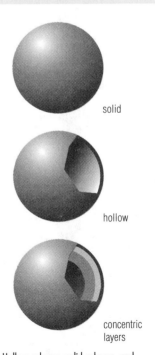

solid

hollow

concentric layers

7-28 Hollow spheres, solid spheres, and concentrically layered spheres can be considered point masses.

The law is not an independent law of motion. It merely quantifies the force of attraction of one object on the other. Gravity is always an attractive force, never repulsive. A repulsive gravitational force might be the result of a particle having a "negative mass." While particles with zero mass are known, no particle with a negative mass has ever been identified, nor is one believed to be possible.

7.17 Cavendish

The gravitational force on an object and its mass can be measured directly, but in order to calculate G, the mass of the earth had to be known as well. For more than a century after Newton discovered the law of universal gravitation, the value of G remained a mystery. Late in the eighteenth century, the eccentric British physical scientist Henry Cavendish was encouraged by a fellow scientist to devise an experiment to determine G directly, independent of the mass of the earth. Then, when the value of G was known, he could "weigh the earth," or more accurately, determine the mass of the earth.

Cavendish was the son of a wealthy family. He was very reclusive, avoiding contact with people, including his servants, most of his life. Biographies indicate that he was a meticulous observer who discovered and described the gas that was eventually called hydrogen, and he determined that water consisted of two gases, thus closing the book on the Greek view that water was one of the four basic elements of the world.

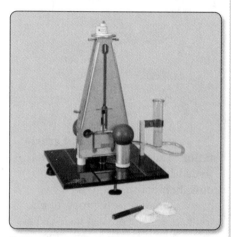

7-29 A modern Cavendish experiment apparatus

7.18 The Cavendish Experiment

Cavendish's experiment measured the gravitational force exerted between known masses in his laboratory. The apparatus consisted of two small lead spheres mounted on the ends of a rod suspended by a quartz thread. This assembly was permitted to rotate freely on the thread, producing a torque in the thread as it twisted. This kind of apparatus is called a **torsion balance.** Another pair of larger lead spheres were symmetrically fixed in such a way that the center of mass of each lay in the same plane with the centers of mass of the lead spheres mounted on the torsion balance. The torsion arm was suspended between the fixed large masses so that as it rotated, the smaller masses would symmetrically approach the larger. See Figure 7-29 for a modern example of Cavendish's apparatus.

Cavendish determined beforehand how much torque was required to twist the quartz thread through various angles. He used a light beam reflected off a mirror attached to the thread to enhance the detection of the tiny deflection angles caused by the thread twisting. By adjusting the distances between the centers of mass (r), he measured the deflection angles, which yielded a torque value. Knowing the physical dimensions of the apparatus permitted him to calculate the gravitational force exerted by the masses on each other.

Solving the universal gravitation formula for G, Cavendish calculated a value equivalent to 6.754×10^{-11} N·m²/kg² in today's SI units. This is in good agreement with the current value of 6.67×10^{-11} N·m²/kg².

7.19 Computing the Mass of Earth

Once Cavendish had the value of G, he could then estimate the mass of the earth. All he had to do was weigh an object of known mass (m) to find its gravitational attraction at the earth's surface (r_E), then solve for M. For example, say he used a

24 lb cannonball ($m \approx 10.9$ kg). Its weight ($F_w = F_g$) would be 107 N. The radius of the earth is approximately 6.37×10^6 m. Solving Equation 7.15 for M:

$$M = \frac{F_w r_E^2}{Gm}$$

$$M = \frac{(107 \text{ N})(6.37 \times 10^6 \text{ m})^2}{(6.754 \times 10^{-11} \text{ N·m}^2/\text{kg}^2)(10.9 \text{ kg})}$$

$$M \approx 5.90 \times 10^{24} \text{ kg}$$

From this mass he was able to calculate the density of the earth (ρ_E), which was of importance to geologists. He assumed that the earth was a sphere and of uniform composition. Density is the mass per unit volume, so he had to calculate the volume of the earth in cubic meters.

$$V_E = \tfrac{4}{3} \pi r_E^3$$

$$V_E = \tfrac{4}{3} \pi (6.37 \times 10^6 \text{ m})^3$$

$$V_E \approx 1.08 \times 10^{21} \text{ m}^3$$

The earth's density can now be calculated from its mass and volume.

$$\rho_E = \frac{M}{V_E}$$

$$\rho_E = \frac{5.90 \times 10^{24} \text{ kg}}{1.08 \times 10^{21} \text{ m}^3}$$

$$\rho_E \approx 5.46 \times 10^3 \text{ kg/m}^3 \text{ or } 5.46 \text{ g/cm}^3$$

This compares well with today's average earth density of 5.52 g/cm³.
Cavendish's result was interesting because the average density of rocks from the crust of the earth is only about 3 g/cm³, which indicates that the interior of the earth must be much denser. Careful measurement of the speeds of seismic waves passing through the interior of the earth indicates that the density of the matter at the center of the earth is as high as 13–15 g/cm³.

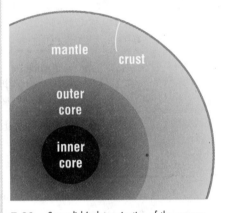

7-30   Cavendish's determination of the average density of the earth revealed that the earth's composition is not uniform.

7C   Section Review Questions

1. What are epicycles? Who proposed their existence?

2. Do the moons of Jupiter have the same Keplerian constant as the moons of Uranus?

3. How is the force of gravity between two objects related to the distance between their centers of mass? What relationships helped Newton understand this?

4. How does Cavendish's experiment disprove the idea that only planet-sized masses cause gravitational attraction?

⊚5. If Venus's orbit has a semi-major axis of 0.723 ua, calculate its orbital period.

⊚6. Calculate the Keplerian constant for Io, a moon of Jupiter, if $T_{Io} = 1.769$ days and $R_{Io} = 421\,600$ km.

⊚7. Calculate the gravitational force that exists between a dog with a mass of 53.0 kg and a cat with a mass of 3.5 kg if they are sitting on opposite sides of the room with their centers of mass 7.5 m apart.

Chapter Review

In Terms of Physics

circular motion	7.2	moment arm (L)	7.12
spin	7.2	rotational equilibrium	7.12
orbital motion	7.2	law of moments	7.12
angular speed (ω)	7.4	geocentric theory	7.13
uniform circular motion	7.4	heliocentric theory	7.13
tangential velocity ($\mathbf{v}_t$)	7.4	Kepler's laws	7.14
centripetal acceleration ($\mathbf{a}_c$)	7.5	semi-major axis (R)	7.14
angular velocity ($\boldsymbol{\omega}$)	7.6	period (T)	7.14
right-hand rule of circular motion	7.6	Keplerian constant (K)	7.14
nonuniform circular motion	7.7	astronomical unit (ua)	7.14
angular acceleration (α)	7.8	universal gravitational constant (G)	7.15
tangential acceleration ($\mathbf{a}_t$)	7.9	Newton's law of universal gravitation	7.15
centripetal force ($\mathbf{F}_c$)	7.11	torsion balance	7.18
torque ($\boldsymbol{\tau}$)	7.12		

Problem-Solving Strategies

7.1 (page 139) Tangential speed can be calculated by the formula $v_t = r\omega$ as well as by $v_t = r\Delta\theta/\Delta t$.

7.2 (page 148) The unit of torque is the m·N. Do not confuse this unit with the N·m, which is a unit of energy and work that is equivalent to the joule (J).

7.3 (page 153) Remember when using Kepler's constant $K = y^2/ua^3$ in his third law equation that R must be in terms of astronomical units and T must be in years.

Review Questions

1. State at least one requirement for uniform circular motion.

2. When the origin of the coordinate system of an orbiting object is at the center of rotation, the object's position vector is parallel to what part of the circle?

3. Why does the complete coordinate system of a rotating object include the z-axis?

4. The quantity "60 rpm" is an example of what kinematic variable?

5. If the radius of an orbiting object increases and its tangential speed remains the same, what must happen to its angular speed? (This would occur if the space shuttle increased its orbital altitude without increasing its speed.)

6. What is the direction of the centripetal acceleration vector in circular motion?

7. What happens to the centripetal acceleration of an orbiting object if the radius of its orbit decreases while its tangential speed increases? (Assume its mass remains constant.)

8. Describe a simple method for determining the direction of the angular velocity vector of a rotating object.

9. What other vector quantity is parallel to angular acceleration? Describe what is happening to the object's motion when the angular acceleration vector points in the negative-z direction.

10. What condition must exist for the three equations of circular motion to hold true?

11. What force is responsible for centripetal acceleration?

12. If an object is experiencing uniform circular motion, is the centripetal force vector constant? Explain your answer.

13. If you swing a lead sinker on the end of a string horizontally over your head in a uniform way, then suddenly shorten the string, what happens to tangential speed if you maintain the centripetal force constant?

14. If an object has unbalanced torques acting on it, what kind of mechanical condition exists?

15. What condition must exist for rotational equilibrium? To what principle of linear motion is this analogous?

16. According to historical records, who was the first to propose that the earth revolves around the Sun?

17. Why did the Greek astronomer Ptolemy believe that epicycles were necessary in his geocentric model of the solar system?

18. What long-held belief about the motions of the planets did Kepler have to discard in order to develop his second law?

19. What is the shape of each planet's orbit around the Sun?

20. At what part of a planet's orbit does it travel fastest?

21. What restriction applies to using the Keplerian constant?

22. Name some of the other scientists who, through their earlier discoveries, contributed to the discovery of the law of universal gravitation by Newton. Which of the scientists mentioned in this section was a contemporary and friend of Newton?

23. How does gravitational attraction depend on the distances between two objects? From which reference points is this distance measured?

24. Why was it important to Cavendish that the diameter of the earth be known accurately when he calculated the earth's average density?

25. The universal gravitational constant can be determined only through experimentation. What are some enhancements to the basic Cavendish apparatus that might permit greater experimental precision compared to what was possible in the eighteenth century? Consider location as well as physical attributes.

True or False (26–35)

26. The tangential velocity of an orbiting object is always perpendicular to its radial position vector.

27. Angular speed is a scalar quantity. It includes both magnitude and the sign indicating its direction of motion.

28. The first and second equations of circular motion can be used with either vector or scalar forms of the variable quantities.

29. Centripetal acceleration is opposite to centripetal force.

30. The tangential speed of a revolving object will increase if its orbital radius increases, and the centripetal force acting on it remains the same.

31. A torque vector is perpendicular to the plane of the applied force and the moment arm.

32. The early heliocentric theories were not much more accurate than the best geocentric theory at predicting the positions of the planets in the heavens.

33. The earth is closer to the Sun during winter in the northern hemisphere. This is why winter seems to drag out so long—the earth is moving more slowly at this point in its orbit.

34. A less massive falling object will not accelerate toward a planet as much as a more massive object.

35. Cavendish had to know the value of G before he could compute the earth's mass.

36. A typical angular speed of an AC induction motor is 3600 rpm. Convert this speed to radians per second.

37. A fishing sinker on a string revolves in a circle of radius 1.00 m with a constant speed of 2.75 m/s. What is its centripetal acceleration?

38. A car drives around a curve with a 125 m radius. The car's maximum centripetal acceleration is 2.60 m/s^2.
    a. What is the car's maximum speed, in km/h, around the curve?
    b. If the car goes twice its maximum velocity for the curve and at the same centripetal acceleration as for part a, what will be the radius of its path?

39. A June bug (*Phyllophaga corrosa*) is clinging to the rim of a 61.0 cm bicycle wheel that is rolling with an angular speed of 1.45 revolutions per second. What is the magnitude of the centripetal acceleration on the insect?

40. If the bicycle in Question 39 started at rest and took 5.0 s to attain the angular speed, what was the magnitude of the June bug's average angular acceleration?

41. If the idle speed of a compact disk in a multi-speed CD-ROM drive is 100 rpm, what is the angular acceleration if it takes 2.2 s for the disk to speed up to 2500 rpm?

42. A 92.2 g blue jay (*Cyanocitta cristata*) wheels through the trees in order to land at a backyard feeding station. If it flies with a constant speed of 24.0 km/h and makes a tight horizontal turn with a radius of 3.50 m, how much centripetal force must its wings exert to make the turn?

43. A 1.40 m thread is tied to a 50.0 g laboratory mass that a physics student is swinging in a horizontal circle around her head. If a force of 4.46 N will break the thread, what is the maximum angular speed at which she can revolve the mass without breaking the thread?

44. If a mechanic exerts 80.5 m·N on a stuck bolt that does not budge, using an 18.0 cm wrench, how much force is the bolt exerting on the mechanic's wrench if the head of the bolt is only 2.0 cm across?

45. Mercury orbits the Sun with an average speed of 4.78×10^4 m/s. About how long does Mercury take to orbit the Sun once?

46. What is the gravitational attraction between a dating couple whose centers of mass are 0.630 m apart if the boy weighs 715 N and the girl weighs 465 N?

47. Jupiter has a mass of 1.90×10^{27} kg. The semi-major axis of its orbit is 7.78×10^{11} m. The radius of the planet is approximately 7.15×10^7 m.
    a. Io, one of Jupiter's Galilean moons, has an orbital radius of 4.22×10^8 m. If Jupiter attracts Io with a force of 6.35×10^{22} N, what is Io's mass?
    b. If Io completes one orbit of Jupiter in 1.53×10^5 s, what is K for Jupiter's moons?
    c. Ganymede, another of Jupiter's Galilean moons, has an orbital radius of 1.07×10^9 m. What is its period?
    d. What is the acceleration due to Jupiter's gravity at its "surface"? (Jupiter is a gaseous planet. Whether it has a distinct surface deep inside its atmosphere is not known.)
    e. Jupiter's closest approach to the earth is 5.89×10^{11} m away. At this distance, what force does Jupiter exert on the earth?
    f. When Jupiter is closest to Earth, what is the acceleration of the earth due to Jupiter's gravity?
    g. When Jupiter is closest to Earth, what force does Jupiter exert on a 55.0 kg person?

48. The diameter of a typical minivan tire is 64.8 cm. A stone is stuck in the tread of the right rear tire. What is the magnitude and direction of the stone's angular velocity vector if the van is traveling at 10.0 km/h?

49. The distance between the centers of Earth and the Moon is 3.84×10^8 m.
    a. At an instant in time, how far from the earth on a line connecting the centers of the earth and the Moon would an object have to be in order to experience equal forces from the earth and the Moon?
    b. Why would this position not be a stable location between the two bodies?
    c. After some research, discuss the locations around two gravitational bodies that permit a third object to experience a stable orbit.

50. Write a brief biography of Sir Edmund Halley. Include in the paper a discussion of the influence of his works on Newton's gravitational theory.

51. Calculate the radius of a "lunasynchronous" orbit. The period of one lunar day, T_M, is equal to 27.3 days (2.36×10^6 s).

52. Prove that the semi-major axis of an ellipse is equal to the average distance of all points in an ellipse from either focus.

$$a_c = \frac{v_t^2}{r} = \frac{60}{10} = 6$$

$$= \frac{70}{5}$$

Circular Motion 159

FACETS of PHYSICS

Roller Coasters

Why are some people addicted to roller coasters? Speed, near weightlessness in the vertical drops, and high acceleration all help produce the adrenaline rush that we associate with thrill rides. Extreme rides achieved new heights in the 1970s with the creation of the looping roller coaster. Those early versions seem tame compared to modern coasters, but the first loop was quite an innovation. If you look closely, you will notice that a loop on a roller coaster is not really a circle but rather an upside down teardrop called a *clothoid loop*. Why was this necessary?

Start by considering a frictionless circular loop. As the coaster train enters the bottom of the loop and climbs to the top, its speed decreases. For the ride to be successful, the initial velocity must be great enough to ensure that the coaster will make it over the top of the loop. However, as the velocity increases, so does the centripetal acceleration force (the *g*-force) experienced by the rider. At the velocity needed to complete the circular loop, the rider would experience a

force of 6 *g*s, or six times the force of gravity, as he enters the bottom of the loop. People not accustomed to such forces start feeling uncomfortable when experiencing 3.5 *g*s and may start blacking out beyond 5 *g*s as the blood is forced out of the brain. Obviously, a circular loop would be neither safe nor popular.

The clothoid loop is a clever way of getting around the difficulty. The coaster train approaches the loop on a linear track and then enters a shallow rising curve with a regularly decreasing radius. Because the radius of curvature at the bottom of the loop is large, the centripetal acceleration exerted on the rider is lower. The upper section of the loop, ±65° from a vertical line bisecting the loop, is approximately a circle. The ride down the loop is the reverse of the ride up. In circular motion, when all other factors are kept constant, as the radius decreases, the tangential speed increases. Now we have two influences at work—the ascension of the loop tending to decrease the train's velocity, and the steadily decreasing radius of curvature, which tends to increase its velocity. The interplay of these two influences ensures that the train has enough energy to go over the top while maintaining accelerations that rarely exceed 3.7 *g*s. As the train falls and its speed increases, the curvature steadily increases so that the acceleration when the train goes through the bottom of the loop is acceptable.

Where should you sit on the roller coaster to get the best ride? Because the train is long, riders in the front, middle and back experience the ride differently. The ride is designed so that the rider traveling at the train's center of mass will have a smooth ride and, at most, feel weightless only briefly. The center of mass of the train is always traveling at the velocity designed for the loop at each point and it will have the lowest velocity at the peak of the loop. Imagine you are in the front of the train. As the train rises to the top of the loop the front cars are pushed over the top faster than the design velocity and, since they are ahead of the center of mass, experience the lowest velocity not directly at the peak, but slightly past the peak. There they feel held back until the center of mass goes over the top and the train begins plunging down. The front cars experience a slight moment of "hang time." The back cars of the train experience their lowest velocity before they reach the peak of the loop. They hang back slightly on the way up. As the center of mass reaches the peak and accelerates, the last cars are whipped over the top. Since riders in the back are going faster than the design velocity for that point, they may feel forced into their seats.

8A	Simplifying Problems	163
8B	Transmitting Mechanical Forces	166
8C	Friction	174
8D	More Applications	179
Facet:	Artificial Gravity	188

New Symbols and Abbreviations

tension (T)	8.1	coefficient of friction (μ)	8.9
normal force (N)	8.5	kinetic friction (f_k)	8.10
friction (f)	8.8	static friction (f_s)	8.11
traction (f_t)	8.8	rolling friction (f_r)	8.12

Sailing ship rigging lines transmit mechanical forces, and pulleys change the direction of those forces.

Applying Newton's Laws

8

> *For our light affliction, which is but for a moment,*
> *worketh for us a far more exceeding and eternal*
> *weight of glory.* II Corinthians 4:17

In this passage, the apostle Paul is writing to the carnal church in the city of Corinth, located in present-day Greece. He was concerned that they were being overcome by the cares of the world—the daily struggle to live, persecution from their pagan countrymen, and the inconsistent testimony of other believers. Paul used the analogy of weight to compare the trivially light burdens of their existence here on Earth to the infinitely greater weight of glory that they would lightly bear in heaven. Weight was a common and easily understood force to the Corinthians. What force will you allow to motivate your life? The relatively light pressures and influences of the world, or the eternally heavy "weight of glory" of God and His Word?

8A   SIMPLIFYING PROBLEMS

Newton's laws are fundamental to mechanics. By applying them systematically, you can solve almost any mechanical problem. But suppose you have a complex problem involving several interacting bodies. How do you apply Newton's laws to such a problem?

8.1   Free-Body Diagrams

The first step in simplifying a complex problem is to isolate each body and consider it as a separate system. For example, Figure 8-1 is a real-world diagram showing three blocks connected by strings. A physics student is pulling upward on the string attached to block 1. To find the net force on block 1, draw a **free-body diagram** of that block. That is, isolate block 1 as the system of interest and determine all the external forces that act *directly* on the system.

What forces act directly on block 1? Gravity pulls down on the block. You can see that the hand does not act directly on the block but on the string, which then exerts the force on the block. Similarly, blocks 2 and 3 do not pull directly on block 1, but the string tied to the bottom of block 1 exerts a downward force on it. The forces of the strings on the block are called **tension (T)** forces because the

> A **free-body diagram** is a vector diagram that isolates a single object in a multi-object system in order to analyze the forces on it.

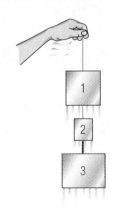

8-1 World diagram of a multi-object system

163

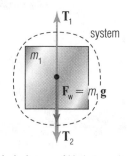

8-2 Free-body diagram of block 1

Problem-Solving Strategy 8.1

Make accurate sketches of multi-body systems that include strings and pulleys. Indicate all known and unknown force vectors and their components in the coordinate system used for each body. Note unknown values such as acceleration, but do not assign a direction until after the analysis is completed.

strings *pull* on the block. These three forces—the force of the upper string, T_1; the force of the lower string, T_2; and the force of gravity (the block's weight), F_w—are the only forces that act directly on block 1. Figure 8-2 shows the free-body vector diagram of block 1.

Keep in mind that a string connecting two objects exerts the same tension on both objects but in different directions.

8.2 Transmitting Forces in a Line: Ideal Strings

The strings attached to objects in this chapter are **ideal strings.** Physicists use the concept of the ideal string when studying multiple-body systems where the objects are not touching but are connected. Ideal strings have the following properties:

- They have no mass; therefore, they do not affect the acceleration of the system of objects.
- They do not stretch when exerting a force.
- They exert only pulling forces, not pushes. A pushing force would crumple the string.
- Forces are exerted only in line with the string.

Real strings and ropes approximate ideal strings as long as the tension forces are limited to avoid stretching or breaking. For example, the force T_1 on block 1 in Figure 8-2 has the same magnitude and direction as the force of the hand on the string. As long as the pull is continuous, an ideal string also holds objects at fixed distances. Therefore, all the objects connected by a string and pulled will have the same speed and acceleration. Although real strings have mass and do stretch somewhat, the mass and stretch are usually small enough to be ignored. So the assumption that a string is ideal is unlikely to introduce noticeable error into your calculations. You can assume that all strings are ideal unless otherwise stated.

8.3 Systems of Connected Objects

There are many approaches to analyzing the dynamics of a system of objects. If all the masses are known, it is generally easier to determine the motion of the system as a whole first and then analyze the forces on the individual objects. You need to identify what information is given in order to determine the approach to solving the problem. There is no "cookbook" method to solving mechanics problems.

Using free-body diagrams, you can specialize the problem-solving method you learned in Chapter 2 to fit problems in mechanics. Most of the changes are additions to step 1, where it was suggested that you sketch the problem. To apply Newton's laws, you should first draw a diagram of the entire arrangement. This is called the "world" diagram. Include in this diagram all connections (such as ideal strings) and an arrow showing the direction of motion if that is given. When you have drawn the overall arrangement, you may have to draw one or more free-body vector diagrams. Then select a coordinate system for each object. This does not necessarily have to be the same for all objects, especially if they are moving in different directions.

After you have drawn free-body diagrams, you can continue with the problem-solving method, applying Newton's laws to each diagram separately. When you have reached a solution, check it to see if it is reasonable. It should have the proper units and should agree with any given information. For example, if you find that the individual accelerations for two objects connected by a string are either of different magnitudes or in opposite directions, then you know that you have made a mistake.

EXAMPLE 8-1

Pulling Strings: Analyzing Systems of Objects

The masses of the blocks in Figure 8-3a are $m_1 = 5.0$ kg and $m_2 = 1.0$ kg. The hand pulls up on the string with a force of 120 N. What is the magnitude of the tension in the string between the blocks?

Solution:

Since the blocks are connected by an ideal string, you should first treat the two blocks as a single 6.0 kg system. The tension in the string connecting the two blocks is not shown because it is a force within the system and cannot affect the system's motion. Only forces from outside a system can affect the motion of the system as a whole.

The acceleration of block 2 equals the acceleration of the system. You can then determine the force on block 2 that will produce that acceleration of the block according to Newton's second law. That force is the tension in the string tied to block 2.

Choose a coordinate system with y vertical.

Calculate the acceleration of the two-block system:

$$\Sigma \mathbf{F}_{sys} = m_{sys}\mathbf{a} = \mathbf{F}_{hand} + \mathbf{F}_w$$

$$m_{sys}a_y = F_{hand\,y} + F_{wy}$$

$$a_y = \frac{F_{hand\,y} + m_{sys}g_y}{m_{sys}}$$

$$a_y = \frac{(+120\text{ N}) + (6.0\text{ kg})(-9.81\text{ m/s}^2)}{6.0\text{ kg}}$$

$$a_y \doteq +1\underline{0}.1\text{m/s}^2 \text{ (up)} \tag{1}$$

Note that all the forces on the system are parallel to the y-axis.

Calculate the tension on block 2 using the acceleration from (1) and Newton's second law for block 2:

$$\Sigma \mathbf{F}_2 = m_2\mathbf{a} = \mathbf{T}_2 + \mathbf{F}_{w\,2}$$

$$\mathbf{T}_2 = m_2\mathbf{a} - \mathbf{F}_{w\,2}$$

$$T_{2y} = m_2 a_y - F_{wy} = m_2 a_y - m_2 g_y$$

$$T_{2y} = m_2\,(a_y - g_y) = (1.0\text{kg})\left[(+1\underline{0}.1\text{m/s}^2) - (-9.81\text{m/s}^2)\right]$$

$$T_{2y} \doteq +1\underline{9}.9\text{ N}$$

$$T_{2y} \approx +20.\text{ N (up)}$$

The tension in the string on block 2 is about 20 N straight up.

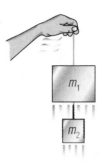

8-3a World diagram of the two-block system

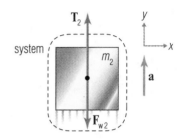

8-3b Free-body diagram of block 2

8A Section Review Questions

1. Explain why a free-body diagram is essential when analyzing the dynamics of an object.

2. Describe the properties of an ideal string. Under what conditions do ropes and real strings act like ideal strings?

3. Compare the tension exerted at one end of a straight string to the tension exerted at the other end of the string.

4. True or False: The force that a system exerts on the surface that supports it is *not* included in a free-body diagram of the system.

5. True or False: If a truck towing two trailers of equal mass in tandem accelerates away from a gas pump, the force exerted by the truck on the first trailer equals the force exerted on the second trailer by the first.

⊙6. Using a diagram, give the location, magnitude, and direction of all forces on a 1 kg brass sphere that is hanging from a string.

⊙7. A family is skating at an ice rink. The 58.2 kg mother is holding the hand of her 35.5 kg daughter. The father grabs his wife's free hand and pulls horizontally with a constant force of 100. N. Assume that the skates glide without friction on the ice and that the family's hands and arms approximate ideal strings. How much force does the daughter experience?

⊙8. Two mountain climbers tied together with an "ideal rope" have slipped on an icy incline. Fortunately, the first climber was belayed to a piton in the ice. The climbers are hanging from their ropes on a frictionless 65.0° incline. The first climber and his gear has a mass of 93.0 kg. The second climber's total mass is 67.0 kg.

   a. What is the tension in the rope tied to the piton?

   b. What is the tension in the rope between the climbers?

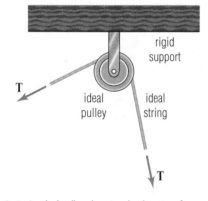

8-4 An ideal pulley changing the direction of a tension force

8B TRANSMITTING MECHANICAL FORCES

8.4 Transmitting Forces Around a Corner: Ideal Pulleys

An ideal string (or a real one, for that matter) cannot exert a force in a direction other than in line with the string. However, one of the main reasons for using a string-like connection between two systems is to exert forces on systems that are not in line with one another. The direction of tension in a string can be changed by using a *pulley*. Physicists employ **ideal pulleys** when considering theoretical systems connected by ideal strings.

An ideal pulley has the following characteristics:

- It consists of a grooved wheel and an axle. The pulley can be mounted to a structure outside the system of interest or attached directly to the system.

- The axle is frictionless.

- The motion of the string around the wheel is frictionless.

- It changes the direction of the tension in the string without diminishing its magnitude.

Actual mechanical systems use pulleys to change the direction of tensional forces as well (e.g., flagpoles, sailboat rigging, etc.). If real systems did not use pulleys, but rather allowed a rope to wrap around an edge of a building or structural beam, for example, the rope would rapidly wear, and the force available at the end of the rope would be greatly reduced because of friction.

In Figure 8-5b, $\mathbf{F}_{wh}$ is the weight of the hanging block. The block exerts this force on the string, which in turn exerts a force $\mathbf{T}'$ on the block on the tabletop. The reaction force of the tabletop block on the hanging block is $\mathbf{T}$. The magnitude $T' = F_{wh}$, but the directions are not the same. The pulley changes the direction of the transmitted force. As you will see in the following example, adjusting the orientation of the coordinate system for each object in a connected system simplifies the analysis.

EXAMPLE 8-2

Getting the Hang of Pulleys: Changing the Direction of a Force

A 1.50 kg block (m_t) rests on a frictionless table. It is attached to a hanging block of mass $m_h = 1.00$ kg by a string passing over an ideal pulley mounted to the table's edge. What is the acceleration of the hanging block?

Solution:

This problem involves linear motion, but each of the elements of the system is moving in a different direction. The tabletop block is moving horizontally, and the hanging block is moving vertically. The tension of the string on each block and the blocks' accelerations have the same magnitudes, but their directions are different.

Choose a coordinate system for each block that maintains the same axis parallel to the motion of the block in its part of the system.

The x-axis for the tabletop block is horizontal and is pointing to the right. The x-axis for the hanging block is *vertical* and is pointing *downward*. This way, whatever direction the acceleration will be, it will have the correct direction relative to the block you are examining.

Perform free-body analysis of the hanging block using Newton's second law:

$$\Sigma \mathbf{F}_h = m_h \mathbf{a}$$

$$\mathbf{T} + \mathbf{F}_{wh} = m_h \mathbf{a}$$

Or in scalar component notation,

$$T_x + m_h g_{h\,x} = m_h a_x. \tag{1}$$

(Note that $g_{h\,x}$ points in the *positive* x-direction due to the rotated reference frame of the hanging block. Therefore, $g_{h\,x} = -g$.)

Find T_x by analyzing the tabletop block. Since the block experiences no vertical motion, any vertical forces must cancel out, and you can work with the scalar components.

$$\Sigma F_x = m_t a_x = T'_x$$

$$T'_x = m_t a_x \tag{2}$$

The acceleration of the tabletop block (a_x) in this equation has the same magnitude and direction relative to the x-coordinate axis as a_x for the hanging block.

You know that $T_x = -T'_x$ because they are an action-reaction force pair transmitted by the ideal string. After replacing T_x in Equation 1, substitute Equation 2 into Equation 1 for T'_x to solve for a_x.

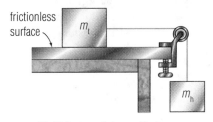

8-5a World diagram of the two blocks

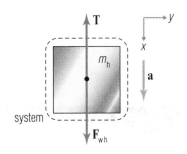

8-5b Forces on the hanging block

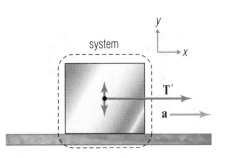

8-5c Forces on the tabletop block

Problem-Solving Strategy 8.2

The notation for tension vectors transmitted by strings connecting objects A and B is as follows: If the string exerts a tension $\mathbf{T}$ on object A, then it exerts a tension $\mathbf{T}'$ on object B. $|\mathbf{T}| = |\mathbf{T}'|$ and $\mathbf{T} = -\mathbf{T}'$ (opposite directions).

$$-T'_x + m_h g_{h\,x} = m_h a_x$$
$$-(m_t a_x) + m_h(-g) = m_h a_x$$
$$-m_h g = m_h a_x + m_t a_x$$
$$-m_h g = a_x(m_h + m_t)$$
$$a_x = \frac{-m_h g}{(m_h + m_t)} = -\frac{m_h}{(m_h + m_t)} \cdot g$$

Substituting the known values from the problem statement gives the following:

$$a_x = -\frac{m_h}{(m_h + m_t)} \cdot g$$

$$a_x = -\frac{1.00 \text{ kg}}{(1.00 \text{ kg} + 1.50 \text{ kg})} \cdot (-9.81 \text{ m/s}^2)$$

$$a_x \doteq +3.9\underline{2}4 \text{ m/s}^2$$

$$a_x \approx +3.92 \text{ m/s}^2$$

The hanging block is accelerating about 3.92 m/s² in the positive-*x* direction in its frame of reference. In the reference frame of the observer, the hanging block is accelerating *downward*.

Problem-Solving Strategy 8.3

It is permissible to use different frames of reference to analyze separate objects in a system as long as the same coordinate axis in each reference frame is parallel to the motion of the system within the reference frame.

Another interesting problem involves two hanging masses on a string over a set of pulleys.

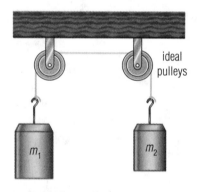

8-6a World diagram of the masses and pulleys in Example 8-3

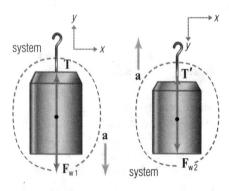

8-6b Free-body diagrams of mass 1 and mass 2

EXAMPLE 8-3

Balancing Act: Another Pulley Problem

Two lab masses, $m_1 = 1.50$ kg and $m_2 = 1.00$ kg, are connected by an ideal string that passes over two ideal pulleys (to keep the masses from interfering with each other). What is the net force on each mass?

Solution:

For mass 1, choose the coordinate axis parallel to the expected vertical motion with the *y*-axis positive upward. Choose the coordinate system for mass 2 to be the vertical with the *y*-axis positive *downward*. Whatever movement occurs in the system will be in the same direction within each mass's reference frame, irrespective of the direction. Consequently, $\mathbf{a}_1 = \mathbf{a}_2 = \mathbf{a}$. The strategy for finding the force sum on each mass involves calculating system acceleration and then solving Newton's second law, $\Sigma \mathbf{F} = m\mathbf{a}$, for each mass.

Analyze the net force on each mass:

Mass 1:

$$\Sigma \mathbf{F}_1 = m_1 \mathbf{a}$$
$$\mathbf{T} + \mathbf{F}_{w\,1} = m_1 \mathbf{a}$$
$$T_y + m_1 g_{1\,y} = m_1 a_y$$
$$T_y = m_1 a_y - m_1 g_{1\,y} \quad (1)$$

Note that $g_{1\,y} = g$.

Mass 2:

$$\Sigma \mathbf{F}_2 = m_2 \mathbf{a}$$
$$\mathbf{T} + \mathbf{F}_{w\,2} = m_2 \mathbf{a}$$
$$T'_y + m_2 g_{2\,y} = m_2 a_y$$
$$T'_y = m_2 a_{2\,y} - m_2 g_{2\,y} \quad (2)$$

Due to the rotated reference frame, $g_{2\,y} = -g$.

The magnitude of the tension in the string is the same throughout the string, but the tension vectors at the ends act on the masses in opposite directions relative to their associated frames of reference. Therefore, T_y and T'_y are an action-reaction force pair and $T_y = -T'_y$.

Combine Equations 1 and 2 and make appropriate substitutions:

$$m_1 a_y - m_1 g_{1\,y} = -T'_y$$

$$m_1 a_y - m_1 g_{1\,y} = -(m_2 a_y - m_2 g_{2\,y})$$

$$m_1 a_y - m_1 g = -\left[m_2 a_y - m_2(-g)\right]$$

$$m_1 a_y - m_1 g = -m_2 a_y - m_2 g$$

$$m_1 a_y - m_2 a_y = m_1 g - m_2 g$$

Solve for a_y:

$$a_y(m_1 + m_2) = (m_1 - m_2)g$$

$$a_y = \frac{(m_1 - m_2)g}{(m_1 + m_2)}$$

$$a_y = \frac{(m_1 - m_2)}{(m_1 + m_2)} \cdot g$$

Substitute the known values from the problem statement:

$$a_y = \frac{(1.50 \text{ kg} - 1.00 \text{ kg})}{(1.50 \text{ kg} + 1.00 \text{ kg})} \cdot (-9.81 \text{ m/s}^2)$$

$$a_y = -1.9\underline{6}2 \text{ m/s}^2$$

The result indicates that mass 1 is dropping (negative-y direction in its reference frame) and mass 2 is rising (also negative-y direction in its reference frame), which is logical, since mass 1 has more mass.

The net force on mass 1 is

$$\Sigma F_{1y} = m_1 a_y$$

$$\Sigma F_{1y} = (1.50 \text{ kg})(-1.9\underline{6}2 \text{ m/s}^2)$$

$$\Sigma F_{1y} \doteq -2.9\underline{4}3 \text{ N}$$

$$\Sigma F_{1y} \approx -2.94 \text{ N (downward)}.$$

The net force on mass 2 is

$$\Sigma F_{2y} = m_2 a_y$$

$$\Sigma F_{2y} = (1.00 \text{ kg})(-1.9\underline{6}2 \text{ m/s}^2)$$

$$\Sigma F_{2y} \doteq -1.9\underline{6}2 \text{ N}$$

$$\Sigma F_{2y} \approx -1.96 \text{ N (upward)}.$$

8.5 Transmitting Forces on Inclines

It is important to realize that coordinate systems are *chosen*. As you have seen, the x-axis does not have to be horizontal, and the positive y-axis does not have to be opposite the direction of gravity. The only restriction is that the x-axis must be perpendicular to the y-axis. Sometimes the forces on an object will be neither horizontal nor vertical, but at an angle. In such cases it is often helpful to define the coordinate system so that one axis is in the same direction as most of the forces. For example, Figure 8-7a shows a block sitting on an incline. Two of the forces on the block are parallel to the incline, and one force is perpendicular to the incline. Therefore, a convenient choice of coordinates would place the x-axis along the incline and the y-axis perpendicular to it, as in the free-body diagram Figure 8-7b.

Figure 8-7b shows one force that is easy to overlook. That force is the **normal force (N),** the force exerted by the incline on the block. This force is called the

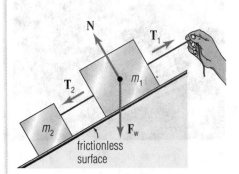

8-7a World diagram of a block on an incline

8-7b Free-body diagram of the block

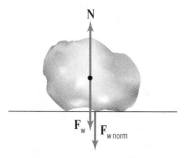

8-8 Relationship between normal force, weight, and the normal component of weight on a surface

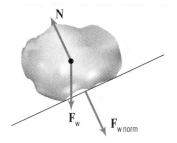

8-9 The relationship between normal force and weight on an incline

Problem-Solving Strategy 8.4

The vector of the force of gravity is always assumed to point straight down. However, the normal component of the object's weight varies with the incline of the surface.

8-11 Because of the angle of the track, the weight of these cars is not perpendicular to the track.

Problem-Solving Strategy 8.5

For inclines, the conventional alignment for the positive x-axis is *up* the slope of the incline. This choice is completely up to you but is consistent with the convention that "up" is positive. If you should choose to have positive directions point down, be sure to change the sign of the negative scalar constant *g*.

normal force because the term *normal* in mathematics means "perpendicular." The normal force of a surface on an object in contact with it is *always* perpendicular to the surface.

To understand the normal force, examine an object resting on a flat table. The force of the earth's gravity on the object pulls it down. This is the object's weight ($\mathbf{F}_w$). Its weight causes the object to exert a force normal to the table that is equal in magnitude and direction to the object's weight. However, this latter force is *not* the weight of the object. Figure 8-8 illustrates the difference between these two forces, where $\mathbf{F}_{w\,norm}$ represents the normal component of the object's weight acting on the table. The table, according to Newton's third law, must exert an opposing force of equal magnitude on the object. This is the normal force ($\mathbf{N}$). Therefore, the normal force can be expressed as the negative of the force exerted by the object's weight vector perpendicular to the table. If the *y*-axis is vertical,

$$\mathbf{N} = -\mathbf{F}_{w\,norm} = -\mathbf{F}_{w\,y}. \qquad (8.1)$$

If the table is tipped, the object's weight no longer presses fully on the table. Instead, the object exerts a force on the table equal to the magnitude of the component of its weight that is perpendicular to the incline, and the normal force decreases accordingly.

EXAMPLE 8-4

The Normal Force on an Incline

A block rests on a 30.0° incline. If the block weighs 50. N, what is the normal force on the block?

Solution:

Choose a rectangular coordinate system so that the *x*-axis is parallel to the inclined surface and the *y*-axis is perpendicular to it. Figure 8-10 (b) is the free-body diagram showing $\mathbf{F}_w$ resolved into a normal component, $\mathbf{F}_{w\,y}$, and a down-slope component, $\mathbf{F}_{w\,x}$. The magnitude of $\mathbf{F}_w$ is 50 N. A right triangle is formed with a hypotenuse of 50. N.

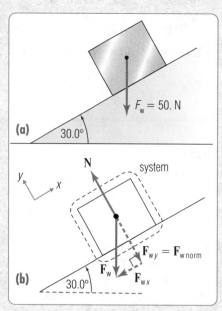

8-10 (a) World diagram and (b) free-body diagram of a block on an incline

The weight vector triangle consists of the block's weight and its two components. The angle between $\mathbf{F}_w$ and $\mathbf{F}_{w\,y}$ equals the incline angle, 30.0°.

Calculate the normal component of weight:

$$|F_{w\,y}| = F_w \cos \theta = (50.\ N) \cos 30.0°$$

$$|F_{w\,y}| \doteq 4\underline{3}.3\ N$$

$$F_{w\,y} \doteq -4\underline{3}.3\ N \text{ (negative-y direction)}$$

Calculate the normal force:

$$N_y = -F_{w\,y}$$

$$N_y \doteq -(-4\underline{3}.3\ N) = +4\underline{3}.3\ N$$

$$N_y \approx +43\ N$$

8.6 Measured Weight

Weights are usually measured with spring scales or some form of electronic scale. The scale exerts an upward force that balances the object's weight. Usually a display indicator shows the amount of force needed to balance the object's weight. For a bathroom scale, this third-law reaction force is the normal force that the scale exerts on the object. In an *unaccelerated reference frame,* the normal force is equal in magnitude but opposite in direction to the weight of an object resting on a horizontal surface.

EXAMPLE 8-5

Weight: Normal Force Read from a Scale

A 25.0 kg crate rests on a package scale. The scale's display shows the magnitude of the normal force of the scale on the crate. The crate and the scale are at rest. What weight does the scale indicate?

Solution:
You want to find *N*, the magnitude of the normal force. Make a free-body diagram for the crate:

$$\Sigma \mathbf{F} = m\mathbf{a}$$

From Newton's first law, you know that the scale-crate system is in translational equilibrium, so $\mathbf{a} = 0$ and

$$\Sigma \mathbf{F} = 0 \text{ N}.$$

The force sum on the crate is

$$\mathbf{N} + \mathbf{F_w} = 0 \text{ N}.$$

Using *y*-components:

$$N_y + F_{w\,y} = 0 \text{ N}$$

$$N_y = -F_{w\,y} = -mg_y$$

$$N_y = -(25.0 \text{ kg})(-9.81 \text{ m/s}^2)$$

$$N_y \doteq +245.2 \text{ N}$$

$$N_y \approx +245 \text{ N}$$

The indicated weight is the magnitude of the normal force exerted by the scale, or about 245 N.

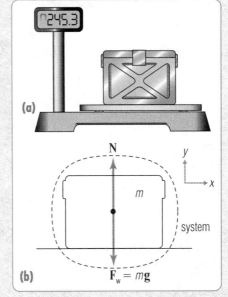

8-12 (a) World diagram of the crate; (b) Free-body diagram of the crate

8.7 Apparent Weight

A spring scale is an accurate measure of weight when the scale and object are not accelerating. But what happens if they *are* accelerating?

EXAMPLE 8-6

Giving Weight a Lift: Accelerated Scales

A 10.0 kg box rests on a bathroom scale in an elevator. The elevator is accelerating upward from rest with a uniform acceleration of +2.00 m/s². What weight does the scale indicate?

Some people think that living a godly life (pure, honest, kind, and good) would be an unbearable burden. This is only an apparent weight, however. Jesus declared "My yoke is easy, and my burden is light" (Matt. 11:30).

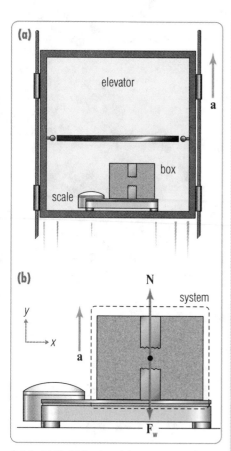

(a)

elevator

a

box

scale

(b)

N

system

y

x

a

$\mathbf{F_w}$

8-13 (a) World diagram of the crate on a scale in an elevator; (b) Free-body diagram of the crate

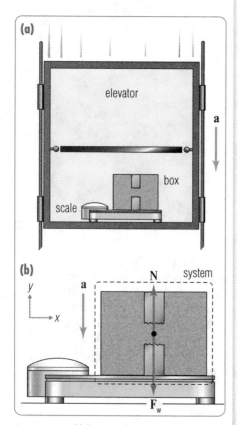

(a)

elevator

a

box

scale

(b)

a **N** system

y

x

a

$\mathbf{F_w}$

8-14 (a) World diagram of the crate on a scale in the elevator; (b) Free-body diagram of the crate

Solution:
The scale reading is the magnitude of the normal force (N) that the scale exerts on the box. The system is the box. The bathroom scale and the elevator are outside the system. You can find the normal force by identifying all of the forces acting on the box.

$$\Sigma\, \mathbf{F} = m\mathbf{a}$$

$$\mathbf{N} + \mathbf{F_w} = m\mathbf{a}$$

Using the y-components,

$$N_y + F_{w\,y} = ma_y. \qquad (1)$$

Solve Equation 1 for the vertical normal force component:

$$N_y = ma_y - F_{w\,y}$$

Since $F_{w\,y} = mg_y$, you now have

$$N_y = ma_y - mg_y, \text{ or}$$

$$N_y = m(a_y - g_y). \qquad (2)$$

Substituting the given and known values into Equation 2 gives you

$$N_y = (10.0 \text{ kg})\big[(+2.00 \text{ m/s}^2) - (-9.81 \text{ m/s}^2)\big]$$

$$N_y \doteq +11\underline{8}.1 \text{ N}.$$

The scale indicates the magnitude of N_y, or about 118 N.

Note that if a_y is in the positive-y direction, as stated in the example problem above, then

$$|ma_y - mg_y| > |mg|,$$

and the bathroom scale indicates an **apparent weight** greater than the weight of the crate (98.1 N).

Will a downward acceleration give the same result?

EXAMPLE 8-7

Weight Loss the Easy Way

The bathroom scale and the box from Example 8-6 are now accelerating downward at -2.00 m/s^2. What weight does the bathroom scale indicate?

Solution:
Again, the scale gives the magnitude of **N**. The scalar values of Equation 2 in the previous example are not specified, so the magnitude of the normal force is still

$$N_y = ma_y - mg_y.$$

Substituting the values into this equation gives you

$$N_y = (10.0 \text{ kg})\big[(-2.00 \text{ m/s}^2) - (-9.81 \text{ m/s}^2)\big]$$

$$N_y \doteq +78.1\underline{0} \text{ N}.$$

The magnitude of the normal force is about 78 N. The crate's apparent weight when the elevator is accelerating downward is less than its actual weight.

The last situation to be evaluated is the elevator dropping in free fall. Real elevators are designed to prevent this, but in this scenario its downward acceleration would be equal to **g**.

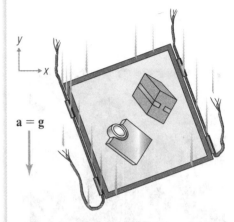

EXAMPLE 8-8

Weight in Free Fall

Suppose the elevator from the previous examples somehow breaks and falls freely down its shaft. What would be the indicated weight of the box?

Solution:

Even if the crate, the scale, and the elevator are all in free fall, the normal force on the crate is still

$$N_y = ma_y - mg_y.$$

But $a_y = g_y$, so

$$N_y = mg_y - mg_y, \text{ and}$$

$$N_y = 0 \text{ N.}$$

The scale exerts no force on the crate. Since you measure weight by the reading of a scale, you might say that the crate is *weightless*. Actually, the weight of the crate has not vanished. It weighs no less than it would if it were stationary; the earth still exerts the same force of gravity on it. However, since both the crate and the scale are in free fall, the scale can't exert a force on the crate. You could say that the crate is "normal force-less!"

8-15 World diagram of the crate, scale, and elevator in free fall

You probably have heard that astronauts in space are weightless. Although the force of gravity lessens with increasing distance from the earth, astronauts are still affected by gravity. However, their vehicle is in orbit, or free fall, around the earth. Since the spacecraft and its occupants fall at the same rate, the astronauts seem to float. Therefore, it only appears that they have no weight.

As you have seen, apparent weight equals real weight when the object being weighed is not accelerating. However, when the object is accelerating, its apparent weight is not the same as the gravitational force on the object.

8-16 A "normal force-less" astronaut

8B   Section Review Questions

1. How can the direction of a tensional force be changed without diminishing the force?

2. Explain why assigning separate coordinate systems to different objects within a moving system simplifies the analysis of the system.

3. What force acting on a system is assumed to always act vertically, regardless of the orientation of the system?

4. True or false: The normal force acting on a system is the third-law reaction force to the system's weight.

◉5. Two masses are suspended by a string placed over a pulley supported by the ceiling ($m_1 = 6.20$ kg and $m_2 = 7.70$ kg).

   a. Which object will descend?

   b. What will its acceleration be?

⊕**6.** A male African elephant (*Loxodonta africana*) has a mass of 5450 kg. Calculate the normal force that supports this critter on a horizontal surface.

⊕**7.** The elephant in Question 6 jumps off of a 10 m cliff into a cool water hole, making an immense splash in the process. Calculate the acceleration of the earth toward the elephant if the earth's mass is 5.974×10^{24} kg.

⊕**8.** Assume that an elevator carrying a bathroom scale supporting a 25.0 kg mass is rising at a uniform +4.00 m/s. It slows to a stop with a uniform acceleration of -2.25 m/s^2. What weight does the bathroom scale display while the elevator is slowing?

8C FRICTION

8.8 Friction Defined

8-17 The normal force and the force of gravity on this woman cannot produce forward motion.

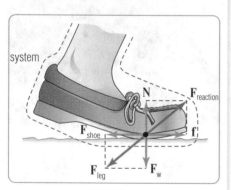

8-18 Friction provides the force needed for walking.

8-19 A large amount of friction is necessary to accelerate a dragster to high speeds.

Have you ever wondered how it is possible for you to walk? According to Newton's first law, if there are no unbalanced external forces on a person, he cannot begin to move. Consider a woman standing in the middle of a driveway. What are the external forces on her? The force of gravity pulls down on her, and the normal force of the ground pushes up. Neither of these forces can propel her forward. The force exerted by her leg muscles is an *internal* force and cannot change her motion directly. How can she create an external force?

The answer lies in Newton's third law. As the woman attempts to walk, her leg muscles, through her feet, exert a force backward and down on the pavement. The pavement therefore exerts a force upward and forward on the foot, thus propelling the woman forward. The upward component of this reaction force is included in the normal force. The forward component of the reaction force is called **friction (f).** Without friction, the woman could not move forward.

Ice is an example of a nearly frictionless surface. Suppose the woman were in the middle of a smoothly frozen, frictionless pond, instead of in a driveway. She could try to exert a backward force on the ice, but her foot would just slip backward. She could still exert a downward force on the ice, and the ice would push back with an equal force, but unless there were some frictional force she could not move forward.

Another kind of motion that is impossible without friction is rolling. As a tire rolls, it pushes backward on the road; consequently, the road pushes forward on the tire. This forward force of friction moves the car forward. When there is little friction (an icy street, for example) and the car's tires spin, they exert little or no force on the ground. The car cannot move forward on level ground when friction is absent. Similarly, a wheel on a frictionless incline will not roll but will just slide without rolling.

The friction that makes walking, rolling, and similar motions possible is called **traction (f$_t$).** Traction is also the name used for friction that prevents unwanted motion. For example, when a car rounds a curve, its tendency is to continue in a straight line. Traction provides the centripetal force to keep the car on the road. Sometimes traction alone would not be adequate to change the car's path. In such cases the curve is often banked so that a component of the normal force also points toward the center of the curve, adding to the centripetal acceleration caused by traction.

Friction does have positive aspects, but sometimes it gets in the way. Friction is, in general, a force between two surfaces that opposes the relative motion of the surfaces. Every time one surface slides across another—for example, a box sliding across a carpet—the sliding object exerts a force on the other surface in the direction of sliding. The surface therefore exerts a reaction force on the sliding object to oppose the sliding. Friction occurs because of the interaction of molecules of the two surfaces. The closer together they get, the stronger the interactions become. If you smooth two surfaces, at first you decrease friction by removing obstacles to the motion, just like removing speed bumps on the road allows a car to travel faster. However, when the surfaces become smooth enough for most of their atoms to touch, the atoms attract each other, increasing friction. Try sliding apart plates of glass that are stacked together.

8.9 Magnitude of Friction

What affects the magnitude of the friction force? Consider two boxes of the same size and material but different weights. Both are pulled across the same surface. Which experiences a greater frictional force? The heavier box does. Apparently weight affects friction. Suppose a rectangular box is dragged across a surface, first on a small side, then on a large side. In which case is the frictional force greater? It is the same in both cases. The frictional force is not affected by the area of contact between the surfaces. Now suppose a box is pulled across a level surface and then across the same surface inclined to the horizontal. In which case is the force of friction greater? It is greater with the level surface.

One force that is related to weight and that decreases on an incline is the normal force. It seems reasonable to assume that the normal force and the frictional force are related. In fact, scientists have found experimentally that

$$f \propto N.$$

This proportion can be written as an equation by adding a proportionality constant:

$$f = \mu N \tag{8.2}$$

The constant, the Greek letter mu (μ), is called the **coefficient of friction.** There is a characteristic coefficient for each pair of materials that must be found by experiment.

8.10 Kinetic Friction

The force of friction on an object depends not only on the object's weight and the material it is made of but also on the object's state of motion. You may have noticed this already. For example, more force is required to start a heavy box moving than to keep it moving. Therefore, for each set of materials there are two coefficients of friction: the coefficient of moving or kinetic friction (μ_k) and the coefficient of static friction (μ_s). Usually, $\mu_s > \mu_k$. For **kinetic friction (f_k),**

$$f_k = \mu_k N. \tag{8.3}$$

Kinetic friction has several properties that generally hold true, although there are exceptions. The kinetic frictional force

- is oriented parallel to the contact surface.
- opposes the motion of the system of interest.
- depends in some ways on the kinds of materials in contact and the condition of the surfaces. This property tends to be highly variable and dependent on other factors.
- is generally *independent* of the relative speed of the sliding surfaces as long as the characteristics of the surfaces remain unchanged (that is, no phase,

Problem-Solving Strategy 8.6

Only the absolute values of friction and normal force may be used in the friction formula. The formula is *not* a vector equation.

The **coefficient of friction** (μ) is the constant of proportionality between the magnitude of a system's normal force and the magnitude of friction. Because friction depends on the materials in contact, the frictional coefficient is unique for each pair of materials.

Problem-Solving Strategy 8.7

The *magnitudes* of variable vectors are represented by the italicized form of the vector symbol, (f and N for example). The *scalar component* symbols for friction and normal force will be f_x and N_y, respectively. These forces have only one nonzero component if they are parallel to the reference frame coordinate axes.

The constant of proportionality between the normal force on a sliding system and kinetic friction is called the *coefficient of kinetic friction,* μ_k.

structural, or chemical change occurs that might alter the surface material properties).

- is generally *independent* of the surface area of contact between two objects.
- is directly *proportional* to the normal force acting on the sliding object.

8.11 Static Friction

However, for **static friction (f_s)** (friction between stationary objects), it is not necessarily true that $f_s = \mu_s N$. When two surfaces attempt to slide against each other due to a shearing force, friction will prevent them from sliding until the force parallel to the surfaces exceeds the static friction. In other words, static friction can vary from zero up to a maximum value, $f_{s\ max}$. As a mathematical expression,

$$0 \le f_s \le f_{s\ max}.$$

As long as the force parallel to the surfaces is less than $f_{s\ max}$, static friction will exactly cancel out the applied force and the surfaces will not slip. Only when $F > f_{s\ max}$ will the surfaces begin to slide. The magnitude for *maximum* static friction is found from Equation 8.2:

$$f_{s\ max} = \mu_s N \tag{8.4}$$

Static friction also has properties that are generally true. Experimentation has shown that static friction

- can be any value between zero and a maximum value characteristic for the two materials in contact.
- is oriented parallel to the contact surface.
- opposes the motion of the system of interest.
- depends on the kinds of materials and condition of the contact surfaces.
- is normally independent of the contact surface area.

EXAMPLE 8-9

Where's the Rub?: Coefficient of Friction

A block of vulcanized rubber weighing 5.00 N rests on a horizontal sheet of clean glass. The rubber block has a screw eye of negligible weight inserted in one side. A spring scale is hooked into the screw eye and the assembly is dragged across the glass at a constant velocity (see Figure 8-20). The spring scale indicates a force of 5.35 N. What is the coefficient of kinetic friction for vulcanized rubber on glass?

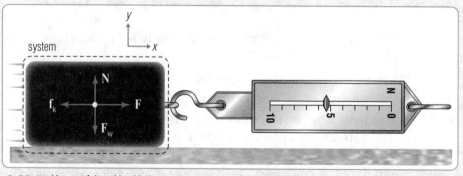

8-20 World view of the rubber block resting on a glass surface

Solution:

Choose a Cartesian coordinate system with the x-axis parallel to the glass surface and the y-axis perpendicular to the surface.

The velocity of the rubber block is constant, so $\mathbf{a} = 0$ m/s^2. Therefore the net force on the block must be zero according to Newton's first law.

The force sum on the block is

$$\Sigma\, \mathbf{F} = \mathbf{F} + \mathbf{f}_k + \mathbf{N} + \mathbf{F}_w = 0 \text{ N}.$$

Analyze the y-components to find N.

$$\Sigma\, F_y = F_y + f_{k\,y} + N_y + F_{w\,y} = 0 \text{ N} \tag{1}$$

In the vertical direction, F_y and $f_{k\,y}$ have zero components. Equation 1 reduces to

$$N_y + F_{w\,y} = 0 \text{ N}$$

$$N_y = -F_{w\,y}.$$

The magnitude of the normal force is just the magnitude of the weight of the block:

$$N = F_w$$

In the x-direction,

$$\Sigma\, F_x = F_x + f_{k\,x} + N_x + F_{w\,x} = 0 \text{ N}. \tag{2}$$

The components N_x and $F_{w\,x}$ are zero, so Equation 2 reduces to

$$F_x + f_{k\,x} = 0 \text{ N, and}$$

$$f_{k\,x} = -F_x.$$

The magnitude of the kinetic friction force equals the magnitude of the pulling force on the block:

$$f_k = F$$

Solve Equation 8.3 for the coefficient of kinetic friction and substitute the known quantities:

$$\mu_k = \frac{f_k}{N} = \frac{F}{F_w} = \frac{5.35 \text{ N}}{5.00 \text{ N}}$$

$$\mu_k = 1.07 \qquad (\mu_k \text{ is dimensionless})$$

The coefficient of kinetic friction between rubber and glass is 1.07.

The next example shows how friction opposes the inception of motion and other forces causing motion.

EXAMPLE 8-10

Giving a Shove: Static Friction

A metal crate having a mass of 100. kg rests on the smooth concrete floor of a warehouse. The coefficients of friction are $\mu_s = 0.94$ and $\mu_k = 0.83$. A worker is able to exert a horizontal force $F = +1250$ N. (a) Can he push the crate across the floor without assistance? In other words, is $F > f_{s\,max}$? (b) If so, what is the crate's acceleration?

Solution:

a. On a horizontal surface, the magnitude of the normal force on a system is just the magnitude of its weight, $F_w = |mg|$. The maximum static friction that can be generated by the crate on the concrete is as follows:

$$f_{s\,max} = \mu_s N = \mu_s |mg|$$
$$f_{s\,max} = (0.94)|(100.\text{ kg})(-9.81 \text{ m/s}^2)|$$
$$f_{s\,max} \doteq 9\underline{2}2 \text{ N}$$

Since the force the man exerts is greater than the maximum static friction, the man can move the crate without help.

b. Once the crate begins to move, kinetic friction takes over. The horizontal second-law force sum on the crate is

$$\Sigma \mathbf{F}_x = F_x + f_{k\,x} = m\mathbf{a}_x. \tag{1}$$

Kinetic friction is determined from Equation 8.3:

$$f_k = \mu_k N = \mu_k |mg|$$
$$f_k = (0.83)|(100.\text{ kg})(-9.81 \text{ m/s}^2)|$$
$$f_k \doteq 8\underline{1}4 \text{ N}$$

Substitute the known values into Equation 1, taking into account the relative directions of the force vectors:

$$F_x + f_{k\,x} = ma_x$$
$$(+1250 \text{ N}) + (-8\underline{1}4 \text{ N}) = (100.\text{ kg})a_x$$
$$a_x = \frac{+4\underline{3}6 \text{ N}}{100.\text{ kg}}$$
$$a_x \doteq +4.\underline{3}6 \text{ m/s}^2$$
$$a_x \approx +4.4 \text{ m/s}^2$$

8C   Section Review Questions

1. What is friction? Discuss several methods for reducing friction between two materials.

2. Write the formula for kinetic friction. Why can the formula *not* be expressed as a vector equation? (*Hint:* Consider the directions and magnitudes of the vector quantities in such an equation.)

3. Why is the coefficient of static friction usually larger than the coefficient of kinetic friction for the same materials?

4. Why is the static friction between two materials not constant?

⊚5. A block with a mass of 9.00 kg is pulled at a constant speed across a horizontal tabletop with a spring scale. The scale reads 61.8 N. Calculate the coefficient of kinetic friction.

⊚6. A warehouse worker is unloading a 240. kg crate from a truck using an "ideal rope" passing over a single ideal pulley above the truck. What is the minimum horizontal frictional force that the floor must exert on the worker to keep him from slipping while slowly lifting the crate if the rope between the pulley and the man makes an angle of 25.0° with the vertical?

8.12 Rolling Friction

Friction for rolling objects is not the same as sliding friction. When an object such as a wheel rolls, every point on the rim of the wheel is in motion relative to the surface on which it is rolling *except* the point where the wheel touches the surface. If the point of contact were completely frictionless, then the wheel would just slide along the road surface and no external torque could be applied to the wheel to roll it. In the real world, some friction always exists between the wheel and the road, causing the wheel to roll. This is traction friction working in reverse.

What is the magnitude of traction on a wheel of a vehicle that is propelled in some way? The applied force (F_{app}) can be either an external force such as a push or gravity, or it can be a force internal to the vehicle, such as a motor or engine. Such forces can be represented as a force vector on the wheel in the direction of application.

When a wheel turns, there is always some friction present as its axle rubs in its bearings. In a more complex vehicle, such as a car, there are many points in the car's *drive train* where friction reduces the force that the engine supplies to the wheels. The sum total of all these points of friction that retard the freedom of motion of the wheel is called **rolling friction (f_r).**

The magnitude of the net propelling force that the wheel applies to the road surface (F_{prop}) is the difference between the magnitudes of the applied force and the rolling friction on the wheel.

$$F_{prop} = F_{app} - f_r$$

The propelling force can be represented diagrammatically by the sum of the external force applied to the wheel by the vehicle, pointing in the direction of motion, and the rolling friction force opposing the applied force, pointing opposite the direction of motion. Since

$$\mathbf{F}_{prop} = \mathbf{F}_{app} + \mathbf{f}_r,$$

the method shown in Figure 8-22 does not introduce errors, and this notation will be useful in later chapters.

The magnitude and direction of the traction force on the wheel depends on the source of the applied force and magnitude of rolling friction. If the applied force is a force external to the vehicle (such as gravity), then the traction vector is equal to just the rolling friction vector. If the applied force includes a torque on the axle of the wheel supplied by the engine, then the magnitude of the traction force is equal to that of the propulsion force in the

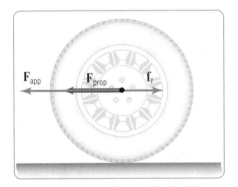

8-22   The net propelling force on the wheel of a vehicle

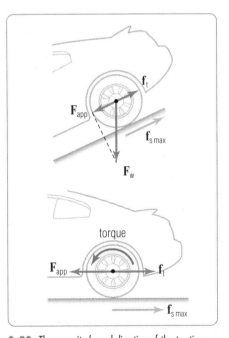

8-23   The magnitude and direction of the traction vector depends on the origin of the forces turning the wheel.

One important source of rolling friction in a car is the car's braking system.

8-21   Brakes in a car purposely increase rolling friction in order to slow the car when required.

Just as rolling friction slows forward motion, the love of money will impede your spiritual growth. "For the love of money is the root of all evil: which while some coveted after, they have erred from the faith, and pierced themselves through with many sorrows" (I Tim. 6:10).

Inexperienced drivers in snow often fail to account for the different orientations of the traction vector when coasting and when applying power to the wheels. A car idling as it coasts down an icy hill can begin to slide if the rolling friction exceeds the static friction available between the tires and the road. Adding a *little* power from the engine while descending the hill can often prevent a skid by reducing the rolling friction to within the traction limits available.

absence of any externally applied forces. No slipping of the wheel will occur as long as the magnitude of the traction force does not exceed the maximum static friction that can exist between the wheel and the road.

EXAMPLE 8-11

On a Roll

A 2.0 kg wheeled laboratory cart rolls without slipping down an inclined plane that makes an angle of 35.0° with the horizontal. If the magnitude of rolling friction between the wheels and axles of the cart is 0.10 N, what is the cart's acceleration?

Solution:

The coordinate system is aligned as in Figure 8-24 (b).

The cart is propelled by gravity. To find the cart's acceleration, you have to know the net force acting on the cart:

$$\Sigma \mathbf{F} = m\mathbf{a}$$

$$\mathbf{f_t} + \mathbf{N} + \mathbf{F_w} = m\mathbf{a}$$

Since motion is in the x-direction, evaluate the x-components:

$$f_{t\,x} + N_x + F_{w\,x} = ma_x \qquad (1)$$

Since $f_{t\,x} = f_{r\,x}$, $N_x = 0$, and $F_{w\,x} = mg \sin \theta$:

$$f_{r\,x} + mg \sin \theta = ma_x$$

$$a_x = \frac{mg \sin \theta + f_{r\,x}}{m}$$

$$a_x = \frac{(2.0 \text{ kg})(-9.81 \text{ m/s}^2)\sin 35.0° + (-0.10 \text{ N})}{2.0 \text{ kg}}$$

$$a_x \doteq -5.\underline{6}7 \text{ m/s}^2$$

$$a_x \approx -5.7 \text{ m/s}^2$$

The cart's acceleration is about 5.7 m/s² down the incline.

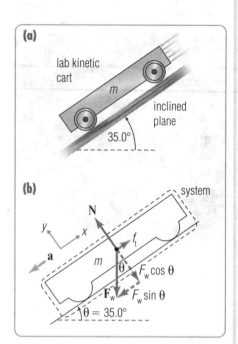

8-24 (a) World diagram of rolling cart on an incline; (b) Free-body diagram of the cart

8.13 Putting It All Together

Now that you have learned the process of free-body analysis, the property of inclines, and the effects of friction, let's look at several examples that combine all of these aspects. The example problems are lengthy, but if you study them carefully you will better appreciate how multiple forces acting on a system determine its motion.

Use the following approach for analyzing inclined-plane dynamics:

a. Assign coordinate systems to each system element so that the x-axis is aligned to the sliding surface and pointing *up* the slope. If other inclined or horizontal surfaces are involved, the x-axes should all be oriented parallel to the respective surfaces and point in the same general direction.

b. Resolve all forces acting on each element of the system into their components relative to the coordinate system for that system element.

c. Determine the maximum static friction possible for the two materials at the angle of incline.

d. Sum the nonfriction forces parallel to the sliding surface for the entire system and compare to the maximum static friction for the system to determine the dynamic state of the system—if the nonfriction forces are less than $f_{s\,max}$, the system remains at rest; if greater, the system is accelerating.

e. If the system is accelerating, calculate the kinetic friction.

f. Sum the x-component forces, including kinetic friction, to find acceleration according to Newton's second law.

EXAMPLE 8-12

Incline, Strings, and Friction

Steel blocks $m_1 = 2.50$ kg and $m_2 = 15.0$ kg rest on horizontal and inclined surfaces as shown in Figure 8–25a. The incline is 38.0° with the horizontal. The blocks are connected by an ideal string passing over an ideal pulley. For this system, $\mu_s = 0.61$ and $\mu_k = 0.50$. If the blocks are initially held at rest, (a) determine whether the blocks will move when released, and (b) if so, calculate the acceleration.

Solution:

a. Assign coordinate systems to each block. The x- and y-axes will be aligned as in Figures 8-25b and c. Note that if the system does slide, the motion of each block will be parallel to the x-direction.

If the system accelerates, both blocks will have the same acceleration (**a**) since they are connected by an ideal string.

b. Define the force sums on each block.

Force sum on block 1:

$$\Sigma\,\mathbf{F}_1 = m_1\mathbf{a}$$

$$\mathbf{T} + \mathbf{f}_1 + \mathbf{N}_1 + \mathbf{F}_{w1} = m_1\mathbf{a}$$

$$\mathbf{T} + \mathbf{f}_1 + \mathbf{N}_1 + m_1\mathbf{g} = m_1\mathbf{a}$$

Force sum on block 2:

$$\Sigma\,\mathbf{F}_2 = m_2\mathbf{a}$$

$$\mathbf{T}' + \mathbf{f}_2 + \mathbf{N}_2 + \mathbf{F}_{w2} = m_2\mathbf{a}$$

$$\mathbf{T}' + \mathbf{f}_2 + \mathbf{N}_2 + m_2\mathbf{g} = m_2\mathbf{a}$$

c. Determine the magnitude of the normal forces acting on the blocks. The normal force is needed in order to calculate maximum theoretical static friction on each block. Rewrite the force sum equations using the y-scalar components.

$T_y + f_{1\,y} + N_{1\,y} + m_1g_{1\,y} = m_1a_y$

The components T_y, $f_{1\,y}$, and a_y are all zero, and $g_{1\,y} = g$. The equation reduces to

$N_{1\,y} + m_1g = 0$ N and

$$N_{1\,y} = -m_1g$$

$$N_1 = |-m_1g|$$

$$f_{s\,max\,1} = \mu_s N_1 = \mu_s|-m_1g|. \quad (1)$$

$T'_y + f_{2\,y} + N_{2\,y} + m_2g_{2\,y} = m_2a_y$

The components T'_y, $f_{2\,y}$, and a_y are all zero, and $g_{2\,y} = g\cos\theta$ because of the rotated reference frame. The equation reduces to

$N_{2\,y} + m_2g\cos\theta = 0$ N

$$N_{2\,y} = -m_2g\cos\theta$$

$$N_2 = |-m_2g\cos\theta|$$

$$f_{s\,max\,2} = \mu_s N_2 = \mu_s|-m_2g\cos\theta|. \quad (2)$$

d. Evaluate the forces parallel to the sliding surface. The down-slope gravitational force on block 2 ($m_2g\sin\theta$) is the only force available to overcome the frictional forces ($f_{1\,x}$ and $f_{2\,x}$) and accelerate the system. If the x-weight component is less than the sum of the theoretical maximum static friction forces on blocks 1 and 2, then the blocks will not move.

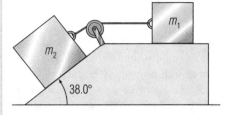

8-25a World diagram of blocks and planes for Example 8-12

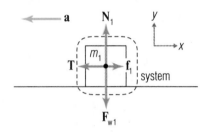

8-25b Free-body diagram for block 1

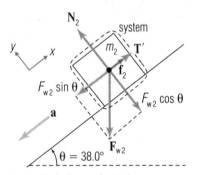

8-25c Free-body diagram for block 2

The total theoretical maximum static friction $f_{s\ max}$ for the system of blocks is

$$f_{s\ max} = f_{s\ max\ 1} + f_{s\ max\ 2}. \tag{3}$$

Substitute Equations 1 and 2 into Equation 3, and solve for $f_{s\ max}$:

$$f_{s\ max} = \mu_s|-m_1 g| + \mu_s|-m_2 g\ \cos\theta|$$

$$f_{s\ max} = \mu_s\big[|-g|\,m_1 + |-g|\,m_2\cos\theta\big]$$

$$f_{s\ max} = \mu_s|-g|\,(m_1 + m_2\cos\theta)$$

$$f_{s\ max} = (0.61)(9.81\ \text{m/s}^2)\big[2.5\ \text{kg} + (15.0\ \text{kg})\cos 38.0°\big]$$

$$f_{s\ max} = (0.61)(9.81\ \text{m/s}^2)(14.\underline{3}2\ \text{kg})$$

$$f_{s\ max} \doteq 85.\underline{6}\ \text{N}$$

The down-slope gravitational force on block 2 is as follows:

$$F_{w2\ x} = m_2 g\ \sin\theta$$

$$F_{w2\ x} = (15.00\ \text{kg})(-9.81\ \text{m/s}^2)\sin 38.0°$$

$$F_{w2\ x} = -90.\underline{5}9\ \text{N}$$

Comparing gravitational force to maximum system static friction, $|F_{w2\ x}| > f_{s\ max}$, so the blocks will accelerate.

e. Determine the *x*-scalar component force-sums on the blocks. These components determine the acceleration of the system:

$$T_x + f_{1\,x} + N_{1\,x} + m_1 g_{1\,x} = m_1 a_x \qquad\qquad T_x' + f_{2\,x} + N_{2\,x} + m_2 g_{2\,x} = m_2 a_x$$

The components $N_{1\,x}$ and $g_{1\,x}$ are zero. The equation reduces to

$$T_x + f_{1\,x} = m_1 a_x$$

$$T_x = m_1 a_x - f_{1\,x}. \tag{4}$$

The component $N_{2\,x}$ is zero. $g_{2\,x} = g\ \sin\theta$ due to the rotated reference frame.

The equation reduces to

$$T_x' + f_{2\,x} + m_2 g\ \sin\theta = m_2 a_x. \tag{5}$$

The tension on each block transmitted by the ideal string is related by the equation

$$T_x' = -T_x.$$

Combine Equations 4 and 5 after making appropriate substitutions:

$$-T_x + f_{2\,x} + m_2 g\ \sin\theta = m_2 a_x \qquad\qquad \text{(Equation 5)}$$

$$-(m_1 a_x - f_{1\,x}) + f_{2\,x} + m_2 g\ \sin\theta = m_2 a_x \qquad \text{(Substitute Equation 4 into 5)}$$

$$-m_1 a_x + f_{1\,x} + f_{2\,x} + m_2 g\ \sin\theta = m_2 a_x \qquad\qquad \text{(Remove parentheses)}$$

$$f_{1\,x} + f_{2\,x} + m_2 g\ \sin\theta = m_1 a_x + m_2 a_x \qquad\qquad \text{(Transpose term)}$$

$$f_{1\,x} + f_{2\,x} + m_2 g\ \sin\theta = a_x(m_1 + m_2) \qquad\qquad \text{(Factor)}$$

$$\frac{f_{1\,x} + f_{2\,x} + m_2 g\ \sin\theta}{(m_1 + m_2)} = a_x \tag{6}$$

The friction terms in Equation 6 are the kinetic friction forces (f_k) on the blocks.

Mass is always positive and $\cos\theta$ is positive between 0° and 90°. Therefore, both terms may be factored from an absolute value expression without changing its value.

f. Calculate f_k for each block.

$$f_{1x} = f_{k\,1} = \mu_k N_1 = \mu_k |-m_1 g|$$

$$f_{k\,1} = (0.50)|-(2.50 \text{ kg})(-9.81 \text{ m/s}^2)|$$

$$f_{k\,1} \doteq 1\underline{2}.2 \text{ N}$$

$$f_{2x} = f_{k\,2} = \mu_k N_2 = \mu_k |-m_2 g \cos \theta|$$

$$f_{k\,2} = (0.50)|-(15.0 \text{ kg})(-9.81 \text{ m/s}^2)\cos 38.0°|$$

$$f_{k\,2} \doteq 5\underline{7}.9 \text{ N}$$

In which direction do the friction force vectors point? Using the same reasoning as in part *d*, we can assume that the blocks will accelerate in the negative *x*-direction. Since friction opposes motion, the values of $\mathbf{f}_k$ will be positive.

g. Calculate the system acceleration. Substitute the friction values into Equation 6 and solve for acceleration.

$$a_x = \frac{(+1\underline{2}.2 \text{ N}) + (+5\underline{7}.9 \text{ N}) + (15.0 \text{ kg})(-9.81 \text{ m/s}^2)\sin 38.0°}{(2.50 \text{ kg} + 15.0 \text{ kg})}$$

$$a_x \doteq -1.\underline{1}7 \text{ m/s}^2$$

$$a_x \approx -1.2 \text{ m/s}^2$$

Whew! The blocks in Figure 8-25a will accelerate to the left at a rate of 1.2 m/s².

EXAMPLE 8-13

Skiing Downhill

A skier and her equipment have a total mass of 63.0 kg. She skis on wet snow down a straight slope that makes a 6.0° angle with the horizontal. She pushes herself off with an initial velocity of 1.00 m/s down the 800. m long slope and reaches the bottom 47.5 s later. (a) What is her average acceleration? (b) What is the coefficient of friction between her skis and the snow?

Solution:

You can use the second equation of motion to find $\bar{a}$, Newton's laws to find the net horizontal and vertical forces, and Equation 8.3 to find μ_k.

Assume that positive-*x* is *up* the slope.

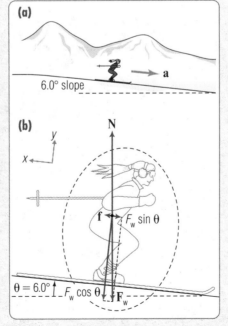

8-26 (a) World diagram of the skier; (b) Free-body diagram of the skier

a. Find $\bar{a}$:

$$d_x = v_{1x}\Delta t + \tfrac{1}{2}\bar{a}_x(\Delta t)^2$$

$$\tfrac{1}{2}\bar{a}_x(\Delta t)^2 = d_x - v_{1x}\Delta t$$

$$\bar{a}_x = \frac{2(d_x - v_{1x}\Delta t)}{(\Delta t)^2}$$

$$\bar{a}_x = \frac{2\big[-800.\ \text{m} - (-1.00\ \text{m/s})(47.5\ \text{s})\big]}{(47.5\ \text{s})^2}$$

$$\bar{a} \doteq -0.66\underline{7}0\ \text{m/s}^2$$

$$\bar{a} \approx -0.667\ \text{m/s}^2 \ (\text{down the slope})$$

b. Determine the coefficient of friction:

$$\Sigma\,\mathbf{F} = m\mathbf{a}$$

$$\mathbf{F_w} + \mathbf{N} + \mathbf{f} = m\mathbf{a}$$

Evaluate the x-component forces:

$$F_{wx} + N_x + f_{kx} = ma_x$$

Since $N_x = 0$ N and $F_{wx} = mg_x = mg\sin\theta$,

$$mg\sin\theta + f_{kx} = m\bar{a}_x$$

$$f_{kx} = m\bar{a}_x - mg\sin\theta$$

$$f_{kx} = m(\bar{a}_x - g\sin\theta)$$

$$f_{kx} = (63.0\ \text{kg})\big[-0.66\underline{7}0\ \text{m/s}^2) - (-9.81\ \text{m/s}^2)\sin 6.0°)\big]$$

$$f_{kx} \doteq +22.\underline{5}8\ \text{N}\ (\text{up-slope}).$$

You know that $f_k = \mu_k N$. To find N, you must evaluate the y-components of the forces acting on the skier:

$$F_{wy} + N_y + f_{ky} = ma_y$$

From the problem's geometry, you know that $f_{ky} = 0$ N, and $a_y = 0$ m/s^2. $F_{wy} = mg_y = mg\cos\theta$. Substitute these values to calculate N_y:

$$mg\cos\theta + N_y = 0\ \text{N}$$

$$N_y = -mg\cos\theta$$

$$N_y = -(63.0\ \text{kg})(-9.81\ \text{m/s}^2)\cos 6.0°$$

$$N_y \doteq 61\underline{4}.6\ \text{N}$$

Solve for μ_k:

$$\mu_k = \frac{f_k}{N} = \frac{|f_{kx}|}{|N_y|} = \frac{22.\underline{5}8\ \text{N}}{61\underline{4}.6\ \text{N}}$$

$$\mu_k \doteq 0.036\underline{7}3$$

$$\mu_k \approx 0.0367$$

The coefficient of friction is a very slippery 0.0367.

8D   Section Review Questions

1. Discuss the steps for dynamically analyzing systems of connected objects.

2. Under what condition will kinetic friction slow a sliding object?

3. What causes a wheel to roll over a surface?

4. For a system of mass *m* sliding on a straight inclined surface, on what does its acceleration ultimately depend?

⊙5. A physics student is experimentally determining the coefficient of static friction for rubber on glass. She places a 150. g rubber stopper on a rectangular sheet of glass, then she slowly raises one end of the glass. She notes that the angle of the incline at the instant the rubber begins to slip is 54.0°. What is the coefficient of static friction for rubber on glass?

⊙6. A cute little 78 g Mongolian gerbil (*Meriones unguiculatus*) is placed on a smooth plastic ramp inclined at an angle of 40.0° to the horizontal. The coefficient of static friction between the gerbil and the ramp is 0.82. Describe the change in motion (if any) that occurs after the animal is released. (Assume that the gerbil does not actively contribute to the motion!)

Chapter Review

In Terms of Physics

free-body diagram	8.1	friction (**f**)	8.8
tension (**T**)	8.1	traction (**f**$_t$)	8.8
ideal string	8.2	coefficient of friction (μ)	8.9
ideal pulley	8.4	kinetic friction (**f**$_k$)	8.10
normal force (**N**)	8.5	static friction (**f**$_s$)	8.11
apparent weight	8.7	rolling friction (**f**$_r$)	8.12

Problem-Solving Strategies

8.1 (page 164)   Make accurate sketches of multi-body systems that include strings and pulleys. Indicate all known and unknown force vectors and their components in the coordinate system used for each body. Note unknown values such as acceleration, but do not assign a direction until after the analysis is completed.

8.2 (page 167)   The notation for tension vectors transmitted by strings connecting objects A and B is as follows: If the string exerts a tension **T** on object A, then it exerts a tension **T**′ on object B. |**T**| = |**T**′| and **T** = −**T**′ (opposite directions).

8.3 (page 168)   It is permissible to use different frames of reference to analyze separate objects in a system as long as the same coordinate axis in each reference frame is parallel to the motion of the system within the reference frame.

8.4 (page 170)   The vector of the force of gravity is always assumed to point straight down. However, the normal component of the object's weight varies with the incline of the surface.

8.5 (page 170)   For inclines, the conventional alignment for the positive *x*-axis is *up* the slope of the incline. This choice is completely up to you but is consistent with the convention that "up" is positive. If you should choose to have positive directions point down, be sure to change the sign of the negative scalar constant *g*.

8.6 (page 175)   Only the absolute values of friction and normal force may be used in the friction formula. The formula is *not* a vector equation.

8.7 (page 175)   The *magnitudes* of variable vectors are represented by the italicized form of the vector symbol (*f* and *N*, for example). The *scalar component* symbols for friction and normal force will be f_x and N_y respectively. These forces have only one nonzero component if they are parallel to the reference frame coordinate axes.

1. What kind of diagram isolates a system so that the forces acting on the system can be analyzed?

2. A bucket is suspended in a well shaft by a rope, which is wound around a windlass. A boy turns the crank of the windlass to lower the bucket. Draw a free-body diagram for the bucket.

3. Why can you not "push" a rope or string?

4. Discuss why it is advisable to assign different coordinate systems to the connected parts of a system that do not move in the same line.

5. How can you change the direction of a force transmitted by a string?

6. What is *always* true about a normal force?

7. A family of three enters an elevator: a 75 kg father, a 55 kg mother, and a 40 kg son.
   a. Compare the weight of the father when the elevator is accelerating upward to his weight when it is at rest.
   b. Compare the normal force on the mother when the elevator is accelerating upward and when it is stationary.
   c. If a difference exists in either *a* or *b,* is the difference the same magnitude for all three family members or is the change proportional to some quantity? Explain your answer.

8. What kind of reference frame does a passenger inside an airliner experience during the plane's takeoff roll? Would you expect Newton's laws of motion to work normally during takeoff? Explain.

9. Explain why dragging a heavy trunk across the floor is harder if you pull horizontally than if you pull at an upward angle.

10. State two situations where kinetic or static friction is beneficial. State two situations where either kind of friction is undesirable.

11. What does friction generally depend on?
    a. surface area of contact
    b. the force pressing two surfaces together
    c. the speed of relative motion

12. What factors contribute most to kinetic friction?

13. Two wooden boxes of mass 500. g each are pulled across a wooden counter with a constant velocity of 10 cm/s. One box is a cube with faces of 64 cm² area. The other box is a rectangle, and the face contacting the counter has a surface area of 90 cm². Which box experiences a greater frictional force?

14. If $f_{s\ max}$ for an object resting on a horizontal surface is 10.0 N, what is the static friction acting on it if the sum of the *other* horizontal forces totals 5.5 N?

15. Which is usually greater for a pair of materials in contact, μ_s or μ_k? State at least one reason for your answer.

16. A concrete block rests on the bed of a pickup truck. The truck slowly accelerates so that the block does not slide relative to the bed. Name the force acting on the block that gives it a horizontal acceleration.

17. When you are rolling down an inclined sidewalk in your Radio Flyer wagon, what factors will determine the magnitude of your acceleration?

18. Compare the magnitudes of propelling force and traction when the tires on a car spin while trying to accelerate on a gravel surface.

True or False (19–28)

19. The total force on the last car in a railroad train is smaller than the total force on the first car when the train speeds up or slows down.

20. The normal force of a surface on a system is *always* perpendicular to the surface.

21. Astronauts aren't weightless in orbit; they just lack normal force. News reports err when they state that astronauts are weightless.

22. You wake up after a one-hour nap on an elevator. While yawning and stretching, you suddenly feel heavier. The elevator must be slowing to a stop.

23. According to Newton's third law of motion, when a skydiver jumps out of an airplane, the earth falls toward him with the same acceleration that he falls toward the earth.

24. A striped bass (*Morone saxatilis*) suspended from a hook at the end of a vertical fishing line experiences a normal force.

25. The formula for determining kinetic friction is *not* a vector equation.

26. The force of static friction on an object always has a magnitude of $f_s = \mu_s N$.

27. Static friction is an "intelligent force." Within certain limits, it "knows" exactly how large it needs to be in order to exactly balance an imposed nonfriction force.

28. Rolling friction on a wheel is exerted by the surface over which it rolls.

⊙29. A sewing thread will break if the tension in it exceeds 5.40 N. What is the maximum unaccelerated mass that this thread can support without breaking?

⊙30. A construction crew is roofing a house. They use a rope connected to an electric winch to pull up packages of shingles from the ground. The rope can transmit only 500. N of force without breaking. Each package of shingles has a mass of 50.0 kg.
   a. Draw a free-body diagram of one package of shingles as it is being lifted by the winch.
   b. With what maximum acceleration can the crew raise the shingles?
   c. If the shingles must be lifted 8.00 m, how long will the lift take at maximum acceleration?

⊙31. A 4.50 kg bowl of potato salad rests on a level picnic table.
   a. What is the normal force of the picnic table on the salad bowl?
   b. The picnic table tips so that it makes a 30.0° angle with the ground. Is the magnitude of the normal force greater than, less than, or equal to the value of the normal force when the table was level?

32. A frictionless playground slide makes an angle of 36.0° with the ground. It is 6.00 m long. A 22.5 kg child holds herself in place at the top of the inclined surface of the slide.

  a. Assume the girl is the system. Draw a free-body diagram of all of the forces acting on the girl-system.

  b. Find the scalar components of the force of gravity on the girl parallel to the slide and perpendicular to the slide.

  c. What normal force does the slide exert on her?

  d. What is the net force on the girl if she stops holding herself back?

  e. What is her resulting acceleration?

  f. How long does it take the girl to slide to the ground?

  g. What is her velocity at the end of the slide?

33. A man on a crowded elevator decides to stand on his newly purchased metric bathroom scale in order to allow more room. As the elevator accelerates downward at 0.200 m/s², the man looks down at the scale to see how well his one-day-old diet is coming. He sees that the scale reads 22.0 N less than he weighed the day before.

  a. If the man's mass is 110.0 kg, how much will the scale read when he steps on it in his house to show his wife his accomplishment?

  b. Will this be the same as the reading in the elevator?

  c. If not, how much weight did he really lose, if any?

34. In yet another elevator situation, a physics student receives permission from the building management to perform an experiment. She suspends a 1.00 kg mass from the ceiling with a spring scale. She hooks an identical mass to the bottom of the first with a second spring scale. The elevator accelerates downward at 2.00 m/s².

  a. Treat each mass as a system and sketch a free-body diagram for each during the interval of the acceleration.

  b. Predict the reading on each of the spring scales. (Ignore the masses of the scales.)

35. A 3.00 kg mass and a 5.00 kg mass are attached to the ends of a string over a pulley. They are both held in place so that they hang 1.00 m below the bottom of the pulley. When they are released, how much time will elapse before the smaller mass hits the pulley?

36. A cook finds that in order to pull his 4.59 kg cast iron skillet across his steel stove top, he must apply more than 18.0 N of force. If he exerts more than 18.0 N of force on the skillet, it moves. If he exerts 18.0 N of force or less, it does not move.

  a. What is the greatest value of the force of static friction?

  b. What is the coefficient of static friction for cast iron on steel?

37. A physics teacher holds a 4.00 kg textbook against a vertical wall with a horizontal force of 180. N so that the book does not slide down the wall. The coefficient of static friction between the wall and the book is 0.280.

  a. Determine the magnitude of the normal force of the wall on the book.

  b. What is the maximum possible static friction force on the textbook?

  c. If the teacher reduces his force on the textbook until it is just about to slip, what is the magnitude of the teacher's force at this point?

38. A 0.100 kg brass mass is placed on a horizontal steel surface. For brass on steel, μ_s is 0.510 and μ_k is 0.440.

  a. If the mass moves with a constant speed of 0.500 m/s, what is the friction on the mass?

  b. If the mass is not moving, what is the maximum theoretical static friction?

39. A car going 32.0 km/h on dry, level pavement jams on its brakes and skids to a stop. The car's mass is 1250 kg. The coefficient of kinetic friction for rubber on asphalt is $\mu_k = 0.400$.

  a. What is the net force on the car?

  b. What is the car's acceleration?

  c. How far does the car travel before stopping?

40. Two ox teams are competing in an ox pull at a state fair. Each must move a loaded sled to remain in the competition. The load on the sled is increased after each team has had an opportunity to try to move it. Team 1 can exert a maximum force of 1.50×10^4 N; team 2 can exert a maximum force of 1.80×10^4 N. The coefficient of friction between the metal sled and the gravel surface is $\mu_s = 0.50$.

  a. What is the mass of the greatest load that team 1 can move?

  b. What is the mass of the greatest load that team 2 can move?

  c. If mass is added in 500 kg increments, what is the mass of the first load that team 2 can pull but team 1 cannot?

41. A 9.50×10^3 kg locomotive is stopped on a track on a slope in the Rocky Mountains. The slope makes an angle of 15.0° with the horizontal. The locomotive's wheels are locked. If the coefficient of static friction between the locomotive's steel wheels and the steel track is 0.57, will the train slip?

42. You are approaching a level curve in the road in your Jaguar XKR. The curve has a uniform radius of 50.0 m. Assuming that the pavement is concrete, what is the maximum constant speed at which you can negotiate the turn without skidding?

43. A 75.0 kg professional dog walker in New York City is out airing two of his charges. All of a sudden, the large 54.5 kg rottweiler decides to head east while the smaller 29.5 kg basset hound heads west exerting an 18.0 N pull west. The human pulls west with 80.0 N in order to stop the rottweiler, but the whole group accelerates east at a uniform 0.30 m/s².

  a. If the rottweiler's leash makes an angle of 30.0° with the ground and the basset hound's leash makes an angle of 40.0° with the ground, what is the magnitude of tension in each leash?

  b. What is the normal force on the dog walker?

44. Prove that, for a system on an incline, the angle between its weight vector and the *y*-component of the weight vector is congruent to the incline angle. Use a formal two-column proof similar to the method discussed in GEOMETRY for Christian Schools, 2nd Edition.

FACETS of PHYSICS

Artificial Gravity

One of the reasons astronauts do so many science experiments in space is that they can study physical and biological processes without the influence of gravity, and the harvest of knowledge from these experiments shows great promise of being rich indeed. Why, then, are scientists interested in creating an artificial gravitational environment in space? There are two main reasons. One focuses on basic science research. By eliminating other variables, scientists can study how varying the simulated gravitational force affects biological and physical processes like plant growth and crystal formation. The International Space Station contains a centrifuge module designed to perform these kinds of experiments, which can generate simulated gravitational forces up to twice that of Earth. The other interest is in human physiology, as scientists try to find ways to keep astronauts healthy during long space flights. Our bodies are designed to operate in a "normal force" world, and in the free fall of space travel, astronauts experience a long list of physiological changes called *deconditioning*. These changes include immune system suppression, decreased blood volume and red blood cell count, loss of bone density, muscle atrophy (including the heart), as well as disorientation and space sickness caused by a lack of normal gravity-related sensory information.

If there is to be any hope of manned exploration to places farther than the Moon, scientists have to know how to keep the astronauts healthy enough to withstand the flight and strong enough to perform the mission when they arrive. There are two approaches to the problem: counteractive measures and preventive measures. Both have limitations. Counteractive measures include exercise and diet. Astronauts have been

exercising in space since the Gemini IV flight in 1965. While astronauts in the cramped quarters of the early space capsules exercised with stretchy bands, residents of the International Space Station spend several hours a day riding a stationary bike or running on a treadmill. This aerobic exercise strengthens the heart muscle as well as both upper and lower body muscle groups, and slows the loss of conditioning.

Many scientists are trying to prevent the ill effects of space travel by providing some sort of artificial gravity. The only way to do this is through circular motion. The centripetal acceleration necessary for circular motion is experienced as a normal force that your body cannot distinguish from the earthly gravitational force. An astronaut can experience simulated gravity if he is spun in a person-sized centrifuge for a few hours a day with his head toward the center and his feet at the perimeter. There are a few problems with this approach, however. The centripetal force experienced by your body depends on its tangential speed. Since each part of the body is traveling at a different tangential speed, there will be a considerable difference in the "gravity" experienced between the head and the feet. This gravity gradient lessens the effectiveness of the treatment. Also, it normally takes astronauts several days of space flight for their bodies to adjust to the disorientation of the sensory nervous system. It is not clear whether they would be able to tolerate repeatedly switching back and forth between artificial gravity and weightlessness.

The other idea, which is still only theoretical, is to generate constant artificial gravity by spinning the entire space station. This could take the form of either a spinning doughnut (called a torus) or two units joined by a tether that spin around the center of mass of the system. This idea has its own problems, however. Artificial gravity would not feel the same or act quite the same as natural gravity. If you are in a torus-shaped station, "down" is toward the rim. Thus, as you walked down the hallway, your head would be pointed toward the center of rotation. If you turned your head suddenly so that you were now looking toward the wall, the fluid in your ears would slosh to one side, causing vertigo similar to seasickness. Any kind of nodding or twisting of the head might cause balance problems.

It might be interesting to take an elevator ride from a docking station at the center of rotation to the outer edge of a rotating station. At the center the tangential speed would be zero, so there would be no normal force. You would have to float into the elevator and make sure that your feet were pointing "down"—at least where "down" would be when you arrived—and that you were facing in the direction of the spin. You will see why in a minute. Two things would happen as you traveled outwards: you would travel from the center to the rim, and you would accelerate from a tangential speed of zero at the center to the tangential speed of the rim. If the elevator had three stages in its outward journey—acceleration, constant velocity, and deceleration—what would you experience during the trip? First, because you would be weightless, as the elevator accelerated toward the edge, the ceiling would exert a force on you. While the elevator was traveling outward at constant velocity, the wall behind your back would be pushing on you, accelerating you to the local tangential speed. During the trip you would float down to the floor and start experiencing a force on your feet. As the elevator decelerated, the force on your feet would increase, and when you stopped, you would be able to walk out the door, having the same tangential speed as the hallway. You can see how important it would be to get yourself oriented properly before the ride. You wouldn't want to arrive upside down with your face pushed against the wall of the elevator!

The rotation of the station would create some other odd effects. Things that fall or are dropped wouldn't appear to fall straight down but would appear to curve in a direction opposite to the spin. If you held a ball four feet off of the floor, its tangential speed would be slower than that of your feet on the floor because the ball would be four feet closer to the center of rotation. If you dropped the ball, it wouldn't gain any tangential speed during the fall, which would really be in the outward direction. Because the floor would be traveling faster than the ball, rather than landing at your feet it would land a short distance away. Raising the release point would increase the difference in tangential speed relative to the floor and thus would increase the amount of deflection. You can see that in a rotating space station any kind of ball-related sports would be difficult to perform.

9A	Work	192
9B	Energy	198
9C	Total Mechanical Energy	207
Facet:	Hydroelectric Dams	212

New Symbols and Abbreviations

work (W)	9.2	watt (W)	9.8
joule (J)	9.3	kinetic energy (K)	9.9
spring constant (k)	9.6	potential energy (U)	9.10
power (P)	9.8	total mechanical energy (E)	9.17

Mount Saint Helens erupting

Work and Energy

> *Ah Lord God! behold, thou hast made the heaven*
> *and the earth by thy great power and stretched out arm,*
> *and there is nothing too hard for thee.* Jeremiah 32:17

When a person creates something, such as a piece of art or furniture, he expends considerable time and effort in order to be pleased with the final results. However, he is merely taking matter that already exists and re-arranging it in a more useful way. In contrast, God created the earth out of nothing and filled it with living things fully formed—an amazing display of His creative power. Yet, as you read this account in Genesis you never get the impression that God has to work very hard to accomplish His purposes. God merely had to speak, and light appeared. In the New Testament we read of Christ healing, or even raising the dead, with a touch or merely a word. Do you ever doubt God's power to work in your life, to help you overcome your sinful habits, or to intervene in any situation? Truly there is nothing too hard for God. Though you may at times not understand how, God is working to accomplish His purpose in your life if you are His child.

9.1 Introduction

On May 18, 1980, an explosion shattered the countryside in southwestern Washington. Mount Saint Helens erupted for the first time in more than one hundred years. Nearly 27 km^3 of solid rock and ice were hurled up to 4 km into the air. The volcano devastated more than 500 km^2 of the surrounding land. To cause the same amount of destruction using nuclear bombs, one Hiroshima-sized bomb would have to be set off every second for seven and one-half hours. The neighborhood of the volcano was changed from a life-filled region to a wasteland. This explosion is an example of the unrestrained release of energy.

Even explosions, however, can be contained and made to do useful work. A gasoline automobile engine, for example, is powered by explosions. Gas and air flow into a cylinder as a piston compresses them into a small volume. When a spark ignites the gas, the mixture explodes, pushing the piston away from the spark plug. This push is the source of the force that turns the wheels of the car.

Both explosions illustrate one attribute of energy: **energy** has the ability to do work. That is, energy can move matter. The work may be useful, as in an engine, or destructive, as in a volcano's eruption. This chapter will discuss the relationship between work and energy.

One description of **energy** is the ability to do work.

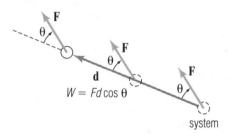

9-1 Relationship of force, displacement, and work

$W = Fd \cos \theta$

system

Problem-Solving Strategy 9.1

In order to determine whether work by a force is positive or negative, find the force's vector component that is parallel to the displacement of the system. If it is in the same direction as the displacement, work is positive. If the force vector component is opposite to the direction of displacement, then work is negative.

The quality of work is determined by the angle between the applied force **F** and the direction of displacement **d**:

If $0° \leq \theta < 90°$, work is positive.

If $\theta = 90°$, work is zero.

If $90° < \theta \leq 180°$, work is negative.

9.2 Defining Work

The word *work* can mean many different things. It can mean exceptionally hard labor, an artistic achievement, a job, or any physical or mental effort. In physics, however, it means the energy required to move an object. You can easily understand that it takes more energy to move a more massive object than it takes to move a less massive object. For example, you must use more energy to carry a 10 kg box 10 m than to carry a 5 kg box the same distance. Also, you use more energy to move the same object a longer distance. It takes more energy to move a box 10 m than to move the same box 5 m.

Considering these facts, it is not surprising that **work (W)** is defined as the product of the force component that is parallel to an object's motion and the distance that the object is moved. Mechanical work is done *by* a force *on* a system. It is important to identify the force for which work is to be calculated and to identify the system on which it acts. In symbols,

$$W \equiv Fd \cos \theta, \tag{9.1}$$

where W is the work done by a force **F** through a displacement **d.** It is not necessary for the force causing the work to be parallel to the displacement. The angle θ is the smallest angle ($\leq 180°$) between the force and displacement vectors when they are placed tail-to-tail. Notice that work itself is a scalar; it is merely a number and can be positive, negative, or zero. Work is the scalar product of the magnitude of the displacement (distance, d) and the magnitude of the force component that is parallel to the displacement ($F \cos \theta$).

The angle between the force on a system and the system's displacement determines the quality of work done. If $\theta < 90°$, Equation 9.1 is positive; therefore, positive work is done because the cosine of angles less than 90° is positive.

If $90° < \theta \leq 180°$, the cosine of the angle is negative, and work is therefore negative. Negative work by the force occurs when the force acting on the system opposes the motion of the system.

If the angle between the force and displacement is 90° (they are perpendicular), there is zero work done by the force on the system because $\cos 90° = 0$.

9.3 Units of Work

The units of work can be found from Equation 9.1. In the SI, F is measured in newtons and d is measured in meters.

$$W = (\text{newtons})(\text{meters}) = \text{N·m}$$

The SI unit of work and energy, the newton-meter, has been given a special derived unit name, the **joule (J).**

$$1 \text{ J} \equiv 1 \text{ N} \times 1 \text{ m}$$

The unit joule is utilized for all forms of energy and work. For mechanical work, the units N·m may be used. Be sure you do not confuse this unit with the unit of torque, the m·N, which is used only for torques.

9.4 Calculating Work

Any kind of force can do work. Contact forces as well as action-at-a-distance forces move objects and therefore do work. For example, a weightlifter exerts a contact force on a barbell that he does work on by lifting it, and the earth exerts the force of gravity on a falling rock that it does work on by pulling it down.

Although any kind of force *may* do work, not every force on an object *will* do work. For example, suppose a rock is at rest on a table. The earth exerts a gravitational force on the rock, but the rock does not move. How much work has the earth done on the rock?

$$W_g = F_g \times 0 \text{ m} = 0 \text{ J}$$

The earth does no work on the rock. Now suppose that the rock is pulled along the horizontal table by a string tied to it. The rock moves a distance of *d*. Now what is the work that the earth does on the rock?

$$W_g = F_{g\,x} d = F_w d \cos \theta$$

$$W_g = F_g d \cos 90°$$

$$W_g = F_g d \times 0 = 0 \text{ J}$$

The earth still does no work on the rock because the component of the force of gravity on the rock in the direction of the rock's motion is zero. The tension in the string, however, does positive work on the rock.

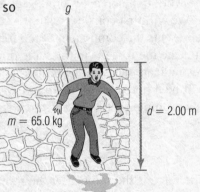

9-2 A rock pulled by a string and various forces that can do work on the rock

Problem-Solving Strategy 9.2

Gravity does positive work when a system loses height, and it does negative work when the system gains height.

EXAMPLE 9-1

Calculating Work: Vertical Force

A 65.0 kg physics student jumps off a 2.00 m high stone wall. How much work does the earth do on the student in his descent to the ground?

Solution:
$\mathbf{F}_g = \mathbf{F}_w$ and is in the same direction as **d**; so

$$W = F_w d \cos 0° = F_w d \times 1 = F_w d$$

$$W = |mg|d$$

$$W = |(65.0 \text{ kg})(-9.81 \text{ m/s}^2)|(2.00 \text{ m})$$

$$W \doteq 12\underline{7}5 \text{ kg·m}^2/\text{s}^2$$

$$W \approx 1280 \text{ N·m} = 1280 \text{ J}$$

g

m = 65.0 kg

d = 2.00 m

9-3 The earth doing work on a student

Problem-Solving Strategy 9.3

Whenever you must calculate the gravitational force on a system in order to compute work done by or against gravity, be sure to use the *magnitude* of gravitational force |*mg*|. The work formula uses the magnitudes of force and displacement, *not* their scalar values.

Recall that the component of the force parallel to the displacement is $F \cos \theta$.

EXAMPLE 9-2

Calculating Work: Force at an Angle to Displacement

A man pushes a lawn mower with a force of 30.0 N directed 60.0° below the horizontal. The lawn mower moves 20.0 m. How much work did the man do on the lawn mower?

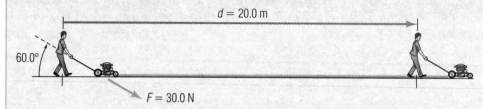

$d = 20.0 \text{ m}$

60.0°

$F = 30.0 \text{ N}$

9-4 A man is doing work with the lawn mower and on the lawn mower.

Solution:

$$W = Fd \cos \theta$$

The angle between the force and the displacement vectors is 60.0°.

$$W = (30.0 \text{ N})(20.0 \text{ m})\cos 60.0°$$

$$W \doteq 300.0 \text{ N·m}$$

$$W \approx 300. \text{ J}$$

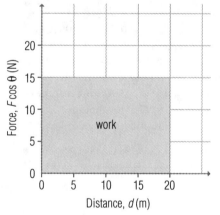

9-5 A force-distance graph

9.5 Determining Work Graphically: Constant Forces

Advanced methods of analysis often use graphs to show how two related variables determine a third. You have seen in previous chapters how the area under a curve approximates a quantity—finding displacement from the area under a velocity-time graph. Similarly, it is sometimes useful to plot the force doing work versus the displacement. As you may suspect, the area under such a graph approximates the work done on a system by the force.

We will first consider a simple case involving a constant force. The vertical axis represents the component of the force parallel to the displacement. Positive values indicate the force is in the same direction as the displacement, while negative values indicate the force is in the opposite direction. For constant forces, this graph will be a horizontal line. Figure 9-5 gives the force-distance graph for the previous example. The constant force is calculated as follows:

$$|F_x| = (30.0 \text{ N})\cos 60.0°$$

$$|F_x| = 15.0 \text{ N}$$

The force is the same after the mower has traveled 0 m, 5 m, 15 m, and any other distance up to $d = 20$ m. The "area" of the shaded block is

$$A = |F_x|d = W.$$

You can use the area under the force-distance curve to represent the work done. This is true for any force-distance curve. If you have trouble recalling the equation used to calculate work, you can use geometry to find the area under the force-distance curve. Remember, the physical size of the graphical area (e.g., in cm²) is not important. It is the product of the dimensions represented by the horizontal and vertical axes that matters.

9.6 Determining Work Graphically: Variable Forces

A force-distance graph is especially helpful for a varying force. For example, the external force required to stretch most springs is proportional to the displacement that the spring has been stretched. That is,

$$\mathbf{F}_{ex} \propto \mathbf{d},$$

where $\mathbf{F}_{ex}$ is the external force and $\mathbf{d} = \Delta x = x - x_0$. The position x_0 is the normal or relaxed length of the spring, called the **equilibrium position.** This observation is known as **Hooke's law.** It can be made into an equation by including a proportionality constant:

$$\mathbf{F}_{ex} = k\mathbf{d} \tag{9.2}$$

The proportionality constant k is called the **spring constant,** which is nearly constant for real springs. In a spring system, the motion and the force are assumed to act along the same one-dimensional coordinate axis, with the origin placed at the spring's equilibrium position. Notice that as a force is applied to stretch or compress the spring, the displacement is in the same direction as the external force, so work done *on* a spring by an external force is ~~always positive~~.

As we have done with other idealized dynamics analyses, we will rely on **ideal springs** in discussions involving Hooke's law. Ideal springs have no mass, and the value of k for the spring is truly constant throughout its range of displacements.

Many systems in nature act very similar to a spring that is displaced from an equilibrium position, particularly systems that tend to vibrate or oscillate. Any force that varies in proportion to the magnitude of displacement from an equilibrium position is called a **Hooke's law force.**

9.7 Work and Ideal Springs

How much work is done to stretch a spring from its equilibrium position by Δx? Plotting F_{ex} versus Δx on a graph, you get Figure 9-6. The shaded triangular area under the graph bounded by x_0 and x is the work on the spring between these two points. The area of the shaded triangle is as follows:

$$A = \tfrac{1}{2}bh$$
$$A = W_{ex} = \tfrac{1}{2}(k\Delta x)(\Delta x)$$
$$W_{ex} = \tfrac{1}{2}k(\Delta x)^2$$

Notice that the height of the triangle is the change in the x-coordinate times the slope of the line. In this case, the slope of the graph is the spring constant, k.

Figure 9-6 shows the work done *by* a force ($\mathbf{F}_{ex}$) *on* the spring. What about the force that a spring exerts? From Newton's third law, you know that the spring exerts a reaction force of

$$\mathbf{F}_s = -\mathbf{F}_{ex}.$$

If the spring is stretched by an external force, the force that the *spring* exerts is

$$\mathbf{F}_s = -k\mathbf{d} \,. \tag{9.3}$$

Notice that the spring force is opposite to the direction of displacements from its equilibrium position caused by an external force. Therefore, the work done *by* a spring on a system connected to it is negative under these conditions. This equation holds true whether the spring is stretched or compressed from its equilibrium position.

Robert Hooke (1635–1703) was an English scientist and inventor who pursued a wide range of interests, from microscopy to architecture. He is best known for first using the term "cell" in biology and for formulating Hooke's law in physics.

The SI units of k are N/m.

An ideal spring has no mass, and its spring constant is truly constant through any displacement of the spring.

Any force that is proportional to the displacement of a system from an equilibrium position is called a **Hooke's law force.** Systems that generate such forces include vibrating strings and bells.

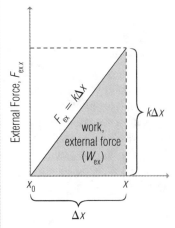

9-6 A force-distance graph for a force acting on a spring

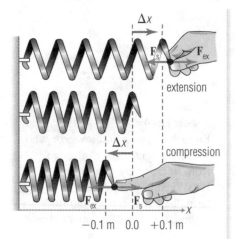

9-7 Comparison of forces on a stretched, relaxed, and compressed spring

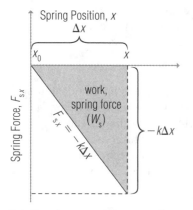

9-8 Force-distance graph for a spring force (Hooke's law)

Problem-Solving Strategy 9.4

You must take care that you calculate the work done by the force that is specified in the problem. If an object attached to a spring is the system of interest, the work done by an external force on the system is not the same as the work done by the spring force.

For example, if an ideal spring is attached to a wall on the left, as in Figure 9-7, and stretched to the right from $x_0 = 0$ m to $x = 0.1$ m, the external force on the spring must be directed to the right. The spring exerts a force toward the left, as you can see from $F_{s\,x}$:

$$F_{s\,x} = -kd_x = -k(x - x_0)$$
$$F_{s\,x} = -k(0.1 \text{ m} - 0 \text{ m}) = -0.1k \text{ N}$$

Since $F_{s\,x} < 0$, the spring force points to the left. If the spring is compressed to the left from $x_0 = 0$ m to $x = -0.1$ m, the external force on the spring is toward the left. The spring, therefore, exerts a force to the right.

$$F_{s\,x} = -kd_x = -k(x - x_0)$$
$$F_{s\,x} = -k(-0.1 \text{ m} - 0 \text{ m}) = +0.1k \text{ N}$$

How much work does an ideal spring do as it is stretched or compressed by an external force? Figure 9-8 gives the force-distance graph for a spring that is stretched. The shaded area represents the work done *by* the spring force.

$$W_s = A = \frac{1}{2}(-k\Delta x)(\Delta x)$$
$$W_s = -\frac{1}{2}k(\Delta x)^2 \tag{9.4}$$

Notice that the negative work region is *below* the horizontal axis of the graph.

Equation 9.4 clearly shows that the work done by the spring on a system in this situation is negative. The spring force acts on the system in the direction opposite to the system's displacement.

EXAMPLE 9-3

Springing to Work: Work by a Variable Force

A spring with a spring constant $k = 0.50$ N/m is stretched from its equilibrium position of $x_0 = 0.00$ m to a position of $x = +0.25$ m. (a) How much work is done on the spring by the external force (W_{ex})? (b) How much work does the spring do (W_s)?

Solution:
a. Work by the external force:

$$W_{ex} = \frac{1}{2}k(\Delta x)^2 = \frac{1}{2}(0.50 \text{ N/m})(+0.25 \text{ m} - 0.00 \text{ m})^2$$

$$W_{ex} \doteq 0.01562 \text{ J}$$

$$W_{ex} \approx 0.0156 \text{ J}$$

b. Work by the spring force:

$$W_s = -\frac{1}{2}k(\Delta x)^2 = -\frac{1}{2}(0.50 \text{ N/m})(+0.25 \text{ m} - 0.00 \text{ m})^2$$

$$W_s \doteq -0.01562 \text{ J}$$

$$W_s \approx -0.0156 \text{ J}$$

9.8 Power

The amount of work required for a warehouse worker to carry, one box at a time, a pallet full of new coffeemakers from the floor to a shelf 4 m above the floor is the same as for a forklift to raise the whole pallet at once. Neglecting the mass of the pallet, the only difference between the two instances is the amount of time needed to do the work. Up to now you have not been concerned with how long it takes to accomplish work, only that it occurs at all. Obviously, your employer would not be happy if your work ethic followed this principle! We use machines because they help us do work faster and with less effort. We will discuss simple machines in Chapter 10.

The time rate of work done on a system by a force is called the **power (P)** of the force. If the total work done over an interval of time is considered, then the *average power* ($\overline{P}$) is computed by

$$\overline{P} = \frac{\text{work done by } \mathbf{F} \text{ during } \Delta t}{\Delta t} = \frac{W}{\Delta t}. \tag{9.5}$$

The rate at which a force does work at any instant is called the *instantaneous power*, or just power.

Customers inquiring about power in a car engine often ask how fast the car can go. They intuitively associate power with speed. This relationship does, in fact, exist. We can expand Equation 9.5 to include the work equation

$$\overline{P} = \frac{W}{\Delta t} = \frac{Fd \cos \theta}{\Delta t}. \tag{9.6}$$

But $d/\Delta t = \overline{v}$ (or just v for uniform motion). Equation 9.6 then becomes

$$P = Fv \cos \theta. \tag{9.7}$$

Power is a scalar quantity that is proportional to the force acting on a system and the speed of the system.

The unit of power is obtained from the dimensions of work per unit time, consisting of the SI units of joules per second. Because power is such a frequently encountered quantity in physics, a derived unit called the **watt (W)** was created in honor of James Watt (1736–1819), a Scottish engineer who was prominent in the development of the first practical steam engine.

$$1 \text{ watt} = 1 \text{ W} = 1 \text{ J/s}$$

More power is needed to work quickly than to work slowly. Thus a belief that God created the universe in six days ascribes far more power to Him than a belief that He created it slowly, over millions of years.

The average power of a source of energy is the work accomplished during a time interval divided by the time interval:

$$\overline{P} = \frac{W}{\Delta t}$$

The SI unit of power is the **watt (W).**
$1 \text{ W} \equiv 1 \text{ J/s}$

EXAMPLE 9-4

Spring Power

How much power is needed to compress a 0.100 m spring with a spring constant $k = 1.25$ N/m to half its length (a) in 10.0 s? (b) in 5.0 s?

Solution:
Work by an external Hooke's law force is calculated from Equation 9.5.

a. Power for the 10 s compression:

$$\overline{P} = \frac{\frac{1}{2}k(\Delta x)^2}{\Delta t} = \frac{(1.25 \text{ N/m})(0.050 \text{ m} - 0.100 \text{ m})^2}{2(10.0 \text{ s})}$$

$$\overline{P} \doteq 1.56 \times 10^{-4} \text{ W}$$

$$\overline{P} \approx 1.6 \times 10^{-4} \text{ W}$$

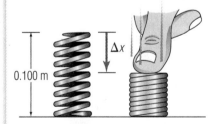

9-9 Compression of a spring

b. Power for the 5 s compression:

$$\bar{P} = \frac{(1.25 \text{ N/m})(0.050 \text{ m} - 0.100 \text{ m})^2}{2(5.00 \text{ s})}$$

$$\bar{P} \doteq 3.\underline{1}2 \times 10^{-4} \text{ W}$$

$$\bar{P} \approx 3.1 \times 10^{-4} \text{ W}$$

9A   Section Review Questions

1. **a.** To what physical quantities is mechanical work proportional?

   b. What kind of quantity is mechanical work?

2. How can you tell whether positive or negative work has been done by a particular force on a system?

3. How can you find the spring constant from a force-displacement graph of an ideal spring?

4. **a.** To what physical quantities is mechanical power proportional?

   b. What kind of quantity is mechanical power?

⊙5. If you slowly lift a 0.95 kg textbook 1.2 m vertically from your desk to a bookshelf, how much work have you done?

⊙6. A 1.95 kg female great horned owl (*Bubo virginianus*) silently swoops directly from a branch 12.4 m high in a tree to capture an unfortunate field mouse (*Apodemus sylvaticus*) on the ground—a 24.0 m distance from the owl's perch. Calculate the work done by gravity on the owl during its downward flight.

⊙7. A crane slowly lifts a 1140 kg load of steel girders vertically 28.0 m.

   a. How much work was done by the crane on the girders?

   b. How much work was done by gravity?

⊙8. A 950. kg car is initially moving at 95.0 km/h. If the car is stopped in 6.50 s with a force of 3860 N opposite to the direction of motion, what is the stopping power of the car's brakes in watts?

9B   ENERGY

9.9   Kinetic Energy

Mechanical work is performed on a system by a force in order to move the system from one place to another. Motion implies that the system has a velocity, and changes of motion imply accelerations. Moving objects can do work when they strike something else. Just consider the damage a flying baseball can do to a windowpane! Evidently, a moving object possesses the ability to do work. This mechanical energy associated with motion is called **kinetic energy (K).** Kinetic energy is a positive scalar quantity that has units of joules in the SI.

The **work-energy theorem** states that the total work done on a system by all the external forces acting on it is equal to the change in the system's kinetic energy.

$$W_{\text{total}} = \Delta K, \text{ or}$$

$$W_{\text{total}} = K_2 - K_1 \qquad (9.8)$$

From Newton's second law you know that

$$\Sigma \mathbf{F} = m\mathbf{a}.$$

Now the work equation can be rewritten as a first-law expression:

$$W_{\text{total}} = \Sigma F_m d = (ma)d, \qquad (9.9)$$

where ΣF_m is the sum of all force components parallel to the movement during the displacement, d. From the third equation of motion, you know that

$$d = \frac{v^2 - v_0^2}{2a}.$$

Multiply both sides by a.

$$ad = \frac{v^2 - v_0^2}{2} = \tfrac{1}{2}(v^2 - v_0^2)$$

Substitute for ad in the right side of Equation 9.9 and distribute the terms.

$$W_{\text{total}} = mad = m\tfrac{1}{2}(v^2 - v_0^2)$$

$$W_{\text{total}} = \tfrac{1}{2}mv^2 - \tfrac{1}{2}mv_0^2 \qquad (9.10)$$

Comparing Equation 9.8 to Equation 9.10, it seems logical to define kinetic energy as

$$K \equiv \tfrac{1}{2}mv^2. \qquad (9.11)$$

The change in kinetic energy is due to the work by *all* external forces acting on the system. Such forces include propelling forces from a rocket engine or a baseball bat, frictional forces, and the force of gravity. You will see how each of these affects kinetic energy in this and later chapters.

The notation F_m is a shorthand way of saying "the force component parallel to the motion that caused the displacement."

Problem-Solving Strategy 9.5

Kinetic energy can only be positive. For any given object, it varies as the square of its speed.

EXAMPLE 9-5

Take a Shot: Bullet Kinetic Energies

Calculate the kinetic energy of two 150 grain (0.00972 kg) 30-caliber rifle bullets loaded with different amounts of powder to produce different muzzle velocities. One bullet has a muzzle velocity of 703 m/s while another has a muzzle velocity of 952 m/s.

Solution:

$$K = \tfrac{1}{2}mv^2$$
$$K_1 = \tfrac{1}{2}(0.00972 \text{ kg})(703 \text{ m/s})^2$$
$$K_1 \doteq 2401 \text{ J} \approx 2.40 \times 10^3 \text{ J}$$
$$K_2 = \tfrac{1}{2}(0.00972 \text{ kg})(952 \text{ m/s})^2$$
$$K_2 \doteq 4404 \text{ J} \approx 4.40 \times 10^3 \text{ J}$$

9-10 A bullet's high speed gives it a high kinetic energy.

Observe that the second bullet has almost twice the kinetic energy of the first bullet even though its muzzle velocity is only a third greater.

9-11 Braking a heavy truck requires a lot of work.

Recall that ΣF_m in this example is the product of the net force and cos θ. Since θ = 180° (ΣF_m opposes motion), cos 180° = −1. Therefore, the value of ΣF_m is negative.

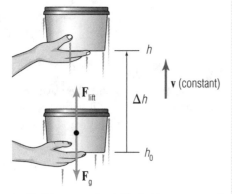

9-12 When a container is lifted with constant speed, the work done on the container does not change its kinetic energy.

Potential energy (U) is the energy associated with the position or condition of a system relative to a reference point of zero potential energy.

We are all born with talents and abilities that represent a potential for doing good, but this potential must be trained, taught, and directed. Common sense says we should cooperate with those willing to spend the time to teach and train us, but "The sluggard is wiser in his own conceit than seven men that can render a reason" (Prov. 26:16). The sluggard can find many foolish excuses to resist learning.

EXAMPLE 9-6

Braking: Changing Kinetic Energy

A 1.00×10^4 kg loaded truck traveling at a speed of 20.0 m/s brakes until its speed is 10.0 m/s. (a) How much work was done on the truck to slow it? (b) If the truck takes 50.0 m to slow, what was the total braking force applied?

Solution:

a. The total work done on the truck to slow it can be calculated from its change of kinetic energy:

$$W_{total} = \Delta K = \tfrac{1}{2}mv^2 - \tfrac{1}{2}mv_0^2$$
$$W_{total} = \tfrac{1}{2}m(v^2 - v_0^2)$$
$$W_{total} = \tfrac{1}{2}(1.00 \times 10^4 \text{ kg})\left[(10.0 \text{ m/s})^2 - (20.0 \text{ m/s})^2\right]$$
$$W_{total} = -1.50 \times 10^6 \text{ J}$$

The work done on the truck is negative because the truck lost energy as it slowed.

b. By definition,

$$W_{total} = \Sigma F_m d.$$

Divide both sides by distance to find the net force on the truck:

$$\Sigma F_m = \frac{W_{total}}{d} = \frac{-1.50 \times 10^6 \text{ kg} \cdot \text{m}^2/\text{s}^2}{50.0 \text{ m}}$$
$$\Sigma F_m = -3.00 \times 10^4 \text{ N}$$

The net braking force is negative because it opposes the motion of the truck.

9.10 Potential Energy

Not all work changes kinetic energy. Suppose you slowly lift a box from the floor at a constant speed to a height that will allow you to place it on a shelf. If the speed of the box is uniform throughout the lift, the kinetic energy of the box does not change. The work simply changes the position of the box. However, before you lifted the box, it had little potential for falling (unless there was a hole in the floor). Once on the shelf, the box has a greater potential for falling to the floor. If it were to fall, its kinetic energy would increase as gravity did work on the box. While it rests on the shelf, therefore, the box is said to have gravitational potential energy. **Potential energy (U)** is energy due to an object's condition or position relative to some reference point assumed to have zero potential energy. As with kinetic energy, the units of potential energy are joules.

There are other forms of potential energy besides gravitational potential energy. They include elastic potential energy, which comes from the forces between atoms, and electrical potential energy, which comes from the electrical forces between charges. Potential energy results from work done against a force. For example, the box was lifted to a shelf, increasing its gravitational potential energy by working against gravity. Similarly, a spring is stretched by working against the elastic (spring) force, thus increasing the elastic potential energy of the system it is connected to.

9.11 Conservative and Nonconservative Forces

Not every force has potential energy associated with it. Only forces that fall into the category of **conservative forces** have their own potential energy. A force is a conservative force if either of the following two conditions is true:

- The work done by the force on a system as it moves between any two points is independent of the path followed by the system.
- The work done by the force on a system that follows a closed path is zero. A system follows a closed path if it begins and ends at the same point in space.

Some examples of conservative forces are

- the gravitational force,
- any central force (a force acting between the centers of two systems),
- any Hooke's law force.

Conservative forces are so named because the energy expended when doing work against them is stored as potential energy and can be regained as kinetic energy.

Other forces do not have these attributes. If the work done by a force on a system over a closed path is *not* zero, or if the work done by a force on a system between two points depends on the path taken, then the force is a **nonconservative force.** Examples of nonconservative forces include

- kinetic frictional force,
- internal resistance forces (e.g., spring internal resistance),
- fluid drag.

When work is done against nonconservative forces, the energy is not stored as potential energy but is converted into other forms of mechanically unusable energy, such as heat, sound, or light.

9.12 Gravitational Potential Energy

All masses, from atoms to galaxies, exert gravitational forces on other masses. If the masses are separated, then the gravitational attraction tends to pull them together. Thus, two masses separated by some distance have the potential to be drawn together by the gravitational force exerted between them. The work required to move these masses apart against the force of gravity is their *gravitational potential energy.*

Near the earth's surface, the force of gravity on an object is nearly constant at $\mathbf{F}_g = m\mathbf{g}$. The work done against gravity in lifting an object at constant speed goes entirely to increasing the object's potential energy. Consider an elevator of mass m that is lifted with constant speed from a height h_1 to a height h_2. Since the elevator's velocity is constant, the acceleration is zero and there is no net force on the elevator. That is, the lifting force just balances the force of gravity, neglecting frictional forces. The lifting force must therefore have the same magnitude, but opposite direction, as the force of gravity:

$$\mathbf{F}_{\text{lift}} = -\mathbf{F}_g = -m\mathbf{g}$$
$$F_{\text{lift}} = |mg| \qquad \text{(upward)}$$

The work that the lift cables do against gravity in lifting the elevator is equal to the product of the force they exert on the elevator and the elevator's change in height.

$$W_{\text{lift}} = F_{\text{lift}}d = |mg|(h_2 - h_1)\cos 0°$$
$$W_{\text{lift}} = |mg|\Delta h$$

Problem-Solving Strategy 9.6

Different forms of potential energy are indicated by a subscript identifying the force to which the potential energy corresponds. For example, U_g is gravitational potential energy and U_s is elastic potential energy of a spring.

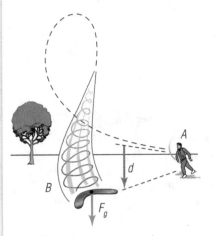

9-13 The work done on a system by a conservative force is independent of the path of the system.

Conservative forces can store work energy as some form of mechanical potential energy.

Nonconservative forces convert work energy into mechanically unusable forms of energy.

Pride sometimes makes us think we are immune to certain temptations. However, we all have the potential for any sin. "Wherefore let him that thinketh he standeth take heed lest he fall" (I Cor. 10:12).

Problem-Solving Strategy 9.7

Recall that the scalar value of gravitational acceleration is $g = -9.81$ m/s^2. Use the absolute value of mg when determining work involving weight or gravitational forces.

Since the force of gravity opposes the lifting force, their components along the direction of the elevator's motion are opposite. Therefore, the work that gravity does on the elevator as it rises is negative.

$$W_g = F_g d = |mg|(h_2 - h_1)\cos 180°$$
$$W_g = -|mg|\Delta h \qquad (9.12)$$

The elevator's potential energy changes by the amount of work the lift cables do on the elevator:

$$\Delta U_g = W_{\text{lift}} = |mg|\Delta h \qquad (9.13)$$

Does it make any difference whether the change in height begins at the earth's surface or at 10 m above the surface? No, as long as the box stays near enough to the earth's surface so that $F_g = |mg|$.

If you compare Equations 9.12 and 9.13, you will notice a fundamental relationship between the work done by a force and the potential energy associated with that force. In the case of the gravitational force,

$$W_g = -\Delta U_g. \qquad (9.14)$$

Equation 9.14 is true for any conservative force. In words, Equation 9.14 states that work must be done *against* a force in order to increase the potential energy of a system with respect to that force. Conversely, as potential energy of a system decreases with respect to a force, the force does positive work on the system.

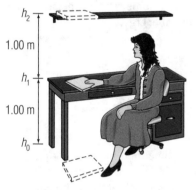

9-14 Lifting the book increases its gravitational potential energy.

EXAMPLE 9-7

Greater Potential: Potential Energy

A woman lifts a 0.750 kg book from the floor to her 1.00 m high desk with a constant velocity. Later she moves the book at a constant velocity to her bookshelf, 2.00 m above the floor. (a) How much work did she do in the first lift? (b) In the second? (c) What is the change in the book's gravitational potential energy for each lift?

Solution:
a. Work for the first lift:

$$W_1 = Fd_1 = |mg|\Delta h_1$$
$$W_1 = |mg|(h_1 - h_0)$$
$$W_1 = |(0.750 \text{ kg})(-9.81 \text{ m/s}^2)|(1.00 \text{ m} - 0.00 \text{ m})$$
$$W_1 \doteq 7.3\underline{5}7 \text{ J}$$

b. Work for the second lift:

$$W_2 = |mg|(h_2 - h_1)$$
$$W_2 = |(0.750 \text{ kg})(-9.81 \text{ m/s}^2)|(2.00 \text{ m} - 1.00 \text{ m})$$
$$W_2 \doteq 7.3\underline{5}7 \text{ J}$$

Notice that $W_1 = W_2$.

c. Since it is assumed that the lifts were made at constant speeds, the change of kinetic energy is zero. No work was expended in changing kinetic energy. All the work went into changing gravitational potential energy.

Therefore:

$$W_{\text{lift}} = \Delta U_g$$

$$\Delta U_{g\,1} = W_1 = 7.36 \text{ J}$$

$$\Delta U_{g\,2} = W_2 = 7.36 \text{ J}$$

The *change* in potential energy is the same as long as the change in height is the same for locations near the earth's surface.

Notice in part *c* of the example that the work by the *lifting force* is being compared to the change in potential energy with respect to the *gravitational force*. In this case, both are positive. The work of one force on a system often causes a change in potential energy with respect to another force also acting on the system. You must be careful when working such problems so that each force is correctly associated with its corresponding work or energy.

9.13 Relative Potential Energy

What is the gravitational potential energy of the book in Example 9-7? The gravitational potential energy of a system is measured by the amount of work gravity can do on it in the absence of any other restraining forces. Since the force of gravity on an object is constant as long as it is near the earth, the work gravity does varies only with the distance the object can fall. So we could say that

$$U_g = |mg|h, \tag{9.15}$$

where *h* is the distance that the object can fall.

How far can an object fall? Figure 9-15 points out the difficulty in answering this question. A rock rests on the edge of a cliff. What is its potential energy? It depends on your *height reference point*—the point from which you measure *h*. If you believe the rock cannot fall at all because it is firmly wedged in at the top of the cliff, then $h = 0$ m and $U_g = 0$ J. If you expect that the rock could fall to the tree, then the tree is the height reference and the rock is at $h = +10$ m; so $U_g = 10|mg|$ J. If you think it's possible that the rock could fall to the surface of the water, then the water's surface is the height reference. The rock's height is $h = +40$ m and $U_g = 40|mg|$ J. If you allow the rock to fall to the bottom of the river, then its height is $h = +50$ m and $U_g = 50|mg|$ J. Which value of the potential energy is "correct"? Each of the values of potential energy is relative to the selected reference point. Since you need to know only the *change* in potential energy, which reference point you choose does not matter. However, after choosing a reference point, you should use it consistently throughout the problem.

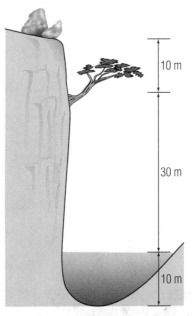

9-15 The potential energy of the rock depends on the point from which you measure *h*.

Problem-Solving Strategy 9.9

When solving gravitational potential energy problems, choose a height reference point that makes sense to you and simplifies the problem.

EXAMPLE 9-8

Loss of Potential Energy: Falling Objects

A 6.50 kg bowling ball falls from an observation deck to the roof of the building, rolls to the edge, falls off, and then makes a bull's-eye dive into an open manhole in the street below (see Figure 9-16). Calculate the ball's change in potential energy for the fall, using as height reference points (a) the roof, (b) the street, and (c) the bottom of the manhole.

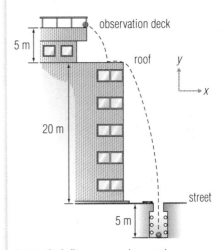

9-16 The ball's gravitational potential energy decreases as it falls.

Solution:

Choose a coordinate system as in Figure 9-16 with y positive upward. The origin will be at the heights specified in each part of the problem.

a. Origin at the roof; deck ($h_0 = +5.0$ m) to bottom of the manhole ($h = -25.0$ m):

$$\Delta U_g = |mg|\Delta h = |mg|(h - h_0)$$
$$\Delta U_g = |(6.50 \text{ kg})(-9.81 \text{ m/s}^2)|\big[(-25.0 \text{ m})-(+5.0 \text{ m})\big]$$
$$\Delta U_g \doteq -1912 \text{ J}$$

b. Origin at the street; deck ($h_0 = +25.0$ m) to the bottom of the manhole ($h = -5.0$ m):

$$\Delta U_g = |(6.50 \text{ kg})(-9.81 \text{ m/s}^2)|\big[(-5.0 \text{ m})-(+25.0 \text{ m})\big]$$
$$\Delta U_g \doteq -1912 \text{ J}$$

c. Origin at the bottom of the manhole; deck ($h_0 = +30.0$ m) to the bottom of the manhole ($h = 0.0$ m):

$$\Delta U_g = |(6.50 \text{ kg})(-9.81 \text{ m/s}^2)|\big[(0.0 \text{ m})-(+30.0 \text{ m})\big]$$
$$\Delta U_g \doteq -1912 \text{ J}$$

The change of potential energy for the entire fall is the same regardless of the height reference point.

9.14 Gravitational Potential

Gravitational potential energy depends on the mass of the object. Suppose two masses, m_1 and m_2, are both at a height h above the surface of the earth. Their potential energies, if $U = 0$ J at the surface, are

$$U_1 = |m_1g|h \qquad \text{and} \qquad U_2 = |m_2g|h.$$

What would be the potential energy of a third mass, m_3, at height h? It would be $|m_3g|h$. The **gravitational potential** for any height h near the earth is defined as $|g|h$ (since mass is always positive). To find the potential energy for an object at a given height, simply multiply the mass of the object by the gravitational potential for that height.

The above expression has been derived for gravitational potential energy near the earth's surface, but what about gravitational energy where $F_g \neq |mg|$? You can find the work done by any gravitational force by using a force-distance curve. You learned in Chapter 7 that the gravitational force between two objects can be calculated from Newton's law of universal gravitation,

$$F_g = G\frac{m_1m_2}{r^2},$$

where r is the distance between the centers of the two masses. The force that one mass exerts on another is plotted versus distance in Figure 9-18.

Suppose that mass m_2 starts at a distance R from mass m_1 and moves away at a constant velocity. A force with the same magnitude as the gravity between the masses but with the opposite direction must move m_2. The work that this force does is equal to the shaded area in Figure 9-18. Calculus shows that the area is equal to

$$G\frac{m_1m_2}{R}.$$

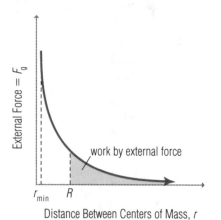

9-17 The gravitational potential for any object is $|g|h$ as long as the same reference point is selected.

The gravitational potential for objects near the surface of the earth is equal to $|g|h$. The potential energy at any given height above the surface is directly proportional to the mass of the object.

work by external force

r_{min} R

Distance Between Centers of Mass, r

9-18 Force-distance graph that defines gravitational potential from a distance R

Since the work was done at a constant velocity ($\Delta K = 0$ J), it all became potential energy. The change in potential energy of m_2 is therefore

$$\Delta U_g = G\frac{m_1m_2}{R}. \tag{9.16}$$

For gravitational potential far from the earth, as well as for potential energy near the earth, the reference point is arbitrary. However, in general gravitational problems it is customary to choose the potential energy at an infinite separation as the zero reference point. For any distance r between m_1 and m_2,

$$\Delta U_g = U_{g\infty} - U_g(r)$$
$$\Delta U_g = 0 - U_g(r),$$

where $U_g(r)$ means "the gravitational potential energy at a separation of r." Therefore, substituting for ΔU_g back into Equation 9.16,

$$-U_g(r) = G\frac{m_1m_2}{r}$$
$$U_g(r) = -G\frac{m_1m_2}{r}. \tag{9.17}$$

Since the constant G, the masses, and distance are always positive, gravitational potential energy is negative for any measurable separation distance. However, increasing r decreases the absolute value of U_g, making it less negative, so potential energy is greater (less negative) the farther apart the two objects are.

9.15 Gravitational Work Function

How does the change in gravitational potential energy relate to the work that gravity does? Well, remember that the external force that pulls m_2 away from m_1 is equal in magnitude but opposite in direction to the gravitational force between the masses. So their components in the direction of motion are opposite. Therefore,

$$W_{ex} = F_{ex}d = -F_gd = -W_g.$$

The change in potential energy is

$$\Delta U_g = W_{ex} = -W_g$$
$$\Delta U_g = -W_g,$$

which is essentially the same as Equation 9.14. Recall that gravitational force is a conservative force, so the work and change of potential energy does not depend on what path the masses take as they move toward or away from each other.

Problem-Solving Strategy 9.10

When finding the change of potential energy from one finite distance to another between two objects, $\Delta U_g(r)$ is just the difference of the final and initial gravitational potential energies at the two positions. Be sure that you use the correct value of r in the denominator of each term.

9-19 The booster rockets provide most of the work necessary to lift the space shuttle into orbit around the earth.

EXAMPLE 9-9

Rocket Science: Gravitational Work

How much work would it take to move a 4.50×10^4 kg rocket from the earth's surface to an infinite distance from Earth? The radius of the earth is 6.38×10^6 m; the mass of the earth is 5.97×10^{24} kg.

Solution:
Assuming that the starting position of the rocket is r_E, the earth's radius, the work necessary to move it to an infinite distance from the earth is

$$W = G\frac{Mm_r}{r_E},$$

where M is the mass of the earth and m_r is the mass of the rocket.

$$W = (6.67 \times 10^{-11} \text{ N·m}^2/\text{kg}^2)\left[\frac{(5.97 \times 10^{24} \text{ kg})(4.50 \times 10^4 \text{ kg})}{6.38 \times 10^6 \text{ m}}\right]$$

$$W \doteq 2.808 \times 10^{12} \text{ J}$$

$$W \doteq 2.81 \times 10^{12} \text{ J}$$

9.16 Elastic Potential Energy

Elastic forces in metals come from the tendency of a metal's atoms to remain at a fixed distance from one another. If the atoms are pushed closer together, they repel each other. If they are pulled farther apart, they attract one another. That is why a spring pulls or pushes back to its equilibrium position if it is stretched or compressed. External forces work against the spring force as the spring is stretched or compressed. Therefore, **elastic potential energy** can be produced with respect to spring-like (elastic) forces, which are conservative forces.

The work done by an external force, W_{ex}, *against* an elastic force to stretch or compress a spring at a constant velocity is equal to the change in the spring's elastic potential energy. The amount of work needed to stretch a spring from a length of x_1 to a length of x_2 is shown in Figure 9-20. Both lengths are *measured from the spring's equilibrium position*. The lightly shaded area is equal to the area of the large triangle minus the area of the small triangle.

$$W_{ex\,1\rightarrow2} = \Delta U_s = \tfrac{1}{2}kd_{2x}^2 - \tfrac{1}{2}kd_{1x}^2$$
$$W_{ex\,1\rightarrow2} = \tfrac{1}{2}k(d_{2x}^2 - d_{1x}^2)$$

If the spring begins at its natural length, then d_{1x} is zero and

$$W_{ex} = \tfrac{1}{2}kd_{2x}^2.$$

Once again, we determine the work done by an external force on the spring by evaluating the change of potential energy with respect to the elastic force.

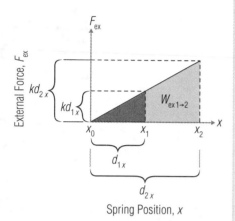

9-20 Work done on a spring between two positions

> Recall that the spring displacement $d_x = \Delta x = x - x_0$.

EXAMPLE 9-10

Change in Spring Potential Energy

A spring with a spring constant of 5.00 N/m is stretched from a length of 0.500 m to a length of 0.700 m. What is its change in potential energy if its rest length (x_0) is 0.450 m?

Solution:
The spring's initial position:

$$x_1 = 0.500 \text{ m}; \; d_{1x} = 0.500 \text{ m} - 0.450 \text{ m} = 0.050 \text{ m}$$

The spring's final position:

$$x_2 = 0.700 \text{ m}; \; d_{2x} = 0.700 \text{ m} - 0.450 \text{ m} = 0.250 \text{ m}$$

$$\Delta U_s = \tfrac{1}{2}k(d_{2x}^2 - d_{1x}^2)$$

$$\Delta U_s = \tfrac{1}{2}(5.00 \text{ N/m})\left[(0.250 \text{ m})^2 - (0.050 \text{ m})^2\right]$$

$$\Delta U_s \doteq 0.150 \text{ J}$$

$$\Delta U_s \approx 0.15 \text{ J}$$

9B   Section Review Questions

1. a. What kind of mechanical energy is specifically related to a system's position or condition, irrespective of its motion?

    b. Is there more than one form of this energy?

2. Discuss why positive work by a force on a system reduces the potential energy of the system with respect to that force.

3. What is the main restriction when using Equation 9.15 for determining relative gravitational potential energy?

4. A daredevil astronaut with a mass of m considers parachuting from orbit to the earth's surface wearing an asbestos space suit. How would the gravitational work on his body during the jump compare to the work done by gravity after he comes to his senses and rides the space shuttle back to earth?

◉5. What is the kinetic energy of a 1.00 g hailstone falling at 8.50 m/s?

◉6. What is the total work required to stop an 85.0 kg football halfback moving at 4.50 m/s?

◉7. What is the change of the gravitational potential energy (with respect to the earth) of a 120. g meteoroid that falls from a distance of 3.80×10^5 km to the surface of the earth?

◉8. A spring with a spring constant of 4.50 N/m is compressed from its rest length of 0.600 m to a length of 0.300 m. What is the change in the potential energy of the spring?

9C   TOTAL MECHANICAL ENERGY

9.17 Defining Total Mechanical Energy

All mechanical work on a system can be subdivided into the work done by conservative forces (W_{cf}) and the work done by all nonconservative forces (W_{ncf}), such as kinetic friction and drag. We can determine their contributions to the change of system kinetic energy with the following equation:

$$W_{total} = W_{cf} + W_{ncf} = \Delta K$$

You know that $W_{cf} = -\Delta U$ because potential energy is associated only with conservative forces. Therefore,

$$-\Delta U + W_{ncf} = \Delta K.$$

Solve for the work done by nonconservative forces.

$$W_{ncf} = \Delta K + \Delta U$$

$$W_{ncf} = (K_2 - K_1) + (U_2 - U_1)$$

Rearranging terms,

$$W_{ncf} = (K_2 + U_2) - (K_1 + U_1), \text{ and}$$

$$W_{ncf} = \Delta(K + U).$$

> The work by nonconservative forces on a system is equal to the change of the sum of kinetic and potential energies of the system.

Therefore, the work-energy theorem (see subsection 9.9) includes the principle that the work done by nonconservative forces is equal to the change of the system's kinetic and potential energies. The term $K + U$ has a special name in physics—**total mechanical energy (E).** The definition of total mechanical energy is

$$E \equiv K + U. \qquad (9.18)$$

You can conclude that the work accomplished by all nonconservative forces on a system during a certain process is equal to the change of total mechanical energy of a system:

$$W_{ncf} = \Delta E \qquad (9.19)$$

9.18 Conservation of Mechanical Energy

This consequence of the work-energy theorem actually helps us to understand how energy exchanges during a process. If we assume that there are no nonconservative forces acting on a system, then the left side of Equation 9.19 becomes 0 J. We routinely do this when we say that no friction or drag occurs during motion. We assume, then, that the change of total mechanical energy is completely due to the changes in the kinetic and potential energies. If there is no change of total mechanical energy, then change of kinetic energy must be equal but opposite to the change of potential energy:

$$\Delta E = \Delta K + \Delta U = 0 \text{ J}$$
$$\Delta K = -\Delta U$$

This special case of the work-energy theorem demonstrates the **conservation of mechanical energy.**

A good problem-solving strategy is to set the sum of the initial kinetic and potential energies of a system equal to the sum of their final values.

$$K_0 + U_0 = K + U \qquad (9.20)$$

If only conservative forces are involved, then Equation 9.20 will represent a true equality. Rearrange terms as necessary to solve for the unknown.

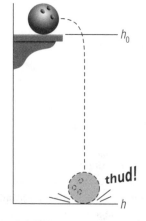

9-21 All the ball's gravitational potential energy is converted into kinetic energy during its fall.

EXAMPLE 9-11

Dropping the Ball: Conservation of Energy

A 5.45 kg bowling ball falls from rest on a shelf 2.0 m above the floor. What is its speed just as it strikes the floor? Assume that no nonconservative forces act on the ball as it falls.

Solution:
Assume that the height reference point is at floor level. Substitute the known values into Equation 9.20.

$$K_0 + U_0 = K + U$$

$$\tfrac{1}{2}mv_0^2 - |mg|h_0 = \tfrac{1}{2}mv^2 + |mg|h$$

$$0 \text{ J} + |(5.45 \text{ kg})(-9.81 \text{ m/s}^2)|(2.0 \text{ m}) = \tfrac{1}{2}(5.45 \text{ kg})v^2 + 0 \text{ J}$$

$$106 \text{ J} = \tfrac{1}{2}(5.45 \text{ kg})v^2$$

$$v = \sqrt{\frac{2(106\ \text{J})}{5.45\ \text{kg}}}$$

$$v \doteq 6.2\underline{3}\ \text{m/s}$$

$$v \approx 6.2\ \text{m/s}$$

The ball is falling at about 6.2 m/s when it hits the floor.

If a nonconservative force was involved in the process, then the two sides of Equation 9.20 will be unequal unless the nonconservative force is accounted for:

$$K_0 + U_0 = K + U + W_{ncf}$$

We will consider the conservation of mechanical energy in greater depth in Chapter 10.

9C Section Review Questions

1. What kinds of forces can change kinetic energy? Potential energy can be changed with respect to what kinds of forces?

2. What kind(s) of mechanical energy does a roller coaster have halfway down its first hill?

3. What kind(s) of energy does a pendulum bob have as it hangs motionless on its string?

4. What kinds of forces could be at work if mechanical energy is not conserved in a process? Describe one such process.

Chapter Review

In Terms of Physics

energy	9.1	kinetic energy (K)	9.9
work (W)	9.2	work-energy theorem	9.9
joule (J)	9.3	potential energy (U)	9.10
equilibrium position	9.6	conservative force	9.11
Hooke's law	9.6	nonconservative force	9.11
spring constant (k)	9.6	gravitational potential	9.14
ideal spring	9.6	elastic potential energy	9.16
Hooke's law force	9.6	total mechanical energy (E)	9.17
power (P)	9.8	conservation of mechanical energy	9.18
watt (W)	9.8		

Problem-Solving Strategies

9.1 (page 192) In order to determine whether work by a force is positive or negative, find the force's vector component that is parallel to the displacement of the system. If it is in the same direction as the displacement, work is positive. If the force vector component is opposite to the direction of displacement, then work is negative.

9.2 (page 193) Gravity does positive work when a system loses height, and it does negative work when the system gains height.

9.3 (page 193) Whenever you must calculate the gravitational force on a system in order to compute work done by or against gravity, be sure to use the *magnitude* of gravitational force $|mg|$. The work formula uses the magnitudes of force and displacement, *not* their scalar values.

9.4 (page 196) You must take care that you calculate the work done by the force that is specified in the problem. If an object attached to a spring is the system of interest, the work done by an external force on the system is not the same as the work done by the spring force.

9.5 (page 199) Kinetic energy can only be positive. For any given object, it varies as the square of its speed.

9.6 (page 201) Different forms of potential energy are indicated by a subscript identifying the force to which the potential energy corresponds. For example, U_g is gravitational potential energy and U_s is elastic potential energy of a spring.

9.7 (page 201) Recall that the scalar value of gravitational acceleration is $g = -9.81 \text{ m/s}^2$. Use the absolute value of mg when determining work involving weight or gravitational forces.

9.8 (page 202) Work by a conservative force on a system is always equal to the negative of the change of the system's potential energy with respect to that force.

9.9 (page 203) When solving gravitational potential energy problems, choose a height reference point that makes sense to you and simplifies the problem.

9.10 (page 205) When finding the change of potential energy from one finite distance to another between two objects, $\Delta U_g(r)$ is just the difference of the final and initial gravitational potential energies at the two positions. Be sure that you use the correct value of r in the denominator of each term.

Review Questions

1. What is the equation that defines work?

2. What is the significance of the cosine of the angle between the force and displacement vectors?

3. Which of the following is the unit for work?
   a. N   **b.** kg   **c.** m   **d.** J   **e.** J/s

4. Explain how negative work on a system occurs.

5. Give an example of a force that acts on a system during a displacement but does no work.

6. What kind of graph can be used to represent mechanical work? How is the graph used to do this?

7. What is the shape of the work region on the graph of a constant force versus displacement?

8. What is the shape of the work region on the graph of a Hooke's law force?

9. For a given object near the surface of the earth, what is the only variable factor that affects gravitational work?

10. Define *power*.

11. Kinetic energy is
   a. energy of motion.   **c.** all of the energy that an object has.
   b. energy of position.   **d.** independent of mass.

12. Define *mechanical potential energy* in your own words.

13. Explain, using an example, how positive work by a conservative force can result in a negative change of potential energy with respect to the same force.

14. What is the main restriction for using the equation $\Delta U_g = |mg|\Delta h$?

15. A brick falls a distance Δh, with a potential energy change of $\Delta U_g = |mg|\Delta h$. If no work is done on the earth-brick system, what is the change in the brick's kinetic energy, ΔK?
   a. 0 J   **b.** $-|mg|\Delta h$ **c.** $+|mg|\Delta h$ **d.** $-mv$

16. What equation expresses the work-energy theorem?

17. What is the sum of the kinetic energy and the potential energy of a system called?

True or False (18–28)

18. Work is a scalar quantity.

19. Every force acting on a system during a displacement does work on the system.

20. The graphical area under a force-displacement curve represents power.

21. Ideal springs follow Hooke's law only when stretched.

22. The force exerted by an ideal spring is a conservative force.

23. External forces that stretch or compress a spring do positive work on the spring.

24. The power delivered by a force can be increased by reducing the distance over which the force works in a given time.

25. The unit of power is J/s, or W.

26. The kinetic energy of a system is directly proportional to its mass and its speed.

27. Only conservative forces can change the kinetic energy of a system.

28. It doesn't matter what height you measure gravitational potential energy from as long as you are consistent within a given problem and gravitational acceleration is essentially constant.

◉29. Calculate the work done by the earth on a 2.00 kg brick that falls from the roof of a 10.0 m building to the ground.

◉30. A 165 g hockey puck moves a distance of 1.00 m while the hockey stick is pushing it with a force of 200. N. How much work does the stick do on the puck?

◉31. A woman pushes on a baby stroller at an angle of 60° below the horizontal. If she does 400. J of work on the stroller while it moves 20.0 m, how much force does she exert?

◉32. An empty spring cart for cafeteria trays has a load of trays placed on it. The spring constant is 490 N/m, and the trays push the spring down 0.50 m.
   a. What force do the trays exert on the spring?
   b. Draw a tray force-distance diagram for the situation.
   c. How much work did the trays do on the spring when they were loaded into the cart?

33. How much power is needed to compress a spring from its rest length of 0.150 m to half that in 10.0 s? The spring constant is $k = 625$ N/m.

34. A 165 g hockey puck resting on a level ice surface is struck by a stick with a force of 225 N. The stick is in contact with the puck over a straight-line distance of 0.840 m.
   a. How much work does the stick do?
   b. What is the change in the puck's kinetic energy?
   c. What is the puck's kinetic energy before it is hit?
   d. What is the puck's speed when the stick breaks contact?

35. An athlete running at a speed of 3.00 m/s decides to speed up on a long, flat stretch. After running 120. m, the runner is at a speed of 4.50 m/s. He has a mass of 75.0 kg.
   a. What is his kinetic energy before he speeds up?
   b. What is his kinetic energy after he speeds up?
   c. How much work was done in speeding up?
   d. What net force changed his speed?
   e. What was his acceleration, from Newton's second law?
   f. From the first equation of motion, how long did it take him to speed up?

36. A 1.50 kg model train initially traveling at 0.50 m/s accelerates to 1.00 m/s in 10. s.
   a. What is the change in the train's kinetic energy?
   b. How much power was expended to accelerate the train?

37. A 2.00×10^3 kg block of granite is lifted from the bottom of a quarry to the top at a constant velocity. The quarry is 176 m deep.
   a. How much work was done to the granite block?
   b. What is the change in potential energy of the block?
   c. If the lift took 70.0 s, how much power was expended?

38. A hydraulic lift raises a 975 kg car 3.00 m to the deck of a ship at a constant speed. What is the change in the car's potential energy?

39. A spring whose rest length is 35.0 cm and whose spring constant is $k = 8.00$ N/m hangs from a beam. The spring is slowly stretched, assuming no acceleration, to a length of 50.0 cm.
   a. How much work was required to stretch the spring?
   b. What is the change in the spring's elastic potential energy?

40. An engine stretches a spring 0.35 m from its rest position at constant speed, doing 75 J of work.
   a. What is the spring's change in elastic potential energy?
   b. Determine its spring constant.

41. An ideal spring is attached to a 0.100 kg block on one end and the wall on the other end, and it rests on a frictionless table. The spring is stretched at constant velocity until it is 2.00 cm longer than its equilibrium position, and then the spring and block are released. The spring constant is 8.00 N/m.
   a. What is the work used to stretch the spring?
   b. What is the potential energy of the block when the spring is stretched?
   c. What is the change in the block's potential energy from the time it is released to the time it reaches the spring's equilibrium position?
   d. What is the change in the block's kinetic energy from the time it is released to the time it reaches the spring's equilibrium position?
   e. What is the block's velocity when it reaches the spring's equilibrium position if it is released from rest?

42. A 0.165 kg ball is thrown vertically upward with a velocity of 10.0 m/s.
   a. What is the ball's initial kinetic energy as it is released?
   b. At the peak of its flight, the ball's velocity is zero. What is the change in its kinetic energy from the time it is thrown to the time it reaches the peak of its flight?
   c. What is the ball's change in potential energy from the time it is thrown to the peak of its flight, assuming that no work is done on the earth-ball system?
   d. How high does the ball rise if it is released at 2.0 m above the ground?

43. A 0.200 kg brick falls from rest a distance of 1.00 m. Find ΔU_g, ΔK, and ΔE.

44. A 1.00×10^6 kg asteroid moves from an infinite distance from the Sun to a distance of 1.75×10^{11} m from the center of the Sun.
   a. What is the force of the Sun on the asteroid at 1.75×10^{11} m?
   b. What is the change in the asteroid's potential energy?
   c. How much work does the Sun do on the asteroid?
   d. If the asteroid returns to its original distance, what will be the change in its potential energy during its return?

45. A 1.00×10^4 kg space probe, after being launched from a space shuttle that is 1.25×10^7 m from the earth's center, has traveled to its orbit, an additional 2.00×10^7 m farther away from the earth.
   a. What is the probe's potential energy with respect to the earth's surface at the time it is launched?
   b. What is the probe's potential energy when it reaches its orbit?
   c. How much work was done against gravity to move the probe to its orbit after its launch from the shuttle?
   d. If the probe falls back to the earth's surface after attaining its orbit, how much potential energy would it lose?
   e. How much work will the earth have done on the probe in such a fall?
   f. By how much would the probe's kinetic energy change during such a fall (ignore air resistance in the atmosphere)?
   g. If the probe has zero radial velocity before it starts its plunge, what is its speed as it reaches the earth's surface (ignore air resistance in the atmosphere)?

Hydroelectric Dams

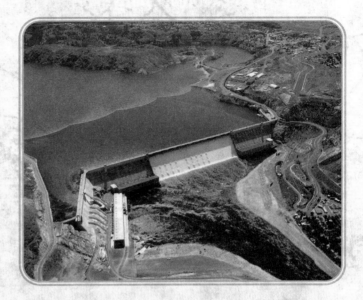

Potential energy is the energy of position—the height of an object or the compression of a spring gives it the potential to perform useful work. The water contained in streams has potential energy that is turned into kinetic energy as it flows downhill. If you have ever waded in a fast-running stream or have been buffeted by ocean waves, you realize that moving water can exert a lot of force. People have been exploiting this natural energy source since the earliest times. In the past, water wheels provided the power for forges, grinders, sawmills, and mills for grinding grain. One of the reasons towns were located along rivers was to take advantage of this natural power source.

Usually a channel was constructed to divert some of the water flow to a containment, often a pond behind a dam. When power was needed, the sluice was opened and the flowing water turned the wheel. As the wheel turned, gears transmitted the power to the machinery. When the water had done its work, it was returned to the stream at a lower elevation than the point at which it was diverted. During the trip from diversion to reentry, some of the water's potential energy was converted into kinetic energy.

Using the same principle, we generate electricity using hydroelectric dams. The dam creates a level ("head") of water at a controlled height. The water is released into an underground inlet tube, called the *penstock,* and is directed against the blades of a turbine, turning the generator shaft. Although the geometry is different, so far this bears a great resemblance to a water wheel. Rather than turning a mechanical device such as a grindstone, however, the shaft turns a large electromagnet (the rotor) within a stationary array of wire coils (the stator). As you will learn in later chapters, a magnet moving past a wire conductor induces a flow of electricity in the wire. The electricity flows to a transformer where the voltage is increased and the electricity is made available through the established power grid. Notice that the energy of the water undergoes several transformations. The potential energy of the water is converted to kinetic energy as the water flows down the penstock. Some of the kinetic energy of the water is transformed to the kinetic energy of the turbines and the generators, which then induce a current in the wire.

The largest hydroelectric facility in North America is the Grand Coulee Dam, built on the Columbia River in the state of Washington. This dam, 5233 feet long, is reported to be the largest concrete structure built to date and raises the water 350 feet above the old riverbed. The three sections of the dam contain 27 individual generators, each fed by its own penstock, the largest being 40 feet in diameter and carrying up to 35 000 ft^3/s of water. The dam's average power generation is 21 billion kilowatt-hours per year.

Some dams, rather than relying simply on the incoming stream, recreate their own heads of water by pumping water back to the upper reservoir after the water has done its work. These pumping stations actually expend more energy pumping the water back up than they gain as it flows down, but they are profitable because they exploit the law of supply and demand. They can use energy for pumping when it is less expensive during times of lower demand (at night) and generate electricity during times of peak demand when they can sell it for high rates. Also, during times of drought, they are able to produce more electricity from the limited supply of water.

The water in the river is guaranteed to lose its potential energy as it flows downstream. This energy in most cases is simply expended in the passing of the water. In order to be useful, the water must be controlled and directed to accomplish a specific purpose. In much the same way, your Creator has blessed you with a sort of potential energy—physical and mental abilities that will allow you to accomplish much throughout your life. How will you use it? You can live a life with no specific purpose, simply reacting to life as you flow the easy way downstream, or you can harness your energy in pursuit of your own will and reap many of the vain material successes available in the world. The righteous choice is fellowship with God found by submitting your will to the Lord and seeking His direction for your life. When your God-given abilities are directed toward His will, you will have real joy and those around you will be blessed as well.

"For whosoever will save his life shall lose it: and whosoever will lose his life for my sake shall find it" (Matt. 16:25).

10A   Total Mechanical Energy   215
10B   Simple Machines   222
Facet: Conserving Energy   221

New Symbols and Abbreviations

theoretical mechanical advantage (TMA)	10.8
actual mechanical advantage (AMA)	10.8
efficiency (η)	10.12

Return to Earth from orbit depends on the fact that mechanical energy is *not* conserved as it is converted into nonmechanical forms, such as heat and light. This is an artist's conception of a replacement for the space shuttle as it prepares to reenter Earth's atmosphere.

Conservation of Energy 10

> *Thus the heavens and the earth were finished, and all the host of them. And on the seventh day God ended his work which he had made; and he rested on the seventh day from all his work which he had made.* Genesis 2:1–2

God created the world in seven days, and then He stopped creating and began preserving what He had made. Since only God can create anything, nothing is now being created. The laws of conservation are a silent but powerful testimony against the hypothetical evolutionary processes that some believe are continuing to produce matter and energy in remote locations in the universe. In fact, the most serious challenges to the big bang hypothesis are the laws of conservation of matter and energy. Scientists have to propose an *ad hoc* explanation that the universe had different laws before the big bang occurred in order for matter and energy to materialize out of nothing. God conserves and preserves His creation through His power. The Bible states that nothing new will be created, except through specific miracles, until He remakes the heavens and the earth as an eternal home for His saints.

Though God may work in different ways at different times, His character never changes. The Bible states that the following attributes of God endure forever:

God Himself—Psalm 102:26
His glory—Psalm 104:31
His mercy—Psalm 106:1
His goodness—Psalm 52:1
His peace—Psalm 72:7
His truth—Psalm 117:2
His righteousness—Psalm 111:3
His name—Psalm 135:13

10A TOTAL MECHANICAL ENERGY

10.1 Conservation Laws

Scientists acknowledge the fact that no truly creative processes can now be observed because of the existence of conservation laws. A **conservation law** states that something is **conserved;** that is, something remains constant under certain conditions. For example, you learned in the last chapter that a system's total mechanical energy remains constant as long as no work is done on the system. Scientists have discovered that mass and electric charge, as well as energy, seem to be conserved in all processes except under the most extreme conditions.

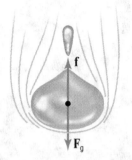

10-1 Forces on a falling raindrop at terminal velocity

10-2 The plasma gases generated by the extremely hot heat shield tiles of the space shuttle bathe the vehicle in a fiery glow during atmospheric reentry. The plasma glow is seen here through the flight deck windows.

10-3 Because there is no air resistance to slow them down, meteors strike the Moon's surface while traveling at great speeds. As a result, they may have produced huge craters such as Clavius, shown here.

When Total Mechanical Energy *Is Not* Conserved

10.2 Air Resistance

Under what conditions is total mechanical energy not conserved? When work is done on the system by *nonconservative forces,* the total mechanical energy changes. This work is often done by frictional forces. One kind of frictional force is *air resistance.*

If you put your hand just outside the window of a moving car, you will feel air resistance. It is a reaction force. Your hand, moving with the car, pushes on air molecules, so the air molecules push back on your hand. If the car goes faster, you must exert more force to keep your hand in the air stream as the force of air resistance increases on your hand.

Consider a raindrop falling through the air. There are two forces on the drop: the force of gravity pulling it down and air resistance pushing upward against it. As the drop's velocity increases, the air resistance on it also increases. Eventually the drop's velocity is high enough that the air resistance balances the force of gravity. The drop stops accelerating and continues to fall with a constant velocity. This velocity is called the drop's **terminal velocity.** We are fortunate that air resistance prevents raindrops from accelerating throughout their fall. Otherwise, a raindrop falling from 6000 m would be traveling at nearly the speed of sound by the time it reached the earth's surface!

When the raindrop reaches its terminal velocity, its kinetic energy remains constant. However, since it falls, its potential energy decreases. What happens to the potential energy that the drop loses? It is changed to thermal energy and lost as heat into the air and within the raindrop. When an object falls through the air, its temperature increases.

In a short fall through the air, the rise in an object's temperature is not measurable. However, in a long or high-speed fall, this temperature increase is important. For example, when a spacecraft falls through the earth's atmosphere, its temperature may exceed 2500 °C. Therefore, all spacecraft intended to return to the earth must be protected by heat shields. The tragic loss of the space shuttle *Columbia* in 2003 is believed to be due to a breach in the integrity of its heat shielding tile system.

Thermal energy is not included in an object's total *mechanical* energy. If potential or kinetic energy is changed to thermal energy, total mechanical energy is not conserved. Thus for an object's fall through the air, total mechanical energy is not conserved. Spacecraft engineers use atmospheric drag to significantly slow the vehicle from orbital speed before it reaches an altitude in the atmosphere where other means are used to slow it to landing speed.

How can energy be conserved for a falling object? By removing air resistance. For example, on a planetary body that has little or no atmosphere, like the Moon, there is no significant air resistance, and total mechanical energy is conserved during a fall.

10.3 Friction

Other frictional forces also change mechanical energy to thermal energy. For example, if you give a wooden block a light push across a table, it will probably stop before reaching the edge. Its kinetic energy changes to thermal energy. The temperatures of the block and the table rise slightly.

The increase in temperature because of friction is especially evident in vehicle brakes. Brakes use friction to stop a vehicle's wheels from spinning. After a car has come to a stop, the brakes are often hot. Overheated brakes can be a problem in loaded trucks or in any vehicle making a long descent down a mountain.

Friction can also change mechanical energy into sound energy. Some of the kinetic energy of a spinning wheel becomes the sound of squealing brakes. A creaking door results when friction at the door's hinges converts kinetic energy to sound. That is why lubricating a squeaking part—decreasing friction—often stops the squeaking.

Not all frictional forces are obvious. A spring, for instance, has internal frictional forces that convert the spring's mechanical energy to thermal energy or sound. The internal forces of friction in the spring eventually stop a spring's oscillations even if no external force acts on it. Most spring problems referred to in this textbook involve springs that have insignificant internal friction. This is another characteristic of an *ideal spring* that was first described in the last chapter. Ideal springs also have no mass, and they obey Hooke's law ($F_x = -k\Delta x$) perfectly. Ideal springs, like ideal strings, only approximate reality.

When Total Mechanical Energy *Is* Conserved

10.4 Conservative Forces

We have discussed frictional forces that do not conserve total mechanical energy. What about *conservative forces,* the forces that do conserve total mechanical energy?

A conservative force between two bodies always pushes or pulls along the line joining the centers of the bodies. You remember from Chapter 7 that this kind of force is called a **central force.** All conservative forces are central forces. One force of this kind is the force of gravity.

Consider a satellite orbiting the earth in a perfectly circular orbit. How much work does the earth do on the satellite? Figure 10-4 shows that the satellite moves in a direction perpendicular to the force of gravity. Therefore, the component of the gravitational force in the direction of the satellite's motion is zero. The earth does no work on the satellite as long as the satellite continues in a circular orbit.

10.5 Path-Independence of Conservative Forces

What if the satellite moves from its orbit? Suppose it gains altitude from point A to point B. How much work must be done on the satellite against gravity? Consider the path followed by the satellite in Figure 10-5a. Assume that the satellite moves from A to C to B. To move it radially (without acceleration) from A to C (Δh), opposite the direction of the force of gravity, will require a force with the same magnitude as the force of gravity, but in the opposite direction ($-\mathbf{F_g}$). The work done on the satellite is

$$W_{AC} = F_g\Delta h.$$

To move the satellite from C to B, no work is needed, since the satellite moves along a constant-radius circular orbit. The total work in moving from A to C to B is

$$W_{AB} = W_{AC} = F_g\Delta h.$$

What if the satellite instead moves from A to D to E to F to B, as Figure 10-5b shows? The satellite moves from A to D and from E to F in circular arcs, so no gravitational work is done on the satellite. To move the satellite from D to E along a radial line, a force of $-\mathbf{F_g}$ is needed for a distance of $\Delta h/2$. The work done against gravity is

$$W_{DE} = F_g\frac{\Delta h}{2} = \tfrac{1}{2}F_g\Delta h.$$

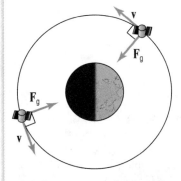

10-4 Because $\mathbf{v} \perp \mathbf{F_g}$, the instantaneous displacement vector $\mathbf{d}$ is always perpendicular to the gravity vector as well. Therefore, the earth does no work on a satellite in a circular orbit.

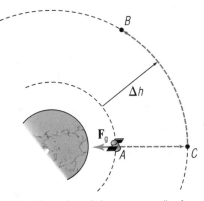

10-5a The work needed to move a satellite from A to C to B equals $F_g\Delta h$.

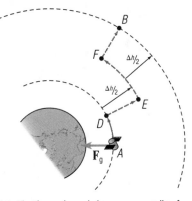

10-5b The work needed to move a satellite from A to B does not depend upon the path taken between A and B.

Conservation of Energy **217**

A force of $-\mathbf{F}_g$ is again needed to move the satellite from F to B, a distance of $h/2$.

$$W_{FB} = F_g \frac{\Delta h}{2} = \frac{1}{2} F_g \Delta h$$

The total work done in moving the satellite from A to D to E to F to B is

$$W_{AB} = \frac{1}{2} F_g \Delta h + \frac{1}{2} F_g \Delta h = F_g \Delta h.$$

This is the same as the work done to move the satellite from A to C to B. In fact, any path from A to B can be thought of as a series of moves in the direction of $-\mathbf{F}_g$ and perpendicular to $-\mathbf{F}_g$. In each case, the work done against gravity is the same. This concept is called *path-independence*. For any conservative force, the work done on an object against the force depends only on the beginning and ending positions of the object. The work done is the same for all paths connecting those points.

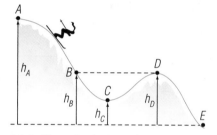

10-6 What is the skier's speed at point B?

EXAMPLE 10-1

The Ups and Downs of Skiing

The skier in Figure 10-6 begins at point A with a speed of 1.00 m/s. If $h_A = 100.0$ m and $h_B = 50.0$ m, what is the skier's speed at point B? His mass is 85.0 kg, and friction and air resistance are negligible.

Solution:

Assume that the height reference point is E. The strategy for this problem involves first finding the change in gravitational potential energy, ΔU_g. Knowing that $\Delta K = -\Delta U_g$ when no work is done on the skier-system by nonconservative forces, you can then solve for the final velocity using the definition of kinetic energy.

a. Find ΔU_g:

$$\Delta U_g = |mg|h_B - |mg|h_A = |mg|(h_B - h_A)$$

$$\Delta U_g = |(85.0 \text{ kg})(-9.81 \text{ m/s}^2)|(50.0 \text{ m} - 100.0 \text{ m})$$

$$\Delta U_g \doteq -41\,690 \text{ J}$$

b. Find ΔK:

$$\Delta K = -\Delta U_g = -(-41\,690 \text{ J})$$

$$\Delta K = +41\,690 \text{ J}$$

c. Solve for v_B:

$$\Delta K = \frac{1}{2} m v_B^2 - \frac{1}{2} m v_A^2$$

$$\Delta K = \frac{1}{2} m (v_B^2 - v_A^2)$$

$$v_B^2 - v_A^2 = \frac{2(\Delta K)}{m}$$

$$v_B^2 = \frac{2(\Delta K)}{m} + v_A^2$$

$$v_B = \sqrt{\frac{2(\Delta K)}{m} + v_A^2} = \sqrt{\frac{2(+41\,690 \text{ J})}{85.0 \text{ kg}} + 1.00 \text{ m/s}}$$

$$v_B \doteq 31.33 \text{ m/s}$$

$$v_B \approx 31.3 \text{ m/s}$$

The skier's speed at point B is about 31.3 m/s.

Suppose the skier in Figure 10-6 continues down the hill and over the hump. What happens to his energy? As he skis to point C, his potential energy changes to additional kinetic energy. Then at point C, he begins to go uphill. His kinetic energy decreases as his potential energy increases. At point D his potential energy is $|mg|h_D = |mg|h_B$. Since his total mechanical energy is conserved, his kinetic energy must be the same as it was at point B, and his speed is again v_B. As the skier continues over the hump and down the hill, his potential energy once again changes to kinetic energy. At the bottom of the hill, point E, his potential energy is zero, having changed completely to kinetic energy.

What is the effect of the hump on the skier's kinetic energy at E? Since his kinetic and potential energies are the same at D as at B, his speed, and thus his kinetic energy at E is the same as it would be if there were no hump and he skied directly from A to B to E. This shows again that an object's energy change between two points is not affected by the path taken between those points.

10.6 Escape Speed

When total mechanical energy is conserved, potential energy can be changed to kinetic energy, and kinetic energy can be changed to potential energy. As an object falls from a shelf, its potential energy changes to kinetic energy. If it is tossed from the floor to the shelf, its kinetic energy changes to potential energy. Similarly, when a meteoroid is pulled from far away in space and impacts the earth's surface, its potential energy changes to kinetic energy; and when a spacecraft at the earth's surface is launched away from the earth, the kinetic energy delivered by its engines changes to potential energy. The **escape speed** is the minimum speed that an object of mass m requires to leave a larger object of mass M so that mass m cannot return due to gravitational attraction alone.

The zero-height reference point for the launch is the center of the earth. Assume that total mechanical energy does not change.

$$\Delta E = \Delta K + \Delta U_g = 0 \text{ J}$$
$$\Delta K = -\Delta U_g$$

Substitute the kinetic and gravitational potential energy formulas for both sides of the equation. Then solve for the escape speed, which is the initial velocity at the radius of the earth, v_R.

$$\tfrac{1}{2}mv_\infty^2 - \tfrac{1}{2}mv_R^2 = -G\frac{Mm}{R}$$
$$\tfrac{1}{2}\cancel{m}(v_\infty^2 - v_R^2) = -G\frac{M\cancel{m}}{R}$$
$$v_\infty^2 - v_R^2 = -2G\frac{M}{R}$$
$$-v_R^2 = -2G\frac{M}{R} - v_\infty^2$$
$$v_R^2 = 2G\frac{M}{R} + v_\infty^2$$
$$v_R = \sqrt{2G\frac{M}{R} + v_\infty^2}$$

In this case, M is the mass of the earth, m is the mass of the rocket, and R is the earth's radius. The minimum escape speed can be calculated when the speed of the spacecraft at an extreme distance is zero.

$$v_R = \sqrt{2G\frac{M}{R}} \tag{10.1}$$

The **escape speed** is the minimum speed that an object of mass m requires to leave a larger object of mass M so that mass m cannot return due to gravitational attraction alone.

Christians have been delivered from God's wrath by Christ. However, there is no amount of good works that allows an unbeliever to escape God's judgment. "He that believeth on him [Jesus Christ] is not condemned: but he that believeth not is condemned already, because he hath not believed in the name of the only begotten Son of God" (John 3:18).

10-7 Since the beginning of the space program, only a scant handful of spacecraft have attained sufficient speed to escape the gravitational tug of our sun. Shown here is an artist's rendering of one of the Voyager probes looking back from interstellar space toward the distant solar system with its sun.

Notice that the rocket's mass did not affect the calculation. The escape speed depends only upon the gravitational force of the planet and its radius. It is the same for all objects.

EXAMPLE 10-2

To Infinity or Even Further: Escape Speed

A 1.00×10^4 kg rocket is launched at a speed of v_R from the surface of the earth, 6.38×10^6 m from the earth's center. It is intended to travel an infinite distance from the earth and arrive "there" with zero speed ($v_\infty = 0$ m/s). Assume that there is no air resistance. What is the smallest initial speed that will launch the rocket to achieve this objective?

Solution:

You know that the final speed is 0 m/s. Substitute the known values for G, M, and R to solve for initial speed.

$$v_R = \sqrt{2G\frac{M}{R}}$$

$$v_R = \sqrt{2(6.67 \times 10^{-11} \text{ N} \cdot \text{m}^2/\text{kg}^2)(5.97 \times 10^{24} \text{ kg})/6.37 \times 10^6 \text{ m}}$$

$$v_R \doteq 11180 \text{ m/s}$$

$$v_R \approx 1.12 \times 10^4 \text{ m/s}$$

The rocket's initial speed would have to be around 11.2 km/s!

10A Section Review Questions

1. Why is the terminal velocity of a 1 g lead pellet higher than for a 1 g piece of paper?

2. An energetic golden hamster (*Mesocricetus auratus*) is running flat out at a constant speed on its frictionless exercise wheel (25.0 cm diameter, weighing 25.0 g). If the hamster is considered part of the exercise wheel system (his efforts are *internal* to the system), how much work is being done on the hamster-wheel system?

3. What is the main difference between a conservative force and a nonconservative one?

4. A physics student removes his textbook from his locker, carries it to the classroom, lays it on the floor under his desk, picks it up at the end of class, and returns it to its place in his locker.

   a. What was the net change in gravitational potential energy on the book?

   b. What was the net work done by gravity on the book?

⦿5. You drop a 100. g scoop of ice cream from a cone 1.20 m above the ground. Oops! Calculate the scoop's gravitational potential energy, its kinetic energy, and its total mechanical energy when it is 1.20 m, 0.80 m, 0.40 m, and 0.00 m from the ground.

6. A 6.80×10^4 kg space shuttle moving at 7.90×10^4 m/s reenters the earth's atmosphere 80.0 km above the earth's surface.

    a. What is the shuttle's total mechanical energy with respect to the earth's surface?

    b. It performs aerodynamic braking to slow sufficiently to land. Assuming landing speed is negligible, how many joules of energy must be dissipated as thermal energy by drag with the atmosphere?

    c. If just 0.1% of the thermal energy generated by reentry is absorbed by the shuttle, what design feature keeps it from melting?

7. A 1.0 kg block sliding along a horizontal frictionless surface at 2.0 m/s encounters an ideal spring attached to a vertical wall. If the spring constant is 80. N/m, how far will the spring be compressed when the block is brought to rest?

8. Calculate the escape velocity for the earth's moon.

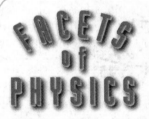

Conserving Energy

Every day we are bombarded by admonitions to "conserve energy." Appliances are proclaimed to be energy efficient, and power companies offer free advice on how to conserve energy. But, as we learned in this chapter, the energy put into a process is equal to the energy leaving the process—no matter what we do. Why, then, do we make a special effort to conserve it?

This seeming contradiction results from our definition of terms. Our statement that "energy into and out of a process is equal" means that the total amount of energy in the universe remains constant. However, not all energy is equally usable. When we use energy to do work in our homes or cars, it becomes less usable. For example, usable electrical energy can become lost thermal energy, such as what occurs in a light bulb.

Those who urge us to "conserve energy" really mean that we should try to keep energy in a usable form. While the total amount of energy in the universe is constant, the amount of usable energy always decreases. We can accelerate this decrease by using energy less efficiently, or we can slow it down by using energy wisely and more efficiently. Therefore, when you hear a plea to "conserve energy," remember that it simply means to use only as much energy as you need (Turn off those lights when you leave the room!), and when you use it, to use it efficiently.

Some modern gas-fueled hot air furnaces are more than 96% efficient.

SIMPLE MACHINES

10-8 Even complicated yard maintenance can be done with simple tools.

10.7 Why Machines?

A large stone rests in the middle of the front yard of your newly built house. Removing the stone will allow you to establish a beautiful level lawn. How will you remove it? You cannot simply lift such a large, sunken stone even if you have several helpers. You may consider using dynamite to shatter the stone, but detonating explosives in a residential area is prohibited. You might think of hiring a crane to remove the rock, but that operation would be prohibitively expensive.

You would probably decide, finally, to remove the stone by using tools you already have or can borrow. To pull the rock out with a lawn tractor, you would first dig around it with a shovel and put a chain around it. On the side of the rock nearest the tractor, you could place boards to form a ramp under the stone to make it slide more easily from its bed. Opposite the ramp, neighbors with pry bars could loosen the rock as the tractor pulls. This strategy would probably remove the stone from its hole in your lawn.

In this illustration several machines have been mentioned. Complex machines such as the crane and the tractor seem indispensable in modern life. However, simple machines, like the shovel, ramp, and pry bars, are not obsolete. Why are these called machines? A **machine** is a device that changes the magnitude or direction (or both) of an applied force. The force may be provided by a human or a nonhuman source of energy like an electric motor. Although machines make some tasks easier by providing an advantage that would not otherwise be available, they do not create energy. In fact, because of friction, the amount of work a machine produces is less than the energy supplied to it.

> Machines in physics are simple devices that change the magnitude or direction of a force.

10.8 Mechanical Advantage

Most simple machines amplify the effect of an applied force. How much does a machine amplify a force? To answer this question, we must define **theoretical mechanical advantage (TMA).** One definition of theoretical mechanical advantage is the force the machine produces in the absence of friction, divided by the force that is applied to the machine. That is,

$$\mathbf{TMA} = \frac{F_{out}}{F_{in}} \text{ (no friction).} \tag{10.2}$$

Mechanical work is accomplished by a force component acting parallel to a displacement. In this discussion of simple machines, we will assume that input and output forces act in the same direction as the displacement of the machine where the force is exerted (e.g., $\theta = 0°$). We make this reasonable simplification to avoid having to resolve the forces into their components. We can then represent the mechanical work function as simply

$$W = Fd, \tag{10.3}$$

since $\cos 0° = 1$.

For a real machine, the work done by the input force, W_{in}, equals the work done by the output force, W_{out}, plus the work done by friction, W_f:

$$W_{in} = W_{out} + W_f$$

You can see that friction reduces the output work in all real machines. However, for ideal frictionless machines, frictional work is zero and

$$W_{in} = W_{out} \text{ (no friction).} \tag{10.4}$$

You can rewrite Equation 10.4 by substituting the definition of work from Equation 10.3,

$$W_{in} = F_{in}d_{in}, \text{ and } W_{out} = F_{out}d_{out}, \text{ so}$$
$$F_{in}d_{in} = F_{out}d_{out} \text{ (no friction)}.$$

Rearranging,

$$\frac{F_{out}}{F_{in}} = \frac{d_{in}}{d_{out}} \text{ (no friction)}. \tag{10.5}$$

The left-hand side of Equation 10.4 is the definition of TMA; so

$$\text{TMA} = \frac{d_{in}}{d_{out}} \text{ (no friction)}. \tag{10.6}$$

EXAMPLE 10-3

Moving Up: Ramps and Theoretical Mechanical Advantage

A ramp is a simple machine called an inclined plane. A 250. kg piano must be moved to the second floor of an apartment building 3.00 m above the ground. The movers intend to use a 10.0 m frictionless ramp to move the piano. (a) What is the theoretical mechanical advantage of the ramp? (b) What input force is needed to move the piano?

Solution:

When considering such a problem, you must ask "What am I basically trying to accomplish?" In this case, you are trying to move the weight of a massive 250. kg piano vertically from street level to an apartment 3.00 m up. These numbers are the "out" quantities in the TMA equations. The "in" quantities are the force exerted and distance covered by the workers.

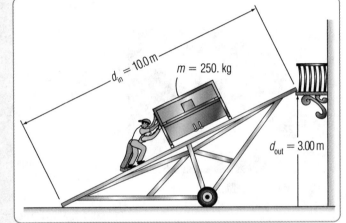

10-9 Simple machines make Herculean tasks easier.

a. Use Equation 10.6 to find the TMA:

$$\text{TMA} = \frac{d_{in}}{d_{out}} \text{ (no friction)}$$

$$\text{TMA} = \frac{10.0 \text{ m}}{3.00 \text{ m}}$$

$$\text{TMA} \doteq 3.3\overline{3}$$

$$\text{TMA} \approx 3.33$$

Notice that TMA is dimensionless—the units always cancel out.

b. Now, solve for the input force or effort. Use Equation 10.2:

$$\text{TMA} = \frac{F_{out}}{F_{in}} \text{ (no friction)}.$$

F_{out} is the weight of the piano. Rearrange to solve for F_{in}:

$$F_{in} = \frac{F_{out}}{\text{TMA}} = \frac{|mg|}{\text{TMA}} = \frac{|(250. \text{ kg})(-9.81 \text{ m/s}^2)|}{3.3\underline{3}3}$$

$$F_{in} \doteq 73\underline{5}.8 \text{ N}$$

$$F_{in} \approx 736 \text{ N}$$

Only 736 N is needed to push the piano up the ramp, compared to about 2450 N to lift the piano vertically. Using the ramp made the job much easier. However, notice that the piano traveled more than 3 m to reach the second floor. The movers "paid" for their easier job by moving the piano farther.

The word *theoretical* in TMA implies that TMA is idealized and unrealistic. In real life, friction is always present. The **actual mechanical advantage (AMA)** of a machine is defined as the output force when friction is present, divided by the input force. The actual mechanical advantage is more easily measured than predicted.

We generally use a machine to make a job easier because we want to deliver more force to the work than the force that we put into the machine or tool. Therefore, in most cases we will want the machine's TMA to be greater than 1. This can mean either that the output force is greater than the input force, or that to accomplish the job we have to move the load farther using the machine than we would if we moved the load directly. We will see how this works as we study various kinds of simple machines.

There are two basic classes of simple machines: inclined planes and levers. Although at first glance shovels, pulleys, and so forth do not seem to fit into either category, closer examination shows that they do.

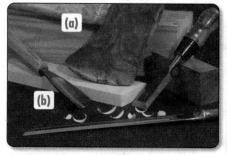

10-10 Types of inclined planes: (a) an axe (a wedge); (b) a gouge (a single inclined plane)

10.9 Inclined Planes

Wedges, screws, and *ramps* are **inclined planes.** Ramps were discussed above. A wedge is usually two inclined planes placed base-to-base. For example, an axe is a wedge. A wedge may also be a single inclined plane, such as a chisel or scalpel, that is sloped on only one side. Wedges are usually used as cutting or splitting tools.

A **screw** is a metal shaft surrounded by a helically coiled inclined plane. It is much like a curving road going up a mountain. Few cars can produce enough force or traction to drive straight up a mountain. However, the slope of a winding road is smaller than that of a straight road; therefore, less force is needed to propel the car forward and up. Figure 10-11 shows the difference between the force needed to counter gravity on a straight road and the force needed on a winding road. Similarly, a screw uses a spiral inclined plane to make it easier to exert great force parallel to the shaft of the screw. With a climbing road, the closer the successive loops are, the less each loop slopes and the easier it is to move the car forward. Of

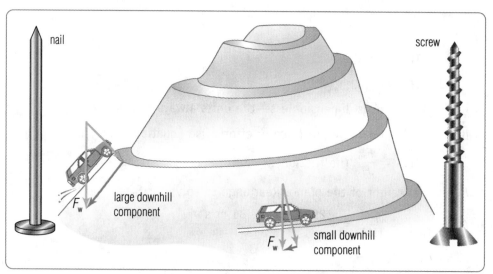

10-11 A screw contains a winding inclined plane. A winding road is easier to ascend than a road going straight up a mountain.

course, the car must travel much farther along the road to reach the mountaintop. Similarly, for a screw, the shorter the distance (the **pitch**) is between two successive threads, the easier it is to turn the screw. However, a screw with a small pitch must be turned more times than a screw with a large pitch.

10.10 Levers

A **lever** is usually a rigid bar that turns around a pivot, the *fulcrum*. In use, an input or *effort force (F_e)* is applied at one end of the lever, called the *effort arm (l_e)*, and the output force (the *resistance* or *load, F_r*) is applied by the *resistance arm (l_r)*. You remember from Chapter 7 that a force applied perpendicular to a line passing through a pivot point is a torque. If the effort and resistance torques acting on the same pivot point are in equilibrium, according to the *law of moments* (see Section 7.13) the lever will be stationary and

$$F_e l_e = F_r l_r. \qquad (10.7)$$

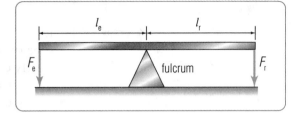

10-12 Parts of a simple lever system

When Equation 10.7 is true, there are no net torques on the lever and the lever just balances on the fulcrum. The theoretical mechanical advantage of a lever system, then, is

$$\text{TMA} = \frac{l_e}{l_r},$$

while the actual mechanical advantage can be determined from the actual forces involved:

$$\text{AMA} = \frac{F_r}{F_e} \qquad (10.8)$$

Problem-Solving Strategy 10.3
Equation 10.8 can be used for determining the AMA of any system by substituting the actual effort and resistance forces into the equation.

There are three kinds of levers. In **first-class levers,** shown in Figure 10.13, the fulcrum is between the resistance and the effort forces. Some common first-class levers are seesaws, crowbars, and scissors. The TMA of a first-class lever varies depending on the relative lengths of the effort arm and the resistance arm. Since

The TMA of first-class levers can be greater than, less than, or equal to 1.

$$\text{TMA} = \frac{l_e}{l_r},$$

TMA > 1 if $l_e > l_r$; TMA = 1 if $l_e = l_r$; and TMA < 1 if $l_e < l_r$.

It is generally desirable for a first-class lever TMA to be greater than 1. If the TMA were less than 1, you would have to exert more effort on the machine than it delivers to the resistance. To illustrate this, try cutting through a thick cord with a pair of scissors by placing the cord between the blades near the pivot point. Then try cutting the same cord using the tips of the blades. You must apply much greater effort to cut the cord using the tips, assuming you are able to do it at all.

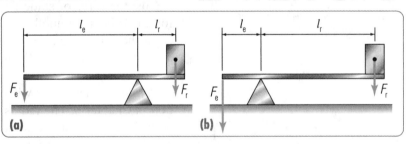

10-13 A first-class lever. (a) TMA > 1; (b) TMA < 1

In **second-class levers,** the resistance is between the fulcrum and the effort force. As you can see in Figure 10-14, the effort arm in a second-class lever system is *always* longer than the resistance arm, so the TMA is always greater than 1.

Some common second-class levers are hinged doors, wheeled tote bags, and nutcrackers.

The TMA of second-class levers is always greater than 1.

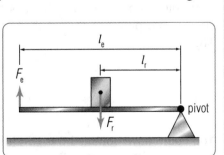

10-14 A second-class lever

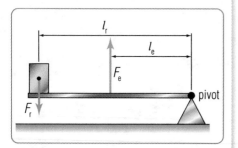

10-15 A third-class lever

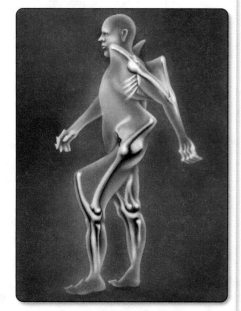

10-16 A human with first-class levers for appendages would be extremely ungainly and inefficient.

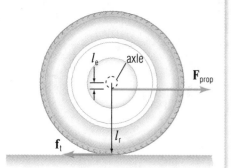

10-17 A vehicle's drive wheels act as a form of third-class lever.

A pulley is a modified first-class lever that can reposition itself indefinitely.

In **third-class levers,** the effort force is between the fulcrum and the resistance. As usual,

$$\text{TMA} = \frac{l_e}{l_r},$$

but here l_e is always *less* than l_r. Therefore, the TMA is always less than 1. A third-class lever diminishes input force. What good is it, then?

A third-class lever moves a resistance a large distance while the effort moves only a small distance. Brooms and tweezers are examples of common third-class levers. But the most common and important application was employed by our Creator in the appendages of every creature with a rigid endo- or exoskeleton. Nearly every major skeletal joint in your body is the fulcrum of a third-class lever. There is a simple biomechanical reason for this design. Muscles develop force at the subcellular level, and this force is applied to our limbs and appendages. Even with millions of cells contracting in unison, the total effort distance a muscle can generate is fairly small. In order to attain maximum possible movement, vertebrates and arthropods were designed with third-class levers in arms, legs, and nearly every other place where a muscle is attached to a rigid support.

10.11 Wheels, Pulleys, and Pulley Systems

A lever has one major limitation: It can move only so far while doing its job. At some point, the effort arm or the resistance arm can no longer move, either because of an obstruction or because the associated force can no longer be applied effectively. It would be ideal if the lever arm could be continually repositioned in order to keep the effort and resistance forces applied. This is exactly what wheels and pulleys do.

Wheels are modified levers. The fulcrum of a wheel-lever is the axle. A force applied at the edge of a steering wheel turns a load near its center. This is a second-class lever, and it makes the load easier to turn. Larger steering wheels, which have longer lever arms (radii), are easier to turn than smaller wheels. For an automobile drive wheel, a force delivered by the engine near the center of the wheel turns it against the force of static friction between the wheel and the road (the resistance force). This is yet another example of a third-class lever. It enables a force moving a short distance to move a load a large distance. Larger diameter wheels move a car farther than smaller wheels for the same angular distance that the force moves.

A *pulley* is a grooved wheel that turns on an axle. You learned how pulleys function in Chapter 8. The radius of the wheel is fixed, of course, but the radius forms both the effort arm and the resistance arm of this special lever. Effort and resistance forces are applied by a rope or string that lies in the groove as it passes over the pulley. The string enters and leaves the pulley tangentially, so the forces are perpendicular to the radii of the pulley. The pulley axle at the center of the radii acts as the fulcrum. Thus, a pulley is a kind of first-class lever.

If you assume an ideal string passes through the pulley, then for every meter the effort force pulls the string, the string also pulls the resistance load one meter. Therefore,

$$d_{in} = d_{out}, \text{ and}$$

$$\text{TMA} = \frac{d_{in}}{d_{out}} = 1.$$

You could have obtained this result by taking the ratio of the lengths of the effort and the resistance arms. Since they are the radii of the same circle, the TMA is 1 using this method as well.

Pulleys can be used in one of two ways. When a pulley is attached to a rigid support, it is called a *fixed pulley*. The advantage of a fixed pulley is that it changes the direction of an applied force. The applied force produces tension in the string that passes undiminished through the pulley. It is often easier to pull down on a rope, putting your weight into it, than to pull up on the load directly. Also, you are not limited by your reach when lifting a load to a greater height.

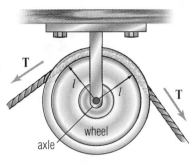

10-18 A simple pulley

For a single fixed pulley, TMA = 1. The main advantage of a fixed pulley is that it changes the direction of the force in a string or rope, thereby allowing the user to pull downward and use his weight to lift an object.

10-19 A mining scoop shovel uses a fixed pulley to redirect its force.

A *movable pulley* works differently. The pulley is directly attached to the load that exerts the resistance force. One end of the string or rope is attached to a rigid support, and a force is applied to the other end. Figure 10-20 (b) shows a free-body diagram for the pulley-load system in the absence of friction. The rope is assumed to be ideal. If all forces on the system in the diagram are in equilibrium, then

$$\Sigma \mathbf{F} = \mathbf{T} + \mathbf{T} + m\mathbf{g} = 0 \text{ N.}$$

You can rewrite this expression using vertical scalar components:

$$T_y + T_y = 2T_y = -mg_y$$

$$T_y = -\frac{mg_y}{2}$$

In other words, each rope carries half the weight of the load. In Figure 10-20 and b, the resistance force (F_r) is the weight of the chest. The man holds only one end of the rope, so only one of the tension forces is the effort force, $T = F_e$. Comparing the magnitudes of the forces, you find that

$$F_r = 2T = 2F_e.$$

The theoretical mechanical advantage of the movable pulley is

$$\text{TMA} = \frac{F_r}{F_e} = \frac{2\cancel{F_e}}{\cancel{F_e}} = 2.$$

You can see then that a movable pulley doubles the effort force. However, it also doubles the distance over which the force must be applied. If the man in Figure 10-20 (a) pulls the rope 1 m, the pulley and storage chest move only $\frac{1}{2}$ m.

A system of fixed and movable pulleys is called a **block and tackle system.** Figure 10-21 (a) shows a block and tackle system with two fixed and two movable pulleys. The effort force equals the uniform tension in the continuous rope that runs through the ideal block and tackle system. The upper fixed pulleys change only the direction of the tensional force. The lower movable pulleys, as

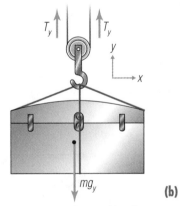

(a)

(b)

10-20 (a) Using a single moveable pulley; (b) Free-body diagram of the chest

A movable pulley is attached to the load to be moved. Movable pulleys individually have a TMA = 2.

Conservation of Energy   **227**

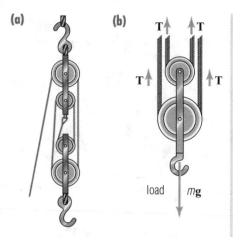

(a)   **(b)**

load | $m\mathbf{g}$

10-21 Block and tackle systems are used to multiply an available lifting force many times.

Problem-Solving Strategy 10.4

The TMA of a block and tackle system is equal to the number of ropes physically supporting the load or resistance force.

figure (b) shows, double the effort force with each pulley. Let's see how this happens.

The net force on the load in figure (b) is the sum of its weight and the four tensional forces in the ropes running through the pulleys attached to the load. If these forces are in equilibrium, then

$$\Sigma\mathbf{F} = m\mathbf{g} + \mathbf{T} + \mathbf{T} + \mathbf{T} + \mathbf{T} = 0 \text{ N}.$$

Once again, the resistance force is the weight of the load and the effort force is just the tension in the rope. Therefore,

$$4\mathbf{T} = -m\mathbf{g}$$
$$4T = |mg|$$
$$4F_e = F_r.$$

The TMA is

$$\text{TMA} = \frac{F_r}{F_e} = \frac{4F_e}{F_e} = 4.$$

Notice that with four ropes pulling up on the load, the TMA is four; with two ropes pulling on the load, as for the movable pulley, the TMA is two; and with one rope pulling on the load, as for the fixed pulley, the TMA is one. This suggests that the theoretical mechanical advantage of a pulley system is equal to the number of ropes supporting the load. This is true for any arrangement of pulleys.

10.12 Mechanical Efficiency

You saw earlier that an ideal simple machine conserves energy. No nonconservative forces act within an ideal simple machine. The work put into the machine equals the work output of the machine. No energy is lost. Real machines do not work this way. Friction between moving parts causes wear. Edges become dull. Axles squeak. Ropes stretch nonelastically. Parts flex and give off heat. The work output of a real machine is never equal to the work put in. It's always less.

Engineers (and consumers) are interested in getting as much energy out of a process as possible compared to the energy available at the beginning of the process. Since energy translates into work, they compare the output work to the work or energy put into a process to determine the **efficiency** (η) of the process. The efficiency is most easily determined by finding the actual mechanical advantage for the process, then dividing the AMA by the TMA.

$$\eta = \frac{\text{AMA}}{\text{TMA}} \times 100\% \qquad (10.9)$$

Other methods exist for determining efficiency that we will consider in later chapters.

EXAMPLE 10-4

The Nearly Perfect Pulley: Mechanical Efficiency

A physics student sets up the block and tackle system shown in Figure 10-22 and attaches a 10 N spring scale to the free end of the string. The load for his experiment is a 1 kg laboratory mass that is hooked to the bottom pulley. He slowly pulls down on the free end of the string until the mass is suspended by the tackle and is hanging motionless. The spring scale indicates 3.30 N. He then lowers the mass to the tabletop, unhooks it from the apparatus, and weighs the mass with the spring scale. The mass weighs 9.82 N. (a) What is the theoretical mechanical advantage of the pulley system? (b) What is the actual mechanical advantage? (c) What is the efficiency of the system?

Solution:

a. From Figure 10-22, you can see that there are three strings supporting the mass and lower pulley, so the TMA for the system is 3.

b. The AMA is the ratio of F_r to F_e. The effort force is $F_e = 3.30$ N. The resistance force is the weight of the mass, $F_r = 9.82$ N. Using Equation 10.8, you can find the AMA:

$$AMA = \frac{F_r}{F_e} = \frac{9.82 \text{ N}}{3.30 \text{ N}}$$

$$AMA \doteq 2.97\underline{5}$$

$$AMA \approx 2.98$$

c. The efficiency is calculated by Equation 10.9:

$$\eta = \frac{AMA}{TMA} \times 100\%$$

$$\eta = \frac{2.97\underline{5}}{3} \times 100\%$$

$$\eta \doteq 99.1\underline{6}\%$$

$$\eta \approx 99.2\%$$

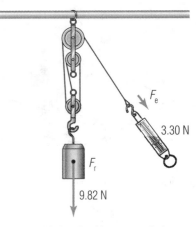

10-22 A block and tackle system with three supporting strings

Stationary pulley systems are very nearly 100% efficient. Their efficiency decreases when the string runs through the pulleys because of the kinetic friction occurring at many points in the system.

Consumers are very interested in the efficiency of the products they buy. Fuel efficiency in cars and furnaces, electrical efficiency in air conditioners and laptop computers, and well-designed hand tools that provide maximum force with minimum effort are all examples of product efficiency. The federal government often mandates that manufacturers prominently post energy efficiency ratings on products prior to their sale.

10B Section Review Questions

1. What force in all its forms do you ignore when determining theoretical mechanical advantage?

2. How does the operator of a simple machine that magnifies force "pay" for this mechanical advantage?

3. Which kind of lever always has a TMA > 1? Why is this so?

4. What quantity would you calculate to determine the overall effect of friction and other nonconservative forces on a process?

⊛5. A frictionless ramp forms a 20.0° angle with the horizontal. A 5.00 kg load rests on the ramp. Calculate the magnitude of the force that will hold the load stationary on the ramp, and then determine the theoretical mechanical advantage of the ramp using this information.

⊛6. A 75.0 kg worker must raise a 2.40×10^3 N shipping crate a few centimeters in order to remove an electrical extension cord that became trapped under the crate. If he intends to use a 2.00 m pry bar and his entire weight to lift the crate, what is the maximum distance from the

crate the fulcrum can be placed that will allow him to just barely move the crate? Assume that the crate's entire weight bears on the crowbar.

⊙7. What is the length of rope that the sailor must pull in order to lift the boat 1.80 m from the ship's deck with the block and tackle shown in the diagram in the margin?

⊙8. A block and tackle pulley system is 96.5% efficient when a load is being lifted. The effort force is 22.6 N. If the theoretical mechanical advantage of the system is 6, how heavy is the load?

SRQ 10B-7 Nineteenth-century sailors relied almost exclusively on block and tackle rigging to operate their ships.

Chapter Review

In Terms of Physics

conservation law	10.1	screw	10.9
conserved	10.1	pitch	10.9
terminal velocity	10.2	lever	10.10
central force	10.4	first-class lever	10.10
escape speed	10.6	second-class lever	10.10
machine	10.7	third-class lever	10.10
theoretical mechanical advantage (TMA)	10.8	block and tackle system	10.11
actual mechanical advantage (AMA)	10.8	efficiency (η)	10.12
inclined plane	10.9		

Problem-Solving Strategies

10.1 (page 218) You can simplify many problems by using the conservation of total mechanical energy. As you learned in the last chapter, $E = K + U$, and $K_1 + U_1 = K_2 + U_2$, or $\Delta K = -\Delta U$.

10.2 (page 220) Escape speed from the surface of a planet does not depend on the mass of the launched object—it depends only on the mass and the radius of the planet.

10.3 (page 225) Equation 10.8 can be used for determining the AMA of any system by substituting the actual effort and resistance forces into the equation.

10.4 (page 228) The TMA of a block and tackle system is equal to the number of ropes physically supporting the load or resistance force.

Review Questions

1. What is a conservation law? How does the existence of conservation laws give evidence for a completed Creation?

2. How can you tell that nonconservative forces are working in a mechanical process?

3. A feather falls through the air. Is its total mechanical energy conserved? Why or why not?

4. An astronaut throws a rock upward while standing on the airless Moon. After the rock leaves the astronaut's hand, is its total mechanical energy conserved?

5. A speeding car screeches to a stop. What happened to its kinetic energy?

6. A mass is attached to a hanging spring that is located far from any source of thermal energy. The temperature of the spring increases as it oscillates. What happens to its total mechanical energy?

7. How much work is done by gravity on an airliner flying level at 30 000 ft between London and New York?

8. Three balls with identical masses are released from rest at the same height, *h,* as shown in the figure, and frictional forces are negligible.
   a. Which ball will be moving fastest at ground level: *A, B,* or *C?*
   b. Which will strike the ground first?

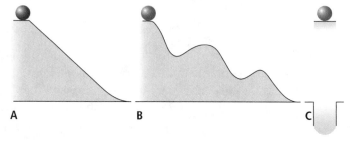

9. Explain why the escape speed from Earth's gravity for a baseball is the same as for a space shuttle.

10. What is the principal difference between a simple machine and a complex machine?

11. Define *theoretical mechanical advantage* (TMA). Why is theoretical mechanical advantage not attainable in real systems?

12. Describe at least three different mathematical ratios that will yield TMA for simple machines. Define each quantity the first time you use it in your answer.

13. Which one of the following is not a simple machine?
   a. bolt
   b. automobile engine
   c. broom
   d. baseball bat
   e. knife
   f. scissors
   g. stairway
   h. crowbar

14. What is the actual mechanical advantage of a machine whose output force is double its input force?

15. Why is it easier to measure actual mechanical advantage than it is to predict it?

16. Name the two basic kinds of simple machines.

17. Which one of the following does not contain an inclined plane?
   a. knife
   b. axe
   c. mountain road
   d. wood plane
   e. broom
   f. screw

18. Which has a greater mechanical advantage, a second-class lever or a third-class lever?

19. What is the theoretical mechanical advantage of a movable pulley?

20. What is the theoretical mechanical advantage of the block and tackle pictured here?

21. Describe the most direct way to calculate the mechanical efficiency of a machine.

True or False (22–31)

22. Gravity is a conservative force.

23. Astronauts beam to Earth TV images of spinning objects floating in the habitable compartments of the space station. Energy is conserved by spinning objects under these weightless conditions.

24. Some thermal energy is transferred to a falling raindrop due to its change of position.

25. Static friction can be both a conservative force and a nonconservative force.

26. The presence of an atmosphere increases the required escape speed of a vehicle launched from a planet's surface.

27. Machines with lower mechanical advantages require more work to do a certain job.

28. A wheel is a modified lever.

29. A screw with a large pitch has a small mechanical advantage.

30. First-class levers always have a mechanical advantage of 1 or more.

31. The theoretical mechanical advantage for any conceivable block and tackle system is equal to the number of ropes or strings physically supporting the load, not including the one exerting the effort force.

⊚32. A 3.00 kg ball is thrown with a velocity of 10.0 m/s upward. How high will it rise before it stops if it is released at a height of 1.60 m?
   a. Solve the problem using the principles of the conservation of energy.
   b. Solve the problem using the equations of motion.

⊚33. A frictionless surface is inclined 35.0° to a horizontal surface. A 1.40 kg ball rests on the incline at a height of 1.00 m above the horizontal surface. An identical ball is suspended 1.00 m above the horizontal surface. If both balls are released at the same instant, how long after release will each ball strike the horizontal surface?

⊚34. A 1250 kg chunk of solid iron is all that is left of a speeding asteroid after it has skimmed the surface layers of the Sun. It has sufficient radial speed to reach a theoretically infinite position from the Sun.
   a. How much does the asteroid's gravitational potential energy change?
   b. If its radial speed at the Sun is v_0 and its speed is 0 m/s at an infinite distance, what is v_0?

⊚35. A 2.00 g penny is dropped from the top of a 110. m building.
   a. How far has the penny fallen when its speed reaches 45.0 m/s?
   b. Neglecting air resistance, how far above the ground is it at that speed?

⊚36. A ramp allows a small force exerted over 10 m to raise a load 1 m vertically. What is the ramp's theoretical mechanical advantage?

⊚37. A sailor exerts 330. N of force to raise a sail at a constant speed using a single fixed pulley. The sail and rope weigh 275 N.
   a. What is the theoretical mechanical advantage of the pulley?
   b. What is the actual mechanical advantage?
   c. What is the efficiency of the pulley?
   d. What would you suggest could be done to improve the efficiency of the pulley?

⊚38. A machine bolt requires 30 complete turns in order to thread it into a hole 6.00 cm deep. What is the pitch of the bolt?

⊚39. Assume that the biceps muscle connects to the radius bone of the forearm 3.00 cm from the elbow. A 10.0 kg mass is held in the hand 30.0 cm from the joint with the forearm horizontal. In addition to holding up the forearm, how much additional force must the muscle exert in order to keep the mass motionless?

⊚40. A 785 kg piano must be lifted 4.50 m to a performance stage in a large auditorium.
   a. How much work must be done to move the piano?
   b. If the piano is moved up a frictionless incline to the stage, how much work is done?
   c. A block and tackle is used instead and an effort force of 2.67×10^3 N is applied to lift the piano at a constant speed. If 13.5 m of rope is pulled through the block and tackle system, what is its efficiency?

⊚⊚41. A spool is a cylinder mounted on a shaft around which a rope or string is wrapped multiple times. Two spools are mounted on the same shaft. One spool has a diameter of 0.300 m, while the smaller spool's diameter is 0.200 m. A string wrapped around the larger spool is tied to a 1.50 kg load. You pull on the string wrapped around the smaller spool, rotating the shaft and lifting the load attached to the large spool. How much force must you exert on the string in order to lift the load at a constant speed? Ignore friction in the axle.

11A	Introducing Momentum	233
11B	Collisions	239
11C	Center of Mass and Angular Momentum	249
Facet:	Gravity Assists	247

New Symbols and Abbreviations	
momentum (p)	11.2
impulse (I)	11.4
angular momentum (L)	11.14

When landing, the space shuttle has so much momentum that it continues to roll for miles before it comes to a stop.

Momentum 11

Cast thy burden upon the Lord, and he shall sustain thee: he shall never suffer the righteous to be moved. *Psalm 55:22*

This chapter deals with another principle of conservation. Conservation implies a sustained condition that is preserved from outside influences. In Psalm 55, one of God's manifold promises is mentioned—the promise to sustain you in the presence of opposing forces. There is but one condition to this promise—you must be among the righteous. Since no one is intrinsically righteous, we must take on the righteousness offered by Jesus Christ. Only then will we be sustained in our walk through life.

11.1 Introduction

"Seven . . . Six . . . Five . . . We have go for main engine start . . . Two . . . One . . . We have liftoff!" These words on April 12, 1981, marked the beginning of the first flight of the space shuttle *Columbia*. Thousands cheered as the *Columbia* blasted into space. Although the space shuttle was called a marvel of modern technology, it used the same principle as its predecessors, the early rockets, to get into space. That same principle also applies to such activities as playing marbles, target riflery, and bowling. The principle is the *conservation of momentum.*

11A INTRODUCING MOMENTUM

11.2 Definition of Momentum

Momentum (p) is a measure of an object's motion. Sir Isaac Newton called this property of an object its "quantity of motion." For example, an object with large momentum is harder to stop than an object with small momentum. What physical properties do you think affect momentum?

The first aspect of momentum should be obvious after a little thought. Which is easier to stop—a 10 000 kg truck moving at 50 km/h or a 0.05 kg tennis ball moving at the same speed? You can stop the tennis ball with one hand, but your whole body would not stop the truck! It seems clear that the mass of an object affects its momentum.

Which is easier to stop—a 0.01 kg bullet moving at 1000 m/s or the same bullet moving at a speed of 1 m/s? You can safely catch the slower bullet in your hand, but the faster bullet can cause great damage before it stops. So, an object's velocity also affects its momentum.

It is not surprising, then, that momentum is defined as the product of mass and velocity:

$$\mathbf{p} \equiv m\mathbf{v} \tag{11.1}$$

Linear momentum is a vector in the same direction as the velocity vector. In the SI, mass is measured in kilograms and velocity is measured in meters per second. Consequently, the unit of momentum is kg·m/s. The unit for momentum does not have a special name.

EXAMPLE 11-1

11-1 Whose momentum is greater?

Law of Gross Tonnage: Momentum

Which has greater momentum, an 85.0 kg halfback running south at 9.00 m/s or a 135 kg defensive lineman running at 6.00 m/s in the same direction?

Solution:

Calculate the magnitude of momenta for each athlete using the scalar values in Equation 11.1:

$$p = mv$$

Halfback:

$$p_{hb} = (85.0 \text{ kg})(9.00 \text{ m/s})$$

$$p_{hb} = 765 \text{ kg·m/s}$$

Lineman:

$$p_{dl} = (135 \text{ kg})(6.00 \text{ m/s})$$

$$p_{dl} = 810. \text{ kg·m/s}$$

The defensive lineman has greater momentum than the halfback.

11.3 Changing Momentum: Constant Mass

Since an object's mass is usually constant, a change in momentum is almost always a change in velocity—that is, an acceleration. Therefore,

$$\Delta \mathbf{p} = \Delta(m\mathbf{v}), \text{ and}$$
$$\Delta \mathbf{p} = m\Delta \mathbf{v}. \tag{11.2}$$

Dividing both sides by the time for the change (Δt) gives

$$\frac{\Delta \mathbf{p}}{\Delta t} = m\frac{\Delta \mathbf{v}}{\Delta t}. \text{ But}$$

$$\frac{\Delta \mathbf{v}}{\Delta t} = \mathbf{a}, \text{ so}$$

$$\frac{\Delta \mathbf{p}}{\Delta t} = m\mathbf{a}.$$

Remember, Newton's second law says that

$$\Sigma\mathbf{F} = m\mathbf{a},\text{ so}$$

$$\Sigma\mathbf{F} = \frac{\Delta\mathbf{p}}{\Delta t}. \tag{11.3}$$

In fact, Equation 11.3 was the way Newton originally stated his second law. A change in momentum results from an unbalanced force.

11.4 Impulse

An unbalanced force on an object changes its momentum. Does it make any difference how long the force acts? Consider a train starting from rest at a station. If its engine exerts a force to turn its wheels for 1 second, will its speed change significantly? Probably not. However, if its engine exerts a force for 1 minute, its speed, and therefore its momentum, will change substantially.

It is easy to see the effect that the time has on the change in momentum by rearranging Equation 11.3:

$$\Delta\mathbf{p} = \Sigma\mathbf{F}\Delta t \tag{11.4}$$

The longer the force is applied (Δt), the more the momentum changes. The product of a force and the time the force is exerted is called the **impulse (I)** of the force.

$$\mathbf{I} \equiv \Sigma\mathbf{F}\Delta t \tag{11.5}$$

By substituting Equation 11.5 into Equation 11.4, you can see that

$$\Delta\mathbf{p} = \mathbf{I}. \tag{11.6}$$

Although any force can have an impulse, we use the term "impulsive force" to refer to the forces on a system that are large compared to the other forces. When a ball is dropped to the ground, gravity is the impulsive force. At the instant that a ball is hit with a bat, the force the bat exerts on the ball is the impulsive force, since it is large compared to the force of gravity.

The following example shows the importance of impulse.

EXAMPLE 11-2

Landing Flat-Footed: Impulse

A 50.0 kg woman jumps to the ground from a wall that is 2.00 m high. At the instant she hits the ground, her velocity is 6.26 m/s downward. After she lands, her velocity is zero. (a) What is the change in her momentum? (b) If she hits the ground flat-footed with her knees locked, taking 0.100 s to stop, what is the force on her feet? (c) If she flexes her knees and lands gradually, taking 0.800 s to stop, what is the force on her feet?

Solution:

Assume all motion takes place in the vertical. Use the y-scalar components for the calculations.

a. Calculate her change in momentum. Use Equation 11.2:

$$\Delta p_y = m\Delta v_y$$

$$\Delta p_y = m(v_2 - v_1)_y = (50.0 \text{ kg})[0 \text{ m/s} - (-6.26 \text{ m/s})]$$

$$\Delta p_y = +313 \text{ kg·m/s (up)}$$

b. Calculate the vertical force on her feet when she lands flat-footed. Use Equation 11.3:

$$\Sigma F_y = \frac{\Delta p_y}{\Delta t}$$

$$\Sigma F_y = \frac{+313 \text{ kg·m/s}}{0.100 \text{ s}} = +3130 \text{ N} \qquad \text{(landing flat-footed)}$$

c. Calculate the vertical force when she lands with flexed legs:

$$\Sigma F_y = \frac{\Delta p_y}{\Delta t}$$

$$\Sigma F_y = \frac{+313 \text{ kg·m/s}}{0.800 \text{ s}}$$

$$\Sigma F_y \doteq +391.2 \text{ N}$$

$$\Sigma F_y \approx 391 \text{ N} \qquad \text{(landing with flexed legs)}$$

The difference in force is substantial. Thus, by spreading out the time of a collision, the force on the colliders can be lessened.

11.5 Conservation of Momentum

Firing a high-caliber rifle causes a powerful recoil. When a rifle fires, the exploding gunpowder exerts a force on the bullet, $\mathbf{F}_{r \to b}$, for a time Δt. The bullet, therefore, exerts a reaction force of $\mathbf{F}_{b \to r}$ on the rifle for the same time, Δt, according to Newton's third law. The impulse of the force on the bullet is

$$\mathbf{I}_b = \mathbf{F}_{r \to b} \Delta t.$$

The impulse of the bullet on the rifle is

$$\mathbf{I}_r = \mathbf{F}_{b \to r} \Delta t = -\mathbf{F}_{r \to b} \Delta t.$$

The change in the bullet's momentum is

$$\Delta \mathbf{p}_b = \mathbf{I}_b = \mathbf{F}_{r \to b} \Delta t.$$

The rifle's momentum also changes.

$$\Delta \mathbf{p}_r = \mathbf{I}_r = \mathbf{F}_{b \to r} \Delta t$$
$$\Delta \mathbf{p}_r = -\Delta \mathbf{p}_b$$
$$\Delta \mathbf{p}_r + \Delta \mathbf{p}_b = 0 \text{ kg·m/s}$$

The change in the rifle's momentum is observed as **recoil.** However, if the rifle and the bullet together are considered the system, the overall momentum of the system does not change.

This discussion reveals the principle of **conservation of momentum:** when no unbalanced outside forces act on a system, the momentum of the system is conserved (stays the same). Notice that the momentum of the bullet and the momentum of the rifle both changed. However, they changed equally in opposite directions so that the momentum change for the system was zero.

Suppose a hunter with a rifle is stranded in the middle of a frozen, frictionless lake. How can he propel himself to the edge of the lake? Because there is no friction between his boots and the ice, he cannot walk. Suppose he fires the gun. The bullet is given momentum in one direction. Since no outside force is assumed to

The magnitude of a decelerating force can be reduced by increasing the amount of time that a given change of momentum takes place. This is the principle behind automobile air bags.

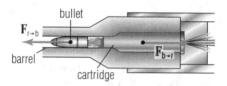

11-2 Reaction forces on a bullet and the rifle as the cartridge fires

Forces between parts within the boundaries of a system are called *internal forces.* All forces that originate outside the boundaries of a system are *external forces.*

Recoil is the velocity given to one object within a system by other parts of the system.

The forces exerted by the exploding powder in the rifle shell come from *within* the system. Internal forces do not change the overall momentum of a system—only external forces do.

Momentum of a system is **conserved** when the net external force acting on the system is zero.

act on the rifle-bullet system, the rifle must gain momentum in the opposite direction to balance the bullet's gain in momentum. Of course, when the rifle moves, the man holding it must move, too, since he is part of the rifle component of the system. Therefore, when the man fires the rifle, he moves in the direction opposite the bullet's direction. He will eventually reach the shore even if he fires only once. If he fires the gun repeatedly, he will reach the shore more quickly than if he fires only once.

11.6 Changing Momentum: Variable Mass

A space vehicle is in the same predicament as the hunter on frictionless ice—there is nothing against which it can push in order to move. The space vehicle's predicament may be solved by a means similar to the hunter firing his gun. Instead of ejecting bullets, a chemical rocket exhausts burned rocket fuel, and an ion propulsion engine ejects charged particles. Such material is called *propellant mass*. As the propellant mass gains momentum in one direction, the spacecraft gains the same amount of momentum in the opposite direction. Therefore, engineers design the combustion chambers of space engines to rapidly eject as much mass as possible in order to develop maximum momentum.

The rocket nozzle shapes and accelerates the flow of exhaust gases to produce the maximum thrust for the mass of exhaust ejected.

Let's examine how rocket propulsion in space works. Assume that the mass of propellant that is ejected as exhaust is Δm_{prop}. The momentum of this propellant exhaust is the product $\Delta m_{prop} v_{exh}$. The exhaust velocity is measured relative to the *rocket*. If we assume that there are no external forces acting on the rocket-propellant system, then the total change in momentum of the system must be zero. Therefore, the momentum of the rocket must change by an amount equal to the change of momentum of the propellant exhaust. The change of momentum of the rocket must then be

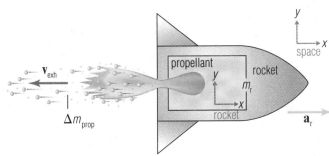

11-3 Conservation of momentum in a rocket-exhaust system

$$\Delta p_{r\,x} = -(\Delta m_{prop})v_{exh\,x}.$$

If we divide both sides by the time (Δt) it takes to exhaust Δm_{prop} of propellant, we obtain the impulsive force of the exhaust on the rocket.

The velocity of the exhaust is in the opposite direction to the momentum of the rocket.

$$F_{exh\,x} = \frac{\Delta p_{r\,x}}{\Delta t} = -\left(\frac{\Delta m_{prop}}{\Delta t}\right)v_{exh\,x}$$

The term $\Delta m_{prop}/\Delta t$ is the rate of propellant mass flow, measured in kg/s. The impulsive force of the exhausted propellant, $F_{exh\,x}$, is called **thrust.** Thrust is the force that accelerates the rocket. Expressing this equation in the form of Newton's second law, we get

$$F_{exh\,x} = m_r a_{r\,x}, \text{ or}$$

$$-\left(\frac{\Delta m_{prop}}{\Delta t}\right)v_{exh\,x} = m_r a_{r\,x}, \qquad (11.7)$$

where m_r is the combined mass of the rocket and the remaining propellant within the rocket. Unlike previous examples, the spacecraft's mass continually changes as long as it ejects propellant mass in its exhaust. The left side of Equation 11.7 normally remains relatively constant until the propellant is gone. Therefore, on the right side of the equation, as the rocket's mass decreases with time, its acceleration must increase. This case of nonuniform acceleration requires more advanced mathematics to analyze the rocket's motion. You *cannot* use the three equations of motion for systems where mass and acceleration vary.

EXAMPLE 11-3

Rocket Science in Deep Space: Conservation of Momentum

A 5.10×10^5 kg space probe is at rest in an assumed fixed reference frame in deep space. In order to maneuver toward a comet nucleus, the probe fires its engine, which exhausts propellant at a mass flow rate of 1.00×10^4 kg/s with a speed of 700. m/s. What is the probe's initial acceleration the instant the engine fires?

Solution:

Assume that all motion is parallel to the x-coordinate axis with the probe moving in the positive direction. Use scalar components for your computations.

Solve Equation 11.7 for acceleration:

$$-\left(\frac{\Delta m_{prop}}{\Delta t}\right) v_{exh\,x} = m_r a_{r\,x}$$

$$a_{r\,x} = -\left(\frac{\Delta m_{prop}}{\Delta t}\right)\frac{v_{exh\,x}}{m_r}$$

Since the mass of the rocket has not changed appreciably, substitute the known quantities into this equation:

$$a_{r\,x} = -(1.00 \times 10^4 \text{ kg/s})\frac{(-700. \text{ m/s})}{(5.10 \times 10^5 \text{ kg})}$$

$$a_{r\,x} \doteq +13.\underline{7}2 \text{ m/s}^2$$

$$a_{r\,x} \approx +13.7 \text{ m/s}^2$$

The initial instantaneous acceleration of the probe is almost 14 m/s².

11A Section Review Questions

1. Define momentum. What must a system be doing in order for momentum to exist?

2. Mathematically, what kind of physical quantity is momentum? How do you determine the direction of momentum?

3. When a dropped object falls toward the ground, is its change of momentum positive or negative? When it hits the ground, is its change of momentum positive or negative? Assume that up is positive.

4. For a given change of momentum, will the force producing the impulse be larger or smaller if the change of momentum takes longer to occur?

5. Calculate the momentum of a 145 g baseball that is moving at 46.9 m/s at an angle of 30.0° to the horizontal.

6. A 4.32 kg rifle is clamped to a frictionless cart in a test firing range, pointing at the distant target. The rifle is fired electronically. The 9.72 g bullet slug exits the muzzle 0.001 80 s later with a speed of 835 m/s. What is the momentum of the bullet?

7. What is the total impulsive force of the burning propellant from the cartridge on the bullet slug in Question 6?

8. What is the recoil velocity of the rifle in Question 6?

⊙⊙9. An astronaut is working around the International Space Station (ISS), attached by a tether cord. The snap on the tether fails and he finds himself adrift, but stationary relative to the ISS. This astronaut was a former physics student and knows that if he can throw some objects away from himself in the right direction, he can propel himself back to the station. His tool kit has a 1.20 kg torque wrench, a 0.70 kg nut driver, and a 0.35 kg bolt. The astronaut, his suit, and all of his gear have an initial mass of 105.00 kg. If he throws away each of the items in his tool kit at 3.0 m/s, one at a time, what will be his final speed relative to the ISS?

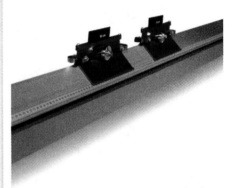

11-4 Gliders on an air track

11B COLLISIONS

11.7 Linear Collisions

We have discussed the conservation of momentum when *internal* forces separate objects within a system. Now let us discuss the conservation of momentum of colliding bodies. In general, the same principle applies as long as we consider the colliding objects as parts of a single system—the momentum of the system is conserved unless an unbalanced *external* force acts on the system.

The air track shown in Figure 11-4 is a traditional physics laboratory apparatus that demonstrates conservation of momentum in collisions. The track is pierced by hundreds of holes through which air flows to form a nearly frictionless cushion of air to support y-shaped gliders. The gliders move in a linear path along the straight track. Suppose glider 1 with mass m_1 on the left part of the track is given a velocity of $\mathbf{v}_1$ toward the right. Glider 2 with mass m_2 is at rest to the right of m_1. We will consider the system to include both m_1 and m_2. There are no external unbalanced forces on the system. Therefore, the total momentum of the system is conserved and

$$\Delta\mathbf{p}_{\text{total}} = 0 \text{ kg·m/s.}$$

What happens when m_1 collides with m_2? We know that the total momentum before the collision is

$$\mathbf{p}_{\text{total bfr}} = \mathbf{p}_{1 \text{ bfr}} + \mathbf{p}_{2 \text{ bfr}},$$

where the subscripts "1" and "2" identify the properties of gliders 1 and 2 respectively, and the subscript "bfr" identifies the properties before the collision. The total momentum of the system after the collision is

$$\mathbf{p}_{\text{total aft}} = \mathbf{p}_{1 \text{ aft}} + \mathbf{p}_{2 \text{ aft}},$$

where the subscript "aft" identifies the properties after the collision.

Since no external force is exerted on the system, the momentum of the system before and after the collision must be the same.

$$\mathbf{p}_{\text{total bfr}} = \mathbf{p}_{\text{total aft}}$$

If we expand these expressions, we can see how the individual momenta change but the overall system momentum is unchanged.

$$\mathbf{p}_{1 \text{ bfr}} + \mathbf{p}_{2 \text{ bfr}} = \mathbf{p}_{1 \text{ aft}} + \mathbf{p}_{2 \text{ aft}}$$

$$\mathbf{p}_{1 \text{ bfr}} - \mathbf{p}_{1 \text{ aft}} = \mathbf{p}_{2 \text{ aft}} - \mathbf{p}_{2 \text{ bfr}}$$

$$-(\mathbf{p}_{1 \text{ aft}} - \mathbf{p}_{1 \text{ bfr}}) = \mathbf{p}_{2 \text{ aft}} - \mathbf{p}_{2 \text{ bfr}}$$

$$-m_1(\mathbf{v}_{1 \text{ aft}} - \mathbf{v}_{1 \text{ bfr}}) = m_2(\mathbf{v}_{2 \text{ aft}} - \mathbf{v}_{2 \text{ bfr}})$$

$$m_1(\mathbf{v}_{1 \text{ aft}} - \mathbf{v}_{1 \text{ bfr}}) = -m_2(\mathbf{v}_{2 \text{ aft}} - \mathbf{v}_{2 \text{ bfr}})$$

In collision problems, the subscripts "bfr" and "aft" refer to properties or conditions before and after the collision.

Problem-Solving Strategy 11.2

Evaluate the effect of all external forces on the system as a whole. If there is no net external force, then the total change of momenta of all parts of the system is zero and momentum is conserved.

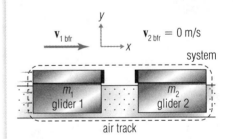

11-5a Before collision

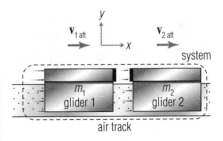

11-5b After collision

This relationship can be summarized in the equation

$$\Delta\mathbf{p}_1 = -\Delta\mathbf{p}_2. \tag{11.8}$$

Before the collision, m_2 is at rest; so $\mathbf{v}_{2\,\text{bfr}} = 0$ m/s. Therefore,

$$m_1\mathbf{v}_{1\,\text{aft}} - m_1\mathbf{v}_{1\,\text{bfr}} = -m_2\mathbf{v}_{2\,\text{aft}}. \tag{11.9}$$

We do not know either $\mathbf{v}_{1\,\text{aft}}$ or $\mathbf{v}_{2\,\text{aft}}$. Since there are two unknowns in Equation 11.9, we cannot solve it. We need another equation involving the unknowns before we can find their values. This requires a closer look at the kind of collision that occurred.

11.8 Collision Modes

Objects may collide in one of three ways or *modes*. In an **elastic collision,** only conservative forces act; therefore, total mechanical energy is conserved. This is a strictly theoretical mode of collision, but many real systems approach this condition, such as the collision of billiard balls rolling on a hard, smooth surface, or the collision of air molecules.

In a **partially elastic collision,** one or more of the colliding masses is deformed (changes its shape), but the masses rebound away from the collision. The material deformation converts some of the system's kinetic energy to thermal energy, sound, or other nonmechanical energy. Most real collisions are partially elastic. In an **inelastic collision,** the colliding masses stick together. The collision converts a large portion of the initial kinetic energy to nonmechanical energy; therefore, kinetic energy is not conserved. Examples of such a collision are a bullet imbedding itself in a tree and a scoop of ice cream falling to the floor.

11.9 Elastic Collisions

Let's suppose that the collision between the gliders on the air track is elastic and x is positive to the right. Kinetic energy is conserved. Therefore, the following relationships are true:

$$K_{\text{total bfr}} = K_{\text{total aft}}, \text{ and}$$

$$\tfrac{1}{2}m_1v_{1\,\text{bfr}_x}^2 + \tfrac{1}{2}m_2v_{2\,\text{bfr}_x}^2 = \tfrac{1}{2}m_1v_{1\,\text{aft}_x}^2 + \tfrac{1}{2}m_2v_{2\,\text{aft}_x}^2$$

In this example, $v_{2\,\text{bfr}_x} = 0$ m/s; so

$$\tfrac{1}{2}m_1v_{1\,\text{bfr}_x}^2 = \tfrac{1}{2}m_1v_{1\,\text{aft}_x}^2 + \tfrac{1}{2}m_2v_{2\,\text{aft}_x}^2. \tag{11.10}$$

Earlier in the discussion we obtained Equation 11.9:

$$m_1\mathbf{v}_{1\,\text{aft}} - m_1\mathbf{v}_{1\,\text{bfr}} = -m_2\mathbf{v}_{2\,\text{aft}}$$

Since the problem is one-dimensional, we can rewrite Equation 11.9 in scalar component notation as

$$m_1v_{1\,\text{aft}_x} - m_1v_{1\,\text{bfr}_x} = -m_2v_{2\,\text{aft}_x}. \tag{11.11}$$

Combining Equations 11.10 and 11.11, we can solve for $v_{1\,\text{aft}_x}$ and $v_{2\,\text{aft}_x}$. We find that

$$v_{1\,\text{aft}_x} = \left(\frac{m_1 - m_2}{m_1 + m_2}\right)v_{1\,\text{bfr}_x}, \text{ and} \tag{11.12}$$

$$v_{2\,\text{aft}_x} = \left(\frac{2m_1}{m_1 + m_2}\right)v_{1\,\text{bfr}_x}. \tag{11.13}$$

Collisions can occur in three modes. **Elastic collisions** conserve both momentum and mechanical energy. **Partially elastic collisions** and **inelastic collisions** do not conserve mechanical energy because some is converted into nonmechanical energy (e.g., the colliding objects become deformed, produce sparks, or make noise). In **inelastic collisions,** objects stick together.

11-6 Car manufacturers design their vehicles so that collisions will be as inelastic as possible—dissipating the initial mechanical energy as nonmechanical energy to reduce injury to the occupants.

Internal, nonconservative forces can change the kinetic energy of internal particles of a system as well as their momenta. While, ideally, momentum is conserved within a system, kinetic energy usually is not.

Problem-Solving Strategy 11.3

Quantities that have multiple subscripts, such as $v_{1\,\text{bfr}_x}$, can be verbally expressed as "vee-one-before." The sub-subscript x is written to indicate a scalar component but may be left unspoken if no confusion will result.

These equations hold for any elastic collision where a moving mass m_1 collides with a resting mass m_2. If $v_{1\,aft_x}$ is negative, it simply means that m_1 rebounds from the collision with a speed $v_{1\,aft}$.

For some people, interaction with spiritual influences is like an elastic collision—they bounce off completely unchanged. Is your heart hard, or are you letting God teach you through His Word, both studied and preached? (See Heb. 4:2.)

11-7 Bumper cars undergo partially elastic collisions.

EXAMPLE 11-4

Tracking Momentum: An Elastic Collision

Two 50.0 g gliders are on an air track. One glider with m_1 is given a velocity of 2.00 cm/s to the right toward the other glider, which is stationary. What are the velocities of the gliders after the collision?

Solution:
Assume x is positive to the right. Use Equations 11.12 and 11.13 to solve for the final velocities of m_1 and m_2, respectively.
Velocity of glider 1 after collision:

$$v_{1\,aft_x} = \left(\frac{m_1 - m_2}{m_1 + m_2}\right) v_{1\,bfr_x}$$

$$v_{1\,aft_x} = \left(\frac{50.0\text{ g} - 50.0\text{ g}}{50.0\text{ g} + 50.0\text{ g}}\right)(+2.00\text{ cm/s})$$

$$v_{1\,aft_x} = 0\text{ cm/s}$$

Velocity of glider 2 after collision:

$$v_{2\,aft_x} = \left(\frac{2m_1}{m_1 + m_2}\right) v_{1\,bfr_x}$$

$$v_{2\,aft_x} = \left(\frac{2(50.0\text{ g})}{50.0\text{ g} + 50.0\text{ g}}\right)(+2.00\text{ cm/s})$$

$$v_{2\,aft_x} = +2.00\text{ cm/s}$$

The moving glider stops, and the stationary glider moves off with the velocity that m_1 originally had. The momentum of each object has changed in the collision, but the total momentum of the system of objects has remained the same.

EXAMPLE 11-5

Irresistible Force: An Unequal Elastic Collision

An out of control 100.0 kg ice skater with a velocity of 10.0 m/s north careens toward a 10.0 kg stationary steel drum that marks the edge of safe ice on the pond. What are the velocities of the skater and the drum after they collide elastically?

Solution:

Assume that y is positive north and that all motions are frictionless.
Ice skater's velocity after the collision:

$$v_{1 \, aft_y} = \left(\frac{m_1 - m_2}{m_1 + m_2}\right) v_{1 \, bfr_x}$$

$$v_{1 \, aft_y} = \left(\frac{100.0 \text{ kg} - 10.0 \text{ kg}}{100.0 \text{ kg} + 10.0 \text{ kg}}\right)(+10.0 \text{ m/s})$$

$$v_{1 \, aft_y} \doteq +8.1\underline{8}1 \text{ m/s}$$

$$v_{1 \, aft_y} \approx +8.18 \text{ m/s}$$

Drum's velocity after the collision:

$$v_{2 \, aft_x} = \left(\frac{2m_1}{m_1 + m_2}\right) v_{1 \, bfr_x}$$

$$v_{2 \, aft_x} = \left(\frac{2(100.0 \text{ kg})}{100.0 \text{ kg} + 10.0 \text{ kg}}\right)(+10.00 \text{ m/s})$$

$$v_{2 \, aft_x} \doteq +18.\underline{1}8 \text{ m/s}$$

$$v_{2 \, aft_x} \approx +18.2 \text{ m/s}$$

Total momentum before the collision:

$$p_{total \, bfr_y} = p_{1 \, bfr_y} + p_{2 \, bfr_y} = m_1 v_{1 \, bfr_y} + m_2 v_{2 \, bfr_y}$$

$$p_{total \, bfr_y} = (100.0 \text{ kg})(+10.0 \text{ m/s}) + (10.0 \text{ kg})(0 \text{ m/s})$$

$$p_{total \, bfr_y} = +1.00 \times 10^3 \text{ kg·m/s}$$

Total momentum after the collision:

$$p_{total \, aft_y} = p_{1 \, aft_y} + p_{2 \, aft_y} = m_1 v_{1 \, aft_y} + m_2 v_{2 \, aft_y}$$

$$p_{total \, aft_y} = (100.0 \text{ kg})(+8.1\underline{8}1 \text{ m/s}) + (10.0 \text{ kg})(+18.\underline{1}8 \text{ m/s})$$

$$p_{total \, aft_y} \doteq +99\underline{9}.9 \text{ kg·m/s}$$

$$p_{total \, aft_y} \approx +1000. \text{ kg·m/s or } +1.000 \times 10^3 \text{ kg·m/s}$$

The moving man barely slows down, while the stationary drum shoots off at nearly twice the man's original velocity. This is true for elastic collisions between a very large moving mass and a relatively small stationary mass. Once again, the momenta of the two objects is very different after the collision, but the total momentum of the system is unchanged.

EXAMPLE 11-6

Exercise in Futility: Another Unequal Elastic Collision

A boy shoots a 0.0300 kg cork from a pop gun with a velocity of 5.00 m/s to the right toward a 6.80 kg stone placed on a fence post. What are the velocities of each after the elastic collision (assume no friction)?

Solution:

Assume that x is positive to the right.

The velocity of the cork after the collision:

$$v_{1\ \text{aft}_x} = \left(\frac{0.0300\ \text{kg} - 6.80\ \text{kg}}{0.0300\ \text{kg} + 6.80\ \text{kg}}\right)(+5.00\ \text{m/s})$$

$$v_{1\ \text{aft}_x} \doteq -4.9\underline{5}6\ \text{m/s}$$

$$v_{1\ \text{aft}_x} \approx -4.96\ \text{m/s (to the left)}$$

The velocity of the stone after the collision:

$$v_{2\ \text{aft}_x} = \left(\frac{2(0.0300\ \text{kg})}{0.0300\ \text{kg} + 6.80\ \text{kg}}\right)(+5.00\ \text{m/s})$$

$$v_{2\ \text{aft}_x} \doteq +0.043\underline{9}2\ \text{m/s}$$

$$v_{2\ \text{aft}_x} \approx +0.0439\ \text{m/s (to the right)}$$

The cork bounces back with nearly its original speed, while the stone hardly budges to the right. This result is typical of an elastic collision of a small moving mass with a large stationary mass. If the stone had been immovable, attached to an extremely large mass, the cork would have bounced back with exactly its original speed. We assume that this is the case when we discuss an elastic collision with a solid wall.

11.10 Inelastic Collisions

We have discussed idealized elastic collisions, but what about the other extreme—inelastic collisions—where the colliding objects stick completely together following the collision? Going back to the air track example, suppose that glider m_2 has tacky glue on the end facing glider m_1. When m_1 collides with it, the gliders stick together. Therefore, both gliders must have the same velocity after the collision:

$$v_{1\ \text{aft}_x} = v_{2\ \text{aft}_x} = v_{\text{aft}_x}$$

Equation 11.9 becomes:

$$m_1 v_{1\ \text{aft}_x} - m_1 v_{1\ \text{bfr}_x} = -m_2 v_{2\ \text{aft}_x}$$

$$m_1 v_{\text{aft}_x} - m_1 v_{1\ \text{bfr}_x} = -m_2 v_{\text{aft}_x}$$

$$-m_1 v_{1\ \text{bfr}_x} = -m_1 v_{\text{aft}_x} - m_2 v_{\text{aft}_x}$$

$$m_1 v_{1\ \text{bfr}_x} = m_1 v_{\text{aft}_x} + m_2 v_{\text{aft}_x}$$

$$m_1 v_{1\ \text{bfr}_x} = (m_1 + m_2) v_{\text{aft}_x} \tag{11.14}$$

Since we know all these quantities except the final speed of the system, we can find v_{aft_x} without using a second equation.

EXAMPLE 11-7

Sticking Together: Inelastic Collisions

A 1.00 kg kinetic cart with a velocity of 1.00 m/s to the right collides with a stationary 2.00 kg cart. Both carts are equipped with mechanisms that firmly latch the carts together when they collide. (a) What is the velocity of the carts after the collision? (b) Is the system's kinetic energy conserved?

Solution:

Assume that x is positive to the right.

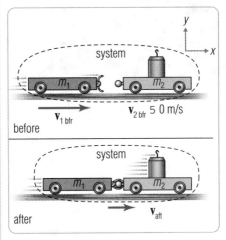

11-8 An inelastic collision

a. Solve for the final velocity of the system:

$$m_1 v_{1\text{ bfr}_x} = (m_1 + m_2)v_{\text{aft}_x}$$

$$v_{\text{aft}_x} = \left(\frac{m_1}{m_1 + m_2}\right)v_{1\text{ bfr}_x} = \left(\frac{1.00 \text{ kg}}{1.00 \text{ kg} + 2.00 \text{ kg}}\right)(+1.00 \text{ m/s})$$

$$v_{\text{aft}_x} = +0.33\overline{3} \text{ m/s}$$

b. Is kinetic energy conserved? If it is, then Equation 11.10 should hold true.

$$\tfrac{1}{2}m_1 v_{1\text{ bfr}_x}^2 \overset{?}{=} \tfrac{1}{2}m_1 v_{\text{aft}_x}^2 + \tfrac{1}{2}m_2 v_{\text{aft}_x}^2$$

$$m_1 v_{1\text{ bfr}_x}^2 \overset{?}{=} (m_1 + m_2)v_{\text{aft}_x}^2$$

$$\left(\frac{m_1}{m_1 + m_2}\right)v_{1\text{ bfr}_x}^2 \overset{?}{=} v_{\text{aft}_x}^2$$

$$v_{1\text{ bfr}_x}\sqrt{\frac{m_1}{m_1 + m_2}} \overset{?}{=} v_{\text{aft}_x}$$

$$(+1.00 \text{ m/s})\sqrt{\frac{1.00 \text{ kg}}{1.00 \text{ kg} + 2.00 \text{ kg}}} \overset{?}{=} +0.33\overline{3} \text{ m/s}$$

$$+0.577 \text{ m/s} \neq +0.33\overline{3} \text{ m/s}$$

Because the two quantities are not equal, kinetic energy is not conserved during an inelastic collision.

Since kinetic energy is not conserved, there must be nonconservative forces acting on the carts. These forces arise from internal friction that converts kinetic energy into other forms of energy.

11.11 Ballistic Pendulums

Another experimental device that depends on completely inelastic collisions is called a **ballistic pendulum.** It is normally used to determine the energy of projectiles such as military and civilian bullets. Indirectly, this information can be used to find the muzzle velocity of guns.

A ballistic pendulum consists of a uniform block of a dense material attached to one or more pendulum arms. Figure 11-9 shows a typical ballistic pendulum. The projectile to be tested impacts the pendulum and embeds itself in the pendulum. The bullet converts essentially all of its kinetic energy into pendulum potential energy as the pendulum rises. An angle indicator records the maximum angle attained by the pendulum arm, and engineers are then able to calculate the kinetic energy and velocity of the projectile at impact.

11-9 A ballistic pendulum

On the Rise: A Ballistic Pendulum

A 10.0 g (0.010 kg) bullet strikes a stationary 2.00 kg ballistic pendulum. The pendulum rises a vertical distance of 12.0 cm (0.120 m). If the bullet remains embedded in the pendulum, what is the bullet's initial speed?

Solution:

The pendulum-bullet system's total mechanical energy is conserved. When the pendulum pauses at the top of its swing, it has gained a potential energy of $\Delta U_g = m_p |g| \Delta h_p$, where Δh_p is the height it rises. At the maximum height, it has no kinetic energy. The pendulum's change in kinetic energy is as follows:

$$\Delta K_p = \tfrac{1}{2}m_p v_{p\,f}^2 - \tfrac{1}{2}m_p v_{p\,i}^2 = 0\text{ J} - \tfrac{1}{2}m_p v_{p\,i}^2$$

$$\Delta K_p = -\tfrac{1}{2}m_p v_{p\,i}^2$$

The velocity of the pendulum and bullet the instant after the bullet imbeds itself ($v_{p\,i}$) is found from the following:

$$\Delta K_p = -\Delta U_{g\,p}$$

$$-\tfrac{1}{2}m_p v_{p\,i}^2 = -m_p |g| \Delta h_p$$

$$v_{p\,i}^2 = 2|g|\Delta h_p$$

$$v_{p\,i} = \sqrt{2|g|\Delta h_p}$$

$$v_{p\,i} = \sqrt{2|-9.81\text{ m/s}^2|(+0.120\text{ m})}$$

$$v_{p\,i} \doteq 1.5\underline{3}4\text{ m/s}$$

Initial pendulum velocity is to the right.
The velocity of the bullet just before impact can be determined from Equation 11.14.

$$m_b v_{b\,bfr_x} = (m_b + m_p)v_{p\,i_x}$$

$$v_{b\,bfr_x} = \frac{m_b + m_p}{m_b} \cdot v_{p\,i_x}$$

$$v_{b\,bfr_x} = \frac{0.0100\text{ kg} + 2.00\text{ kg}}{0.0100\text{ kg}} \cdot (+1.5\underline{3}4\text{ m/s})$$

$$v_{b\,bfr_x} \doteq +30\underline{8}.3\text{ m/s}$$

$$v_{b\,bfr_x} \approx +308\text{ m/s}$$

The bullet's speed just before impact is about 308 m/s.

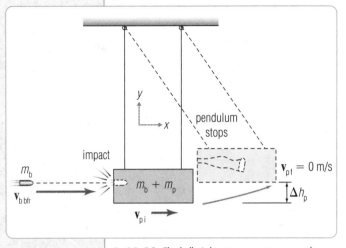

11-10 The bullet's kinetic energy is converted to the pendulum's potential energy.

11.12 Two-Dimensional Collisions

Most collisions occur in two or three dimensions. Momentum is conserved in two- and three-dimensional collisions as well as in one-dimensional collisions, but the equations are more complicated for more dimensions.

An excellent apparatus for demonstrating two-dimensional collisions is an air table, very similar to an air hockey table. Its surface has numerous holes through which air flows to suspend pucks of various masses and to permit nearly friction-less motion.

In Figure 11-11a, puck m_1 approaches the stationary puck m_2 from the left with velocity $\mathbf{v}_{1\,bfr}$. Puck 1 elastically strikes puck 2 off-center and caroms with a velocity $\mathbf{v}_{1\,aft}$. At the same time, m_2 rebounds with velocity $\mathbf{v}_{2\,aft}$. Momentum is conserved, and Equation 11.8 holds true.

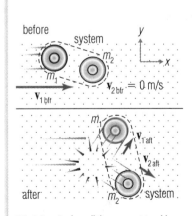

11-11a Pucks colliding on an air table

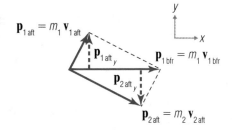

$\mathbf{p}_{1\,\text{aft}} = m_1\mathbf{v}_{1\,\text{aft}}$

$\mathbf{p}_{1\,\text{bfr}} = m_1\mathbf{v}_{1\,\text{bfr}}$

$\mathbf{p}_{2\,\text{aft}} = m_2\mathbf{v}_{2\,\text{aft}}$

11-11b Resolving the momenta of two-dimensional collisions

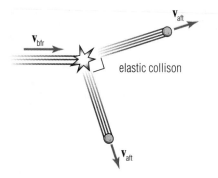

elastic collison

11-12a The rebound angle for an off-center, perfectly elastic collision is always 90°.

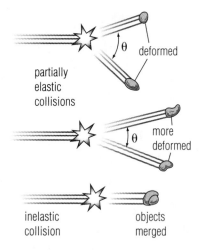

partially elastic collisions

deformed

more deformed

inelastic collision

objects merged

11-12b Rebound angles decrease as collisions become less elastic. The rebound angle is 0° for an inelastic collision.

$$\Delta\mathbf{p}_1 = -\Delta\mathbf{p}_2$$

$$m_1\mathbf{v}_{1\,\text{aft}} - m_1\mathbf{v}_{1\,\text{bfr}} = -(m_2\mathbf{v}_{2\,\text{aft}} - m_2\mathbf{v}_{2\,\text{bfr}}), \text{ or}$$

$$m_1\mathbf{v}_{1\,\text{aft}} - m_1\mathbf{v}_{1\,\text{bfr}} = -m_2\mathbf{v}_{2\,\text{aft}},$$

since $\mathbf{v}_{2\,\text{bfr}} = 0$ m/s. Collecting terms for momenta before and after the collision,

$$-m_1\mathbf{v}_{1\,\text{bfr}} = -m_1\mathbf{v}_{1\,\text{aft}} - m_2\mathbf{v}_{2\,\text{aft}}, \text{ and}$$

$$m_1\mathbf{v}_{1\,\text{bfr}} = m_1\mathbf{v}_{1\,\text{aft}} + m_2\mathbf{v}_{2\,\text{aft}}.$$

In order for momentum to be conserved, the *x*-component of the momentum before collision must equal the sum of the *x*-components afterward. The same is true of the *y*-components. But $m_1\mathbf{v}_{1\,\text{bfr}}$ is only in the *x*-direction—it has no *y*-component—while $m_1\mathbf{v}_{1\,\text{aft}}$ and $m_2\mathbf{v}_{2\,\text{aft}}$ both have *y*-components. This difficulty is resolved by Figure 11-11b. The *y*-components of the two momenta after the collision exactly cancel each other. To find the final velocities of objects involved in a two-dimensional collision, obtain the components of their velocities from the *x*- and *y*-components of their momenta.

When two identical particle-like objects collide elastically, the post-collision angle between the objects' paths is exactly 90°. As the amount of deformation in one or both objects increases, and the collision becomes more inelastic, the angle becomes smaller. For completely inelastic collisions, the rebound angle is zero (see the illustrations in Figures 11-12a and b).

11B Section Review Questions

1. Under what conditions will momentum be completely conserved in a collision between two objects?

2. True or false: If the only forces at work on a system are internal, momentum is conserved.

3. Under what conditions will kinetic energy be conserved during a collision?

4. Following an off-center elastic collision between two objects, what is the vector sum of the objects' momenta?

⊚5. An air track glider of mass $m_1 = 1.80$ kg with an initial velocity of 1.20 m/s to the right collides elastically with a stationary glider of mass $m_2 = 5.00$ kg. Calculate the speed and direction of the two gliders after the collision.

⊚6. An air track glider ($m_1 = 1.80$ kg) moving to the right collides with a stationary glider (m_2). What must the mass of the second glider be in order for its velocity after the collision to be 1.2 times glider 1's initial velocity?

⊚7. A 6.80 g bullet traveling at 1010 m/s to the right imbeds itself in a 3.00 kg cube of oak sitting on a fence post.

   a. What is the velocity of the block and bullet mass after the impact?

   b. How much mechanical energy was lost to thermal and other forms of energy at impact?

⊚8. A 0.250 kg air table puck traveling at 1.00 m/s to the right rebounds elastically off of a 2.00 kg stationary puck. The first puck rebounds with a velocity of 0.870 m/s at an angle of +60.° from the positive *x*-axis. What is the velocity of the second puck after the collision?

Gravity Assists

Ever since the first space flight, men have dreamed of exploring the farthest reaches of our solar system. This dream was considered impossible using typical chemical rocket launchers and propulsion systems. The spacecraft loses much of its initial velocity just escaping from Earth orbit. If it is to explore the solar system, it has to use fuel in order to counteract the constant pull of the Sun's gravity and regain an appropriate velocity. The amount of fuel that would have to be carried for such a mission is enormous. What great things could be accomplished if only we could get free energy! The scope of exploration was greatly expanded when aeronautical researchers found a source of virtually free energy available to spacecraft if they can only escape Earth orbit. What is this new technology? Well, it is not new. In fact, this force has been around since God formed the universe. It's not really even technology. Energy can be imparted to a spacecraft, increasing its speed, by the exploitation of a planet's gravity and angular momentum to "slingshot" the craft out into space as it passes by the planet. This technique is called a gravity assist (or flyby) and has two main advantages. First, because it is not using fuel for acceleration, the amount of fuel the craft must carry is greatly reduced, decreasing its weight. Second, greater speed means it will take less time to reach the destination. The discovery of this method of trajectory design has made possible the launch of unmanned probes to planets that were considered unreachable by traditional methods.

We usually think of the launch vehicle sitting on the pad as a stationary object. That vehicle, though not in Earth orbit, is already orbiting the Sun along with Earth. When the rocket launches, it travels far enough away from the earth so that, while Earth's gravity is still present, the Sun's gravity is the predominant influence, and the spacecraft begins to go into its own orbit around the Sun. What happens when that spacecraft approaches a planet that is traveling in the same direction? The planet has a large angular momentum in its orbit around the Sun. As the spacecraft comes within the planet's gravitational influence, it begins to fall toward the planet. However, since it is traveling so fast, it doesn't go into orbit

around the planet but escapes, traveling in a different direction from when it came in. Relative to the planet, the spacecraft's velocity is the same as before the interaction. Because the planet is traveling so much faster, however, while the spacecraft is "stuck" to the planet (under the gravitational influence), the spacecraft accelerates. There is an exchange of angular momentum between the planet and the spacecraft. The spacecraft gains angular momentum and the planet loses angular momentum. The planet is so massive that the loss doesn't appreciably affect its motion, but because the spacecraft's mass is so small, it can accelerate greatly. Relative to the Sun, the spacecraft is going much faster when it leaves

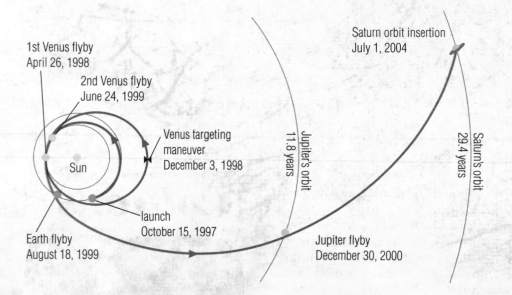

1st Venus flyby
April 26, 1998

2nd Venus flyby
June 24, 1999

Venus targeting
maneuver
December 3, 1998

Sun

launch
October 15, 1997

Earth flyby
August 18, 1999

Jupiter's orbit
11.8 years

Saturn's orbit
29.4 years

Saturn orbit insertion
July 1, 2004

Jupiter flyby
December 30, 2000

the planet than before it arrived. Gravity assists can also be used to decelerate a spacecraft by flying by in a direction opposite to the planet's orbit.

The trajectory designed for the mission must take into account not just the position of the destination planet, but the positions of the intermediate planets as well. This is why there is often only a certain time that a mission can be launched. The heavier planets, because of their greater mass, are ideal for gravity assists. The trajectory of the *Cassini* mission to Saturn (launched in 1997) included two flybys of Venus and one each of Earth and Jupiter. Notice that the spacecraft was first launched toward the Sun even though its final destination was in the opposite direction. This trajectory imparted an acceleration which would have taken 68 000 kg (75 tons) of fuel to achieve in a direct trajectory. The mission was calculated to reach Saturn in July 2004. That may seem like a long time to wait, but without gravity assists, it probably couldn't get there at all.

11.13 Explosions and Center of Mass

A fireworks rocket streaks into the sky. The initial burst throws out sprays of light. It isn't your imagination if it seems like the whole pattern of color continues to climb with the same direction and speed as the original rocket. If you do not look carefully, the parts of the explosion just seem to scatter randomly. However, if it were possible to plot the positions of all the particles of the rocket during the burst, you would find that their average positions would correspond to the original trajectory of the rocket's center of mass.

If you could run the explosion in reverse, it would seem that all of the particles converge in one great simultaneous inelastic collision. This is a reasonable approximation of the situation if we ignore the mass of the explosive.

If we consider the entire collection of particles resulting from an explosion to be the system, then the most significant forces acting on the particles during the instant of the explosion are internal forces. Because internal explosive forces are usually very large compared to external forces such as gravity and air resistance, they will predominate when determining the particle momenta at the instant of the explosion. Internal forces do not affect the total momentum of the system. Therefore, the sum of the momenta of all particles immediately after an explosion or disintegration is equal to the initial momentum of the object (assuming external forces are insignificant),

$$\mathbf{p}_{\text{sys before}} = \Sigma \mathbf{p}_{i \text{ after}},$$

where $\mathbf{p}_i$ is the momentum of each individual particle.

Although momentum is conserved to a good approximation, the kinetic energy of the particles is not. The forces that accelerate particles in explosions and other kinds of disintegrations come from chemical, mechanical, or other nonconservative sources.

11-13 To a first approximation, the center of mass of the glowing particles of an exploding fireworks rocket initially follows the same path that the rocket would have followed.

Momentum is a vector quantity that can be oriented in any direction in three dimensions. Each component of the momentum vector has its own vector equation.

The momentum of an object before an explosion is equal to the sum of the momenta of its parts after the explosion if external forces acting on the system are insignificant compared to the internal forces involved.

In the physical realm, internal forces do not affect the total momentum of the system, but internal forces can greatly affect the "momentum" of your spiritual life. You must seek victory over sin (the most detrimental internal force) rather than look for excuses to keep on sinning. In Psalm 4:2 the Holy Spirit, through David, says, "O ye sons of men, how long will ye turn my glory into shame? how long will ye love vanity, and seek after leasing?" (*Leasing* is an archaic English term meaning all the different kinds of excuses, alibis, and rationalizations.)

EXAMPLE 11-9

Breaking Up: Conservation of Momentum and Explosions

A 12.0 kg blob-like object is rising vertically at a speed of 20.0 m/s. At some point, it explodes apart into three smaller parts. The 1.0 kg part continues straight up at 40.0 m/s. The 3.0 kg part flies away horizontally to the right at 10.0 m/s. Find the direction and speed of the remaining 8.0 kg piece.

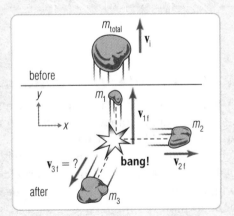

11-14 The explosion remnants

Solution:

The coordinate system and velocities are as indicated in Figure 11-14. Masses $m_1 = 1.0$ kg, $m_2 = 3.0$ kg, and $m_3 = 8.0$ kg.

From the conservation of momentum, you know the following equation relationships are true:

$$\mathbf{p}_{\text{total } i} = \mathbf{p}_{\text{total } f}$$

$$m_{\text{total}}\mathbf{v}_i = m_1\mathbf{v}_{1\,f} + m_2\mathbf{v}_{2\,f} + m_3\mathbf{v}_{3\,f} \qquad (1)$$

Rewrite Equation 1 in terms of its x- and y-components:

$$m_{\text{total}}v_{i_x} = m_1v_{1\,f_x} + m_2v_{2\,f_x} + m_3v_{3\,f_x}, \text{ and}$$

$$m_{\text{total}}v_{i_y} = m_1v_{1\,f_y} + m_2v_{2\,f_y} + m_3v_{3\,f_y}$$

Solve each of these equations for the unknown horizontal and vertical components.

x-component:

$$v_{3\,f_x} = \frac{m_{\text{total}}v_{i_x} - m_1v_{1\,f_x} - m_2v_{2\,f_x}}{m_3}$$

$$v_{3\,f_x} = \frac{(12.0 \text{ kg})(0 \text{ m/s}) - (1.0 \text{ kg})(0 \text{ m/s}) - (3.0 \text{ kg})(+10.\text{ m/s})}{8.0 \text{ kg}}$$

$$v_{3\,f_x} = -3.\underline{7}5 \text{ m/s}$$

y-component:

$$v_{3\,f_y} = \frac{m_{\text{total}}v_{i_y} - m_1v_{1\,f_y} - m_2v_{2\,f_y}}{m_3}$$

$$v_{3\,f_y} = \frac{(12.0 \text{ kg})(+20.\text{ m/s}) - (1.0 \text{ kg})(+40.\text{ m/s}) - (3.0 \text{ kg})(0 \text{ m/s})}{8.0 \text{ kg}}$$

$$v_{3\,f_y} = +25 \text{ m/s}$$

Determine the velocity of the third piece:

$$v_{3\,f} = \sqrt{v_{3\,f_x}^2 + v_{3\,f_y}^2} = \sqrt{(-3.\underline{7}5 \text{ m/s})^2 + (+25 \text{ m/s})^2}$$

$$v_{3\,f} \doteq 2\underline{5}.2 \text{ m/s}$$

$$\alpha = \tan^{-1}\frac{|v_{3\,f_y}|}{|v_{3\,f_x}|} = \tan^{-1}\frac{|+25 \text{ m/s}|}{|-3.\underline{7}5 \text{ m/s}|}$$

$$\alpha \doteq 81.5°$$

The third piece moves away at about 25 m/s at 81.5° above the horizontal to the left.

11-15 The secret to a successful hammer throw is developing maximum angular momentum.

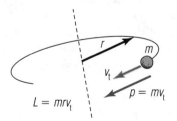

11-16 Derivation of Equation 11.15

11.14 Angular Momentum

A mass being whirled around on a string or a spinning wheel has momentum, but its momentum is somewhat different from the momentum of a mass traveling in a straight line. Mass moving in a circle is said to have **angular momentum (L)**, which means momentum associated with circular motion.

Let's look at a dimensionless particle of mass m orbiting a point at a distance r with tangential speed v_t. The magnitude of the particle's instantaneous linear (tangential) momentum, p, is

$$p = mv_t.$$

Angular momentum is the product of the linear momentum and the moment arm of the particle. (Recall that the moment arm is the shortest distance from the center of rotation *perpendicular* to the *force* on the particle.) In this case, the linear momentum is always perpendicular to the radius, so the moment arm is the distance from the center of rotation to the particle's center of mass. Therefore,

$$L = rp = mrv_t. \tag{11.15}$$

The units of angular momentum are kg·m²/s. There are no derived SI units for angular momentum.

Recall that tangential speed is related to the radius of the circular path and the angular speed by the formula

$$v_t = r\omega.$$

So, angular momentum for a *single orbiting particle* is

$$L = mr^2\omega. \qquad (11.16)$$

Similar to linear momentum and external forces, angular momentum is conserved unless an external torque is applied to the rotating system. For example, when a mass is whirled on a string of radius r with an angular speed of ω, if the string's length is reduced to half the radius, its tangential speed will double and its angular speed will increase as the square of the change in tangential speed. This is the same principle that a competition ice skater uses when performing a spin. His rotational speed is slower with his arms extended and much faster when the mass of his arms is held close to his body.

Assuming that no external torques act on the system:

$$\Delta L = m_2 r_2^2 \omega_2 - m_1 r_1^2 \omega_1 = 0 \text{ kg·m}^2/\text{s}$$
$$m_2 r_2^2 \omega_2 = m_1 r_1^2 \omega_1$$

Mass doesn't change, so $m_1 = m_2 = m$. The radius reduces to half, so

$$r_2 = \frac{r_1}{2}.$$

Substituting:

$$m r_2^2 \omega_2 = m r_1^2 \omega_1$$
$$\cancel{m}\left(\frac{r_1}{2}\right)^2 \omega_2 = \cancel{m} r_1^2 \omega_1$$
$$\frac{\cancel{r_1^2}}{4}\omega_2 = \cancel{r_1^2}\omega_1$$
$$\omega_2 = 4\omega_1$$

The angular speed of the mass quadruples.

Angular momentum of a particle is a vector quantity that is technically the product of two vectors—tangential momentum and the radial position vector. While its magnitude can be found from either Equation 11.15 or 11.16, its direction is found using the rotational right-hand rule (see Chapter 7). With the fingers of the right hand curled around the rotational axis in the direction of rotation, the extended thumb points in the direction of the angular momentum vector. Counterclockwise rotation has positive angular momentum.

11C Section Review Questions

1. When analyzing the momentum of a system of particles resulting from an explosion, discuss at least two assumptions that are made to simplify the analysis.

2. If a trap shooter shatters a flying clay pigeon with a blast of shot from his shotgun, will the particles follow the original trajectory of the intact clay pigeon? Explain your answer.

11-17 The flywheel of a heavy machine stores angular momentum.

The magnitude of angular momentum of a revolving particle is equal to the product of the orbit radius and the linear momentum of the particle. Units of angular momentum are kg·m²/s.

11-18 A skater draws his arms in to his axis of rotation to increase his rate of spin without changing his angular momentum.

3. What must occur for the angular momentum of a rigid spinning system to increase?

4. Six children are riding on the perimeter of a playground merry-go-round. If they all lean outward simultaneously, will the merry-go-round speed up, slow down, or continue spinning as before? Explain what happens.

⊛5. Two identical air track gliders of mass 0.125 kg approach each other at equal speeds of 0.500 m/s. Assume that the two gliders are a system. Since the center of mass of a system of objects is the point at which all of the mass can be assumed to be concentrated, choose a convenient coordinate system and determine the location of the center of mass for the following times:

   a. 0.5 s before the collision

   b. collision

   c. 0.5 s after the collision

   What is the momentum of the center of mass for this system?

⊛6. Determine the speed of the center of mass of a system of the two gliders described in problem 5 if glider 1 is stationary and glider 2 approaches glider 1 at 0.500 m/s to the right. What is the momentum of the system in this case?

Chapter Review

In Terms of Physics

momentum ($\mathbf{p}$)	11.2	elastic collision	11.8
impulse ($\mathbf{I}$)	11.4	partially elastic collision	11.8
recoil	11.5	inelastic collision	11.8
conservation of momentum	11.5	ballistic pendulum	11.11
thrust	11.6	angular momentum ($\mathbf{L}$)	11.14

Problem-Solving Strategies

11.1 (page 234) You will be introduced to many equations in this chapter. You do not need to memorize them. So many conditions apply to using most of the equations that you probably could derive them from the principles you will learn before you could recall them or find them in your notes. Physics is not just about memorizing equations.

11.2 (page 239) Evaluate the effect of all external forces on the system as a whole. If there is no net external force, then the total change of momenta of all parts of the system is zero and momentum is conserved.

11.3 (page 240) Quantities that have multiple subscripts, such as $v_{1\ bfx}$, can be verbally expressed as "vee-one-before." The sub-subscript x is written to indicate a scalar component but may be left unspoken if no confusion will result.

1. What is linear momentum? How does it differ from any other kind of momentum?

2. How do you determine the direction of the linear momentum vector?

3. What is the momentum of a 10 000 kg truck at rest?

4. Define *impulse*. For a constant net external force on a system, what determines the magnitude of the force's impulse?

5. How does a gymnast reduce or eliminate injury from a fall by using the principle of impulse?

6. At the instant it is struck, a baseball has the following forces acting on it: the bat, the earth's gravitational force, the Sun's gravitational force, atmospheric drag, and the pseudo-force due to the Coriolis effect. Which of these forces could be considered an impulsive force?

7. State the law of conservation of momentum.

8. What basic condition must exist for a system's momentum to remain unchanged?

9. Why is momentum not conserved when a rock is dropped?

10. A wad of soft clay is thrown against a wall and sticks in place. Is the clay's momentum conserved? Explain your answer.

11. For a given rocket engine at a fixed throttle setting, the force exerted by the engine is constant. Which rocket will achieve the higher final velocity with the same thrust setting—a rocket with a large propellant-to-payload ratio or one with a small ratio?

12. Which of the following are elastic collisions?
    a. A piece of clay collides with a baseball and sticks to it.
    b. A golf ball strikes an identical resting golf ball, which moves away with the original ball's speed while the original ball stops.
    c. A polystyrene block strikes a table edge and bounces off, squashed.
    d. A bullet strikes and lodges in a ballistic pendulum.
    e. A steel ball bearing, dropped on a steel plate, rebounds with the same speed it had before the impact.

13. Which of the following is *not* conserved in an inelastic collision?
    a. kinetic energy   c. mass
    b. momentum   d. total energy

14. Under which situation will a moving object (m_1) completely transfer its mechanical energy to a stationary object (m_2) during a one-dimensional collision?
    a. $m_1 \ll m_2$ c. $m_1 \gg m_2$
    b. $m_1 = m_2$ d. m_1 sticks to m_2

15. A physicist observes a subatomic particle passing through a special apparatus called a bubble chamber that makes the tracks of particles visible. It decays at one point into at least two particles that also leave tracks in the chamber. Assuming that the tracks in the figure all lie in the same plane, how can he conclude that there must have been at least one additional undetected particle resulting from this disintegration?

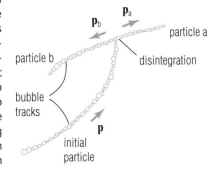

16. Using your knowledge of angular momentum, explain why aerial ropewalkers use long bars to maintain balance during their stunts.

17. A planet orbiting a star moves closer to the star. If no outside forces act on the planet-star system, what would have to happen to the planet's orbital speed?

18. A competition high diver often tucks in order to obtain as many revolutions as possible before entering the water. Explain why timing the point at which he straightens out is critical to successfully completing the dive. Does the diver actually stop rotating?

19. The engine of an Alaskan bush pilot's plane sputters and dies, and the pilot bails out with a parachute and a survival kit. He lands in the middle of a perfectly smooth frozen lake. There is no wind, no friction, and no wood on the lake. The survival kit contains two weeks' supply of canned food, a can opener, a hatchet, and matches. If he stays in the middle of the lake, he will probably freeze to death before help comes. However, if he can reach the lake's wooded shore, he can survive the week needed to reach town. Using the principles of momentum conservation he learned in high school, how can he reach the lake's shore before he freezes to death?

20. A motorcyclist collides with a bumble bee (*Bombus griseocollis*) squarely in the middle of his helmet visor while driving 113 km/h (around 70 mi/h). Assuming the collision is elastic (*not* a good assumption), which receives the larger impulse?
    a. the motorcyclist
    b. the bumble bee
    c. neither—the impulses are equal

True or False (21–30)

21. The object with the greatest velocity always has the greatest momentum.

22. Under normal circumstances, the momentum of an object changes because its velocity changes.

23. Any unbalanced external force changes an object's momentum.

24. A car that maintains a constant speed around a corner has constant momentum.

25. Gravity cannot have an impulse because it does not act suddenly like an explosive force or a baseball bat striking a ball.

26. An inelastic collision is a collision in which momentum is not conserved.

27. Most collisions are partially elastic as long as no external forces act on the system of colliding objects.

28. Ideally, the motion of the center of mass of an object is not affected if the object explodes.

29. Object 1 collides elastically with stationary object 2. Both move off in the same direction. From this scenario, you can conclude that object 1 had more mass than object 2.

30. The angular momentum vector points in the same direction as the motion of the rotating object.

31. What is the momentum of a 0.50 kg kitten running with a velocity of 1.00 m/s to the left?

32. Compare the momentum of a 1500. kg car traveling at 20. km/h north with the momentum of a 900. kg car traveling at 40. km/h north.

33. A 1.00 kg ball traveling at a velocity of 3.00 m/s west strikes a wall and rebounds with a velocity of 3.00 m/s east. The ball contacts the wall for 0.0100 s.
  a. What is the ball's momentum before it hits the wall?
  b. What is the ball's momentum after it hits the wall?
  c. What is the ball's change in momentum?
  d. What force must the wall exert on the ball in order to change its momentum?

34. A 0.400 kg cart traveling 0.0750 m/s to the right encounters a spring. In the next 0.125 s, the spring exerts an average force of 0.480 N to the left on the cart.
  a. What is the cart's momentum before it reaches the spring?
  b. What is the change in the cart's momentum as a result of the encounter?
  c. What is the cart's momentum after it leaves the spring?
  d. What is the cart's velocity after it leaves the spring?

35. A 975 kg car travels northeast at 8.00 m/s.
  a. What is its momentum?
  b. What impulse is needed to stop the car?
  c. What force is needed to stop the car in 10.0 s?
  d. How long would it take a 7.80×10^3 N force to stop the car?

36. A 10.0 g bullet leaves the muzzle of a 4.00 kg rifle with a velocity of 8.50×10^2 m/s east.
  a. What is the bullet's momentum?
  b. What is the rifle's momentum?
  c. What is the recoil velocity of the rifle?
  d. If the explosion took 0.00200 s, how much force does the rifle exert on the rifleman's shoulder?

37. A 1.50×10^3 kg freight dolly travels north at a speed of 1.50 m/s. A 125 kg box drops vertically onto the dolly in an inelastic collision.
  a. What is the dolly's momentum before the box drops onto it?
  b. What is the momentum of the box and dolly together after the box lands on the dolly?
  c. Is this an example of conservation of momentum within a system? Explain your answer.

38. A 25.0 g golf ball traveling to the right with a speed of 10.0 m/s collides elastically with a stationary 145 g baseball. Calculate the velocity of each ball after the collision.

39. A 176 g billiard ball (m_1) collides elastically with an identical stationary billiard ball (m_2). With a speed of 1.50 m/s, m_2 moves off in the original direction of m_1.
  a. What was the initial speed of m_1 ($v_{1\ bfr}$)?
  b. What is the final speed of m_1 ($v_{1\ aft}$)?

40. A croquet ball with a mass of 0.369 kg moves to the left with a speed $v_{1\ bfr}$ of 11.9 m/s toward a stationary golf ball. Assuming that both balls are particles and they collide elastically, the golf ball moves off with a speed $v_{2\ aft}$ of 21.3 m/s in the direction of $\mathbf{v}_{1\ bfr}$.
  a. What is the mass of the golf ball?
  b. What is the final velocity of the croquet ball?

41. A 10.0 g bullet traveling at 850. m/s embeds itself in a stationary 3.60 kg block of wood resting on a frictionless surface. Calculate the final speed of the block of wood.

42. A 0.0150 kg bullet strikes a stationary 2.485 kg ballistic pendulum. The pendulum rises 0.200 m before stopping.
  a. How much does the pendulum-bullet system's potential energy change?
  b. How much does its kinetic energy change?
  c. What is the pendulum-bullet system's speed after the bullet strikes the pendulum?
  d. What is the bullet's original speed?

43. Is it possible for two objects within the same system to collide elastically and both be at rest after the collision? Explain your answer.

44. Research the theory and operation of spacecraft ion propulsion systems. Discuss the advantages and disadvantages of ion propulsion versus conventional combustion rocket engines.

45. If a nuclear particle at rest decays (splits apart) into two identical pieces, explain why the two parts *have* to rebound in exactly opposite directions.

12A	Simple Harmonic Motion	258
12B	Periodic Motion and the Pendulum	262
12C	Oscillations in the Real World	269
12D	Waves	272
Facet:	Foucault Pendulum	266
Facet:	Earthquakes	276

New Symbols and Abbreviations

restoring force (F_r)	12.2	moment of inertia (I)	12.9
simple harmonic motion (SHM)	12.3	natural oscillation frequency (f_0)	12.11
amplitude (A)	12.3	wavelength (λ)	12.14
frequency (f)	12.3	wave speed (v)	12.14
hertz (Hz)	12.3	intensity (I_s)	12.15
pendulum arm (L)	12.6	decibels (dB)	12.15
total acceleration (a_{total})	12.7		

The Egg Nebula (CRL-2688) is an unusual celestial object located approximately 3000 light-years from Earth. In this false-color image, the shells of ejected material are illuminated by strange narrow beams of light that illuminate the wavelike structure of the nebula surrounding the hidden star

Periodic Motion

> *And Hezekiah answered, It is a light thing for the shadow to go down ten degrees: nay, but let the shadow return backward ten degrees. And Isaiah the prophet cried unto the Lord: and he brought the shadow ten degrees backward, by which it had gone down in the dial of Ahaz.* II Kings 20:10–11

This passage describes one of the many miracles God used throughout Israel's history to authenticate His Word. The length of a day is the same every day to within a few seconds. The regular, periodic character of diurnal time was recognized by people thousands of years ago. The daily rotation of the earth on its axis causes the apparent rising and setting of the Sun, which is the most obvious periodic motion in the heavens. This apparent motion is often referred to in the Bible when the word *day* occurs. The doubtful king Hezekiah requested that this natural, predictable motion be miraculously changed to prove that the prophet's words were true. Today, Christians have the Holy Spirit to authenticate God's Word, giving them the faith that it is true.

12.1 Introduction

Vibrations are all around us. Vibrating air carries sound to our ears. Vibrating electromagnetic fields provide light for our eyes. Vibrating atoms give a feeling of heat through our sense of touch. We depend on vibrations in our everyday life, yet many people are oblivious to them.

We also use vibrations to make our lives easier. Precisely controlled electronic vibrations in the form of radio waves speed communication. Microwave ovens generate electromagnetic vibrations that force water molecules in food to vibrate, heating the food quickly. Clocks depend on regular vibrations to show the passing of time. The repetitive motion of an engine's pistons produces the energy needed to move a car.

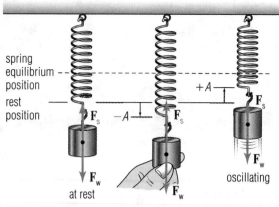

spring
equilibrium
position
rest
position

$+A$

$-A$

F_s

F_s

F_s

F_w

F_w

F_w

at rest

oscillating

12-1 A laboratory mass oscillates on the end of a spring.

The force sum on the mass is expressed in the equations

$\Sigma \mathbf{F} = \mathbf{F}_w + \mathbf{F}_s$ and

$\Sigma F_y = mg_y + (-k\Delta y)$,

where the y-axis is vertical and positive upward.

Friction that opposes the restoring force in oscillating systems produces **damping,** the gradual reduction of the amplitude of oscillations.

Hooke's force law may be summarized by the equation $\mathbf{F}_{\text{Hooke}} = -k\Delta x$. The force exerted by the spring is proportional to the displacement of the spring and opposite in direction to its displacement.

12.2 Describing Periodic Motion

Vibrations often repeat themselves. For example, if you pluck a violin string, it will vibrate many times. Motion that repeats at a constant rate is called **periodic motion.**

Figure 12-1 shows a laboratory mass attached to an ideal spring hanging from a support. The oscillation of the hanging spring-mass system is an example of periodic motion. Initially, the mass is attached to the spring and allowed to hang at rest. At this position, the force of the spring pulling up on the mass equals the force of gravity pulling down on the mass. Since the mass is at rest, this position is called the *rest position,* or *equilibrium position,* of the system. Note that this is *not* the equilibrium position of the spring alone. When the mass is attached, the spring is stretched downward until an upward force equal to the weight of the mass is developed by the spring according to Hooke's law.

To begin the oscillations, the mass is pulled down to $y = -A$. Here the upward force of the spring is greater than the force of gravity on the mass, so when the mass is released, it is accelerated upward. When the mass reaches the system equilibrium position, the forces on it are again balanced, so at this instant it experiences no acceleration (Newton's first law). Velocity is maximum as the mass passes through the equilibrium position. After it rises above the equilibrium position, the spring force continues to decrease while gravity maintains a constant force on the mass. The net force on the mass increases in the downward direction and decelerates the mass until it comes to rest at $y = +A$. The net force then propels the mass downward. The mass again travels through its equilibrium position at maximum velocity, and the increasing spring force finally draws the mass to a temporary halt at $y = -A$. At this point the cycle starts over again.

The mass will continue moving up and down forever in the absence of frictional forces. However, in the real world frictional forces are always present. A spring may oscillate for a long time, but it will eventually stop. The effect of friction is called **damping.** See Section 12C for a discussion of damped oscillators.

One reason the mass keeps moving is that the net force on it always tends to return it to its equilibrium position. Below the equilibrium position, the upward pull of the spring is stronger than the downward pull of gravity. Above the equilibrium position the spring's pull weakens, and if the mass passes the spring's rest length, it even changes to a downward push. Thus the net force on the mass always tends to restore the mass to its equilibrium position. The force is therefore called a **restoring force ($\mathbf{F_r}$).** Often two or more forces must be summed to find the restoring force.

12.3 Characteristics of Simple Harmonic Motion

There are many kinds of periodic motion, but periodic motion that is controlled by a restoring force proportional to the system displacement from its equilibrium position is called **simple harmonic motion (SHM).** The restoring force in simple harmonic motion is described by the formula

$$F_r = k\Delta x, \qquad (12.1)$$

where k is a constant and Δx (or d_x) is the displacement of the system from the equilibrium position. Notice that Equation 12.1 is identical in form to Hooke's law, so a spring-mass system can approximate simple harmonic motion in the absence of friction. Many natural systems involving periodic motion approximate simple harmonic motion.

The following table summarizes the relationships during simple harmonic motion between restoring force, system velocity, and acceleration.

TABLE 12-1									
Characteristics of an Oscillating System									
Position	$x = +A$	$0 < x < +A$	$x = 0$	$-A < x < 0$	$x = -A$	$-A < x < 0$	$x = 0$	$0 < x < +A$	$x = +A$
Restoring Force	$F_{x\,max}\,(-)$	$\lvert F_x \rvert \downarrow$	$F_x = 0$	$\lvert F_x \rvert \uparrow$	$F_{x\,max}\,(+)$	$\lvert F_x \rvert \downarrow$	$F_x = 0$	$\lvert F_x \rvert \uparrow$	$F_{x\,max}\,(-)$
Acceleration	$a_{x\,max}\,(-)$	$\lvert a_x \rvert \downarrow$	$a_x = 0$	$\lvert a_x \rvert \uparrow$	$a_{x\,max}\,(+)$	$\lvert a_x \rvert \downarrow$	$a_x = 0$	$\lvert a_x \rvert \uparrow$	$a_{x\,max}\,(-)$
Velocity	$v_x = 0$	$\lvert v_x \rvert \uparrow$	$v_{x\,max}\,(-)$	$\lvert v_x \rvert \downarrow$	$v_x = 0$	$\lvert v_x \rvert \uparrow$	$v_{x\,max}\,(+)$	$\lvert v_x \rvert \downarrow$	$v_x = 0$

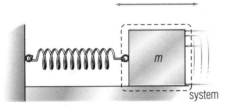

12-2 Diagram of horizontal oscillating system on frictionless surface

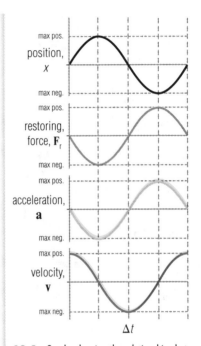

12-3 Graphs showing the relationships between position, restoring force, acceleration, and velocity of an oscillating system

Simple harmonic motion is different from the straight-line or circular motions that you have studied. One characteristic of simple harmonic motion is that there is a maximum displacement from the equilibrium condition. The mass on the spring never goes farther from its equilibrium position than it was originally pulled, and a pendulum never rises above the height from which it was originally released. This maximum displacement is called the **amplitude (A)**. An object undergoing simple harmonic motion goes from a distance A on one side of its equilibrium point to a distance A on the other side of its equilibrium point and back again.

The mass in Figure 12-1 repeatedly goes through the same set of motions: from $-A$ to 0 to A to 0 to $-A$. This set of motions is called a **cycle.** The time that the mass takes to complete a cycle is constant. For example, if the mass moves from $-A$ to A to $-A$ in 1 s, the next time it moves from $-A$ to A to $-A$ will also take 1 s. This time is called the mass's *period*. You studied the concept of period in Chapter 7 in association with planetary orbits—another form of periodic motion. Sometimes you want to know how many cycles the mass will complete in a unit of time. This is called the mass's **frequency (f)** and is usually measured in cycles per second, or **hertz (Hz)**. A hertz is defined as follows:

$$1 \text{ cycle/s} = \text{s}^{-1} \equiv 1 \text{ Hz.}$$

Since frequency is the number of cycles completed per unit of time, and period is the amount of time needed to complete a cycle, frequency and period are reciprocal quantities. In symbols,

$$f = \frac{1}{T} \quad \text{and} \quad T = \frac{1}{f}.$$

The unit of frequency is cycles per second, or s^{-1}. It is also given the unit *hertz* (Hz).

Heinrich Rudolf Hertz (1857–94) was a German physicist whose experiments laid the foundation for the wireless telegraph and radio.

Problem-Solving Strategy 12.1

Remember that period and frequency are reciprocals of each other.

12.4 Reference Circle

Circular motion is similar to simple harmonic motion. Some aspects of an object's position in simple harmonic motion are related to position in circular motion. Figure 12-4 shows a ball rotating in a circle at a constant speed on the end of a thin rod. Light shines directly on the rotating ball from the left, casting the ball's shadow on the wall at the right. The shadow moves like the mass in Figure 12-1 moved, from a central "equilibrium" position down to a maximum "amplitude," then back up through the central position to an upward "amplitude," back down, and so on, as long as the ball is rotating. In fact, the motion of the ball's shadow and the motion of the spring-mass system can be synchronized. This helps in analyzing the mass's motion. Assume that the mass of the ball equals the mass of the spring-mass system.

When the motions of the ball's shadow and the mass are synchronized, the mass's period equals the ball's period. If the ball's speed around the circle is v_t and its orbital radius is r,

$$T = \frac{2\pi r}{v_t}. \tag{12.2}$$

In Figure 12-5, the "restoring force" that seems to act on the shadow is equal to the x-component of the centripetal force (F_c) on the ball (assuming the x-axis is oriented vertically and no other external forces act on the ball). The triangle formed by $\mathbf{F}_r$ and $\mathbf{F}_c$ is similar to that formed by Δx and r. Therefore,

$$\frac{F_r}{\Delta x} = \frac{F_c}{r} \text{, or}$$

$$\frac{F_r}{F_c} = \frac{\Delta x}{r}. \tag{12.3}$$

But $F_r = k\Delta x$, and

$$F_c = ma_c = \frac{mv_t^2}{r}.$$

So, combining these formulas in Equation 12.3,

$$\frac{k\Delta x}{mv_t^2/r} = \frac{\Delta x}{r} \text{, or}$$

$$\frac{kr}{mv_t^2} = \frac{1}{r}. \tag{12.4}$$

By rearranging Equation 12.4, you can see that

$$\frac{r^2}{v_t^2} = \frac{m}{k} \text{, or}$$

$$\frac{r}{v_t} = \sqrt{\frac{m}{k}}. \tag{12.5}$$

Substitution of the right side of Equation 12.5 for r/v_t in Equation 12.2 gives

$$T = \frac{2\pi r}{v_t} = 2\pi \frac{r}{v_t} \text{, or}$$

$$T = 2\pi\sqrt{\frac{m}{k}}. \tag{12.6}$$

You can now calculate the period of the spring-mass system's motion without constructing a corresponding circular motion diagram. The period is proportional to the square root of the oscillating system's mass.

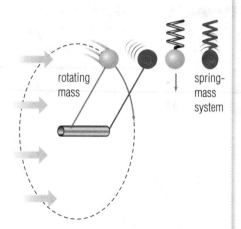

12-4 Circular motion and spring-mass system oscillations are mathematically similar.

(a)

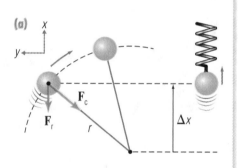

(b)

12-5 The geometry showing the similarity between circular motion and oscillations

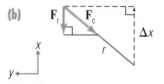

Problem-Solving Strategy 12.2

A longer period means that the spring system moves slower and takes longer to complete one cycle. The larger the mass, the longer the period. The "stiffer" the spring, the shorter the period.

EXAMPLE 12-1

SHM in a Spring-Mass System

A 0.190 kg mass is hung from an ideal spring with a spring constant $k =$ 76.0 N/m. It is set in motion by displacing it 3.00 cm downward. (a) What is the period of its motion? (b) What is the frequency?

Solution:

a. Use Equation 12.6 to solve for period:

$$T = 2\pi\sqrt{\frac{m}{k}} = 2\pi\sqrt{\frac{0.190 \text{ kg}}{76.0 \text{ N/m}}}$$

$$T \doteq 0.31\underline{4}1\sqrt{\frac{\text{kg}}{(\text{kg}\cdot\text{m})/(\text{m}\cdot\text{s}^2)}} \text{ (unit cancellation)}$$

$$T \approx 0.314 \text{ s}$$

b. Take the reciprocal of the period to find the frequency:

$$f = \frac{1}{T} = \frac{1}{0.31\underline{4}1 \text{ s}}$$

$$f \doteq 3.1\underline{8}3 \text{ s}^{-1}$$

$$f \approx 3.18 \text{ Hz}$$

Notice that the initial displacement, or amplitude, of the block has no effect on period (T) or frequency (f).

12A Section Review Questions

1. How does the equilibrium position of a mass hanging on a spring compare to the rest position of the spring alone?

2. What is the property of all periodic motion that tends to return the system to its equilibrium position?

3. What differentiates simple harmonic motion from other periodic motion?

4. If the mass of a system in simple harmonic motion is increased, will the period of the system increase or decrease?

•5. An ideal spring with a spring constant of 55.0 N/m is suspended from a support. The spring is 12.0 cm long. What is the stretched length of the spring if a 500. g laboratory mass is hung from the spring? .5kg

•6. What is the period of the spring-mass system in Question 5?

•7. What is the frequency of the spring-mass system in Question 5?

•8. Convert the frequency of a sound at 261.6 cycles/s (middle C) to Hz.

12-6 A replica of Huygens's pendulum clock based on Galileo's design

All the mass of an ideal pendulum is assumed to be centered at the end of the pendulum arm. This is a reasonable approximation as long as the mass of the arm is negligible compared to the mass at the end of the arm.

It is important that we specify the time that we inventory the forces acting on the pendulum as *the instant of release*. As the pendulum begins to fall, other forces appear that affect the acceleration of the pendulum.

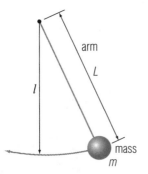

12-7 Parts of a simple pendulum

12B PERIODIC MOTION AND THE PENDULUM

12.5 Historical Overview

The origin of the **pendulum** is unclear. There is evidence in cave paintings that pendulums were used in some occult way to dowse for water. The pendulum was a mystery to ancient Greeks because it defied the notion that a heavy weight would single-mindedly seek its natural place at the center of the earth. To have a weight on a rope continuously swing up and down in an arc did not fit with Aristotle's natural order of things.

It was not until the sixteenth century that the pendulum was first scientifically investigated by Galileo Galilei. Although educated in Aristotelian science, he concluded early in his life as a scientist that much of what he had learned was often contradicted by simple observation. For example, while still a student, he observed in the cathedral at Pisa, Italy, that the periods of swinging chandeliers having equal length chains were independent of the amplitudes of their swing. This was an important discovery that eventually led to research in pendulum regulators for clocks.

The periods of both pendulums and spring-mass systems in simple harmonic motion are independent of the amplitude of their initial displacements. In this section, we will investigate the properties of pendulums and how they approximate simple harmonic motion.

12.6 Description of Pendulum Motion

A simple or ideal pendulum consists of a mass suspended from an ideal string or massless rod of length L, called the **pendulum arm.** All of the pendulum's mass (m) is assumed to reside at a point at a distance l from the point of suspension. Therefore, for an ideal pendulum, $l = L$. In this section, the symbol l specifically means the distance between the pendulum's pivot point and its center of mass. When a pendulum swings, its center of mass follows an arc of a circle with radius l. The need to emphasize this point will be clear at the end of this section when we discuss physical (nonideal) pendulums.

The forces acting on a pendulum at rest are its weight ($m\mathbf{g}$) and the tension in the pendulum arm ($\mathbf{T}_p$). When the pendulum is hanging still, these two forces are in equilibrium. This hanging position is the equilibrium position for a pendulum. Suppose an experimenter pulls the pendulum to the side so that the pendulum arm makes an angle θ with its equilibrium position. At the instant he releases it, the only forces acting on the mass are the weight of the mass and tension in the arm.

12.7 Pendulum Restoring Force

What is the restoring force that causes the pendulum to return to its original equilibrium position? The only freedom of movement that the pendulum mass has is tangential along a circular path of radius l. What force acts in a tangential direction?

The weight of the pendulum mass is always directed vertically downward. Therefore, weight alone cannot move the mass back to its equilibrium position. If the string were cut, the pendulum could only move downward. The tension applied by the pendulum arm on the mass is oriented radially along the arm, so it cannot cause tangential motion by itself either. Only the sum of these two forces acting at an angle forms the restoring force that returns the pendulum to its equilibrium position.

The reaction force of the mass acting on the pendulum arm ($\mathbf{T}'_p$) is a component of the pendulum's weight parallel to the tension vector and in the opposite direction (see Figure 12-8). The magnitude of this component is

$$T'_p = |mg| \cos \theta.$$

The other component of the pendulum's weight is perpendicular to $\mathbf{T}'_p$. This is the component of weight tangent to the circular path. Its magnitude is calculated by

$$|mg| \sin \theta.$$

This is the force that produces the tangential acceleration responsible for the motion of the pendulum. We will call this vector sum the "restoring force" ($\mathbf{F}_r$) because it is the net force sum on the pendulum mass.

$$F_r = |mg| \sin \theta \qquad (12.7)$$

At the instant of release, the tangential or restoring force is equal to the vector sum of the tension vector $\mathbf{T}_p$ and the weight vector $m\mathbf{g}$. The vector equation defining the restoring force is

$$\mathbf{F}_r = \mathbf{T}_p + m\mathbf{g}.$$

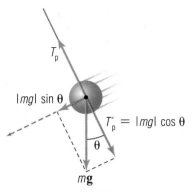

12-8 Forces on a pendulum

We will see shortly that this force does not always correspond to the simple harmonic motion restoring force except under certain conditions.

After it is released, the pendulum accelerates as it falls toward its equilibrium position. Several interesting things happen. As the pendulum's tangential speed increases, it is being accelerated toward the center of its circular path according to the principles of circular motion. The centripetal force ($\mathbf{F}_c$) that causes this centripetal acceleration is added to the tension in the pendulum arm supporting the weight of the mass. In equation form,

$$\mathbf{T}_{total} = \mathbf{T}_p + \mathbf{F}_c,$$

where $T_p = |mg| \cos \theta$ and $F_c = mv_t^2/r$. The vector sum of $\mathbf{T}_{total}$ and $m\mathbf{g}$ is no longer tangent to the pendulum path but tends to rotate toward its interior (see Figure 12-9). The tangential acceleration vector ($\mathbf{a}_t$) and the centripetal acceleration ($\mathbf{a}_c$) sum together to yield the **total acceleration ($\mathbf{a}_{total}$)** of the pendulum mass. The total acceleration vector points into the interior of the circular path ahead of the system's radial position vector if the system is accelerating and behind the position vector if the system is slowing. The force that causes this total acceleration is the pendulum's "restoring force."

Tangential acceleration decreases to zero as the pendulum reaches the bottom of its arc because the tangential component of gravitational acceleration disappears at the equilibrium position. As the pendulum rises in the second half of its swing, tangential acceleration begins to increase in the opposite direction to reduce tangential speed. The total acceleration vector points behind the pendulum, indicating that the vector sum of total tension and weight (the "restoring force") acts in the same direction. Examine Figure 12-10 to see how this interplay of forces and accelerations controls a pendulum's movement. It should be evident that the force restoring a pendulum to its equilibrium position is not proportional to the magnitude of the pendulum's displacement, nor does it act in the opposite direction to the displacement. So it seems that a pendulum's motion does not conform to the rules for simple harmonic motion.

The "restoring force" in a pendulum is the vector sum of the pendulum's weight and the tension in the pendulum arm.

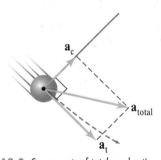

12-9 Components of total acceleration

In nonuniform circular motion, such as that of a pendulum, the vector sum of tangential and centripetal accelerations is its **total acceleration ($\mathbf{a}_{total}$).**

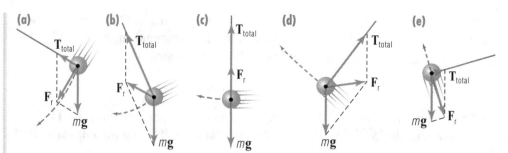

12-10 Relationships between forces and position of a pendulum swinging with a large amplitude

A pendulum experiencing large-amplitude swings does not follow the rules for simple harmonic motion.

You can see that as the pendulum's initial displacement (its amplitude) from its vertical equilibrium position increases, the tangential speed at the bottom of its swing increases. Centripetal force on an orbiting object is proportional to the *square* of its tangential speed, so the "restoring force" on the pendulum is significantly affected by its speed—much more so than by its displacement for large initial angles ($\theta > \pi/8$ rad, or more than 22.5°).

12.8 Small Amplitude Pendulum Motion

If we consider a pendulum starting its swing from a small displacement angle ($\theta < \pi/8$ rad from the vertical), then we can make some helpful approximations to analyze the periodic motion of a pendulum. With small initial displacement angles, the pendulum does not fall very far before it reaches its equilibrium position, nor does it acquire any appreciable tangential speed during its swing. Thus centripetal force is negligible and tension force in the pendulum arm is essentially dependent only on the weight of the mass and the displacement angle. The restoring force acting on the pendulum is then equal to the vector sum of just the pendulum arm tension and the pendulum's weight. The restoring force vector is also essentially parallel to the displacement of the pendulum mass from equilibrium—a key requirement for simple harmonic motion.

The restoring force of a pendulum oscillator (for small angles) is dependent on the weight of the pendulum and the pendulum's displacement angle.

The magnitude of the pendulum's initial perpendicular displacement (d) is the length of the side opposite to the initial angle θ in the right triangle formed by the pendulum arm l and the vertical. You can see that

$$d = l \sin \theta.$$

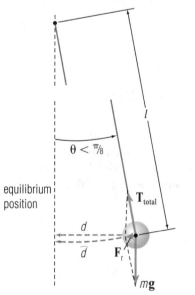

12-11 Forces on a pendulum with a small-amplitude swing

The semi-arc length of the pendulum's path ($\widehat{d}$)—half the pendulum swing—is

$$\widehat{d} = r\theta = l\theta$$

(θ is in radians). However, for small angles, $\sin \theta \approx \theta$. Therefore,

$$d = l \sin \theta \approx l\theta.$$

But $l\theta = \widehat{d}$, so for small θ,

$$d \approx \widehat{d}.$$

The initial perpendicular displacement is nearly the same as the arc length of the pendulum's path from equilibrium. It follows that

$$\sin \theta = \frac{\widehat{d}}{l}.$$

If you substitute this into the restoring force equation (Equation 12.7), you obtain the following:

$$F_r = |mg|\sin\theta$$

$$F_r = |mg| \cdot \frac{\widehat{d}}{l}$$

$$F_r = \frac{|mg|}{l} \cdot \widehat{d} \qquad (12.8)$$

For any given pendulum, the first factor in Equation 12.8 is a constant (k). We have previously shown that $d \approx \widehat{d}$, so Equation 12.8 has a form of the Hooke's force law equation,

$$F = kd,$$

where $k = |mg|/l = m|g|/l$. If you substitute this value for k into Equation 12.6 (simple harmonic motion period formula), you obtain:

$$T = 2\pi\sqrt{\frac{m}{k}} \qquad \text{(Equation 12.6)}$$

$$T = 2\pi\sqrt{\frac{m}{m|g|/l}}$$

$$T = 2\pi\sqrt{\frac{\cancel{m}l}{\cancel{m}|g|}}$$

$$T = 2\pi\sqrt{\frac{l}{|g|}} \qquad (12.9)$$

Therefore, for small initial displacement angles, a pendulum's motion closely approximates simple harmonic motion. Longer pendulum arms produce longer periods of swing (pendulums appear to swing more slowly). With greater gravitational acceleration, the swing period is shorter (a pendulum appears to swing more rapidly).

The pendulum's period is *not* related to the mass of the pendulum, for the same reason that a heavier mass does not fall with greater acceleration than a smaller mass. Although the larger mass weighs more and has a greater force acting on it, its larger inertia neutralizes the larger weight.

Problem-Solving Strategy 12.3
When calculating the period of a pendulum, be sure to use the *magnitude* of g. If you use the scalar value, the result will be an imaginary number.

For small initial displacements, the period of a pendulum is proportional to the square root of pendulum arm length and inversely proportional to the square root of gravitational acceleration.

EXAMPLE 12-2

Determining Gravitational Acceleration

A physics student discovers from his studies that he can determine the acceleration of gravity at his location using just a pendulum and a stopwatch. He carefully measures the period of his 0.750 m pendulum 20 times and calculates an average period of 1.739 s. (a) What is gravitational acceleration at his location? (b) Based on what you know about the accepted value of *g* near the earth's surface, what could you deduce about the student's location relative to sea level?

Solution:

a. Rearrange Equation 12.9 to solve for the magnitude of *g*:

$$T = 2\pi\sqrt{\frac{l}{|g|}}$$

$$|g| = \frac{4\pi^2 l}{T^2}$$

$$|g| = \frac{4\pi^2(0.750 \text{ m})}{(1.739 \text{ s})^2}$$

$$|g| \doteq 9.7\underline{9}0 \text{ m/s}^2$$

$$|g| \approx 9.79 \text{ m/s}^2$$

b. You know that gravitational acceleration decreases with distance from the center of the earth. Therefore, assuming his measurements are accurate, you can deduce that the student lives at a high elevation in a mountainous region, since his value of $|g|$ is less than the accepted sea-level value of 9.81 m/s².

Foucault Pendulum

12.9 Physical Pendulums

The pendulums considered up to this point could be called "ideal pendulums." All of their mass is concentrated at a point, and the pendulum arm is assumed to be massless. Real systems that swing as a pendulum have their mass distributed to some extent along the length of the pendulum arm. In fact, any object that can swing from a pivot point acts like a pendulum. These systems are called **physical pendulums.** The period of physical pendulums must take into account the distribution of mass and its distance from the center of rotation. This characteristic is constant for any given pendulum and is represented by a quantity called the **moment of inertia (*I*)** of the swinging mass. A table of moments of inertia for various-shaped objects is provided in Appendix F.

The period of a physical pendulum is determined experimentally to be

$$T = 2\pi \left(\frac{l}{|g|}\right)^{1/2} \left(\frac{I}{ml^2}\right)^{1/2},$$

where the second factor containing the moment of inertia (*I*) for the physical pendulum "corrects" the basic period formula for the mass distribution of the physical pendulum. When this equation is simplified, we have

$$T = 2\pi \sqrt{\frac{I}{|mg|l}}. \tag{12.10}$$

The **moment of inertia (*I*)** of an object quantifies the distribution of its mass around its rotational center.

Problem-Solving Strategy 12.4
The larger the moment of inertia for a physical pendulum, the farther its center of mass is from the center of rotation. Thus, larger moments of inertia correspond to longer pendulum oscillation periods.

Do you believe the earth rotates once a day? If so, can you prove it? You cannot feel the rotation, since the angular speed is about 0.0007 rpm. One strong proof that the earth rotates is the Foucault pendulum, an invention of the French physicist Jean Bernard Leon Foucault.

In 1851 Foucault hung a pendulum from the dome of the Pantheon in Paris. The pendulum consisted of a 28-kilogram iron ball on the end of an approximately 67.5 meter wire, which was attached to its support so that it could swing in any direction. The period of this extra-long pendulum was about 16.5 seconds—it would swing only four times each minute.

Since the pendulum's only contact with the earth was its support, which exerted only a vertical force on it, the pendulum would not follow the rotation of the earth. If the earth truly rotated, it would move under Foucault's pendulum so that the pendulum's orientation would appear to change. Foucault set the pendulum in motion by pulling the bob to one side and releasing it. He then drew a line on the floor to mark the direction of the pendulum's swing. At first the pendulum seemed to continue to follow the line, but eventually the change was noticeable. After 9 hours, the pendulum was swinging perpendicular to the line. After nearly 32 hours, the pendulum's swing had rotated through a full circle. Since there were no unbalanced forces on the pendulum, Foucault proved that the earth must rotate.

You might wonder why the pendulum took 32 hours to complete one cycle. This effect is due to the location of the pendulum. At one of the earth's poles, the pendulum would take exactly 23.93 h to complete one full *precession* cycle as the earth rotates under the pendulum. This was actually done in 2001 at the South Pole. At the equator, the pendulum would not precess at all. However, at any in-between latitudes, the period of precession is equal to 23.93 h divided by the sine of the latitude angle. So at the latitude of Paris (48.8N), the precession period should be

$$T_{\text{precession}} = \frac{23.93 \text{ h}}{\sin 48.8°} = 31.8 \text{ h}.$$

Not only is the precession period affected by pendulum location, but the *direction* of precession is affected as well. In northern latitudes, the precession is clockwise, while it is counterclockwise in the southern hemisphere. This is related to the same Coriolis effect that determines the global cyclical wind and ocean current motions.

Much research has gone into understanding the properties of physical pendulums. Beginning with Christiaan Huygens, the inventor of the first pendulum-regulated clock, scientists and inventors have attempted to improve the periodicity and accuracy of pendulums. The most notable of these efforts was the development of a chronometer that would withstand the rigors of the sea so that mariners could determine their longitude. Within the last century, pendulum-like devices were developed to control countless machines and activate safety systems, such as seatbelts and airbags in cars.

In order to understand the walking gait of people, you must know how to determine the natural period of the human leg. This knowledge directly affects the design of athletic footwear and of prosthetic limbs for amputees. The military's research into augmenting human limbs in "powered combat suits" also involves these principles.

12-12 A smart prosthetic leg that works similarly to a natural leg

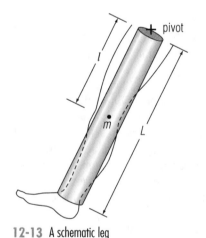

12-13 A schematic leg

EXAMPLE 12-3

Period of a Pendulum—Your Leg

What is the natural period of a human leg? This is an important consideration for "smart" prosthetic legs that adjust to the motion of the user.

Solution:

First, model an average human leg as a thin, solid rod of length $L = 1.5$ m and mass $m = 16.0$ kg. From Appendix F the moment of inertia for a thin, solid rod is $I = {}^{mL^2}/_3$. The pendulum arm l is the distance from the pivot point to the center of mass. Assume for this example that the mass is uniformly distributed throughout the leg; then $l = {}^L/_2$.

Use Equation 12.10 to determine the period of a human leg.

Substitute the known information into the formula and solve for T.

$$T = 2\pi \sqrt{\frac{I}{m|g|l}} = \sqrt{\frac{mL^2/3}{m|g|(L/2)}}$$

$$T = 2\pi \sqrt{\frac{2L}{3|g|}} = 2\pi \sqrt{\frac{2(1.5 \text{ m})}{3|-9.81 \text{ m/s}^2|}}$$

$$T \doteq 2.00 \text{ s}$$

$$T \approx 2.0 \text{ s}$$

The natural period for an entire human leg is about 2 seconds.

12B Section Review Questions

1. Why did the study of pendulums present a problem to Greek natural philosophers?

2. a. When a pendulum is pulled to the side and then released, what forces are acting on the pendulum mass at the instant of release?
   b. What is the effect of the vector sum of these forces?

3. a. What additional force is acting on the pendulum at the bottom of its swing?
   b. Why does this force exist?

4. What limitation exists for considering pendulum motion as simple harmonic motion?

5. True or false: The period of a pendulum is directly proportional to the length of the pendulum arm.

⊚6. a. What forces must you consider when determining the "restoring force" that causes a pendulum to return to its equilibrium position?

b. For large amplitude swings, how does the "restoring force" vary with the position of the pendulum arm?

⊚7. a. What is the period of an ideal pendulum that has a mass of 0.250 kg centered at the end of a pendulum arm that is 0.650 m long? Assume that it is swinging at the earth's surface.

b. If the same pendulum were taken to the moon, what would its period be?

⊚8. What is the length of an ideal pendulum that swings with a period of 1.40 s near the earth's surface?

⊚9. Find the period of a pendulum consisting of a thin solid rod weighing 8.50 kg that is 3.50 m long.

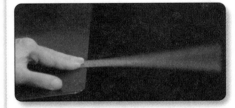

12-14 Harmonic oscillators naturally experience damping.

12C OSCILLATIONS IN THE REAL WORLD

12.10 Damped Oscillations

When an ideal spring-mass system is set to oscillating, it will theoretically continue its simple harmonic motion indefinitely in the absence of any frictional effects. If we hang a real mass from a vertical physical spring in a laboratory and set it to oscillating, in a relatively short time the amplitude of the oscillations is reduced to a slight bobbing that eventually stops. Resistance within the spring and the drag of the air on the mass produce forces that slow the motion of the oscillating mass. An oscillator that experiences such forces is called a **damped harmonic oscillator.** We can find such oscillators in everything from hand bells to diving boards.

A **damped harmonic oscillator** would experience simple harmonic motion in the absence of all frictional forces.

The medium surrounding the oscillator has a large effect on its behavior. Consider a mass hanging from a vertical spring suspended in a thick, viscous liquid such as heavy motor oil or a gooey syrup. If the mass is displaced and released, it slowly moves back to its equilibrium position and no further. In fact, the system is not even an oscillator since it cannot complete even a small fraction of a cycle before its motion is damped out. This motion is called an **overdamped oscillation.**

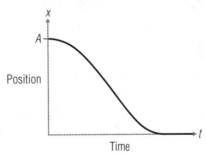

12-15 An overdamped oscillator does not complete even one cycle.

In some ways our spiritual effectiveness is like a harmonic oscillator, with the world system constantly trying to damp the oscillations. Several passages in the Bible outline tactics that alert us to this dampening effect. Paul wrote, "Beware lest any man spoil you through philosophy and vain deceit, after the tradition of men, after the rudiments of the world, and not after Christ" (Col. 2:8).

If we gradually thin the liquid by adding an appropriate solvent, we will eventually be able to get the system to just barely overshoot the equilibrium point before it comes to a stop—one truncated cycle, if you will. When the oscillator and medium viscosity are matched so that the mass hardly overshoots the equilibrium point, then the system is **critically damped.** If we continue to thin the liquid medium, then the system behaves like a damped oscillator as described earlier in this subsection.

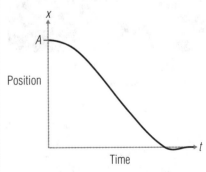

12-16 A critically damped oscillator barely completes one cycle before damping stops all motion.

The mathematical modeling of damping is somewhat complex, and the derivation of the applicable equations is beyond the scope of this textbook. However, we can consider some of the characteristics of damped oscillators. Damping in an oscillator that is modeled by an ideal spring occurs because of frictional drag from the medium. The magnitude of the frictional force is approximately proportional to the velocity of the system through the medium. This can be written in equation form as

$$\mathbf{f}_x = -\beta\mathbf{v}_x,$$

where β is a friction proportionality constant. The friction force opposes the restoring spring force on the system. The net force sum acting on the system is then

$$\Sigma F_x = F_{\text{spring } x} + f_x, \text{ or}$$
$$\Sigma F_x = (-k\Delta x) + (-\beta v_x).$$

Rewriting the second equation in the form of Newton's second law, we have

$$-k\Delta x - \beta v_x = ma_x. \tag{12.11}$$

In Equation 12.11, velocity, and thus the frictional damping force, is dependent on the system's displacement from equilibrium. Velocity is zero at the maximum system displacement and maximum when displacement is zero. When the system is initially released, the spring force accelerates the mass and there is little damping friction. But as system velocity increases, the damping force increases proportionately, so the mass does not accelerate as much as it otherwise would. A lower acceleration means that there is less speed to carry the system beyond the equilibrium point against the increasing restoring force of the spring on the other side. The displacement at the far side of the oscillation is less than the initial displacement. When the system begins to return, there is less initial acceleration because of the smaller displacement. As velocity increases toward the middle of the oscillation, frictional force again reduces the net force accelerating the mass. The net effect is that the amplitude for each half-cycle is smaller than the one before it. Eventually, the motion is completely damped out.

12.11 Driven Oscillations and Resonance

Although all real vibrations will be stopped by friction if left alone, they can be sustained by the application of an additional force to the system. This is known as *driving* the system. The force cannot be applied in a haphazard manner, however. To illustrate this, let us look at a child on a swing.

A swing can be considered a simple pendulum. If you give the child a small push, the swing will begin to move in damped harmonic motion and with a small amplitude. The child will probably want to be pushed "faster." Changing tactics, if you perform five very rapid pushes in a short period of time, you note that the amplitude is still not very large. What is the correct way to push the swing so that you can increase the amplitude without overexerting yourself and causing uncomfortable accelerations to the child?

You've learned from experience that giving a small push when the swing is at its maximum height (amplitude) produces the desired effect. This works because you are applying a force at the frequency of the swing-oscillator that is assisting the restoring force of the system. The frequency at which the force is most effective in increasing the swing amplitude is called the **natural oscillation frequency** (f_0). As you push the child in this way, your pushes are in *phase* with the swing's oscillations. Engineers call such pushes in an oscillating system *pulses*.

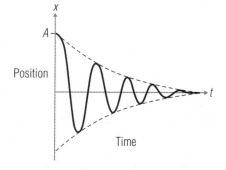

12-17 The amplitude of a damped oscillator gradually diminishes until motion stops.

12-18 Pushing a child on a swing is most effective if the pushes are given at the natural frequency of the child-swing system.

Objects tend to vibrate at a characteristic frequency called the **natural oscillation frequency (f_0),** also called the **resonant frequency.**

If the added force exceeds the frictional damping forces and is in phase with the restoring force on the system, then the oscillator's amplitude will increase. When the oscillations increase in amplitude due to the force supplied at the natural frequency of the system, they are called **driven oscillations.** Driven oscillations can result in very large amplitude oscillations—a condition called **resonance.** Resonance sometimes has a good effect, such as the pure resonating tone in a musical instrument, but sometimes it causes damage.

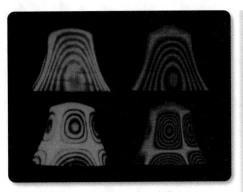

12-19 The resonating regions of the bell are not uniformly distributed over the bell's surface, as seen in these photos taken in polarized light.

One example of the damage resonance can cause is its effect on buildings during an earthquake. Earthquakes are low-frequency vibrations that travel through the earth. Tall buildings tend to suffer greater damage than shorter buildings because their natural frequencies are lower than those of smaller buildings. You can demonstrate this fact by clamping long and short rulers to a desktop and comparing their vibrations. The long ruler vibrates at a lower frequency than the shorter one does. The lower natural frequency of taller buildings is closer to the vibrational frequency of the ground during an earthquake, and resonant conditions are more easily induced, causing much greater damage.

Complex structures such as buildings and bridges have many natural frequencies. Engineers expend a lot of effort and use computer modeling extensively in order to ensure that their designs do not develop resonance under anticipated wind, road traffic vibration, or even earthquake conditions.

One well-known demonstration of the catastrophic potential for resonance was the collapse of the Tacoma Narrows Bridge near Tacoma, Washington. This bridge, spanning a narrow river gorge, was often buffeted by winds. It was known by local residents to sway noticeably on windy days. One day in November of 1940, the way the winds blew applied a force at the natural frequency of most of the bridge's components, particularly the road-deck. The internal resistance of the metal and other materials was not sufficient to dampen the pulses provided by the wind; as a result, the bridge swayed with greater and greater amplitudes until it finally fell apart. This collapse would not have occurred if the bridge had been constructed of proper materials and with component shapes that had different natural frequencies.

A driven oscillator has three forces acting on it:

- The restoring force, or Hooke's law force;
- The damping resistance force that opposes the velocity of the system;
- The pulsed force that is applied in the same direction as the restoring force.

The applied force may occur as discrete pulses at intervals of T or $T/2$ seconds, but many natural sources of driving forces often vary sinusoidally with time.

Oscillation amplitude is a function of the frequency of the system. Maximum amplitude occurs when the frequency of the applied force matches the natural frequency of the oscillator. Theoretically, the maximum resonant amplitude is infinite in the absence of damping. However, in every real system there is always some damping, so the maximum amplitude is limited by the damping effects on the system. These effects become more significant with increasing amplitude, which eventually limits the resonant amplitude to a maximum value.

The condition of **resonance** is indicated by large amplitude oscillations. These **driven oscillations** result from the periodic application of a force to the system.

12-20 The collapse of the Tacoma Narrows bridge

Natural sources of oscillation-driving forces include sound, water, and earthquake waves. Each of these can be modeled by a sinusoidal function.

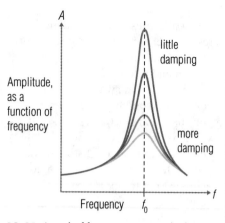

12-21 A graph of frequency versus amplitude showing how damping limits the peak amplitude of a resonating system

12C Section Review Questions

1. What eventually happens to a real oscillating system without any outside assistance?

2. Assuming an ideal spring restoring force acts on an oscillating system, what is the major source of the damping force?

3. To which property of an oscillating system is the magnitude of the external damping force proportional?

4. What kind of condition exists if the damping force is so great that no overshoot of the equilibrium position occurs after the system is released?

5. The escapement system in a pendulum-regulated clock applies a small force to the pendulum arm every cycle. What are these momentary forces called, and what kind of oscillations are occurring?

6. At what frequency should force be applied to increase the amplitude of an oscillator?

7. During World War II, army field commanders noticed that floating pontoon bridges quickly fell apart when large numbers of troops marched across them in step. The marching cadence was very close to the natural frequency of the pontoon structures, producing damaging resonance in the bridges. Discuss one way the commanders could (and did) solve the problem without redesigning the bridges.

12D WAVES

12.12 Defining Waves

Simple harmonic motion pertains mostly to oscillations of isolated objects, such as a mass attached to a spring or a bob on a string. **Waves,** on the other hand, are oscillations of extended bodies. For example, water waves are oscillations of a body of water.

The material through which the waves travel is called the **medium.** The oscillations in the medium are called a **disturbance.** The disturbance usually begins at one point in the medium and travels from there. If you drop a pebble into a smooth pond, the water immediately around the pebble begins to move, and soon the water at the edge of the pond is vibrating. The disturbance has traveled to the edge of the medium.

One thing to notice about a wave is that the *disturbance* travels; the *medium* does not move far. A cork floating in the pond mentioned above bobs up and down but stays in the same area, as the disturbance caused by the pebble passes by. The disturbance traveled, but the water supporting the cork did not.

A wave is complicated. Suppose you tie one end of a rope to a tree and hold the other so that the rope is taut. Then you shake the end of the rope. In this example, the rope is the medium for the wave.

How can you describe the wave in the rope? A graph can help. Refer to Figure 12-22. Figure (a) shows the rope 0.5 s after it was shaken. The disturbance has pulled knot 1 down, but it has not yet reached knot 2. Figure (b) shows the string 1 s after it was shaken. The disturbance has now passed knot 1 and has dropped knot 2. The shape of the disturbance has not changed, but the wave has moved. The graphs in figures (a) and (b) are called *waveform graphs*. A waveform graph is essentially a snapshot of the entire wave at an instant in time. Figures (c) and (d) show a different way of graphing the wave. Figure (c) is a graph of knot 1's vertical position versus time. Figure (d) is the graph for knot 2. These latter graphs are called *vibration graphs*. Vibration graphs show the position of a single point in the medium over a span of time.

12.13 Types of Waves

There are two basic kinds of waves. A **longitudinal wave** is a disturbance that moves the medium along its line of travel. A spring, for example, vibrates longitudinally when it expands and compresses along its length. A **transverse wave** is a disturbance that moves the medium perpendicular to its line of travel. For example, the wave along a snapped string is a transverse wave. Many waves have both longitudinal and transverse components.

Any physical medium can carry a longitudinal wave. Figure 12-23 shows an air-filled tube with a closed end. The hammer strikes one end of the tube, causing it to vibrate. The end of the tube pushes the air molecules next to it, and these in turn push other molecules. This sets up a longitudinal disturbance traveling down the tube.

As the wave travels, the air molecules in some regions are forced together while the molecules in other regions are pushed apart. The regions where the molecules are pushed together have higher density and pressure than the average for the medium and are called **compression zones.** The regions where the molecules are forced apart have lower density and pressure and are called **rarefaction zones.** These zones move with the wave.

A longitudinal wave travels more quickly in a medium whose molecules are closer together (more dense). Therefore, longitudinal waves such as sound travel more rapidly through solids than through gases. For this reason, nineteenth-century Indians in the American West could put their ears to a railway track and detect approaching trains long before the trains could be heard through the air.

Some mechanical transverse waves, unlike longitudinal waves, cannot pass through gases or liquids. However, solids can transmit all forms of mechanical transverse waves. You will learn in later chapters that light is a transverse wave that does not require a medium for propagation.

An example of a wave with both transverse and longitudinal components is a water wave. The transverse and longitudinal forces combine to make a force that causes the particles in a wave to travel in a vertical circle. Earthquakes are also "combination" waves.

12.14 Periodic Waves

When a disturbance is repeated regularly, it produces a periodic wave. Periodic waves, like the waves resulting from a single disturbance, can carry information and energy from one place to another. Waves carry information like a code carries information. For example, if soldiers are told that one whistle means "move ahead," while two whistles means "stay put," then when they hear one whistle, they think "move ahead." Similarly, your ears and eyes regard waves as a code. When the cells in your eyes detect a light wave that vibrates 3.8×10^{14} times each second, they report "red."

Waves also carry energy. For example, tsunamis—huge water waves caused by earthquake shocks—carry an earthquake's energy far from the quake's origin. The 1883 eruption of Krakatoa, a volcano in the Sundra Strait between Java and Sumatra, killed 36,000 people with the waves it produced.

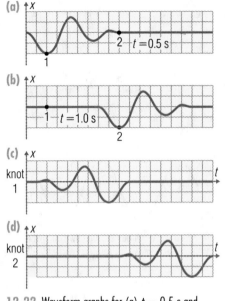

12-22 Waveform graphs for (a) $t = 0.5$ s and (b) $t = 1.0$ s. Figures (c) and (d) are vibration graphs for knots 1 and 2, respectively.

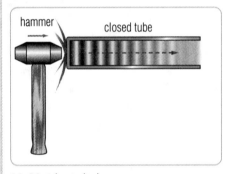

12-23 A longitudinal wave

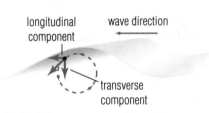

12-24 Velocity components of a particle in a water wave

Strange things happen during earthquakes, but perhaps the strangest thing that ever happened is recorded in Numbers 16:31–32. Korah, Dathan, Abiram, and their followers tried to take the leadership away from Moses. God caused the ground to open up and they all fell into the abyss. The ground then closed up again, leaving no evidence of the rebels or the fissure in the ground.

To describe waves, some of the same terms are used as were used for harmonic motion. For a transverse wave, the *amplitude, A,* is the greatest distance the wave displaces a particle. A longitudinal wave's physical amplitude is usually half the distance between the maximum and minimum pressure differences it produces:

$$A = \frac{1}{2}(y_{peak} - y_{trough}) \text{ (transverse), or}$$

$$A = \frac{1}{2}(x_{max} - x_{min}) \text{ (longitudinal)}$$

One **wavelength** (λ) is the distance from one peak (or compression zone) to the next, or from one trough (or rarefaction zone) to the next. The wave completes one *cycle* as it moves through one wavelength. The wave's *frequency* is the number of cycles a wave completes per unit of time.

Each wave that we will study moves at a constant speed called the **wave speed** (*v*). This is the speed of the *disturbance,* not of the medium. For example, suppose a traffic light turns red when a careless driver is not looking. The other cars stop, but this driver continues until he strikes the car in front of him. That car hits the next car, which hits the next, and soon the car at the front of the line is struck. It may be 1 s from the time the careless driver rams the last car to the time the first car, 20 m ahead, is struck. The wave has traveled at 20 m/1 s = 20 m/s, or 72 km/h. None of the cars approaches this speed.

For a periodic wave, this speed is related to wavelength and frequency. Since wavelength is the number of meters per cycle and frequency is the number of cycles per second, the product of wavelength and frequency is the number of meters the wave travels per second—the wave speed. In equation form,

$$\lambda f = v, \tag{12.12}$$

where *v* is the wave's speed through the medium. By analyzing the units on the left side of Equation 12.12, you can see that the appropriate units for speed appear on the right:

$$\lambda f \Rightarrow \frac{m}{\cancel{cycle}} \cdot \frac{\cancel{cycle}}{s} = \frac{m}{s} \Rightarrow v$$

Wave speed is a physical property of the medium. If the medium is uniform, λ and *f* are inversely proportional.

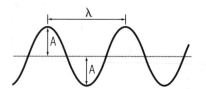

12-25 A wave can be described by its amplitude (*A*) and its wavelength (λ).

Wave speed is a physical property of the *medium*—it is *not* due to the shape or frequency of the wave passing through the medium.

Problem-Solving Strategy 12.5

For a uniform medium, the wave speed is constant for all frequencies. Therefore, frequency and wavelength are inversely proportional. For example, if frequency doubles, then wavelength is halved.

EXAMPLE 12-4

Working With Waves: Determining Wavelength

A photon of red light has a speed of 3.00 × 10⁸ m/s and a frequency of 3.80 × 10¹⁴ Hz. What is its wavelength?

Solution:

$$\lambda f = v$$

$$\lambda = \frac{v}{f} = \frac{3.00 \times 10^8 \text{ m/s}}{3.80 \times 10^{14} \text{ s}^{-1}}$$

$$\lambda \doteq 7.8\underline{9}4 \times 10^{-7} \text{ m}$$

$$\lambda \approx 789 \text{ nm}$$

EXAMPLE 12-5

Analyzing Waves

Figure 12-26a shows a wave graph, and Figure 12-26b shows a vibration graph for a wave. Find the wave's (a) amplitude (A), (b) wavelength (λ), (c) frequency (f), (d) period (T), and (e) speed (v).

Solution:

a. Amplitude:

$$A = \tfrac{1}{2}(d_{peak} - d_{trough}) = \tfrac{1}{2}\left[(+10 \text{ cm}) - (-10 \text{ cm})\right]$$

$$A = 10 \text{ cm}$$

b. Wavelength: $\lambda = 20$ cm (from Figure 12-26a)

c. Frequency:

$$f = \frac{n \text{ cycles}}{\Delta t} = \frac{1 \text{ cycle}}{2 \text{ s}} \text{ (from Figure 12-26b)}$$

$$f = 0.5 \text{ s}^{-1}$$

d. Period:

$$T = \tfrac{1}{f} = \frac{1}{0.5 \text{ s}^{-1}}$$

$$T = 2 \text{ s}$$

e. Speed:

$$v = \lambda f = (0.2 \text{ m})(0.5 \text{ s}^{-1})$$

$$v = 0.1 \text{ m/s}$$

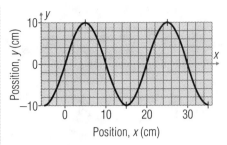

12-26a Wave graph

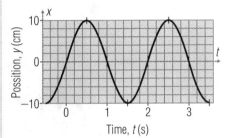

12-26b Vibration graph

12.15 Sound Waves

Sound waves are an important part of everyday life. Sound waves are longitudinal pressure waves that come from a vibrating body, such as a tuning fork, and are detected by your ears. When you strike a tuning fork, it vibrates, and the vibration alternately compresses and rarefies the air around it. This compression and rarefaction sets up a pressure wave that reaches your eardrum. Your eardrum vibrates in response to the wave and stimulates nerves in your inner ear through a mechanical linkage of tiny bones in your middle ear. Your brain interprets the neural impulses from this vibration as sound.

Because sound waves travel by compressing and rarefying their medium, they must have a material medium to travel through. Sound cannot travel through a vacuum. Like other longitudinal waves, sound travels most rapidly through solids and most slowly through gases. Thus, you can hear a sound sooner under water than in air, and even sooner if your head is in contact with a solid. You may have noticed that you hear sounds originating on your head bones differently from sounds only a short distance away. For example, if you scratch a pencil with your fingernail as you hold it a few inches from your ear, you can hardly hear the sound. If you instead hold the pencil between your teeth as you scratch it, you can hear the sound clearly. Your solid skull conducts sound much better than the air next to your ear.

TABLE 12-2	
Medium (at 0 °C)	**Speed** (m/s)
air	332
hydrogen	1284
glycerol	1849
water	1450
aluminum	6420
Pyrex glass	5640

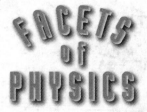

Earthquakes

Earthquakes demonstrate the ability of shock waves to pass through solids and to transport energy. While the causes of earthquakes are still not completely understood, the characteristics of earthquakes are well known. Earthquakes usually consist of three kinds of waves: P, S, and L.

The P, or primary, waves travel fastest and arrive first at an observing station. These waves are longitudinal waves and can travel through solids, liquids, and gases. The second waves to arrive at a seismic station are the S, or secondary, waves. These transverse waves can travel only through solids. The last waves to arrive at a seismic station are called L waves. These waves have much longer wavelengths than P or S waves, and they travel around the earth at its surface rather than through it. Two types of waveforms compose L waves—translational and elliptical.

Analyzing the waves from earthquakes can tell scientists about the earth's structure. For example, when scientists discovered that P and S waves change speed when they travel more than about 50 kilometers below the earth's surface, they concluded that the earth's composition changes at that depth. Similarly, seismologists noticed that S waves do not travel directly through the earth's center, but P waves do. This evidence suggests that portions of the earth's center, or core, are liquid.

Large earthquakes can be disastrous, but all kinds of earthquakes help us to learn about the interior of the earth. In fact, following the cessation of underground nuclear weapons testing, earthquakes are the only source of information we have about the deep structure of the earth.

The sound you hear depends not only on the characteristics of the wave but also on the characteristics of your hearing. Three basic characteristics, influenced by both the sound wave and your hearing, describe what you hear. *Loudness* is the interpretation your hearing gives the intensity of the wave. Sound **intensity (I_s)** is the amount of power transported by the wave per unit area and is measured in watts per square meter (W/m²). It is proportional to the square of the wave's amplitude.

The intensity of the sound does not directly equal perceived loudness. A sound must be ten times as intense as another in order to sound twice as loud. The least intense sound a human ear can detect is about 10^{-12} W/m². A sound with an intensity of 10^{-11} W/m² seems only twice as loud as the least intense sound. Therefore, scientists have devised a scale of loudness that reflects this characteristic of hearing. Loudness was originally measured in *bels,* where an increase of 1 bel doubles the loudness of the sound and increases the sound intensity by a factor of 10. Now scientists measure loudness in tenths of bels, or **decibels (dB).** A loudness of 10 dB is barely audible to most people. A sound that is 20 dB is twice as loud as a 10 dB sound.

A second characteristic of sound is its *frequency,* which you interpret as **pitch,** or how high or low the sound is. A sound with high frequency has a high pitch, and a sound with low frequency has a low pitch. The average human can detect sound frequencies from 20 Hz to 20 000 Hz. Some animals detect sounds of higher and lower frequencies. Dogs, for instance, detect sounds from 15 Hz to 50 000 Hz, and bats and porpoises detect sounds with frequencies as high as 120 000 Hz.

Most sounds are actually combinations of waves with several frequencies. The particular combinations of frequencies can cause two sounds with the same pitch and loudness to sound different, thus giving sound a third characteristic—**quality.** For example, a trumpet playing middle C at 40 decibels sounds much different from an oboe playing middle C at 40 decibels. Actually, each instrument emits multiples of the frequency of middle C as well as the frequency itself. The frequency of the dominant note is called the **fundamental;** the multiples of this frequency are called **harmonics.** Each instrument produces different harmonics. The oboe tends to emit odd harmonics, while the trumpet emits both odd and even harmonics. Thus, the quality of the sounds from the instruments is different even when their pitch and loudness are the same.

The three characteristics of sound—loudness, pitch, and quality—affect one another. Sounds with high pitch tend to seem louder than sounds of low pitch. Also, the loudness of a sound affects its quality. You are less likely to hear the harmonics in a barely audible sound than in a sound that is comfortably loud. Therefore, you cannot fully describe one characteristic of sound without also describing the other characteristics.

The loudness of a sound (β) of intensity, I_s, is computed by the formula

$$\beta = (10\ \text{dB})\log\left(\frac{I_s}{10^{-12}\ \text{W/m}^2}\right).$$

Problem-Solving Strategy 12.6

The common logarithm (log) of a number is just the exponent of 10 that gives the argument x in log (x). The ratio of I_s over 10^{-12} W/m² yields the relative strength of the sound compared to a reference intensity. Taking the log of this ratio gives the magnitude of the ratio.

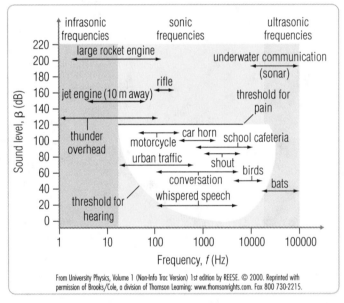

12-27 Comparison of sound levels for various sources of sound and the range of human hearing

12.16 Doppler Effect

So far in this discussion of waves, it has been assumed that the observer and the wave source are not moving with respect to each other. Relative motion changes the observer's perception of the waves. Imagine that you are standing beside a long conveyor belt with milk bottles on it. Two bottles pass you each second. If you begin to walk away from the end of the belt that the bottles are coming from, perhaps only one bottle will pass you each second. If you turn around and walk toward the source of the bottles, three bottles may pass you each second. The

Christian Johann Doppler (1803–53) was an Austrian physicist who first described the effect of relative motion on perceived frequencies in the electromagnetic spectrum.

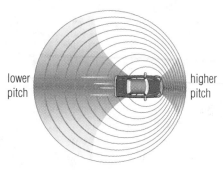

12-28 The Doppler effect

bottles are coming from the source at a constant rate, but the frequency of the bottles passing you depends on the relative velocity of you and the source of bottles. Similarly, the perceived frequency of any sound waves depends on the relative velocity of the observer and the wave source. This is called the **Doppler effect.**

The Doppler effect is responsible for the change in frequency you hear when a car passes you—first an increase in pitch as it moves toward you, then a decrease in pitch once it is past. When the car, the source of the sound waves, is moving toward you, the crests of the waves are hitting your ear more frequently. This is perceived as a higher pitch. When the source of the sound waves is moving away from you, the crests of the waves hit your ear less frequently, and this is perceived as a lower pitch. Notice that this is a perceptual phenomenon. The actual sound emitted by the car may not change—it is only perceived differently. We are so used to this phenomenon that we form a perception of how fast a car is going by the intensity of the Doppler effect as it drives by.

12D Section Review Questions

1. Explain how a physical wave moves through a medium. What actually moves?

2. Discuss the difference between waveform graphs and vibration graphs.

3. Describe longitudinal and transverse waves. Which category do sound waves fall under? Light waves?

4. Discuss the relationship between intensity and perceived loudness of sound waves.

5. **a.** Why does the pitch of a car's engine seem to increase as the car approaches you and decrease as it travels away from you?

   b. What is this phenomenon called?

⊛6. The speed of sound in air at 20 °C is 343 m/s. What is the wavelength of a 440. Hz sound wave?

⊛7. What is the period of the sound wave in Question 6?

⊛8. If a sound has an intensity of 10^{-7} W/m^2, how loud is the sound in dB?

In Terms of Physics

periodic motion	12.2		resonance	12.11
damping	12.2		wave	12.12
restoring force ($\mathbf{F}_r$)	12.2		medium	12.12
simple harmonic motion (SHM)	12.3		disturbance	12.12
amplitude (A)	12.3		longitudinal wave	12.13
cycle	12.3		transverse wave	12.13
frequency (f)	12.3		compression zone	12.13
hertz (Hz)	12.3		rarefaction zone	12.13
pendulum	12.5		wavelength (λ)	12.14
pendulum arm (L)	12.6		wave speed (v)	12.14
total acceleration ($\mathbf{a}_{total}$)	12.7		intensity (I_s)	12.15
physical pendulum	12.9		decibels (dB)	12.15
moment of inertia (I)	12.9		pitch	12.15
damped harmonic oscillator	12.10		quality	12.15
overdamped oscillation	12.10		fundamental	12.15
critically damped oscillation	12.10		harmonics	12.15
natural oscillation frequency (f_0)	12.11		Doppler effect	12.16
driven oscillations	12.11			

Problem-Solving Strategies

12.1 (page 259) Remember that period and frequency are reciprocals of each other.

12.2 (page 260) A larger period means that the spring system moves slower and takes longer to complete one cycle. The larger the mass, the longer the period. The "stiffer" the spring, the shorter the period.

12.3 (page 265) When calculating the period of a pendulum, be sure to use the *magnitude* of g. If you use the scalar value, the result will be an imaginary number.

12.4 (page 267) The larger the moment of inertia for a physical pendulum, the farther its center of mass is from the center of rotation. Thus, larger moments of inertia correspond to longer pendulum oscillation periods.

12.5 (page 274) For a uniform medium, the wave speed is constant for all frequencies. Therefore, frequency and wavelength are inversely proportional. For example, if frequency doubles, then wavelength is halved.

12.6 (page 277) The common logarithm (log) of a number is just the exponent of 10 that gives the argument x in log (x). The ratio of I_s over 10^{-12} W/m^2 yields the relative strength of the sound compared to a reference intensity. Taking the log of this ratio gives the magnitude of the ratio.

Review Questions

1. What is periodic motion?

2. What is the name for periodic motion that is controlled by a restoring force $F_r = k\Delta x$ and is not subject to friction?

3. What is the net force on a simple harmonic oscillator at its equilibrium position?

4. Which of the following is *not* used to describe simple harmonic motion?
   a. frequency, f
   b. wavelength, λ
   c. amplitude, A
   d. period, T

5. Name two motions that approximate simple harmonic motion.

6. Mass m_1 undergoes simple harmonic motion on a spring. It has a period T_1. If it is replaced by a mass $m_2 = 2m_1$, is the new period double the old ($T_2 = 2T_1$)?

7. Under what conditions can a pendulum's motion be considered simple harmonic motion?

8. A clock contains a pendulum that swings from one side to the other in 1.00 s near the earth's surface. Assume that it exhibits simple harmonic motion. What must its period be?

9. What is the term for the friction that naturally opposes periodic oscillations?

10. Which kind of damping would be most comfortable in a car shock absorber?
   a. overdamped
   b. critically damped
   c. slightly damped

11. What is the name of the frequency at which energy is most efficiently added to an oscillating system?

12. What condition occurs when the energy added to a driven oscillator exceeds the natural damping of the system?

13. A cork floats in a smooth pond. A pebble is dropped next to it and creates a wave with a wave speed of 10.0 m/s. If the wave travels to the shore 1000 m away, will the cork be significantly closer to the shore after 10 s?

14. What is the difference between a waveform graph and a vibration graph?

15. What are the two major types of waves?

16. Classify the following waves according to type (one of the two major types or a combination).
    a. wave in an iron bar that is struck by a hammer parallel to its length
    b. wave in a taut rope whose end is snapped
    c. sound wave
    d. wave of collisions when a car in traffic stops suddenly
    e. waves from an earthquake
    f. light wave

17. Why can you hear sounds sooner when your ear is to the ground than when you are standing up?

18. Which one of the following do waves not transmit?
    a. information
    b. matter
    c. energy

19. Sound travels faster in
    a. gases.
    b. liquids.
    c. solids.

20. Which of the following determines a sound's pitch?
    a. frequency
    b. intensity
    c. quality

21. Decibels measure a sound's
    a. intensity.
    b. loudness.
    c. pitch.
    d. quality.

22. On what does the quality of a sound depend?

True or False (23–34)

23. A laboratory mass hanging from an ideal spring can oscillate in simple harmonic motion.

24. The acceleration experienced by an oscillating system is directly proportional to the system's displacement from its equilibrium point.

25. Simple harmonic motion is analogous to uniform circular motion.

26. An oscillating system's maximum displacement from one extreme position to the other is its amplitude.

27. The position of a vibrating system at time $t + T$ is the same as it was at time t.

28. A weaker spring (smaller k) implies a higher frequency of oscillation.

29. Ignoring air resistance and other forms of friction, there are only two forces that act on a pendulum at any point in its swing—its weight and the total tension in the pendulum arm.

30. Assuming an initial pendulum angle of $\pi/4$ when it is released, as the pendulum swings through the lowest position in its arc, the weight vector of the mass is larger because of centrifugal force.

31. A pendulum will probably swing more slowly at the top of a high mountain than at its base.

32. Damping tends to oppose the acceleration of a harmonic oscillator.

33. A physical pendulum is a damped oscillator.

34. In order to cause resonance to occur in a system with a higher natural frequency, you must apply a larger force than for a system with a lower natural frequency.

⊚35. A mass is attached on the end of a spring. Then it is displaced 4.00 cm from its equilibrium position. In the next 0.500 s, the mass moves through its equilibrium position, stops 4.00 cm on the other side, then returns to its original displacement. It continues to vibrate at this rate. What are its amplitude, period, and frequency?

⊚36. A 1.00 kg mass hangs from a spring. It is set in motion with a period of 6.28 s. What is the spring constant, k?

⊚37. A mass on a spring with a spring constant $k = 5.00$ N/m has a frequency of $f = 0.250$ Hz when undergoing simple harmonic motion.
    a. What is its period?
    b. What is the mass?

⊚38. A spring stretches by 0.0500 m when a 2.55 kg mass is hung from it. The mass is then pulled down 0.100 m and released. It undergoes simple harmonic motion.
    a. What is the force that the mass exerts on the spring?
    b. What is the spring constant?
    c. What is the amplitude of the simple harmonic motion?
    d. What is the period of the mass's motion?
    e. What is the frequency of the mass's motion?

⊚39. A spring-mass system oscillates with an amplitude of 0.100 m. The spring constant is 1.00×10^2 N/m. Take the potential energy of the system to be 0 J at the equilibrium position ($x = 0$ at equilibrium).
    a. What is the mass's potential energy at $x = A$?
    b. What is the mass's total mechanical energy at $x = A$?
    c. What is the mass's kinetic energy at $x = 0$?
    d. If the mass is 1.00 kg, what is its speed at $x = 0$?
    e. Find the period and frequency of the motion.

⊚40. A pendulum has a period of 1.28 s on the earth's surface. What period will the same pendulum have on the surface of the Moon ($|g_{Moon}| = 1.62$ m/s²)?

⊚41. What is the period of a 1.00 m long pendulum on Mars ($|g_{Mars}| = 3.71$ m/s²)?

⊚42. If a given pendulum has a length l and period T, what would be the length of a second pendulum if its period were $T/3$?

⊚43. A blue light has a wavelength of 4.55×10^{-7} m and a speed of 3.00×10^8 m/s. What is its frequency?

44. Refer to the wave graph to answer the questions below.

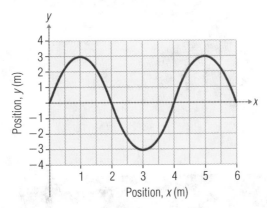

a. What is the wavelength?
b. What is the amplitude?
c. The wave's speed is 64 m/s. What is its frequency?
d. What is the wave's period?

45. A 5.50 kg laboratory mass rests on a surface with coefficients of friction $\mu_s = 0.30$ and $\mu_k = 0.25$. A horizontal spring is attached to the mass and is stretched 0.160 m. The mass remains motionless. However, if the spring is stretched 0.161 m, the mass slips. What is the spring constant?

46. A device is built to shoot a 150. g puck across a frictionless surface using a triggered spring launcher. The spring is compressed 8.0 cm and has a spring constant of 45 N/m. What is the initial acceleration of the puck when the spring trigger is released? Describe the puck's acceleration as a function of position throughout the entire launch. For students familiar with calculus, describe the acceleration of the puck with respect to time. (*Hint:* A graph would be useful in your answer.)

UNIT
3

Thermodynamics and Matter

The first faltering steps to harness the energy of steam in the eighteenth century provided the impetus for a totally new area of science—thermodynamics, the second main area of classical physics. Scientists soon discovered that thermodynamic principles underlie every process where energy is liberated or absorbed, from a ship's snorting steam engine to the great heat engine of the earth's oceans. The laws of thermodynamics also provide the strongest evidence against a naturalistic origin to the universe.

In Unit 3, you will learn how the current model of matter became established as scientists gained a better understanding of thermodynamic principles. You will investigate the phases or states of matter and why phase changes take place. You will also learn about thermal energy and how it can produce useful work according to the laws of classical mechanics.

13A	Theories of Matter	286
13B	States of Matter	290
Facet:	Ice Skating	294
Facet:	Relative Humidity	298

New Symbols and Abbreviations

Avogadro's constant (N_A)	13.4	strain (ε)		13.8
mole (mol)	13.4	elastic (Young's) modulus (E)	13.8
stress (σ)	13.8	shear modulus (G)		13.8

The majority of the matter in the visible
universe is in the bulk of stars, such as our
sun. Therefore, the majority of visible matter
in the universe is in the plasma state.

Properties of Matter

> *Thou, even thou, art Lord alone; thou hast made heaven, the heaven of heavens, with all their host, the earth, and all things that are therein, the seas, and all that is therein, and thou preservest them all; and the host of heaven worshippeth thee.* Nehemiah 9:6
> *And he is before all things, and by him all things consist.* Colossians 1:17

The Old Testament author Nehemiah understood through inspiration and his knowledge of the Scriptures that the world and everything in it, living and inanimate, was the product of God's creative acts. However, God didn't just create the universe and then abandon it as the deists of the eighteenth century believed. Rather, He continuously sustains and upholds the matter of the entire creation moment by moment. This is the infinitely powerful God who is the object of our worship. Never let the political or social influences of this world cause you to view Him in any lesser role.

13.1 Introduction

The seemingly infinite ways that matter is manifested in nature can range from awe-inspiring to mundane. The continental mountain ranges and vast canyons and river valleys give evidence of tremendous forces that operated in the past to move immense quantities of matter in their formation.

13-1 It is likely that cataclysmic forces during and after the Flood were responsible for the building of these mountains.

13-2 Vehicle Assembly Building, Kennedy Space Center

Man's efforts seem puny by comparison, but huge structures such as the Vehicle Assembly Building at the Kennedy Space Center, extraordinary water diversion and hydroelectric projects such as the Three Gorges Dam in China, and the massive cleanup of the debris from the destruction of the World Trade Center towers in New York City showcase man's considerable ability to manipulate all forms of matter for his purposes.

The advancement of science and technology depends on a thorough understanding of the nature of matter. Without this understanding, we would not be able to obey the dominion mandate, given to Adam in the first chapters of the Book of Genesis, to subdue the earth. Beginning with this chapter, you will investigate some of the properties of matter and how the principles of classical mechanics determine those properties.

> Man manipulates matter, but God created it all: "In the beginning God created the heaven and the earth" (Gen. 1:1).

13A THEORIES OF MATTER

13.2 Nature of Matter

Mankind has been curious about the nature of matter since the dawning of historical times. Matter is the tangible stuff of the universe that surrounds us and is commonly separated into four states or phases: solids, liquids, gases, and plasmas.

Solids

Solids are normally considered to be fairly rigid, having definite shape and volume. Examples of solids are bowling balls, granite, and steel girders. Solids are usually thought to be massive and dense, but they also can be finely divided and have little density (for example, dust).

Liquids

The liquid state has about the same density as the solid state and has a definite volume. However, liquids assume the shape of their containers even if they do not fill them. Liquids also can flow from one point to another under the influence of a force. Examples of pure liquids include water, mercury, and ethanol.

Gases

The gaseous state is characterized by low density (under normal circumstances), and it can flow just like the liquid phase. When gas is placed in an evacuated container, it completely fills it. Gases can also be easily compressed and rarified. The atmosphere contains a mixture of gases.

Plasmas

The plasma phase exists where the particles of matter have enough kinetic energy that at least some of their electrons have been stripped away. This occurs at very high temperatures. Physically, plasmas act as a gas or liquid (they are fluidlike) and consist of a neutral mixture of electrons and positively charged particles. Plasmas exist in the Sun and most stars, fluorescent light bulbs, and neon lights.

13-3 The three phases of water

The first three states have been well understood for centuries. The fourth was not even suspected until late in the nineteenth century. The states of matter are difficult to define with any precision. Many common materials can have properties of two or more states depending on conditions, and some materials in one state have properties normally associated with other states. For example, a finely ground solid material can partially fill its container and flow from one place to another like a liquid (e.g., the sand in an hourglass). Water may exist in any of three states depending on temperature and pressure. It can even exist in all three states at once if conditions are right. Therefore, associating a state of matter with a particular material is not entirely correct, even though we tend to do it all the time.

In order to understand the reason for the states of matter, we will begin with a review of the background of the current theory of the structure of matter.

13-4 A fluidlike solid

13.3 Particles of Matter

The basic chemical building blocks of matter are called **elements.** These are substances that cannot be broken down into simpler substances by ordinary chemical means. In the early nineteenth century, John Dalton noticed that elements always combine in definite ratios of masses. For example, for every gram of hydrogen in water there are always 8 g of oxygen. To explain this observation, Dalton suggested that matter is made up of small particles that he called **atoms.** All atoms of an element have the same mass, which is unique to that element. Thus, as a first guess Dalton suggested that oxygen atoms are eight times as massive as hydrogen atoms. (Actually, oxygen atoms are sixteen times as massive as hydrogen atoms, but there are two hydrogen atoms for each oxygen atom in water.) He further theorized that atoms are never created, destroyed, or even changed. Dalton's atoms were hard, indivisible spheres. His theory was useful in correlating many observations.

Since Dalton's time, other researchers have discovered that atoms can be subdivided. The major components of an atom are *protons, neutrons,* and *electrons.* Positively charged protons and neutral particles called neutrons, which are similar in mass to protons, are bound together to form the atom's nucleus. Electrons—negatively charged entities about 1/1860 of the mass of a proton—occupy a region of space around the nucleus.

These major constituents of the atom are made of even smaller components called *elementary particles* and *quarks.* What exactly these even more fundamental things are made of scientists do not fully understand. To say they are physical entities composed of energy merely moves the mystery back one level to the question "What is energy?" To answer that question as we did in Chapter 9—"Energy is the ability to do work"—is inadequate here. The truth is that scientists do not really know what the matter of elementary particles is made from.

A pure sample of an element is composed of atoms that all have the same number of protons. For instance, every oxygen atom has eight protons each, while every gold atom has seventy-nine protons. Atoms can exist independently, but they are usually found bound to one or more other atoms in order to establish a more advantageous arrangement of electrons around their nuclei. The group may consist of atoms of one element or it may include atoms of two or more elements. Such groups are called **molecules** or **formula units,** depending on the kind of bonding between the atoms. If the atoms are of different elements, the combination is called a **chemical compound.** Atoms and groups of atoms can gain or lose electrons. The charge imbalance between the nuclei and the remaining electrons results in charged particles called **ions.**

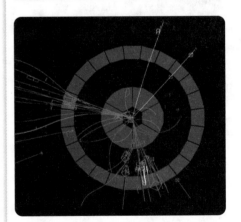

13-5 Model of a helium atom

An **atom** is the smallest particle that exhibits the chemical properties of its element.

John Dalton (1766–1844) was an English scientist whose atomic theory led to the table of atomic weights and a mathematical basis for chemistry.

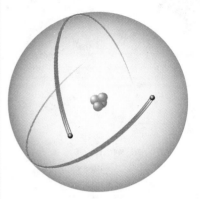

13-6 A computer-generated image of the spray of elementary particles from the destroyed nucleus of an atom in a particle accelerator

An **ion** is a charged particle consisting of a single atom or two or more atoms bonded together that have a mismatch in the total number of electrons compared to protons in the atoms' nuclei.

An Atom or group of atoms that has gained or lost electrons, resulting in a net charge on the particle

13-7 Each sample contains 1 mole of a substance.

The *atomic mass unit (u)* is a convenient unit of mass equal to one-twelfth of the mass of a carbon-12 atom.

Avogadro's constant or **number** (N_A) is the number of particles in one mole of a substance. The mole is the SI unit for quantity. There are 6.022×10^{23} particles in a mole.

Pressure is defined as force per unit area:

$$P \equiv \frac{F}{A}$$

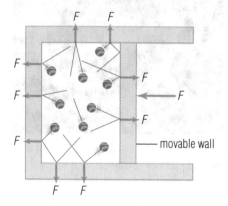

13-8a The total force the marbles exert on the sides of the container is a function of number of marbles, their speed, and the frequency of collision.

13.4 Atomic Mass

Scientists sometimes need to measure the number of atoms in a physical sample. The conversion between kilograms, atomic mass units, and the number of particles depends on a conversion factor using **Avogadro's constant (N_A).** Avogadro's constant (or *number*) is defined as follows: The mass, in grams, of Avogadro's number of particles is numerically the same as the number of atomic mass units in one particle of the substance. For example, exactly 12 g of carbon-12 contains Avogadro's number of carbon-12 atoms, because the atomic mass of a carbon-12 atom is exactly 12 u. The number N_A has been experimentally determined to be about 6.022×10^{23} particles. This number is also the numerical value of the SI unit **mole (mol)** that is used to quantify large numbers of particles.

13.5 An Application of the Kinetic Theory: Gas Pressure

One area of observation that Dalton's atomic model explains well is the behavior of gases. Using his model, scientists formed a good approximation of real gases that is known as the *ideal gas model.* You will study this model in greater detail in Chapter 14. This model assumes that the particles of a gas have no volume, interact with each other only through perfectly elastic collisions, and are in constant random motion. These assumptions make their behavior much easier to study and are usually good approximations to reality.

Imagine a box with low sides, one of which can be moved. The bottom of the box contains a number of marbles that are free to move in constant random motion. The marbles collide elastically with each other and with the sides of the container. In these two-dimensional collisions, the marbles exert impulsive forces on each other and on the sides of the container. The collision forces on the container are opposed by the rigid sides except on the movable side of the box. The movable side will tend to be pushed back, enlarging the space in which the marbles are confined, unless the side is supported by an opposing force. The sum of the impulsive forces on the sides of the box at any instant divided by the length of the sides (the "area" of the side in this two-dimensional example) defines the *pressure* that the marbles exert on the container. Since their motion is random, the marbles collide with equal areas of the box's walls with about equal force and frequency; so they exert the same pressure on each wall of the container.

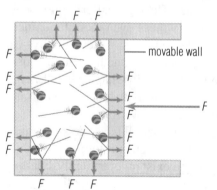

13-8b More marbles means more total force on the sides

What factors affect the "pressure" that the marbles exert on the box? Whatever affects either the frequency of collisions with the box or the force of these collisions will affect the pressure. For a given average kinetic energy (which equates to average velocity if the marbles are identical), the number of marbles in the box affects the frequency of collision. Logically, the more marbles that are in the box, the more marbles there are to collide with the box; so collisions will be more frequent if their average speed is constant.

The "area" of the box sides also affects the collision frequency. If the box becomes smaller, the marbles will have less room to travel before striking a wall and will therefore collide with walls more frequently. The average speed of the marbles ($\bar{v}$) also affects the frequency of collision. If the marbles are moving faster, they will collide with the walls of the box more often. Combining these factors, the frequency of collision (f) is proportional to the number of marbles (N) and the average speed of the marbles and is inversely proportional to the area of the box (A). In symbols,

$$f \propto \frac{N \cdot \bar{v}}{A}. \tag{13.1}$$

The main factor affecting the force of the collisions is the average speed of the marbles. Remember from Chapter 11 that the force a marble exerts in a collision is equal to its change in momentum divided by the time required for the collision. If we assume that the average collision is perpendicular to the wall of the box, then the final momentum of the gas particle is equal to but opposite of its initial momentum before impact. Thus, the magnitude of the force of an average collision is

$$|F_x| = \left|\frac{m\Delta\bar{v}_x}{\Delta t}\right| = \left|\frac{m(\bar{v}_{\text{aft }x} - \bar{v}_{\text{bfr }x})}{\Delta t}\right|.$$

But $\bar{v}_{\text{bfr }x} = -\bar{v}_{\text{aft }x}$, so

$$|F_x| = \left|\frac{m[\bar{v}_{\text{aft }x} - (-\bar{v}_{\text{aft }x})]}{\Delta t}\right|, \text{ and}$$

$$F = \left|\frac{2m\bar{v}_x}{\Delta t}\right| = \frac{2m\bar{v}}{\Delta t}. \tag{13.2}$$

The force of each collision is affected by the average speed of the marbles.

The "pressure" of the marbles on the sides of the box is proportional to both the frequency of collisions (Equation 13.1) and the force of each collision (Equation 13.2). In symbols:

$$P \propto f \cdot F$$

$$P \propto \frac{N\bar{v}}{A} \cdot \frac{2m\bar{v}}{\Delta t}$$

$$P \propto \frac{2Nm\bar{v}^2}{A\Delta t}$$

Therefore, gas pressure is proportional to $m\bar{v}^2$, which is proportional to the average kinetic energy of the gas particles.

According to the ideal gas model, gas molecules behave as the marbles in a box behave, but in three dimensions. The gas molecules exert pressure on their container by colliding with it, and this pressure is proportional to their average kinetic energy. Average kinetic energy of a gas is measured indirectly by a thermometer as temperature. Thus, adding energy to a gas increases the average kinetic energy of its particles, which appears as an increase in gas temperature. This topic will be discussed in more detail in Chapters 14 and 15.

13.6 Solids, Liquids, and the Kinetic Theory

The particles of solids and liquids also possess kinetic energy, but the ideal gas model does not apply to them. Their range of motion in a given time interval is extremely limited compared to that of a gas particle. Separate particles in solids and liquids are very close together and thus *do* exert noticeable forces on each other. Such forces are called **cohesion** and **adhesion.** The forces binding solid and liquid particles restrain their motion to such a degree that these particles can move only short distances from an equilibrium position, being pulled back and forth in

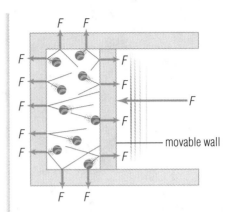

13-8c In a smaller box, the marbles do not travel as far before another collision occurs. The result is a higher frequency of collisions.

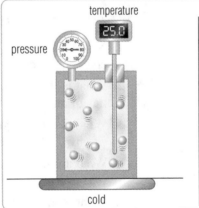
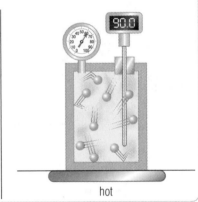

13-9 Pressure of a confined gas increases as its temperature increases.

Cohesion is the nonchemical bonding of identical particles through intermolecular forces. **Adhesion** is the nonchemical bonding of dissimilar substances through attraction at their surfaces.

a manner similar to simple harmonic motion. The velocity of the vibrations determines the particles' kinetic energy. Even though their range of motion is limited, solid and liquid particles may still possess a large amount of kinetic energy indicated by a high temperature.

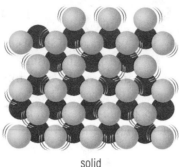
solid

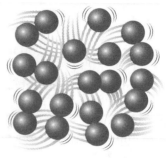
liquid

13-10 The particles of solids and liquids are close together and exert significant forces on each other.

The concept of matter as a collection of numerous, extremely tiny particles in continuous motion forms the classical **kinetic theory of matter.** The kinetic theory does a very good job of predicting the behavior of matter under many conditions. It also explains to a good first approximation how matter changes from one of the standard three states to another. As with any model, it is not perfect. We will note its limitations as we discuss various aspects of the properties of matter in this and later chapters.

13A Section Review Questions

1. Name the four states of matter and briefly describe their distinguishing characteristics.
2. What is the smallest particle of matter that determines the characteristics of an element?
3. What are the major components of atoms? Identify their charges (if any), relative masses, and their locations within the atom.
4. What are the names given to the groups of atoms that determine the characteristics of their compounds?
5. What is the atomic mass unit? Why did scientists establish such a unit?
6. What theory was most effective in describing the properties of gases?
7. Discuss the factors that determine the pressure a gas will exert in any instant in time.
8. What is the relationship between the temperature of a gas and the kinetic energy of the gas particles?

13B STATES OF MATTER

13.7 Arrangement of Particles in Solids

Particles in a solid have little kinetic energy compared to the cohesive forces or chemical bonds between them. Therefore, the particles of a solid have little motion and are held in fixed arrangements. In most solids, this arrangement is a repeating pattern. Such solids are called **crystalline solids.** The smallest pattern of particles that is repeated throughout the solid, the *unit cell,* is composed of particles in a three-dimensional pattern. Figure 13-11 shows the geometric arrangement of ions

a solid in which the atoms or molecules occur in a regular, repeating manner

The **kinetic theory of matter** is an interpretation of the physical properties of matter that relies strictly on classical Newtonian dynamics. It states that all matter consists of tiny particles in constant, random motion.

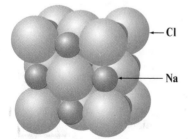

—Cl

—Na

13-11 The face-centered cubic unit cell of sodium chloride—table salt

in sodium chloride (NaCl). The distances between the atoms and the angles between planes of atoms are fixed. For sodium chloride, the angles at the corners of its unit cell are ninety degrees. Table salt is a good example of a crystalline compound. Most metals are also crystalline arrays of atoms.

Solids whose molecules do not form repeating patterns are called **amorphous solids.** The atoms may be in any arrangement, although the particular arrangement a solid assumes does not change unless the solid is melted. Glass is probably the most familiar amorphous solid.

13.8 Solids Under Pressure: The Elastic Modulus

We usually assume that a solid moving in response to a force does not change its shape—it's perfectly rigid. As with many concepts in an introductory physics course, this ideal condition doesn't really exist. Solids *can* change shape, sometimes significantly, in response to certain forces.

When we discussed the transmission of forces by an ideal string, we assumed the string did not stretch and tension was transmitted undiminished from end to end. A *tensile* force tends to pull particles of the string apart. Since atoms in a solid *can* move a limited amount, the particles will tend to spread out and the material will stretch.

A wire is structurally similar to a string. It can be considered a long, thin cylinder of solid metal. Suppose we apply a tension to the wire by pulling on both ends. How much will the wire stretch? The answer depends on the following three factors: stress, strain, and the wire's elastic modulus.

Stress (σ) (also called *tensile* or *normal stress*) is related to the tension force normal to the cross-sectional area of the wire. In this context, stress is defined as the force per unit area:

$$\text{normal stress} = \text{perpendicular force} \div \text{area}$$

$$\sigma = \frac{F_\perp}{A}$$

The symbol $F_\perp$ means the "force perpendicular to" something—the cross-sectional area in this case.

Strain (ε) (also called *tensile* or *linear strain*) is the amount that the wire stretches (Δl) divided by the original length of the wire (l_0):

$$\text{linear strain} = \text{change in length} \div \text{original length}$$

In an object subject to mechanical stress, the ratio of the

$$\varepsilon = \frac{\Delta l}{l_0}$$

amount of change in a dimension to the original dimension.

Strain is usually expressed as a simple decimal with no units or as a percent of the initial length.

Finally, if we take the ratio of the normal stress to the linear strain we obtain the wire's **elastic (Young's) modulus** (E):

$$\text{elastic modulus} = \frac{\text{normal stress}}{\text{linear strain}}$$

$$E = \frac{\sigma}{\varepsilon}$$

The SI units of the elastic modulus reduce to newtons per square meter, or N/m^2. These are the same units as pressure. The approved SI derived unit for pressure and related quantities is the pascal (1 Pa = 1 N/m^2). Elastic moduli (plural for modulus) are determined experimentally for many solids and tabulated in standard references. For brevity, the values of elastic moduli for most solids are usually listed in terms of GPa (10^9 Pa).

The elastic modulus is a measure of the material's resistance to change in shape. For example, it takes more force to change the shape of a strip of carbon steel (E = 203 GPa) than to proportionately change the shape of a strip of rubber

Elastic moduli are determined under specified conditions. Variations in temperature, humidity, and moisture content can radically alter the elastic moduli.

($E = 0.0017$ GPa). If a wire's elastic modulus, its cross-sectional area, its original length, and the tension exerted on it are known, we can estimate its change in length.

$$\Delta l = \frac{F_\perp \cdot l_0}{AE} \qquad (13.3)$$

EXAMPLE 13-1

Stressing: Strain and Elastic Modulus

Two men pull on the ends of a 10.00 m section of copper wire ($E = 130$ GPa) with a cross-sectional area of 2.50×10^{-6} m². They exert a force of 500. N. How long is the wire as they pull on it?

Solution:

You must find the total length of the wire as the men are pulling on it, or

$$L = l_0 + \Delta l,$$

where L is the stretched length of the wire.

$$\Delta l = \frac{F_\perp \cdot l_0}{AE} = \frac{(500.\ \text{N})(10.00\ \text{m})}{(2.50 \times 10^{-6}\ \text{m}^2)(130 \times 10^9\ \text{N/m}^2)}$$

$$\Delta l \doteq 0.015\underline{3}\ \text{m}$$

$$\Delta l \approx 0.015\ \text{m}\ \left(1.5 \times 10^{-2}\ \text{m}\right)$$

$$L = 10.0\underline{0}\ \text{m} + 0.01\underline{5}\ \text{m} \approx 10.02\ \text{m}$$

Thomas Young (1773–1829) was an English physician and physicist best known for Young's modulus and for his experiments with interference that proved the wave theory of light.

Mechanical forces can cause other kinds of deformations in solids. *Compressive forces* tend to crush or push the particles of matter together. *Shearing forces* tend to cause layers of particles within the solid to slide parallel to each other like a deck of cards. Each of these forces has its own moduli. The compressive elastic modulus is usually the same as the tensile elastic modulus. Both elastic moduli together are often called the Young's modulus (after Thomas Young).

→ like a dragging force ←

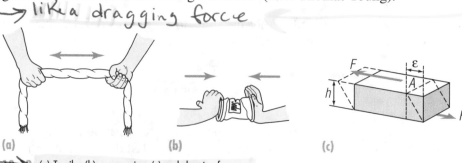

3 basic forces acting on solids →

(a)   (b)   (c)

13-12 (a) Tensile, (b) compressive, (c) and shearing forces

Some materials, such as spider silk, exhibit strain hardening with increased stress. Silk's elastic modulus rapidly increases with stress until it breaks.

Shear stress and strain occur when the lines of action of forces are parallel but do not act through the same point on a system or object. Consider the rectangular prism (parallelepiped) in Figure 13-12 (c). The forces act on opposite surfaces in opposite directions. The shear stress equals the force exerted parallel to the surface, divided by the surface area. The shear strain is the ratio of deformation of the object parallel to the force, divided by the separation of the two surfaces of the object. The **shear modulus (G)** is the ratio of shear stress to shear strain:

$$G = \frac{\text{shear stress}}{\text{shear strain}}$$

The shear modulus has no special name. It usually has a smaller magnitude than the Young's modulus for a given material.

13.9 Stress-Strain Graph

If the tensile stress and corresponding strain are plotted together, a graph similar to Figure 13-13 is obtained. Each kind of material has a characteristic stress versus strain graph. Note that there are several key regions along the graph that are indicative of how the material responds to stresses. Where the graph is linear, the Young's modulus is the slope of the line and the solid acts as a perfectly elastic material—the deformation is completely reversible. In other words, it will return to its original shape when the stress is removed. The maximum strain that can be experienced without permanent deformation is called the *proportional limit*. If the material is deformed beyond this limit, some permanent change in shape occurs but it still behaves somewhat elastically. The limit of reversible deformation is called the *elastic limit*. Beyond the elastic limit, the material permanently deforms with little elastic character. Comparatively little stress is required beyond the elastic limit to cause a large change of shape. Eventually, after enough force is applied, the solid breaks.

A number of material properties determine the shape of the curve in Figure 13-13. The hardness, brittleness, and number of cycles of stress all contribute to the shape of the graph for that material. For example, some materials *stress harden*. In other words, when stress is applied in a cyclic way, the material becomes harder or more brittle. This has the effect of making the stress-deformation curve steeper and moving the curve closer to the vertical axis. Farriers use this property of iron to harden a horse shoe by repeatedly striking the cooling shoe with a hammer as they form it. Similarly, vibrations are responsible for causing cracks in the metal alloy framework of aircraft.

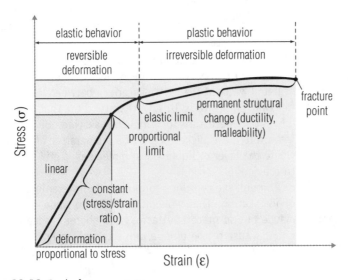

13-13 Graph of stress vs. strain

TABLE 13-1

Approximate Elastic Moduli for Selected Materials

Material	Young's Modulus E (GPa)	Shear Modulus G (GPa)
Aluminum alloy	70	25
Bone		
tension	18	
compression	9	
Brass	100	36
Brick	15	
Copper (element)	130	42
Glass (Pyrex)	70	23
Gold (element)	78	
Iron (cast)	90	
Iron (element)	211	
Lead	13	5.6
Lucite (PMMA)	3.0	
Nickel	200	77
Nylon	2.8	
Pine		
tension	14	
compression	0.042	
Polyethylene	0.22	
Rubber	0.0017	
Silk (silk worm)	16	
Silk (spider, max.)	0.28	
Steel	203	84
Tungsten (element)	41.1	150

Note: The values given in Table 13-1 are generalized from various sources. They represent typical values rather than actual test results.

If we can consider trials as moral stress and patience as the stress hardening of our spiritual fortitude, James 1:2–4 becomes clearer: "My brethren, count it all joy when ye fall into diverse temptations; Knowing this, that the trying of your faith worketh patience. But let patience have her perfect work, that ye may be perfect and entire, wanting nothing."

13.10 Transition Between the Solid and Liquid States

In most materials, the change from solid to liquid takes place at a predictable temperature—its melting point. At the melting point, the molecules of a solid gain enough kinetic energy to break out of their rigid arrangements and move more freely with respect to each other. The temperature at which a liquid becomes a solid is less predictable. When a liquid freezes, it gives up energy so that its molecules are constrained to stay in a rigid arrangement.

A solid absorbs energy when it melts. Molecules held in their solid phase positions require additional kinetic energy in order to break out of their rigid patterns. The absorption of energy by melting ice is the process that cools an iced drink. The ice simply absorbs thermal energy from the drink.

The melting point of a solid also depends on pressure. Most substances contract when they freeze. For these, higher pressure helps to force the molecules together and makes possible freezing at higher temperatures. There is an important exception to this general rule—water. Water expands when it freezes; so higher pressure hinders freezing. Water contracts when it melts; so higher pressure allows melting at lower temperatures. Simply adding pressure to ice at temperatures near—but below—0 °C can cause it to melt. This, however, requires *a lot* of pressure—approximately 120 atmospheres to lower the melting temperature 1 °C.

It *is* possible to cool a liquid to a temperature below its freezing point and maintain a liquid state. Such a liquid is said to be supercooled.

The phrase "120 atmospheres" of pressure means 120 times the normal atmospheric pressure at sea level.

FACETS of PHYSICS

Ice Skating

Ice skating is a popular winter sport, especially in regions where the ponds and lakes freeze in winter. Every skating enthusiast knows that this sport requires well-designed skates and a smooth ice surface. But not all skaters may be aware of the most important requirement for skating: the unique properties of water.

Unlike any other liquid, water expands just before it freezes. The ice takes up more space than the same mass of cold water; so the ice is less dense. Therefore, it floats and forms a smooth covering on the surface. If freezing water contracted and sank to the bottom, ice skaters would have a hard time skating (unless they wanted to wear scuba gear)!

It was believed for many years that the tendency of ice to melt under pressure (regelation) was the principle method for lubricating ice skates so that they glide easily. However, this belief ignores the fact that the average skater cannot generate more than about 6 atmospheres of pressure on the ice—even with a sharp blade—which is insufficient to cause regelation.

A more plausible explanation was first proposed by Michael Faraday in the nineteenth century. Ice naturally has a thin film of liquid water at the ice-air interface between –10 °C and the freezing point. It is this thin water film that provides the lubrication necessary for enjoyable skating. Because warmer ice is softer, the slickest ice is at a temperature of about –7 °C, where the layer of liquid water molecules is still thick enough to provide lubrication.

Through the centuries young people have enjoyed the sport of ice skating. That wonderful ability to glide gracefully (or maybe not so gracefully) across the glistening ice is possible only because of the unique properties of water.

This property of water can be demonstrated with a block of ice, a wire, and two very heavy weights (see Figure 13-14). The apparatus should be at the same temperature as the ice in order to eliminate the effects of heat conduction along the wire from the surroundings. The block of ice must be supported at its ends. The weights are fastened to the ends of the wire and the wire is placed over the ice block. The small diameter of the wire coupled with the force exerted by the attached weights produces many atmospheres of pressure on the ice directly under the wire. If the setup is observed over a period of days, the wire will slowly move vertically through the ice. The pressure exerted by the wire is sufficient to melt a thin layer of water molecules below the wire that flow around it and freeze on the upper, low-pressure side of the wire. Eventually, the wire and attached weights will fall free of the ice block, which will remain intact. This process of melting ice under pressure and refreezing it is called **regelation.** The movement of glaciers over the ground is lubricated in part with liquefied ice (water) from regelation due to the massive weight of the overlying glacier ice.

Other factors besides temperature and pressure can affect the melting and freezing points of liquids, including the presence of solutes. Investigating the freezing and melting points of chemical solutions is usually a topic covered in basic chemistry courses. In this course we will concern ourselves only with physical factors affecting phase changes.

13-14 Experiment demonstrating regelation

Don't be like a liquid, conforming to whatever social setting you find yourself in. Rather, let God shape and form you to His will, and then keep your shape like a solid. "And be not conformed to this world: but be ye transformed by the renewing of your mind, that ye may prove what is that good, and acceptable, and perfect, will of God" (Rom. 12:2).

13.11 Phase Diagrams

Figure 13-15 is a special graph called a **phase diagram.** It shows the relationship of the phases of a substance compared to controlling factors such as pressure and temperature. Other factors such as solute concentration may also be plotted on a phase diagram. This phase diagram of pure water shows regions where water is a solid, liquid, and gas. The boundaries between these regions involve specific kinds of phase changes depending on the direction of the change.

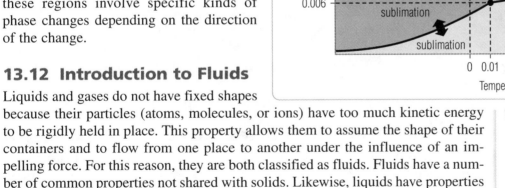

13-15 Phase diagram of pure water

13.12 Introduction to Fluids

Liquids and gases do not have fixed shapes because their particles (atoms, molecules, or ions) have too much kinetic energy to be rigidly held in place. This property allows them to assume the shape of their containers and to flow from one place to another under the influence of an impelling force. For this reason, they are both classified as fluids. Fluids have a number of common properties not shared with solids. Likewise, liquids have properties that distinguish them from gases.

13.13 Liquids and Surface Tension

Surface tension is a distinctly liquid property. Although liquid particles can slide freely over one another, cohesive forces between them are still significant. The strength of the cohesive forces depends on the liquid. Let's assume that the liquid is a molecular compound. Some nonpolar organic liquids have very low cohesive forces. Polar liquids, such as water, have much higher cohesive properties. Cohesion has little effect on the motion of molecules within the bulk of a liquid sample because the forces on each particle are balanced in all directions. However,

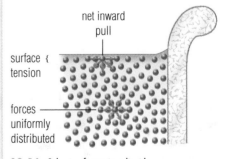

13-16 Cohesive forces in a liquid

(a)

(b)

13-17 Examples of surface tension: (a) Droplets of water; (b) Meniscus of liquid mercury

<u>surface tension</u>
the tendency for the surface of a liquid to present a barrier to penetration due to unbalanced forces on the surface particles

molecules at the air-liquid interface (the surface) are not completely surrounded by symmetrical cohesive forces. The external forces on these particles are not balanced, and the net force on a surface molecule is inward toward the bulk of the liquid. Consequently, the layer of molecules at the surface of a liquid acts like a skin that resists penetration from outside or from within. This is why it is possible for small but dense objects like pins, paper clips, and insects to float on the water's surface and why water droplets are nearly spherical. This phenomenon is called **surface tension.**

Figure 13-17 shows different ways in which surface tension is manifested. Figure (a) is a photograph of small droplets of water. The surface tension is so strong compared to the weight of the drop that it tends to pull the water molecules together into a volume having the smallest possible surface area. The resulting geometric shape is a sphere. Surface tension also explains the characteristic convex meniscus of mercury in a tube (figure (b)) and why you can overfill a glass with water.

In some situations, adhesion to an adjoining surface attracts a liquid's surface molecules more than the inner liquid molecules do. When adhesion is strong, the surface molecules are affected by this force. The net force, the sum of adhesion and cohesion, causes the liquid molecules to spread out over the other surface, even against gravity. The liquid can move against gravity until the weight of the rising liquid balances the adhesive force. This effect is called **capillarity.** The concave meniscus made by water in a test tube is a result of capillarity. Capillarity also permits liquids to flow into fibrous and porous materials. This property is what makes cloth and paper towels so absorbent. Ground water moves through porous rock by a combination of capillarity and gravity.

The movement of a liquid through a narrow passage or tube caused by the adhesion of the molecules in the liquid to the surface with it's contact.

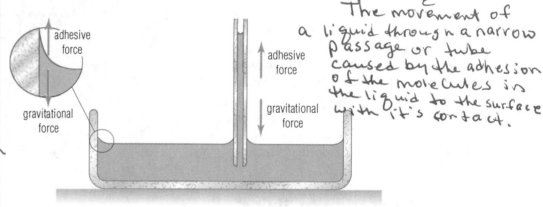

13-18 Adhesion causes a meniscus to form.

13.14 Liquids and Vapor Pressure

Surface tension prevents liquid molecules with average or less-than-average kinetic energies from escaping from the body of liquid through the surface interface. How, then, do liquids evaporate without being heated? The term *average* implies that some liquid molecules have less and some have more energy than the average kinetic energy. When one of the molecules with more kinetic energy than average reaches the surface, the cohesive forces are not enough to hold it at the surface. Such molecules escape from the liquid, forming a gaseous vapor of energetic molecules above the liquid. This process is known as **evaporation.** Evaporation is reversible. A gaseous phase molecule can just as easily reenter the liquid phase.

As energetic liquid molecules escape to the gaseous phase, the *average* kinetic energy of the remaining liquid molecules decreases, since only less energetic molecules remain. Since temperature is proportional to the average kinetic energy of the liquid, the liquid cools as it evaporates.

Evaporation is the escape of individual energetic molecules from the liquid phase to the gaseous phase. Evaporation can occur at any temperature below the boiling point.

If the liquid is enclosed in a container, then the vapor, like all gases, exerts pressure on its boundaries, including the liquid. As more liquid molecules escape to the gaseous phase, the pressure increases until the number of gaseous molecules returning to the liquid phase equals those evaporating, and net evaporation stops. This is an example of *dynamic equilibrium*—a static condition resulting from two opposing processes occurring at the same rate. The pressure of the gas when the closed system has reached equilibrium is called the **vapor pressure.**

> *the equilibrium pressure exerted by the gaseous phase of a substance in contact with its liquid or solid phase in an enclosed container.*

equilibrium
higher vapor
pressure

not in
equilibrium

equilibrium

warm

13-19 Vapor pressure is measured when a substance's gaseous phase is in equilibrium with its liquid phase in a closed container. Liquids in open containers generally do not reach equilibrium.

The equilibrium vapor pressure required to stop evaporation in a closed container increases with increasing liquid temperature. Thus, vapor pressure depends on the kind of liquid and its temperature. Comparing the tabulated vapor pressures of liquids at the same temperature gives a measure of their volatility. Liquids with low cohesive forces have high vapor pressures. We call such liquids *volatile*. If the cohesive forces are strong, then vapor pressure at a given temperature tends to be lower, and we call these liquids *nonvolatile*. A liquid in an open container will continue to evaporate, even if its temperature is in equilibrium with its surroundings, because of thermodynamic principles you will study in Chapter 16.

13.15 The Gaseous State

Unlike either solids or liquids, gas molecules have enough kinetic energy to overcome the cohesive forces between them. This is why we can assume that gas molecules interact during collisions only when temperatures and pressures are close to normal. Since gas molecules are moving at high speeds and have little effect on other molecules, they occupy much more space at a given temperature and pressure than liquid molecules. Therefore, a substance always significantly expands, if not contained, when it changes from a liquid to a gas. As you learned earlier in this chapter, rapidly moving molecules collide frequently with their container, causing gas pressure.

Most substances change from a liquid to a gas at a predictable temperature that varies according to the pressure of the gaseous phase above the liquid. Higher pressures raise the **vaporization point.** The vaporization point is the temperature at which *all* the liquid molecules reach a kinetic energy high enough to escape from the liquid form. At lower temperatures, as we have seen, *some* liquid molecules will have enough kinetic energy to escape to the gaseous phase regardless of the temperature.

The **vaporization point** is the temperature at which *all* the particles of the liquid have sufficient energy to enter the gaseous phase. It is also known as the *boiling point.*

When the liquid and vapor are confined together in a container, the evaporated molecules exert pressure. The liquid will continue to evaporate until the gas phase pressure in the container reaches the characteristic vapor pressure for that liquid at that temperature. If any more molecules evaporate, the excess molecules will tend to reenter the liquid phase, or *condense,* in order to maintain the gas dynamic equilibrium at that temperature. When the temperature drops, fewer molecules are evaporating and the gas-phase molecules have less kinetic energy, so they are more easily captured by the liquid phase molecules. The fewer remaining gas-phase molecules exert a lower vapor pressure at the lower temperature. On a phase diagram, condensation occurs when temperature and pressure coordinates move from the gas region into the liquid region of the graph.

If the liquid is in an open container, its vapor pressure is negligible unless there is a huge volume of liquid such as that found in the earth's bodies of water. A lake, for instance, experiences some evaporation during the day. Since its surroundings are so large, the vapor is not likely to reach the vapor pressure of water at daytime temperatures. However, at night air temperature drops nearer to the vapor pressure temperature of the water's surface. The water vapor molecules begin to stick together because they have insufficient kinetic energy to rebound during collisions. Eventually, visible droplets of water condense. The condensation is called dew and fog. When the temperature drops quickly below the freezing point, the vapor skips the liquid phase and changes directly to ice. The phase change that goes directly from gas to solid or solid to gas is called **sublimation.** Frost forms by sublimation. On a phase diagram, sublimation occurs when the temperature and pressure coordinates move from the gas region into the solid region of the graph.

without ← ← passing through the liquid State

FACETS of PHYSICS

Relative Humidity

On a hot, damp summer day you might have complained to your friends about your discomfort: "It's not the heat—it's the humidity!" Perhaps the weatherman said that the humidity was 95% that day. Have you ever wondered, "95% of what?"

Absolute humidity is simply the amount of water vapor per unit volume of air. A desert in midsummer and a seaport on a cold, clammy winter day may have the same absolute humidity. The difference in comfort lies in the temperature.

Relative humidity is the ratio of the absolute humidity to the amount of water vapor that could be contained per unit volume in the air at that temperature with no condensation. This quantity is expressed in percent. The relative humidity is what your body responds to. We need some humidity to live. However, the cooling mechanism of the human body depends on evaporation. When relative humidity is near 100%, little water or perspiration can evaporate. This blocks our cooling mechanism and makes us much warmer than we would be at the same temperature with lower relative humidity. So the cliché is true: It's not the heat—it's the *relative* humidity.

13.16 The Triple Point

You probably noticed the point on Figure 13-15 labeled **triple point,** where the three phase regions come together on the graph. This point represents the combination of temperature and pressure at which all three phases of water coexist. The triple point for water occurs at 0.01 °C and 611.73 Pa. Other substances have triple points as well. You will learn the significance of this point when you study temperature scales in Chapter 14.

13B Section Review Questions

1. Name the different kinds of pure solids and discuss the differences between the arrangements of particles in each.

2. What is the significance of a material's elastic modulus? Would a larger or smaller elastic modulus be desirable for metal used in bridge supports?

3. What is the slope of the linear portion of a stress-strain graph equal to? Would the graph for a block of hard rubber be steeper or shallower in this region compared to the graph for a block of steel?

4. What is the purpose of a phase diagram?

5. What property of liquids is not shared by gases? What kind of force is responsible for this property?

6. Is vapor pressure a property of the liquid phase or the gaseous phase? Explain.

7. Compare the volume occupied by a certain mass of liquid to the volume occupied by the same mass of gas at similar conditions.

⦿8. A section of a rubber band having a cross-sectional area of 3.14×10^{-6} m^2 is stretched 2.00 cm. If the original length of the rubber band is 6.00 cm, how much force is required to stretch it?

Chapter Review

In Terms of Physics

element	13.3	stress (σ)	13.8
atom	13.3	strain (ε)	13.8
molecule	13.3	elastic (Young's) modulus (E)	13.8
formula unit	13.3	shear modulus (G)	13.8
chemical compound	13.3	regelation	13.10
ion	13.3	phase diagram	13.11
Avogadro's constant (N_A)	13.4	surface tension	13.13
mole (mol)	13.4	capillarity	13.13
cohesion	13.5	evaporation	13.14
adhesion	13.5	vapor pressure	13.14
kinetic theory of matter	13.6	vaporization point	13.15
crystalline solids	13.7	sublimation	13.15
amorphous solids	13.7	triple point	13.16

Problem-Solving Strategies

13.1 (page 292) When calculating Δl using the elastic modulus, E, remember that you are finding the *change* in length of the object, not the final length. You must add the change in length to the original length to find the stretched (or compressed) length.

Review Questions

1. List and describe the four states of matter.

2. What are the chemically simplest substances?

3. What did Dalton think atoms were?

4. How is the modern idea of atoms different from Dalton's?

5. A silver atom has a mass of 107.9 u. What is the mass, in grams, of 6.022×10^{23} silver atoms?

6. Discuss the dependence of gas pressure on the number of gas particles, their kinetic energy, and the surface area of the container.

7. What happens to the pressure of a gas in a sealed container when its temperature rises?

8. What properties of matter are stated in the kinetic theory of matter?

9. What are the two common ways that the particles of pure solids can be arranged?

10. Which state of matter (solid, liquid, or gas) has the fastest-moving molecules at a given temperature?

11. Will a wire snap back to its original length if it is stretched past its elastic limit?

12. Why are surface molecules of a liquid more affected by cohesion than interior molecules are?

13. Why is energy required to melt a solid?

14. What happens to the melting temperature of ice when a lot of pressure is applied to it?

15. According to a phase diagram, such as Figure 13-15, what phase tends to exist under conditions of low pressure and high temperature?

16. What kind of force dominates capillarity?

17. Discuss the difference between evaporation and vaporization.

18. Describe *vapor pressure*. How does the vapor pressure of ethyl alcohol compare to that of water at the same temperature?

19. Will more dew form on a cool night or on a warm night? Why?

20. Between which two states does sublimation occur?

True or False (21–30)

21. Man has a biblical commandment to study and use matter in all of its forms.

22. Based on the descriptions of the four states of matter given in this chapter, plasma must be the most common state of matter in the visible universe.

23. The existence of gas pressure is explained well by the kinetic theory of matter.

24. Gas molecules exert pressure on a container by colliding with the container.

25. Strain is related to force, and stress is related to the material deformation caused by a force.

26. The elastic modulus has units of pressure.

27. A large elastic modulus implies that the material is "stretchy."

28. Some soft metals can be hardened by someone striking them repeatedly with hammerlike blows.

29. An ice skater depends on regelation for smoothly sliding skates.

30. The adhesion of water molecules to each other results in surface tension.

◉31. How many grams of iron metal are contained in 1.00×10^{12} atoms of iron (atomic mass = 55.85 u)?

◉32. A student pushes down on a polystyrene cube with a force of 10.0 N. The 7.50 cm high cube is compressed to a height of 6.00 cm.
  a. What is the stress on the cube?
  b. What is the strain on the cube?
  c. What is the cube's Young's modulus?

◉33. A wooden two-by-four stud (3.81 cm × 8.89 cm) in an exterior wall of your house supports a load of 1340 N. Describe the kind of stress applied, and determine its magnitude.

◉34. If the originally 2.36 m (93 in.) long stud in Question 33 shortens by 1.65 cm under load, what is the strain of the piece of lumber?

◉35. Assuming a stud experiences a normal stress of 3.00×10^5 N/m² and a strain of 0.0072, what is the modulus of elasticity for the wood?

◉◉36. A 65.0 kg physics student jumps stiff-legged from a 1.00 m wall vertically to a concrete sidewalk. He stops in 0.050 s as he lands. If the cross-sectional area of his femur (thigh bone) is 5.00 cm², calculate the stress on the bone as he lands on the pavement. (*Hint:* Calculate his landing speed and the resulting impulse of the force as he stops, then calculate the stress on the bone. Also, consider that he has *two* legs.)

14A	Thermal Properties	303
14B	Measuring Temperature	309
14C	Gas Laws	314
Facet:	Ball and Ring	309
Biography:	Count Rumford	324
Biography:	James Joule	325

New Symbols and Abbreviations

coefficient of linear expansion (α)	14.2
coefficient of volume expansion (β)	14.3
Boltzmann's constant (k)	14.14
universal (ideal) gas constant (R)	14.14
standard temperature and pressure (STP)	14.14

At the same pressure, a given mass of hot air occupies more volume and is less dense than cool air. Thus, hot-air balloons can float in the surrounding cool air.

Expansion and Temperature

14

> *I know thy works, that thou art neither cold nor hot: I would thou wert cold or hot. So then because thou art lukewarm, and neither cold nor hot, I will spue thee out of my mouth.* Revelation 3:15, 16

In the above passage, the aged apostle John faithfully records the words of the glorified Lord Jesus Christ as He describes the saints of the Laodicean church. He is grieved because the Christians of this church are not spiritually at one end or the other of the temperature scale. Early readers of this letter would have understood the significance of these references, for Laodicea was a significant urban center situated between the geothermal hot springs of Hierapolis several miles to the north and the cold springs of Colosse a similar distance to the south. These bodies of water promoted health and were refreshing. However, the Laodiceans were lukewarm, perhaps like the water supply of their city, which caused one to vomit in disgust. In a similar way, Christ does not want His followers to be self satisfied and spiritually useless as were the Laodiceans. What is the temperature of your spiritual life?

cold

14A THERMAL PROPERTIES

14.1 Basics of Thermal Expansion

How does a hot bar of copper differ from a cold bar of copper? Other than the temperature, what other properties can be measured to compare the two objects? If we could observe individual copper atoms, how could we distinguish between a "hot atom" and a "cold atom"? According to the kinetic theory of matter, "hot atoms" have more kinetic energy than similar "cold atoms" do. "Hot atoms" are moving or vibrating faster than "cold atoms."

To visualize the difference, it helps to portray the copper atoms as tiny solid spheres held in position by springs attached to adjacent atoms. A displacement of the atom from its equilibrium position will generate a restoring force that will tend to push it back toward its equilibrium position. The faster it moves, the more momentum it

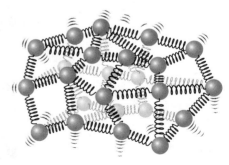

hot

14-1 The atoms of solids can be considered as particles connected to each other with springs. The more kinetic energy they have, the harder they push on their neighbors.

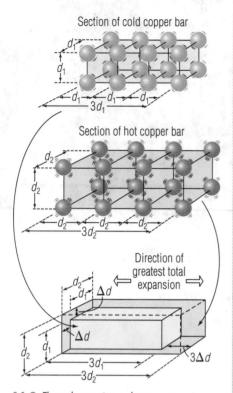

Section of cold copper bar

Section of hot copper bar

Direction of greatest total expansion

14-2 Thermal expansion—the more atoms in a row, the greater the expansion in that direction.

14-3 Thermal expansion can buckle railroad tracks on hot days.

has, and the harder it pushes against its neighbors. Since no matter is perfectly rigid, the atoms tend to spread out as they are forced apart. Consequently, vibrating atoms occupy more space at higher temperatures than at lower ones.

Imagine a copper bar made up of sixteen atoms (Figure 14-2). When it is cold, its atoms vibrate with small amplitude and low velocity, so they take up little space. However, if the bar's temperature rises, each atom will vibrate faster and move farther from its equilibrium position. As a result, each atom will occupy a greater volume, and the bar will expand in every direction.

Since each atom occupies a larger interatomic space, the bar will expand more in the directions where more atoms are lined up in a row. In the illustration, the bar is two atoms high (one interatomic space), two atoms wide (one interatomic space), and four atoms long (three interatomic spaces). As the interatomic distance increases with temperature, the bar expands the same amount in height and width, but three times that amount in length. For this reason, when a wire is heated it expands much more in length than in diameter because its length is proportionately so much greater (more atoms in a line). This effect can be easily observed in the larger amount of sag experienced by an overhead high-voltage power line in the summer compared to its sag in the winter. However, a micrometer caliper would be required to detect the wire's seasonal change in diameter.

The expansion of a heated material produces noticeable and potentially undesirable effects. For example, railroad tracks that are laid too close end-to-end in cold weather will sometimes buckle in hot weather. Railroad builders try to leave a small gap between track sections for the track to expand, or they use alloys that expand only a little with high temperatures. Even with these precautions, some tracks occasionally buckle on very hot days, causing hazardous conditions for trains. Other structures have similar expansion joints or special materials. Bridge road surfaces, for example, are usually connected by interlocking joints that prevent buckling when the bridge expands or contracts. The bridge's road-support structures themselves often rest on rollers. Even concrete sidewalks have fibrous expansion joint fillers or gaps between sections to prevent buckling.

Glassware is also affected by thermal expansion. Ordinary bottle glass expands significantly when it is heated. However, it is a poor heat conductor. If hot water is poured into a cold, thick glass bottle, the inner glass surface quickly warms and expands. The middle and outer thickness of glass, however, is not heated as much and expands little. Because glass is brittle, this difference in expansion rates can cause the bottle to crack. Two ways to avoid this problem are to use thin glass so that the entire thickness heats up and expands almost simultaneously, or to use glass that expands very little. The second choice is more practical for normal use since thin glass breaks easily. Pyrex brand glass is a kind of strong glass that does not expand significantly when heated.

14-4 Engineered solutions can prevent expansion-related damage in structures.

14.2 Linear Thermal Expansion of Solids

Suppose a 100 cm aluminum rod is uniformly heated along its entire length. Is there any way to predict how much it will expand? Physicists determined the answer to this question by experimentally measuring the change of length of a metal rod with a corresponding change of temperature and plotting the results on a graph. Since the change of length is caused by the change of temperature, temperature change is the independent variable and length is the dependent variable. Figure 14-5 is a typical plot of experimental data.

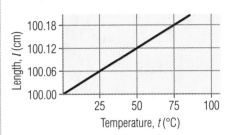

14-5 Thermal expansion graph

Mathematically, a straight line graph is described by the familiar equation

$$y = mx + b.$$

Similarly, the equation for the line in Figure 14-5 could be expressed as

$$l = m\Delta t + l_0, \qquad (14.1)$$

where l is the length of the rod as a function of its temperature, l_0 is the rod's original length at the reference temperature, and Δt is the difference from the reference temperature. The slope (m) is related to the original length of the rod (the slope is steeper for a longer rod and less steep for a shorter rod) and includes a constant denoted α.

$$m = \alpha l_0$$

Replacing m in Equation 14.1 gives

$$l = (\alpha l_0)\Delta t + l_0.$$

The rod's length has changed by

$$\Delta l = l - l_0, \text{ so}$$

$$\boxed{\Delta l = \alpha l_0 \Delta t.} \qquad (14.2)$$

This equation applies to the thermal expansion of any solid material. The constant α, called the **coefficient of linear expansion,** is different for each material. Table 14-1 gives the linear expansion coefficients for some common solids.

The units of the coefficient of linear expansion (α) are $°C^{-1}$ or K^{-1}. The original length l_0 may be given in any unit of length, and Δl will be expressed in the same unit. Of course, Δt must be expressed in the same temperature unit as that of α. The "length" of an expanding object may actually be its linear dimension in any direction. Since each dimension increases in proportion to its size, α is the same for any direction.

TABLE 14-1

Approximate Coefficients of Expansion for Selected Solids

Material	Linear Expansion (α) ($\times 10^{-6}\ °C^{-1}$)	Volume Expansion ($\beta = 3\alpha$) ($\times 10^{-6}\ °C^{-1}$)
Aluminum	23.1	69.3
Brass	20.3	60.9
Copper	16.5	49.5
Glass (ordinary)	8.7	26.1
Glass (Pyrex)	3.2	9.6
Gold	14.2	42.6
Ice	51	153
Iron	11.8	35.4
Lead	28.9	86.7
Pine wood (along grain)	5	15
Pine wood (across grain)	30	90
Silver	18.9	56.7
Steel (mild)	11.7	35.1
Steel (stainless)	17	51

EXAMPLE 14-1

A Bridge for Every Season: Thermal Expansion

A steel bridge is 200.00 m long on a winter day with a temperature of −15.0 °C. How many centimeters longer is it on a warm summer day with a temperature of 35.0 °C? ($\alpha_{steel} = 11.7 \times 10^{-6}\ °C^{-1}$)

Solution:
Compute the change in temperature:

$$\Delta t = 35.0\ °C - (-15.0\ °C) = +50.0\ °C$$

Problem-Solving Strategy 14.1
The thermal expansion equation can be simplified a bit if the length of the object, l_0, is measured at the reference temperature of 0.0 °C. Then the change of length is just

$$\Delta l = \alpha l_0 t,$$

where l_0 is the reference length and t is the new Celsius temperature.

Compute the change of length:

$$\Delta l = \alpha l_0 \Delta t$$

$$\Delta l = (11.7 \times 10^{-6} \ °C^{-1})(200.00 \ m)(+50.0 \ °C)$$

$$\Delta l = 0.117 \ m, \ or$$

$$\Delta l = 11.7 \ cm$$

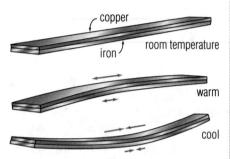

14-6 A bimetallic strip bends when heated or cooled.

14-7 A common application for a bimetallic strip—a household thermostat

Different materials expand different amounts with equal temperature change. This fact can be an advantage for some applications. For example, suppose strips of copper and iron metal are bonded along their entire lengths. When the resulting **bimetallic strip** is heated, the copper expands more than the iron does and the strip curves with the iron toward the inside. Bimetallic strips are commonly used as the temperature-sensitive components of a thermostat. A long, thin strip of the two metals is fabricated in a spiral shape with the copper on the inside of the spiral to maximize the effect of the temperature change on the sensor. As temperature rises, the spiral expands, and as temperature drops, the spiral contracts. Movement of the end of the bimetallic strip controls an electrical contact that in turn operates the heating or cooling system.

14.3 Thermal Volume Expansion of Solids

What is the volume of a solid after it is heated? For regular solids such as cubes and spheres, we can predict new volumes by calculating new dimensions based on the linear expansion relationship and using these dimensions in the appropriate volume formula. But what if the solid has an irregular shape? A formula similar to the linear thermal expansion formula (Equation 14.2) has been developed that predicts the change in volume,

$$\Delta V = \beta V_0 \Delta t, \tag{14.3}$$

where β is called the **coefficient of volume expansion** and is a constant for each material. For each material, $\beta = 3\alpha$. See Table 14-1 for a list of some representative coefficients of volumetric expansion.

EXAMPLE 14-2

Bulking Up: Volumetric Expansion

A solid sphere of ordinary glass has a volume of 7250.0 cm³ at 20.0 °C. Find its volume at 80.0 °C.

Solution:
Calculate the change of temperature:

$$\Delta t = 80.0 \ °C - 20.0 \ °C = +60.0 \ °C$$

Compute the change of volume. Obtain the value of β from Table 14-1.

$$\Delta V = \beta V_0 \Delta t$$

$$\Delta V = (26.1 \times 10^{-6} \; °C^{-1})(7250.0 \; cm^3)(+60.0 \; °C)$$

$$\Delta V \doteq 11.\underline{3}5 \; cm^3$$

Compute the final volume:

$$V = V_0 + \Delta V$$

$$V = 7250.0 \; cm^3 + 11.\underline{3}5 \; cm^3$$

$$V \doteq 7261.\underline{3}5 \; cm^3$$

$$V \approx 7261.4 \; cm^3$$

14.4 Thermal Expansion of Liquids

Liquids respond much more to temperature changes than do solids. Because the molecules in a liquid are bound less tightly to each other than in a solid, an increase in kinetic energy leads to a proportionately greater increase in volume in a liquid than it does in a solid. Thus, coefficients of volume expansion for liquids are generally greater than those for solids, as a comparison of Tables 14-1 and 14-2 shows. When measuring the volume of a liquid after heating, you can generally disregard any effects that expansion of the container might have on the accuracy of your results.

EXAMPLE 14-3

Gasoline Spill: A Consequence of Liquid Expansion

A car's owner tops off his car's 75.7 L stainless steel gas tank in the morning when the gas temperature is 10.0 °C. He drives a short distance to work. The car then stands in the sun until mid-afternoon, when the gas temperature is 23.9 °C. How much gas spills out of the tank's overflow vent because of thermal expansion? Calculate this (a) by ignoring the expansion of the tank, and then (b) by including this expansion. The volumetric expansion coefficient of the stainless steel tank is $\beta_t = 51 \times 10^{-6} \; °C^{-1}$.

Solution:

a. Calculate the volume of expansion of the gasoline alone (ΔV_g):

$$\Delta t = 23.9 \; °C - 10.0 \; °C = +13.9 \; °C$$

$$\Delta V_g = \beta V_{g\,0} \Delta t$$

$$\Delta V_g = (960. \times 10^{-6} \; °C^{-1})(75.7 \; L)(+13.9 \; °C)$$

$$\Delta V_g \doteq +1.0\underline{1}0 \; L$$

Ignoring the expansion of the gas tank, slightly more than a liter of gasoline spills out of the tank.

b. Considering the expansion of the tank as well, the tank can hold ΔV_t more gas, and this additional volume reduces the gas spilled. Determine the volumetric expansion of the tank:

$$\Delta V_t = \beta V_{t\,0} \Delta t$$

Problem-Solving Strategy 14.2

Be sure to use the appropriate coefficient of expansion for solid materials. Use α for linear expansion and β for volumetric expansion.

The volumetric coefficients of expansion of liquids tend to be as much as several orders of magnitude larger than for solids because molecules in the liquid state are bound together much less firmly.

Problem-Solving Strategy 14.3

There is no linear coefficient of expansion for fluids because, unlike solids, fluids do not have fixed dimensions.

TABLE 14-2

Approximate Coefficients of Expansion (β) for Selected Liquids Near Room Temperature ($\times 10^{-6} \; °C^{-1}$)

Acetone	1490
Carbon disulfide	1150
Ethyl alcohol	1120
Gasoline	960
Glycerin	485
Mercury	183
Methyl alcohol	1134
Turpentine	900
Water (20°)	207

$$\Delta V_t = (51 \times 10^{-6} \ °C^{-1})(75.7 \ L)(+13.9 \ °C)$$

$$\Delta V_t \doteq +0.05\underline{3}6 \ L$$

The amount of gas that vents off in this case is as follows:

$$\Delta V_g - \Delta V_t = 1.0\underline{1}0 \ L - 0.05\underline{3}6 \ L \doteq 0.9\underline{5}6 \ L$$

$$\Delta V_g - \Delta V_t \approx 0.96 \ L$$

Slightly less than a liter of gas is spilled.

Example 14-3 illustrates why you should stop fueling your car when the automatic delivery nozzle shuts off. The nozzle setpoint is designed to leave expansion room in your tank so that gas does not spill out of the vent due to temperature changes. Spilling gas this way is poor stewardship because it wastes money, can damage the pavement, and pollutes the ground. It also creates a fire hazard.

14.5 Thermal Expansion of Gases

Gas particles are so far apart compared to solids at normal pressures and temperatures that they exert little or no force on each other. For this reason, gas particles in an evacuated space will continue to spread apart until they encounter the container boundaries, if any exist. In fact, the volume occupied by a given number of gas particles will expand to fill any empty (evacuated) container. The larger the container, the lower the pressure that will be exerted by the fixed number of particles. However, if the container is designed so that it can change volume in order to maintain a constant gas pressure, gases will then have temperature-dependent volume like liquids and solids. Most gases have volumetric expansion coefficients around $3700 \times 10^{-6} \ K^{-1}$. Table 14-3 lists the expansion coefficients of several common gases. These coefficients are quite similar because gas particles interact very little since they are so far apart. Therefore one gas behaves much like another. You will study the response of gases to temperature and pressure changes in more detail in Section 14C.

14A Section Review Questions

1. Briefly explain from the basis of the classical kinetic theory why most materials expand when heated.

2. Give some examples of engineered solutions for problems stemming from expansion and contraction of materials.

3. What three factors determine the amount of thermal expansion by a solid in a given dimension?

4. How is the volume expansion coefficient (β) related to the linear expansion coefficient (α) for solids?

●5. Aluminum and copper are two metals that were commonly used for electric wiring in homes in the U.S. during the 1960s and '70s. If a large electrical current is carried by the wire, its temperature can increase significantly. What is the difference in linear expansion experienced by a 1.00 m length of aluminum wire versus 1.00 m of copper wire for a $+30.0 \ °C$ change in temperature?

●6. How much will $1.000 \ cm^3$ of mercury expand in volume for a $+1.00 \ °C$ temperature change?

In physics, an "empty container" is one that is completely evacuated, having no particles of any kind inside when the gas is released into it.

Gases are far more compressible than either solids or liquids. Therefore, the coefficients of volumetric expansion for gases are all determined and reported at a constant standard pressure. Standard pressure is atmospheric pressure measured at mean sea level.

TABLE 14-3	
Coefficient of Volume Expansion (β) for Selected Gases at Standard Pressure ($\times 10^{-6} \ K^{-1}$)	
Air	3670
Carbon dioxide	3740
Helium	3665
Hydrogen	3660

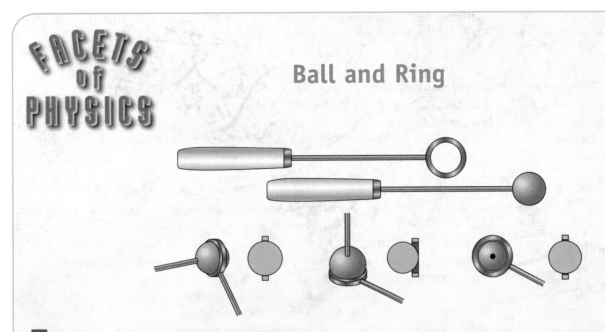

FACETS of PHYSICS

Ball and Ring

If you heat a metal washer, does the hole in the middle become larger or smaller? One way to find the answer is to use a ball-and-ring demonstrator. The ball and ring are made of the same metal so that they expand at the same rate. When they are both at room temperature, the ball just fits through the ring, but what happens if one or the other is heated?

If only the ball is heated, obviously it will no longer fit through the ring. If only the ring is heated, experiments show that the ball fits loosely through the ring. Therefore, the hole expands when it is heated, but by how much? You can find out by heating both the ball and the ring. Experiments show that a heated ball just fits through a heated ring. That is, the hole expands the same distance that the ball expands. A hole in a metal ring expands and contracts exactly as much as a disk of the same material and size would expand or contract.

14B MEASURING TEMPERATURE

14.6 Thermometers

Since materials expand in proportion to a temperature change, they provide a way of measuring temperature. Galileo used this property of expansion to make the first known **thermometer.** His *thermoscope,* as it was called, was based on the expansion of air and consisted of an air-filled glass bulb connected to a tube immersed in an open dish of water. When the temperature of the air in the glass bulb rose, it expanded, forcing the water down the tube. When the air temperature dropped, it contracted, allowing the water level to rise.

The drawback to the thermoscope is that it responds to changes in atmospheric pressure as well as temperature. When the pressure of the surrounding air is high, it forces the water up the tube even when the temperature has not changed. Thus, without

14-8 Galileo's thermoscope

A **thermometer** is any device that uses a thermometric property to indicate temperature. A *thermometric property* is any characteristic of a material or instrument that varies directly with temperature.

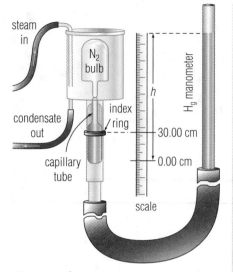

14-9 A gas thermometer

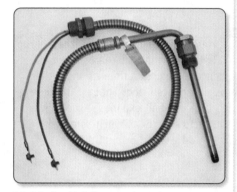

14-10 One kind of remote electrical temperature sensor is called an RTD (resistive temperature detector).

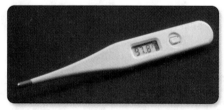

14-11 Digital thermometers are common household items for monitoring fever.

additional instrumentation, we can never be sure whether a thermoscope is showing a change in temperature or a change in pressure.

A gas thermometer can be a better tool for measuring temperature if the gas volume is maintained constant. This can be done by connecting the gas column with a column of mercury. When the gas thermometer is read, the mercury column position is adjusted so that the left-hand column height is returned to its original position, thus restoring the gas volume to the reference volume condition. The difference in height between the left- and right-hand mercury columns represents the pressure difference required to maintain a constant volume. The height difference is read on a calibrated scale as temperature.

To avoid the problems of thermoscopes and the inconvenience of gas thermometers, modern thermometers use liquids such as glycol (an alcohol) or mercury sealed in fine glass tubes under vacuum. The liquids expand so much more than their glass containers that the expansion of the glass usually has a negligible effect on the temperature reading. For very sensitive measurements, the thermometer is carefully calibrated to take into account the expansion of the glass at each temperature. *Digital thermometers,* such as those used for household and medical purposes, measure temperature using a thermometric electrical resistance property, while weather balloons, equipment temperature sensors, and other remote temperature sensing devices rely on the potential difference at a **thermocouple** junction.

To be useful for measuring temperature, a liquid-filled thermometer must have several features:

- The thermometer must be calibrated so that the length of the column of liquid corresponds to a specific temperature.

- A thermometer must be sensitive enough that a small change in temperature produces a noticeable change in liquid level. This qualification is met by making the glass envelope sufficiently narrow so that a small change in the liquid's volume makes a large change in the column's length.

- A thermometer should respond quickly to changes in temperature. A liquid thermometer responds quickly if the glass of the bulb is thin enough.

- The mass of the thermometer must be small in relation to the bulk of the object whose temperature is being measured. If you were to attempt to measure the temperature of a few milliliters of a liquid with a standard laboratory thermometer, the thermal energy transferred between the mass of the thermometer and the liquid sample would materially affect the reading.

- The medium used in the thermometer must be appropriate for the range of temperatures being measured. Since mercury freezes at –39 °C, mercury thermometers cannot be used to measure temperatures at or below that point.

Electronic thermometers have corresponding requirements for their electrical components.

14.7 Temperature Scales

Since there are instruments to measure temperature, there must be units in which to measure it. A thermometer may have a column of liquid that rises to a particular height when it reaches a certain temperature, but what should that height be called? Historically, this decision was not settled until the nineteenth century. The issue concerned choosing appropriate reference points to establish a temperature scale.

The Fahrenheit Scale

Probably the most familiar temperature scale to people in the United States is the Fahrenheit scale, devised by Daniel Gabriel Fahrenheit. He initially made his zero point the lowest temperature he could obtain with a water-ice-salt solution, and his other fixed point was the normal temperature of the human body. Following a suggestion from Newton, he set the temperature of the human body as 12° and used this as the upper reference temperature. These choices resulted in degrees that were too large to be practical, so he subdivided them into eighths, and renumbered, which then gave a human body temperature of 96°.

After recalibrating his thermometer in the smaller units, he discovered that pure water boiled near 212° and froze near 32°. There are 180° between these two points, so he decided to make these physical properties of pure water his new reference points. Once again he re-marked his scale, and as a result the average human body temperature, since it was no longer a fixed reference point, was measured as 98.6 °F. These freezing and boiling temperatures of pure water are valid at one atmosphere of pressure.

Gabriel Fahrenheit (1686–1736) was a German scientist who invented the mercury thermometer, established the Fahrenheit temperature scale, and discovered the effect of atmospheric pressure on the boiling points of liquids.

Reference standards for various kinds of measuring scales are also called *fiducial* points. The fiducial points for the Fahrenheit and Celsius scales are the freezing and boiling points of pure water.

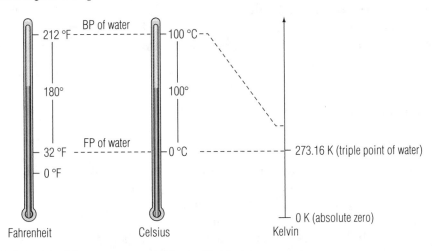

14-12 Comparison of three temperature scales showing fiducial points

The Celsius Scale

Anders Celsius also used the freezing and boiling points of water as reference points for calibrating his scale. However, Celsius, following the growing metric movement in Europe, chose to have 100 degrees between the freezing and boiling points of water. Consequently, the freezing point of pure water is 0 °C and the boiling point is 100 °C. This is the temperature scale used by most of the world other than the U.S. and is especially convenient for the scientific, technical, and engineering communities.

The Absolute or Kelvin Scale

Both the Fahrenheit and the Celsius scales have negative temperatures. For many years, no one knew how large the negative temperatures could get. For this reason, it was known that 60° was *not* twice the temperature of 30° in either temperature scale. What was needed was a single absolute reference temperature from which all others could be measured so that physical properties that depended on proportional temperatures could be correctly determined. After experimenting with the temperature-volume relationship of gases, scientists extrapolated the minimum possible theoretical temperature to be −273.15 °C. British physicist Lord Kelvin suggested in 1854 that this temperature be established as the **absolute zero** for measuring all temperatures. This temperature is not directly measurable because it cannot be reached. However, another single reference point was known

Anders Celsius (1701–44) was a famous Swedish astronomer who is most noted for developing the metric temperature scale for meteorological studies. He was part of the expedition that confirmed Newton's prediction that the earth was an oblate spheroid.

Absolute zero is the temperature of matter with no thermal energy. This is not achievable in the real world because there must be a colder location for the thermal energy to flow to in order to cool the substance to absolute zero.

and experimentally measurable—the triple point of water—where all three phases of water coexist. This point occurs at $0.01\ °C$ and 611.73 Pa. So scientists define the triple point temperature to be *exactly* 273.16 Celsius degrees above the absolute zero point, which provides the *measurable* reference point for the absolute temperature scale.

The temperature unit on the absolute or Kelvin scale is called the **kelvin (K)** in honor of Lord Kelvin's many contributions to physical science. The kelvin is the same temperature interval as the Celsius degree on a thermometer. The kelvin unit is never used in conjunction with the word "degree," nor with the degree symbol. Therefore, when spoken, an absolute temperature is followed by the word "kelvins" rather than "degrees kelvin," and when written the unit symbol is just "K" and *not* "°K."

14.8 Conversions Between Temperature Scales

Conversions between temperature scales are straightforward. The Celsius and Kelvin scales differ only by an added constant. Kelvin temperatures are *always* larger than Celsius, so converting from Celsius degrees to kelvins involves adding the constant:

$$T = t + 273.15$$

Converting from kelvins to Celsius is just the opposite:

$$t = T - 273.15$$

The constant is usually rounded to the decimal place of the given temperature before conversion. According to the significant digit addition rules, the result cannot have any more decimal places than the number with the fewest decimal places.

Between the freezing and boiling points of pure water, 180 Fahrenheit degrees correspond to 100 Celsius degrees. Therefore, Fahrenheit degrees are "smaller" than Celsius degrees. The two scales have different zero points as well (on the Celsius scale, $0\ °C$ is the freezing point of water; on the Fahrenheit scale, $0\ °F$ is $32°$ below the freezing point of water). For these reasons, the conversion between Celsius and Fahrenheit temperatures is a bit more complicated.

When converting from Celsius to Fahrenheit, the number of Celsius degrees needs to be multiplied by the appropriate conversion factor (greater than 1) to give the corresponding larger number of Fahrenheit degrees:

$$t_F = \frac{180}{100} \cdot t_C = \frac{9}{5} \cdot t_C \qquad \text{(uncorrected)}$$

Then, the appropriate number of Fahrenheit degrees ($32°$) are added to shift the temperature reference point from $0°$ on the Celsius scale to $32°$ on the Fahrenheit scale.

$$t_F = \frac{9}{5} \cdot t_C + 32° \qquad \text{(corrected)}$$

To convert from Fahrenheit to Celsius, undo these operations:

$$t_C = \frac{5}{9}(t_F - 32°)$$

To convert from Kelvin to Fahrenheit, simply convert from Kelvin to Celsius and then to Fahrenheit.

EXAMPLE 14-4

Converting from Celsius to Kelvin

Convert 78 °C to kelvins.

Solution:

$$T = t + 273.15$$

$$T = 78 \text{ °C} + 273°$$

$$T = 351 \text{ K}$$

$$78 \text{ °C} = 351 \text{ K}$$

The given information is provided to the nearest degree, so the conversion constant is also rounded to the nearest degree.

EXAMPLE 14-5

Converting from Kelvin to Celsius

The boiling point for elemental oxygen is 90.2 K. What is this temperature in degrees Celsius?

Solution:

$$t = T - 273.15$$

$$t = 90.2 \text{ K} - 273.2°$$

$$t = -183.0 \text{ °C}$$

$$90.2 \text{ K} = -183.0 \text{ °C}$$

Note that the conversion constant was rounded to tenths of a kelvin before subtraction to match the decimal places of the given information.

EXAMPLE 14-6

Converting from Fahrenheit to Celsius

The temperature of water in your water heater at home is 131 °F. What is its Celsius temperature?

Solution:

$$t_c = \frac{5}{9}\left(t_F - 32°\right)$$

$$t_c = \frac{5}{9}\left(131 \text{ °F} - 32°\right)$$

$$t_c = 55 \text{ °C}$$

$$131 \text{ °F} = 55 \text{ °C}$$

Problem-Solving Strategy 14.4

It is recommended that you express temperatures mainly in the Celsius scale. The Celsius scale is more convenient, and you will become more familiar with the metric system as you use it.

For most processes, a given change in temperature may not be as important as where on the absolute temperature scale the change takes place. In a similar way, *what* we pray is more important to God than just the fact that we did pray. "Hearken unto the voice of my cry, my King, and my God: for unto thee will I pray. My voice shalt thou hear in the morning, O Lord; in the morning will I direct my prayer unto thee, and will look up. For thou art not a God that hath pleasure in wickedness: neither shall evil dwell with thee" (Ps. 5:2–4).

14B Section Review Questions

1. What is a *thermometric* property? Give an example of one discussed in this section.

2. What was the key historical issue affecting the establishment of scientifically useful temperature scales?

3. What is the *measurable* reference temperature for the absolute temperature scale? What physical property establishes this temperature?

4. What two aspects complicate the conversion between Celsius and Fahrenheit temperatures?

⊛5. Assume that 1.000 cm³ of mercury expands 1.80×10^{-4} cm³ for a +1.00 °C temperature change. If the degree scale marks on a thermometer are 1.00 mm apart, what must be the diameter of the cylindrical capillary tube in the thermometer to accommodate this volume change?

⊛6. What is the average human body temperature in kelvins?

⊛7. For a comfortable room temperature setting on a thermostat, 72 °F is often chosen. What is this temperature in Celsius?

⊛8. Calculate absolute zero in degrees Fahrenheit.

Jacques Charles (1747–1823) was a French physicist who, after hearing of the Montgolfier brothers' airship balloon experiment, built his own balloon using hydrogen. His balloon stayed aloft for 45 minutes before landing in a field, where it was set upon and destroyed by superstitious farmers.

Joseph Louis Gay-Lussac (1778–1850) was a French chemist and physicist credited with discovering the relationship between temperature and pressure. He conducted many experiments involving a gas balloon. During one such experiment, he reached a height of over 21,000 feet above sea level.

14C GAS LAWS

14.9 Ideal Gases

The gaseous phase of a substance has no definite volume or shape. In order to measure how much of a gas is present (how much mass or how many particles), you must establish a standard pressure and temperature. This, in turn, establishes a certain known volume for the gas under these standard conditions. As you study the interaction of a gas's temperature, pressure, and volume, you will use the idea of an **ideal gas** for the same reason that you used an ideal spring and other ideal concepts in physics—to make the basic properties of gases easier to understand.

An ideal gas has the following properties:

• Ideal gas particles are infinitesimally small—they have no volume.

• Ideal gas particles do not interact with each other except when they collide. Their collisions are perfectly elastic.

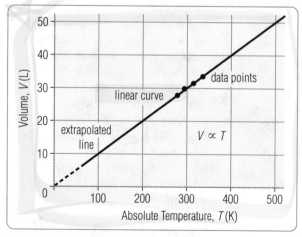

14-13 Graph showing the relationship between temperature and volume of an ideal gas in Charles's law

14.10 Charles's Law

In the late 1700s, Jacques Charles and Joseph Gay-Lussac studied the effect of temperature on the volume of a gas at constant pressure. Both men concluded that all gases expand at nearly the same rate, about 1/273 of their volume at 0 °C for each Celsius degree rise in temperature. Figure 14-13 shows this relationship. Extending the graph by extrapolation back to zero volume gives a temperature around –273 °C. More accurate experimentation in the nineteenth and twentieth centuries defined the theoretical minimum temperature to be –273.15 °C.

A gas theoretically has zero volume at absolute zero (0 K). (Actually, gases condense to the liquid phase above this temperature; so the line in Figure 14-13 does not apply for very low temperatures.) The graph represents a simple proportion:

$$V \propto T, \text{ or}$$
$$V = kT,$$

where k is an appropriate proportionality constant and T is absolute temperature measured in kelvins. This equation can be rearranged as

$$\frac{V}{T} = k.$$

The volume of a gas (at constant pressure) divided by its *Kelvin* temperature is a constant. If the temperature changes from T_1 to T_2,

$$\frac{V_1}{T_1} = k \text{ and } \frac{V_2}{T_2} = k, \text{ so}$$

$$\frac{V_1}{T_1} = \frac{V_2}{T_2}. \qquad (14.4)$$

Equation 14.4, called **Charles's law,** is true for any fixed quantity (mass) of an ideal gas *as long as Kelvin temperature is used* and pressure is the same when measuring volume and temperature.

14-14 The volume of air in this balloon has contracted significantly with a reduction of absolute temperature.

EXAMPLE 14-7

Using Charles's Law

What volume will a gas occupy at 277 °C if it occupies 0.500 L at 27 °C? Assume pressure remains constant.

Solution:
The "initial" conditions are $V_1 = 0.500$ L and $T_1 = 27\ °\text{C} + 273° = 300.$ K. The "final" condition is $T_2 = 277\ °\text{C} + 273° = 550.$ K, and V_2 is the unknown.

Solve Equation 14.4 for V_2 and substitute the known information to find the "final" volume:

$$\frac{V_1}{T_1} = \frac{V_2}{T_2}$$

$$V_2 = V_1 \left(\frac{T_2}{T_1} \right)$$

$$V_2 = (0.500 \text{ L}) \left(\frac{550.\text{ K}}{300.\text{ K}} \right) \leftarrow (\text{"correction factor"} > 1)$$

$$V_2 \doteq 0.916\underline{6} \text{ L}$$

$$V_2 \approx 0.917 \text{ L}$$

Notice that the appropriate ratio of the given temperatures (>1) "corrects" the given volume to yield the unknown volume. Temperature increased; therefore volume must increase according to Charles's law.

Problem-Solving Strategy 14.5

When solving gas-law problems, the solution process can be greatly simplified by evaluating beforehand what effect the changing property has on the property of the unknown quantity, assuming the other variables are held constant. Multiply the given value by an appropriate ratio of the known variables to yield the unknown value.

When could you see Charles's law in effect? Try heating a bowl of water tightly wrapped with plastic food wrap in a microwave oven for several minutes. If the wrapping doesn't leak, the bowl will soon have a bulging plastic dome covering it. The water vapor and gases within heat up and expand. The plastic wrap maintains approximate atmospheric pressure in the trapped gases so you can see their change in volume with temperature.

14.11 Boyle's Law

For Charles's law to be true, pressure must be constant. What if pressure is allowed to vary and temperature is held constant? Robert Boyle, the brilliant Christian physicist and chemist, investigated this possibility. To carry on his experiments, Boyle invented a J-shaped glass tube, called the Boyle's law tube. The short end of the J was melted shut and the long end remained open. When Boyle poured mercury into the open end of the tube, he trapped some air in the closed end. He then measured this initial volume of air. By pouring more mercury into the tube and measuring the differences in the metal's height between the long and short ends of the tube, he could determine the increase of pressure on the trapped air while being careful to keep its temperature constant. With each addition of mercury he measured the air's volume. He found that when the pressure was doubled, the air's volume was halved.

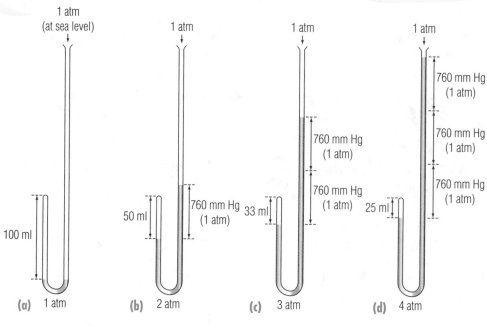

14-15 Boyle's law experiment. Note that the level of mercury in the open tube is measured relative to the level in the closed end.

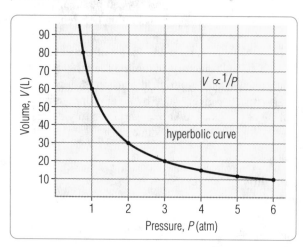

14-16 The volume of the balloon in differing air pressures illustrates Boyle's law.

14-17 Graph showing the inverse relationship of pressure and volume according to Boyle's law

After experimenting thoroughly, Boyle established that the volume of a gas is inversely proportional to the pressure imposed on the gas as long as its temperature remains constant. Mathematically, this is equivalent to saying that the product of a gas's volume and pressure is constant. That is,

$$VP = k.$$

If the pressure and volume change while temperature remains constant,

$$V_1 P_1 = k, \text{ and } V_2 P_2 = k, \text{ so}$$

$$V_1 P_1 = V_2 P_2. \tag{14.5}$$

Equation 14.5, called **Boyle's law,** is true for any fixed quantity of an ideal gas at constant temperature.

EXAMPLE 14-8

Using Boyle's Law

A chemist collects 2.50 L of oxygen from a chemical gas generator at a pressure of 800. torr. What volume will the oxygen occupy at a pressure of 760. torr? Assume that temperature remains the same.

Solution:

You know that pressure changes from 800. to 760. torr. According to the inverse relationship of Boyle's law, a pressure decrease means the volume must be greater at the lower pressure if temperature remains constant. The "correction factor" for volume must be greater than 1. This is confirmed if we solve Boyle's law equation for the unknown V_2:

$$V_1 P_1 = V_2 P_2$$

$$V_2 = \frac{V_1 P_1}{P_2} = V_1 \left(\frac{P_1}{P_2} \right)$$

$$V_2 = (2.50 \text{ L}) \left(\frac{800. \text{ torr}}{760. \text{ torr}} \right) \leftarrow (\text{"correction factor"} > 1)$$

$$V_2 \doteq 2.6\underline{3}1 \text{ L}$$

$$V_2 \approx 2.63 \text{ L}$$

Pressure decreased; therefore volume must increase.

The *torr* is a non-SI unit of pressure commonly used by chemists for routine laboratory work. The derivation of the torr will be discussed in Chapter 17.

Problem-Solving Strategy 14.6

It is not necessary to memorize the equations for Charles's or Boyle's laws if you can remember how gas volumes respond to changes of temperature and pressure. Use the appropriate "correction factor" and the given volume to obtain the required volume.

You probably have experienced a painful reminder of Boyle's law if you have rapidly descended or ascended a mountain or flown in an airplane when you had a head cold. If the air pressure change external to your ear cannot be equalized quickly enough, the volume of the air trapped in your middle ear changes, stretching your eardrum (tympanic membrane). The pain can usually be relieved by swallowing, chewing gum, or yawning to open the Eustachian tube to your middle ear.

14.12 Gay-Lussac's Law

Suppose the volume of a gas is held constant while the pressure and temperature are allowed to vary. What will happen? Gay-Lussac experimented to answer this question and found that the gas's pressure divided by its Kelvin temperature is constant as long as its volume is constant. That is,

$$\frac{P}{T} = k.$$

If the gas's pressure or temperature changes,

$$\frac{P_1}{T_1} = k \text{ and } \frac{P_2}{T_2} = k, \text{ so}$$

$$\frac{P_1}{T_1} = \frac{P_2}{T_2}. \qquad (14.6)$$

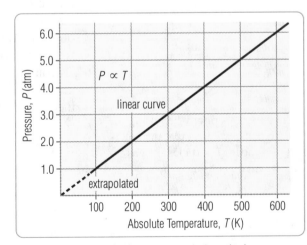

14-18 Graph showing the direct proportional relationship between temperature and pressure for a sample of an ideal gas

Equation 14.6, called **Gay-Lussac's law**, is true for any fixed quantity of an ideal gas at constant volume when T is expressed in kelvins.

EXAMPLE 14-9

Using Gay-Lussac's Law

Air in a closed container is initially at a temperature of 27 °C and a pressure of 760. torr. What is the pressure of the air if it is cooled to −78 °C?

Solution:

The temperature decreases from 300. K to 195 K; therefore pressure must drop in a constant-volume container according to Gay-Lussac's law. The pressure "correction factor" must be less than 1.

$$\frac{P_1}{T_1} = \frac{P_2}{T_2}$$

$$P_2 = \frac{T_2 P_1}{T_1} = P_1 \left(\frac{T_2}{T_1}\right)$$

$$P_2 = (760. \text{ torr}) \left(\frac{195 \text{ K}}{300. \text{ K}}\right) \leftarrow (\text{"correction factor"} < 1)$$

$$P_2 = 494 \text{ torr}$$

The factor that "corrects" the initial pressure for the temperature change is less than 1.

Gay-Lussac's law can be observed in action when a soccer ball that is full and rigid on a warm summer day becomes soft and flaccid on a cold winter day.

14.13 The Combined Gas Law

The three gas laws that we have studied thus far had a constraint on one gas property—either pressure, temperature, or volume. What if all three properties are permitted to change without any constraint? Can we still predict the value of an unknown property? The answer, of course, is yes. Ideal gases behave in such a way that if the temperature and pressure of a given quantity of gas change simultaneously, its final volume is the same as if temperature and pressure changed stepwise, one property at a time.

You could use two steps to calculate an answer when both temperature and pressure vary, but it is not necessary. The **combined gas law** equation shows that the same effect can be achieved in one mathematical operation:

> The only constraint when using the combined gas law is that the quantity (number of particles) of gas must remain constant.

$$\frac{P_1 V_1}{T_1} = \frac{P_2 V_2}{T_2} \tag{14.7}$$

Equation 14.7 is true as long as the quantity of gas is constant.

Any of the three gas laws can be derived from the combined gas law equation by holding the appropriate property constant. However, it should not be necessary to memorize all four (or even three) gas law equations if the effects of the basic laws are understood. Just use the appropriate "correction factors" on the required property to find the answer according to the respective gas laws.

EXAMPLE 14-10

The Combined Gas Law

A 1.00 L sample of gas initially at 27 °C and 760. torr is cooled to −123 °C at a pressure of 400. torr. What is its volume at the new conditions?

$$\frac{P_1V_1}{T_1} = \frac{P_2V_2}{T_2}$$

$$V_2 = V_1\left(\frac{T_2}{T_1}\right)\left(\frac{P_1}{P_2}\right)$$

| given volume | temperature correction (< 1) | pressure correction (> 1) |

$$V_2 = (1.00\ \text{L}) \left(\frac{150.\ \text{K}}{300.\ \text{K}}\right) \left(\frac{760.\ \text{torr}}{400.\ \text{torr}}\right)$$

$$V_2 = 0.950\ \text{L}$$

Note that since temperature decreased, volume should decrease according to Charles's law (temperature "correction" < 1). Pressure decreased; therefore volume should increase according to Boyle's law (pressure "correction" > 1).

14.14 The Ideal Gas Law

If Robert Boyle's experiments with a J-tube or similar device were performed at different temperatures, we would discover that the product of pressure and volume is directly proportional to the absolute temperature. In an equation,

$$PV = KT, \tag{14.8}$$

where K is the temperature-dependent proportionality constant.

If we take two samples of the gas having the same volume, temperature, and pressure and combine them so that their temperature and pressure are unchanged, the gas volume doubles. This means that K must double as well to maintain the equality of Equation 14.6. Therefore, K is directly proportional to the *amount* of gas, or the number of gas particles present. We can express this proportionality as

$$K = kN, \tag{14.9}$$

where N is the number of gas particles in the sample and k is the proportionality constant known as **Boltzmann's constant.** This constant has the value

$$k = 1.381 \times 10^{-23}\ \text{J/K}.$$

To avoid using the extremely large numbers associated with individual particles in a gas sample, we normally express the amount of a gas in moles (mol). Recall from Chapter 13 that a mole of any kind of particle (atoms, molecules, etc.) is Avogadro's number (N_A) of particles.

The number of particles is

$$N = nN_A,$$

where n is the number of moles of gas. Equation 14.9 can now be rewritten

$$K = nkN_A.$$

Ludwig Boltzmann (1844–1906) was an Austrian physicist and mathematician best known for his development of statistical mechanics, which states that the visible properties of matter are determined by the properties of the atoms and molecules of which it is made.

The product of Boltzmann's and Avogadro's constants, kN_A, is conveniently combined into one constant called the ideal or **universal gas constant** (R):

$$R = kN_A = (1.38066 \times 10^{-23} \text{J/K})(6.022137 \times 10^{23} \text{ mol}^{-1})$$
$$R \doteq 8.31452 \text{ J/(K·mol)}$$
$$R \doteq 8.315 \text{ J/(K·mol)} \quad \text{(to four SDs)}$$

Equation 14.8 can be rewritten as

$$PV = nRT, \tag{14.10}$$

which is known as the **ideal gas law.** Observe that rearranging Equation 14.10 to isolate the constant R places all of the possible gas variables in one ratio:

$$R = \frac{PV}{nT}$$

The ideal gas law predicts that this ratio should hold true for any pressure (pressure determines how close the particles are to each other). Real gases hold closely to this relationship up to several atmospheres of pressure.

The ideal gas constant as given thus far is expressed in SI units. In order to use it in this form, volume must be expressed in cubic meters, temperature in kelvins, and pressure in pascals. Since gas-law experiments may be conducted using other units, either the experimental units must be converted to SI units or the constant R must be converted to the units being used. Table 14-4 provides values of R in commonly used volume and pressure units. Note that temperature must *always* be expressed in kelvins regardless of the value of R used.

The ideal gas law is useful because we can determine one of a gas's four properties if the other three are known. In order to compare the properties of one gas to another, data for both must be obtained under the same conditions. The scientific community has established **standard temperature and pressure (STP)** to be 273.15 K and 1.013×10^5 Pa respectively. Obviously, their equivalents in other unit systems may be used if the situation requires.

TABLE 14-4

Units for Use in Ideal Gas Law

V	P	T	n	R
m^3	N/m^2	K	mol	8.315 J/(K·mol)
L	atm	K	mol	0.08207 L·atm/(K·mol)
L	torr	K	mol	62.36 L·torr/(K·mol)
m^3	Pa	K	mol	8.315 m^3·Pa/(K·mol)

EXAMPLE 14-11

Using the Ideal Gas Law: Find Volume

Find the volume of 1.00 mole of hydrogen in liters at STP.

Solution:

Solve Equation 14.10 for volume and substitute the known values for the other quantities. Note that we may use either value of R that includes L in its unit. The other appropriate parameters must be used as well.

$$V = \frac{nRT}{P} = \frac{(1.00 \text{ mol})(0.08207 \text{ L·atm/K·mol})(273.15 \text{ K})}{1.00 \text{ atm}}$$

$$V \doteq 22.\underline{4}1 \text{ L}$$

$$V \approx 22.4 \text{ L}$$

One mole of an ideal gas at STP occupies a volume of 22.4 L.

EXAMPLE 14-12

Using the Ideal Gas Law: Find Pressure

If 2.00 moles of ammonia occupy 50.0 L at 350. K, what is the pressure of the gas in pascals?

Solution:

In this case you will solve Equation 14.10 for pressure. Since you need to find pressure in pascals, $R = 8.315$ m³·Pa/K·mol.
In order to use this value of R, you must convert liters to cubic meters:

$$\frac{50.0 \text{ L} \mid 1 \text{ m}^3}{\mid 1 \times 10^3 \text{ L}} = 0.0500 \text{ m}^3$$

Solve for pressure:

$$P = \frac{nRT}{V} = \frac{(2.00 \text{ mol})(8.315 \text{ m}^3 \cdot \text{Pa/K} \cdot \text{mol})(350. \text{ K})}{0.0500 \text{ m}^3}$$

$$P \doteq 1.164 \times 10^5 \text{ Pa}$$

$$P \approx 1.16 \times 10^5 \text{ Pa}$$

14.15 Real Gases

The ideal gas model assumes that particles of a gas have no volume and interact only through collisions. Of course, real gas particles *do* occupy space and attract each other. These properties can usually be ignored if the particles are so far apart that their volume is insignificant and are moving so rapidly that the forces between them have little effect. Thus, the ideal gas model is accurate for gases at high temperatures and low pressures.

At higher pressures or lower temperatures, a gas begins to deviate from the predictions of the ideal gas model. At first, volumes at higher pressures or lower temperatures are *less* than the ideal gas law predicts. The attraction between molecules reduces their impulsive collision forces with the container. As a result, the real gas cannot occupy as much volume at a given pressure as predicted by the ideal gas model. At extremely high pressures or extremely low temperatures, the gas molecules are so close together or have so little kinetic energy that their individual volumes are no longer insignificant compared to the volume of the gas sample as a whole. In this state, the gas begins to resist further compression, and the volume is *more* than the ideal gas law would predict for that pressure and temperature.

One of the first men to realize the limitations of the ideal gas model and to devise an experimentally corrected equation was Johannes Diederik van der Waals. His gas law is

$$\left(P + a \cdot \frac{n^2}{V^2}\right)(V - bn) = nRT.$$

The factors a and b are constant for each gas. The constant a represents a pressure correction for the force between gas molecules. The factor b is a volumetric correction based on the moles of molecules present. These factors must be found experimentally for each gas. This equation holds for most gases over a wide range of temperatures and pressures. Those gases that obey the van der Waals equation are called **van der Waals gases.**

As with other real systems in the world, real gases behave in more complex ways than ideal gases. In religious matters, it seems that just the opposite is true. Man creates great and complex religious works in unbelieving worship where God demands only the simplest evidences of a saving faith. "Pure religion and undefiled before God and the Father is this, To visit the fatherless and widows in their affliction, and to keep himself unspotted from the world" (James 1:27).

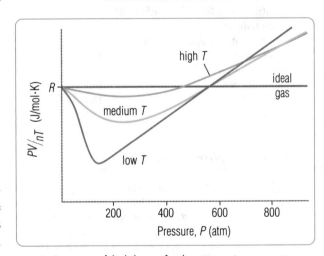

14-19 Comparison of the behavior of real gases at various temperatures and pressures to an ideal gas. Note that only at higher temperatures and lower pressures do real gases approximate ideal gases.

More complicated equations have been devised to describe more gases over wider ranges of temperatures and pressures. However, the simplest equation, the ideal gas equation, is still the most useful approximation for the behavior of gases under typical conditions.

14C Section Review Questions

1. What are the characteristics of an ideal gas?

2. Describe in your own words Charles's law. Under what conditions is this principle true?

3. Describe in your own words Boyle's law. Under what conditions is this principle true?

4. What does the ideal gas law take into account that the other gas laws assume to be constant?

⊙5. An ideally elastic balloon contains 0.650 L of air at 25.0 °C. What is the volume of the balloon after the sun warms it to 42.0 °C? Assume that pressure within the balloon remains constant.

⊙6. A hot air balloon contains 120. m^3 of air while moored to the ground. Surrounding air pressure is 1.00 atm and temperature is 60.0 °C. The balloon launches and rises until the air pressure is 0.95 atm. What is the new volume of the hot air in the balloon, assuming that the hot air temperature remains constant and the amount of hot air does not change?

⊙7. What is the new volume of hot air in the balloon in Question 6 if the hot air temperature decreases to 48.0 °C at the final altitude?

⊙8. How many moles of an ideal gas would it take to occupy 1.00 L at STP?

Chapter Review

In Terms of Physics

coefficient of linear expansion (α)	14.2	Boyle's law	14.11
bimetallic strip	14.2	Gay-Lussac's law	14.12
coefficient of volume expansion (β)	14.3	combined gas law	14.13
thermometer	14.6	Boltzmann's constant (k)	14.14
thermocouple	14.6	universal (ideal) gas constant (R)	14.14
absolute zero	14.7	ideal gas law	14.14
kelvin (K)	14.7	standard temperature and	
ideal gas	14.9	pressure (STP)	14.14
Charles's law	14.10	van der Waals gases	14.15

Problem-Solving Strategies

14.1 (page 305) The thermal expansion equation can be simplified a bit if the length of the object, l_0, is measured at the reference temperature of 0.0 °C. Then the change of length is just $\Delta l = \alpha l_0 t$, where l_0 is the reference length and t is the new Celsius temperature.

14.2 (page 307) Be sure to use the appropriate coefficient of expansion for solid materials. Use α for linear expansion and β for volumetric expansion.

14.3 (page 307) There is no linear coefficient of expansion for fluids because, unlike solids, fluids do not have fixed dimensions.

14.4 (page 313) It is recommended that you express temperatures mainly in the Celsius scale. The Celsius scale is more convenient, and you will become more familiar with the metric system as you use it more.

14.5 (page 315) When solving gas-law problems, the solution process can be greatly simplified by evaluating beforehand what effect the changing property has on the property of the unknown quantity, assuming the other variables are held constant. Multiply the given value by an appropriate ratio of the known variables to yield the unknown value.

14.6 (page 317) It is not necessary to memorize the equations for Charles's or Boyle's laws if you can remember how gas volumes respond to changes of temperature and pressure. Use the appropriate "correction factor" and the given volume to obtain the required volume.

Review Questions

1. How are molecules of hot water and molecules of cold water different?

2. What factors cause the particles in solid matter to require more space as the material becomes hotter?

3. What features are designed into a vehicle bridge to accommodate the expansion of materials as they are heated?

4. What two factors determine how much a metal rod will expand in length for a given temperature change?

5. Suppose you want to buy gold *by volume*. Would you get more for your money if you heat the gold or cool it before measuring it? (*Hint:* Temperature does not affect the gold's weight.)

6. Why is a thermoscope alone a poor instrument for measuring temperature?

7. How does a mercury or alcohol thermometer indicate temperature?

8. Which one of the following is *not* the freezing point of water?
   a. 32 °F
   b. 0 K
   c. 0 °C
   d. ~273 K

9. Which one of the following is the boiling point of water?
   a. 100 °F
   b. 273 K
   c. 100 °C
   d. 373 °C

10. Which temperature scale is based on a single reference point? What is the *measurable* reference point for this scale?

11. Which temperature unit represents the smallest change in temperature: the Celsius degree, the kelvin, or the Fahrenheit degree?

12. Under what conditions does $V_1/T_1 = V_2/T_2$ for a gas?

13. When using the gas laws, what temperature scale must be used?

14. For every gas law involving ideal gases except the "ideal gas law," what is assumed about the sample of gas?

15. Write the ideal gas law.

16. What is the product of Boltzmann's constant and Avogadro's constant called?

True or False (17–25)

17. Materials expand when heated because the atoms physically become larger, taking up more space.

18. The coefficient of volume expansion is the cube of the coefficient of linear expansion.

19. The coefficients of expansion for gases are very similar because gas particles are so far apart that they do not affect each other at normal temperatures and pressures except when they collide.

20. A thermometric property must be proportional to temperature.

21. It is best if a thermometer's mass is large compared to the mass of the material whose temperature it is measuring so that it will be more sensitive.

22. The accepted fiducial point of the absolute temperature scale is 0 K.

23. Charles's law states that the volume of an ideal gas is directly proportional to its absolute temperature.

24. Boyle's law states that the volume of an ideal gas is proportional to the reciprocal of the gas's pressure, assuming temperature is held constant.

25. Real gases adhere closely to the ideal gas law at high pressures and low temperatures.

26. A bar of gold is 10.0 cm × 5.00 cm × 5.00 cm at room temperature, 25.0 °C. What is its change of volume at 50.0 °C?

27. An uncapped 1.00 L jar is filled to the brim with turpentine early in the morning, when the temperature is 10.0 °C. How much turpentine spills over the brim that day if the temperature reaches 30.0 °C? Ignore the jar's expansion.

28. At 100.0 °C, what will be the volume of mercury that has a volume of 1.00 cm³ at 0.00 °C?

29. The pine mast of a New England charter sailing ship is 20.000 m tall from deck to peak on a frigid (−40.0 °C) winter day. The following summer, the mast expands to 20.007 m on a hot summer day. Assuming the mast is at a uniform temperature for each measurement, what is the mast's temperature on the summer day? (The grain of the wood runs lengthwise in the mast.)

30. Convert the following temperatures to Fahrenheit.
   a. 25 °C
   b. −15 °C
   c. −40 °C
   d. 273 K
   e. 373 K

31. Convert the following temperatures to Celsius.
   a. 0 K
   b. 503 K
   c. 328 K
   d. 113 °F
   e. 59 °F

32. Convert the following temperatures to Kelvin.
   a. 12 °C
   b. 217 °C
   c. 122 °C
   d. −73 °C
   e. 104 °F

33. A gas occupies a volume of 2.00 L at 612 K and 760. torr. What volume does it occupy at 306 K and 760. torr?

34. A gas is confined to a 1.50 L expandable container at 300. K and 380. torr. What is its pressure if it is allowed to expand to a volume of 2.00 L at constant temperature?

35. A 2.5 L airtight, rigid container full of carbon dioxide at 325 K is heated to 525 K, where its pressure is 150 torr. What was its original pressure?

36. Find the volume of 3.5 moles of oxygen at 20.0 °C under 760.0 torr.

37. A total of 1.75 L of acetylene gas is collected from a chemical generator at 735 torr and 35.0 °C. How many moles of acetylene were collected?

38. Write a brief report on the development of the temperature scales used in scientific research. Discuss the current advances and standards for determining the key fiducial points.

39. Write a brief report on the work of Jacques Charles.

Count Rumford
(1753–1814)

Scientists try to explain what they observe with theories. From time to time, new experiments show that an old theory must be discarded and a new theory devised. A striking example of this process is Count Rumford's cannon-boring experiment.

Benjamin Thompson, who later became Count Rumford, was born to poor parents in a small town in Massachusetts. His family could not afford to keep him in school. At the age of 19, Thompson improved his lot by marrying a rich widow from Rumford (now Concord), New Hampshire. When the colonies began to plot war against England, Thompson sided with the British as a spy. But his sedition was discovered, and he had to flee to England. There his knowledge of colonial affairs won him a military position.

Following the war, Thompson's services were no longer needed in England. He had no hope of returning to America, where he was considered a traitor. Of the few choices remaining, he accepted a minor position in the Bavarian government; yet before Thompson went he persuaded George III to knight him by promising to work undercover in Germany. Thompson's skill in politics soon earned him a position as the minister of war and police. He introduced social reforms, promoted the potato plant, reorganized the outdated Bavarian army, and supervised the production of modern weapons. In 1792 the Bavarian government awarded him the title Count Rumford.

Along with everything else Thompson was doing, he found time to conduct scientific experiments. While watching munitions plants bore their brass cannons, he noticed that the searing heat never died down no matter how long the brass was bored. His contemporaries believed that heat was a fluid locked in the metal's interstices and released by friction. But if that were the case, then the liquid would be used up in the process of boring. Thompson theorized that heat is not a fluid, but motion.

Rumford investigated other aspects of heat, especially convection. He was the first to detect convection currents and the expansion of cooling water. Using this knowledge, he invented an efficient stove, a double boiler, and a drip coffee-maker. He also improved the common chimney and fireplace.

Rumford's plans to return to the United States were frustrated when President John Adams would not allow him to direct a military academy he had designed. So Rumford returned to England in 1798 and founded the Royal Institution of London, an organization to promote science among the common people. Its employees included Sir Humphry Davy and Michael Faraday. In 1804 Rumford left London for Paris, where he was warmly received by Napoleon Bonaparte. The count remained in Paris until his death ten years later.

James Joule
(1818–89)

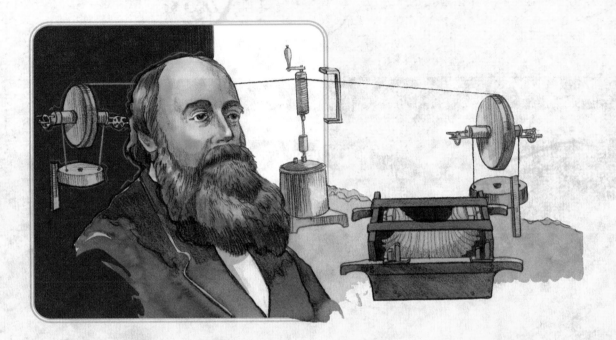

Afundamental law of physics is the conservation of energy. It is surprising, therefore, that no one discovered this law (or even defined energy) until the nineteenth century. The major experiments that prove this law were first performed by James Prescott Joule.

A sickly child of wealthy parents, James had to be educated at home until he was fourteen. When he went to Manchester University, he studied under the famous chemist John Dalton, who had formulated the theory that atoms are indivisible spheres. Always experimenting, young James had to settle for makeshift equipment from the kitchen until his father could build him a laboratory.

Joule was especially interested in electricity and the heat that an electric current produces. His early experiments showed that the heat generated per unit time by electricity in a wire is proportional to the product of the wire's resistance and the square of the current ($P \propto RI^2$). This fact is now called Joule's law.

Whether the electricity was produced by chemical energy (in a voltaic cell) or by mechanical energy (in an electromagnetic generator he had invented), the resulting heat was always the same. Joule contrived an experiment to see if mechanical energy could produce heat directly without first becoming electricity. The success of his experiment proved that any form of energy, whether electrical, chemical, or mechanical, produces an equivalent amount of heat.

Joule went a step further and devised four experiments to determine the precise mechanical equivalent of heat. In his most famous experiment, a paddle wheel, driven by the mechanical energy in a falling weight, heated the water in a tub just enough to be measured by a sensitive thermometer. With more careful experiments he refined his results until, by 1850, he found that 772 foot-pounds of work always raise the temperature of one pound of water one Fahrenheit degree. He proved that heat is not a fluid as was popularly believed, but a form of energy. He showed that this energy is conserved during work. His careful experiments paved the way for general acceptance of the conservation of energy.

Joule's work went unnoticed until William Thomson, Lord Kelvin, discovered it. Kelvin then asked Joule to collaborate with him in several experiments, the most famous of which demonstrated the "Joule-Thomson effect." This discovery, that the temperature of gas falls when it is allowed to expand, ushered in the huge refrigeration industry later in the century. With Thomson's help, Joule received acclaim as a scientist. In his last years, the British government granted him a pension in honor of his achievements.

15A	Theories of Heat	327
15B	Thermal Energy and Matter	330
15C	Mechanisms for Heat Transfer	339
Biography:	Nicolas Sadi Carnot	344

New Symbols and Abbreviations

calorie (cal)	15.7	latent heat of fusion (L_f)	15.11
heat capacity (C)	15.8	latent heat of vaporization (L_v)	15.11
heat (Q)	15.8	Stefan-Boltzmann constant (σ)	15.14
specific heat (c_{sp})	15.9		

The Sun is the source of most of the
thermal energy at the earth's surface

Thermal Energy and Heat 15

> *But the day of the Lord will come as a thief in the night; in the which the heavens shall pass away with a great noise, and the elements shall melt with fervent heat, the earth also and the works that are therein shall be burned up.* II Peter 3:10

The apostle Peter writes in this passage of the future time after the gathering of the believers and the judgment of the wicked, when God purges the universe of all remnants of sin in preparation for the establishment of the eternal abode of Christ and His saints. The "fervent heat" indicates a source of energy of such intensity that even the elements melt. A standard reference of physical properties shows that carbon is the element with the highest melting point—3800 K. However, the verse gives the sense that more than a mere phase change is taking place—the very atoms of the elements are being destroyed and remade, completing the sin-purging process. The source of thermal energy that can remake the entire universe can be only God Himself.

15A THEORIES OF HEAT

15.1 Introduction

What makes one object hot and another cold? What exactly is the "thing" called heat? Scientists have studied these questions for centuries, and their conclusions have changed as they have learned more about how heated and unheated objects behave.

15.2 The Kinetic Theory

Modern scientists agree that what we call "heat" is really a form of energy called thermal energy. According to the kinetic theory of matter, all substances contain tiny, constantly moving particles. The particles in hot materials move faster than the particles in colder materials. Thermal energy consists of the kinetic energy of the random motion of the particles. The average kinetic energy of the particles is proportional to the temperature of the material. However, temperature alone does not indicate how much thermal energy is present. We will look into this aspect later. For now, let's review some important observations that give evidence for the kinetic-particle theory of matter.

The **kinetic theory** is also called the **kinetic-molecular theory** of matter because the two concepts of tiny discrete particles in constant motion are inseparable.

327

15-1 The spreading of this dye throughout the water illustrates the principle of diffusion.

15-2 Diffusion of gases is clearly seen in the mixing of bromine vapor and air.

Robert Brown (1773–1858) was an English botanist who first described Brownian motion while studying pollen grains under a microscope.

15.3 Matter Can Be Subdivided

Suppose that you place a crystal of blue dye in a beaker of water and stir it until it is completely dissolved and all the water has become dark blue. The crystal has somehow divided itself and spread throughout the water in the beaker. If you pour the contents of the beaker into a much larger container of water and stir the solution again, the color becomes less intense but is still completely distributed throughout the larger container. The process may be repeated many times. The color becomes progressively fainter but is still present. The material from the crystal seems to keep dividing indefinitely. For this to be true, there must be a multitude of minuscule particles. These particles are called molecules, atoms, or ions.

15.4 Diffusion

You have seen evidence that molecules exist. Now let's consider evidence that these particles are in motion. Suppose you drop a crystal of the same blue dye into a beaker of water *without* stirring. If you wait patiently, you see that the crystal dissolves and that eventually the dye and the water completely mix (though this may take a very long time). This observation can be explained by assuming that the water molecules are in constant motion. Their motion distributes molecules of the dye throughout the water.

Diffusion takes place much faster in gases than in liquids. This observation leads us to believe that gas molecules move faster than liquid molecules. For example, when a few drops of household ammonia are spilled from a bottle on one side of a room, the scent of ammonia can soon be smelled on the far side of the room. Judging by the speed at which the ammonia drifts, air molecules must move at high speeds (around 800 m/s at room temperature).

Scientists can visually demonstrate diffusion in gases, using bromine vapor and air. Bromine vapor is brown and dense; it is about five and one-half times as dense as air. (It is also poisonous—that is why only qualified people should demonstrate this.) Suppose that a container of air is placed upside down over a similar container of bromine vapor that has its open end facing up but covered by a glass plate (see Figure 15-2). When the glass plate is slid out from between the containers, the brown bromine vapor will start diffusing upward into the transparent air. Air will also move downward into the bromine vapor. This cannot be explained by gravity, because the lighter air goes downward while the heavier bromine vapor moves upward. After a period of time, the two gases will be evenly mixed. The molecules of the air and the bromine must each have motion of their own to mix themselves so quickly and thoroughly. This process occurs even more rapidly at higher temperatures.

15.5 Brownian Motion

In 1827 Robert Brown discovered more direct evidence that matter is in constant motion. Brown used a microscope to examine plant spores floating in water. He saw the spores jostling back and forth as if they were being struck repeatedly from different sides. Yet nothing was touching them except the water in which they were floating. The explanation for his observation is that water molecules are in constant random motion, and they collide with the spores. This effect, seen only in tiny particles, is called **Brownian motion.**

It is also possible to see Brownian motion in smoke particles. While looking through a microscope, introduce a small amount of smoke before the objective lens and illuminate it with a bright light. You will be able to clearly see the smoke particles jostling about in vibratory motion. This motion is caused by collisions with high-speed air molecules.

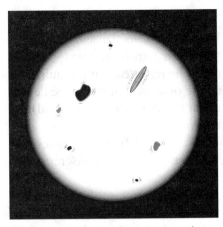

15-3a Brownian motion of microscopic particles suspended in a liquid or gas can be observed using a light microscope.

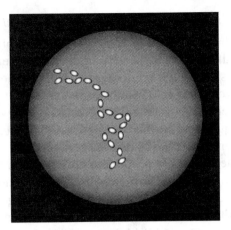

15-3b Time-lapse microphotography shows the position of an inanimate particle experiencing Brownian motion at given time intervals.

liquid molecules striking particle (size exaggerated)

microscopic particle

15-3c Brownian motion of microscopic particles is caused by the random, asymmetrical collisions of liquid or gas molecules against the particles.

15.6 The Caloric Theory

Many eighteenth-century scientists believed the **caloric theory**—the view that thermal energy is a material fluid that flows from hot bodies to cold bodies. Earlier scientists, including Francis Bacon, Robert Boyle, and Robert Hooke, had previously advocated the rival kinetic theory—the theory that thermal energy is a property of an object arising from the motion of its particles. Still, such notable scientists as Joseph Black, who discovered carbon dioxide, enthusiastically endorsed the caloric theory. Antoine Lavoisier, a famous French chemist, preferred not to take a stand on which theory, caloric or kinetic, was superior.

A major blow to the caloric theory came with the famous cannon-boring experiment conducted by Count Rumford. According to the caloric theory, the amount of thermal energy produced in drilling a metal should be proportional to the volume of the material the drill removed. The caloric was believed to reside between the most elementary particles of the metal and be released when the particles were separated.

However, as you learned in Chapter 1, Rumford observed no correlation between the volume of the material removed and the thermal energy produced. On the contrary, he observed that some of the greatest heat was produced when a dull drill removed only a little metal after a great amount of mechanical work was expended. An almost unlimited supply of thermal energy could be produced under such conditions. In a report to the Royal Society in 1798, he theorized that thermal energy was the result of motion in the elementary particles of the metal. Humphry Davy continued this line of attack on the caloric theory in the early 1800s.

In 1840 James Prescott Joule of Manchester, England, began a series of experiments designed to unite thermal and mechanical energy under a single conservation law. In a lecture he delivered in 1847, Joule administered the *coup de grâce* to the faltering caloric theory. He showed that since caloric can be converted to kinetic or potential energy, it cannot possibly be a material substance. He explained the source of the heating in Rumford's experiment as a change from mechanical to thermal energy. The caloric theory became scientific history.

Robert Boyle (1627–91) was an Irish-born English physicist and devout Christian most noted for his critique of Aristotelian chemistry and alchemy, which opened the door to true chemistry. He also investigated the properties of gases and discovered the importance of air to life and combustion.

Joseph Black (1728–99) was a Scottish physician who conducted research in the thermal properties of matter and discovered carbon dioxide. His work was foundational in the establishment of the science of chemistry.

Antoine Lavoisier (1743–94) was a brilliant French chemist often called the "Father of Modern Chemistry." He discovered the conservation of mass, oxygen as the active agent in combustion and respiration, and many new elements. He was beheaded in the French Revolution.

Humphry Davy (1778–1829) was an English chemist who specialized in discovering many new elements and investigating their properties. He discovered the law of combining proportions that supported Dalton's atomic theory of matter. He also conducted qualitative research in many areas of science.

15.7 The Mechanical Equivalent of Heat

The first results of Joule's research, published in 1843, described a device in which falling weights turned an electric generator, which in turn heated water with a resistance wire. For thirteen separate measurements he reported that it required an average of 4.50 N·m of work to raise the temperature of a gram of water one degree Celsius. This quantity of thermal energy was given a name—the **calorie (cal).** Its equivalent in the usual energy units, called the *mechanical equivalent of heat,* varied somewhat from experiment to experiment because of frictional losses, but the values obtained were close enough to support the idea that the discrepancies would disappear in the absence of friction.

Joule repeated his attempts to change other forms of energy to thermal energy with other devices. His experiments did not agree exactly. For the equivalent of 1 cal, Joule reported 4.14 N·m of work for heating water with a perforated piston, 4.27 N·m for the temperature drop of rapidly expanding air, and 4.16 N·m for experiments involving water moving through narrow tubes. By 1850 he had settled on a value of 4.15 N·m for 1 cal. The currently accepted value of the mechanical equivalent of thermal energy is 4.186 N·m = 1 cal (at 15 °C). In honor of the fundamental work Joule accomplished in thermodynamics, the N·m was renamed the joule, the SI derived unit of energy and work.

> The **calorie (cal)** was the unit of thermal energy for many years. However, the joule is now the primary unit of energy in most scientific research. The calorie, or more specifically, the dietary Calorie (1 Calorie = 1 kcal or 1000 cal), is still in use for quantifying the energy content of foods.

15A Section Review Questions

1. Describe how thermal energy is manifested in matter according to the kinetic theory.

2. How is temperature related to thermal energy?

3. Discuss three evidences that support the kinetic-molecular theory of matter.

4. What observation produced a key piece of evidence that contradicted the caloric theory of heat and led to its eventual demise?

5. What logical argument proved that heat could not be the material fluid called *caloric?*

6. State the SI equivalent for 1 cal.

7. Ignoring losses to the surroundings, if 10.0 J of work is done stirring 2.0 L of water, how much thermal energy is transferred to the water?

15B THERMAL ENERGY AND MATTER

15.8 Heat Capacity

Recall from earlier chapters that momentum and velocity are related but different quantities. It is not always the fastest moving object that has the greatest momentum. For example, a slow-moving truck may have more momentum than a speeding motorcycle because the truck has far more mass than the motorcycle. Momentum is the product of mass and velocity.

Similarly, it is not always the hottest object that contains the greatest amount of thermal energy. There are two other factors involved here—the mass of the object and its ability to hold thermal energy. Both of these factors are included in the object's **heat capacity (C),** which is the amount of thermal energy required to raise the temperature of the entire object one degree Celsius. The amount of thermal energy added to or taken from a system is called **heat (Q).** The SI unit for heat capacity is joules per degree Celsius (J/°C). Heat capacity of an object is to

> The **heat capacity (C)** of an object is the quantity of thermal energy required to raise the temperature of the *entire object* 1 degree Celsius. Units are J/°C. Heat capacity is a unique property of each object.

> **Heat (Q)** is the quantity of thermal energy that is transferred to or from a system. It is not correct to say that an object contains a certain amount of heat.

thermal energy as mass is to momentum. Heat capacity may be calculated by the formula

$$C = \frac{Q_{object}}{\Delta t},$$ (15.1)

where Δt is the change of temperature of the object,

$$\Delta t = t_{final} - t_{initial}.$$

Heat is analogous to mechanical work. Mechanical work on or by a system involves the transfer of mechanical energy to or from the system. Heat involves the movement of thermal energy. So it is technically incorrect to say a system *has* a certain amount of heat, just as it is improper to say a system *has* a certain amount of work. Heat is not a property of a system as thermal energy is. It takes great discipline to use the two terms correctly.

Just as momentum is a function of a fixed property of an object (its mass) and a variable property (its speed), the thermal energy of an object is dependent on its heat capacity (a fixed property) and its temperature (a variable property).

Recall that lowercase *t* refers to Celsius temperature as well as time. This is another case in physics where a single symbol is used for more than one quantity. The context of the problem will indicate what dimensional property is meant.

15.9 Specific Heat

Mass is a property of a specific object, not of a kind of substance. Similarly, heat capacity is a property of an object. To compare the relative "heaviness" of one substance to another, scientists divide the mass of a sample of the substance by its volume, which yields the *specific density* of the material—its mass per unit volume. Similarly, to find the thermal energy capacity of a substance (not an object), scientists divide the heat capacity of an object by its mass to find its *specific heat capacity*—its heat capacity per gram. The **specific heat (c_{sp})** of a substance is the amount of thermal energy required to raise the temperature of 1 g of the substance one degree Celsius. It is properly expressed in SI units of joules per gram-degree Celsius (J/g·°C). Older textbooks sometimes refer to units of calories per gram-degree Celsius. Comparing the definitions of specific heat and calorie, we can see that the specific heat of water is, by definition, exactly 1 cal/g·°C (at 15 °C). In SI units, the specific heat of water is about 4.18 J/g·°C (near room temperature). The specific heats of some common substances are listed in Table 15-1 on page 332. Specific heat varies somewhat with temperature and pressure. However, the specific heats of most solids are nearly constant for temperatures between 20 °C and 100 °C and pressures near 1 atm.

Specific heat capacity, or just **specific heat (c_{sp}),** is the amount of heat required to raise the temperature of 1 g of a substance 1 °C.

The definition of the calorie, dependent as it is on the heat capacity of water (which varies with temperature), must be given with an associated temperature. Typically, the calorie is defined at 15 °C, although other temperatures are used as well.

How can we find the specific heat of a material? Using the definition of specific heat, if we add a known amount of thermal energy (Q) to an object of known mass (m) and note the corresponding temperature change, Δt, in degrees Celsius, then we can find the specific heat of the material from the equation

$$c_{sp} = \frac{Q}{m \Delta t}.$$ (15.2)

EXAMPLE 15-1

Taking Some Heat: Determining Heat Capacity

When 89.7 J of thermal energy is added to a 10.0 g aluminum block at 20.0 °C, the block's temperature rises to 30.0 °C. Compute (a) the specific heat of aluminum and (b) the heat capacity of the block.

Solution:
The system is the aluminum block.

TABLE 15-1

Specific Heat of Selected Substances

Substance	c_{sp} (J/g·°C)
air (27 °C/300 K)	1.16
aluminum	0.897
benzene	1.75
brass	0.375
carbon (graphite)	0.709
chlorine	0.479
copper	0.385
ethyl alcohol	2.44
glass (crown)	0.670
gold	0.129
iron	0.449
lead	0.129
mercury	0.140
oil (olive)	1.79
oxygen	0.918
platinum	0.133
silver	0.235
tin	0.228
tungsten	0.132
water (solid/ice)	2.09
water (liquid)	4.18
water (gas/steam)	2.01
wood	1.70
zinc	0.388

An **adiabatic** process is one that does not gain or lose thermal energy to its surroundings.

a. Computing specific heat:

$$c_{sp\ Al} = \frac{Q}{m\Delta t} = \frac{89.7\ \text{J}}{(10.0\ \text{g})(10.0\ °\text{C})}$$

$$c_{sp\ Al} = 0.897\ \text{J/g·°C}$$

b. Computing heat capacity:

$$C = mc_{sp\ Al}$$

$$C = (10.0\ \text{g})(0.897\ \text{J/g·°C})$$

$$C = 8.97\ \text{J/°C}$$

The amount of heat that is needed to raise the temperature of an object a certain number of degrees is released when the object is cooled the same number of degrees. If, in the previous example, the aluminum block were cooled from 30 °C to 20 °C, the cooling process would release 89.7 J of heat.

15.10 Conservation of Thermal Energy

When an object gains heat, its surroundings lose the same amount of heat. Thermal energy is conserved if it is the only form of energy exchanged in a process. In this chapter we will assume that thermal energy is conserved, so the heat gained by a system is lost by its surroundings and vice versa. Symbolically,

$$Q_{\text{system}} = -Q_{\text{surroundings}},\ \text{or}$$
$$Q_{\text{system}} + Q_{\text{surroundings}} = 0\ \text{J}.$$

Such expressions are called *heat-balance equations*.

EXAMPLE 15-2

Conservation of Thermal Energy: The Calorimeter

A 10.0 g aluminum block has an initial temperature of 20.0 °C. It is placed in an **adiabatic** vessel—one that allows no heat to enter or leave its contents—that contains 20.0 g of water at 31.1 °C. The two parts of the system come to equilibrium at a final temperature of 30.0 °C. Compare the heat gained by the metal block to the heat lost by the water.

Solution:

The system consists of the water and the aluminum block. It does *not* include the vessel at this time. You are seeking the quantity of heat lost and gained by the respective parts of the system. Equation 15.2 must be rearranged to solve for Q:

$$Q = mc_{sp}\Delta t$$

Calculate the heat gained by the water:

$$Q_{H_2O} = (mc_{sp}\Delta t)_{H_2O} = [mc_{sp}(t_f - t_i)]_{H_2O}$$
$$Q_{H_2O} = (20.0\ \text{g})(4.18\ \text{J/g·°C})(30.0\ °\text{C} - 31.1\ °\text{C})$$

$$Q_{H_2O} \doteq -91.9 \text{ J}$$

$$Q_{H_2O} \approx -92 \text{ J} \tag{1}$$

The heat "gained" by the water is negative, so thermal energy is actually lost from the water. There is a net flow of heat out of the water and into the aluminum block.

Calculate the heat gained by the block:

$$Q_{Al} = (mc_{sp}\Delta t)_{Al} = [mc_{sp}(t_f - t_i)]_{Al}$$

$$Q_{Al} = (10.0 \text{ g})(0.897 \text{ J/g·°C})(30.0 \text{ °C} - 19.7 \text{ °C})$$

$$Q_{Al} \doteq +92.39 \text{ J}$$

$$Q_{Al} \approx +92.4 \text{ J} \tag{2}$$

Notice that $Q_{Al} = -Q_{H_2O}$ or $+92 \text{ J} = -(-92 \text{ J})$ expressed to 2 SDs.

In the previous example an adiabatic container was specified. There are no truly adiabatic containers. Therefore we must take into account the thermal energy lost or gained by the container and the surrounding air. A **calorimeter** is a container designed to minimize the loss of thermal energy to the surroundings. It consists of an insulated outer cup, a lighter inner cup, and an insulated lid with small holes for a stirrer and a thermometer. Figure 15-4a is a photograph of a typical school laboratory calorimeter. Figure 15-4b is a research-quality calorimeter. It is possible to determine the heat capacity of a calorimeter by using the principle of the conservation of energy.

EXAMPLE 15-3

Finding the Little Foxes: Calorimeter Heat Capacity

10.0 g of water at room temperature is placed into a calorimeter. The equilibrium temperature of the calorimeter and water is 20.0 °C. Then 10.0 g of water heated to 40.0 °C is added to the water in the calorimeter. The new equilibrium temperature is 27.0 °C. Find the heat capacity of the calorimeter.

Solution:
The system is the calorimeter, the cool water, and the additional hot water. You measure the water temperature and assume that the calorimeter's temperature is the same at equilibrium. The essential concept at work here is this: After the hot and cool water mix and you determine the heat gained and lost by the two quantities of water, any difference in heat *must* be due to the heat transferred to or from the calorimeter based on the conservation of thermal energy.

$$Q_{hw} + Q_{cw} + Q_{cal} = 0 \text{ J} \tag{1}$$

Calculate the heat transfer of the cool water:

$$Q_{cw} = (mc_{sp}\Delta t)_{cw} = [mc_{sp}(t_f - t_i)]_{cw}$$

$$Q_{cw} = (10.0 \text{ g})(4.18 \text{ J/g·°C})(27.0 \text{ °C} - 20.0 \text{ °C})$$

$$Q_{cw} \doteq +2\underline{9}2 \text{ J} \tag{2}$$

15-4a Components of a calorimeter

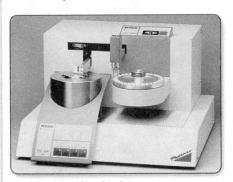

15-4b A research-grade calorimeter

The title of Example 15-3 refers to Song of Solomon 2:15: "Take us the foxes, the little foxes, that spoil the vines"—the little foxes are an allegorical reference to the small iniquities in our lives that rob us of our joy in our Lord just as the calorimeter affects, if only in a small way, the thermal energy transferred between its contents.

Calculate the heat transfer of the hot water:

$$Q_{hw} = (mc_{sp}\Delta t)_{hw} = [mc_{sp}(t_f - t_i)]_{hw}$$

$$Q_{hw} = (10.0 \text{ g})(4.18 \text{ J/g·°C})(27.0 \text{ °C} - 40.0 \text{ °C})$$

$$Q_{hw} \doteq -543.4 \text{ J} \tag{3}$$

Substitute results (2) and (3) into Equation (1) and solve for the heat transfer of the calorimeter:

$$Q_{hw} + Q_{cw} + Q_{cal} = 0 \text{ J}$$

$$Q_{cal} = -(Q_{hw} + Q_{cw})$$

$$Q_{cal} = -(-543.4 \text{ J} + 292 \text{ J})$$

$$Q_{cal} \doteq +251 \text{ J} \tag{3}$$

The calorimeter heat capacity is calculated from Equation 15.1:

$$C_{cal} = \frac{Q_{cal}}{\Delta t} = \frac{+251 \text{ J}}{(27.0 \text{ °C} - 20.0 \text{ °C})}$$

$$C_{cal} \doteq 35.8 \text{ J/°C}$$

$$C_{cal} \approx 36 \text{ J/°C}$$

Using a calorimeter with a known heat capacity, the heat capacity or specific heat of an unknown sample can be easily determined.

EXAMPLE 15-4

Experimentally Determining Specific Heat

A calorimeter with a heat capacity of 36.0 J/°C contains 20.0 g of water at 50.0 °C. A 10.0 g metal alloy sample is at 20.0 °C. When the sample is placed into the calorimeter, the final equilibrium temperature is 49.0 °C. What is the specific heat of the alloy?

Solution:

The system is the calorimeter, the alloy sample, and the hot water. Rearranging Equation 15.2 permits you to calculate specific heat of the alloy.

$$c_{sp \text{ alloy}} = \frac{Q_{alloy}}{(m\Delta t)_{alloy}} \tag{1}$$

You need to determine the heat transfer of the alloy sample when it is placed into the calorimeter in order to solve Equation (1). From the conservation of energy principle, you know that

$$Q_{hw} + Q_{alloy} + Q_{cal} = 0 \text{ J}. \tag{2}$$

First, determine the heat lost by the hot water:

$$Q_{hw} = (mc_{sp}\Delta t)_{hw} = [mc_{sp}(t_f - t_i)]_{hw}$$

$$Q_{hw} = (20.0 \text{ g})(4.18 \text{ J/g·°C})(49.0 \text{ °C} - 50.0 \text{ °C})$$

$$Q_{hw} \doteq -83.6 \text{ J}$$

Next, determine the heat lost by the calorimeter:

$$Q_{cal} = (C\Delta t)_{cal} = [C(t_f - t_i)]_{cal}$$

$$Q_{cal} = (36.0 \text{ J/°C})(49.0 \text{ °C} - 50.0 \text{ °C})$$

$$Q_{cal} \doteq -36.0 \text{ J}$$

Solve Equation (2) for the heat gained by the alloy (Q_{alloy}):

$$Q_{alloy} = -(Q_{hw} + Q_{cal}) \tag{3}$$

Substitute Equation (3) for Q_{alloy} into Equation (1) and determine $c_{sp \; alloy}$:

$$c_{sp \; alloy} = \frac{-(Q_{hw} + Q_{cal})}{(m\Delta t)_{alloy}} = \frac{-(Q_{hw} + Q_{cal})}{\left[m(t_f - t_i)\right]_{alloy}}$$

$$c_{sp \; alloy} = \frac{-\left[(-83.6 \text{ J}) + (-36.0 \text{ J})\right]}{(10.0 \text{ g})(49.0 \text{ °C} - 20.0 \text{ °C})}$$

$$c_{sp \; alloy} \doteq 0.412 \text{ J/g·°C}$$

$$c_{sp \; alloy} \approx 0.41 \text{ J/g·°C}$$

The procedure just described in this example is similar to many such heat balance calculations performed by scientists and engineers for a variety of reasons. For example, the latent heats required to bring about various phase transitions are determined using calorimeter procedures.

15.11 Heat and Phase Transitions

Every material has a characteristic specific heat capacity. You would expect that any thermal energy gained by an object would cause an increase in temperature, but this is not always the case.

Consider an ice cube that is forming in a freezer ice cube tray. We have taken the liberty of inserting a thermocouple junction (an electrical temperature sensor) in the water before it freezes. When we are ready for the demonstration, we quickly remove the instrumented ice cube from the tray, connect it to a temperature display instrument, and immerse the cube in a beaker of ice and water. A laboratory burner is lighted under the beaker.

The initial temperature is around −10 °C, the temperature of the freezer. We observe the temperature every ten seconds as the ice begins to warm. At first, the temperature rises rapidly. At 0 °C, we note that the cube's temperature stops rising. The ice cubes in the beaker, including our "wired" cube, are all shrinking in size as they melt. When the ice fragments are nearly gone, we observe the temperature starts rising again, but more slowly than before. The water's temperature continues to rise at a steady rate until the water nears the boiling point (numerous small bubbles form and detach from the bottom of the beaker). Then the rate of temperature rise drops off and temperature becomes constant when the water reaches a rolling boil.

You may wonder why the temperature does not rise continuously during the entire period. The flame has been adding thermal energy to the beaker at the same rate the whole time. The answer lies in the water's changes of state. The thermal energy the water absorbed without changing temperature was used to break the bonds between water molecules in the ice. The same energy is released when the phase change is reversed. That is, the amount of thermal energy absorbed when ice melts is released when water freezes. Similarly, the amount of thermal energy absorbed when water vaporizes is released when steam condenses. This amount of

15-5 Engineers use calorimetric procedures in order to determine the rate of heat transfer and the efficiency of their power plants.

thermal energy per unit mass is a constant for a given phase change of each substance. The **latent** (hidden) **heat of fusion (L_f)** of a substance is the amount of thermal energy required to melt 1 kg of the substance at its melting point. The **latent heat of vaporization (L_v)** is the amount of thermal energy required to vaporize 1 kg of a substance at its boiling point. The values of these quantities are determined at standard conditions so that they may be compared. Table 15-2 lists some heats of fusion and vaporization for selected substances.

Heats of fusion and vaporization are determined by the following relationships:

$$L_f = \frac{Q_{melt}}{m} \tag{15.3}$$

$$L_v = \frac{Q_{boil}}{m} \tag{15.4}$$

Units of L_f and L_v are kJ/kg (or J/g).

TABLE 15-2

Heats of Fusion and Vaporization with Associated Phase Transition Temperatures for Selected Substances

Substance	Melting Point t_m (°C)	Heat of Fusion L_f (kJ/kg)	Boiling Point t_b (°C)	Heat of Vaporization L_v (kJ/kg)
Acetone	−94.8	98.0	56.1	501
Aluminum	660.3	397.0	2520	10 900
Carbon Dioxide	−56.6	205	—	—
Copper	1084.6	208.7	2563	4726
Ethanol	−114.1	78.4	78.3	837
Gold	1064.2	63.7	2857	1645
Iron	1538	247.3	—	—
Lead	327.5	23.0	1750	866
Mercury	−38.8	1.4	356.7	295
Nitrogen	−210.0	25.3	−195.8	198
Oxygen	−218.8	13.8	−183.0	213
Silver	961.8	104.7	2163	2323
Tungsten	3422	284.5	5900	4820
Water	0.0	333.5	100.0	2256

EXAMPLE 15-5

Determining Heat of Fusion: A Single Phase Change

When 10.0 g of ice is placed in a calorimeter with a heat capacity of 36.0 J/°C, the equilibrium temperature is 0.0 °C. Then 20.0 g of water at 50.0 °C is added, and the new equilibrium temperature is 5.2 °C. The ice has melted. Calculate the heat of fusion (L_f) of ice.

Solution:
The heat of fusion is calculated from Equation 15.3:

$$L_f = \frac{Q_{melt}}{m} \text{ or}$$

$$Q_{melt} = mL_f \tag{1}$$

Note which quantities are changing and which are constant. The ice melts, but its temperature does not change. The cold meltwater from the ice (subscript cw) warms from 0.0 °C to 5.2 °C. The warm water (subscript ww) cools from 50.0 °C to 5.2 °C. The calorimeter experiences the same temperature change as the cold water. Therefore, heat flows from the warm water into the ice to melt it and into the resulting cold water and the calorimeter.

From the conservation of energy in an adiabatic container you know that

$$Q_{cal} + Q_{cw} + Q_{ww} + Q_{melt} = 0 \text{ J}.$$

Solve for Q_{melt}:

$$Q_{melt} = -\left(Q_{cal} + Q_{cw} + Q_{ww}\right)$$

Substitute Equation (1) for Q_{melt} and solve for L_f:

$$m_{ice}L_f = -\left(Q_{cal} + Q_{cw} + Q_{ww}\right)$$

$$L_f = \frac{-\left(Q_{cal} + Q_{cw} + Q_{ww}\right)}{m_{ice}} \tag{2}$$

Now calculate each of the heat expressions in Equation (2):

Calorimeter: $\quad Q_{cal} = (\Delta t C)_{cal} = \left(+5.2 \text{ °C}\right)\left(36.0 \text{ J/°C}\right)$

$$Q_{cal} \doteq +18\underline{7} \text{ J}$$

Cold water: $\quad Q_{cw} = (mc_{sp}\Delta t)_{cw} = (10.0 \text{ g})(4.18 \text{ J/g·°C})(+5.2 \text{ °C})$

$$Q_{cw} \doteq +21\underline{7} \text{ J}$$

Warm water: $\quad Q_{ww} = (mc_{sp}\Delta t)_{ww} = (20.0 \text{ g})(4.18 \text{ J/g·°C})(-44.8 \text{ °C})$

$$Q_{ww} \doteq -374\underline{5} \text{ J}$$

Solve Equation (2) for L_f:

$$L_f = \frac{-[18\underline{7} \text{ J} + 21\underline{7} \text{ J} + (-374\underline{5} \text{ J})]}{10.0 \text{ g}}$$

$$L_f \doteq 3\underline{3}4 \text{ J/g}$$

$$L_f \approx 330 \text{ J/g}$$

This result obtained in a typical high school laboratory calorimeter agrees reasonably well with the accepted value of 333.5 J/g.

EXAMPLE 15-6

From Ice to Steam: Thermal Energy Needed for Two Phase Changes

How much thermal energy is required to raise the temperature of 10.0 g of water from −10.0 °C to 110.0 °C?

Solution:

This problem involves five parts. We must determine the thermal energy changes for the following:

(1) Heating the ice from -10.0 °C to 0.0 °C (Q_{ice});

(2) Converting the ice to liquid water at 0.0 °C (Q_{melt});

(3) Heating the liquid water from 0.0 °C to its boiling point at 100.0 °C (Q_w);

(4) Converting the liquid water to gaseous water (steam) at 100.0 °C (Q_{boil});

(5) Heating the steam from 100.0 °C to 110.0 °C (Q_s).

$$Q_{total} = Q_{ice} + Q_{melt} + Q_w + Q_{boil} + Q_s \qquad (1)$$

Heats of fusion and vaporization can be obtained from Table 15-2. Specific heats can be obtained from Table 15-1.

Determine the heat transferred for each term on the right side of Equation (1):

Heat ice: $\quad Q_{ice} = (mc_{sp}\Delta t)_{ice} = (10.0 \text{ g})(2.09 \text{ J/g·°C})(+10.0 \text{ °C})$

$\qquad\qquad Q_{ice} = +20\underline{9}.0 \text{ J}$

Melt ice: $\quad Q_{melt} = m_{ice}L_f = (10.0 \text{ g})(333.5 \text{ J/g})$

$\qquad\qquad Q_{melt} = +33\underline{3}5 \text{ J}$

Heat water: $\quad Q_w = (mc_{sp}\Delta t)_w = (10.0 \text{ g})(4.18 \text{ J/g·°C})(+100.0 \text{ °C})$

$\qquad\qquad Q_w = +41\underline{8}0 \text{ J}$

Boil water: $\quad Q_{boil} = m_w L_v = (10.0 \text{ g})(2256 \text{ J/g})$

$\qquad\qquad Q_{boil} = +22\,\underline{5}60 \text{ J}$

Heat steam: $\quad Q_s = (mc_{sp}\Delta t)_s = (10.0 \text{ g})(2.01 \text{ J/g·°C})(+10.0 \text{ °C})$

$\qquad\qquad Q_s = +20\underline{1}.0 \text{ J}$

Sum all of the heat quantities to find the total thermal energy transferred:

$$Q_{total} = 20\underline{9} \text{ J} + 33\underline{3}5 \text{ J} + 41\underline{8}0 \text{ J} + 22\,\underline{5}60 \text{ J} + 20\underline{1} \text{ J}$$

$$Q_{total} \doteq 30\,4\underline{8}5 \text{ J}$$

$$Q_{total} \approx 30\,500 \text{ J} \text{ (Allowable precision is to nearest 100 J.)}$$

Problem-Solving Strategy 15.3

Calculating the thermal energy required to change a solid substance to a vapor, or vice versa, can be broken down into five heat-transfer steps. The key to success is to use the proper specific heat value for the corresponding phase and the correct latent heat for each phase change. Be sure to observe proper significant digit rules when summing the final result.

15B Section Review Questions

1. Discuss why, when evaluating heat processes involving water, it might be more convenient to use the unit calorie.

2. Define and clearly differentiate the terms temperature, heat, and thermal energy.

3. Discuss the difference between the heat capacity of a cylinder of steel and its specific heat.

4. As thermal energy is added to a solid, why does its average temperature stop rising when it begins to melt?

5. What is the quantity of heat required to melt a gram of a substance at its melting point called? What do we call the quantity of heat per gram that must be removed to freeze the substance at its freezing point?

⊛6. In a laboratory exercise, you are required to tentatively identify a 65.0 g sample of metal from Table 15-1 by its specific heat, using a calorimeter. The calorimeter and the 50.0 g of distilled water that it contains are both at the temperature of 24.0 °C. The initial temperature of the metal just before it is placed into the calorimeter is 98.5 °C. The final temperature of the calorimeter and its contents is 26.5 °C. The calorimeter's heat capacity is 39.5 J/°C. What metal is the sample made of?

⊛7. How much heat must be lost in order to completely condense 100.0 g of nitrogen at its condensation point?

⊛8. Compare the heat loss required to condense 1.0 g of steam to water at 100.0 °C and the heat lost to cool 1.0 g of water 1.0 °C at 100.0 °C. The specific heat of water at 100.0 °C is 4.22 J/g·°C. Which, gram for gram, is more damaging to the unprotected skin—water at its boiling point or condensing steam?

15C MECHANISMS FOR HEAT TRANSFER

There are three mechanisms for the transfer of thermal energy from one system to another: conduction, convection, and radiation.

15.12 Conduction

Hold a stainless steel spoon in a pot of boiling soup. You will begin to feel the spoon handle growing warm. Eventually, you will either have to drop the spoon or withdraw it from the soup to prevent being burned. This is an example of **conduction.** The hot soup particles accelerate the motion of the metal atoms in the bowl of the spoon. These rapidly vibrating molecules collide with the adjacent, slower-moving molecules within the metal, transmitting kinetic energy to the atoms farther up the handle. The rapid motion and associated kinetic energy is transmitted from atom to atom along the handle to your fingers.

Some materials conduct thermal energy easily, while others act as thermal insulators. For example, if you had used a wooden spoon to stir the soup, you could have held onto the spoon indefinitely. To understand the difference, you must return to the atomic model. Recall that the major subatomic particles are protons, neutrons, and electrons. The protons and neutrons are in the nucleus, but the electrons surround the nucleus. Some substances hold their electrons firmly. Other substances, especially metals, hold some of their electrons so loosely that it takes little energy to remove them. In fact, some electrons are so independent of their nuclei that they are called **free electrons.** These free electrons readily transmit kinetic energy between atoms throughout a substance. Substances with free electrons are therefore good thermal conductors. Substances lacking free electrons do not conduct thermal energy well, so they are thermal insulators.

15.13 Convection

Air is a poor thermal conductor. However, a hot water radiator surrounded by air can heat an entire room. This is possible because air transports thermal energy by

—ceramic cooktop

heating element

15-6a Conduction

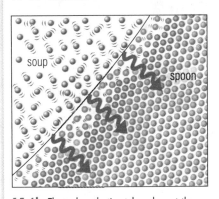

soup

spoon

15-6b Thermal conduction takes place at the molecular level.

Substances that contain free electrons, such as metals, generally are good conductors of heat. Substances that do not contain free electrons, such as most solid ionic and molecular compounds, are generally poor thermal conductors (good insulators).

→ this makes medals good thermal conductors

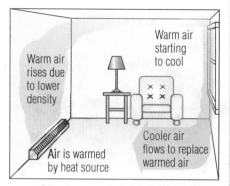

15-7 Convection

convection. **Convection** moves thermal energy by physically moving material that has thermal energy. Gravity is the main force that causes this motion. The hot radiator transmits heat to the air mass directly around it through conduction and *radiation* (see subsection 15.14). This warmer air expands, becoming less dense, and is displaced upward by the colder, denser air around it, which replaces the warmer air next to the radiator. The cold air is soon heated and replaced with other cold air from the room. The cycle continues, setting up a convection current of rising and falling air in the room. Eventually the warmed air cools, falls, and returns to the radiator to be heated again. In order to make most efficient use of this convection process, a well-designed house has its heat sources near the floor.

Water, as well as air, transmits thermal energy by convection. In bodies of water, the cooling process is especially interesting. Water is densest at 4 °C. A body of water—a lake, for example—cools from the top. When the top layer cools to 4 °C, it sinks and is replaced by warmer water. Since water at 4 °C is densest, no water can cool to below 4 °C until all the water in the lake has reached that temperature. When all the water is 4 °C, the surface layer cools to lower temperatures, and the lake freezes over from the top down. Since ice is less dense than liquid water at the same temperature, it floats. The net effect of this phenomenon is that freezing is delayed throughout the depth of the lake as long as possible. In deep lakes, even after the surface freezes, some unfrozen water exists under the ice all winter. This phenomenon, ordained by the Creator, preserves aquatic life through the winter months.

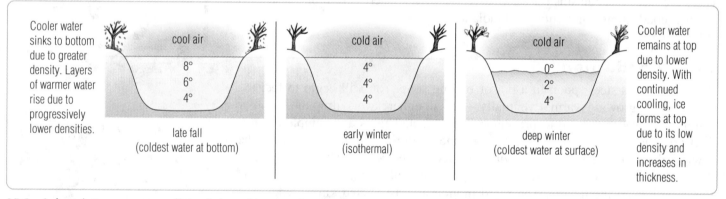

15-8a Surface-to-bottom temperature profile in a freshwater lake as it cools in winter

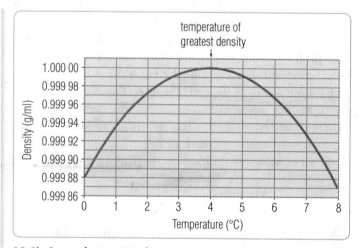

15-8b Density of water near its freezing point

15.14 Radiation

The earth is warmed by thermal energy from the Sun. However, since the density of the particles between the Sun and the earth is so low, the Sun's thermal energy cannot be transmitted by conduction or convection. The Sun's electromagnetic energy reaches the earth by **radiation**, which is converted to thermal energy when it is absorbed by matter. The Sun is not the only object that transfers thermal energy by radiation. Every object with a temperature greater than absolute zero transfers thermal energy in this way. "Cold" objects emit less radiation than "hot" objects. **Stefan's law** gives the correspondence between absolute temperature (T) and the power of its radiant energy (S),

$$S = \sigma T^4, \tag{15.5}$$

where σ is a proportionality constant, called the **Stefan-Boltzmann constant** ($\sigma = 5.67 \times 10^{-8}$ W/(m^2·K^4)), and T is the Kelvin temperature of the object's surface. Because S is proportional to temperature to the fourth power, an object radiates sixteen times as much energy at a given absolute temperature as it radiates at half that temperature.

Radiant energy, of which visible light is one form, is different from thermal energy, which can exist only in matter. Radiant energy is transformed into thermal energy when it is absorbed into matter. It is important to distinguish between these two forms of energy.

Josef Stefan (1835–93) was an Austrian physicist known for discovering the relationship between the radiant energy of an object and the fourth power of its absolute temperature (Stefan's law).

EXAMPLE 15-7

Thermal Energy and Radiation: Stefan's Law

What is the ratio of the radiant power of an iron bar at 600. K to its radiant power at 200. K?

Solution:

$$S = \sigma T^4$$

Find the radiant power at 600. K.

$$S_{600\ K} = \sigma(600.\ K)^4\ W/m^2$$

Find the radiant power at 200. K.

$$S_{200\ K} = \sigma(200.\ K)^4\ W/m^2$$

Find the ratio of the radiant energies.

$$\frac{S_{600\ K}}{S_{200\ K}} = \frac{\sigma(600.\ K)^4}{\sigma(200.\ K)^4} = \frac{\sigma(3)^4 \cdot (200.\ K)^4}{\sigma(200.\ K)^4}$$

$$\frac{S_{600\ K}}{S_{200\ K}} = (3)^4 = 81$$

A threefold increase in the absolute temperature of the bar yields an eighty-onefold increase in the rate of thermal energy radiated.

Help your Christian brothers and sisters to radiate Christ by noticing and commending their Christian virtues—not for their glory but for Christ's. "Hearing of thy love and faith, which thou hast toward the Lord Jesus, and toward all saints; That the communication of thy faith may become effectual by the acknowledging of every good thing which is in you in Christ Jesus." (Philem. 5–6)

Black objects absorb radiant energy, including visible light. Therefore a black object is also the best absorber of radiant energies that are the most easily converted to thermal energy. (That is why solar collectors on rooftops are painted black.) For similar reasons, a black object is the best radiator. Since lighter objects do not absorb energy as easily, there is less energy to re-radiate. A perfect (ideal) radiator and absorber is called a **blackbody.**

15C Section Review Questions

1. Which mechanism of heat transfer involves the physical movement of matter?
2. Which mechanism(s) of heat transfer require(s) a material medium?
3. Why is it *not* correct to say that the Sun is radiating thermal energy?
4. What seems to be the mechanism for the rapid transmission of thermal energy *within* metals?
5. How is the radiant energy of an object related to its Kelvin temperature?
⊙6. How much radiant power (in W/m^2) does a person's skin emit if its temperature is 32.0 °C?

Chapter Review

In Terms of Physics

kinetic-molecular theory	15.2	latent heat of fusion (L_f)	15.11
Brownian motion	15.5	latent heat of vaporization (L_v)	15.11
caloric theory	15.6	conduction	15.12
calorie (cal)	15.7	free electrons	15.12
heat capacity (C)	15.8	convection	15.13
heat (Q)	15.8	radiation	15.14
specific heat (c_{sp})	15.9	Stefan's law	15.14
adiabatic	15.10	Stefan-Boltzmann constant (σ)	15.14
calorimeter	15.10	blackbody	15.14

Problem-Solving Strategies

15.1 (page 333) Heat-balance problems often result in slightly different calculated values of heat lost versus heat gained. This is due to the different precisions of the various measured quantities and the rounding conventions involved in the calculations. Differences of a few percent are acceptable for the purposes of this course.

15.2 (page 335) Solving heat-balance problems requires finding the change in heat (both magnitude and sign) of every part of the system within the adiabatic boundary. The conservation of heat within an adiabatic boundary assures you that the sum of all heat transfers will be zero.

15.3 (page 338) Calculating the thermal energy required to change a solid substance to a vapor, or vice versa, can be broken down into five heat-transfer steps. The key to success is to use the proper specific heat value for the corresponding phase and the correct latent heat for each phase change. Be sure to observe proper significant digit rules when summing the final result.

Review Questions

1. Give two evidences that matter is in constant motion.
2. What is the name of the discredited theory that heat is an invisible fluid?
3. Describe briefly one experiment that cast doubt on the caloric theory of heat.
4. What theory states that thermal energy is energy due to the motion of particles of matter?
5. What is the difference between heat capacity and specific heat?
6. Discuss the factors that determine the heat capacity of an object.
7. What laboratory device can thermally isolate a process from its surroundings?
8. Briefly describe the steps you would take to determine the specific heat of a small sample of a metal.
9. What assumption must you make in order to apply the principle of the conservation of thermal energy to a process?
10. Which contains more thermal energy, 1 g of liquid water at 100 °C or 1 g of steam at 100 °C? Explain your answer.
11. A glass of ice water rests on a table in a 20.0 °C room. From the time the ice begins to melt to the time the last bit of ice melts, how much does the temperature of the water change? Why?

12. You place a cold frying pan on a hot electric burner. A minute later, when you touch the pan's metal handle, you get burned. By what process did thermal energy travel from the burner to your hand?

13. How is thermal energy transferred through a vacuum?

14. If your home is heated by convection, where in a room should your source of thermal energy be located?

15. Is the rate that an object radiates energy twice as much at 2000 K as it is at 1000 K? If not, what is the relative rate at the higher temperature?

True or False (16–31)

16. Temperature alone does not determine the thermal energy of an object.

17. The dissolving of a sample of a solid substance is a key evidence for particle motion in matter.

18. Gas particles move faster than liquid particles.

19. According to the caloric theory, cold objects absorb caloric from hotter objects.

20. Joule appealed to the logic that energy cannot be converted to matter, and vice versa, by purely mechanical means, to discredit the caloric theory of heat.

21. The mechanical equivalence of heat is demonstrated where 1 N·m of work on a gram of water will raise its temperature 1 °C.

22. It is possible for two objects made of different materials to have the same heat capacity.

23. Speed is to momentum as temperature is to thermal energy.

24. The quantity of heat lost or gained by an object is inversely proportional to its temperature change.

25. Under normal conditions, the specific heat of most solids is nearly constant.

26. The term *adiabatic* means "constant thermal energy."

27. The thermal energy that must be lost in order to freeze a kilogram of a substance at its freezing point is its latent heat of fusion.

28. In order to transfer heat between two points by conduction, there must be a continuous path of matter between the points.

29. In Chemistry you learned that compounds of nonmetal elements usually form bonds where their electrons are closely shared between the atoms. You would expect such compounds to easily conduct heat.

30. Convection can take place only in fluids.

31. Only glowing objects radiate energy.

⊙32. A 20.0 g lump of chromium has a heat capacity of 9.03 J/°C. What is the specific heat of chromium?

⊙33. What is the heat capacity of a 100. g bar of silver?

⊙34. If a spool of copper wire has a heat capacity of 13.6 J/°C, what is the spool's mass?

⊙35. A 30.0 g gold coin and a 30.0 g silver coin sitting on a windowsill each receive 63.0 J of energy from the Sun. If they were originally at the same temperature, which will be warmer after the energy is absorbed?

⊙36. A 425 g glass jar has a temperature of 20.0 °C. If c_{sp} glass = 0.837 J/g·°C, what is the jar's temperature after it absorbs 3560 J of heat?

⊙37. A 27.5 g asbestos pad's temperature rises from 25.0 °C to 50.0 °C when 1.05×10^5 J of heat is absorbed.
   a. What is the pad's heat capacity?
   b. What is the specific heat of asbestos?

⊙38. A 10.0 g lump of lead is placed in boiling water (100.0 °C) until its temperature is constant. Then the lead is removed, quickly dried, and placed into 10.0 g of water at 20.0 °C in an adiabatic calorimeter. The final temperature of the lead and water is 22.4 °C.
   a. How much heat does the water gain?
   b. How much heat does the lead lose?
   c. What is the lump's heat capacity?
   d. What is the specific heat of lead?

⊙39. 25.0 g of boiling water is added to a calorimeter near room temperature (20.0 °C). The final temperature of the calorimeter and water is 81.5 °C.
   a. How much heat does the water lose?
   b. What is the calorimeter's heat capacity?

⊙40. How much heat does 1.00 g of steam/water release as it cools from 105.0 °C to 55.0 °C?

⊙41. A 10.0 g lump of metal at 77.6 °C is dropped into a calorimeter containing 10.0 g of ice and 10.0 g of water at 0.0 °C. 1.0 g of the ice melts before thermal equilibrium is reestablished.
   a. What is the final temperature of the mixture?
   b. How much heat does the ice that melts gain?
   c. How much heat do the water and calorimeter gain?
   d. What is the lump's heat capacity?
   e. What is the metal's specific heat?

⊙42. In order to double the rate of radiant energy output of a hot plate, how much hotter must the plate be? Express your answer to three significant digits.

⊙⊙43. 50.0 g of liquid oxygen at its boiling point is sealed in a 50.0 g adiabatic aluminum calorimeter cup initially at room temperature (25.0 °C). Is there enough thermal energy in the cup to vaporize the entire quantity of liquid oxygen at its boiling point? Assume pressure in the calorimeter remains constant. (*Hint:* Converting all temperatures to kelvins will make this problem easier to solve.)

Nicolas Sadi Carnot
(1796–1832)

Early steam engines were too inefficient to be practical, except in a few rare cases. The first engines were used in mines to pump out water, but the engines' massive, coal-fired burners converted only 1% of the heat energy into mechanical energy. After James Watt invented the first practical steam engine in 1769 (with a 2% efficiency), other men tried to make further improvements. Among them was Sadi Carnot, a French military engineer whose theories proved more useful to science than to engineering.

Nicolas Léonard Sadi Carnot was actively involved in the French government with the rest of his family. His father served on the five-man Directory that ruled France before Napoleon's reign, and Napoleon later appointed him minister of war. Years later, Carnot's nephew became president of France's Third Republic. Carnot himself was an engineer in the French army reserves most of his life.

Carnot became intrigued with the problem of designing good steam engines. Instead of becoming lost in the mechanical detail, as other engineers had done, Carnot reduced the operation of a steam engine to its bare essentials: a source of heat (the boiler), a receiver of unused heat (the condenser), and a working substance (water). He realized that the working substance does not matter in an ideal steam engine; what matters is the difference in temperature between the source of heat and the receiver. In 1824 he published his theories in a short, nontechnical paper entitled *Reflections on the Motive Power of Fire*.

Many details of the work are mistaken because Carnot followed the caloric theory of heat. But the theorems that he deduced about the "Carnot cycle" were significant. Yet Carnot's discoveries were ignored until long after his death, when Lord Kelvin and Rudolf Clausius rediscovered his work. His ideas helped them to discover the second law of thermodynamics.

Although Carnot continued his scientific investigations after 1824, he never published his later findings. He died in relative obscurity at the age of 36, when a cholera epidemic struck Paris. Years later his brother published Carnot's old notes, proving that Joule was not the first man to discover that a given amount of mechanical work is equivalent to a certain amount of heat. However, Carnot's reputation still rests on the insight and clarity of his classic paper of 1824.

16A   The Zeroth and First Laws   347
16B   The Second and Third Laws   355
16C   Entropy and Its Consequences   361
Facet: Growth and the Second Law   364

New Symbols and Abbreviations

internal energy (U)	16.1
thermal efficiency (ε)	16.11
performance coefficient (K)	16.12
entropy (S)	16.14

Steam is a convenient medium for transforming thermal energy into other useful forms of energy. The efficient use of steam requires knowledge of thermodynamics.

Thermodynamic Laws

16

The apostle Peter relates several important truths in this verse. First, corruption (biological disorder, decay, disease) is in the world because the sin of Adam and Eve brought the penalty of death onto the entire creation. The word *corruption* describes the inevitable end of all natural processes. You will learn in this chapter that the preceding statement is the broadest statement of the second law of thermodynamics. If God were not to intervene, as promised in His Word, the entire universe would eventually succumb to this principle. However, the promise in II Peter 1:4 is that those who are saved will escape this judgment and will have eternal bodies that are not subject to this law. Do you have the assurance of this promise?

16A THE ZEROTH AND FIRST LAWS

16.1 Introduction

You have so far studied two forms of energy—mechanical and thermal. Mechanical energy is a property of visible objects. Two types of mechanical energy are potential energy and kinetic energy. Potential energy, one result of work, can be changed to kinetic energy—energy of motion. A motionless ball poised at the top of an inclined plane has potential energy. As the ball rolls down the inclined plane, it gains kinetic energy as it loses potential energy. When the ball reaches the bottom of the plane, the last of its gravitational potential energy disappears and it rolls along the flat surface with only kinetic energy. Total mechanical energy (E) is the sum of the kinetic and potential energy due to the motion and position of physical objects.

Once the appropriate reference points for determining the system's motion and position are established, the system's total mechanical energy can be known. Subsequently, only the changes of the system's total mechanical energy (ΔE) are important from a physics perspective.

16-1 Mechanical potential energy can change to mechanical kinetic energy.

Thermal energy is due to the rapid, random motion of the atomic, molecular, and subatomic particles of matter. Like mechanical energy, thermal energy can be subdivided into potential energy and kinetic energy. The particles of a substance are constantly moving; therefore, they have kinetic energy. Their average kinetic energy is proportional to the temperature of the substance.

The innumerable particles of a substance also have potential energies, but it is extremely complicated to determine the reference points to establish these energies. The sum total of the particle kinetic and potential energies is called **internal energy (U).** As with total mechanical energy, once the internal energy of a system is established or assumed for a given state of the system, only the change in internal energy becomes important. The changes of internal energy (ΔU) are usually much easier to calculate than the internal energy states at the beginning and end of a process.

16.2 The Zeroth Law of Thermodynamics

Some materials conduct thermal energy well, while other materials are good insulators. An ideal insulating wall through which no thermal energy can pass is called an *adiabatic* wall. An ideal conductor of thermal energy is called a **diathermic** wall. Although no real adiabatic or diathermic materials exist, the concepts are useful for simplifying discussions.

As you have probably noticed, when you place a hot object in contact with a cold object, the cold object becomes hotter and the hot object becomes colder. In time, the two objects reach the same temperature. That is, they reach a state of **thermal equilibrium.** Thermal energy has flowed from the hot object to the cold object. As you learned in Chapter 15, the flow of thermal energy is called *heat.*

Figure 16-2 depicts what is called the **zeroth law of thermodynamics.** This law states that two systems that are in thermal equilibrium with a third system must be in thermal equilibrium with each other. The law received its name because it is more basic than either the first or second law but was formulated after the other laws had been named.

In figure (a), system A is separated from system B by an adiabatic wall, and both are linked to system C by diathermic walls. Although A and B cannot exchange thermal energy with each other, both can exchange thermal energy with C. In figure (b) the walls have been switched so that A and B can exchange thermal energy with each other, but C is thermally insulated from both A and B. According to the zeroth law, if no energy exchange occurs in figure (a), then none will occur in figure (b).

The principle becomes still more understandable when you view it in terms of temperature. In figure (a), systems A and B are at the same temperature as C. (You must, of course, assume that every part of a given system is at the same temperature as every other part.) System A must therefore be at the same temperature as System B. When the adiabatic wall between them is replaced with a diathermic wall, neither system will warm (that is, transfer thermal energy to) the other. Moreover, since all three systems are at the same temperature, no thermal energy can flow in any direction, regardless of how the walls are placed.

16.3 The General Law of Conservation of Energy— The First Law

Before progressing further, we need to gather together what we know about energy transfers in order to establish a fundamental or general energy conservation principle. Energy can be added to a system by either of two processes: (a) mechanical

Just as hot and cold objects have the potential to change each other's temperature, Proverbs 22:24–25 cautions against friendship with an angry man, "lest thou learn his ways and get a snare to thy soul." We are influenced by our friends for good and bad, so we must pick our friends wisely.

The **zeroth law of thermodynamics** states that if two systems are in thermal equilibrium with a third, then they are in thermal equilibrium with each other.

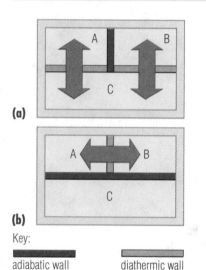

(a)

(b)

Key:

adiabatic wall (insulating material)

diathermic wall (conducting material)

16-2 Three systems in thermal equilibrium demonstrating the zeroth law of thermodynamics. (a) If systems A and B are individually in thermal equilibrium with system C, then (b) they must be in thermal equilibrium with each other.

work (W') through the application of external, nonconservative forces or (b) heat transfer (Q) through temperature differences with the system's surroundings. The energy transferred to the system by these means is seen as a change of the internal energy of the system (ΔU) or as a change in the system's total mechanical energy (ΔE) or both. The *general energy conservation law* can be written

$$Q + W' = \Delta U + \Delta E. \quad (16.1)$$

When we discussed basic mechanics, we assumed that no heat transfers took place and no changes of internal energy occurred ($\Delta U = 0$ J). When discussing basic thermodynamics principles, we will make the assumption that the total mechanical energy of the system is constant ($\Delta E = 0$ J). Therefore, the only forces doing work on the system are nonconservative forces. Equation 16.1 can then be rewritten as

$$Q + W' = \Delta U. \quad (16.2)$$

Rather than being concerned about the work done on a system, as we were in our study of mechanics, it is more useful in thermodynamics to evaluate the effects of a system on its **surroundings;** therefore the work notation in Equation 16.2 needs to be modified slightly. Recall from our study of Newton's laws that every force on a system is paired with an opposite force acting on another system. It follows that, if a system's surroundings do mechanical work on it through the application of a nonconservative force, then the system must also do work on its surroundings via the reaction force. The work by the system on its surroundings is therefore the negative of the work on the system:
Equation 16.2 then becomes

$$Q - W = \Delta U, \text{ or}$$
$$Q = \Delta U + W. \quad (16.3)$$

Equation 16.3 is the mathematical statement of the **first law of thermodynamics.** The first law extends the principle of the conservation of total mechanical energy to internal energy as well. We will investigate the applications and limitations of the first law in the following discussion of thermodynamic processes.

16.4 Heat Engines

Remember from Chapter 15 that heat refers to thermal energy moving from one system to another. Thermal energy can be changed into mechanical energy. An engine called a **heat engine** can do mechanical work by absorbing and discharging heat.

The simplest "machine" that converts thermal energy to work is an expanding gas. For a gas to expand usefully, it must be confined in an expandable container. One such container is a cylinder fitted with a gas-tight piston. Assuming that the piston is massless and frictionless simplifies calculations. As the gas expands, it is no longer in thermal equilibrium. It heats or cools unevenly. To avoid this problem, we allow the gas to expand in extremely minute steps, letting it return to thermal equilibrium between steps. Consequently, the gas expands without ever being far from thermal equilibrium. A process that proceeds in this way is called a *quasi-static process.*

For quasi-static processes, the gas pressure inside the cylinder is at all times in equilibrium with the external pressure. When the gas is compressed from a volume V_0 to a volume V by an external pressure, P, work is done on the gas. (You can verify that energy is added to the gas because it warms when it is compressed.)

The symbol W' represents the mechanical work done *on* a thermodynamic system by external nonconservative forces. This symbol is different from that used in earlier chapters because in studying thermodynamics, it is usually more important to analyze the work a system does on its surroundings (W).

The **surroundings** of a system are everything in the universe that is outside the system's boundaries.

$$W = -W'$$

The work done on a thermodynamic system is the negative of the work done on the system's surroundings. This is a direct consequence of Newton's third law of motion: every force is balanced by an equal but opposite force on a different system.

The **first law of thermodynamics** states that the heat transferred to or from a system is equal to the sum of the change of the system's internal energy and the work the system does on its surroundings.

$$Q = \Delta U + W$$

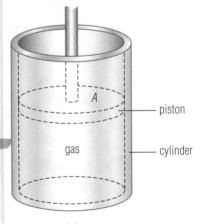

16-3 A simple heat engine

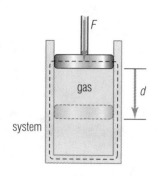

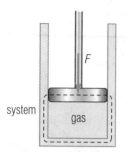

16-4 A gas is compressed by a pressure, P.

The simplified formula for work is

$$W = Fd. \qquad (16.4)$$

Pressure is force per unit area, so force exerted by the gas equals pressure times the cross-sectional area of the piston:

$$F = P_{gas}A$$

Substituting for force in Equation 16.4,

$$W = (P_{gas}A)d.$$

But the cross-sectional area of the cylinder times the displacement distance of the piston is just the volume change (ΔV) of the gas in the cylinder; so the work done *by* the gas as it is compressed by a constant external pressure is

$$W = P_{gas}\Delta V.$$

The change of volume when a gas is compressed is negative. Therefore, the work done by the gas *on its surroundings* (see subsection 16.3) is negative when it is compressed. This result makes sense, since the surroundings are losing energy and the gas is gaining energy.

When gas expands against the applied pressure, the gas does *positive* work on its surroundings. (A gas that does work by expanding loses thermal energy, and it cools.) In general, for expansion or contraction against a constant pressure, the work done *by* the gas is

$$W = P(V - V_0).$$

The pressure against the gas is not always constant. When the pressure varies during an expansion or a contraction, the work is the sum of the work done in each step. In a quasi-static process, you can assume that for a small volume change the pressure is constant; so

$$W_{total} = (P\Delta V)_1 + (P\Delta V)_2 + (P\Delta V)_3 + \ldots + (P\Delta V)_n. \qquad (16.5)$$

16.5 *P-V* Diagrams

Equation 16.5 becomes easier to solve if you plot the pressure versus the volume on a graph called a ***P-V* diagram.** For instance, in Figure 16-5 the gas is originally at 0.50 L and 1.00×10^5 Pa. It expands to 1.00 L ($\Delta V = 0.50$ L) at 1.00×10^5 Pa. Then while the volume remains constant the pressure changes to 0.75×10^5 Pa. The gas expands to 1.50 L at 0.75×10^5 Pa. The pressure is again lowered to 0.50×10^5 Pa. Finally, the gas expands to 2.00 L. The total work done by the gas is as follows:

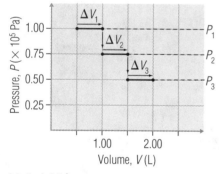

16-5 A *P-V* diagram

$$W_{total} = (P\Delta V)_1 + (P\Delta V)_2 + (P\Delta V)_3$$
$$W_{total} = (1.00 \times 10^5 \text{ Pa})(+0.50 \text{ L}) + (0.75 \times 10^5 \text{ Pa})(+0.50 \text{ L})$$
$$+ (0.50 \times 10^5 \text{ Pa})(+0.50 \text{ L})$$
$$W_{total} = 0.50 \times 10^5 \text{ Pa} \cdot \text{L} + 0.375 \times 10^5 \text{ Pa} \cdot \text{L} + 0.25 \times 10^5 \text{ Pa} \cdot \text{L}$$
$$W_{total} = 1.125 \times 10^5 \text{ Pa} \cdot \text{L}$$
$$W_{total} \approx 1.13 \times 10^5 \text{ Pa} \cdot \text{L}$$

Since 1 Pa = 1 N/m^2, and 1 L = 0.001 m^3,

$$\frac{1.13 \times 10^5 \text{ N}}{\text{m}^2} \left| \frac{0.001 \text{ m}^3}{1 \text{ L}} \right. = 113 \text{ N} \cdot \text{m} = 113 \text{ J}$$

Notice that each term in the work equation is equal to the block of area under the corresponding step of the graph of the process on the *P-V* diagram (Figure 16-6). That is, the area under the curve representing the process on a *P-V* diagram is equal to the absolute value of the work done by the gas during the process. This result is true for any confined gas process (expansion or contraction). The sign of the work depends on whether the gas gains or loses energy. If the gas expands, it does work on its surroundings, so the sign is positive. If the gas contracts or is compressed, the surroundings do work on it; the sign is negative.

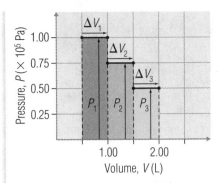

16-6 An area under a curve on a *P-V* diagram indicates the work done by a process between the two states represented by the endpoints of the curve.

EXAMPLE 16-1

Work of an Expanding Gas

A gas originally at 1.00×10^5 Pa and 1.00 L expands until its volume is 2.00 L and its pressure is 0.50×10^5 Pa. How much work is done by the gas?

Solution:

The gas in the cylinder is the system. Work is determined relative to the system's surroundings.

The work by the gas is related to the numerical value of the shaded area of Figure 16-7. The easiest way to find the area is to calculate the area of the triangle above the dotted line and add it to the area of the rectangle below the dotted line.

The area of the triangle is

$$A_\triangle = \tfrac{1}{2}bh, \text{ where } b = \Delta V = 1.00 \text{ L and } h = \Delta P = 0.50 \times 10^5 \text{ Pa}$$

$$A_\triangle = \tfrac{1}{2}(1.00 \text{ L})(0.50 \times 10^5 \text{ Pa})$$

$$A_\triangle = 0.25 \times 10^5 \text{ Pa·L} = 25 \text{ N·m}$$

$$A_\triangle = 25 \text{ J.}$$

The rectangle's area is

$$A_\square = bh, \text{ where } b = \Delta V = 1.00 \text{ L and } h = P_2 = 0.50 \times 10^5 \text{ Pa}$$

$$A_\square = (1.00 \text{ L})(0.50 \times 10^5 \text{ Pa})$$

$$A_\square = 0.50 \times 10^5 \text{ Pa·L} = 50. \text{ N·m}$$

$$A_\square = 50. \text{ J.}$$

The total area is then

$$A_{total} = A_\triangle + A_\square$$

$$A_{total} = 25 \text{ J} + 50. \text{ J}$$

$$A_{total} = 75 \text{ J.}$$

The total area under the curve equals the absolute value of the work done, so

$$|W| = 75 \text{ J.}$$

Since the gas expanded, the work done by the gas is positive.

$$W = +75 \text{ J}$$

The gas does about 75 J of work on its surroundings as it expands.

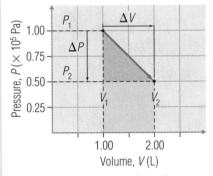

16-7 A graph showing the work done by an expanding graph for Example 16-1

Thermodynamic Laws   **351**

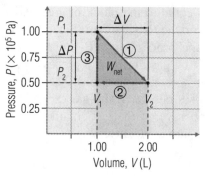
16.6 Expansion Cycles

If a gas is to be useful as a machine, it must be able to expand repeatedly, following a cycle. When the gas from the previous example returns to its original state, as Figure 16-8 shows, it is ready to expand again. For the first step in the cycle, ①, the gas expands as it did in the example, and the same amount of work is done *by* the gas on its surroundings:

$$W_1 = 7.5 \times 10^4 \text{ Pa·L} = +75 \text{ J}$$

For the second step, where the volume is reduced at constant pressure, the area under curve ② is

$$A_2 = (1.00 \text{ L})(5.0 \times 10^4 \text{ Pa})$$
$$A_2 = 5.0 \times 10^4 \text{ Pa·L}$$
$$A_2 = 50. \text{ J}.$$

Since the gas is compressed (contracts), the work *by* the gas is negative.

$$W_2 = -50. \text{ J}$$

For the third step, ③, the volume is constant; therefore, no work is done.

$$W_3 = 0 \text{ J}$$

The total work for the cycle is

$$W_{total} = W_1 + W_2 + W_3$$
$$W_{total} = (+75 \text{ J}) + (-50 \text{ J}) + 0 \text{ J}$$
$$W_{total} = +25 \text{ J}.$$

Notice that the net work by the gas is the same as the area enclosed by the path of the cycle (the triangular area in Example 16-1). This is a general principle: *For a cycle, the absolute value of the total work done is equal to the area enclosed by the path of the cycle on a P-V diagram.* A cycle that has a clockwise path on the diagram does positive work. Such a system is a heat engine. A cycle that has a counterclockwise path on the diagram does negative work. An example of a system that follows such a path is a refrigerator.

The work done by a gas, unlike work done against gravity, *depends on the path of the process on a P-V diagram.* The thermal energy added to the gas also depends on the expansion path. Experiments show, however, that for systems whose total mechanical energy is constant, the quantity $(Q - W)$ is the same regardless of the thermodynamic path between two states. This quantity is the change in the internal energy of the gas, ΔU, according to the first law of thermodynamics. Since the change of internal energy does not depend on the way the energy is added (the path between two thermodynamic states), the internal energy is called path-independent. Path-independent quantities are called **state variables** since they depend only on the thermodynamic state of the gas at a given time and not on how it got there. Internal energy, temperature, pressure, and volume are some state variables. They define the state of a gas.

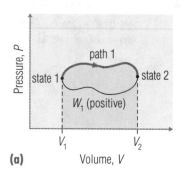

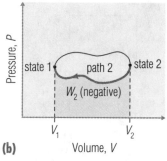

16-9 (a) The expanding gas does positive work while expanding from V_1 to V_2. (b) The gas does negative work as it is compressed from V_2 to V_1.

One state variable that you are familiar with is the balance of your bank account. It doesn't really matter to you what form of money (coins, bills, checks, ATM transfers, etc.) adds value to the account balance, as long as your balance increases. The state of your account is the cash balance that it has at any given moment.

16.7 Systems

We have mentioned in our discussion an entity called a **thermodynamic system.** As you know, a system is a piece of the universe isolated, at least mentally, for study. Other examples of possible thermodynamic systems are a calorimeter containing hot water; an ice cube, and an internal combustion engine. The boundaries of the system are set by the person studying the system. For an expanding gas, the cylinder walls and the piston are the boundaries of the system. Anything that is not a part of a system is a part of its surroundings.

Systems are classified into three categories. An **open system,** such as an ice cube resting on a kitchen counter, can exchange both energy and matter with its surroundings. A **closed system** can exchange energy, but not matter, with its surroundings. An expanding gas in a thermally-conducting cylinder with a gas-tight piston is a closed system. An **isolated system,** such as a liquid in a perfectly insulated vacuum flask, cannot exchange either matter or energy with its surroundings. In practice, there are no truly isolated systems except for the universe itself.

In an isolated system, energy is conserved. That is, the amount of energy in an isolated system is constant. Energy may be converted from one form to another, but no energy enters or leaves the system, nor does it appear or disappear. Since the universe is an isolated system, the total amount of energy is constant. When energy is exchanged between an open or closed system and its surroundings, all the energy lost by one must be gained by the other, since the system and its surroundings ultimately include the entire universe. This fact is the reason for the following equation in Chapter 15:

$$Q_{system} + Q_{surroundings} = 0$$

Although no practical system is isolated, a system enclosed in a calorimeter or other vacuum-insulated container approximates an isolated system for short periods of time. The first law of thermodynamics is one of the great conservation laws of science. It states that in a closed system, the total quantity of energy (in the forms of thermal energy and mechanical work) is constant, neither being created nor destroyed. This law is consistent with numerous biblical references proclaiming the finished work of God at the end of the Creation week.

16.8 Thermodynamic Processes

A *thermodynamic process* is a change in the thermodynamic state of a system. Although it is possible for all state variables to change at the same time, analysis and calculations are easier if at least one variable is held constant throughout the process. Processes are named to indicate which variables are held constant.

An *adiabatic process* exchanges no thermal energy between the system and its surroundings. For adiabatic processes, Q is zero; therefore, the first law becomes

$$0 \text{ J} = \Delta U + W, \text{ or}$$

$$\Delta U = -W.$$

Since internal energy of a gas is directly proportional to temperature, a change in internal energy is indicated by a change in temperature. If the gas expands, doing positive work, then internal energy decreases and the gas temperature drops; if the gas is compressed so that it does negative work, temperature and internal energy increase.

For an **isothermal** process, the temperature of the system is constant. If you assume that the system is not in the process of melting, freezing, boiling, or

Thermodynamically, neither the piston nor the cylinder walls are parts of the gas system.

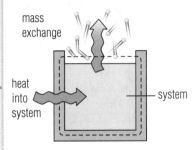

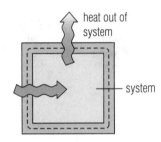

16-10a An open system

16-10b A closed system

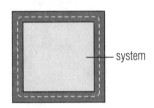

16-10c An isolated system

The first law of thermodynamics is one of the most thoroughly tested and confirmed laws in physical science. Anytime a scientist proposes that energy or mass (a different form of energy) spontaneously appeared in the universe, you know his idea violates this law of science.

In *adiabatic* processes, thermal energy is constant ($Q = 0$ J).

Recall that W is work *on the surroundings* by the system and is the negative of the work *on the system* ($W = -W'$).

In **isothermal** processes, temperature is constant ($\Delta U = 0$ J).

Thermodynamic Laws   **353**

condensing (phase changes involving the exchange of latent heats at constant temperatures), the internal energy is also constant ($\Delta U = 0$ J). The first law becomes

$$Q = 0 \text{ J} + W, \text{ or}$$
$$Q = W.$$

This means that all of the heat transferred is converted to work.

During an **isochoric** (ī′ so-kôr′-ik) process, the volume is constant. Remember, work is done on or by a gas only when its volume changes. Therefore,

$$Q = \Delta U + 0 \text{ J, or}$$
$$Q = \Delta U.$$

All the thermal energy added to the system increases the internal energy of the system.

An **isobaric** process is a process for which the pressure is constant. Isobaric processes give the simplest equation for work (Equation 6.3),

$$W = P\Delta V.$$

The first law of thermodynamics for this process is

$$Q = \Delta U + P\Delta V.$$

The calculations for work and energy are much simpler when it is possible to use the ideal gas relationships in the general gas law. At low pressures, real gases behave much like ideal gases, and the approximation is valid. A process that allows the use of ideal gas relationships is known as an **ideal gas process.**

16A Section Review Questions

1. What is a barrier that conducts no thermal energy called?

2. On a cold winter morning, if the air outside a car is in thermal equilibrium with the glass in the car's window and the window glass is in thermal equilibrium with the air inside the car, what can be said about the temperatures of the inside and outside air? What is your authority for your answer?

3. How is the work by the surroundings on a thermodynamic system related to the work the system does on its surroundings? What principle of mechanics allows you to answer this question?

4. What is the area under a *P-V* curve equivalent to?

5. What kind of thermodynamic engine follows a counterclockwise path on a *P-V* diagram?

6. What kind of thermodynamic system is a sealed plastic container full of steaming hot soup in a refrigerator?

7. For a closed gas thermodynamic system, what kind of process exchanges thermal energy for change in internal energy?

⊙8. A gas cylinder system similar to that described in this section originally has a volume of 1.80×10^{-3} m³ at 1.01×10^5 Pa. If the volume and pressure of the gas changes along a linear path on the *P-V* diagram to 1.22×10^{-3} m³ and 2.02×10^5 Pa, what is the work done by the gas in joules?

THE SECOND AND THIRD LAWS

16.9 Heat Engines and the Second Law

All thermodynamic processes except adiabatic processes involve exchanges of energy between a system and its surroundings. The surroundings must therefore contain either a source of thermal energy, a sink (or receiver) for thermal energy, or both. An easy way to visualize a thermal energy source or sink is to imagine a **heat reservoir** at a specific temperature. The reservoir is so large that no addition or subtraction of energy can change its temperature significantly. A reservoir at a higher temperature than the system, called a *hot reservoir,* is a source of thermal energy for the system. A reservoir at a lower temperature than the system, called a *cold reservoir,* is a thermal energy sink for the system. Both a hot reservoir and a cold reservoir are used to operate a heat engine.

16-11 The Sun, a vast heat reservoir, occasionally releases huge amounts of energy in the form of solar flares.

A source of thermal energy for a system must be "hotter" than the system. The reason is that thermal energy flows from a "hotter" body to a "colder" body. This principle agrees with human experience. If you place a glass of hot water against a glass of cold water, the cold water warms while the hot water cools. What would you think if, after a few minutes of contact, the cold water froze and the hot water boiled? One form of the **second law of thermodynamics** is the principle that energy flows from an area of higher concentration to an area of lower concentration.

A typical heat engine requires a hot reservoir, a cold reservoir, and a working fluid (a liquid or gas). Thermal energy absorbed from the hot reservoir causes the fluid to expand against a piston or some other movable part. The expansion forces the part to move, performing mechanical work. Then the fluid gives up thermal energy to the cold reservoir and contracts. The fluid is then ready to expand again.

16.10 Early Steam Engines

A steam engine is a heat engine using steam as a working fluid. For centuries men have known that steam can do work. The first steam engine, the *aeolipile,* was described by Hero of Alexandria, who lived about the time of Christ. The aeolipile is a sphere or cylinder with bent outlet pipes (Figure 16-12) that is connected by pipes to a container of boiling water. The connecting pipes are attached in a way that allows the sphere to rotate. When the water in the container boils, steam is forced through the connecting pipes into the sphere and finally out the outlet pipes.

Probably the greatest heat engine that affects us daily is created by the interaction of the atmosphere and the hydrosphere—our weather. "The sun also ariseth, and the sun goeth down. . . . The wind goeth toward the south, and turneth about unto the north; it whirleth about continually, and the wind returneth again according to his circuits. All the rivers run into the sea; yet the sea is not full; unto the place from whence the rivers come, thither they return again" (Eccles. 1:5–7). Do you think the Author of this passage understands the underlying physics that causes weather?

There are several ways to state the second law of thermodynamics. One statement of the second law is that energy flows from an area of higher concentration to an area of lower concentration. You observe the second law in action when thermal energy flows from hotter locations to colder ones.

Hero of Alexandria (first century A.D.) was a mathematician and engineer whose writings were an early contribution to mechanics and related sciences.

16-12 The aeolipile

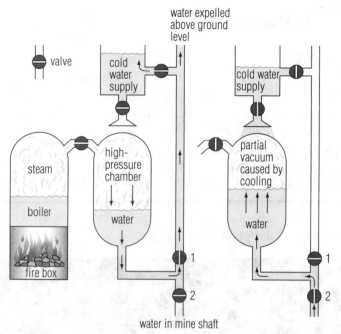

16-13 Schematic of a Savery engine (1698)

Thomas Savery (1650–1715) was an English engineer and inventor.

Thomas Newcomen (1663–1729) was an English iron magnate and inventor. His was the first practical engine to use a piston in a cylinder.

James Watt (1736–1819) was a Scottish engineer who devoted most of his life to designing and building improved steam engines.

The escaping steam exerts a reaction force on the nozzles that produces a torque, causing the sphere to rotate. The aeolipile is not a cyclic engine; the engine uses only a single irreversible thermodynamic process between two states until its water boils away. All the thermal energy (and the water) that enters the process is lost.

Because it is not cyclic, the aeolipile is not a practical steam engine. The first practical steam engine was a water pump invented by Thomas Savery. A better steam engine, which could be used as a pump or for raising loads, was invented by Thomas Newcomen. Newcomen's engine dominated the market until James Watt invented an even more efficient design. Watt's engine was the first to use separate chambers to heat and cool the steam to avoid the wasteful process of repeatedly heating and then cooling the cylinder walls. Watt's engine was able to power factories and transportation as well as pump water from deep mines. The Industrial Revolution began with Watt's steam engine.

The basic idea behind the steam engine is that steam expands when it is heated and contracts when it is cooled. In the Savery engine (Figure 16-13), the steam expands to push water out of a chamber and through the pipe. Valve 1 is open; valve 2 is closed. When all the water is out of the chamber, valve 1 is closed and valve 2 is opened. The steam contracts, and the decrease in pressure draws water up through the pipe into the chamber.

The Newcomen engine, Figure 16-14 (a), has a movable piston that goes up when the steam expands and down when the steam condenses. It uses a first-class lever to lift the load, so the condensing stroke does the lifting work. The condensed steam is drained from the cylinder before the next compression begins. One problem that existed with this engine was that the cylinder was cast iron. The rough interior surface made it difficult to obtain a good seal with the piston. Another problem was that the cylinder was cooled by the condensation step; the steam entering at the start of the next cycle condensed prematurely, before any work was done.

The Watt engine, Figure 16-14 (b), is similar to the Newcomen engine in that it uses a piston. However, the Watt engine exerts its force using steam pressure rather than steam condensation. Also, in later Watt engines, the piston is *double-acting—*

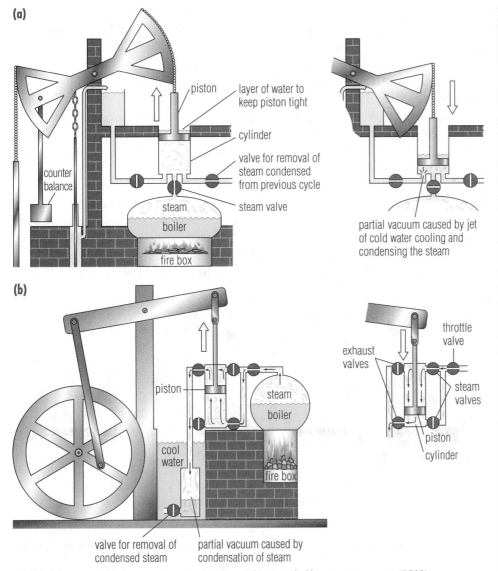

16-14 Schematics of (a) Newcomen's engine (1712) and (b) Watt's double-action rotary engine (1782)

force is exerted on both the forward and reverse strokes of the piston through the automatic operation of appropriate steam supply and exhaust valves. The exhausted steam is directed to a second chamber cooled by water to condense the steam (the condenser), forming a partial vacuum that increases the net pressure on the piston applied by the steam. The Watt engine also had the advantage of a machine-bored cylinder. The extra-smooth walls allowed a good seal with the piston, preventing the wasting of steam.

The valves indicated in the figures are all mechanically linked to the piston in order to ensure they operate at the proper times in the cycle.

16.11 An Ideal Heat Engine Cycle—The Carnot Cycle

A **reversible** process is a quasi-static process that leaves the system in exactly the same state after occurring twice, once normally and once in reverse. The processes that you study here can be assumed to be reversible unless the text states that one is not. However, no real processes are completely reversible due to the presence of nonconservative effects, although some are more nearly reversible than others.

A reversible *cycle* leaves the system in the same state as before the entire process occurred. The pressure, volume, and temperature of an ideal gas are the same at the end of a reversible cycle as they were at the beginning. Reversible

A thermodynamic process is a thermodynamic change between two states. A cycle is made up of several processes that enclose an area on a *P-V* diagram. The starting and ending points of a cycle are identical.

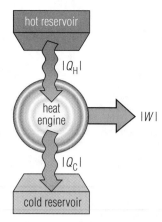

The **Carnot cycle** is the most efficient ideal heat engine cycle.

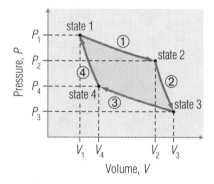

16-17 *P-V* diagram for the Carnot cycle

The curves that represent isothermal Steps 1 and 3 follow *P-V* curves called *isotherms* on the graph. There are an infinite number of isotherms in any given diagram. The isotherms followed by an ideal Carnot cycle are the temperatures of the hot and cold reservoirs.

cycles are the most efficient means of converting thermal energy to mechanical work. Cyclic processes are practical because they are very efficient and they can repeat many times without the operating conditions of the engine having to be reset. The engine doesn't continually gain heat until it melts or lose heat until it freezes. There are several idealized cycles that convert thermal energy to work, with varying efficiency.

The most efficient cycle that can operate between two temperatures is the **Carnot cycle,** a four-step reversible cycle (see Figure 16-16). Step 1 in the Carnot cycle, figure (a), is an isothermal expansion from V_1 to V_2 at temperature T_H. The system absorbs a thermal energy of Q_H from the hot reservoir. Step 2, shown in figure (b), is an adiabatic expansion from V_2 to V_3. The temperature changes from T_H, the temperature of the hot reservoir, to T_C, the temperature of the cold reservoir. The system absorbs no heat in this step. In the third step, figure (c), an isothermal compression from V_3 to V_4 occurs at T_C. The system gives up a thermal energy of Q_C to the cold reservoir. Step 4, figure (d), is an adiabatic compression to the original conditions, T_H and V_1. The system absorbs no thermal energy in this step.

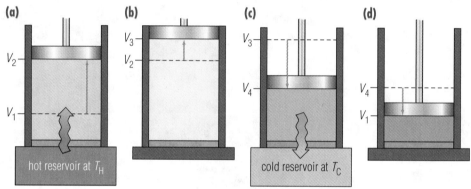

16-16 A Carnot engine completes one cycle. (a) Step 1—isothermal expansion (b) Step 2—adiabatic expansion (c) Step 3—isothermal compression (d) Step 4—adiabatic compression to original conditions

The work done by the system during the cycle can be found from the first law of thermodynamics:

$$Q_{\text{cycle}} = \Delta U_{\text{cycle}} + W_{\text{cycle}}$$

For a cycle, ΔU_{cycle} is zero.

$$Q_{\text{cycle}} = 0 \text{ J} + W_{\text{cycle}}$$
$$Q_{\text{cycle}} = W_{\text{cycle}}$$

Since the gas does work on its surroundings, W is positive. The sum total of the thermal energy exchanged is Q_{cycle}:

$$|Q_{\text{cycle}}| = |Q_H| - |Q_C|$$

The absolute value signs are useful but not necessary in order to avoid confusion with signs. By assuming that the gas heat engine does positive work, we assume that Q_H added to the system from the hot reservoir is positive and Q_C lost from the system to the cold reservoir is negative.

Substituting work for the net heat gained by the cycle,

$$|W_{\text{cycle}}| = |Q_H| - |Q_C|.$$

The portion of the heat absorbed by the heat engine from the hot reservoir that is converted into useful work is a measure of the efficiency of the engine. The **thermal efficiency (ε)** of a Carnot engine is defined as

$$\varepsilon = \frac{|W_{cycle}|}{|Q_H|}, \text{ or}$$

$$\varepsilon = \frac{|Q_H| - |Q_C|}{|Q_H|}. \tag{16.6}$$

This result is true for any heat engine cycle that operates between two heat reservoirs. For the portions of the cycle involving isothermal processes at T_H and T_C,

$$\frac{|Q_H| - |Q_C|}{|Q_H|} = \frac{T_H - T_C}{T_H}, \text{ or}$$

$$\varepsilon = \frac{T_H - T_C}{T_H}. \tag{16.7}$$

You can increase the efficiency of a Carnot engine by raising the temperature of the hot reservoir, lowering the temperature of the cold reservoir, or doing both. However, even in theory the engine is not 100% efficient. The efficiency of a real engine will be lower than that of a theoretical cycle because of friction and other losses.

16.12 Heat Pumps

A heat engine that runs in the counterclockwise direction around the P-V diagram, with work added to move thermal energy from the cold reservoir to the hot reservoir, is called a **refrigeration engine** or **heat pump.** The Carnot cycle can be reversed to work as a heat pump. A Carnot heat pump would first adiabatically expand so that the gas's temperature drops below the cold reservoir temperature, then isothermally expand as heat is transferred from the cold reservoir to the heat pump, then adiabatically contract, raising the gas's temperature above the hot reservoir temperature, and finally contract isothermally while it discharges heat to the hot reservoir, thus returning to its original conditions. A refrigerator is a heat pump used to cool a volume. It uses work to move thermal energy from a cold place (inside the refrigerator) to a warm place (the room). An air conditioner is a kind of refrigerator. It cools a room by pumping thermal energy outside. That is the reason that some air conditioners are installed in windows. Heat pumps that function as air conditioners in the summer and as heaters in the winter (by pumping thermal energy from the outside to the room) are also available.

16.13 Various Statements of the Second Law and the Third Law

It was stated earlier in this chapter that thermal energy naturally flows from hot bodies to cold bodies. This is one statement of the second law of thermodynamics, a law governing the direction of natural processes. Another statement of the second law is that thermal energy cannot be completely converted to work in a cyclic process. A cycle always loses some energy to a cold reservoir. No exceptions to the second law have been observed.

According to the second law, no engine can be completely efficient. Efficiency for an ideal heat engine is expressed by Equation 16.6,

$$\varepsilon = \frac{|Q_H| - |Q_C|}{|Q_H|}.$$

If no energy were lost to the cold reservoir, Q_C would be zero, and efficiency would be 100%. Complete efficiency is ruled out by the second statement of the second law because some energy is always lost to the cold reservoir.

The ocean is a vast reservoir of thermal energy. Why can't we use this energy to drive a ship across the ocean? The ship could extract energy from the ocean to

16-18 Domestic heat pumps transfer heat from the cooler space to the warmer space in summer and winter.

drive its engine, leaving the ocean a trifle colder than before. Unfortunately, the second law of thermodynamics says that thermal energy does not flow from a cold reservoir—like the ocean—to a hot reservoir—like the ship's engine—unless work is done to force this flow. The work done to extract energy from the ocean would be more than the work the energy could do, so the ship would go nowhere.

The first law says that when converting energy into work you can never get out of a transaction more than you put into it. The second law says that you cannot get as much work out a machine as you put into it. Both laws forbid perpetual motion machines. The first law forbids perpetual motion machines of the first kind, which produce more energy than they consume. The second law forbids perpetual motion machines of the second kind, which are 100% efficient. Many inventors have tried to invent a perpetual motion machine; none has succeeded.

A heat engine could be 100% efficient only if Q_C were zero. For an isothermal process,

$$\frac{|Q_C|}{|Q_H|} = \frac{T_C}{T_H}. \qquad (16.8)$$

If $|Q_C|$ is zero, then T_C must be zero if the proportion is to hold true. Equation 16.8 is valid only for Kelvin temperatures. Therefore, for a heat engine to be 100% efficient, its cold reservoir must be at absolute zero. Scientists have tried to reach absolute zero. They have reached temperatures lower than 10^{-9} K. However, the closer they get to absolute zero, the harder it becomes to reach lower temperatures. Therefore, scientists are convinced that it is impossible to reach a temperature of absolute zero. This principle is included in the **third law of thermodynamics.**

16B Section Review Questions

1. What are the essential parts of a thermodynamic heat engine?

2. a. What did the Savery engine lack that the Newcomen and Watt engines had?

   b. What was the major improvement of the Watt engine over the Newcomen engine?

3. What two basic reversible thermodynamic processes occur at various points in a Carnot cycle?

4. A steamship's propulsion plant uses the ocean water to absorb its waste heat according to the Carnot cycle. What happens to the propulsion plant's efficiency if seawater temperature becomes warmer (as in the tropics)?

5. a. What condition would have to exist for an ideal heat engine to be 100% efficient?

   b. What principle states that this condition is not possible?

6. a. Why does your car engine have a radiator?

   b. What principle predicts that a gasoline engine would need one?

7. a. Sketch a P-V diagram of a typical Carnot cycle for a heat engine.

   b. Shade and label the area representing work done during the adiabatic and isothermal expansions.

   c. Shade with a contrasting color and label the work done during the adiabatic and isothermal contractions.

   d. Identify the area that represents the net work done by the engine.

8. Sketch a P-V diagram Carnot cycle for a heat pump. What does the area enclosed by the graph represent?

Statements of the second law:

• Heat flows from a place of higher temperature to one of lower temperature.

• Thermal energy cannot be completely converted to work in a cyclic process.

• No heat engine is 100% efficient.

The expression of Equation 16.8 for heat pumps using the Carnot cycle is

$$\frac{|Q_C|}{|Q_H|} = \frac{T_C}{T_H}.$$

As the difference between hot reservoir temperature and cold reservoir temperature increases, the performance of a Carnot refrigeration engine decreases.

The third law of thermodynamics states that absolute zero is unattainable. Therefore, cold reservoirs at $T_C = 0$ K are not possible.

ENTROPY AND ITS CONSEQUENCES

16.14 What Is Entropy?

We have already discussed three statements of the second law of thermodynamics. A fourth statement of the second law is that entropy increases in all natural processes. **Entropy (S)** is another state variable of a system, like volume, temperature, or pressure. Because entropy is related to the microscopic properties of the particles of a system, it is similar in some ways to internal energy. Just as with internal energy, the change of entropy (ΔS) is more important and more easily measured than the system's entropy at any particular state of thermodynamic equilibrium.

Some define entropy as a measurement of the randomness, or disorder, of the particles in a specific part of the universe. The second law says that every *natural* process makes the universe more disorderly. Disorder implies unusable energy, such as the thermal energy rejected by a heat engine. The energy still exists, but it can no longer do useful work in a particular process. Mathematically, entropy for a reversible process is defined as

$$\Delta S \equiv \frac{\Delta Q}{T}, \qquad (16.9)$$

where ΔQ is the change of the system's thermal energy at isothermal conditions and T is the absolute temperature of the system. The units for entropy are J/K. For quasi-static processes, the changing quantities in Equation 16.9 experience infinitesimally small changes at each step.

In reversible processes, the entropy of the universe remains constant. The entropy change in the system is exactly balanced by the entropy change in the surroundings:

$$\Delta S_{\text{system}} = -\Delta S_{\text{surroundings}}$$

In *natural* processes—irreversible processes—entropy increases (ΔS is positive). There may be local decreases in entropy for the system, but these are more than offset by an increase in entropy of the surroundings. Therefore, the entropy of the universe increases.

The entropy of a system is a state variable because the entropy change depends only on the initial and final states of the system. Therefore, the entropy change for a reversible process between two states is the same as the entropy change for an irreversible process between the same two states. Entropy calculations for a free expansion of a gas use this principle.

In a free expansion, an ideal gas is allowed to expand adiabatically into a vacuum. The gas originally occupies a volume of V at a temperature T. It expands into a vacuum also of volume V, for a final volume of $2V$. This is an irreversible process. Since the gas is adiabatically isolated, there is no heat exchanged with its surroundings. Because the gas is expanding into a vacuum, it does no work on its surroundings, so $W = 0$ J as well. Therefore, according to the first law, internal energy does not change either. Then the temperature remains constant at T throughout the process. A direct entropy calculation for this process using Equation 16.9 is not possible, because that equation applies only to reversible processes. However, you know that entropy has increased during the process because the particles of the gas are farther apart (less ordered), even if they don't have more energy.

The entropy change for the system in the free expansion is the same as that for the system during the isothermal (reversible) process. However, since the free expansion is adiabatic, the system has no contact with its surroundings, so the

The second law of thermodynamics also states that **entropy** increases in all natural processes.

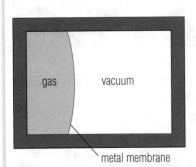

16-19 A gas trapped by a membrane expands rapidly into the evacuated volume of the vessel when the membrane ruptures.

entropy of the surroundings does not change. The total entropy change of the gas is greater than zero. Intuitively, this increase in entropy makes sense because the gas particles are farther apart and are therefore more disorganized.

In the kinetic theory, entropy is a measure of the likelihood that a system will be in a particular state. Because of the second law, a statistically more probable state for a system has a high entropy, and a statistically unlikely state has low entropy. The most likely state of a system is one of disorder. An example of disorder is diffusion of a dye throughout a beaker of distilled water. Even if the system is maintained at isothermal conditions, the random motion of the particles will eventually distribute the dye throughout the water. The dye particle distribution is more spread out and random with time compared to the original arrangement of particles in a solid crystal of the dye. This randomness is the natural consequence of diffusion, so the entropy of the dispersed dye is greater than its entropy as a crystal. The orderliness of an isolated system tends to naturally deteriorate to maximum disorder. Another way of stating this result is to say that the entropy of an isolated system is at its maximum.

Entropy can also be related to the encoding of information. Information with high entropy contains meaningless pieces, which in communications are called *static*. Information with low entropy is exact, with few meaningless parts (for example, when a message is communicated clearly from one place to another). Experts try to minimize the static and keep the message's entropy low. Another example of highly ordered information is the biological code stored in our genes. Entropy in this genetic information is a measure of its randomness. Because this information is highly ordered, it has low entropy. A random change, known as a mutation, destroys some information and harms a body. Such a change increases the organism's genetic entropy. When an organism dies, the cellular processes that maintain the DNA code in good, meaningful condition no longer function, and the natural random destruction of the molecules eventually erases the genetic information. Organized infusion of energy is required to maintain low entropy.

16.15 Conservation and Degeneration in Nature

You may wonder why the universe still has energy available for work if the second law is true. The answer is that God created the universe with an immense supply of usable energy. Obviously, the laws of thermodynamics were not valid during the very acts of Creation. The first law is a result of the fact that only God can create or destroy. The only exceptions to this law are divine actions. Since the universe began with relatively little entropy, it will be a long time before it comes to a state of maximum entropy.

The first law says that energy and mass are conserved. If something is to be conserved, it must exist. Some scientists believe that the physical laws we know now have *always* been in effect. This belief is called **uniformitarianism.** Uniformitarian scientists must believe either that matter does not now exist or that it has always existed. Most uniformitarians believe that matter is eternal. Some believe that before this universe was formed, matter existed in the form of subatomic particles. These particles eventually moved close together and then exploded (the so-called big bang). With the energy from the explosion, the particles united to form atoms. Eventually, the early universe developed to its present state. The big bang is only one of the theories of evolutionistic cosmology.

The first law, by itself, neither supports nor refutes this theory of evolution. However, assuming that the primordial universe, highly compacted and ordered, exploded and expanded in an irreversible process, the universe's entropy would have had to rapidly increase during the early period of its supposed expansion. Furthermore, in order for galaxies, stars, and planets to form, a *decrease* of

16-20 The immense amount of information stored in DNA is maintained by complex cellular mechanisms that require energy.

The Bible predicts many of the heresies and pseudosciences of the last days. Does it predict uniformitarianism? Peter affirmed that this worldview would be prevalent in the last days when he wrote, "There shall come in the last days scoffers, walking after their own lusts, And saying, Where is the promise of his coming? For since the fathers fell asleep, all things continue as they were from the beginning of the creation" (II Pet. 3:3–4).

entropy is required. Scientists have never been able to identify any natural ordering process that would tend to gather particles of dust together into these celestial structures. Such an ordering process could be accomplished only through an intelligent infusion of energy. Therefore, naturalistic evolutionary cosmology is thermodynamically impossible.

Evolutionists point out that, according to kinetic theory, the second law is based on probability. Although unlikely, they say, a violation of the second law is possible. Given enough time—ten billion years or so—a violation is almost inevitable. It is true that the kinetic theory suggests that an entropy decrease is possible. However, there are two reasons that evolutionists are in error when they say that the second law allows evolution. First, no violation of the second law has ever been observed. The second law is based on observations. Until a violation of it is observed, the idea that such a violation can occur is only an idea. Second, the theory of evolution requires not *one* decrease in entropy but almost *continual* decreases in entropy whose effects are preserved by the system. If the second law can be ignored as many times as evolution requires, it is not an accurate reflection of nature. Evolution implies that there is a tendency for a system to become more ordered, but the second law says that the tendency is for a system to become disordered. The theory of evolution directly contradicts the second law.

The second law states that energy is becoming less available for use. Eventually, if the universe is left to itself, all energy will be unavailable for use. The universe will be at a uniform low temperature. Nothing will be able to live. This effect is called the **heat death** of the universe. This heat death is far in the future because there is still much energy available for use. Actually, the heat death will never happen, because the universe will not be left to itself. The Lord will return and replace this universe with a new one. With Him ruling and sustaining the new universe, you will not have to worry about exhausting available energy.

> The second law is supported by experiments; evolution is merely an unsupported theory.

16C Section Review Questions

1. What property of a system does entropy describe?

2. According to the second law of thermodynamics, what happens to the entropy of a system subject to natural processes?

3. A mixture of ice and water is contained in a near-perfect adiabatic container. As the entropy of the melting ice increases, what must happen to the entropy of the water surrounding it?

4. Briefly discuss how the free expansion of a gas in an adiabatic container increases the entropy of the gas.

5. a. What would be the likelihood of all the oxygen in your classroom clumping together in a 1 L volume in one corner?

   b. If it did, would the entropy of the gas be higher or lower than is normally the case?

6. The belief that matter is eternal is often a tenet of what system of philosophy?

7. As one follows the progress of total entropy in the expanding universe after the supposed big bang, what direct violation of the second law of thermodynamics must have occurred for the universe to have acquired its present form?

8. What is the flaw in some evolutionists' logic when they appeal to the statistical nature of entropy over immense periods of time as a justification for evolution?

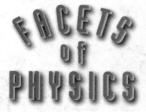

Growth and the Second Law

The universe is filled with countless galaxies containing billions of stars each.

The concept of "heat death"—a cold, lifeless universe—is both frightening and hard to accept. Man is naturally repelled by the second law of thermodynamics, a law that implies that all things tend to break down and that the present universe is forever losing usable energy that can never be restored. Common sense seems to indicate that the law is not true. After all, with the coming of each winter, the warm breezes of spring are sure to follow. Also, human life is not becoming less complex but is becoming increasingly more complex; each year men invent new and better machines. Every day in nature, photosynthesis converts sunlight into complex organic matter that animals transform into more complex tissues such as nerves and brain cells.

Thus it is not surprising that scientists did not even discover the second law of thermodynamics until Kelvin's famous paper was published in 1852, and only after much heated debate was it accepted in the scientific community. Yet God's Word had declared this law millennia before scientists stumbled upon it. "The earth shall wax old like a garment, and they that dwell therein shall die in like manner" (Isa. 51:6).

"For the creature was made subject to vanity. . . . The whole creation groaneth and travaileth in pain together until now" (Rom. 8:20, 22). "All go unto one place; all are of the dust, and all turn to dust again" (Eccles. 3:20).

The most apparent contradiction of the law of entropy is the growth of plants and animals from single cells into complex life forms. For example, acorn kernels expand into mighty one-hundred-foot oak trees. Even more amazing than oak trees is the life of a man. One microscopic cell subdivides and specializes until in nine months there appears the most complex collection of matter in the universe. Within a few years, this mass of molecules can think, speak, and commune intelligently with God. Social Darwinists use human progress to prove that man is evolving into a higher state and becoming perfect. But how does the second law refute their arguments, and how does it explain the growth of complex organisms?

When you consider the growth of cells, particularly of human cells, you should realize that adult organisms are no more complex than the initial fertilized egg. Every major cell function takes place in the first cell, and the first cell's DNA contains all the information needed by future cells. A baby's DNA is only as complex as his parents' DNA, and the parents' DNA ultimately came from Adam. Again we see that God intervened, in this case forming a complex body "of the dust of the ground" (Gen. 2:7).

Human "progress" after Adam strongly supports the second law of thermodynamics. Adam's healthy body lasted 930 years before it returned to dust, but the longevity of his descendants decreased, especially after the Flood, until Moses said of his generation "The days of our years are threescore years and ten" (Ps. 90:10). Adam's brain needed only one day to register and name every beast of the field and every fowl of the air. Scientists estimate, however, that people today can use only a small part of their brain's capacity.

Thermodynamics considers the result of natural processes. What is a man's life? "It is even a vapour, that appeareth for a little time, and then vanisheth away" (James 4:14). The human body never evolves; it merely battles to preserve the life that it was given for an appointed time. During that brief period, the body consumes massive amounts of energy to maintain itself; yet eventually it dies and decays. This process fits perfectly into the laws of thermodynamics.

Fortunately, the second law of thermodynamics is not eternal, so "heat death" will never occur. Even though God cursed the earth and the human body after Adam's sin, He promised that a Messiah would come who would one day restore the human body and replace this universe with a new heaven and a new Earth. The creation, being "made subject to vanity," was given hope that it would "be delivered from the bondage of corruption into the glorious liberty of the children of God" (Rom. 8:20, 21). The earth shall again be a place where "there shall be no more death, neither sorrow, nor crying, neither shall there be any more pain: for the former things are passed away" (Rev. 21:4).

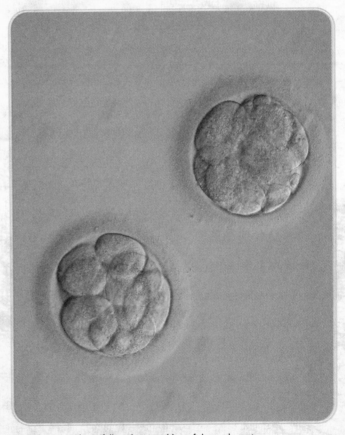
Even human embryos follow the second law of thermodynamics.

Chapter Review

In Terms of Physics

internal energy (*U*)	16.1	isobaric	16.8
diathermic	16.2	ideal gas process	16.8
thermal equilibrium	16.2	heat reservoir	16.9
zeroth law of thermodynamics	16.2	second law of thermodynamics	16.9
surroundings	16.3	reversible	16.11
first law of thermodynamics	16.3	Carnot cycle	16.11
heat engine	16.4	thermal efficiency (ε)	16.11
P-V diagram	16.5	refrigeration engine	16.12
state variable	16.6	heat pump	16.12
thermodynamic system	16.7	performance coefficient (*K*)	16.12
open system	16.7	third law of thermodynamics	16.13
closed system	16.7	entropy (*S*)	16.14
isolated system	16.7	uniformitarianism	16.15
isothermal	16.8	heat death	16.15
isochoric	16.8		

Problem-Solving Strategies

16.1 (page 352) If a cycle is followed clockwise on a *P-V* diagram, then the system is a heat engine doing positive work. If the cycle is followed counterclockwise, then work is being done on the system (as in a refrigerator).

16.2 (page 352) If a process moves from left to right along any path on a *P-V* diagram, then the system is doing work. If the process moves from right to left along any path, then work is being done on the system.

Review Questions

1. What thermodynamic state variable is dependent on the temperature of the system?

2. What is the zeroth law of thermodynamics?

3. Discuss the two ways that energy can be exchanged between a system and its surroundings.

4. State the first law of thermodynamics. What principle of mechanics does it extend?

5. Define *heat engine*. Give an example.

6. Describe the three kinds of thermodynamic systems and give an example of each.

7. Classify each of the following systems as open, closed, or isolated.
    a. flour in a sifter
    b. ice cube in an adiabatic calorimeter
    c. a living human being
    d. hot coffee in a mug in a cold room
    e. a leakproof steam engine

8. Why does a heat engine require a fluid to operate?

9. What advantage did the Newcomen engine have over the Savery engine?

10. As a heat engine completes a cycle, how does its final internal energy compare to its starting internal energy?

11. Discuss two ways to thermodynamically (rather than mechanically) improve the efficiency of a Carnot heat engine.

12. What law gives the reason that a heat engine cannot be 100% efficient?

13. Which of the following is *not* a statement of the second law?
    a. Thermal energy can flow from cold to hot if work is done.
    b. Thermal energy cannot be converted entirely to work in a cyclic process.
    c. The entropy of an isolated system tends to decrease.
    d. Every natural process makes the universe more disorderly.

14. How does the theory of evolution contradict the second law of thermodynamics?

15. Discuss several examples that illustrate it is necessary to do work on a system in order to prevent an increase of entropy.

16. Discuss why the heat death of the universe will not occur.

True or False (17–32)

17. Mechanical kinetic and potential energies are properties of the macroscopic system, but internal energy is a property arising from the microscopic particles of the system.

18. The zeroth law of thermodynamics establishes the "transitive property of thermal equilibrium"—if A and B are both in thermal equilibrium with C, then they are in equilibrium with each other.

19. Fiberglass insulation approximates a good diathermic material.

20. In order to simplify the statement of the first law of thermodynamics, it is assumed that the total mechanical energy of the system does not change during a thermodynamic process.

21. A quasi-static process proceeds smoothly without any steps or stops.

22. Thermodynamics concerns the work done on the system more than the work the system does on its surroundings.

23. The work done during a thermodynamic process graphed on a *P-V* diagram does not depend on the path the process takes.

24. A steaming cup of coffee on a tabletop is an open system.

25. Pushing down on the plunger of a tire air pump causes the air in the pump to undergo an isochoric process.

26. Engines that heat the working fluid only once and then discharge it have low efficiencies compared to those that heat the working fluid over and over again.

27. James Watt invented the first practical steam-powered engine.

28. A Carnot engine is more efficient if the temperature difference between the hot and cold reservoirs is increased.

29. A heat engine would be 100% efficient if it discharged no heat to the cold reservoir.

30. Entropy is a form of energy.

31. As with some other state variables, it is simpler to determine the change of entropy than to measure it at a given state.

32. The fact that your bedroom gets messier with time is an example of the second law of thermodynamics in action.

⦿33. A pressure of 1.52×10^5 Pa forces a gas to compress from a volume of 5.00×10^{-3} m³ to a volume of 2.50×10^{-3} m³. How much work is done on the gas, in N·m?

⦿34. A gas in a cylinder with a cross-sectional area of 60.0 cm² holds the piston at a height of 5.00 cm when a force of 18.0 N is applied. Then, under the same force, the gas expands in a quasi-static process to a height of 7.50 cm.
    a. What is the gas's original volume in m³?
    b. What is the original pressure on the gas in N/m² (Pa)?
    c. What is the final volume (in m³) of the gas?
    d. How much work does the gas do, in J?

⦿35. A gas expands as Figure (a) shows.
    a. How much work does the system do in step 1?
    b. How much work does it do in step 2?
    c. How much work does it do overall?

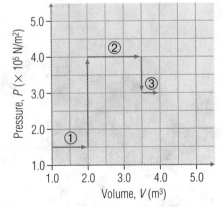

(a)

⦿36. Find the work done by a gas following one cycle as shown in Figure (b) for
    a. each step.
    b. the entire cycle.

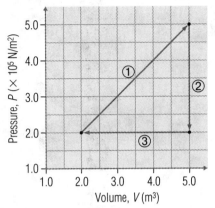

(b)

⦿37. A Carnot engine has a hot reservoir at a temperature of 490 K and a cold reservoir at a temperature of 290 K. What is its efficiency?

⦿38. If the hot reservoir of a Carnot engine is at room temperature (24 °C), what must be the temperature of the cold reservoir for the engine to have an efficiency of 50%?

⦿⦿39. If a ship's steam propulsion boilers produce 1.20 MW of thermal energy at full power, and the propulsion plant is 15% efficient, how much power is discharged to the sea via the seawater cooling system?

⦿⦿40. A gas expands and contracts in the cycle in Figure (c). How much work does each part of the cycle do? ($A_{\bigcirc} = \pi r^2$)

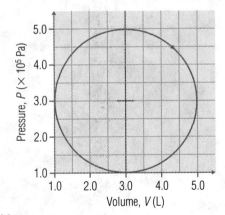

(c)

17A Hydrostatics: Fluids at Rest 369
17B Hydrodynamics: Fluids in Motion 382
Facet: Maple Syrup Hydrometers 381
Facet: Aerodynamically Designed Cars 389

New Symbols and Abbreviations

density (ρ)	17.2	gauge pressure (P_g)	17.3
specific gravity (s.g.)	17.2	absolute pressure (P)	17.3
atmosphere (atm)	17.3	coefficient of viscosity (η)	17.19
millibar (mb)	17.3		

Water in motion can be put to useful work in hydroelectric dam generators. This view is of a penstock discharge at Hoover Dam.

Fluid Mechanics 17

> *And the flood was forty days upon the earth; and the waters increased, and bare up the ark, and it was lift up above the earth. And the waters prevailed, and were increased greatly upon the earth; and the ark went upon the face of the waters. And the waters prevailed exceedingly upon the earth; and all the high hills, that were under the whole heaven, were covered. Fifteen cubits upward did the waters prevail; and the mountains were covered.* Genesis 7:17–20

God's judgment by the Genesis Flood was without question the greatest demonstration of the mechanics of fluids in motion in the history of the earth. If it were not for the existence of the buoyant force, the ark would never have floated, and the lives of Noah and his family would not have been preserved by this method (although God would have provided another means, according to His purpose). Some geologists that do research using a biblical framework for interpretation believe that the moving floodwaters during the abatement phase of the Flood were responsible for the major erosional landforms that remain visible in many places of the world today. Such features, including the mesas and buttes in the western United States, could have been formed only by the movement of continent-sized sheets of water flowing off of the land as the land rose out of the Flood and the waters abated.

God promised mankind that he would never judge the world by a flood again. The seal of His promise is the rainbow. However, every person faces a personal judgment at the end of time. Only those sealed by the Holy Spirit are judged to be worthy of eternal life. This worthiness is obtained only by accepting Jesus Christ as your Savior. Have you made this decision?

17-1 An erosional remnant in the western U.S.

17A HYDROSTATICS: FLUIDS AT REST

17.1 The Science of Fluids

Objects can be immersed in fluids, and fluids can pass through objects. Fluids can flow around objects and structures, as well as flow through pipes and orifices. For these reasons, fluids exhibit some interesting mechanical properties that have great significance to the proper functioning of life processes as well as to strictly physical phenomena. In earlier chapters we discussed nonfluid mechanics—the mechanics of rigid objects. In this chapter you will study how Newtonian principles apply to fluids, a field called **fluid mechanics.** This area of mechanics can be broadly subdivided into **hydrostatics**—the study of stationary fluids where all forces are in equilibrium—and **hydrodynamics**—the study of fluids in motion.

17.2 Density

A major difference between the study of nonfluids and the study of fluids is that fluids are continuous. If you pour twenty cupfuls of water into a bucket, the water will assume one unified structure, but if you place twenty wooden blocks in a bucket, they will retain their own identities. Therefore, it is impractical to talk about the water and the wood in the same way. One consequence of this difference is that the mass of a fluid sample is less useful than the mass of a solid. The **density** (ρ) (mass per unit volume) of a fluid is usually more important in a physics context. Table 17-1 lists the densities of some common materials.

The densities in Table 17-1 are given in grams per cubic centimeter (g/cm^3). This unit is the common or CGS unit for density. The SI unit of density is kg/m^3. Water's density is 1000 kg/m^3. Since water's density is not numerically 1 in the SI, the densities of other substances in these units do not indicate their relative density directly as they do in the CGS system. However, if the densities of other substances are divided by the density of water, then we obtain a relative density called the **specific gravity (s.g.).** The specific gravity of a substance is a dimensionless number that is numerically equal to the density of the substance in g/cm^3.

TABLE 17-1
Representative Densities
(Liquids in Color)

Substance	Density (g/cm³)
Hydrogen	0.00009
Nitrogen	0.00126
Air (dry)	0.00129
Oxygen	0.00143
Wood (balsa)	0.12
Wood (cork)	0.24
Wood (pine)	0.50
Gasoline	0.68
Wood (oak)	0.72
Ethyl alcohol	0.79
Oil (crude)	0.88
Ice	0.92
Oil (olive)	0.92
Water	1.00
Seawater	1.03
Blood (whole)	1.05
Table salt	2.17
Sand	2.32
Glass	2.40–2.80
Aluminum	2.70
Bromine	3.12
Iron/Steel	7.86
Copper	8.92
Silver	10.50
Lead	11.30
Mercury	13.60
Gold	19.30
Platinum	21.45

17.3 Units of Pressure

We have mentioned the concept of pressure in several contexts in earlier chapters. Fluid pressure is a distinctive property of liquids and gases. Perhaps you have experienced a change of pressure in your sinuses during a rapid descent or ascent on a mountain road or in an airplane. Skin and scuba divers do not have to go very deep before they experience significant pressure effects. How can we characterize pressure in a fluid?

Pressure is defined as the force exerted perpendicular to a unit area. If you select a small hypothetical planar area dividing two regions of fluid anywhere within the volume of the fluid (see Figure 17-2), the force exerted by the fluid on one side of the planar area is equal in magnitude but opposite to the force exerted by the fluid on the other side of the area. The two forces acting in opposite directions on different parts of the fluid form an action-reaction force pair and cancel each other according to Newton's

TABLE 17-2			
Densities and Specific Gravities for Some Selected Substances			
Substance	Density		Specific Gravity
	g/cm³	kg/m³	
Air	0.001 29	1.29	0.001 29
Ethyl alcohol	0.79	791	0.791
Water	1.00	1000	1.00
Iron	7.86	7860	7.86
Mercury	13.60	13 600	13.60
Gold	19.30	19 300	19.30

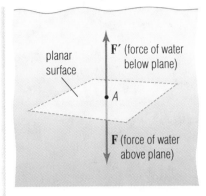

17-2 The pressure on one side of a hypothetical plane surface is exactly equal to the pressure on the opposite side when the fluid is in mechanical equilibrium.

third law. Consequently, at every point in a sample of fluid at rest, the pressure at that point is equal in all directions. At the boundaries of the fluid, the container exerts a pressure on the fluid identical to the pressure the fluid exerts on the container.

The SI unit of pressure is the pascal (Pa). Some other units of pressure are commonly used by different professions. The **atmosphere (atm)** was probably the earliest reference pressure. It is the average pressure of the atmosphere at sea level. In pascals,

$$1 \text{ atm} = 1.013 \times 10^5 \text{ Pa.}$$

While this is the accepted average value for atmospheric pressure, the actual value at a given location will vary due to factors such as elevation, weather, and temperature.

Chemists and other scientists often use the **torr,** named after Evangelista Torricelli, the inventor of the mercury barometer. This non-SI pressure unit is defined as $^1/_{760}$ of the pressure exerted by a column of mercury 760 mm high at sea level. This value was chosen because one atmosphere of pressure will support a 760 mm column of mercury in an evacuated glass tube, which was the basis for the first barometer. It should not be surprising that pressure may also be measured in mm of mercury or even inches of mercury. Atmospheric pressure is often reported in the news media in these units.

Meteorologists are more likely to report atmospheric pressure in non-SI units called **bars** or **millibars (mb)**. A bar is 10^5 Pa, so

$$1 \text{ atm} = 1.013 \text{ bar.}$$

Millibars permit the convenient reporting of atmospheric pressures in whole numbers (or with only a single decimal place). A list of pressures equivalent to 1 atm reported in various SI and non-SI units is provided in Appendix C.

Engineers often report pressures in piping systems in **gauge pressure.** Mechanical pressure gauges are designed so that one side of the pressure-detecting mechanism senses atmospheric pressure and the other side senses system pressure. When the gauge indicates zero, it still experiences 1 atm. Therefore, the actual pressure, or **absolute pressure (P)**, in the system is system gauge pressure (P_g) plus 1 atm.

17.4 Pressure in Incompressible Fluids

We have seen that pressure at a point in a fluid is the same in all directions. But we also know that pressure varies with the vertical position in the fluid. How does this occur? Let's first look at how pressure changes with depth in a liquid.

Some news services and other sources of weather reports use inches of mercury for atmospheric pressure. 1 atm = 29.92 in. Hg

The **torr** has no unit abbreviation. Some references give the unit for torr as 1 torr = 1 mm Hg.

Evangelista Torricelli (1608–47) was an Italian mathematician and physicist. His many works included deriving principles in plane and solid geometry, describing ballistic motion (with Galileo), and demonstrating that a vacuum could be formed.

Most problems in this textbook will involve absolute pressure. However, in everyday life you will more likely read gauge pressure. The common system pressure unit in the U.S. is pounds per square inch (gauge), or psig.

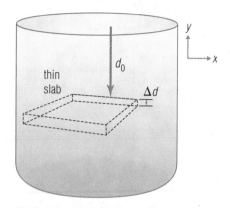

17-3 A thin slab of fluid at equilibrium

The particles in liquids are very close together, so they cannot be forced much closer together than they are already. For this reason, most liquids are essentially incompressible to a first approximation, so their densities can be assumed to be constant throughout the bulk of the liquid.

Let's choose a vertical coordinate system with the origin at the surface of the liquid so that $y_0 = 0$ m and positive is upward. Consider a small volume of the liquid, a thin rectangular slab with a square horizontal face of area A and thickness Δd. The top of the slab is at depth $y = d_0$ (a negative value) beneath the surface of the liquid. The liquid has a density of ρ, which is constant throughout the liquid. The pressure at the surface of the liquid is P. The pressure at depth d_0 is P_{d_0}. The pressure of the liquid at the bottom of the slab is P_d, where the depth is

$$y = d = d_0 + \Delta d_y.$$

(Note that Δd_y is a change of depth in the $-y$ direction.)

Assuming that the liquid is not in motion, then the slab is not moving, which means that the net force on the slab of liquid (the system) is zero. According to Newton's first law,

$$\Sigma \mathbf{F} = 0 \text{ N}.$$

Recall that

$$P = \frac{F}{A}, \text{ so}$$

$$F = PA.$$

The forces on the equal areas of the opposing sides of the slab are opposite and equal to each other, so they cancel and do not need to be considered further. The vertical forces acting on the slab of liquid must be in equilibrium as well. The downward force on the upper surface at depth d_0 is

$$\mathbf{F}_{d_0} = -P_{d_0}A, \tag{17.1}$$

and the upward force on the lower surface of the slab at depth d is

$$\mathbf{F}_d = +P_dA. \tag{17.2}$$

There is one more vertical force to consider—the weight of the liquid in the slab itself,

$$\mathbf{F}_w = m\mathbf{g}.$$

The mass of the liquid is determined by the product of the liquid's density and its volume:

$$m = \rho V$$

But the volume of the slab is the product of the area and the magnitude of the change in depth (Δd_y is negative):

$$V = A|\Delta d_y| = A(-\Delta d_y)$$

So, the weight of the mass of liquid is the expression

$$\mathbf{F}_w = \rho A\mathbf{g}(-\Delta d_y) = -\rho A\mathbf{g}\Delta d_y, \tag{17.3}$$

which is a downward force on the slab of liquid. Referring the forces to the vertical coordinate system where upward is positive, we have the force sum in the vertical direction.

$$\Sigma \mathbf{F}_y = \mathbf{F}_{d_0} + \mathbf{F}_d + \mathbf{F}_w = 0 \text{ N} \tag{17.4}$$

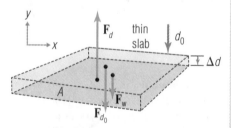

17-4 Forces on the hypothetical slab of liquid

Substituting Equations 17.1, 17.2, and 17.3 for each force in Equation 17.4, we have

$$(-P_{d_0}A)_y + (+P_dA)_y + (-\rho Ag\Delta d)_y = 0 \text{ N}.$$

This is a one-dimensional problem, so we can dispense with the vector component notation. Setting the pressure terms equal to the weight term,

$$-P_{d_0}A + P_dA = \rho Ag\Delta d.$$

Dividing both sides by the area A and rearranging the left side yields

$$P_d - P_{d_0} = \rho g\Delta d. \tag{17.5}$$

Equation 17.5 indicates that the change of pressure from the top to the bottom of any slab of a liquid is a function of only the change of the depth, since liquid density and gravitational acceleration are assumed to be constant throughout the liquid (a reasonable assumption). Therefore, we can calculate the pressure at any depth d ($\Delta d = d - 0$) in a liquid using the following equation:

$$P_d = P_0 + \rho gd, \tag{17.6}$$

where d is expressed in *negative* scalar distances below the reference level at the surface of the liquid, and $g = -9.81 \text{ m/s}^2$. As you can see, pressure increases with fluid depth. If the container with the liquid is open to the atmosphere, the reference level d_0 is the liquid's surface, and the pressure $P_0 = P$ (the atmospheric pressure). For a closed container, P_0 is often greater than the atmospheric pressure.

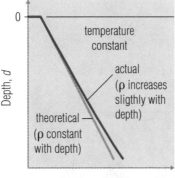

17-5 Pressure varies as an essentially linear function with depth. Compression of liquids at great depths causes the graph to depart from a truly linear relationship.

EXAMPLE 17-1

Ear Squeeze: A Practical Pressure-at-Depth Problem

Most people's eardrums can withstand a difference of up to 3.2×10^4 Pa between the external pressure and middle ear pressure. This limitation can be a factor when free diving in a body of water. If a freshwater lake has a density of $1.00 \times 10^3 \text{ kg/m}^3$, how deep can an average person dive without fear of rupturing his eardrums? Assume that middle ear pressure remains at atmospheric pressure during the dive.

Solution:
We know that the maximum difference in pressure allowed (ΔP_{max}) is 3.2×10^4 Pa (1 Pa = 1 N/m^2). Let $P_0 = P$ = atmospheric pressure. Equation 17.6 will permit us to calculate d_{max}.

$$\Delta P_{max} = P_d - P_0$$

$$\Delta P_{max} = (P + \rho gd_{max}) - P$$

$$\Delta P_{max} = \rho gd_{max}$$

$$d_{max} = \frac{\Delta P_{max}}{\rho g} \tag{1}$$

Substitute the known values into Equation (1) to solve for the maximum depth:

$$d_{max} = \frac{3.2 \times 10^4 \text{ N/m}^2}{(1.00 \times 10^3 \text{ kg/m}^3)(-9.81 \text{ m/s}^2)}$$

$$d_{max} \doteq -3.\underline{2}6 \, \tfrac{\text{N/m}^2}{(\text{kg/m}^3)(\text{m/s}^2)} = -3.\underline{2}6 \, \left(\tfrac{\text{kg·m}}{\text{m}^2 \cdot \text{s}^2}\right)\left(\tfrac{\text{m}^3}{\text{kg}}\right)\left(\tfrac{\text{s}^2}{\text{m}}\right)$$

Change newtons to base SI units in order to complete the cancellation. Careful cancellation yields the expected unit for depth.

$$d_{max} \doteq -3.\underline{2}6 \text{ m}$$

$$d_{max} \approx -3.3 \text{ m}$$

The average person can free dive to a depth of about 3.3 m, or about 10 feet, without equalizing pressure in his ears and not rupture his eardrums. As a diver approaches this depth, a painful condition called *ear squeeze* occurs. Divers avoid this condition by forcing air into their middle ears via their eustachian tubes in order to equalize the pressure across their eardrums.

The shape and volume of a container has no bearing on the pressure at a given depth in a liquid. The apparatus in Figure 17-6 demonstrates this principle. The variously shaped containers have the same external pressure—the pressure of the surrounding air. When the liquid levels are the same in each container, the pressure is the same at the same depth.

17-6 Pascal's vases, illustrating the principle that fluids at the same depth have the same pressure

17.5 Pressure in Compressible Fluids

We saw in the previous discussion that the pressure at a given depth in an incompressible fluid is proportional to the depth and the density of the liquid above that depth. This is true for pressure in a gaseous fluid as well. However, the density is not constant with height (or depth, depending on the location of the reference point) in a gas because of a gas's inherent compressibility. In fact, gas density decreases exponentially with height, so gas pressure varies the same way as long as other factors remain constant. (Recall from Chapter 14 that the density of an unconfined gas is also dependent on temperature as well as pressure.) In equation form, this relationship is

> In Equation 17.7, e is the base of the natural logarithm. Your scientific calculator should provide a function that facilitates this calculation.

$$P = P_0 e^{-\frac{\rho_0}{P_0}|g|h}, \tag{17.7}$$

where P is the gas pressure at height h above the reference height, P_0 is the pressure at the reference height, and ρ_0 is the gas density at the reference height. Note that only the magnitude of g is used in the exponent.

EXAMPLE 17-2

Mountain Sickness: Pressure in a Compressible Fluid

What is atmospheric pressure at the top of Mt. Everest (8850 m)? Assume that pressure at sea level is 1.013×10^5 Pa, and atmospheric density does not depend on temperature (not a realistic assumption).

Solution:

You already have the information necessary to solve Equation 17.7 for pressure at 8850 m, or you can look it up.

$$P = P_0 e^{-\frac{\rho_0}{P_0}|g|h}$$

$$P = (1.013 \times 10^5 \text{ Pa}) \exp\left[-\frac{1.29 \text{ kg/m}^3}{1.013 \times 10^5 \text{ N/m}^2}(9.81 \text{ m/s}^2)(8850 \text{ m})\right]$$

$$P = (1.013 \times 10^5 \text{ Pa}) \exp\left[-1.105 \left(\frac{\text{kg}}{\text{m}^3}\right)\left(\frac{\text{m}^2 \cdot \text{s}^2}{\text{kg} \cdot \text{m}}\right)\left(\frac{\text{m}}{\text{s}^2}\right)\text{m}\right]$$

$$P = (101\,300 \text{ Pa})\, e^{-1.105}$$

$$P \doteq 33\,550 \text{ Pa}$$

$$P \approx 3.36 \times 10^4 \text{ Pa}$$

Problem-Solving Strategy 17.1

Complex exponential expressions are common in physics. In order to prevent confusion, the term "exp" is inserted before the expression in order to alert the reader that the following parenthetical statement is the exponent of the natural exponential base e. All units in the exponential expression must cancel.

Ignoring the effect of temperature on the density of the atmosphere, the pressure at the top of Mt. Everest is about one-third of sea level pressure. This lack of air pressure at high altitudes causes *anoxia,* lack of oxygen, in mountain climbers who climb without supplemental oxygen tanks. It is even a problem for athletes who have trained at elevations near sea level but must compete at high-altitude locations.

Observe that all units in the complicated exponent in the exponential expression cancel out—exponents are dimensionless numbers.

17.6 Hydraulic Devices

Fill a sports bottle completely to the top. After tightening the cap with the pop top open, squeeze the bottle. What happens? The stream of water squirting from the cap indicates that the force of your hand was somehow transmitted directly to the mass of the water being ejected from the bottle.

In the seventeenth century, Blaise Pascal studied the effect of pressure on incompressible fluids (liquids) that completely filled their containers. His experiments led him to the principle that is now called **Pascal's principle:** *the external pressure applied to a completely enclosed incompressible fluid is distributed in all directions throughout the fluid.* Consider Equation 17.6 again. If we shut the cap of the sports bottle and then squeeze on the bottle to change P_d, the reference pressure P_{d_0} must increase by the same amount in order to maintain the equality.

Pascal's principle provides the theoretical basis for **hydraulic devices,** which are machines that transmit forces via enclosed liquids. Figure 17-9 shows how a small input force can generate a large output force. Since we are considering the *change* in pressure caused by an applied force, we can ignore the static pressure due to liquid depth throughout the

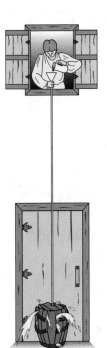

17-8 In a famous experiment, Blaise Pascal attached a long, narrow pipe to a tightly sealed cask. After he filled the cask with water, the relatively small amount of water needed to fill the tube provided enough pressure to break the cask.

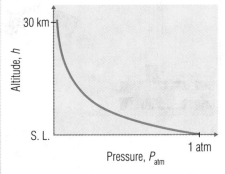

17-7 Atmospheric pressure varies exponentially with altitude. Blaise Pascal predicted that a vacuum existed above the atmosphere after he measured a decrease of atmospheric pressure as he ascended a mountain with a barometer.

Blaise Pascal (1623–62) was a French mathematician, physicist, and philosopher known for his theory of probability and his work on pressure. He invented one of the first calculating machines. An older computer language was named after him.

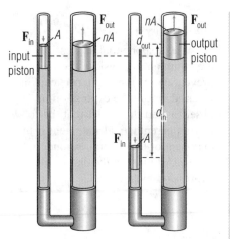

17-9 The principle behind hydraulic devices

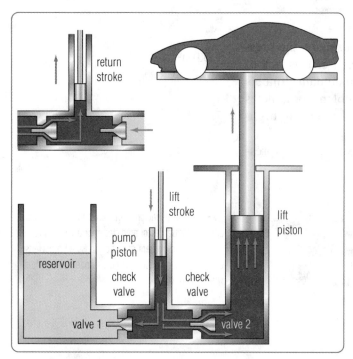

17-10 A hydraulic lift

device. Static pressure is the same at equal depths throughout the device and does not affect the change of output pressure and the associated force.

The pressure created in the fluid by the force on the small piston is transmitted to the large piston. Assume that the cross-sectional area of the small piston is A and that of the large piston is nA. The pressure at the small piston is

$$P = \frac{F_{in}}{A}. \tag{17.8}$$

Since the large piston is at the same height as the small piston, there is no effect on pressure due to height differences. Therefore, the pressure at the large piston is P. The force on the large piston is

$$F_{out} = PnA.$$

Substituting Equation 17.8 for P, the areas cancel, giving the equation

$$F_{out} = nF_{in}. \tag{17.9}$$

A small force on the small piston can lift a heavy load on the larger piston if the larger piston is big enough. For example, a small force on a hydraulic brake pedal can stop the wheels of a speeding car. Similarly, a child can lift a car by using a hydraulic jack.

It seems, at first glance, that a hydraulic device produces more energy than it consumes. In reality, the device simply transmits work. Mechanical work is the scalar product of force and displacement. The small piston experiences a smaller force, but it travels a greater distance than the larger piston. This is the distance principle you observed in simple mechanical machines. The work done on the large piston is equal to or smaller than the work done on the smaller piston. In real hydraulic devices, some work is used to slightly compress the fluid and overcome friction in the device, so less work is produced than is consumed.

The hydraulic jack has a spiritual analogy. A little pressure from family or friends can have a major impact on our decisions. Unfortunately, Satan knows well how to manipulate this pressure. In Job 2:9, Job's wife says, "Dost thou still retain thine integrity? curse God, and die." Fortunately Job was able to resist her pressure, but the account illustrates how careful we should be as we choose our friends and especially our spouses.

EXAMPLE 17-3

Multiplying Force: A Hydraulic Jack

A hydraulic jack uses water as its working fluid. The face of its small input piston is 10.0 cm above the bottom of the water reservoir in the jack. The area of the input piston is $A_{in} = 2.00$ cm^2. The output piston's area is $A_{out} = 80.0$ cm^2. (a) If the magnitude of the net force applied to the input piston is 25.0 N, what is the pressure at the *bottom* of the water reservoir? (b) What is the magnitude of the force exerted on the output piston?

Solution:
The pressure exerted by the input piston is computed from Equation 17.8:

$$P = \frac{F_{in}}{A_{in}} = \frac{25.0 \text{ N}}{2.00 \text{ cm}^2} \left| \frac{10^4 \text{ cm}^2}{1 \text{ m}^2} \right.$$

$$P = 1.25 \times 10^5 \text{ N/m}^2 \tag{1}$$

(a) Taking result (1) as P_0, calculate the pressure at the bottom of the reservoir using Equation 17.6:

$$P_d = P_0 + \rho g d$$

$$P_{d-10\text{ cm}} = 1.25 \times 10^5 \text{ N/m}^2 + (1.00 \times 10^3 \tfrac{\text{kg}}{\text{m}^3})(-9.81 \tfrac{\text{m}}{\text{s}^2})(-0.100 \text{ m})$$

$$P_{d-10\text{ cm}} \doteq 1.259 \times 10^5 \text{ N/m}^2$$

$$P_{d-10\text{ cm}} \approx 1.26 \times 10^5 \text{ N/m}^2 \qquad (2)$$

(b) Pressure at the same height in a closed container is equal everywhere. Therefore, no pressure adjustment for height is required. The ratio of the output piston's surface area to the input piston's is

$$n = \frac{A_{out}}{A_{in}} = \frac{80.0 \text{ cm}^2}{2.00 \text{ cm}^2}$$

$$n = 40.0.$$

Use Equation 17.9 to calculate the output force:

$$F_{out} = nF_{in} = (40.0)(25.0 \text{ N})$$

$$F_{out} = 1.00 \times 10^3 \text{ N}$$

The output force exerted by the jack is 40 times the input force.

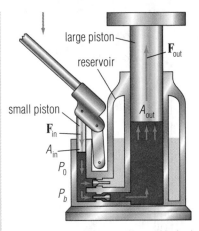

17-11 Schematic of a typical manual hydraulic jack

> **Problem-Solving Strategy 17.2**
>
> Under normal circumstances, the difference of pressure due to height differences between input and output pistons in hydraulic devices is negligible compared to the force transmitted by the hydraulic system.

17.7 Liquid-Medium Pressure Indicators

It is often important to know a fluid's pressure within a system, especially in engineering applications. For example, the crankcase of a diesel engine is a sealed chamber at the bottom of the engine, containing the crankshaft and the engine's lubricating oil. Operators need to know if a significant pressure is building up in the crankcase, which would indicate that fuel and exhaust vapors are accumulating and could cause an explosion. This pressure buildup can be caused by worn piston rings. A device called a **manometer** is connected to the crankcase to indicate a positive or negative pressure compared to atmospheric pressure.

A manometer is a U-shaped transparent tube that contains a dense liquid such as mercury or some other nonvolatile liquid. One end is attached to the container of the fluid whose pressure is unknown, and the other end is open to the atmosphere. A flexible tube connects the two ends of the instrument. Before the fluid pressure is applied, the movable end of the manometer is adjusted until the level of mercury in both ends is the same. Then the fluid pressure is applied to the closed end of the manometer. Assuming that the fluid pressure is greater than atmospheric, the mercury level will drop in the closed end and rise in the open end of the manometer. The difference in the levels of the mercury on the two sides of the tube is proportional to the difference between the pressure in the fluid and that in the atmosphere. This difference of levels is read as a pressure on an attached scale. Obviously, if the monitored fluid's pressure is less than atmospheric pressure, then the mercury will rise in the closed end and drop in the open end of the manometer.

The first instrument to accurately measure atmospheric pressure was the mercury **barometer.** This is essentially a long glass tube sealed at one end, filled with mercury, and carefully overturned into an open reservoir of mercury so that no air enters the tube. If the tube is sufficiently long, the mercury meniscus at the top of the tube will drop until atmospheric pressure exerted on the reservoir just balances

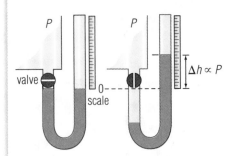

17-12 A manometer measures relative pressure.

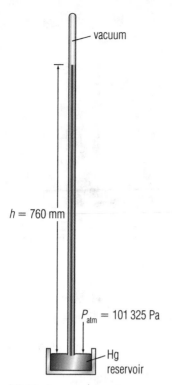

17-13 A mercury barometer

the weight of the column of mercury. Greater atmospheric pressure supports a higher column of mercury. Standard atmospheric pressure, 101 325 Pa, supports a 760 mm column of mercury. As mentioned earlier in this chapter, these relationships provide various systems of units for reporting atmospheric pressure.

17.8 Buoyancy

A story goes that more than two hundred years before the birth of Christ, the scientist and mathematician Archimedes was given an assignment from his king, Hieron II. The king had ordered a goldsmith to make him a crown from the gold the king gave him. The goldsmith made the crown and delivered it to the king. Then the king began to doubt the craftsman's honesty. Had he used some silver in the crown and kept some of the gold for himself? The king immediately weighed the crown and found that it weighed the same as the gold the king had given the goldsmith. But what if the craftsman had replaced the gold with an equal weight of silver? Was there any way to find out? This was the problem the king took to Archimedes.

Archimedes thought hard about the gold crown. In what way do silver and gold differ, other than in value? Of course, their densities are different, but Archimedes did not have instruments to measure density accurately. Then one day the scientist was relaxing at the local bath. As he lowered himself into the full bath, some water spilled over the edge. Suddenly Archimedes had the answer to the crown problem. If he placed the crown in a full container of water, the amount of water that spilled out would depend on the crown's volume. A given weight of gold would force out less water than would a gold-silver mixture. When Archimedes tested the crown, he found that it indeed was a gold-silver mixture. Thinking about the solution led Archimedes to the principle that the buoyant force of water on an object equals the weight of the water it displaces.

17.9 Submerged Objects

What happens to an object placed in a fluid? Let's consider a volume of fluid to be the system. Its shape doesn't matter as long as it is in equilibrium with the surrounding fluid. The forces acting on the fluid system are its weight (F_{w-f}) and its equilibrant force, which we shall call the **buoyant force** (F_b). The buoyant force results from the fluid pressure on the system. Because the fluid system is in equilibrium, these two forces are equal in magnitude but opposite to each other.

If the fluid system is now replaced by an object that is exactly the same shape, the gravitational force on the new object-system (F_{w-o}) will be different. However, the buoyant force that is dependent on the pressure of the rest of the fluid on the *volume* of the object remains the same. The object experiences an upward force that is equal to the weight of the fluid it displaces. The magnitude of the buoyant force is determined from the equation

$$F_b = \rho |g| V,$$ (17.10)

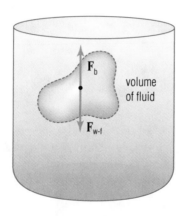

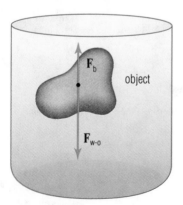

17-14 Buoyancy is the force a fluid exerts on an immersed object.

where ρ is the density of the *displaced fluid*. The fluid-system experienced the same force, but it was balanced by its own weight. Equation 17.10 is the mathematical expression for **Archimedes' principle.** This principle states that *any system that is submerged or floats in a fluid is acted on by an upward buoyant force equal in magnitude to the weight of the fluid it actually displaces.*

Archimedes' principle holds for a system of any shape. The buoyant force may be greater than a system's weight, less than its weight, or equal to its weight. If it is equal to the system's weight, the system behaves exactly like the fluid it displaces. It is not accelerated at all because the forces acting on it are balanced. This implies that the overall density of the system must equal the fluid it displaces, because its volume and weight (and mass) are the same.

Similarly, if the weight of a system is greater than that of the displaced fluid, its density is greater than the fluid's. Its downward weight will exceed the upward buoyant force, and the object will begin to sink. However, the net force accelerating the system downward is the difference of the system's weight and the buoyant force. Consequently, the system appears to weigh less in the fluid. You can verify this fact by lowering a heavy stone into a pond. As it submerges, the stone seems to become lighter.

If the weight of a system is less than that of the displaced fluid, its density is less than the fluid's. An object less dense than the fluid will not have enough weight to balance the buoyant force. The net force will accelerate the system upward in the fluid. If the fluid is a liquid, the system will eventually break through the surface. When this happens, the buoyant force on the system rapidly decreases as the volume of liquid that is displaced decreases. Eventually, the buoyant force equals the weight of the system, and the system stops rising with part above the liquid's surface and part below.

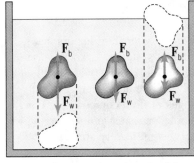

$$\rho_o > \rho_f \qquad \rho_o = \rho_f \qquad \rho_o < \rho_f$$

17-15 Knowing the comparative densities of the immersed object and the fluid will determine whether the object sinks, floats, or is neutrally buoyant.

EXAMPLE 17-4

Floating Like a Brick: Buoyancy

A 1.00×10^3 cm^3 brick ($\rho = 1.74$ g/cm^3) is placed in mercury ($\rho = 13.6$ g/cm^3). (a) What is the buoyant force on the brick? (b) Will it float?

Solution:
(a) Determine the magnitude of the buoyant force on the brick fully submerged in the mercury:

$$F_b = \rho |g| V$$
$$F_b = (13.6 \text{ g/cm}^3)(9.81 \text{ m/s}^2)(1.00 \times 10^3 \text{ cm}^3)$$
$$F_b \doteq 133\,400 \text{ g·m/s}^2$$

Convert grams to kilograms in order to obtain newtons:

$$F_b \approx 133 \text{ kg·m/s}^2 = 133 \text{ N}$$

(b) The magnitude of the brick's weight is as follows:

$$F_w = m|g| = (\rho V)|g|$$
$$F_w = (1.74 \text{ g/cm}^3)(1.00 \times 10^3 \text{ cm}^3)(9.81 \text{ m/s}^2)$$
$$F_w \doteq 17\,060 \text{ g·m/s}^2$$
$$F_w \approx 17.1 \text{ N}$$

The brick's weight is less than the maximum possible buoyant force of 133 N; so the brick will float in the mercury.

17-16 Because the density of icebergs ($\rho_{ice} = 0.92$ g/cm^3) is less than that of seawater ($\rho_{sw} = 1.03$ g/cm^3), icebergs float on the ocean. The visible portions are often huge compared to oceangoing vessels, yet only 11 percent of their entire volume is above the ocean's surface.

Problem-Solving Strategy 17.3

If you need to know if an object will float in a liquid, compare the densities of the object and the liquid. If the object's density is less than the liquid's density, it will float. Otherwise it will submerge or even sink.

If the fluid is a gas and the system can rise without restriction, such as a balloon in the atmosphere, the gas density decreases with altitude, and thus the buoyant force decreases as well. Eventually, an altitude is reached at which the densities of the balloon and the displaced gas are equal and the balloon stops rising.

There are other fluid factors that can affect system buoyancy: variations in fluid density with depth and height, temperature changes, and changes of fluid composition. For example, seawater temperature and salinity significantly affect the buoyancy of a ship or submarine.

17.10 Buoyancy of Real Objects

We have assumed that the submerged systems we have considered are rigid and have constant densities. These ideal systems do not respond to changes in fluid pressure as their depths or heights change. Real systems, such as helium balloons and submarines, *do* respond to pressure changes. For example, a submarine's crew can adjust the sub's weight at a specific depth so that its weight exactly equals its buoyant force—it is *neutrally buoyant*. If the submarine then moves to a shallower depth, the sea pressure becomes less and its hull expands slightly as the metal cylinder of the hull is less compressed. The submarine's volume then increases but its weight remains the same, so its density decreases. As a result, the buoyant force exceeds the weight of the submarine, and the net force acting on the ship is upward—it is *positively buoyant*. The opposite effect occurs if the ship goes deeper—it becomes *negatively buoyant*. Balloons and "rigid" airships experience similar effects due to the variation in atmospheric pressure with altitude.

17.11 Center of Buoyancy

Every object submerged in a fluid has both a center of gravity (center of mass) and a center of buoyancy. The **center of buoyancy** is the center of mass of the fluid that would occupy the submerged space that the object occupies. If the object has uniform density and is completely submerged, its center of mass and center of buoyancy are the same. The same is true if the object's mass is distributed symmetrically horizontally. However, if an object's mass is not symmetrically distributed horizontally, its center of gravity will not coincide with the center of buoyancy. Therefore, the buoyant force and the gravitational force will not be on the same vertical line. A torque will exist on the object, and the object will rotate as it sinks, rises, or stays in place until its center of gravity is on a vertical line with its center of buoyancy.

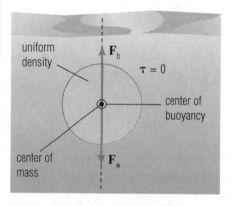

17-17a In a submerged object of uniform density and shape, the centers of mass and buoyancy are located at the same point.

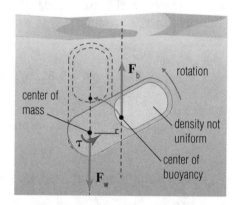

17-17b When the mass is not distributed in a horizontally symmetrical position, the resulting torque will rotate the object until its centers of mass and buoyancy fall on the same vertical line.

17.12 The Hydrometer

The relationship between density and buoyancy provides a method for measuring fluid density. One instrument used to measure density is a **hydrometer.** The hydrometer is a cylinder with a weighted end. When it is placed in a low-density liquid, the hydrometer sinks more than it does in a high-density liquid. The volume of the liquid that the hydrometer must displace to float is inversely proportional to the density of the liquid. Each hydrometer is calibrated for use with a specific liquid. The scale is placed so that the surface of the liquid intersects it at the value corresponding to the liquid's density or its specific gravity. Hydrometers are

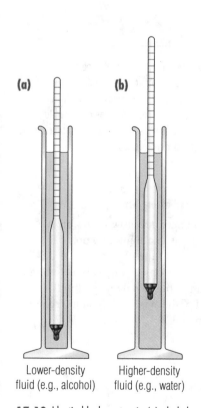

(a) (b)

Lower-density fluid (e.g., alcohol)   Higher-density fluid (e.g., water)

17-18 Identical hydrometers in (a) alcohol and (b) water commonly used to test the amount of antifreeze in a radiator and the concentration of ions in a battery. Scientists use hydrometers to measure the amount of suspended sediment in lake and river water as well.

Do not confuse *hydrometers* with *hygrometers*. The latter instrument is used to determine the relative humidity of the atmosphere.

FACETS of PHYSICS

Maple Syrup Hydrometers

Maple syrup can cost more than twenty-five dollars per gallon. Since the syrup is valuable, sensible manufacturers are careful to preserve it properly. An important part of this preservation is making syrup the desired density.

To make the syrup, producers gather maple sap and boil it. Some of the water in the sap evaporates, leaving a concentrated sugary solution that is much denser than before. The syrup is then filtered to remove impurities, poured into containers, and sealed.

If the syrup is not dense enough, it will spoil easily; but if it is too dense, the sugar will precipitate out of the solution and crystallize. The proper density for hot syrup is about 1.33 g/cm³.

Vermont, which normally produces the most maple syrup in the United States, requires that each gallon labeled "pure Vermont maple syrup" have the right density. A maple syrup hydrometer determines the density of the syrup. This instrument is an ordinary hydrometer with a colored mark at 1.33 g/cm³. A syrup producer pours boiling syrup into a tall metal cup and then places the hydrometer in the cup. If the mark is level with the surface of the syrup, the syrup has boiled enough. If the mark is below the surface, the syrup is not dense enough and must be boiled more. If the mark is above the surface, the syrup is too dense and must be diluted with fresh sap and boiled again.

The sugar industry (also including corn, cane, and sugar beet syrup, as well as candy producers) uses a variety of traditional density units such as degrees brix, degrees baumé, degrees twaddle, and other whimsical names. Each relates a "degree" of density to a specific mass of sugar per unit volume. Maple sugar producers must ensure a minimum density of 66.0 degrees brix at 68 °F.

In recent decades, the maple syrup industry has been producing up to five different grades of syrup based primarily on color. The highest quality, fancy grade syrups (lightest in color) are often produced near the minimum of 1.33 s.g. Fuller bodied, darker syrups are considered to be lower in grade and are usually somewhat higher in density, since the sugar is more concentrated.

17A Section Review Questions

1. **a.** Name and briefly describe the two subsections of fluid mechanics.

   b. In what fundamental way does fluid mechanics differ from the mechanics you have studied prior to this chapter?

2. **a.** Using qualitative terms, how does fluid pressure vary horizontally through a fluid?

   b. How does it vary vertically?

3. What three factors determine the static pressure in any fluid at a given depth (or height)?

4. What determines the force a fluid exerts against a surface?

5. What two factors determine the buoyant force exerted on an object submerged in a fluid?

⊛6. How deep (in meters and feet) must a diver go (in fresh water) in order to experience two atmospheres of pressure? (Recall that a diver experiences 1 atm at the water's surface.) State your answer to two significant digits.

⊛7. Assuming that atmospheric pressure at sea level is 1.00 atm, what would be atmospheric pressure at the level of the Dead Sea (altitude: −400. m)? Ignore difference in pressure due to variation of temperature.

⊛8. A block of solid oak (30.0 cm × 40.0 cm × 50.0 cm) floats in fresh water. What percentage of the block's volume is *not* under water?

17B HYDRODYNAMICS: FLUIDS IN MOTION

17.13 Ideal Fluids

The concepts that you have been studying in hydrostatics are fairly simple. Hydrodynamics, in contrast, is one of the most complicated areas of physics. To simplify the study, we must establish the concept of the **ideal fluid.** In the case of an ideal fluid, it is assumed that

- the fluid flows smoothly;
- the velocity of the fluid does not change with time at a fixed location in the fluid path;
- the density of the fluid is constant (the fluid, whatever its density, is incompressible);
- friction has no effect on fluid flow.

These assumptions are only approximately true for real fluids, whether liquids or gases.

The first assumption, the idea that ideal fluids flow smoothly, means that every particle passing through a given point follows the same path. Such a path is called a **streamline.** Streamlines have no physical reality, but they are useful in visualizing fluid flow. If the streamlines are continuous, then the flow is smooth, or **laminar.** If the streamlines have discontinuities (bumps, twists, or turns), then the flow is **turbulent** and thus requires much more complex modeling. Several streamlines can define a **flow tube.** A flow tube may be an actual cylindrical pipe, or it may be a small volume within a larger volume of flowing fluid. The fluid particles do not cross the boundaries of a flow tube.

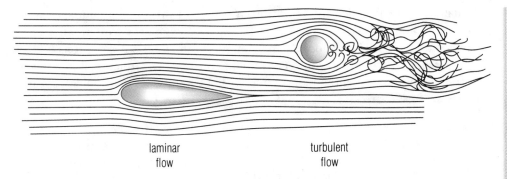

laminar
flow

turbulent
flow

17-19 Streamlines around various objects showing laminar and turbulent flow

The second assumption establishes that the velocity of the fluid at any one location is constant with time. However, it does *not* mean that fluid velocity cannot be different at various points within a body of flowing fluid at the same time.

The third assumption is that fluids are incompressible. This, combined with the law of conservation of mass, requires that the rate of volume flow into a segment of a flow tube equal the rate of volume flow out of the flow tube segment. That is, if 1 L of fluid flows into one end of a flow tube in 1 s, 1 L of fluid must flow out the other end in the same second. If this were not true, then either some mass would disappear in the tube or some mass would be created in the tube.

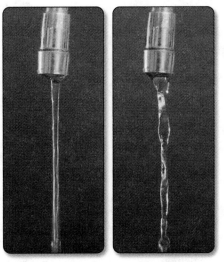

17-20 Laminar and turbulent flow in a real fluid

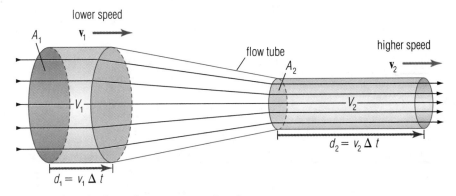

17-21 A flow tube showing how a constriction affects flow rate

17.14 Flow Continuity

The volume of fluid flowing into a flow tube segment is shaped roughly like a cylinder. The base of the cylinder is the cross section of the flow tube at the beginning of the segment, which has area A_1. The "height" of the cylinder is the length traveled by a particle of the fluid in time Δt. This distance is equal to the velocity in that segment of the flow tube times the time interval. The volume flowing into the flow tube segment is

$$V_1 = A_1 v_1 \Delta t.$$

Similarly, the volume of fluid flowing out of the segment in time Δt is

$$V_2 = A_2 v_2 \Delta t.$$

These volumes must be equal.

$$V_1 = V_2$$
$$A_1 v_1 \Delta t = A_2 v_2 \Delta t$$
$$A_1 v_1 = A_2 v_2 \qquad (17.11)$$

Equation 17.11 is called the **equation of flow continuity.** Notice that in order for this equation to hold true, lengths of flow tube segments with smaller cross sections must have higher velocities than lengths with larger cross sections. Otherwise the product of *A* and *v* would not remain constant. This relationship between area and velocity is evident in nature: a river flows rapidly in areas where it is narrow; but as the river widens, its flow rate decreases.

EXAMPLE 17-5

Slow with the Flow: Flow Continuity

A fluid flows through a level pipe at a speed of 4.00 m/s at a point where the pipe's cross-sectional area is 10.0 cm². How fast will it flow where the pipe's area is 20.0 cm²?

Solution:

$$A_1v_1 = A_2v_2$$

$$v_2 = \frac{A_1v_1}{A_2} = \frac{(10.0 \text{ cm}^2)(4.00 \text{ m/s})}{20.0 \text{ cm}^2}$$

$$v_2 = 2.00 \text{ m/s}$$

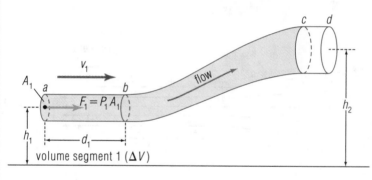

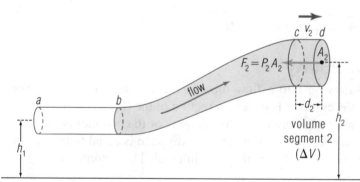

17-22 A fluid gains potential energy by moving against gravity.

17.15 Bernoulli's Principle

Consider the energy and work involved in moving a segment of fluid between *a* and *c* (the colored portion) to the volume between *b* and *d* in Figure 17-22. This is equivalent to transferring a volume of fluid, ΔV, from a height h_1, velocity v_1, and area A_1 to a height h_2, velocity v_2, and area A_2. The change in kinetic energy is

$$\Delta K = \frac{1}{2}mv_2^2 - \frac{1}{2}mv_1^2.$$

Since the density of the fluid is the mass divided by the volume of the fluid, $\rho = m/\Delta V$, then $m = \rho \Delta V$ and

$$\Delta K = \tfrac{1}{2}\rho \Delta V v_2^2 - \tfrac{1}{2}\rho \Delta V v_1^2. \tag{17.12}$$

The change in mechanical potential energy is due in this case to the change in height of the fluid sample above an arbitrary reference height,

$$\Delta U = m|g|h_2 - m|g|h_1, \text{ and}$$

$$\Delta U = \rho \Delta V|g|h_2 - \rho \Delta V|g|h_1. \tag{17.13}$$

Since we assumed that the flow is not turbulent and there are no friction or viscosity effects in or on the fluid, all the work done on the segment of fluid by all other nonconservative forces just changes the total mechanical energy of the fluid. In other words, according to the work-energy theorem,

$$W_{ncf} = \Delta K + \Delta U. \tag{17.14}$$

The total work done *on* the fluid is equal to the work done by the fluid force on the left side of the first volume segment ($\mathbf{F}_1$) plus the work done by the fluid force on the right side of the second volume segment ($\mathbf{F}_2$). The work of each force is calculated from the product of the force and its displacement through each segment of fluid. So the work of the first force is

$$W_1 = F_1 \Delta d_1,$$

since the force and displacement of the fluid are oriented in the same direction. The fluid in the second segment is still moving to the right, but the fluid force exerted on it is oriented to the left. Therefore, the work on the second fluid segment is

$$W_2 = -F_2 \Delta d_2.$$

From the definition of pressure, we know that

$$F_1 = P_1 A_1, \text{ and}$$

$$F_2 = P_2 A_2.$$

Substituting these expressions for force into the respective work equations gives us

$$W_1 = P_1 A_1 \Delta d_1, \text{ and}$$

$$W_2 = -P_2 A_2 \Delta d_2.$$

The arbitrarily small fixed volume ΔV is the product of the area of each fluid segment and the arbitrarily small fixed displacement of the fluid in the flow tube at each segment.

$$\Delta V = A_1 \Delta d_1 = A_2 \Delta d_2$$

We can now substitute ΔV for $A_1 \Delta d_1$ and $A_2 \Delta d_2$ in the work equations. The work done by each of the fluid forces is then as follows:

$$W_1 = P_1 \Delta V$$

$$W_2 = -P_2 \Delta V$$

The work done by nonconservative forces is

$$W_{ncf} = W_1 + W_2, \text{ and}$$

$$W_{ncf} = P_1 \Delta V - P_2 \Delta V. \tag{17.15}$$

In the context of work and potential energy, ΔU means the change of mechanical potential energy rather than the change of internal energy as in Chapter 16.

In this discussion, the symbol Δd is used to represent an arbitrarily small fixed displacement rather than a specific distance.

Substituting Equations 17.15, 17.12, and 17.13 for W_{nc}, ΔK, and ΔU in the work-energy theorem (Equation 17.14), respectively, gives the following:

$$W_{ncf} = \Delta K + \Delta U$$

$$P_1 \Delta V - P_2 \Delta V = (\tfrac{1}{2}\rho \Delta V v_2^2 - \tfrac{1}{2}\rho \Delta V v_1^2) + (\rho \Delta V |g| h_2 - \rho \Delta V |g| h_1)$$

Dividing each term by ΔV gives

$$P_1 - P_2 = \tfrac{1}{2}\rho v_2^2 - \tfrac{1}{2}\rho v_1^2 + \rho |g| h_2 - \rho |g| h_1.$$

Rearranging to group terms associated with each fluid segment, we obtain

$$P_1 + \tfrac{1}{2}\rho v_1^2 + \rho |g| h_1 = P_2 + \tfrac{1}{2}\rho v_2^2 + \rho |g| h_2. \qquad (17.16)$$

Equation 17.16 is called **Bernoulli's equation.**

17.16 Bernoulli's Principle and Constant Velocity

There are two special cases of Bernoulli's equation. One is the case where the cross section of the pipe, and therefore the velocity, does not change. Then

$$v_1 = v_2 = v.$$

Substituting this into Bernoulli's equation:

$$P_1 + \tfrac{1}{2}\rho v^2 + \rho |g| h_1 = P_2 + \tfrac{1}{2}\rho v^2 + \rho |g| h_2$$

$$P_1 + \rho |g| h_1 = P_2 + \rho |g| h_2$$

The difference in pressure ($\Delta P = P_2 - P_1$) depends only on the difference in depth in the fluid ($\Delta h = h_2 - h_1$). This is the same relationship that determines the change of hydrostatic pressure.

17.17 Bernoulli's Principle and Constant Height

The other special case is one in which the fluid remains at the same elevation. Then

$$h_1 = h_2 = h.$$

Substituting this into Equation 17.16:

$$P_1 + \tfrac{1}{2}\rho v_1^2 + \rho |g| h = P_2 + \tfrac{1}{2}\rho v_2^2 + \rho |g| h$$

$$P_1 + \tfrac{1}{2}\rho v_1^2 = P_2 + \tfrac{1}{2}\rho v_2^2$$

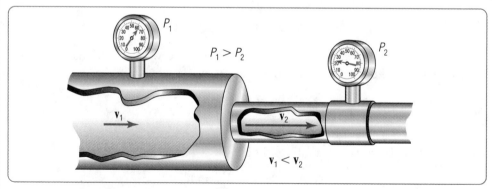

17-23 The faster flow has lower pressure.

Daniel Bernoulli (1700–82) was a Swiss mathematician who studied many different scientific topics. He is best known for his work in hydrodynamics, vibrating strings, and probability.

Problem-Solving Strategy 17.5

In an ideal flow tube, if the flow cross section remains constant, then fluid pressure depends only on the height of the fluid above the reference height according to Bernoulli's equation.

Problem-Solving Strategy 17.6

In an ideal flow tube, if the height remains constant, fluid pressure will increase when fluid speed decreases and vice versa, according to Bernoulli's equation.

This equation shows that in order to maintain the equality, if the velocity of the fluid increases in a segment of flow tube, the corresponding pressure must decrease, and vice versa.

You can verify this result by a simple experiment. Hold two sheets of paper about an inch apart, as in Figure 17-24. Blow steadily between the sheets. You might expect them to fly out from the force of your blowing, but instead they draw together. The moving air between them exerts less pressure on them than the still air around them, so they are pushed together. The photographs on the right demonstrate how lift is created by the Bernoulli effect.

17-24 These two demonstrations show how unbalanced forces across a fluid boundary created by the Bernoulli effect can produce work.

17-25 The moving air lowers the pressure, drawing the liquid up the tube.

EXAMPLE 17-6

Pressure and Speed: Bernoulli's Principle

Water flowing through a level pipe has a pressure of 2.50×10^4 Pa when its speed is 1.00 m/s. The water flows through a constriction into a smaller pipe. What is its pressure when its speed increases to 3.00 m/s?

Solution:

$$P_1 + \tfrac{1}{2}\rho v_1^2 = P_2 + \tfrac{1}{2}\rho v_2^2$$

$$P_2 = P_1 + \tfrac{1}{2}\rho v_1^2 - \tfrac{1}{2}\rho v_2^2 = P_1 + \tfrac{1}{2}\rho(v_1^2 - v_2^2)$$

$$P_2 = 2.50 \times 10^4 \text{ Pa} + \tfrac{1}{2}(1.00 \times 10^3 \text{ kg/m}^3)\,[(1.00 \text{ m/s})^2 - (3.00 \text{ m/s})^2]$$

$$P_2 = 2.50 \times 10^4 \text{ Pa} - 4000 \text{ Pa}$$

$$P_2 = 2.10 \times 10^4 \text{ Pa}$$

$$P_2 < P_1$$

As expected, fluid pressure decreases with an increase of speed.

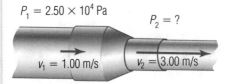

$P_1 = 2.50 \times 10^4$ Pa $P_2 = ?$

$v_1 = 1.00$ m/s $v_2 = 3.00$ m/s

17-26 Pipe restrictions often occur in real plumbing systems.

You can tell what parts of a fluid are moving faster by studying the streamlines. Faster portions have smaller areas; therefore, the streamlines are diagrammatically closer together. Since faster portions have lower pressure, closely spaced streamlines also indicate lower pressure.

17.18 Lift

Airplanes obtain some of their lift by using a modified version of Bernoulli's principle. Some airplane wings have a special shape called an **airfoil** that forces the air

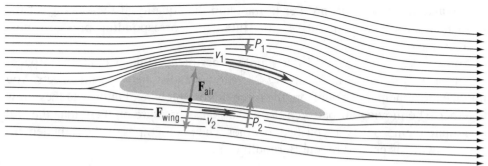

17-27 Streamlines over an airplane wing

passing over the top of the foil to travel a longer path than the air under the foil. Because of the principle of flow continuity, the particles of air that start out at the front of the wing and take the two paths must meet at the back if turbulence is kept to a minimum, so the air moving over the top surface must flow faster than the air underneath. According to Bernoulli's principle, the faster moving air will generate lower pressure above the wing. The higher pressure underneath lifts the wing and the airplane. The most efficient wing designs use a combination of Bernoulli's principle coupled with the reaction force generated by air on the underside of the wing as the wing deflects the air downward.

17-28 Helicopters use a combination of Bernoulli's principle and the third law of motion for lift.

17.19 Real Fluids

In real-world aerodynamics and hydrodynamics, designers and engineers must take into account friction, fluid viscosity, and turbulence, all of which greatly complicate the analysis of fluid flow around immersed objects.

The cohesive forces between particles of a fluid produce a type of internal friction called **viscosity.** Viscosity determines how freely a fluid flows. A **coefficient of viscosity** (η) indicates the resistance of a fluid to flow. Lower viscosity coefficients mean the fluids flow more easily than those with higher viscosity coefficients. For instance, water has a lower viscosity than molasses. Viscosity is familiarly known as the "thickness" of a fluid.

The adhesive forces between a flowing fluid and its confining walls (such as a pipe) cause the outermost particles of fluid to be essentially stationary. Particles farther into the fluid have small velocities that increase with distance from the wall of the pipe. The fluid particles at the center have the highest velocities. The farther the particle is from the wall of the pipe, the faster it goes. Therefore, the flow of fluid at the center of a large pipe is faster than the fluid at the center of a smaller pipe at the same supply pressure. This effect is a combination of friction with the pipe and fluid viscosity. Because of fluid friction, pressure drops with distance along a pipe. This is why the shower in your house farthest from the water pipe coming into your house has the slowest flow (least pressure).

17-29 Viscosity determines the ease with which objects move through a fluid. Both marbles were dropped at the same instant. The left liquid is water; the right liquid is a thick syrup.

The **coefficient of viscosity** (η) is given the SI unit of pascal-second (Pa·s). An older unit of viscosity is the *poise.* Viscosity coefficients are tabulated at a specified temperature, since temperature significantly affects cohesive forces.

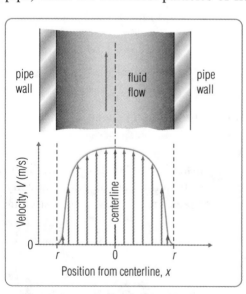

17-30 Viscosity and friction with the wall surface reduce fluid velocity near the margins of a pipe.

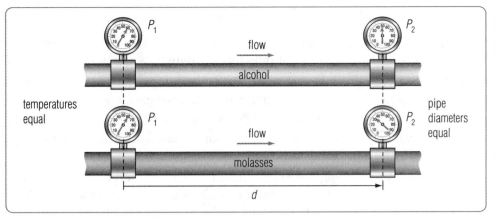

17-31 Viscosity lowers pressure with distance as a fluid moves through a pipe.

Aerodynamically Designed Cars

In order to conserve energy and reduce emissions, automobile designers are doing their best to make cars more fuel-efficient. Although increasing engine efficiency and making cars lighter are important, much of the fuel savings comes from designing the body to move smoothly through the air.

Actually, the idea of an aerodynamically efficient car is not entirely new. As early as 1934, Chrysler produced the Airflow, a sleek vehicle with smooth, rounded curves and fenders. The idea did not become popular, however, until the fuel shortage crises of the 1970s forced manufacturers and consumers to explore every avenue of cutting fuel consumption.

At highway speeds, a car uses more than half its power to overcome air resistance. Therefore, reducing air resistance reduces fuel consumption. The main goal is to make the flow of air around the car as nearly laminar as possible. Models of new designs are first tested using sophisticated computer modeling. Fractional-scale and full-scale testing is done in wind tunnels, where smoke streams around the models, revealing turbulence and potential drag.

If you compare newer passenger cars (including SUVs) with older models, you will see that the latest are designed to let air flow smoothly over, under, and around them. Headlights are molded into the fenders, front hood lines flow smoothly into the front bumper structure, windows are flush with the body, and the overall profile promotes the laminar flow of air over the vehicle. Even the underside of the car is engineered to remove as many obstructions to the flow of air as possible. Even recreational vehicles, with their boxy profiles, are designed to minimize wind resistance.

These efforts, combined with newer, more efficient engines, have dramatically improved the fuel efficiency of passenger vehicles.

17B Section Review Questions

1. Describe the properties of an ideal fluid. What kinds of fluids approach the description of an ideal fluid most closely?

2. When you are taking a shower, where is the water moving fastest—in the ½-inch diameter pipe connected to the shower nozzle or the ¾-inch diameter pipe connected to the house water meter? Explain your answer.

3. Explain why most municipal water supply tanks are located on top of nearby hills or on tall structural steel supports.

4. The Alyeska oil pipeline connecting the Alaskan North Slope oil fields to the oil terminals on the coast extends 800 miles over land. Why does the pipeline have pumping stations placed at regular intervals along its length?

⊙5. If water flows at 1.00 m/s in a 2.54 cm diameter garden hose, how fast is it moving when it exits the 0.25 cm diameter orifice at the end of the hose nozzle?

⊙6. Assume that water pressure supplied to your house is 4.14×10^5 Pa (about 60 psi). If the second floor bathroom shower fixture is 5.00 m above the water supply connection to the house, what is the (static) water pressure (in Pa) at the shower head?

⊙7. Assume that your parents' shower is on the opposite side of the wall from the shower you use. Using Bernoulli's equation, explain what happens to your shower pressure if your father turns on his shower and doubles the flow rate of water supplying both shower fixtures.

Chapter Review

In Terms of Physics

fluid mechanics	17.1	buoyant force ($\mathbf{F}_b$)	17.9
hydrostatics	17.1	Archimedes' principle	17.9
hydrodynamics	17.1	center of buoyancy	17.11
density (ρ)	17.2	hydrometer	17.12
specific gravity (s.g.)	17.2	ideal fluid	17.13
atmosphere (atm)	17.3	streamline	17.13
torr	17.3	laminar	17.13
bar	17.3	turbulent	17.13
millibar (mb)	17.3	flow tube	17.13
gauge pressure (P_g)	17.3	equation of flow continuity	17.14
absolute pressure (P)	17.3	Bernoulli's equation	17.15
Pascal's principle	17.6	airfoil	17.18
hydraulic devices	17.6	viscosity	17.19
manometer	17.7	coefficient of viscosity (η)	17.19
barometer	17.7		

Problem-Solving Strategies

17.1 (page 375) Complex exponential expressions are common in physics. In order to prevent confusion, the term "exp" is inserted before the expression in order to alert the reader that the following parenthetical statement is the exponent of the natural exponential base e. All units in the exponential expression must cancel.

17.2 (page 377) Under normal circumstances, the difference of pressure due to height differences between input and output pistons in hydraulic devices is negligible compared to the force transmitted by the hydraulic system.

17.3 (page 379) If you need to know if an object will float in a liquid, compare the densities of the object and the liquid. If the object's density is less than the liquid's density, it will float. Otherwise it will submerge or even sink.

17.4 (page 384) It is permissible to use different metric systems in one calculation as long as the units cancel. This often occurs in simple proportional calculations.

17.5 (page 386) In an ideal flow tube, if the flow cross section remains constant, then fluid pressure depends only on the height of the fluid above the reference height according to Bernoulli's equation.

17.6 (page 386) In an ideal flow tube, if the height remains constant, fluid pressure will increase when fluid speed decreases and vice versa, according to Bernoulli's equation.

Review Questions

1. What is specific gravity?

2. What is the SI unit of pressure?

3. The containers in the figure contain the same fluid to the same height. Which has greater pressure at its base?

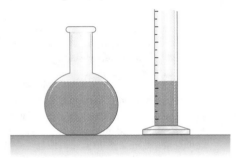

4. The pressure of a liquid at the middle of a sealed container, 10 cm below its surface, is 1.0 atm. What pressure does the liquid exert on the sides of its container at 10 cm below its surface?

5. Explain why atmospheric pressure varies with height in a different way from how liquid pressure varies with depth.

6. Two spheres are lowered into a pool of water to a depth of 3.0 m. One sphere has twice the surface area of the other. How does the pressure on the larger sphere compare to the pressure on the smaller sphere?

7. Which of the following instruments are designed to measure pressure?
   a. thermometer
   b. barometer
   c. hydrometer
   d. manometer

8. State whether the following will float in mercury, in water, in both, or in neither.
   a. gold
   b. cement ($\rho = 2.8$ g/cm^3)
   c. iron
   d. wood (oak)
   e. lead

9. How can a ship made of steel float on water?

10. Many bony fish have an organ called a swim bladder that assists them in maintaining proper buoyancy. This gas-filled bag within the fish's organ cavity varies in volume as necessary. If the fish is negatively buoyant, how must the swim bladder change in volume in order to make the fish neutrally buoyant?

11. Discuss how streamlines are used in a flow diagram to indicate the movement of fluid (include ideas such as spacing, direction, etc.).

12. In what section of the pipe will water run fastest: *A, B,* or *C?*

13. What assumptions are made when deriving Bernoulli's equation?

14. Using Bernoulli's principle, explain why a shower curtain tends to billow into the shower when the water is running.

15. Some backyard swimming pool owners remove the water from their pools using devices called aspirators attached to their garden hoses. Explain how such a device might work.

16. What is the effect of friction in fluids called?

True or False (17–31)

17. A fluid is a liquid.

18. Specific gravity is the ratio of the density of a substance to the density of water.

19. One torr is $^1/_{760}$ of normal atmospheric pressure.

20. Fluid pressure increases with depth from the reference height.

21. The pressure in a fluid is dependent only on the depth (or height) in the fluid and the pressure at the reference height.

22. Atmospheric pressure decreases at a nearly linear rate with height.

23. A hydraulic device works because a small force applied to a large diameter piston transmits the pressure via hydraulic fluid and produces a large force on a small diameter piston.

24. The buoyant force on an object is proportional to the density of the fluid and inversely proportional to the volume of fluid displaced.

25. As a helium balloon rises in the atmosphere, it becomes heavier and eventually stops rising.

26. A floating object is stable if the torque produced by the misalignment of the centers of gravity and buoyancy tends to move them back into the same vertical line rather than away from it.

27. It is valid to assume that fluids are incompressible when discussing liquids, but it is not valid when discussing gases.

28. The equation of flow continuity shows that the rate at which a mass of fluid enters a flow tube must equal the rate at which it exits.

false 29. Doubling the speed of a fluid reduces its pressure to half the original value. *The change of pressure due to flow rate is related to the square of the change of fluid velocity.*

30. An airfoil works because air exerts a drag on the lower surface of the foil as it moves through the air.

31. Larger pipes experience a higher flow rate for a given inlet pressure than smaller pipes.

⊙32. The density of sea water is about 1030 kg/m³. The pressure at the top of the ocean is usually 101 325 Pa. The Marianas Trench is 10 924 m below sea level. What is the pressure there?

⊙33. A mercury-containing glass tube sealed at its upper end is inverted with its open end in a dish of mercury. Above the mercury in the tube is a vacuum ($P = 0$ Pa). The mercury in the tube settles 76.0 cm above the surface of the mercury in the dish. What is the pressure at the surface of the mercury in the dish? ($\rho_{Hg} = 13.6$ g/cm³)

⊙34. A hydraulic jack has a small piston with an area of 3.50 cm² and a large piston with an area of 31.5 cm².
   a. What force is needed on the small piston to lift a 250. kg engine with the large piston?
   b. If a man can exert a force of 175 N on the small piston, how much lifting force can he generate using the jack?
   c. Suppose the jack is an ideal hydraulic device. (An ideal hydraulic device is one in which the fluid does not compress and there are no frictional losses.) How far must the small piston move to move a load 1.00 m?
   d. What is the theoretical mechanical advantage of the jack? (TMA = A_{out}/A_{in})

⊙35. The figure shows an open-tube manometer connected to a gas sample. Atmospheric pressure (P_0) is 1.01×10^5 Pa. The mercury in the open end of the manometer stands 0.450 m above position A.

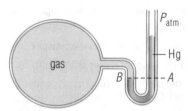

   a. What is the pressure in the column of the mercury at point A?
   b. What is the pressure of the gas at point B?

⊙36. A rock with a mass of 1.53 kg weighs 13.5 N when it is immersed in water.
   a. What is the buoyant force on the rock?
   b. What is the rock's volume?
   c. What is the rock's density?

⊙37. A 20.0 g mass, when fully immersed in a liquid, exerts a force of 1.00×10^{-1} N on a supporting spring scale. The mass has a volume of 2.75 cm³.
   a. What is the buoyant force on the mass?
   b. What volume of liquid does the mass displace?
   c. What is the liquid's density?

⊙38. A level pipe narrows from a cross-sectional area of 11.5 cm² to a cross-sectional area of 5.00 cm². If a liquid moves at a speed of 0.50 m/s through the wider section, how fast will it go through the narrow section?

⊙39. Water runs through a hose with a cross-sectional area of 11.5 cm² at a speed of 1.50 m/s. A child playing with the hose has a hand over its end, leaving a 0.200 cm² opening for the water. At what speed does the water pass the child's hand?

⊙40. A 35.0 km/h wind blows around a house that contains still air at a pressure of 1.01×10^5 Pa. Assume the air is an ideal fluid.
   a. What is the difference in pressure between the air inside the house and the air at the same elevation outside the house?
   b. Based on your answer to part *a*, discuss some reasons why it is inappropriate to use Bernoulli's equation with a low-density, compressible gas.

⊙⊙41. The force exerted by air pressure can be dramatically demonstrated using a pair of simple laboratory devices called *Magdeburg hemispheres*, named after the town in Germany where Otto von Guericke demonstrated his air pump invention to Emperor Ferdinand III in 1654. He constructed two hollow metal hemispheres with a gasket between them that were then evacuated using his air pump. Two teams of 15 horses each could not pull them apart. Assuming that the hemispheres were 1.00 m in diameter, that they were completely evacuated, and that standard atmospheric pressure existed on that day,
   a. what was the magnitude of the force exerted on each hemisphere by air pressure?
   b. what was the minimum force each horse would have had to exert in order to pull the hemispheres apart?

Electromagnetics

For centuries, natural philosophers and scientists were aware of electrical and magnetic phenomena in nature. However, it was not until the nineteenth century that scientists realized that the two phenomena were aspects of the same entity—electromagnetism, the third major area of classical physics. We are the beneficiaries of a host of phenomena that depend on electromagnetic energy, from the very nature of matter and light to the technology that powers modern-day civilizations. For our study and enjoyment, the heavens contain radiant electromagnetic sources in the form of the Sun, stars, and the awe-inspiring aurora displays.

In Unit 4, you will be introduced to electrostatic forces, electric fields, and the law of charges. You will learn why and how charges flow and how moving charges create and are affected by magnetic fields. The properties of magnets and magnetic fields will be discussed. You will also become acquainted with the basics of electricity and electrical schematics as you analyze the flow of current through circuits.

18A	Electrification	397
18B	Detecting Electric Charge	402
Biography:	William Gilbert	399
Facet:	Static Electricity	402
Biography:	Michael Faraday	410

New Symbols and Abbreviations

fundamental charge (*e*)	18.4
electrostatic unit (esu)	18.11
coulomb (C)	18.11

Lightning is nothing more than the discharge of static electricity as charges move between the ground and the clouds.

Electric Charge

Wherefore the Lord said, Forasmuch as this people draw near me with their mouth, and with their lips do honour me, but have removed their heart far from me . . .   Isaiah 29:13

In the above verse, the prophet Isaiah reports God's lament over His people's backslidden condition. God knows that when the hearts of His children draw away from dependence on His Word and commandments, His influence and protection rapidly diminish, similar to the way in which the force between two electric charges rapidly decreases as the distance increases between them.

Do you feel sometimes that your prayers go unanswered and that your spirit is burdened with the cares of this world? Perhaps there is too much distance between you and your Lord. Who has moved? James wrote, "Draw nigh to God, and he will draw nigh to you. Cleanse your hands, ye sinners; and purify your hearts, ye double minded" (James 4:8). The closer you are to Jesus, the greater the effect He has on your life.

18A ELECTRIFICATION

18.1 Introduction

What is electricity? You probably use electricity every day, but what do you know about it? Most Americans realize that electricity is carried by wires; that it allows stoves to heat, refrigerators to cool, and computers to function; that it costs money; and that a modern city is crippled without it. Beyond such simple ideas, many know little about electricity.

18-1 Electricity is essential to the existence of a modern city.

The word "static" in "static electricity" means fixed or motionless. In static electricity, charges are usually fixed but are capable of moving certain distances for brief moments (even though those distances might be great, such as with lightning). In contrast, the charges in current electricity are continuously moving.

18-2 A beautiful sample of amber containing a trapped insect

Amber is fossilized resin that oozed naturally from the bark of trees thousands of years ago. Because the majority of amber samples found today are obtained from buried deposits, it is likely that these trees were inundated in the Noachian flood. Amber has traditionally been used for personal jewelry.

18.2 History

Six hundred years before Christ, a Greek philosopher recorded that a piece of amber rubbed with fur would attract dust and bits of dry leaves. The Greek word for amber, *elektron,* became associated with this attraction, and today it is called **static electricity.**

An Elizabethan physician, William Gilbert, was among the first to experiment extensively with *electrification.* He discovered that glass and other materials besides amber would attract small objects after being rubbed with fur or cloth. Gilbert called a material that can be electrified an *electrick.*

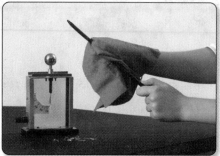

18-3a A student charging an "electrick"

18-3b The charged "electrick" attracts bits of paper.

Many different materials are "electricks." As scientists experimented with these materials, they discovered that they can affect other similar materials as well as small objects. For example, an electrified amber rod repelled another electrified amber rod, and an electrified glass rod repelled another electrified glass rod. Also, an electrified glass rod attracted an electrified amber rod. Scientists divided electricks into two categories: materials that acted like amber were called **resinous** (amber is fossilized resin) electricks, and materials that acted like glass were called **vitreous** electricks (*vitro* is Latin for "glass").

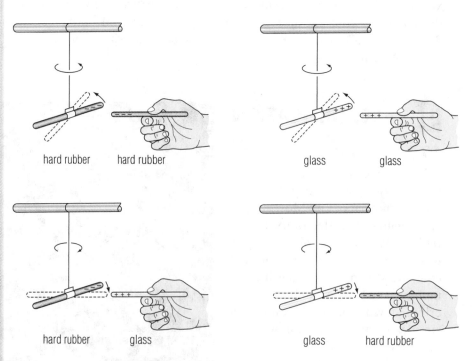

hard rubber   hard rubber   glass   glass

hard rubber   glass   glass   hard rubber

18-4 Charged rods of different materials demonstrate the law of charges. No charge attracts or repels *both* of the known kinds of charge. Therefore, it can be concluded that there are only two kinds of charge.

After Gilbert, Charles Du Fay proposed an explanation for the existence of resinous and vitreous electricks. Du Fay thought that each electrick contained an electrical fluid. Since there were two kinds of electricks, Du Fay decided that there must be two kinds of fluids: resinous electricks, which contained resinous fluid, and vitreous electricks, which contained vitreous fluids.

Benjamin Franklin, on the other hand, thought that only one kind of electrical fluid existed. He believed that vitreous electricks had an excess of fluid, while resinous electricks had a deficiency of fluid. He called vitreous electrification *positive* and resinous electrification *negative*. (This designation will become important in the discussion of electric current in later chapters.)

Franklin and Du Fay agreed about the cause of the electrification. They said that when an electrick was rubbed, some fluid was transferred between the electrick and the cloth. The resulting imbalance of fluid produced an electric *charge* on the electrick. The electric charge enabled the electrick to act according to the **law of charges:** unlike charges attract, and like charges repel.

Charles-François de Cisternay Du Fay (1698–1739) was an extremely gifted and intelligent French academician. Originally an army officer, he devoted his life to investigating many areas of physical and biological science, including electrostatics and the properties of crystalline solids.

Benjamin Franklin (1706–90) was an American printer, philosopher, statesman, scientist, and inventor.

The **law of charges** states that opposite charges attract each other, whereas like charges repel each other.

William Gilbert (1544–1603)

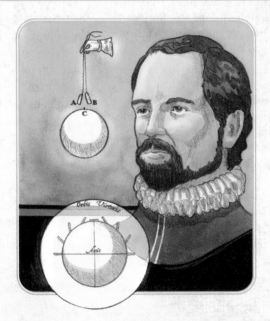

From ancient times, people have been fascinated by mysterious phenomena such as lightning, electric eels, and sparkling glow on ship's spars and mastheads (called St. Elmo's fire). Two other materials have received special attention: natural magnets from the fields of Magnesia (called lodestones) and fossilized tree sap (called amber). When amber was rubbed vigorously, it would attract particles of straw and wheat. The Greeks attributed this force to *sympathy;* in other words, amber was being friendly. The lodestone, on the other hand, attracted heavy metals such as iron tools. When people noticed that a lodestone needle (a compass) always pointed north, they proposed many bizarre theories to account for this magnetic effect. None of the theories were investigated scientifically until the time of William Gilbert.

Gilbert was a trained physician who served in the court of Queen Elizabeth I. In his spare time, he experimented with the effects of amber and lodestone. He felt contempt for earlier thinkers who studied these effects only by reading what ancient authorities had said. He himself would accept only those theories that he could support with experiments.

Gilbert noticed that a lodestone always has two poles and that the poles are not points but regions. He distrusted the popular explanations for the behavior of the compass needle. He offered a dramatic new theory that the earth itself is a lodestone with two poles, one north and one south. To support his idea, he made a huge spherical lodestone and placed a tiny magnetized needle on it. The needle behaved exactly like a compass would on the earth.

While studying amber, he discovered that many other common materials can become charged by friction. He called this force that attracted objects *electrick*. His experiments showed for the first time how electricity is different from magnetism: magnetism attracts few materials to two poles and is unaffected by intervening materials, whereas the electrostatic force attracts unlimited materials to one center and is greatly affected by intervening materials. Gilbert published his findings in *De Magnete* (1600), which established him as one of the foremost English scientists of his day.

Gilbert's book is important not only because it summarizes everything known in his day about magnetism but also because it encouraged others to test their ideas with experiments. The science of electromagnetism eventually rose from this beginning.

18.3 Discovery of the Electron

The idea of an electric fluid implied that electric charge was continuous; that is, there could be any amount of electric charge. In 1897 the English scientist J. J. Thomson disproved this idea when he demonstrated that all matter contains tiny negatively charged particles with the same charge-to-mass ratio. He called these particles **electrons.** As a consequence, an object could have only a charge that was an integer multiple of the charge of an electron. The magnitude of this charge remained unknown for more than two decades.

18.4 Determination of the Fundamental Charge

In 1921 Robert Millikan devised an experiment to find the charge on each electron. He sprayed oil drops into a chamber that could be irradiated with x-rays. He expected the oil drops to acquire electrons from gas molecules bombarded by the x-rays. Observing the oil drops through a calibrated microscope, Millikan allowed them to fall under the influence of only gravity and air resistance. The drops soon reached a terminal velocity, which Millikan calculated by averaging the results of seventeen trials. He then applied an electric field and adjusted it until some of the oil drops were stationary. Since he knew the strengths of the forces of gravity and air resistance on the drops and the strength of the electric field, he could calculate the charge on the drops. After many trials, Millikan discovered that all the charges he calculated were multiples of a **fundamental charge (e).** This charge he assumed to be the charge on a single electron. In the many years since Millikan's experiment, his results have been vindicated. All charges are multiples of *e*; no charges have been found that are fractions of *e*.

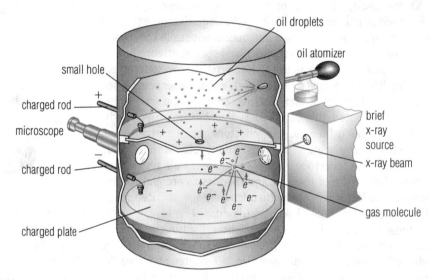

18-5 Millikan's apparatus

Experiments have shown that there are only two kinds of charges, so in this respect modern scientists agree with Du Fay. However, Franklin's terms were used to identify the charges, although any terms identifying opposites could have been used. The charge on an electron was termed negative ($-e$) because in Thomson's experiment the electrons were attracted to the electrode that was positively charged according to Franklin's designation.

The proton was discovered several years after the electron and was found to have a positive charge. Further experimentation determined that it possessed a single fundamental charge ($+e$). In solids, the positively charged protons in an atom do not move appreciably because the atoms themselves are in fixed positions in solids. The outer electrons in the atoms of some materials are loosely held and can be easily removed. Electrons moving from one region to another in or on the object cause the electrical charge on a solid. When electrons leave an area, the remaining protons produce a net positive charge. In the areas to which the electrons move, the extra electrons produce a net negative charge. Franklin's theory was close to modern theories for solids, but modern theories substitute electrons for an electrical fluid.

18.5 Classification of Substances by Electrical Properties

Not all solids have atoms with mobile electrons. Those solids that *do* have free electrons are called **conductors;** metals are a good example. Materials that hold their electrons tightly are called **insulators;** examples of good insulators are porcelain, rubber, and glass. Insulators can be used to confine electricity, as with the insulation around a wire. **Semiconductors** are materials that usually act like insulators but can act like conductors under certain conditions.

Gilbert's electricks were insulators that gained or lost electrons when rubbed. When a neutral object, such as a speck of dust, comes near such a charged material, its electrons are attracted or repelled by the charge. The side of the dust speck nearest the charged insulator then has a charge opposite to the insulator's charge, and the dust speck is attracted to the insulator. When the dust speck touches the insulator, electrons move to give the speck some of the insulator's excess charge. The speck then has the same charge as the insulator and is therefore repelled. The same process happens with water molecules present in the air. As a result, the charge on the insulator does not last long but leaks away into the air. A conductor also gains or loses electrons when rubbed, but because its electrons are much more mobile, it conducts its charge to the air almost immediately.

18A Section Review Questions

1. What was the substance that was first reported to have electrical properties? How is its name preserved and used today?

2. Briefly summarize the scientists and discoveries leading to the current understanding of electrical charges.

3. State the law of charges. Explain why, based on this law, scientists believe that there are only two kinds of electrical charges.

4. Why was the electron assigned the negative charge?

5. Of what category of electrical materials would an effective lightning rod system be constructed?

6. The "electricks" (test rods) that are used in school labs are made of what category of electrical material?

●7. List the variables that Millikan had to control in his famous oil-drop experiment.

Conductors are usually materials composed of the *metal* elements, principally from the middle and left side of the periodic table (see Appendix H), and their alloys. **Insulators** are usually composed of *nonmetal* elements from the upper right-hand corner of the table and their compounds. **Semiconductors** are the group of elements, called *semimetals* or *metalloids,* that occurs between the other two kinds.

You experience more static cling in your clothes and static shocks from doorknobs during the winter because the relative humidity of the air is so much lower during that season. During the humid summer months, the water vapor in the air carries away the static charges before they can build up.

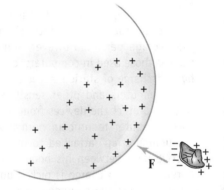

18-6 A charged rod induces a charge on a particle of dust.

18.6 An Early Detection Device

To study electrical charge, you must be able to detect it. The ability to measure charge is also important. Early scientists used lightweight balls made from cork or pith (material from inside plant stems) to detect charges. Two balls were hung side-by-side from silk threads, close to each other but not touching. If both balls were charged, they would either attract or repel each other. The scientists tested to see if a material had a charge by touching it to both balls. The charge would pass to the balls, and, since it would be the same kind of charge, the balls would separate.

FACETS of PHYSICS

Static Electricity

To most people, static electricity is a minor annoyance. An occasional shock or static cling does not greatly bother the average person. However, sensitive electronic devices, such as those used in computers, can be damaged or destroyed by a discharge of static electricity. Therefore, those who work with computers and other sensitive circuits have developed ways to protect the devices from static electricity.

A computer technician normally wears what is called a "grounding strap" attached to his wrist that provides a conductive path between his body and the chassis of the computer. Then, if a charge is picked up by the worker, it quickly bleeds away to the grounded chassis, preventing any possibility of a charge jumping to the computer components from the technician's hands.

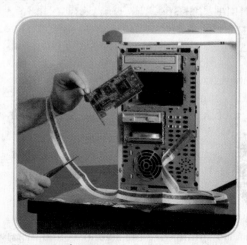

Using a grounding strap

Another method is to keep easily charged materials away from the circuit. Surrounding the sensitive circuit with a conductor insures that a charge will not build up around it.

Antistatic materials are used to prevent electrical charge buildup. Plastics have been developed that quickly drain off charges. Sensitive circuits are often wrapped in such plastics during shipping and may be surrounded by an antistatic plastic after being installed.

Carpets and sprays used around computers contain a special antistatic chemical that absorbs water. Charges cannot accumulate on the carpet or sprayed material because the water conducts away the charges. Consumers can also buy this chemical to minimize annoying static shocks and static cling.

18.7 The Electroscope

An **electroscope** is an improved method of detecting charge. You can make an electroscope from a glass flask or bottle, a rubber stopper, a metal rod that acts as an electrode, and two thin leaves of metal foil. Put the electrode rod through the rubber stopper, attach the foil to the L-hook end of the rod so that the foil pieces can move, and insert the stopper into the container so that the leaves hang freely inside the container.

When you bring a charged test rod near the electrode of the electroscope, the leaves separate. Why? Suppose the test charge is negative. The electrons in the electrode are repelled by the negative test charge and travel down the rod to the leaves. Because both leaves receive excess electrons, both have net negative charges. Since the leaves can move, they repel each other. If the charged test rod is positive, it attracts electrons from the leaves to the top of the electrode, leaving both leaves with a net positive charge. Again, both leaves have the same charge and thus repel each other. When you move the charged test rod away from the electroscope, the leaves return to their original positions.

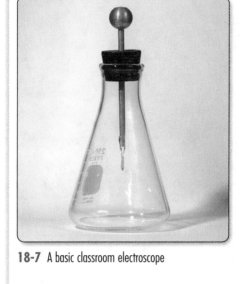

18-7 A basic classroom electroscope

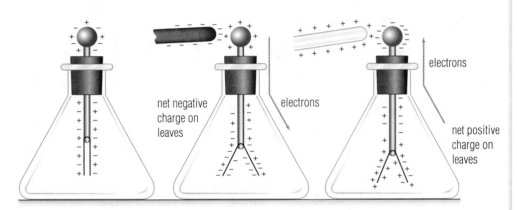

18-8 Testing for charge with an electroscope

Suppose you *touch* an electroscope with a positively charged test rod. What happens? The electrons in the electroscope are attracted to the test rod. If the test rod is removed, the electrons go with it, leaving the electroscope positively charged. If you move another positively charged test rod near the charged electroscope, it will attract more electrons from the leaves. The leaves become even more charged, and they separate more. However, if you bring a negatively charged rod near the electroscope, the electrons in the test rod will force the electrode's electrons down to the leaves, reducing their positive charge. The leaves will therefore fall closer together. With a charged electroscope, you can not only detect charge but also distinguish between positive and negative charges, if you know what kind of charge is initially present.

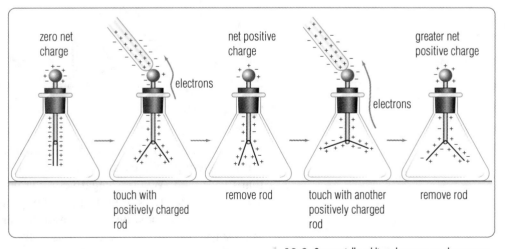

18-9 Sequentially adding charge to an electroscope

18.8 Conduction Tester

You can use a pair of electroscopes to see how well materials conduct. Begin with one neutral electroscope and one that is charged negatively. Connect the electroscopes by bridging their electrodes with the material you want to test. If the connector is a good conductor, like a metal, electrons will flow through the conductor from the charged electroscope to the neutral electroscope, charging the neutral electroscope. Its leaves will separate almost immediately. If you connect the electroscopes with an insulator, electrons cannot easily travel to the neutral electroscope. Its leaves will separate slowly, if at all.

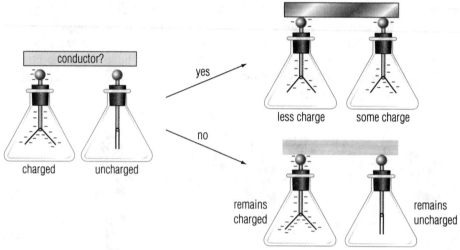

18-10 Testing for conduction

18.9 Methods of Establishing a Charge

Electroscopes illustrate two ways to charge a body: by **contact**—when a charged object touches an uncharged object and electrons are transferred from one to the other—and by **induction**—when the electrons on the surface of an uncharged object are redistributed by a charged object that is brought near.

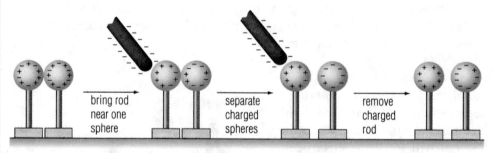

18-11 Charging by induction and separation of objects

It is possible to induce a permanent charge on a neutral body. Consider two neutral conducting spheres that are mounted on insulating stands and positioned so that they touch. When a negatively charged test rod is brought near one of the spheres, the electrons of both spheres migrate away from the rod, leaving the near sphere with a net positive charge and the far sphere with a net negative charge. If the test rod is removed, the electrons will return to their places and the spheres will again be neutral. However, if the spheres are separated while the test rod is near, they will remain charged. The rod's removal will have no effect on this charge since the spheres can no longer exchange electrons. This is charging by induction.

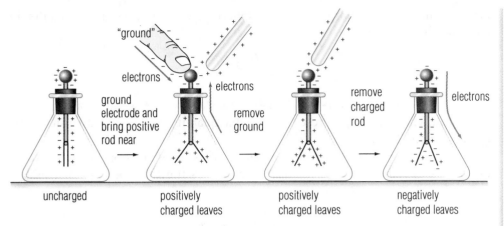

18-12 Charging a grounded electroscope by induction

You can also charge an object by induction using a neutral body connected to an **electrical ground.** An electrical ground is a large body whose charge is unaffected when electrons are added or removed. Suppose you connect a neutral electroscope to an electrical ground and bring a positively charged rod near the electroscope. The positive charge attracts electrons from the electroscope and the ground. If the connection to the ground is now broken, the electrons from the ground in the electroscope cannot return to the ground. The electroscope is left with extra electrons and, therefore, a negative charge.

18.10 Coulomb's Law

Although good for detecting charges, the electroscope is not very useful for measuring electric force. The instrument that *can* measure **electrostatic force** accurately is very similar to the one that was first used to accurately measure gravitational force—the torsion balance. The instrument for measuring electrostatic force consists of a fiber supporting a rod with an insulated sphere on each end. A fixed sphere is mounted near one of the hanging spheres. The fixed sphere and the hanging sphere nearest it are usually charged so that the fixed sphere repels the hanging sphere.

When the hanging sphere moves, it produces a torque on the fiber through the rod. As the fiber twists, a torsional force develops in the fiber that resists twisting like a spring resists compression. The amount of torsional force the fiber exerts against the torque exerted by the charged spheres is related to the angle through which it has twisted. Eventually, the torsion in the fiber balances the mechanical torque produced by the spheres. The torsion in the fiber can be calculated from the angle through which the fiber has twisted. From this the force on the movable sphere can be determined.

The first man to report having done this experiment was Charles Augustin de Coulomb, a French physicist. He studied the electrostatic force as it relates to the distance between the spheres and discovered that when he doubled the distance between the spheres, the force between them was one-fourth the original. When he tripled the distance, the force was one-ninth. Thus, he concluded that electrostatic force is proportional to the reciprocal of the square of the distance between the charges, or

$$F_{\text{elect}} \propto \frac{1}{r^2},$$

where *r* is the distance between the spheres. He then tried doubling the charge on one sphere, which doubled the force. Doubling the charge on both spheres quadrupled the force. Coulomb concluded that the electrostatic force is directly

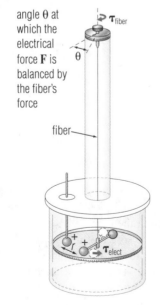

angle θ at which the electrical force **F** is balanced by the fiber's force

18-13 An electrostatic torsion balance

proportional to the charge on each sphere. He summarized his results in **Coulomb's law:**

$$F_{elect} = k\frac{|q_1||q_2|}{r^2},$$ (18.1)

where the constant k is a proportionality constant whose value depends on the units used. The symbol q represents a variable quantity of charge that has no physical dimension (a *point charge*).

The electrostatic force on each charged object is equal and opposite to the force exerted on the other object, according to Newton's third law of motion. This law is true even if the charges are unequal.

$$\mathbf{F}_{q_1 \to q_2} = -\mathbf{F}_{q_2 \to q_1}$$

Note that the form of Coulomb's law is identical to Newton's law of universal gravitation that predicts the force of two masses M and m on each other:

$$F_{grav} = G\frac{Mm}{r^2}$$

Two key differences exist between the gravitational force and electrostatic force: (1) There are two kinds of charges that produce the electrostatic force, but there is only one kind of mass that produces the gravitational force. Even antiparticles have "positive" mass. (2) Electrostatic forces may be attractive or repulsive, but gravity is only attractive.

You will discover additional similarities between gravitational and electrostatic force in later chapters. You should expect this, since both of these forces are central forces—they act along a line connecting the centers of the objects producing the force. They also are both conservative forces in that the work done by the force between two positions is independent of the path taken.

18.11 Units of Charge

To use Equation 18.1, you must specify the unit of charge. One experimental unit of charge is the **electrostatic unit (esu).** In the Gaussian or CGS system of units, one esu of charge on each of two dimensionless objects exactly 1 cm apart exerts a force of exactly 1 dyne on each object ($1 \text{ dyne} = 1 \times 10^{-5}$ N).

$$1 \text{ dyne} = k_{esu}\frac{(1 \text{ esu})(1 \text{ esu})}{(1 \text{ cm})^2}$$

The constant k_{esu} is

$$k_{esu} = 1\frac{\text{dyne} \cdot \text{cm}^2}{\text{esu}^2}.$$

The esu was chosen because it allows the constant to be 1.

The SI unit of charge is the **coulomb (C),** which for now will be defined as the magnitude of the charge on 6.24×10^{18} electrons. For this unit, the constant k_C is defined as

$$k_C = 8.99 \times 10^9 \frac{\text{N} \cdot \text{m}^2}{\text{C}^2}, \text{ so}$$

$$F_{elect} = 8.99 \times 10^9 \frac{\text{N} \cdot \text{m}^2}{\text{C}^2} \cdot \frac{q_1 q_2}{r^2}.$$

In this textbook, we will normally evaluate electrical problems using the coulomb for the unit of charge.

EXAMPLE 18-1

Distance and Charge: Coulomb's Law

Two point-charges of 1.00×10^{-6} C each are separated by a distance of 0.100 m. (a) What is the force on each particle? (b) If the distance is 0.0100 m, what is the force?

Solution:

$$F_{elect} = k\frac{q_1 q_2}{r^2}$$

a. Force at 0.100 m apart:

$$F_{0.1m} = 8.99 \times 10^9 \tfrac{N \cdot m^2}{C^2} \cdot \frac{(1.00 \times 10^{-6}\ C)(1.00 \times 10^{-6}\ C)}{(0.100\ m)^2}$$

$$F_{0.1m} = (8.99 \times 10^9\ N)(1.00 \times 10^{-10})$$

$$F_{0.1m} = 8.99 \times 10^{-1}\ N$$

$$F_{0.1m} = 0.899\ N$$

b. Force at 0.0100 m apart:

$$F_{0.01m} = 8.99 \times 10^9 \tfrac{N \cdot m^2}{C^2} \cdot \frac{(1.00 \times 10^{-6}\ C)(1.00 \times 10^{-6}\ C)}{(0.0100\ m)^2}$$

$$F_{0.01m} = (8.99 \times 10^9\ N)(1.00 \times 10^{-8})$$

$$F_{0.01m} = 8.99 \times 10^1\ N$$

$$F_{0.01m} = 89.9\ N$$

As expected, the smaller distance between charges results in a much larger force between them.

18B Section Review Questions

1. Describe how one or more pith balls hanging from a support can be used to determine the kind of charge, if any, on an object.

2. a. Describe how an electroscope works.

   b. Why should the foil leaves be as thin and light as possible?

   c. Why must the stopper supporting the electrode be made of an insulator?

3. What is the consequence of touching a neutral electroscope electrode with

   a. a positively charged test rod?

   b. a negatively charged test rod?

4. To what quantities is the electrostatic force between charged objects proportional?

●5. Sketch a diagram of the force vectors on negative charge q_1 and positive charge q_2 if $|q_1| = 2|q_2|$.

●6. a. What is the electrostatic force between two point charges $q_1 = +1.00$ C and $q_2 = -1.00$ C if the distance between them is 1.00 m?

b. What can you conclude about the ability of opposing electrostatic charges to do work?

Chapter Review

In Terms of Physics

static electricity	18.2	electroscope	18.7
resinous	18.2	contact (charging)	18.9
vitreous	18.2	induction (charging)	18.9
law of charges	18.2	electrical ground	18.9
electrons	18.3	electrostatic force	18.10
fundamental charge (e)	18.4	Coulomb's law	18.10
conductors	18.5	electrostatic unit (esu)	18.11
insulators	18.5	coulomb (C)	18.11
semiconductors	18.5		

Problem-Solving Strategies

18.1 (page 406) When using the Coulomb's law equation, be sure that units used for charge, force, and distance match those in the Coulomb constant. The SI units are coulombs, newtons, and meters, respectively.

Review Questions

1. What did William Gilbert mean by the word *electrick?*

2. Why does amber attract dust and dry leaves after it has been rubbed with fur?

3. What happens when an electrified amber test rod is placed
   a. near an electrified glass rod?
   b. near an electrified amber rod?

4. How did the two different "electrical fluid" theories explain the differences between electrified amber and electrified glass?

5. What happens to a neutral electroscope when you hold a negatively charged object near it? Why?

6. When you bring a charged rod near a negatively charged electroscope, its leaves separate farther. Is the charge on the rod negative or positive?

7. You have two neutral conducting spheres. How can you give one a negative charge and the other an equal positive charge without touching either sphere with a charged object?

8. Explain what will happen when particles of paper picked up by a negatively charged test rod gain electrons from the rod.

9. Explain why a charged electroscope will discharge with time if left to itself.

10. What happens, if anything, when you touch the electrode of a charged electroscope with your finger?

11. Why is it advisable on a cold, dry winter day to touch the metal case of your computer before starting it or touching any other part of the system?

12. Many appliances and hand tools have at least three wires in their electrical cords. The center prong of the plug is connected to the ground wire. Where do you suppose this wire leads to when it is plugged into an electrical outlet?

13. To what other kind of force is the electrostatic force similar? How is it different?

14. Is a coulomb of electrons a large charge or a small charge? In other words, is a charge of this size dangerous—can it do a lot of damaging work in a short time? (*Hint:* Review the results of Question 6 in the section review for 18B.)

True or False (15–24)

15. An "electrick" is any substance that can easily conduct electricity.

16. Glass, amber, rubber, and Lucite are "electricks."

17. If two objects repel each other, you know they are both negative charges.

18. The magnitude of the fundamental charge is the same for positive and negative charges.

19. When a charge is established on an object, the charge exists because electrons moved to or from the object.

20. On humid days, charged objects lose their charges more quickly because there are more conductive water molecules in the air.

21. Electroscopes can indicate the presence, sign, and magnitude of electrostatic charge on an object.

22. Induction of an electrostatic charge requires only the presence of a charged object, not actual contact, to establish a charge in another object.

23. If the distance between two charged particles doubles, the force exerted between them will be halved.

24. If two charges exert an electrostatic force on each other and one charge is twice the other, then the larger charge will exert twice the force of the smaller charge and the smaller charge will exert half the force of the larger.

◉25. Two charges separated by a distance r exert a force of 1.00 N on each other. If one charge is tripled without anything else being changed, what force will the charges exert on each other?

◉26. One thousand grams of water (about 1 L) contains 55.5 moles of water molecules. (One mole is 6.022×10^{23} particles.) One electron is removed from every hundredth water molecule and placed in a container 100. m from the water.
   a. How many water molecules are there in the water container?
   b. How many electrons have been removed from the sample of water?
   c. What is the charge, in coulombs, of the electrons that were removed?
   d. What is the force of attraction between the container of water and the container of electrons?

◉27. Two conducting spheres, when given a charge of 2.50×10^{-6} C each, repel each other with a force of 2.50×10^{-4} N at a certain distance apart. How far apart are the balls?

◉28. The charged tips of two test rods are suspended 10.0 cm from each other. One rod has a charge of 2.50×10^{-5} C. The rods attract each other with a force of 5.00 N. What is the charge on the other rod?

◉29. Three conducting spheres hang as shown in the figure, in a straight line. Sphere A has a charge of -2.00×10^{-6} C. Sphere B, 5.00 cm to the right, has a charge of $+1.00 \times 10^{-6}$ C. Sphere C, 10.00 cm to the right of sphere B, has a charge of $+3.00 \times 10^{-6}$ C.

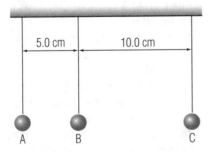

5.0 cm   10.0 cm

A   B   C

   a. What is the force between spheres A and B?
   b. What is the force between spheres B and C?
   c. What is the total force on sphere B?

◉◉30. Write a brief biographical report on Charles Du Fay or Benjamin Franklin.

◉◉31. Research and give an oral report on the origins and properties of natural lightning.

◉◉32. Build and demonstrate a mechanical static electricity generator, such as a Wimshurst machine. A good reference for this exercise is HOMEMADE LIGHTNING by R. A. Ford (TAB Books, The McGraw-Hill Companies, 1996).

Michael Faraday
(1791–1867)

Some scientists have special talent for devising theories to explain the observations of others; other scientists are unusually good at experimenting to test and develop theories. Michael Faraday was good at both. He is undoubtedly one of the greatest experimental geniuses in history.

Faraday was born into a poor London family. After a scanty education, he was apprenticed to a bookbinder. Faraday supplemented his education by reading the science books that came into the shop. When a customer gave him tickets to a series of lectures by Sir Humphry Davy, Faraday attended, listened attentively, and took careful notes. Later he sent the notes with illustrations to Davy as evidence of his eagerness to "enter into the service of science." Flattered, Davy hired him as a laboratory assistant.

With Davy's help, Faraday soon became a skilled experimenter. He started with gases and successfully liquefied chlorine and isolated benzene. When Hans Christian Oersted showed that electric currents affect magnets, Faraday became interested in electromagnetism. In his first great experiment in electricity, he transformed electrical energy directly into mechanical energy by letting a current-carrying wire revolve around a magnet. These successes brought Faraday widespread fame. He was appointed superintendent and later lab director at the Royal Institution. He was also elected a fellow of the Royal Society.

Next, Faraday tried to produce an electrical current from a magnet. He was convinced that there is a divine unity in the forces of nature. Undismayed by ten years of fruitless experiments, he was finally rewarded for his efforts when he discovered that changing magnetism induces a current. The principles behind Faraday's generator were essential to the development of our modern electric power industry.

To explain his magnetic generator, Faraday proposed that a field made up of magnetic lines of force surrounds a magnet. When a conductor cuts this field, the magnetic field induces a current in the conductor. He later extended his idea to electricity: electric field lines surround electric charges. His field theory formed the basis of Maxwell's classic field theory, and it provided a strong foundation for Einstein's theory of relativity.

While experimenting with chemical solutions that conduct electricity, he arrived at the basic laws of electrolysis, called Faraday's laws. Unfortunately, the sustained effort led to a breakdown from which Faraday never fully recovered. He returned to the laboratory long enough to discover that magnetism can affect light, but his failing mind could no longer solve any more of the scientific riddles that fascinated him.

Faraday was well respected for his speaking ability as well as his genius. His captivating lectures drew large crowds to the Institution every Friday evening. He was popular among children as well; he started a series of Christmas lectures for them that became a tradition.

One reason for Faraday's success was undoubtedly his love for God. A devout Christian, he believed that science is simply a search for knowledge about God's creation. In all his experiments, he was dedicated to finding the truth. He continually checked his work for errors and abandoned any theories that did not fall in line with the facts. Faraday's life was a testimony to his fellow scientists, who realized that his strength of mind came from his worship of God. His life is evidence that a good Christian can be a good scientist.

19A	Modeling the Electric Field	413
19B	Capacitors	419
Facet:	Kinds of Capacitors	425

New Symbols and Abbreviations

test charge (q)	19.2	dielectric constant (κ)	19.8
electric field strength (E)	19.2	capacitance (C)	19.9
electrical potential (V)	19.4	permittivity of free space (ε_0)	19.9
potential difference (ΔV)	19.4	permittivity of material (ε)	19.9
volt (V)	19.4	farad (F)	19.9

Tesla coils generate extremely intense electric fields that permit the discharge of powerful electrical sparks.

The return of our Lord in the heavens at the end of the Tribulation is compared to lightning—the gigantic electrostatic discharges between clouds and the earth and among clouds in the sky. In this chapter you will learn that the atmosphere is an insulator called a dielectric. Lightning "sparks" occur when the dielectric of the atmosphere breaks down. In a similar way, the Lord will break down the barrier between heaven and Earth at His Second Coming. He will then establish His millennial kingdom upon Earth, and all of us who are believers will reign with Him for a thousand years.

19A MODELING THE ELECTRIC FIELD

19.1 Lines of Force

Michael Faraday, a nineteenth-century English scientist, had little formal education. He gained most of his knowledge through personal study and experimentation. As a result, Faraday's mathematical skills were somewhat limited. However, this did not keep him from taking a theoretical approach (rather than a mathematical one) to explain certain phenomena that he had observed. Faraday proposed a model of forces that has proved fruitful in many areas of physics: the **field model.**

Faraday considered each charged object to be surrounded by an **electric field.** Through this field, the charged object exerts a force on any other charged object that ventures into the field. Faraday represented an electric field by **lines of force,** which give the direction of the force that the charged object exerts on a small positive charge. For example, Figure 19-1 shows the lines of force that represent the electric field of a positively charged object (*A*). A small positive charge (*B*) will experience a force in a direction directly away from the positively charged body.

The **field model** is a theoretical approach to accurately explaining such three-dimensional phenomena as magnetic, electric, and gravitational forces, as well as sound propagation and subatomic particle fluxes.

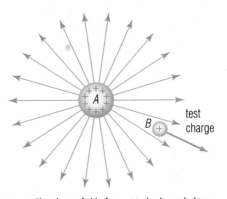

19-1 The electric field of a positively charged object

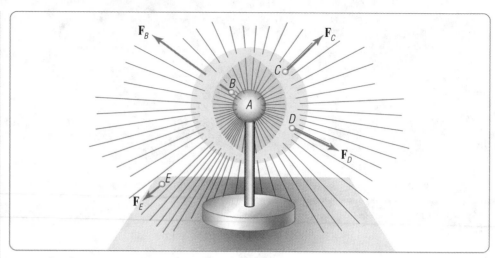

19-2 A three-dimensional view of the electrostatic field of a positively charged object

Lines of force can also show the relative magnitude of the force at each point. Closely spaced lines represent strong force. Widely spaced lines represent weak force. For example, in Figure 19-2 the lines are close together near the charged object and far apart away from the object. This shows that the object exerts a greater force on a charge near it than on a more distant charge.

19.2 The Test Charge and Field Strength

Suppose there is more than one charged body, as in Figure 19-3. What is the electric field like between the objects? How can you measure its magnitude and direction? There are two ways you can find the lines of force. First, you can use Coulomb's law to find the force that a positive **test charge** (q) would experience between the bodies. In this calculation, it is helpful to define the **electric field strength (E).** The electric field strength is the force per unit of positive charge at each point in the electric field. The magnitude of **E** generated by a charge Q is defined as

$$E \equiv \frac{F}{q}. \tag{19.1}$$

From Coulomb's law,

$$F = k\frac{Qq}{r^2},$$

where r is the distance between Q and q. Therefore,

$$E = \frac{kQq}{r^2} \cdot \frac{1}{q}, \text{ and}$$

$$E = \frac{kQ}{r^2}. \tag{19.2}$$

The direction of **E** at any location in the field is tangent to the electric field line through that point. The vector points radially *away* from Q if the charge is positive and radially *toward* Q if it is negative. (The direction of **E** is the same direction as the arrows representing the lines of force.)

To calculate the electric field strength between the charged objects in Figure 19-3, you must combine the electric field strength of the negatively charged object with the electric field strength of the positively charged object at each point, using vector addition. Figure 19-4 shows the result. This method is accurate but difficult and tedious to use.

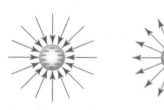

19-3 What is the shape of the electric field between two charged objects?

A **test charge** is a *positive* charge small enough that it exerts negligible force on other charges in the problem.

Equation 19.2 permits the calculation of the *magnitude* of the field strength due to a single charge Q. The orientation of the vector **E** depends on the sign of the charge. The sign associated with a positive Q is positive, and **E** points radially away from Q. If Q is negative, **E** points radially toward Q.

19-4 Calculating the electric field strength between two charged objects. Figure (a) shows the vector addition method for determining the electric field at a single point, *P*, in the electric field. This method must be applied to each point in the electric field. Figure (b) shows the electric field lines superimposed upon the individual field strengths throughout the grid. Figure (c) shows that the direction of the electric field strength vector is tangent to the field lines.

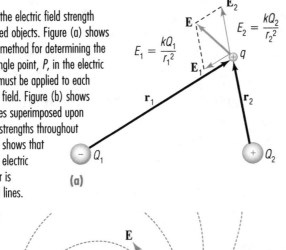

$E_1 = \dfrac{kQ_1}{r_1^2}$

$E_2 = \dfrac{kQ_2}{r_2^2}$

(a)

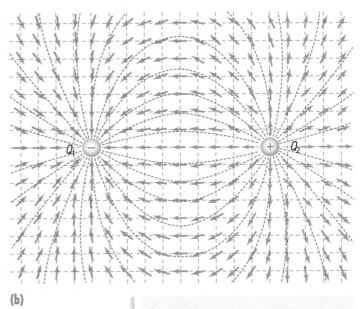

(b)

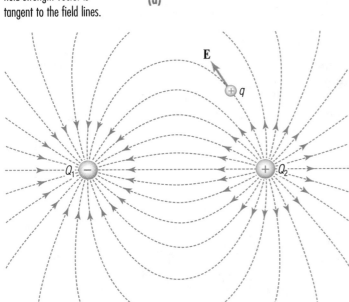

(c)

Diagrams of electric fields have the following properties:

- Lines lie parallel to the electric field strength vector **E.**
- Lines point from positive charge to negative.
- A stronger field is shown by lines that are closer together, and a weaker field is shown by lines that are farther apart.
- Field lines do not cross other field lines.

19.3 Electric Field Orientation

Notice in Figure 19-4 (c) that the lines of force between the objects begin on the positive object and end on the negative object. That is because a test charge (which is positive) would move from the positive object to the negative object along a field line. All field lines are assumed to begin on a positive charge and end on a negative charge. However, in the figure some field lines begin on the positive object and appear to go nowhere, whereas others appear to begin nowhere and end on the negative object. We assume that they are too far from the objects to be shown in the diagram, but they still originate at some point that has a different charge from the ones shown.

The second way to show the electric field is to set up an experiment. Set two charged objects in a dish containing grass seeds suspended in a liquid. Grass seeds are oblong in shape. The electrons in the grass seeds will migrate to one end of each seed so that each seed will align itself with the electric field. The seeds thus form a visible picture of the electric field. The illustrations in Figure 19-5 show the electric fields of several combinations of charged objects.

If you know the electric field strength, *E*, and the magnitude of the test charge, *q*, but not *F*, you can find *F* by rearranging Equation 19.1:

$$F = Eq$$

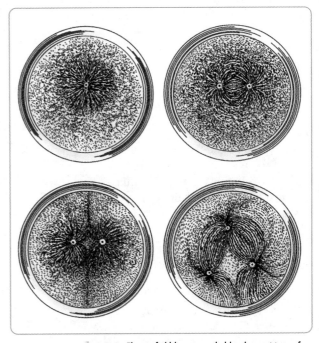

19-5 Electric field lines revealed by the positions of grass seed

If the test charge q is large, it can distort the field so that E is no longer due to the original charged object(s) alone. However, for electrostatic problems in this textbook, the magnitude of q is assumed to be so small that it has no effect on the electric field. The electrostatic force on the test charge (F_e) is due only to the field strength derived from the charged object(s) producing the field.

19.4 Electrical Potential and Electrical Potential Energy

All forces, including electrical forces, can do work. When an "electrified" piece of amber attracts a dry leaf, it does work on the leaf. Pushing like charges together or pulling unlike charges apart requires work, just as lifting a mass against gravity requires work. A lifted mass has gravitational potential energy, since it can fall. Similarly, a positive charge pushed near another positively charged object has potential energy with respect to the electrostatic force, since it will be repelled and will move away if allowed.

Electrical potential energy (U_e) is a measure of the amount of mechanical work that is necessary to move a charge from the location of a point charge to a radial location **r** from the point charge. The **electrical potential (V),** in contrast, is the amount of work *per unit charge* that is needed to move a charge the same distance r from the point charge. In other words,

$$V \equiv \frac{U_e}{q}.$$

Electrical potential is analogous to gravitational potential, which equals U_g/m, or $|g|h$.

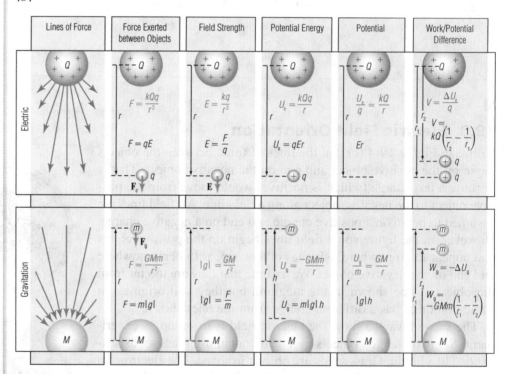

19-6 Comparison of the electrostatic force and the gravitational force

Electrostatic force has the same form as gravitational force:

$$F_g = G\frac{Mm}{r^2}$$

$$F_e = k\frac{Qq}{r^2}$$

Similarly, electrical potential energy has essentially the same form as gravitational potential energy (the difference in sign is due to the repulsion of like charges):

$$U_g = -G\frac{Mm}{r}$$

$$U_e = k\frac{Qq}{r}$$

The electrical potential of a test charge q is U_e/q; so

$$V = k\frac{Q}{r},$$

where Q is a point charge and r is the radial distance from the position of the point charge.

Like gravitational potential energy, electrical potential energy's zero point is usually taken to be an infinite separation between the charge source and the test charge. However, as we have seen elsewhere, only the difference in electrical potential energy, the **potential difference (ΔV),** is important. The potential difference between any two positions in an electrical field is the electrical work needed to move a unit charge between the positions, in the same way that mechanical work is the change in potential energy of a system in a gravitational field.

Since potential energy of all kinds is measured in joules, and charge is usually measured in coulombs, electrical potential and potential difference (ΔV) are both measured in joules per coulomb:

$$\Delta V = \frac{\Delta U_e}{q} \Rightarrow \frac{J}{C}$$

This unit of potential difference is called a **volt (V).**

$$1 \text{ volt} \equiv \frac{1 \text{ J}}{1 \text{ C}}$$

The magnitude of the electric field strength vector (E) is related to the potential difference in the same way that force is related to the difference in gravitational potential energy. That is, for the simple case where force and displacement are in the same direction,

$$F_g \times d = \Delta U_g,$$

it follows that

$$F_e \times d = \Delta U_e.$$

Therefore,

$$\frac{F_e}{q}d = \frac{\Delta U_e}{q}.$$

Since $F_e/q = E$ and $\Delta U_e/q = \Delta V$,

$$Ed = \Delta V. \qquad (19.3)$$

But $d = |\Delta \mathbf{r}|$, so Equation 19.3 can be rearranged to give

$$E = \frac{\Delta V}{d} = \frac{\Delta V}{|\Delta \mathbf{r}|}. \qquad (19.4)$$

Equation 19.4 shows that the field strength is related to the rate of change of the potential difference with position in an electric field.

Problem-Solving Strategy 19.1

Do not confuse electrical potential and electrical potential energy. Electrical potential is the property of a *position* in an electric field. It is measured in joules per coulomb (J/C) or volts (V). Its scalar value can be positive, negative, or zero depending on the sign of the charge producing the field.

Electrical potential energy is something a charged object has based on its location in an electric field. This quantity is expressed in joules (J).

The unit of electrical potential difference is the **volt (V).**

Problem-Solving Strategy 19.2

Do not confuse the derived unit **volt (V)** with the electrical potential variable symbol, *V*.

Problem-Solving Strategy 19.3

The direction of the electric field is always from the more positive (less negative) values of electrical potential toward lower (more negative) values of electrical potential.

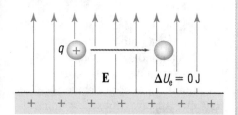

19-7 As charge q moves parallel to the charged surface, no work is done on the charge.

19.5 Equipotential Surfaces

Consider a flat, charged plate on which the charges are stationary (that is, they don't appear to be moving around). If the electric field had a component parallel to the plate's surface, the surface charges would move in response to the field strength component. However, since the surface charges do not move, the electric field must be perpendicular to the plate's surface.

Suppose you move a test charge parallel to the plate's surface (perpendicular to the lines of force). A force does no work on a system that moves at right angles to the line of action of the force; so you do no work with respect to the electrical potential by moving the charge parallel to the surface of the plate. Since no work is done on the test charge, its potential energy remains the same as it moves along the plate's surface. Therefore, every point on the surface of the plate must have the same electrical potential. For this reason, the surface of the plate is called an **equipotential surface.**

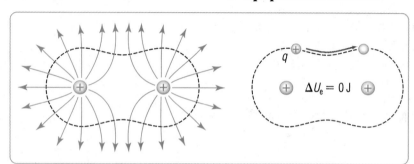

19-8 An equipotential line representing the cross section of a three-dimensional equipotential surface surrounding two charges

You can construct imaginary equipotential surfaces in any electric field by positioning the surfaces so that they are perpendicular to the electric field lines at every point on the surface. For example, the dotted line in Figure 19-8 is a two-dimensional cross section of a three-dimensional surface that is perpendicular to the electric field at every point surrounding the two charges. Therefore, it is an equipotential line. It is imaginary because there is nothing physically where the dotted line is. However, a test charge will have the same electrical potential everywhere along the line.

Every charged object is surrounded by an infinite series of three-dimensional equipotential surfaces. Since it is difficult to make three-dimensional drawings, we will represent these surfaces as two-dimensional equipotential lines. Figure 19-9 shows equipotential lines (dotted lines) around a spherical positively charged object.

The shape of the charged object affects the shape of its electric field. You have already seen the shape of a spherical object's field. What about the field of an irregular object? Figure 19-10 shows the electric field of an object containing a cavity. The dotted lines are equipotential lines around the object. The equipotential surfaces around the object are drawn so that they are always perpendicular to the field lines representing the orientation of **E.**

Although the technique of drawing an electric field perpendicular to a conductor's surface is useful, it has limitations. For example, consider a charged cube, which Figure 19-11 represents as a square. It is easy to find a perpendicular for the sides, but what about the corners? What direction is perpendicular to a corner? Experimentation has shown that the electric field is very strong at a sharp point like the corner of a cube. In fact, if the potential difference between the cube's surface and the surrounding air is large enough, electrons will fly from the corners of the cube into the air, creating a spark.

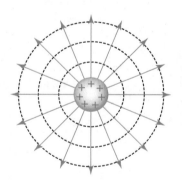

19-9 A charged sphere

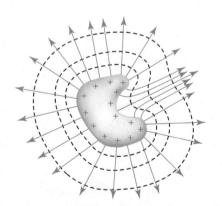

19-10 A charged object having an irregular shape

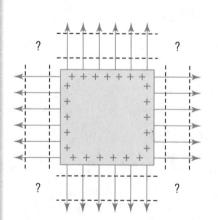

19-11 A charged cube

19A Section Review Questions

1. **a.** What model accounts for the spatial distribution and orientation of the electrostatic force associated with a charge?

   b. What graphical aid is used to represent this force in a diagram?

2. Why must the test charge that allows you to define the electric field strength vector be as small as possible?

3. Where do lines of force begin and end according to the current model of electric fields?

4. If you move a test charge from position $\mathbf{r}_1$ to position $\mathbf{r}_2$ on the opposite side of a point charge Q, and $r_1 = r_2$, how much total work did you do on the test charge? Explain your answer.

5. The rate of change of what quantity determines the field strength in an electric field?

⊚6. Draw three pairs of small circles on your paper. The circles represent point charges. Position them approximately 3 cm apart. Pair (a) are both positive charges. Pair (b) consists of a positive and a negative charge. Pair (c) are both negative charges.

   a. For each pair, sketch enough field lines to define the behavior of the electric field between and around the two charges. Make sure to show the direction of the field.

   b. For each pair, draw at least three lines representing equipotential surfaces surrounding the charges.

19B CAPACITORS

19.6 Storing Charge

One hindrance to early investigators of electricity was that they could not store charge. They could accumulate charges by using friction, but these charges soon leaked away. The investigators did discover that it is easier to store charge using two conductors separated by an insulator—positive charge on one and negative on the other—than on an isolated conductor.

In the mid-eighteenth century, scientists at the University of Leyden, Netherlands, devised a device called a **Leyden jar** to hold charge. The inside and outside of the glass jar had a lead coating. A conducting rod inserted through an insulating stopper was connected to the inner surface of the jar by a metal chain. The outside of the jar was grounded. When a charged object touched the conducting rod, it charged the inner lead surface. The glass molecules then aligned themselves so that the inner surface of the glass had a net charge opposite to the inner lead coating, and the outer glass surface had the same charge as the inner lead. In this way, the insulating glass maintained a neutral charge overall. The net charge of the outer glass surface induced a charge on the outer lead coating by forcing electrons to or from the ground. When the charged object was removed, the Leyden jar was charged and stayed charged for a reasonable time.

19.7 Structure of a Capacitor

Modern descendants of the Leyden jar are called **capacitors.** A capacitor consists of two conductors, called **plates,** separated by a **dielectric.** The dielectric may be a vacuum, air, glass, or one of many other materials. Dielectrics are usually

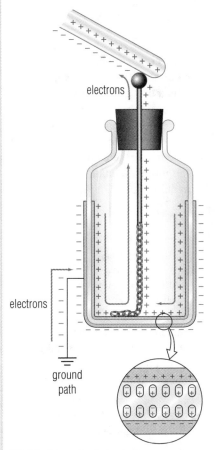

19-12 Schematic of a Leyden jar

Electric Fields **419**

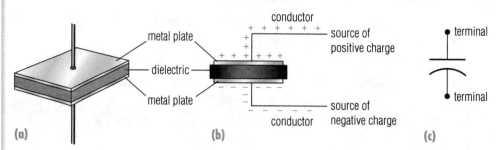

19-13 (a) shows the internal structure of a capacitor; (b) shows how a charge is established on a capacitor; and (c) is the electrical schematic symbol for a capacitor.

insulators. Depending on the construction of the capacitor, the leaves of metal foil that form the plates may be interleaved with each other in a flat, parallel configuration or rolled into a cylinder. Each conductor's surface is at a constant potential, but when the capacitor is charged, the potential is opposite on each plate. Therefore, there is a potential difference between the plates of a capacitor. For the sake of simplicity, we will limit our discussion to capacitors containing two flat plates arranged parallel to each other, which are called *parallel plate capacitors*.

If a conductor is connected between one plate of a capacitor and the other, the excess electrons on the negative plate travel to the positive plate, and the charge is neutralized. This process is called discharging the capacitor. The potential difference between the plates seems to produce a force that moves the electrons, although this is not a truly accurate explanation based on the present model of electron flow through a conductor. The more charge there is on each plate, the greater the potential difference is between the plates, and the greater the number of electrons that can move between the plates.

19.8 Dielectric Constant

The field lines that pass between the conducting plates of a parallel plate capacitor with a *vacuum* dielectric have a field strength that is proportional to the charge on the plates and is inversely proportional to the square of the distance between them. Through experimentation, it can be shown that when other substances are inserted between the plates of the capacitor, the field strength is reduced. This reduction of the vacuum field strength (E_0) is defined by a quantity called the **dielectric constant** (κ). The magnitude of the field strength in the dielectric ($E_{\text{die.}}$) is found by the equation

$$E_{\text{die.}} = \frac{E_0}{\kappa}. \tag{19.5}$$

The dielectric constant for vacuum is $\kappa_0 = 1$, so in a parallel-plate vacuum-dielectric capacitor, Equation 19.5 reduces to an identity. The dielectric constant for paper is about 4, so the field strength in a capacitor with a paper dielectric is one-fourth that of a vacuum. Therefore, for any dielectric, the vacuum field strength is reduced by a factor of κ in a capacitor of identical dimensions. The dielectric constant for each material is found by experiment.

Consequently, the dielectric reduces the potential difference between the plates compared to a vacuum capacitor. From Equation 19.3, we know that

$$V = Ed.$$

The **dielectric constant** is represented by the Greek letter kappa (κ). It is a unitless, positive real number (≥ 1) that is a measure of the ease with which electrostatic fields can permeate a volume of space or matter. The larger the number, the more attenuation of the electrostatic field.

The symbol ΔV could be used here for potential difference, but it is assumed that one of the plates of the capacitor is at zero relative potential. Therefore, it is usual to represent potential difference in such a case as just V. You saw this same principle in Chapter 9, where change in relative height determined the change in potential energy.

Substituting Equation 19.5 for the field strength in the presence of a dielectric, we have

$$V_{die.} = \left(\frac{E_0}{\kappa}\right)d = \frac{E_0 d}{\kappa} = \frac{V_0}{\kappa}.$$

Thus, a charged capacitor with a paper dielectric will have approximately one-fourth the potential difference between its plates compared to a vacuum capacitor with the same amount of charge. Next, you will see how this affects the performance of a capacitor.

19.9 Capacitance and the Farad

How much charge can a capacitor store? That depends on the potential difference between the plates. In this discussion, Q represents the entire charge on a conductor. The ratio Q/V is experimentally found to be a constant for each capacitor. This ratio is called its **capacitance (C):**

$$C = \frac{|Q|}{|V|} \tag{19.6}$$

A larger dielectric constant effectively increases the capacitance of a capacitor by reducing the potential difference required to store a given amount of charge. The capacitance of a nonvacuum dielectric capacitor is

$$C = \frac{|Q|}{|V_0/\kappa|} = \frac{\kappa|Q|}{|V_0|} = \kappa C_0,$$

where C_0 is the capacitance of a parallel plate capacitor with a vacuum dielectric.

Because the capacitance of a capacitor is always positive regardless of the kind of charge and the sign of the potential difference, the absolute value signs in Equation 19.6 are required. Since the capacitance for any given capacitor is a constant, the values of Q and V cannot vary independently of one another. That is because C is also dependent on the geometry of the capacitor. The formula for capacitance using geometric quantities is

$$C = \frac{\kappa \varepsilon_0 A}{d}, \tag{19.7}$$

where A is the area of each plate, ε_0 is called the **permittivity of free space,** and d is the distance between the plates. The constant ε_0 is defined as

$$\varepsilon_0 \equiv \frac{1}{4\pi k} = 8.854 \times 10^{-12} \text{ C}^2/(\text{N} \cdot \text{m}^2).$$

The constant k in the denominator is the Coulomb's law constant, 8.988×10^9 N·m²/C². The product $\kappa\varepsilon_0$ is related to the dielectric and is called the material's **permittivity (ε).**

The **farad (F)** is the SI derived unit of capacitance. It is defined as one coulomb per volt:

$$1 \text{ F} \equiv \frac{1 \text{ C}}{1 \text{ V}}$$

The farad is a very large unit; so the microfarad (μF $= 10^{-6}$ F), the nanofarad (nF $= 10^{-9}$ F), or the picofarad (pF $= 10^{-12}$ F) are usually used in actual electronic circuits involving discrete **electrical components.**

TABLE 19-1

Selected Dielectric Constants

Substance	κ
Vacuum	1
Air	1.0005
Waxed Paper	2.2
Mylar	3.1
Paper	3.7
Pyrex Glass	5.6
Bakelite	~5
Mica	5.4
Porcelain	6.5
Water	80

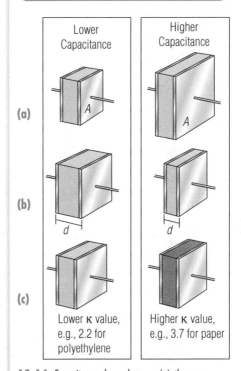

19-14 Capacitance depends upon (a) the area of the plates, (b) the separation of the plates, and (c) the dielectric constant (κ) of the medium between the plates.

The **farad** was named after Michael Faraday.

An **electrical component** is any device that is used in an electrical circuit. Examples of electrical components include capacitors, resistors, switches, batteries, lamps, inductors, transistors, and a host of other devices.

EXAMPLE 19-1

Capacitance and Charge

A parallel plate capacitor consists of two 1.00×10^{-4} m^2 plates separated by a distance of 1.00×10^{-3} m. It has air for a dielectric ($\kappa = 1.00$). (a) What is its capacitance? (b) If there is a potential difference of 1.00 V between the plates, what charge is stored on them?

Solution:

a. Find the capacitance using Equation 19.7:

$$C = \frac{\kappa \varepsilon_0 A}{d}$$

$$C = \frac{(1.00)(8.854 \times 10^{-12} \text{ C}^2/\text{N}\cdot\text{m}^2)(1.00 \times 10^{-4} \text{ m}^2)}{1.00 \times 10^{-3} \text{ m}}$$

$$C = 8.854 \times 10^{-13} \text{ C}^2/\text{N}\cdot\text{m}$$

$$C = 8.854 \times 10^{-13} \text{ C}^2/\text{J}$$

$$C = 8.854 \times 10^{-13} \text{ C/V} \quad (1 \text{ V} = 1 \text{ J/C})$$

$$C \approx 0.885 \text{ pF}$$

b. Find the charge using Equation 19.6:

$$C = \frac{|Q|}{|V|}$$

$$|Q| = C|V| = (8.854 \times 10^{-13} \text{ F})|1.00 \text{ V}|$$

$$|Q| = 8.854 \times 10^{-13} \text{ (C/V)(V)}$$

$$Q \approx 8.85 \times 10^{-13} \text{ C}$$

19.10 Connected Capacitors

Capacitors are electrical components that store charges in an electrical circuit. In order to begin understanding some of the basic electrical properties of capacitors, we will examine how they function when connected to other capacitors. They may be connected in either of two ways. The first is a *parallel* connection. Two or more capacitors are arranged such that the wires from the plate at one end of each capacitor are all attached together (with solder or an electrical terminal), and the wires to the other plates are all attached together (see Figure 19-15). In this way, the corresponding plates of parallel connected capacitors are at the same potential. Therefore, the potential difference between the plates of all capacitors connected in parallel is the same.

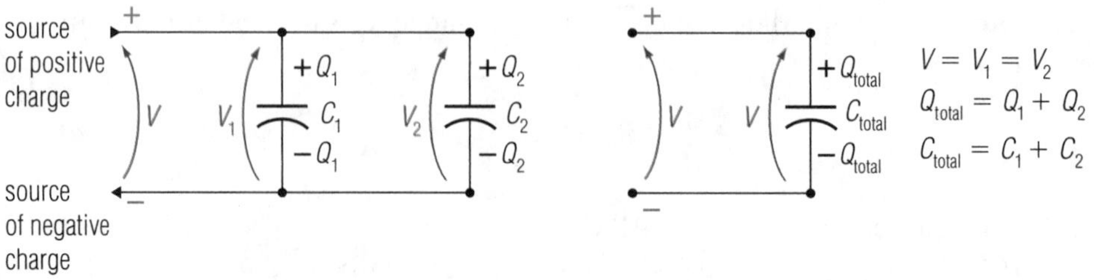

19-15 Capacitors in parallel

The total charge stored in the circuit segment is the sum of the charges on the capacitors:

$$Q_{\text{total}} = Q_1 + Q_2 \qquad (19.8)$$

Since $Q = CV$, and the potential difference between the plates is the same for each capacitor, Equation 19.8 can be rewritten as

$$C_{\text{total}}V = C_1V + C_2V.$$

Canceling the Vs gives us

$$C_{\text{total}} = C_1 + C_2. \qquad (19.9)$$

The total capacitance of parallel connected capacitors is just the sum of the individual capacitances.

The other way to connect capacitors is to connect one plate of one capacitor to one plate of another capacitor in a chain, as Figure 19-16 shows. The unconnected plates of the end capacitors are attached to the source of potential difference. This arrangement is called a *series* connection. The potential difference between plates A and D induces equal and opposite charges on these plates. Plate A then induces a charge of $-Q$ on plate B, and plate D induces a charge of $+Q$ on plate C, since that is how capacitors work. Note that the net charge along any continuous **conductor** containing series capacitors is zero. The total potential difference in the circuit segment is the sum of the potential differences across the capacitors.

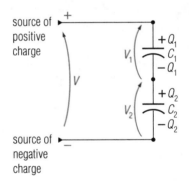

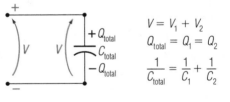

19-16 Capacitors in series

For series connected capacitors, you can see that

$$V_{\text{total}} = V_1 + V_2. \qquad (19.10)$$

Since $C = Q/V$, then $V = Q/C$, and Equation 19.10 can be rewritten as

$$\frac{Q}{C_{\text{total}}} = \frac{Q}{C_1} + \frac{Q}{C_2}.$$

Canceling the Qs,

$$\frac{1}{C_{\text{total}}} = \frac{1}{C_1} + \frac{1}{C_2}. \qquad (19.11)$$

EXAMPLE 19-2

Total Capacitance of Connected Capacitors

What is the total capacitance of a 2.00 μF capacitor and a 4.00 μF capacitor if they are (a) connected in parallel? (b) connected in series?

Solution:

(a) Total capacitance in parallel using Equation 19.9:

$$C_{total} = C_1 + C_2$$

$$C_{total} = 2.00 \ \mu F + 4.00 \ \mu F$$

$$C_{total} = 6.00 \ \mu F$$

(b) Total capacitance in series using the shortcut given above:

$$C_{total} = \frac{C_1 C_2}{C_1 + C_2} = \frac{(2.00 \ \mu F)(4.00 \ \mu F)}{2.00 \ \mu F + 4.00 \ \mu F}$$

$$C_{total} = \frac{8.00 \ (\mu F)^2}{6.00 \ \mu F}$$

$$C_{total} = 1.3\overline{3} \ \mu F$$

Note that the total capacitance for parallel-connected capacitors will always be higher than for the same capacitors connected in series.

If the potential difference between a capacitor's plates becomes too high, even a good insulating dielectric will break down and conduct charge between the plates. The damage is usually permanent, and the capacitor's function is degraded or destroyed. Some dielectrics will even explode if they are exposed to a potential difference that is too high. For this reason, most capacitors are marked with a potential difference rating as well as a capacitance. The capacitor will probably cease to function if the potential difference rating is exceeded.

Some capacitors are marked with a positive sign on one terminal. These components are called *polarized capacitors* because they are designed to be connected in a circuit only one way—with the positive plate at the higher potential. If improperly connected, even normal voltages can cause the internal insulation to break down, resulting in failure.

19B Section Review Questions

1. Describe the construction of a Leyden jar and compare it to the structure of a modern capacitor.

2. a. How does electric field strength between capacitor plates in a vacuum compare to field strength when an insulating material replaces the vacuum?

   b. What electrical quantity accounts for this fact?

3. In an isolated charged capacitor of given dimensions, how does the potential difference between the plates with a vacuum for a dielectric compare to the potential difference when some other insulating material is present?

4. What factors determine the capacitance of a capacitor?

⊛5. What is the capacitance of a capacitor that has a plate area of $2.3 \times 10^{-4} \ m^2$ and a distance of $1.8 \times 10^{-3} \ m$ between the plates? The dielectric is a mylar-like material.

⊛6. If the capacitor in Question 5 is connected to the terminals of a 9.0 V battery, how much charge can be stored in the capacitor?

⊚ **7.** Describe the arrangement of capacitor connection in the adjacent figure. Determine the total capacitance of these capacitors.

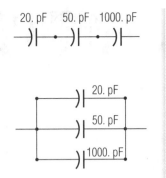

⊚ **8.** Describe the arrangement of capacitor connection in the adjacent figure. Determine the total capacitance of these capacitors.

FACETS of PHYSICS

Kinds of Capacitors

Many different kinds of capacitors do many different jobs. Air capacitors, whose dielectric is air, do not need to be enclosed. However, the plates must be far enough apart that they will not touch by accident. Vacuum capacitors can tolerate a much higher potential difference without allowing a current between their plates. Of course, vacuum capacitors must be enclosed in an airtight container.

Many capacitors have solid dielectrics. Ceramic capacitors have high capacitance for small sizes, but they also become less reliable with age. Mica dielectrics are especially useful in powerful radio transmitters. Paper dielectrics are inexpensive and work well where they are unlikely to get wet.

Liquid dielectrics are also common. Electrolytes (conducting solutions) form capacitors by depositing an insulator on a conducting plate. Since they operate properly when the current goes in one direction but not when the current reverses, electrolytic capacitors are called polarized capacitors. They must be connected in the proper direction or they will not work. Modern capacitors are often polarized.

Capacitors need not have a fixed capacitance. Variable air capacitors usually consist of many semicircular plates that may or may not overlap. The more plates that are overlapping, the greater the capacitance is. These capacitors are essential for radio frequency tuning circuits.

Chapter Review

In Terms of Physics

field model	19.1	capacitor	19.7
electric field	19.1	plate	19.7
lines of force	19.1	dielectric	19.7
test charge	19.2	dielectric constant (κ)	19.8
electric field strength (**E**)	19.2	capacitance (C)	19.9
electrical potential (V)	19.4	permittivity of free space (ε_0)	19.9
potential difference (ΔV)	19.4	permittivity of material (ε)	19.9
volt (V)	19.4	farad (F)	19.9
equipotential surface	19.5	electrical component	19.9
Leyden jar	19.6	conductor	19.10

Problem-Solving Strategies

19.1 (page 417) Do not confuse electrical potential and electrical potential energy. Electrical potential is the property of a position in an electric field. It is measured in joules per coulomb (J/C) or volts (V). Its scalar value can be positive, negative, or zero depending on the sign of the charge producing the field.

Electrical potential energy is something a charged object has based on its location in an electric field. This quantity is expressed in joules (J).

19.2 (page 417) Do not confuse the derived unit **volt (V)** with the electrical potential variable symbol, V.

19.3 (page 417) The direction of the electric field is always from the more positive (less negative) values of electrical potential toward lower (more negative) values of electrical potential.

19.4 (page 423) The term "total capacitance" can also be called "equivalent capacitance," since you can replace all of the capacitors connected together with a single one of an equivalent value.

19.5 (page 423) Review the process for summing the reciprocals of quantities. Recall that

$$\frac{1}{C_{total}} = \frac{1}{C_1} + \frac{1}{C_2} \neq \frac{1}{C_1 + C_2}.$$

Once you have found the sum of the reciprocal capacitances, you must take the reciprocal of *that* sum to find the total capacitance.

19.6 (page 423) In the case of just two series-connected capacitors, a much simpler formula to use is

$$C_{total} = \frac{C_1 C_2}{C_1 + C_2}.$$

Review Questions

1. What are lines of force?

2. In the figure, in what region is the electrical force the strongest?

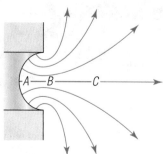

3. Draw the lines of force surrounding a spherical, negatively charged body.

4. Describe a test charge.

5. What is electric field strength?

6. An object with a charge of 2.00 mC rests far from other charges. Write an equation that gives electric field strength at a given distance r from the charge.

7. Where does an electric field line end? Why?

8. What is electrical potential?

9. The dotted lines in the figure are equipotential lines. If the potential difference between them at A is 3 V, what is the potential difference between them at B? Why?

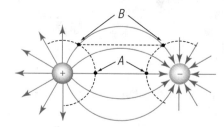

10. What does a capacitor do?

11. Name the predecessor of the capacitor.

12. List the parts of a capacitor.

13. Name three ways to increase a capacitor's capacitance.

14. During a lab period, you and your lab partner charge a capacitor with a 6 V battery. You say that the capacitor has a charge of $|Q|$. Your lab partner says that there is zero charge on the capacitor. You are both right. Explain why this is so.

15. How does charge vary with potential difference applied to a capacitor? Sketch a graph of charge versus potential difference for a capacitor with charge on the vertical axis and potential difference on the horizontal.

16. For a group of series-connected capacitors, which capacitor will have the greatest effect determining the total capacitance—the one with the largest capacitance or the one with the smallest? Explain your answer from either a mathematical or an electrical viewpoint.

True or False (17–30)

17. Lines of force can actually cross each other where electrical fields are exceptionally strong.

18. Field lines point in the direction that a small, positively charged object would move in response to the electric field.

19. The field strength vector **E** is tangent to the field's lines of force and points toward the positive charge.

20. Every field line originates at a positive charge and terminates at a negative charge, even if the line is not visible beyond the edge of a diagram.

21. Electrical potential is measured with reference to an infinite position from a charge.

22. Field strength is the rate of change of electrical potential with distance in an electric field.

23. The sharper the curvature of an equipotential surface, the more concentrated the lines of force.

24. The thicker the lead coating in a Leyden jar, the greater the amount of charge that could be stored, assuming all other dimensions and materials were held the same.

25. Within limits, the greater the potential difference supplied to a capacitor, the greater the amount of charge that can be stored.

26. A large dielectric constant implies that more charge can be stored for a given potential difference between the plates.

27. The capacitance of a vacuum dielectric capacitor is greater than for a capacitor of the same dimensions having a glass dielectric.

28. Doubling the plate area of a capacitor doubles its capacitance.

29. Capacitance in SI base units is C^2/J.

30. Doubling the number of identical capacitors connected in series doubles the amount of charge that can be stored by the entire circuit.

◉31. A large charged object exerts a force of 10^{-3} N on a 1 μC charge. What is the electrical field strength at the small charge's position?

◉32. A 3.00 μC charge has an electrical potential energy of 9.00×10^{-5} J. What is the electrical potential of its position?

◉33. A tiny spherical object with a charge of +1.00 mC rests far from all other charges.
   a. What is the electrical potential 0.0100 m from the object?
   b. What is the electrical potential 1.100 m from the object?
   c. What is the potential difference at a point 1.100 m from the object to 0.0100 m away?

◉34. Draw an equipotential line for the electric field in the adjacent figure.

◉35. A parallel-plate vacuum-dielectric capacitor has a plate area of 3.00×10^{-4} m² and has a distance of 3.00×10^{-2} m between the plates. What is the capacitance?

◉36. A round capacitor with a radius of 1.50×10^{-2} m has a vacuum dielectric. The distance between the parallel plates can be manually varied. The maximum distance apart is 1.00×10^{-2} m and the closest they can be is 5.00×10^{-4} m.
   a. What is the area of each plate?
   b. What is the capacitance at maximum plate separation?
   c. What is the capacitance at minimum plate separation?
   d. When the charge on the capacitor is 2.00×10^{-11} C, what is the potential difference across the capacitor at maximum separation?
   e. When the charge on the capacitor is 2.00×10^{-11} C, what is the potential difference across the capacitor at minimum separation?

◉37. A capacitor with square plates 5.00 cm on each side and a Pyrex glass dielectric has a separation distance of 5.00×10^{-3} m. What is its capacitance?

◉38. When a capacitor has a charge of 6.00×10^{-11} C, the potential difference between its plates is 20.0 V. What is its capacitance?

◉39. Two capacitors, one with $C_1 = 2.00$ μF and one with $C_2 = 3.00$ μF, are connected in parallel. What is the equivalent capacitance of the arrangement?

◉40. Suppose you have a box full of only 1.50 pF capacitors and you need a capacitance of 1.00 pF. How could you arrange capacitors to get the desired capacitance? (*Hint:* Try various combinations of series and parallel capacitors.)

◉41. Two 6.00 μF capacitors are arranged in series. What is the arrangement's capacitance?

◉42. You have three capacitors: two 4.00 pF and one 2.00 pF.
   a. What is the capacitance if you connect them in parallel?
   b. What is the capacitance if you connect them in series?
   c. What is the capacitance if you connect the two 4.00 pF capacitors in series?
   d. What is the total capacitance if you connect the series arrangement of the 4.00 pF capacitors in parallel with the 2.00 pF capacitor? (*Hint:* treat the two 4.00 pF capacitors as a unit.)

◉◉43. Write a brief report, including diagrams, of the construction and usage of 1+ farad capacitors.

◉◉44. Investigate and report on what happens when too high a voltage is applied to a capacitor. How can this be quantified?

◉◉45. Lightning is an example of breakdown of a dielectric. Prepare a report on lightning and its associated upper atmospheric effects (such as lightning sprites).

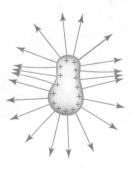

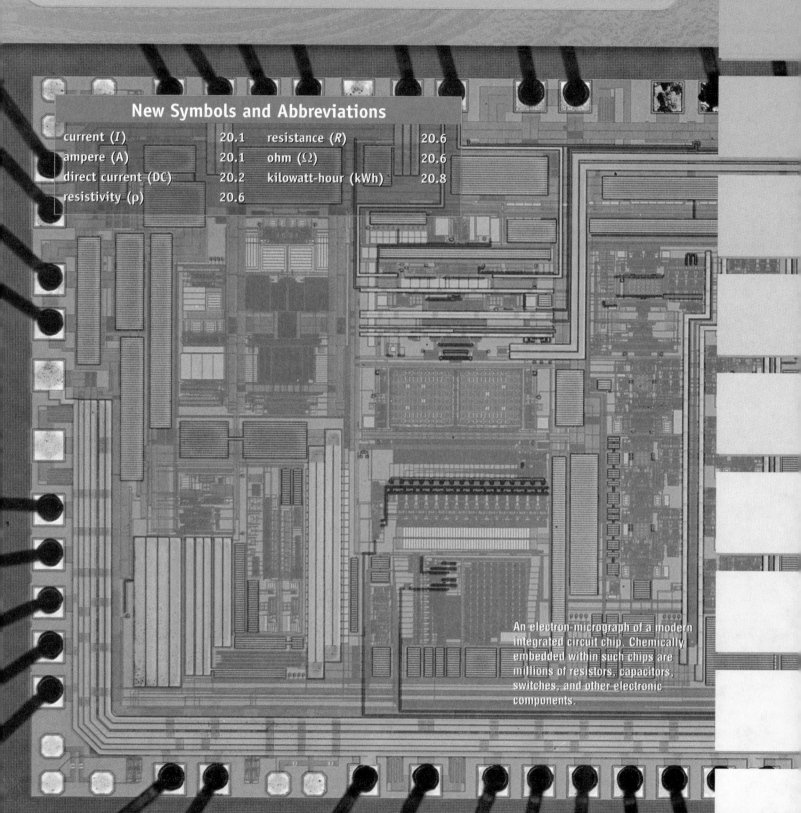

20A	Current, Voltage, and Resistance	429
20B	Electrical Circuits	436
20C	Electrical Safety	445
Facet:	Fuel Cells	452

New Symbols and Abbreviations

current (I)	20.1	resistance (R)	20.6
ampere (A)	20.1	ohm (Ω)	20.6
direct current (DC)	20.2	kilowatt-hour (kWh)	20.8
resistivity (ρ)	20.6		

An electron-micrograph of a modern integrated circuit chip. Chemically embedded within such chips are millions of resistors, capacitors, switches, and other electronic components.

Electrodynamics 20

The Apostle Paul wrote this solemn warning to Christians who were not obeying the commandments of God. Just as a resistor decreases electrical potential, spiritual resistance decreases the power of God in a person's life. The spiritual allegory gets even more drastic. One who completely rejects the ordinances of God is like an infinite resistance. There is no power of God in his life and he risks severe judgment. Don't be a resistor, but rather be a perfect conductor—pass on the potential of God's power and grace to others undiminished.

20A CURRENT, VOLTAGE, AND RESISTANCE

20.1 Current

The last two chapters dealt with stationary charged objects (electrostatics). However, charged objects do move when they are attracted or repelled by other charged objects. They may set up a **current (I),** which is defined as a continuous flow of charge. The study of the causes and effects of current electricity is called **electrodynamics.**

The motion of electrons creates a current in metals. This electrical conduction is similar to thermal conduction, which is the transmission of kinetic energy by collisions of free electrons. Electrical conduction is the transmission of electrical potential energy by repulsions and attractions of free electrons. Free electrons, remember, are electrons that are loosely held by their nuclei.

Electrical conduction in metals begins with a potential difference in the metal. For instance, if a positively charged object approaches one end of a wire, that end of the wire has a higher electrical potential than the other end. The wire's free electrons move toward the positively charged object. When the electrons move, they leave a space with a net positive charge. Therefore, new free electrons are attracted to the region that other free electrons vacated.

Free electrons do not move far from their original positions; yet an impulse of motion spreads throughout the metal. To illustrate how this occurs, consider a line of dominoes with many dominoes standing on end less than one domino-length

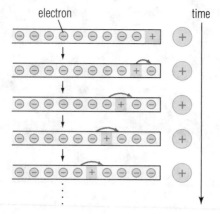

20-1 Electrons moving in a metal-like solid

Equation 20.5 is the classical empirical definition of the ampere. The present-day SI definition is given in Appendix A.

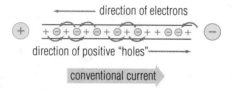

direction of electrons

direction of positive "holes"

conventional current

20-2 The direction of conventional current flow

The flow of current is in the direction that tends to cancel the potential difference causing the current. This textbook adopts the convention that current is the flow of positive charges from higher (positive) potentials to lower potentials.

The potential difference between a point at 10 V and a point at 5 V is 10 V − 5 V = +5 V. The potential difference between a point at 5 V and a point at 10 V is 5 V − 10 V = −5 V. The second number is always taken as the reference potential.

apart. If you push the first domino over, the last one will eventually fall. The dominoes have not moved far from their original positions, but the impulse of falling down has spread through all the dominoes.

The velocity of the electrons, called the *drift velocity,* is slow—on the order of millimeters per second. The free electrons in a conductor tend to move in random directions, even when an electrical potential is applied, but more electrons move toward the higher potential than in any other direction. This net movement with time is the drift velocity. Each electron needs to move only far enough to attract or repel another electron. As soon as the electron moves even a small distance, the free electrons nearest it are affected. The impulse moves much faster than the electrons do—at nearly the speed of light.

In solid conductors, electrons usually carry the current. In *electrolytic* (conducting) solutions, though, electrons are not the sole current-carriers. *Electrolytes* are materials that are able to conduct electricity when they are dissolved. They separate into positive ions (atoms or molecules that have a deficiency of electrons) and negative ions (atoms or molecules that have an excess of electrons). In a solution, the current consists of negative ions and positive ions traveling in opposite directions.

Since current is the flow of charge, it is natural to define the unit of current, the **ampere (A),** as 1 coulomb of charge per second.

$$1 \text{ A} = \frac{1 \text{ C}}{1 \text{ s}}$$

20.2 Current Direction

These facts raise a question: What is the current direction? The choice of current direction is arbitrary. The first scientists to study electricity assumed that current was the flow of positive charge, after Ben Franklin's naming convention. Therefore, they decided that current flows from a higher potential (positive charge) to a lower potential (negative charge). This text will follow this convention. When considering current in a solid such as a wire, you need to be careful to avoid confusion. Although both positive and negative charges move through solutions, in solids only the electrons move.

Since the *conventional current* direction is the direction that positive charges would move, this direction is opposite the direction of the electron flow. It makes little difference, since the flow of positive charge ("holes") in one direction is equivalent to the flow of negative charge in the opposite direction. In either case, the charges move to cancel the potential difference. The motion of charges to cancel a *constant-direction* potential difference is called **direct current,** or **DC.**

20.3 Potential Difference and Voltage

Potential difference is a difference in electrical potential between two positions. The potential difference between point A and point B is $V_A - V_B$, while the potential difference between point B and point A is $V_B - V_A$. The second number is always assumed to be the reference potential. Obviously, one value will be the negative of the other:

$$V_A - V_B = -(V_B - V_A)$$

Consequently, it is important to identify how potential difference is measured. Potential difference can be positive, negative, or zero. In order to identify the location of higher and lower potentials on a source of potential difference, **polarity markings** are used—positive (+) for the higher potential and negative (−) for the lower. These points do not necessarily have to be at positive or negative potentials, respectively. It just means that the point with the $+$ is at a higher potential than the point with the $-$.

Any device that creates a difference in potential between two points is a source of potential difference. If the potential difference is created by a self-contained device, especially by a chemical reaction such as in a battery, it is sometimes called *emf*. *Emf (electromotive force)* is a term scientists gave to potential difference before they realized that it is not a force. Even though emf is not a force as we normally use the word in physics, it is sometimes useful to resort to this term for brevity's sake.

Differences in electrical potential can be produced in several ways. Most processes involve a change from another form of energy to electrical potential energy. A *battery* changes chemical energy to electrical potential; a *photovoltaic cell* changes light energy to electrical potential; a *Van de Graaff generator* (static electricity generator) and *piezoelectric crystals* convert mechanical energy to electrical potential; and a *thermocouple* converts thermal energy to electrical potential.

Potential difference is measured in volts (V), or joules of energy per coulomb of charge.

$$1 \text{ V} = \frac{1 \text{ J}}{1 \text{ C}}$$

One coulomb (C) is defined as the total charge of a specific number of fundamental charges:

$$1 \text{ C} = 6.24 \times 10^{18} \ e,$$

where e is the absolute value of the charge of an electron. Since a specific number of a large quantity of fundamental charges is impossible to measure with present technology, scientists resort to the classical definition of the ampere and define the coulomb in SI units as an ampere-second.

$$1 \text{ C} \equiv 1 \text{ A·s}$$

20.4 The Voltaic Cell

Probably the most convenient source of steady potential difference for laboratory use is one based on the **voltaic cell,** an electrochemical device that spontaneously changes chemical energy to electrical energy. The conversion of chemical energy to electrical energy was first observed by Luigi Galvani as he was studying the anatomy of a frog. He discovered that if he touched a nerve in a frog's leg with a metal scalpel while the frog was on a metal tray, the touch established an electric current, which contracted the frog's leg muscles. Galvani touched the nerve with different metals and discovered that some metals caused a stronger contraction that others. He thought that the frog's tissues were the source of the current.

Considering Galvani's observations, Alessandro Volta correctly concluded that the source of the current is the contact of two different metals (the tray and the scalpel) in a conducting medium (the frog's body fluids). The difference in chemical activities of the two metals causes electrons and positive ions to flow between the metals if both metals contact an electrolytic solution. Volta made a cell from pieces of copper and zinc separated by seawater-soaked cardboard. This cell is called a *voltaic pile,* after Volta.

The metal conductors in voltaic cells are called **electrodes.** One electrode (the *cathode*) is made of a metal that has a stronger attraction for electrons than the other electrode (the *anode*). The cathode pulls electrons from the anode through the wire. The accumulation of electrons on the cathode attracts positive ions (cations) from the solution that accept the electrons to form a neutral plating on the cathode. The anode, which lost electrons to the cathode via the wire, is a site

20-3 A Van de Graaff generator can safely produce large static electrical potentials. A static charge can be the beginning of a bad-hair day!

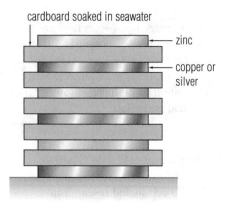

cardboard soaked in seawater
zinc
copper or silver

20-4 A voltaic pile

An electrochemical cell consists of a single anode and cathode immersed in an electrolyte. A **battery** is two or more cells connected together.

of positive charge immersed in the solution. It attracts electrons from the negative ions (anions) in the solution. This process continues as long as electrolytic ions remain in the solution and the metal electrodes remain in contact with the solution.

Although the electrons flow from the anode to the cathode, their charge is negative; so the external *conventional current* from a voltaic cell is, by our convention, directed from the cathode to the anode. With the wire in place, the current has a complete path: positive ions go from the anode to the cathode in the *solution,* and positive "holes" flow back to the anode through the *wire*. The closed path of electrical current thus described is called a **circuit.**

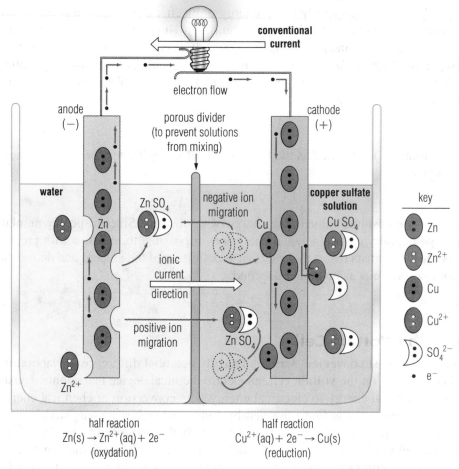

20-5 A voltaic or electrochemical cell

20.5 Types of Electrochemical Cells and Batteries

Modern voltaic cells function in a similar way to Volta's cell. There are several different types of cells. **Primary cells** simply convert chemical energy to electrical energy. Since a primary cell depends on irreversible chemical reactions, it cannot be recharged. The standard battery cells you use in a flashlight are primary cells.

Unlike primary cells, **storage cells** use only reversible chemical reactions. Therefore, storage cells can be recharged by the addition of electrical energy, which is converted to chemical energy. Probably the most familiar storage cell is found in an automobile battery. These batteries usually consist of a group of connected storage cells. (Technically, a **battery** is a series of cells. Flashlight "batteries" are really single voltaic cells.) Rechargeable storage cells that use lithium ion and nickel-metal hydride technology are found in batteries used in cellular phones, digital cameras, and laptop computers.

Another term you may hear applied to voltaic cells is the **dry cell.** A dry cell is a primary or storage cell that has an electrolytic paste instead of a liquid solution. Since there is no chance of spilling the solution, a dry cell is easier and safer to use than a "wet" cell. The common D-cell is a dry primary cell, whereas a car battery consists of wet cells.

20.6 Resistance

Every material tends to impede the flow of charge. In wires, the free electrons collide with the electric fields of the inner atoms when these atoms are in the electrons' path. Such collisions convert the electric potential energy of the free electrons to kinetic energy of the inner atoms. At higher temperatures, the inner atoms have more kinetic energy and collide more forcefully with the conducting electrons. Therefore, both the material and the temperature of an object affect how well it conducts electricity.

The tendency to impede current is represented by **resistivity (ρ).** Resistivity (a different quantity from resistance) is a property of a material that varies with temperature but is not dependent on its size and shape. Good conductors such as copper have lower resistivities than poor conductors such as glass. Only **superconductors**—certain materials at temperatures near absolute zero—appear to have zero resistivity.

Any circuit component that is designed to convert electrical potential energy to thermal energy (and produce a potential difference in the process) is called a **resistor.** Electrons in a long, thin wire suffer more collisions than those in a short, thick wire, since they must travel past more inner atoms. Thus a long, thin wire is a better resistor than a short, heavy wire. The geometry of a circuit component also affects how well it conducts or resists. A resistor must be made of materials of the proper resistivity and be formed into the right size and shape for its purpose.

Resistance (R) is a quantity that takes into account both the resistivity of the material and the geometry of the resistor. Objects with long current paths resist more than short objects of the same material, and thin objects resist more than wide objects of the same material. The following equation represents this relationship:

$$R = \rho \frac{L}{A},$$ (20.1)

where L is the length of the object's current path and A is the cross-sectional area of the current path. A good conductor has a low resistance, and a poor conductor has a high resistance. The symbol $-\!\!\bigwedge\!\!\bigwedge\!\!-$ represents a resistor in a circuit diagram. The unit for resistance is the **ohm (Ω),** which is further defined below.

Resistivity is the temperature-dependent property of a material that is the measure of the way it impedes current flow. The symbol for resistivity is the Greek letter rho (ρ). The unit of resistivity is the ohm-meter ($\Omega\cdot$m).

What is the spiritual equivalent of resistivity? Probably a combination of unbelief, disobedience, and sin. "Take heed, brethren, lest there be in any of you an evil heart of unbelief, in departing from the living God. But exhort one another daily, while it is called To day; lest any of you be hardened through the deceitfulness of sin. For we are made partakers of Christ, if we hold the beginning of our confidence stedfast unto the end" (Heb. 3:12–14).

TABLE 20-1
Resistivities at Room Temperature

Material	ρ ($\Omega\cdot$m)
Conductors	
Silver	1.6×10^{-8}
Copper	1.7×10^{-8}
Aluminum	2.6×10^{-8}
Tungsten	5.5×10^{-8}
Nickel	7.8×10^{-8}
Lead	21×10^{-8}
Nichrome	100×10^{-8}
Semiconductors	
Carbon	3.5×10^{-5}
Germanium	0.5
Silicon	2.3×10^{3}
Insulators	
Glass	$10^{10} - 10^{14}$
Hard Rubber	10^{16}
Quartz	7.5×10^{17}

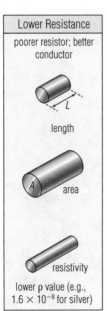
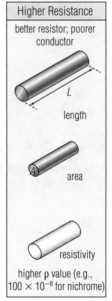

Lower Resistance	Higher Resistance
poorer resistor; better conductor	better resistor; poorer conductor
length	length
area	area
resistivity	resistivity
lower ρ value (e.g., 1.6×10^{-8} for silver)	higher ρ value (e.g., 100×10^{-8} for nichrome)

20-6 Factors influencing magnitude of resistance

20.7 Ohm's Law

Every electrical component converts some electrical potential energy to thermal energy; each circuit component causes a decrease of potential in the direction of current flow. How much potential energy is lost? In 1827, Georg Ohm found the answer to this question. He began experimenting by measuring the current in a circuit made from uniform wire while he changed the potential difference, and he found that current is directly proportional to potential difference. Then, keeping the potential difference the same, Ohm measured the current while he changed the length of the current path by adding or removing identical wires. He found that current is inversely proportional to the length of wire, which represents the resistance. Finally, Ohm expressed his findings in an equation known as **Ohm's law,** which says that the change of potential across any circuit component is

$$V = IR,$$ (20.2)

where V is the potential difference across the component, I is the current through the component, and R is the resistance of the component.

The unit of resistance, the ohm (Ω), is defined according to Ohm's law: One volt of potential difference across one ohm of resistance produces one ampere of current through the resistance.

$$1 \, \Omega \equiv \frac{1 \text{ V}}{1 \text{ A}}$$

20.8 Electrical Work and Power

The purpose of most electrical components is to convert electrical energy to some other form of energy. That is, electrical components are intended to do work. Since potential difference is measured in volts, or joules per coulomb, and work is expressed in joules,

$$\text{potential difference} = \frac{\text{work}}{\text{charge}}.$$

From this equation you can see that

$$\text{work} = \text{potential difference} \times \text{charge, or}$$
$$W = V \times q.$$ (20.3)

Equation 20.3 is true for a circuit as well as for individual components in the circuit. The rate at which work is done (work per unit time) is the definition of power:

$$P = \frac{W}{\Delta t}$$

The electrical power consumed by a circuit is

$$P = \frac{Vq}{\Delta t} = V \frac{q}{\Delta t}.$$

But $q/\Delta t$ is the flow of charges per unit time, or the current, I. Therefore, the electrical power used is

$$P = VI.$$ (20.4)

The electrical power used in the component is the product of the potential difference and the current.

Each resistor absorbs a specific amount of electric power when it changes the electrical energy of the free electrons going through it to thermal energy. Joule discovered that the thermal energy produced by a resistor each second (thermal power) is related to electrical power by the expression

$$P = I^2R.$$ (20.5)

Equation 20.5, known as **Joule's law,** looks different from the first equation for power. But is it? Ohm's law says that

$$V = IR.$$

Joule's law, expanded a bit, is

$$P = IR \times I.$$

So Joule's law can be written

$$P = VI,$$

which is Equation 20.4.

The electrical power used in a circuit or a circuit component can be expressed in three ways:

$$P = VI$$
$$P = I^2R$$
$$P = \frac{V^2}{R} \qquad (20.6)$$

You can prove that Equation 20.6 is equivalent to the other two by using Ohm's law.

Since power is energy or work per unit time, it is expressed in joules per second, or watts (W). The kilowatt (kW) is 1000 W. The **kilowatt-hour (kWh),** a unit of energy, is equal to

$$1 \text{ kWh} = 1000 \text{ W} \times 3600 \text{ s, or}$$
$$1 \text{ kWh} = 3\ 600\ 000 \text{ J.}$$

20.9 Electrical Units

Table 20-2 sums up the relationships between the various electrical variables and their units.

20A Section Review Questions

1. Describe how electrical conduction occurs in a conductor.

2. Discuss the main difference between conduction in a solid conductor and in an electrolytic solution.

3. How can you tell which electrical connection on a device is supposed to be at the higher potential?

4. List at least three sources of electrical potential.

5. Which wire would impede the flow of current less, copper or aluminum? Explain your answer.

⦿6. What is the resistance of a silver wire 10.0 m long and 0.050 cm in diameter?

⦿7. How much power is used in a coil of copper wire 100. m long with a wire diameter 0.010 cm if there is 1.00 A of current flowing through it?

⦿8. a. How many kilowatt-hours do you waste if you forget to turn off your 100. W bedroom lamp in the morning before school and it remains on for 9.0 hours?

   b. How many joules of energy does this quantity represent?

⦿⦿9. How many grams of water could you heat from 0.0 °C to 100.0 °C with the thermal equivalent of the energy used in Question 8?

Electrical "power" companies sell electrical *energy,* not power. They care for only how much energy you use, not how fast you use it.

TABLE 20-2
Some Basic Electrical Dimensions

Dimension	Unit (Symbol)	Definition	Circuit Symbol
current	ampere (A)	1 C/s	I
potential	volt (V)	1 J/C	V
resistance	ohm (Ω)	1 V/A	R
charge	coulomb (C)	1 A·s	q or Q
power	watt (W)	1 J/s	P

Problem-Solving Strategy 20.3

Remember that the unit symbol indicates the kind of dimension of a property, while the variable (circuit) symbol represents its numerical value.

20.10 Terminology and Symbols

Recall from Chapter 19 that there are two ways of connecting multiple electrical components together—series and parallel. A **series** circuit has a single path for current. The components are linked together like a train or chain. The current, denoted *I*, passes through each circuit component in turn. A **parallel** circuit has more than one path for current. The current therefore divides and goes through all the paths simultaneously. The direction of current in a circuit diagram is represented by an arrow, and a value may be written by the arrow if it is known. A source of potential difference (*V*) is represented by a symbol of one or more voltaic cells ⊣|ı|⊢. A curved arrow points from the lower potential to the higher potential alongside the symbol.

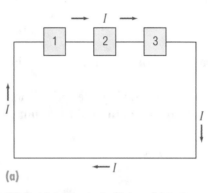
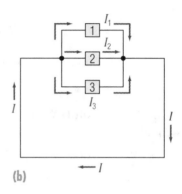

20-8 (a) A series circuit; (b) A parallel circuit

Electrical components must be connected to each other in order to create a circuit. Their conductors are either permanently connected by soldering (using a type of low-melting-point metal), or they are temporarily clamped together using screws, spring clamps, or some other method.

In a series circuit, the current flows through the circuit components sequentially. Since the circuit does not produce or consume electrons, the current must remain constant throughout the circuit. (That is, electrical current does not speed up or slow down in a series circuit.) The potential difference changes throughout a series circuit because each circuit component converts some electrical potential energy to kinetic energy of the inner atoms (thermal energy). Therefore, the potential is higher just before the component than just after the component. This decrease in potential is known as a **voltage drop.** Calling a potential difference "voltage" is similar to talking about the "footage" of film, meaning its length. The magnitude of the potential difference across the voltage source is equal to the magnitude of the sum of the voltage drops in the circuit.

$$|V_{\text{circuit}}| = |V_1 + V_2|$$

This is similar to gravitational potential energy. Imagine a forklift that raises a box to a platform 2 m high. The box must fall down two steps in order to return to floor level, where the forklift can raise it again. The sum of the drops through

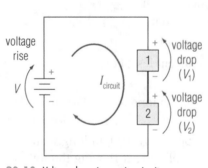

20-10 Voltage drops in a series circuit

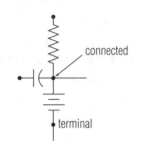

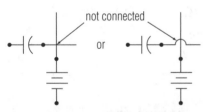

20-9 The points where components are joined to other components or to conductors are indicated on an electrical circuit schematic as a black dot. If lines on a schematic cross without connecting, this fact is indicated by a lack of a dot or an arch in the crossing line.

which the box falls is equal to the height at which the box was originally. Similarly, the sum of the voltage drops in a series circuit is equal to the original potential difference.

In a parallel circuit, there is more than one path for the current to follow. Since there are no sources or sinks to create or consume electrons, the sum of the currents in each parallel branch is equal to the current entering or leaving the parallel portion of the circuit. In Figure 20-12

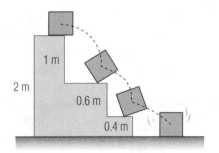

20-11 Stepwise drops in gravitational potential energy are analogous to voltage drops in a series circuit.

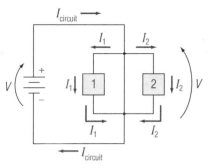

20-12 Current paths in a parallel circuit

$$I_{circuit} = I_1 + I_2.$$

The potential difference is the same across all parallel paths. Returning to the forklift situation, imagine that the forklift raises several boxes 2 m to a platform. There are two ramps from the platform to the floor. Some boxes slide down the steep ramp, and the other boxes slide down the shallow ramp. Both ramps span a 2 m gap between the platform and the floor. Therefore, the two ramps have the same gravitational potential. Similarly, all parallel circuit paths have the same potential difference at the points where they are connected together.

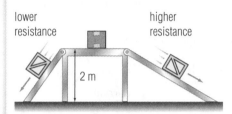

20-13 Boxes sliding down different inclined ramps from the same height are analogous to current falling through the same voltage drop via different paths.

20.11 Plumbing Analogy

If the principles governing electrical circuits are unfamiliar, an analogy may make them clearer. Electrons flowing through a circuit behave somewhat like water flowing through connected pipes. (This analogy is not perfect, since electrons are not identical to water in pipes, but the similarities can clarify the new ideas.) Current, the flow of charges, is similar to the flow of water. A source of potential difference resembles a pump, and pressure is analogous to voltage. Pressure is highest at the discharge of the pump and lowest at its inlet. Resistance is like the fluid friction that occurs when a constriction is encountered in the pipe. Electrical switches are like valves in a pipe. They open and close the circuit to control current flow.

Consider a "series" water system. The pump raises the pressure of the water at the pump outlet pipe. As the water encounters constrictions, friction causes drops in pressure (see Chapter 17). Since no water is added or released, all the water must be moving at the same volumetric rate (volume per unit time). The pressure drops are similar to voltage drops. The pipes decrease the water pressure by friction just as wires decrease potential difference by resistance.

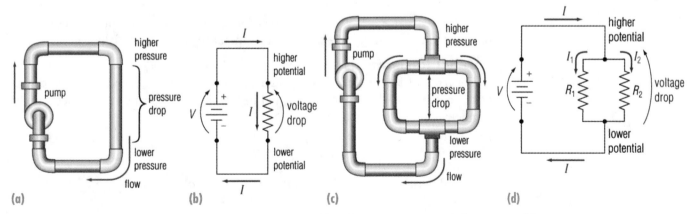

20-14 Similarities between water circuits and electrical circuits. Figures (a) and (b) show series circuits; (c) and (d) compare parallel circuits.

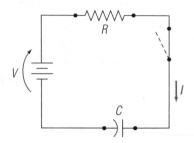

20-15 A simple resistor-capacitor (R-C) circuit

Symbol	Meaning
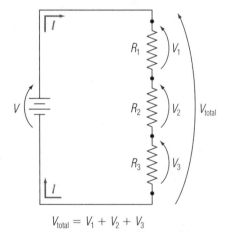	battery
$\underrightarrow{I}$	current
⟋⟍⟋⟍	resistor
⊣⊢	capacitor
⟋	switch

The total resistance of series-connected resistors is the simple sum of all individual resistances:

$R_{\text{total}} = R_1 + R_2 + \ldots + R_n$

$V_{\text{total}} = V_1 + V_2 + V_3$

20-16 Resistors in series

A "parallel" water system is also possible. In this system, T-joints allow the water flow to split into several paths. Some water flows in each pipe. The total volumetric flow rate remains the same, so the water must have the same rate of flow after the pipes rejoin as it had before the pipes divided. Therefore, the sum of the flow rates in the parallel pipes is equal to the rate of flow coming out of the pump. The pressure drop in the two arms is the same, since both pipes are connected at the T-joints. This is true even if the parallel pipes are of different diameters and have different flow rates. As was noted before, this is how electricity behaves in a parallel circuit.

20.12 Analyzing Circuits

Figure 20-15 is a diagram of a simple circuit. Each circuit component has a symbol that represents it on a diagram. Direct-current potential difference is ⊣⊢; current is $\underrightarrow{I}$; resistance is ⟋⟍⟋⟍; and capacitance is ⊣⊢. A switch, ⟋, is used to start and stop current flow by opening and closing the circuit. Circuit schematic symbols can represent properties as well as discrete components. The values of components' properties are often shown adjacent to the components on the diagram.

20.13 Equivalent Resistance

Series Resistances

For series connections, current is uniform and the sum of the voltage drops across each component equals the total voltage drop across all of the series components. Resistance, from Ohm's law, is

$$R = \frac{V}{I}.$$

For resistors in series, the total voltage drop is

$$V_{\text{total}} = V_1 + V_2 + \ldots + V_n.$$

The total resistance is therefore:

$$R_{\text{total}} = \frac{V_{\text{total}}}{I}$$

$$R_{\text{total}} = \frac{V_1 + V_2 + \ldots + V_n}{I}$$

$$R_{\text{total}} = \frac{V_1}{I} + \frac{V_2}{I} + \ldots + \frac{V_n}{I}$$

But V_1/I is R_1, V_2/I is R_2, and so forth, so

$$R_{total} = R_1 + R_2 + \ldots + R_n . \qquad (20.7)$$

The total resistance of two or more resistors in a series is the sum of the individual resistances. Adding resistors in series increases the total resistance.

Parallel Resistances

For parallel connections, the voltage drop is the same in each branch. The sum of the currents in the branches is equal to the current entering the divided circuit segment. From Ohm's law,

$$R_1 = \frac{V}{I_1}; R_2 = \frac{V}{I_2}; \ldots ; R_n = \frac{V}{I_n}.$$

The total current flowing through the parallel branches is

$$I_{\text{total}} = I_1 + I_2 + \ldots + I_n.$$

Therefore, the combined resistance of the resistors is

$$R_{\text{total}} = \frac{V}{I_{\text{total}}}, \text{ or}$$

$$R_{\text{total}} = \frac{V}{I_1 + I_2 + \text{p} + I_n}.$$

Taking the reciprocal of each side gives

$$\frac{1}{R_{\text{total}}} = \frac{I_1 + I_2 + \ldots + I_n}{V}, \text{ or}$$

$$\frac{1}{R_{\text{total}}} = \frac{I_1}{V} + \frac{I_2}{V} + \text{p} + \frac{I_n}{V}.$$

The terms I_1/V, I_2/V, and I_n/V are the reciprocals of R_1, R_2, and R_n, respectively.

$$\boxed{\frac{1}{R_{\text{total}}} = \frac{1}{R_1} + \frac{1}{R_2} + \ldots + \frac{1}{R_n}} \tag{20.8}$$

The total resistance of two or more resistors in parallel is the reciprocal of the sum of the reciprocals of the individual resistances. Resistors connected in parallel have less total resistance than any of the individual resistors. Adding more resistors in parallel decreases the total resistance.

A technique used to simplify circuits for analysis is that of replacing several resistors with an equivalent resistor (R_{eq}). For example, circuit (a) in Figure 20-18a looks complicated, but it can be theoretically replaced by circuit (h), which has only one equivalent resistor. The replacement must be done in steps. It uses only the two resistance rules just discussed.

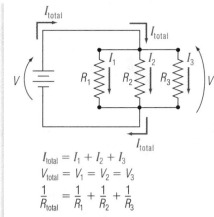

$$I_{\text{total}} = I_1 + I_2 + I_3$$
$$V_{\text{total}} = V_1 = V_2 = V_3$$
$$\frac{1}{R_{\text{total}}} = \frac{1}{R_1} + \frac{1}{R_2} + \frac{1}{R_3}$$

20-17 Resistors in parallel

The total resistance of a group of resistances connected in series is the reciprocal of the sum of the reciprocals of the individual resistances:

$$\frac{1}{R_{\text{total}}} = \frac{1}{R_1} + \frac{1}{R_2} + \ldots + \frac{1}{R_n}$$

Problem-Solving Strategy 20.5

After summing the reciprocals of the parallel resistances, remember to take the reciprocal of the sum to find the total resistance.

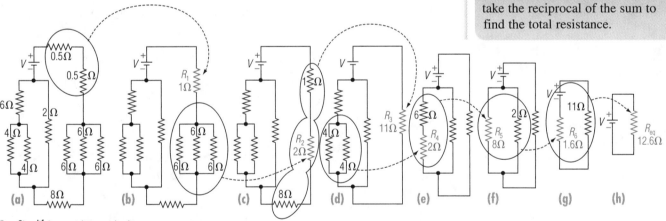

20-18a Simplifying a resistance circuit

EXAMPLE 20-1

Uncomplicate Your Life: Equivalent Resistance

Find the resistance of the equivalent resistor for circuit (a) in Figure 20-18a. (The circuits in Figure 20-18a appear again as Figure 20-18b on the next page.)

Solution:
Begin at the positive terminal of the battery, and proceed in the direction of the current. You first come to two resistors in series. Replace them with an equivalent resistor of resistance R_1:

$$R_1 = 0.5 \text{ } \Omega + 0.5 \text{ } \Omega = 1 \text{ } \Omega$$

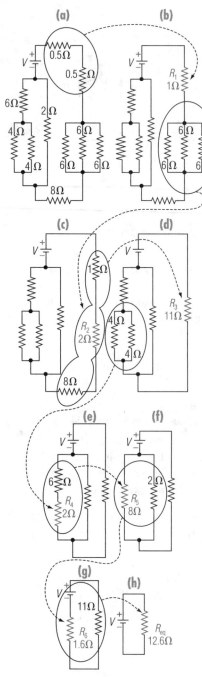

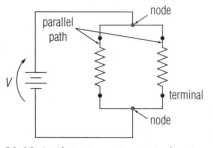

20-18b Simplifying a resistance circuit, from Figure 20-18a

20-19 A node is a junction in a circuit where current divides or converges.

Circuit (b) is the result. Now notice that there are three resistors in parallel. Replace them with R_2:

$$\frac{1}{R_2} = \frac{1}{6\ \Omega} + \frac{1}{6\ \Omega} + \frac{1}{6\ \Omega}$$

$$\frac{1}{R_2} = \frac{3}{6\ \Omega} = \frac{1}{2\ \Omega}$$

$$R_2 = 2\ \Omega$$

Circuit (c) is the result. Notice the three resistors in series; replace them with R_3:

$$R_3 = 1\ \Omega + 2\ \Omega + 8\ \Omega$$

$$R_3 = 11\ \Omega$$

Now you have circuit (d). You have reached the complicated parallel arrangement. Simplify the left branch first. The parallel resistors can be replaced with R_4:

$$\frac{1}{R_4} = \frac{1}{4\ \Omega} + \frac{1}{4\ \Omega}$$

$$\frac{1}{R_4} = \frac{2}{4\ \Omega} = \frac{1}{2\ \Omega}$$

$$R_4 = 2\ \Omega$$

The branch now has two series resistors. Replace them with R_5:

$$R_5 = 2\ \Omega + 6\ \Omega = 8\ \Omega$$

The parallel circuit segment is now two parallel resistors. Replace them with R_6:

$$\frac{1}{R_6} = \frac{1}{8\ \Omega} + \frac{1}{2\ \Omega}$$

$$\frac{1}{R_6} = \frac{1}{8\ \Omega} + \frac{4}{8\ \Omega} = \frac{5}{8\ \Omega}$$

$$R_6 = 1.6\ \Omega$$

The circuit (g) shows the result of these operations. Since the two remaining resistors are in series, you can replace them with R_{eq}:

$$R_{eq} = 11\ \Omega + 1.6\ \Omega$$

$$R_{eq} = 12.6\ \Omega$$

The circuit (h) contains one equivalent resistor. Circuit simplification may take a long time, but it is not hard if you work consistently and apply the proper rules.

In complicated circuits, it is sometimes difficult to tell where the parallel part of a circuit begins. The place where a current divides or converges is called a **node** and is represented on a circuit diagram by a connection dot. Parallel paths begin and end at a node, as in Figure 20-19.

20.14 Kirchhoff's Rules

The rules for current and potential difference in series and parallel circuits can be summed up in two general rules called **Kirchhoff's rules:**

1. The sum of voltage drops in a simple *closed* path equals the sum of voltage rises in the path. A simple closed path does not contain any cross-connecting paths.

2. The sum of currents entering a node equals the sum of currents leaving the node. In other words, the algebraic sum of the currents entering and leaving a node is zero.

For these laws to be valid, the rules for positive and negative signs must be consistent. To be consistent, you should begin the analysis of any circuit diagram by assigning positive and negative sides to each circuit component (Figure 20-20). Start with the positive side of the voltage source (the battery in this instance), and proceed in the direction of the current. Mark the first side of any circuit component you reach positive, and mark the other side negative. In this way current always flows from positive to negative.

Notice in Figure 20-20 that the current crosses all voltage drops (the resistors) from positive to negative. For all components in a circuit other than the source, the potential is higher at the current inlet of the component and lower at the outlet. This is understandable since current flows down the potential "hill." The potential difference, ΔV, for these components is therefore negative. However, the current crosses the source of voltage (the battery) from negative to positive. This is consistent with what you know already. The purpose of voltage sources is to supply a high electrical potential. The potential difference across a source is therefore positive in the direction of current flow, since the outlet potential is higher than the inlet potential:

$$\Delta V = V_{\text{out}} - V_{\text{in}},$$

where $V_{\text{out}} > V_{\text{in}}$.

Gustav Kirchhoff (1824–87) was a Prussian physicist known for his work on electrical currents and spectral analysis.

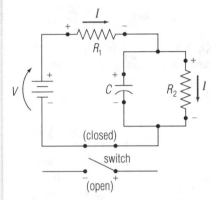

20-20 Assigning positive and negative polarities in a circuit

---

EXAMPLE 20-2

Using Kirchhoff's Current Rule

Figure 20-21 shows a circuit segment. If all the resistors have a resistance of 10 Ω, and the total current (I) is 1 A, find the current through each resistor.

Solution:
Use Kirchhoff's current rule to find the relationships among the currents. At node 1,

$$I = I_1 + I_4$$
$$1 \text{ A} = I_1 + I_4.$$

At node 2,

$$I_2 + I_3 = I$$
$$I_2 + I_3 = 1 \text{ A}.$$

You obtained no new information from node 2. That will be the case as long as there are no nodes between nodes 1 and 2. You can get more specific information from Kirchhoff's potential-difference rule. The only *closed* path in the figure is the path that goes through R_1, R_2, R_3, and R_4 in that order

Problem-Solving Strategy 20.6

When applying Kirchhoff's voltage rule, remember that voltage rises across voltage sources (ΔV is positive) and drops across other circuit components (ΔV is negative) for conventional current flow. The opposite is true for negative current flow.

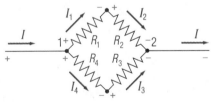

20-21 Current leaving a node must equal the current entering a node.

or the reverse order. Kirchhoff's rule gives

$$V_1 + V_2 + V_3 + V_4 = 0 \text{ V},$$

since there is no voltage source along the path. Note that V_1 and V_2 are negative and the other two voltages are positive, according to the convention we are using. Apply Ohm's law to find these voltage drops.

$$(-I_1 R_1) + (-I_2 R_2) + I_3 R_3 + I_4 R_4 = 0 \text{ V}. \qquad (1)$$

Since R_1 and R_2 are in series, and R_3 and R_4 are in series, then the currents through the respective pairs of resistors are equal. Therefore,

$$I_1 = I_2 = I_{1,\,2} \text{ and } I_3 = I_4 = I_{3,\,4}.$$

Substitute for currents and rearrange terms in Equation (1):

$$I_{3,\,4}(R_3 + R_4) = I_{1,\,2}(R_1 + R_2)$$
$$I_{3,\,4}(10 \ \Omega + 10 \ \Omega) = I_{1,\,2}(10 \ \Omega + 10 \ \Omega)$$
$$I_{3,\,4} = I_{1,\,2}$$

Then, $I_1 = I_2 = I_3 = I_4$.

To find the values of the currents, use the results of the current rule at node 1.

$$1 \text{ A} = I_1 + I_4 = I_{1,\,2} + I_{3,\,4}$$
$$1 \text{ A} = I_{1,\,2} + I_{1,\,2} = 2\,I_{1,\,2}$$
$$I_{1,\,2} = 0.5 \text{ A}$$
$$I_1 = I_2 = I_3 = I_4 = 0.5 \text{ A}$$

The current through every resistor is 0.5 A.

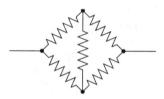

20-22 A circuit segment that cannot be simplified into an equivalent resistance

This example could have been done as easily by using equivalent resistances to calculate the voltage drop across the segment and then using that value in Ohm's law to find the current in each branch. However, some circuits cannot be simplified using equivalent resistances because there are multiple paths that the current can follow. For example, the circuit segment in Figure 20-22 would have to be analyzed using Kirchhoff's rules.

20.15 Electrical Instruments

To calculate any quantity associated with a circuit, you must have some information. You will usually obtain this information by using instruments to detect the basic quantities. An **ammeter** measures current. A **galvanometer** is a sensitive ammeter for detecting very small currents. Devices to measure current must be connected to the energized circuit in series with the current to be measured.

An **ohmmeter** measures resistance. It is not connected in a circuit because it contains its own source of potential difference. An ohmmeter can be connected to a single component, a circuit segment, or an entire circuit, as long as there is no outside source of voltage present that could damage the instrument.

A **voltmeter** measures the voltage drop or rise across an energized circuit component. In some instruments, the energy of the potential difference being measured is used to produce the instrument reading. The instrument is connected to the circuit in parallel with the component for which the voltage is being measured.

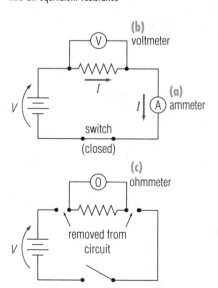

20-23 Electrical schematics showing the connections and symbols for using (a) an ammeter, (b) a voltmeter, and (c) an ohmmeter.

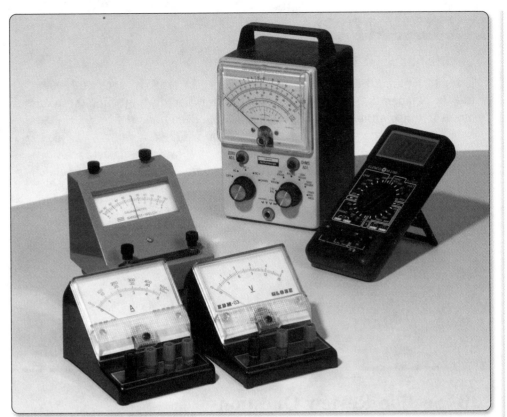

20-24 An ammeter, a voltmeter, and a galvanometer; the analog and digital multimeters (in back and right) can act as an ohmmeter as well as perform the other three functions.

20.16 Resistance Bridges

Figure 20-25a demonstrates the analysis of a somewhat complicated resistance circuit. Such a circuit is called a **bridge circuit,** which is commonly used to accurately measure temperatures using a *resistance temperature detector (RTD).* This circuit segment consists of two standard resistors, one precise variable resistor, and a galvanometer. The unknown resistor, labeled R_x (e.g., the RTD), is connected as Figure 20-25a shows. The *variable resistor* is adjusted until the galvanometer detects no current. Now the circuit is analyzed using Kirchhoff's rules. In figure 20-25b, the potential difference rule for the left-hand path gives

$$-I_1R_x + I_2R_3 = 0 \text{ V}$$
$$I_1R_x = I_2R_3. \tag{20.9}$$

For the right-hand path in Figure 20-25b, the potential difference rule gives

$$-I_1R_1 + I_2R_2 = 0 \text{ V}$$
$$I_1R_1 = I_2R_2. \tag{20.10}$$

Now divide Equation 20.9 by Equation 20.10:

$$\frac{I_1R_x}{I_1R_1} = \frac{I_2R_3}{I_2R_2}$$
$$\frac{R_x}{R_1} = \frac{R_3}{R_2}$$
$$R_x = \frac{R_1R_3}{R_2} \tag{20.11}$$

Equation 20.11 can give the unknown resistance precisely, and it requires only simple calculations.

A more detailed discussion of the various electrical meters and their uses is provided in Appendix E of the Laboratory Manual.

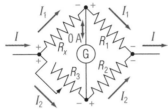

20-25a A resistance bridge circuit segment

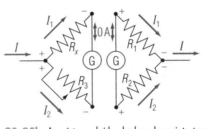

20-25b A resistance bridge broken down into two equivalent closed paths for analysis using Kirchhoff's current rule

A *variable resistor,* ⌐\/\/\/⌐, is capable of providing any resistance between two preset values by changing the length of the current path through the resistor. Typically, a variable resistor is constructed of a very long length of wire coiled around a support. A metal "wiper" is connected to one end of the circuit, and one end of the resistor's wire coil is connected to the other end of the circuit. As the current flows through the wiper resting on the wire coil, the current path length varies depending how far the wiper is from the connected end of the coil. Such a variable resistor is called a **rheostat.**

Electrodynamics   **443**

TABLE 20-3

Temperature [°C]	RTD [Ω]†
0	1000.0
10	1039.0
20	1077.9
30	1116.7
40	1155.4
50	1194.0
60	1232.4
70	1270.8
80	1309.0
90	1347.1
100	1385.1

†URL: http://www.weedinstrument.com/
pdf/rvt.PDF

Bridge circuits are useful for measuring very minute resistances and the total resistance of complex circuits. The following verse tells how to detect and remove minute resistance (subtle unbelief) in your own life. "Take heed unto thyself, and unto the doctrine; continue in them: for in doing this thou shalt both save thyself, and them that hear thee" (I Tim. 4:16).

EXAMPLE 20-3

Circuit Analysis: The RTD Bridge

Refer to Figure 20-25a on the previous page. In order to accurately measure the temperature in a cooled-down nuclear reactor plant pipe, an RTD is monitored using the following bridge resistance readings: $R_1 = 1500.\ \Omega$ and $R_2 = 2500.\ \Omega$. The variable resistor, R_3, is adjusted to a value of $1796.5\ \Omega$ in order to zero the galvanometer reading. (a) What is the resistance of the RTD (R_{RTD}) for the temperature in the reactor pipe? (b) What is the temperature in the pipe?

a. Substituting the values of the known resistances into Equation 20.11, you have

$$R_{RTD} = \frac{R_1 R_3}{R_2}$$

$$R_{RTD} = \frac{(1500.\ \Omega)(1796.5\ \Omega)}{2500.\ \Omega}$$

$$R_{RTD} = 1077.9\ \Omega.$$

b. According to Table 20-3, the piping temperature is 20 °C.

20B Section Review Questions

1. Is the switch for your kitchen ceiling light arranged in series or parallel with the light fixture? Explain your answer.

2. Which point is at the higher potential in the figure? Explain.

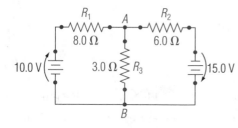

3. State Kirchhoff's voltage rule and his current rule in your own words.

4. List the three basic electrical instruments, how they are connected when in use, and the condition of the circuit/component when a measurement is being taken (i.e., energized/de-energized, connected/disconnected to a circuit, etc.).

5. What kind of circuit is used to find the value of an unknown resistance in the circuit by balancing the currents flowing through its two branches?

⊛6. What is the equivalent resistance of the resistors 1 MΩ, 100. Ω, and 1.00 Ω

    a. in series? (Assume all resistors have a 5% tolerance for rounding purposes. See p. 291 in the Laboratory Manual for discussion of manufacturer's tolerances.)

    b. in parallel?

⊛7. For each case in Question 6, is the order of magnitude of the equivalent resistance closer to the largest or the smallest resistance?

⊛⊛8. Using Kirchhoff's rules, determine the current through and voltage drop across each resistor in the accompanying figure.

ELECTRICAL SAFETY

20.17 Overcurrent Hazards

Most houses in the United States use a voltage of around 110 V for supplying electricity to lights and most small appliances in the home. This electricity is furnished as alternating current (AC) instead of the direct current that you studied earlier in this chapter. Alternating current continually changes magnitude and direction. In the United States, these changes occur 60 times per second (60 Hz). In many other parts of the world, alternating current frequency is 50 Hz. While both AC and DC electricity each have their own electrical safety issues and hazards, for the discussion here it will be assumed that both are equally hazardous under the appropriate circumstances.

Electrical devices connected to most house circuits are in parallel. That way, when a light bulb burns out, for example, the rest of the circuit still receives current. Similarly, you can shut off one device without turning them all off at the same time. Another advantage of parallel wiring is that all of the current required by the entire circuit does not have to pass through every device. Connecting many items in parallel presents a problem, however. Remember, when resistors are connected in parallel, their combined resistance is *less* than any of the individual resistances. In fact, adding a resistor in parallel always reduces the total resistance of the circuit. The current supplying a circuit increases when the circuit's resistance decreases. The supply lines for the circuit carry all the current. Their resistance is small, but if the lines carry a large current, the thermal energy produced becomes significant. This thermal energy can start an electrical fire.

Another possible cause of excess current is a **short circuit.** A lamp cord usually contains two wires. The wires are coated with an insulating material. If the coatings are broken or cut, the wires can make electrical contact. The current flows from one wire to the other through the break, since that path has much less resistance than the intended path through the lamp bulb. A short circuit provides little resistance to current compared to a normal electrical load, so high currents result.

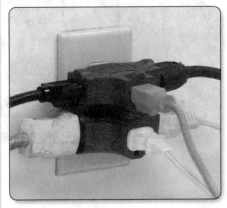

The condition that exists when a circuit is carrying more current than its conductors can safely handle is called an **overcurrent** condition.

20-26 An overloaded circuit occurs when too many electrical devices are connected to it, drawing more current than the circuit is designed to carry. This outlet will experience a very high current if all of these loads are energized at the same time.

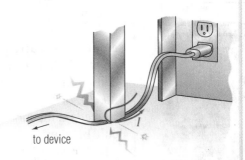

to device

20-27 Short circuits occur when some condition causes a direct conducting path between the supply and ground wire in a supply cable or extension cord, bypassing the electrical device (and its resistance) completely.

20.18 Fuses

Early in the past century, devices called **fuses** were invented to protect homes and facilities from circuit overloads and short circuits. A fuse is placed in series with the voltage supply and the electrical loads in the circuit. A fuse consists of a durable container electrically attached to a strip of metal inside having a low melting point. When the current exceeds the maximum safe level for that circuit, it heats the metal strip or wire in the fuse to its melting point. The fuse breaks, opening the circuit, and the current can no longer flow anywhere in the circuit. In household fuses, a "blown" fuse is indicated by a blackened, smoky appearance inside the glass window. Fuses without windows must be tested with an ohmmeter to determine whether they have "blown."

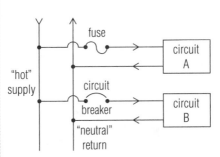

20-28 The location of a fuse and a circuit breaker in a circuit

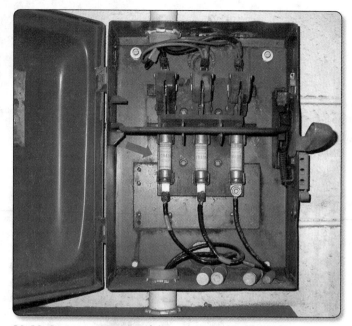

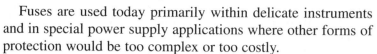

Fuses are used today primarily within delicate instruments and in special power supply applications where other forms of protection would be too complex or too costly.

Fuses are not used for protecting homes and businesses much anymore. The standardized national and international electrical codes for builders require that new houses be built with mechanical, reusable safety devices called *circuit breakers* for overcurrent protection.

20-29 Fuses come in a variety of shapes and sizes, depending on their use.

20.19 Circuit Breakers

A **circuit breaker** is a switch that automatically opens when the current through it is too high. Circuit breakers are more convenient than fuses because circuit breakers need only be switched back on to restore current. Fuses must be replaced, which means replacements must be available. Circuit breakers are located in a panel mounted on a wall, usually in a garage, in the basement, or near the kitchen or a utility room. They should always be accessible in order to permit quick action to turn off a circuit. Never pile objects in front of the panel. When a circuit breaker is shut or open, the lever handle is fully in the left/right or up/down position on the breaker module. Modern breaker boxes are designed to give you a visual indication that the breaker has tripped—either the handle is in an in-between position or there is a colored indicator of some kind. After correcting the overcurrent condition, you must push the handle to the full off position before shutting the breaker to restore electricity to the circuit.

Fuses and circuit breakers are designed to interrupt a certain amount of current. Typical ratings for circuits supplying mainly lights are 15 amps and 20 amps. Circuits supplying air conditioners and other heavier appliances may have ratings of 20, 30, or 50 amps. Ratings for the circuit breakers are clearly printed on the modules. Never replace a lower-rated fuse or circuit breaker with a higher-rated one in the electrical service panel in your home. This is another way that wires in a circuit can become unprotected. You should also never replace a fuse with a coin or other conductor. Do not replace a fuse or reset the switch on a circuit breaker until you have corrected the problem that caused the high current. Ignoring such a problem could allow a fire to start and destroy your home.

20-30 The circuit breaker on the top is tripped. The breaker on the bottom is shut.

20.20 Electrical Shock

Electrical fires are dangerous, but electric shock is a greater danger to you personally. The seriousness of the injury from electric shock depends on the current passing through the body. You can usually feel a current of 0.001 A (1 milliamp, or mA). A current of 0.01 A (10 mA) can contract your hand muscles. If you grasp a wire that allows 10 mA to pass through your body, you may not be able to let go of the wire. A current of 0.1 A (100 mA) can upset the rhythm of your heart so much that your heart cannot maintain a steady flow of blood. This effect is called *ventricular fibrillation,* and it can kill you very quickly. A current twice as large, 0.2 A (200 mA), paralyzes your heart. Surprisingly, a current between 0.1 A and 0.2 A is more likely to be fatal than a current greater than 0.2 A. It is easier to restart a stopped heart than to stabilize the fluttery beat of a fibrillating heart.

The amount of current passing through your body depends on the potential difference across your body and the resistance of your body. Your skin has more resistance than the interior tissues of your body because the rest of your body is mostly water containing electrolytes in solution. The resistance of your skin varies from a few hundred ohms for wet skin to over $10\ 000\ \Omega$ for dry skin. If you are touching a bare 120 V wire with your hand and are standing barefoot on the ground, the potential difference across your body is 120 V. Will you be electrocuted? If your skin is dry, the current through your body will not exceed about 10 mA. That is probably enough to "freeze" your hand to the wire if you are holding it. If you are just touching it with the back of your hand, your hand will contract and fall away from the wire. However, if you have wet skin, your resistance would be less than $1000\ \Omega$. The current would be over 100 mA, which is enough to electrocute you. You can never be sure that your skin is dry enough to be a good resistor. Sweat makes your skin wet as much as water does, and sweat is a good electrolyte. It is therefore best to avoid touching a bare wire and to replace cracked wires or damaged insulation.

20.21 Ground Fault Protection

You are probably saying, "I would never touch a bare, energized wire on purpose." But many electrocutions occur every year because of electrical faults inside hand-held tools and appliances. Vibration or breakage from dropping can cause electrical wires to come loose inside the equipment and energize conducting parts that are accessible to your touch. The hazard of this kind of problem can be minimized by inserting special devices in the circuit that detect extremely small currents between the conductor and ground. These devices are called **ground fault interrupters (GFI)** or ground fault circuit interrupters (GFCI).

Building codes require GFI protection for 120 V electrical receptacles in specific locations, including garages, rooms containing plumbing fixtures, and outdoors. GFI devices can be individual receptacles in walls, portable, extension cord–like receptacles, or special circuit breakers in an electrical service panel. Most GFI devices installed in homes are designed to open all the conductors supplying a circuit if an imbalance of more than 5 mA between the current-carrying conductors is detected, indicating that the current is following a path other than through the circuit wires (e.g., through you). Thus, if you are using a curling iron over a sink full of water and accidentally drop it into the sink, the current flowing to ground through the water will immediately trip the GFI near the sink and turn off the curling iron before you could react and pick it up, eliminating a shock hazard.

When someone experiences an electrical shock, he tends to drop everything (if he can) and back away. If you are a Christian, sooner or later you will encounter someone who will hate you just because you are a Christian. Don't reflexively drop everything and back away from Christ's Great Commission. Hang in there. Just remember that they hated Jesus, also. "And ye shall be hated of all men for my name's sake: but he that endureth to the end shall be saved" (Matt. 10:22).

ground potential

20-31 Most domestic circuits contain a "hot" wire and one or more wires at ground potential. If you touch the hot wire, your body becomes a second path to ground, and some current will flow through you, possibly causing an electrocution.

20-32 Individual GFI receptacles are provided with test buttons to verify they are operating properly. They should be tested at least once a month.

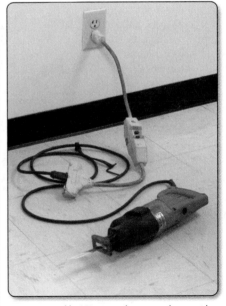

20-33 Portable GFI receptacles are used to provide electrical service at work sites that may not otherwise have GFI protection or that present an increased shock hazard.

Electrodynamics   **447**

20C Section Review Questions

1. What is the greatest electrical hazard to personal safety?
2. **a.** What is a circuit breaker designed to prevent?

   b. Why can it not prevent an electrical shock?
3. How can an electrical short circuit cause a fire?
4. Why will a shock current passing between the thumb and little finger of one hand not likely kill a person?
5. Discuss the steps to take in order to restore power to a circuit after its breaker has been tripped.
6. What should you do if you accidentally drop your running, corded shaver into a bathtub full of water?

Chapter Review

In Terms of Physics

current (I)	20.1	Ohm's law	20.7
electrodynamics	20.1	Joule's law	20.8
ampere (A)	20.1	kilowatt-hour (kWh)	20.8
direct current (DC)	20.2	series	20.10
polarity markings	20.3	parallel	20.10
voltaic cell	20.4	voltage drop	20.10
galvanic cell	20.4	Kirchhoff's rules	20.14
electrode	20.4	ammeter	20.15
circuit	20.4	galvanometer	20.15
primary cell	20.5	ohmmeter	20.15
storage cell	20.5	voltmeter	20.15
battery	20.5	bridge circuit	20.16
dry cell	20.5	rheostat	20.16
resistivity (ρ)	20.6	overcurrent	20.17
superconductor	20.6	short circuit	20.17
resistor	20.6	fuse	20.18
resistance (R)	20.6	circuit breaker	20.19
ohm (Ω)	20.6	ground fault interrupter (GFI)	20.21

Problem-Solving Strategies

20.1 (page 432) It is not normally necessary to keep track of which direction the moving charges are going. Just remember that current flows from the highest potential to the lowest potential in the circuit.

20.2 (page 434) Ohm's law is true for a single circuit component, a segment of a circuit, or an entire circuit.

20.3 (page 435) Remember that the unit symbol indicates the kind of dimension of a property, while the variable (circuit) symbol represents its numerical value.

20.4 (page 436) The symbol for a potential difference source has parallel lines of different lengths. The shorter line is conventionally the negative end of the source. Think of it as a minus sign.

20.5 (page 439) After summing the reciprocals of the parallel resistances, remember to take the reciprocal of the sum to find the total resistance.

20.6 (page 441) When applying Kirchhoff's voltage rule, remember that voltage rises across voltage sources (ΔV is positive) and drops across other circuit components (ΔV is negative) for conventional current flow. The opposite is true for negative current flow.

20.7 (page 442) Kirchhoff's rules are used to generate as many current equations as there are unknown currents in the simple closed paths in the circuit. Solving these equations simultaneously using the techniques you learned in algebra classes permits you to find the individual currents. After the current values are known, you can find the voltage drops and the power absorbed for each resistance.

Review Questions

1. What is an electrical current?

2. What carries current in metals?

3. What is the difference between conventional current and the physical movement of charges in a conductor such as a metal?

4. In what kind of current do the charges move in only one direction?

5. When installing a battery in your calculator, how do you know which end has the electrode with the higher potential?

6. List the kinds of devices that convert the following forms of energy to electrical potential:
   a. light   **c.** thermal
   b. chemical   **d.** mechanical

7. What carries the current in electrolytic solutions?

8. List three sources of potential difference.

9. What is the difference between a primary cell and a storage cell?

10. Discuss the difference between resistance and resistivity.

11. **a.** What class of materials appear to lack resistance?
    b. What environmental condition limits the usefulness of this property?

12. Why do you think that scientists chose to define the ampere in the SI on the basis of a force between two conductors rather than using the classical definition that involved the rate of the flow of charge? (Review the Chapter 2 discussion of the SI.)

13. This series circuit has a current I_A at point A. At point B, three circuit components farther along, is I_B greater, less than, or equal to I_A?

14. A circuit has a current I_A at point A just before the circuit branches. There is one circuit element in each branch. Is the current in each branch greater than, less than, or equal to I_A?

15. Using the plumbing analogy of electricity, what electrical properties or components correspond to the following plumbing concepts?
    a. piping
    b. piping constriction
    c. pump
    d. pressure drop
    e. volumetric flow rate
    f. in-line flow meter
    g. pressure gauge
    h. shutoff valve

16. In Kirchhoff's voltage rule, what restriction applies when summing the voltage drops and rises along a path?

17. Name the instrument that you would use to measure a circuit's current and describe how you would connect it.

18. Name the instrument you would use to measure a resistor's resistance and describe how you would connect it.

19. What does a resistance bridge circuit measure?

20. Are most items on household circuits connected in series or in parallel? Why?

21. Why does adding another item to a household circuit increase the circuit's current?

22. What is the primary electrical hazard
    a. for a dwelling or other structure?
    b. for a human?

23. What is the function of a fuse?

24. Why should you not replace a fuse with a coin?

25. Why should you be especially careful of electrical circuitry when your skin is wet?

26. Which is more likely to harm you, a 100 V shock in which your body draws 0.005 A or an 80 V shock in which your body draws 0.1 A?

True or False (27–36)

27. Charged particles carrying a current move at speeds near that of light.

28. The work done on an electron can be calculated from the equation $W = (emf)(d)$, where *emf* is the electromotive force and *d* is the distance that the electron moved.

29. Every material has some resistance at room temperature.

30. If you applied a certain voltage across the 1 cm thickness of a 1 m² copper plate, you would discover that it had the same resistance as a 1 cm long copper wire.

31. Current in parallel branches must be equal in order for Kirchhoff's rules to be valid at the two nodes connected by the branches.

32. Current is constant throughout a series circuit.

33. Conventional current flowing through a resistor travels from the positive end to the negative end.

34. Voltage measured from the positive end of a resistor to the negative end is negative.

35. It is safe, if not reliable, to replace a fuse of a certain rating with one of a smaller rating.

36. Electrocution should not occur when one is operating electrical equipment while standing on damp concrete as long as the equipment is plugged into a GFI-protected outlet.

⊚37. What is the resistance of a 1.00 m long copper wire with a cross-sectional area of 1.00×10^{-6} m²?

⊚38. A certain kind of glass has a resistivity of 1.54×10^{11} Ω·m. What is the resistance of a 1.00 m long strand with a cross-sectional area of 1.00×10^{-6} m²?

⊚39. A sample of silver has a volume of 2.70×10^{-5} m³. Find the resistance along the longest dimension if
  a. the sample has the shape of a cube.
  b. the sample is a bar with a cross-sectional area of 3.00×10^{-4} m².
  c. the sample is a wire with a cross-sectional area of 1.00×10^{-6} m².
  (Remember, $V = Bh$ or $V = Al$.)

⊚40. A current of 1.50×10^{-2} A flows through a 30.0 Ω resistor. What is the voltage drop across the resistor?

⊚41. A 1.50 V cell is connected to a 1.50 Ω light bulb in a simple circuit. How much current flows through the bulb?

⊚42. A 12.0 V battery produces a current of 1.00 mA when connected in a circuit to resistor R. What is R's resistance?

⊚43. A 55.0 W bulb is connected in a circuit to a voltage source of $V_{max} = 110.$ V.
  a. What current does it draw?
  b. What is its resistance?

⊚44. A 40.0 Ω resistor in a circuit draws a current of 0.100 A.
  a. What is the voltage drop across the resistor?
  b. How much electrical power does it absorb?

⊚45. Find the resistance of two 1.50×10^3 Ω resistors
  a. in series.
  b. in parallel.

⊚46. Find the resistance of the circuit segment shown.

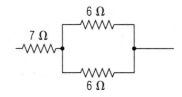

⊚47. In the figure below, $R_1 = R_2 = 20.0$ Ω and $V = 10.0$ V. Find I.

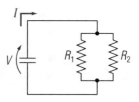

⊚48. What is I in the figure?

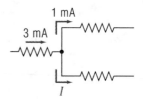

⊚49. What is V in the figure?

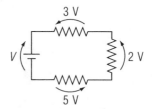

⊚50. **a.** Find the equivalent resistance of the circuit.

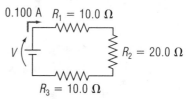

  b. Find the voltage drop across R_1.
  c. Find the voltage drop across R_2.
  d. Find the voltage drop across R_3.
  e. What is V?

⊚51. The voltage drop across R_1 is 2.00 V. $R_1 = 20.0$ Ω; $R_2 = 30.0$ Ω; $V = 6.00$ V.

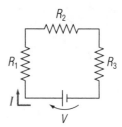

  a. What is the current through R_1?
  b. What is the current through R_3?
  c. What is the voltage drop across R_2?
  d. What is the total voltage drop in the circuit external to the voltage source?
  e. What is the voltage drop across R_3?
  f. What is R_3?

52. $V = 6.00$ V; $R_1 = 30.0$ Ω; $R_2 = 60.0$ Ω

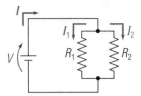

a. Find the equivalent resistance of the circuit.
b. Find the total current I through the circuit.
c. What is the voltage drop across R_1?
d. What is the voltage drop across R_2?
e. What is the current through R_1?
f. What is the current through R_2?

53. $V = 10.0$ V; $I_1 = 0.100$ A; $I_2 = 0.0500$ A.

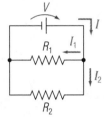

a. What is the voltage drop across R_1?
b. What is the voltage drop across R_2?
c. Calculate R_1.
d. Calculate R_2.
e. Calculate the circuit's equivalent resistance.
f. Calculate the current through the voltage source.
g. Compare R_1/R_2 with I_1/I_2.

54. The resistor bridge in the figure contains fixed resistors $R_1 = R_2 = 10.0$ Ω, variable resistor R_3, and unknown resistance R_x. Find R_x if, when the galvanometer (G) reads 0 A, and $R_3 = 50.0$ Ω.

Fuel Cells

In the history of the United States, most electricity has been generated by mechanically rotating a coil of wire (called an armature) in a magnetic or electromagnetic field. The mechanical power used to rotate the coil has come from a variety of sources including water wheels, wind, steam engines, hydraulic turbines, steam turbines, gas turbines, diesel engines, gasoline engines, and sterling engines. The fuels used in the engines (both internal and external combustion) have included charcoal, wood, coal, crude oil, natural gas, gasoline, kerosene, diesel fuel, and atomic fuel. But all of this will probably change soon because of a device called a *fuel cell*.

Fuel cells generate electricity without the use of mechanical power and rotating parts. There are five main types of fuel cells:

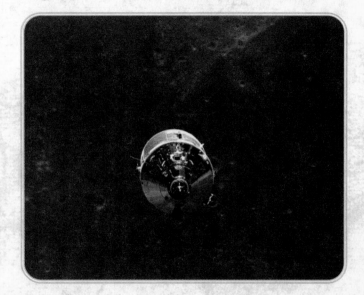

The *alkaline fuel cell* was the first type to be used extensively, largely because it was developed and ready for use before the others. It was used in the Apollo Command Module and the space shuttle, but it required pure hydrogen and oxygen and was too expensive for commercial use.

Three other types, using *phosphoric acid, solid oxide,* and *molten carbonate,* have high operating temperatures and may be suitable only for stationary power plants and marine applications.

The *proton exchange membrane fuel cell* shows the most promise for consumer applications. It consists of three parts: (1) a fuel chamber, part of which also functions as an anode; (2) a proton exchange membrane (PEM), which may have a catalytic film on the cathode side; (3) an air chamber (part of which also functions as the cathode), and an exhaust duct (see the schematic below).

The fuel, hydrogen, enters the fuel chamber under pressure. As each atom of hydrogen reaches the PEM, its electron is stripped from the nucleus but the nucleus passes through the membrane. The anode collects the electrons and delivers them as 0.7 volts of direct current. By connecting several cells in series, higher voltages are obtained. In the air chamber, the nuclei of the hydrogen atoms combine with oxygen to form pure water.

The PEM fuel cell operates at 70 °C and requires no cooling. It generates no pollutants, requires no lubricants, pumps, or filters, and has no moving parts to wear out. The only by-product is pure water, suitable for drinking, cooking, washing, irrigating, etc.

The major problem with fuel cells is the fuel, hydrogen. Because the hydrogen atom is so small, it is difficult to make a tank that will contain hydrogen under pressure. The atoms

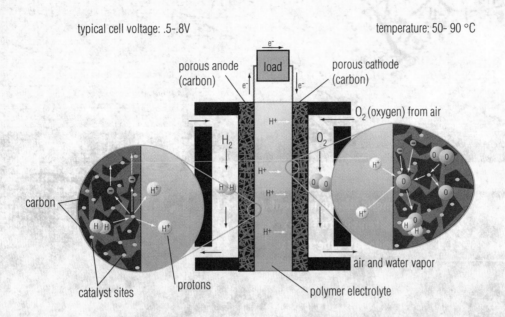

typical cell voltage: .5–.8V temperature: 50– 90 °C

porous anode (carbon)

load

porous cathode (carbon)

e⁻

e⁻ e⁻

H⁺

H₂

O₂ (oxygen) from air

O₂

H⁺

H⁺

H⁺

air and water vapor

carbon

catalyst sites

protons

polymer electrolyte

leak out between the atoms or molecules of the tank material. Hydrogen is extremely flammable, so leaks create fire and explosion hazards. It is hazardous to transfer liquid or gaseous hydrogen from one vessel to another. And there are no filling stations where you can pull in and purchase ten gallons for your fuel tank. In spite of these problems, seven major cities around the world were operating PEM fuel cell–powered buses in 2002, and ten more cities expect to begin in 2003. Major car manufacturers are planning PEM fuel cell–powered passenger cars for mass production in the near future. By the time you read this, PEM fuel cell–powered vehicles may be a reality.

A device called a *reformer* offers one possible solution to the problems associated with the hydrogen fuel. The reformer can be tailored to work on any hydrocarbon fuel, such as gasoline, kerosene, propane, ethane, methane, butane, or natural gas. It strips the hydrogen from the fuel and supplies it on demand to the PEM fuel cell.

Another possible solution uses a compound with sodium borohydride. Drivers of PEM fuel cell–powered vehicles will be able to pull in to a filling/reprocessing station, buy a container of this fuel, and drive away. When the hydrogen is used up, they bring the container back and exchange it for a recharged one.

Other possible solutions are being explored, but it is not yet clear which concept will be adopted. One thing is clear: the use of fuel cells has the potential to reduce or eliminate dependence on an ever-diminishing oil resource.

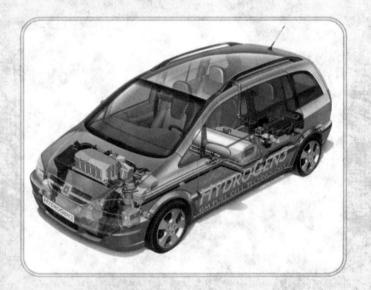

21A   Describing Magnetism   455
21B   Electromagnetism and Charges   464
21C   Electromagnetism and Conductors   469
Facet: Paleomagnetism and the Age of the Earth   462

New Symbols and Abbreviations

magnetic field vector (B)	21.2	magnetic dipole moment (μ)	21.4
tesla (T)	21.2	Curie temperature (T_C)	21.4
relative permeability (κ_m)	21.3	cathode ray tube (CRT)	21.10

Earth's magnetic field is magnificent evidence of God's protection of life on Earth.

Magnetism 21

In describing the manner of His impending death on the cross, Jesus made this profound statement. Did He really mean that every person would come to Him—to acknowledge Him as Savior? Obviously that is not the case. The unfortunate reality is that not everyone will come to Christ (Rom. 3:11). C. H. Spurgeon explained Christ's meaning by pointing out that all *kinds* of people would be saved—Jew, Gentile, cleric, lay person, scientist, and student. The source of this magnetlike force is the Father, who through His Holy Spirit draws us to Himself (John 6:44–45). A compass always seeks magnetic north. Are you continuously seeking Christ?

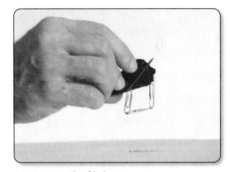

21-1 A sample of lodestone

21A DESCRIBING MAGNETISM

21.1 Historical Background

Before the time of Christ, people knew that certain stones would attract other similar stones and some metals. The Romans called such stones *magnets* after the city of Magnesia, where legend says that natural magnets were first found. During the third century B.C., the Chinese discovered that a natural magnet suspended by a string would rotate until one end pointed north. As early as the eighth century A.D., Chinese sailors were navigating using stone magnets, which the Europeans called **lodestones** ("leading stones"). By the end of the twelfth century, the navigational compass equipped with a magnetized metal needle was in common use.

In the thirteenth century, Petrus Peregrinus de Maricourt performed several experiments to learn more about lodestones. He discovered that two regions at opposite ends of a lodestone, which he called **poles,** had greater attraction for metals than the surface between them. He also discovered that there are two kinds of poles, north and south. Poles of the same kind repelled each other, while poles of different kinds attracted each other. This principle is called the **law of magnetic poles.** This property is very similar to the law of charges. There are two kinds of electrical charge, positive and negative. Like charges repel, whereas opposite charges attract. Early investigators believed that static electricity and magnetism had much in common.

The ancient city of Magnesia was located in the coastal region of what is now Thessaly in Greece on the western shore of the Aegean Sea.

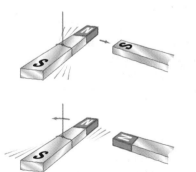

21-2 Opposite poles attract, whereas like poles repel.

Petrus Peregrinus de Maricourt (thirteenth century) was a French scholar and scientist.

455

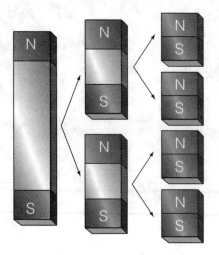

21-3 Breaking a magnet in half repeatedly just produces smaller dipole magnets.

21.2 Description of the Magnetic Field

Magnetic poles come in north-south pairs. Realizing that long magnets have little magnetic force at the center of their lengths, early experimenters tried to isolate a magnetic pole by breaking a magnet in half. They invariably got shorter, weaker magnets, each with a north and a south pole. No matter how small the magnetic pieces are made, two poles are always present. No example of a single, isolated magnetic pole has ever been observed. This is a significant difference from electrical charges.

A magnet with two poles is called a *dipole,* while a single magnetic pole is called a *monopole.* While no magnet monopole has been found, electric monopoles are common. The negative and positive unit charges (*e*) on electrons and protons are monopoles.

The effect of a magnet on another magnet, metals, and moving electric charges is best described by a *magnetic field.* Like an electric field, a magnetic field is a convenient description of how the space surrounding a magnet is affected by it. A magnetic field line gives the direction that an (imaginary) magnetic north monopole would travel if released along the line. Consequently, magnetic lines of force are curved arrows pointing toward south magnetic poles and away from north magnetic poles. If the magnetic monopole is replaced by a tiny dipole magnet and suspended in a large magnetic field, it will align itself so that it lies along the field with its north pole pointing in the direction of the arrows.

You can find the general shape of the magnetic field of a magnet easily. Simply sprinkle some iron filings onto a piece of paper and hold the paper over the magnet. The iron filings will act like small magnets and align themselves along the magnetic field. Another way to map a magnetic field is to place a compass in it. The magnetized compass needle will point in the direction of the magnetic field. These methods are limited because they are able to portray only a two-dimensional cross section of the field surrounding a magnet. Like electric and gravitational fields, magnetic fields are three-dimensional.

Notice in Figure 21-4 that the magnetic field lines are closer together at a magnet's pole than anywhere else. The number of field lines passing through an area of space represents the strength of the field at a point. More lines indicate a stronger field. Both the strength and the direction of the magnetic field at a point are described by the **magnetic field vector (B),** whose magnitude is called the *magnetic flux density.* The direction of **B** at any point in space is tangent to the magnetic field line through that point. The strength of the field, and therefore the magnitude of **B,** is measured in **teslas (T)** after Nikola Tesla. One tesla is equal to one volt-second per square meter:

$$1 \text{ T} = 1 \text{ V·s/m}^2$$

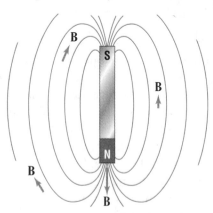

21-4 Orientation of the magnetic field vector in a magnetic field

Nikola Tesla (1856–1943) was a brilliant Serbian-American electrical engineer and inventor. He invented the methods for the practical production and use of alternating current that became the modern commercial power industry.

Another convenient unit of magnetism is the **gauss,** often used in research involving small magnetic fields as well as in geophysics.

$$10^4 \text{ gauss} = 1 \text{ T}$$

The earth's magnetic field strength at its surface is approximately 1 gauss.

21.3 Magnetic Properties of Matter

When a magnet attracts a metal, it first transforms the metal into a temporary magnet. This process is called **magnetic induction.** A north pole induces a south pole; a south pole induces a north pole. Most induced magnets lose their magnetization as soon as the permanent magnet is removed.

Materials differ in their responses to magnetic fields. A magnetic field passing through a material is usually not the same as in a vacuum. We saw this with electrical fields and capacitors. The dielectric constant κ is the ratio of the material's permittivity to the permittivity of free space (vacuum) (ε_0). In a similar way, scientists describe the way a material affects a magnetic field that is within it by a property called the material's *magnetic permeability* (μ). A more easily interpreted quantity is the material's **relative permeability** (κ_m), which is defined as the ratio of the permeability of the material and the permeability of free space (μ_0):

$$\kappa_m = \frac{\mu}{\mu_0} \qquad (21.1)$$

This is equivalent to the ratio of the strength of a magnetic field within a material to the strength of the same field in a vacuum.

Materials can be classified by their magnetic properties according to the magnitude of their relative permeabilities.

One group of materials is readily attracted to magnets. Their relative permeabilities are much greater than 1. Therefore, a magnetic field is much stronger within one of these materials than in a vacuum. These materials can be induced to become permanent magnets. A material with these properties, such as iron, nickel, cobalt, and certain alloys, is called **ferromagnetic.**

Another group of materials, **paramagnetic** materials, is not noticeably attracted to small magnets. The relative permeability of these materials is slightly greater than 1. The magnetic field in a paramagnetic material is only slightly greater than it would be in a vacuum. Aluminum, platinum, and sodium are examples of paramagnetic materials.

TABLE 21-1 Relative Permeabilities	
Material	κ_m
Ferromagnetic Materials	
Iron (Fe)	200 – 5000
78 Permalloy (21% Fe, 79% Ni)	5000 – 100 000
Supermalloy (5% Mo, 79% Ni, 16% Fe)	100 000 – 1 000 000
Paramagnetic Materials	
Aluminum (Al)	$1 + 2.2 \times 10^{-5}$
Sodium (Na)	$1 + 7.2 \times 10^{-6}$
Platinum (Pt)	$1 + 2.8 \times 10^{-4}$
Uranium (U)	$1 + 4.0 \times 10^{-4}$
Diamagnetic Materials	
Copper (Cu)	$1 - 9.8 \times 10^{-6}$
Lead (Pb)	$1 - 1.8 \times 10^{-5}$
Mercury (Hg)	$1 - 3.0 \times 10^{-5}$
Silver (Ag)	$1 - 2.6 \times 10^{-5}$

A third group of materials is those that actually diminish a magnetic field. Their relative permeabilities are slightly less than 1. They are repelled by magnets. Materials with these properties, such as copper, gold, lead, and sodium chloride, are called **diamagnetic.** Table 21-1 lists some common metals and their relative permeabilities.

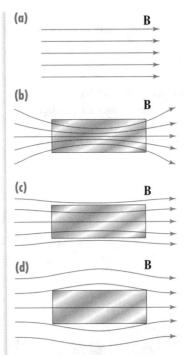

21-5 Magnetic field line density within (a) a vacuum, (b) a ferromagnetic metal, (c) a paramagnetic material, (d) a diamagnetic material

The magnetic character of materials can be described by their relative permeabilities:

Ferromagnetic $\kappa_m \gg 1$
Paramagnetic $\kappa_m > 1$
Diamagnetic $\kappa_m < 1$

Hans Christian Oersted (1777–1851) was a Danish physicist who first demonstrated the relationship between electricity and magnetism—by accident—during a classroom demonstration.

André Marie Ampère (1775–1836) was a French physicist and mathematician. He was largely self-taught and eventually became the Inspector General of the French system of universities.

21.4 Causes of Magnetism

What makes a magnet magnetic? In 1820 Hans Christian Oersted discovered that a wire carrying a current moves a nearby compass needle. From that discovery, André Marie Ampère concluded that all magnetic fields result from currents. Today most scientists agree with Ampère. Yet permanent magnets need not be connected to a source of potential difference. Where, then, does the current come from?

Remember, an atom consists of a positively charged nucleus orbited by negatively charged electrons. The electrons are moving charges. Therefore, each atom contains a current. The magnetic effect of an atomic current is represented by its **magnetic dipole moment (μ)** vector. (Do not mistake this vector quantity for the scalar magnetic permeability, μ.) The magnetic dipole moment of an atom is a vector; its direction is the direction of the atom's magnetic field, and its magnitude is related to the strength of the magnetic field. The electrons in an atom seem to "spin" as well as orbit. Each electron therefore has its own magnetic moment.

Pairs of electrons "spin" in opposite directions so that their magnetic moments cancel each other within the atom. Only those substances that have unpaired electrons in certain atomic energy levels may be magnetic. Even then, in most materials the magnetic moments of their atoms point in random directions. On the average, the random magnetic moments sum to cancel each other. That is why most materials have no net magnetic field. In ferromagnetic materials, however, the atomic magnetic dipole moments are strong enough to influence each other. The magnetic moments of all the atoms within tiny crystalline regions of the material are aligned in one direction. These regions are called **domains.**

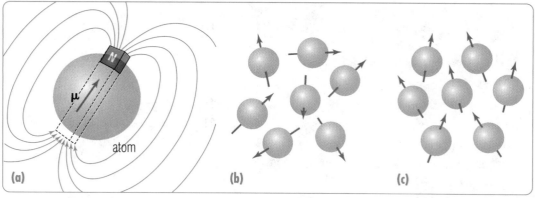

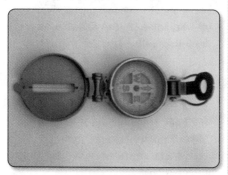

21-6 The atoms of some materials act as tiny bar magnets (a). If they are randomly oriented, the material is completely unmagnetized (b). More commonly, atoms within individual crystals of the material have the same orientation, producing domains of magnetism (c).

Pierre Curie (1859–1906) was a French physicist who discovered the piezoelectric effect and showed that magnetic properties of a substance change at a given temperature. He and his wife, Marie, discovered the elements radium and polonium and were awarded half of the 1903 Nobel Prize in Physics for their studies in radiation.

Usually, the alignment of the domains varies randomly, so there is no overall reinforcement of the individual domain dipole moments. In the presence of a strong external magnetic field, however, the domains tend to align with the field, and the material becomes magnetized. This magnetization becomes permanent if the individual atomic dipole moments are strong enough to overcome the tendency for the moments to become disoriented due to the thermal motion of their atoms (ferromagnetic materials). Otherwise, the magnetism lasts only until the field is removed (paramagnetic materials).

Diamagnetic materials start out with no net magnetic moment. When an external magnetic field is applied, a magnetic dipole moment is induced in the atoms in the same way that electrostatic charge is induced in a neutral object. North is attracted to south. The resulting induced field opposes the external field and diminishes its strength within the material. A diamagnetic object can actually repel a standard magnet.

The ordered state of magnetic domains is greatest very close to absolute zero, when there is little thermal motion. At higher temperatures, atomic thermal motion interferes with the alignment of magnetic moments. When it exceeds a characteristic temperature, called its **Curie temperature (T_C),** a ferromagnetic material loses its ferromagnetic properties and becomes paramagnetic.

21.5 Terrestrial Magnetism

Why does the north pole of a compass point north? A magnetic compass is a small, light magnet mounted on a pivot so that it can rotate freely in a horizontal plane. The poles of a magnet are called the north magnetic pole and the south magnetic pole. A magnet's *north magnetic pole* points toward the earth's north magnetic pole (which is actually a *south* magnetic pole). This confusing state of affairs exists because opposite poles attract, and a north pole is defined as one that points north.

21-7 A magnetic compass that a hiker might use

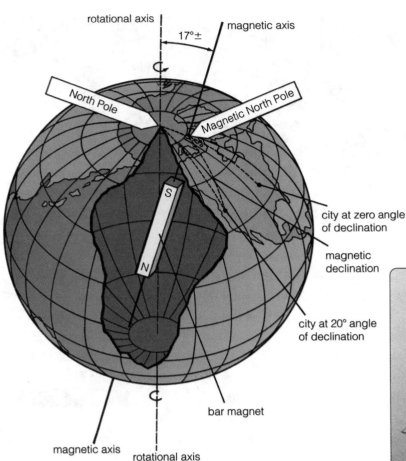

21-8 The poles of the earth's magnetic field are not aligned with its geographic poles. This is the main reason for magnetic declination.

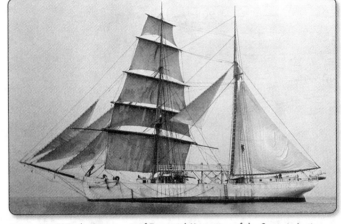

21-9 In 1905, the Department of Terrestrial Magnetism of the Carnegie Institute, Washington, D.C., commissioned the first-ever detailed magnetic survey of the Pacific Ocean using a nearly nonmagnetic wooden sailing vessel, the *Galilee*.

William Gilbert reasoned that magnets are attracted by magnets. Therefore, the earth must be a magnet. To test his conclusions, Gilbert milled a large piece of lodestone into a sphere. When he placed a compass at various positions on the surface of the lodestone, the compass reacted in the same way as it would when responding to the earth's magnetic field. The earth behaves like an immense lodestone, or as if it has a bar magnet in its center oriented with its south magnetic pole near the earth's north pole.

The earth's magnetic axis is not in the same place as its rotational axis. The magnetic pole is inclined about 17° from the rotational axis. A compass points to the magnetic pole. Assuming the earth's magnetic field is symmetrically distributed around the line passing through the north and south magnetic poles, a compass would not point to geographic north unless it was aligned with a magnetic field line that coincides with the meridian containing both the geographic north pole and the magnetic north pole. When the magnetic field is not aligned with this meridian, it points a certain angular distance away from the geographic pole. The difference between true north and magnetic north is called **magnetic declination.** It is usually given in degrees east or west.

Determining magnetic declination is far more complicated than a mere exercise in spherical geometry. The distribution of the material in the earth's core that produces the terrestrial magnetic field is not uniform. In addition, the crust's thickness and local conditions also affect the orientation of the earth's field at its surface. The locations of the magnetic poles continuously change, as well. Therefore,

Problem-Solving Strategy 21.1

The direction of magnetic declination at your location can be determined by imagining the meridian line connecting your position and the geographic north pole. If the magnetic north pole is to the left (west) of this line, then the declination is west; if to the right (east) of the line, then declination is east.

The annual change of direction of the earth's magnetic field vector at any given point is fairly constant over a span of several years. This rate is called the **annual variation,** and it is usually printed in the legend on better quality maps. It is given in units of degrees per year east or west, depending on the direction of change of local declination.

Magnetism   **459**

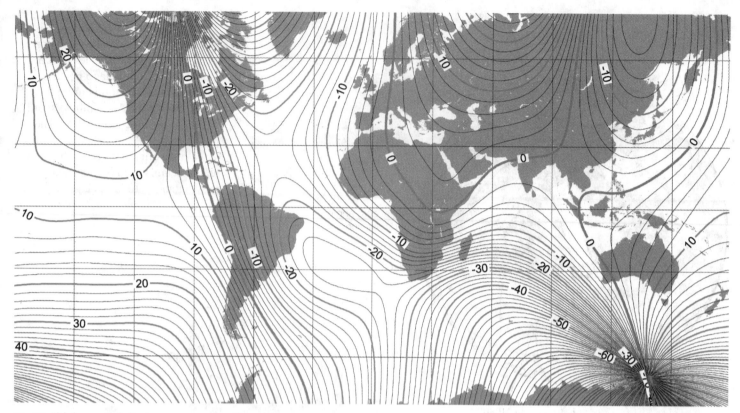

21-10 Global magnetic declination

magnetic field data must be periodically taken and plotted on maps such as Figure 21-10.

21.6 The Magnetosphere

21-11 The earth's magnetic field, or magnetosphere, distorted by the solar wind

"And all the days of Methuselah were nine hundred sixty and nine years: and he died" (Gen. 5:27). Why did people live so much longer before the Flood? Nobody knows for sure, but one theory is that the earth's magnetic field could have been much stronger, giving better protection from the solar wind and cosmic radiation. With the weakening magnetic field, more mutation-inducing radiation has penetrated to the earth's surface, destroying the natural vitality and longevity that God originally created in living things.

The volume of space containing the earth's effective magnetic field is called the **magnetosphere.** The shape of the magnetosphere is not exactly like the symmetrical field of a laboratory bar magnet. It is distorted by the **solar wind,** the continuous stream of charged particles emitted by the Sun. These charged particles act like a fluid (plasma) and are deflected by the magnetic field. The field, in turn, is carried along with the plasma flow. Therefore, the magnetosphere facing the solar wind is not as thick as the side away from the Sun, which tends to be drawn out like a magnetic tail (Figure 21-11).

Some particles are trapped in two areas of the magnetosphere. These belts of charged particles are called the **Van Allen belts** after Dr. James Van Allen, who discovered them in 1958 during the flight of America's first satellite, *Explorer I.* If the earth's magnetic field did not trap or deflect the radiation of the solar wind, it would reach the earth and damage all living things on the earth's surface. The existence of the strong magnetosphere is essential to preserve life on the earth and is evidence for God's protection of His creation.

The particles trapped in the Van Allen belts spiral back and forth from one end of the earth to the other in a few seconds. Sometimes, when the solar wind is especially strong, a high electrical potential (>10 000 V) is established between

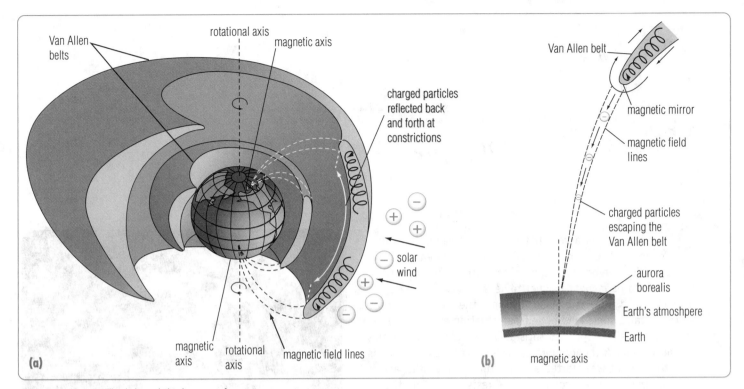

21-12 (a) The Van Allen belts and (b) the cause of auroras

the lower ends of the belts and the ionosphere. High-energy electrons rain down and collide with the atmosphere molecules about 100 km above the earth's surface. The energy of the collisions is dissipated with the emission of light. The light as observed from the ground (and space) is known as an **aurora** (from the Latin word for *dawn*). The aurora in northern latitudes is called the *aurora borealis* or the *northern lights*. The southern aurora is called the *aurora australis* or the *southern lights*.

21A Section Review Questions

1. **a.** What is a magnetic pole?

   b. How many kinds of poles are there?

   c. State the law of magnetic poles.

2. How is the magnitude and direction of the magnetic field represented *at a point* in a diagram?

3. What is the standard unit of magnetism and its SI equivalent?

4. **a.** What is the process that makes magnets out of nonmagnetic metals?

   b. What kind of metals retains some magnetism after this process?

5. Ultimately, what two factors result in magnetism in a given material?

6. What condition can destroy ferromagnetism in even the strongest magnets?

7. For a point on the surface of the earth, what is the horizontal angle between the directions to magnetic and true north called?

8. Describe the safety mechanism God provided to protect life on Earth from the effects of the solar wind.

Paleomagnetism and the Age of the Earth

Most scientists agree that the earth's magnetic field is the result of some kind of current in the earth's core, but they disagree on the source of the current. The core is far too hot for a permanent magnet to be the cause of the earth's magnetic field. There are many "dynamo" theories that say the earth contains some kind of self-perpetuating electric generator. None of these theories satisfactorily explains the earth's magnetic field. A theory more consistent with observations is that the current was started at some time in the past, and it has not yet been overcome by flow resistance. This theory would predict that the current, and therefore the earth's magnetic field, is decreasing. The earth's magnetic field has decreased by a significant amount since it was first measured in 1839 by Carl Friedrich Gauss.

The idea that the earth's magnetic field is consistently decreasing bothers those who believe that the earth is billions of years old. The measurements of the earth's magnetic field show that its strength is decreasing exponentially. The field's half-life—the time in which the magnetic field strength halves—is about fourteen hundred years. (More recent research has shown that the half-life may be as low as one thousand years, making the strength of the field in the past even more than the values presented here.) Today the strength of the earth's magnetic field is less than 10^{-4} T. Based on these facts, if the earth's magnetic field has been decreasing in the same way since it started, then one million years ago it would have had a strength of about 10^{250} T! Some evolutionists believe that the recorded magnetic field data indicates that the field is decreasing in a linear fashion. If this were the case, then only twenty thousand years ago the magnetic field of 1.8 T would have been stronger than most manmade magnetic fields. Such a strong field would require a current so large that it would produce enough heat to damage the earth's core. If the field continues decreasing at the same linear rate, it will disappear in just a few thousand years and life on Earth will be seriously endangered.

The rapidly decreasing magnetic field creates a significant problem for evolutionists. Most of those who believe that the earth is billions of years old believe that its magnetic field must be about as old because the solar wind would destroy or

A model of the magnetically active core of the earth. The larger wireframe structure is the surface of a spherical outer core, and the smaller wireframe is the bounds of a spherical inner core to compare with the actual shapes obtained from seismic data.

damage life on the earth if the magnetic field did not deflect it. Therefore, since a constant field would already have disappeared, they have to theorize that the earth's magnetic field oscillates. According to this theory, we are observing only the decrease since the last maximum. For evidence, they cite rocks containing magnetic materials that show magnetism with a polarity opposite to the earth's present magnetism.

When a hot ferromagnetic material cools, it often becomes magnetized to align with any external magnetic field. When a volcano erupts, any magnetic materials in the lava usually tend to orient themselves along the earth's magnetic field. Geologists have noticed that igneous rocks bordering the mid-Atlantic ridge contain bands of materials magnetized in opposite directions, as the figure on the next page shows. Assuming that (according to the uniformitarian plate tectonics model) as the continental plates slowly spread apart and new magma wells up into the crack, the rocks "record" the orientation of the earth's magnetic field at the time of their formation, geologists conclude that the earth's magnetic north and south poles have exchanged places several times in the earth's history.

Scientists working within a Creation-Flood model believe that the magnetic banding detectable near the mid-ocean ridges was caused by the devastating breakup of the earth's crust during and immediately following the Genesis Flood. Except for this short disruption, the earth's overall magnetic field has been decaying continuously in an exponential way since Creation about 6000 years ago.

Some scientists, especially those who believe that the earth is relatively young, do not find the evidence for changing magnetic fields convincing for several reasons. One reason is that the magnetization in rocks is so small that it is difficult to measure. Some measuring devices can even change the magnetic field they measure. In addition, the evolutionist model assumes that the decay and recovery of the earth's magnetic dipole field is accomplished by storing energy in the weaker 4-pole and 8-pole fields that are components of the much stronger dipole field. However, recent data shows that the slight gain in "poly-pole" field intensity does not account for the loss of energy in the main dipole field during the past 30 years. The theory also assumes that the energy transfer between the fields is 100% efficient. Nowhere else in the universe is energy conversion 100% efficient. In short, the data "proving" that the earth's magnetic field has reversed are not very reliable.

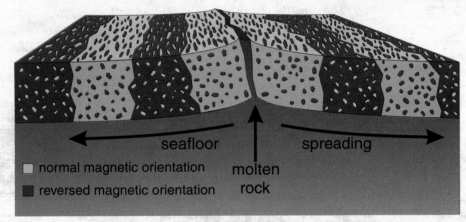

seafloor spreading

molten rock

- ☐ normal magnetic orientation
- ■ reversed magnetic orientation

Recent research has shown that the magnetic banding along midocean ridges has been confirmed. However, drill core samples from the bands reveal that the magnetic orientation in rocks beneath the surface is highly variable with depth, showing no coordinated pattern. Rather than being evidence for a slow, gradual spreading of the seafloor according to uniformitarian geology, the data supports a catastrophic model of rock formation that occurred quickly along with rapid local changes of magnetic field orientation during the time of the Genesis Flood.

21.7 Magnetic Force on a Moving Charge

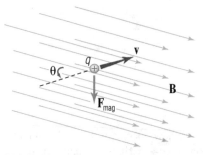

21-13 A point charge moving through a magnetic field

A current-carrying wire can produce a magnetic force that can attract or repel a magnet. When Oersted discovered this fact, scientists wondered, "Do magnets also affect electrical charges?" The answer is yes—if the charges are moving. A magnet has no effect on a stationary charge, and a stationary charge has no effect on a magnetic field. Scientists have made the following observations regarding the interplay of magnetic fields and electric charges:

1. A point charge of either polarity at rest in a magnetic field experiences zero force.

2. A charge of either polarity that moves parallel to magnetic field lines in either direction experiences zero force.

3. If any point charge moves so that it crosses magnetic field lines at an angle, a *magnetic force* (F_{mag}) is exerted on the charge. The magnitude of the force is proportional to the speed of the charge and a function of the angle with which it crosses the field lines. For a given speed, the force is maximum when the charge's velocity is perpendicular to the magnetic field lines.

4. The force on a moving charge is proportional to the density of the magnetic field lines, or field magnitude (B).

5. The direction of the force vector on a charge depends on the polarity of the charge.

The magnitude of the magnetic force on a point charge is calculated by the formula

$$F_{mag} = |q|vB \sin \theta, \tag{21.2}$$

where θ is the smallest angle ($\leq 180°$) between the velocity vector and magnetic induction vector (**B**). The absolute value sign is required because the charge q may be either positive or negative.

Surprisingly, the force of a magnetic field on a moving charge (or moving field on a stationary charge) is *not* in the direction of the charge's direction of motion, nor is it in the direction of the field; it is perpendicular to both the magnetic induction vector and the velocity vector of the charge. Because the magnetic force on the charge has no component in the direction of the charge's motion, it does no work on the charge. This interaction is similar to the force of gravity on an orbiting object. Earth's gravity keeps the Moon in its orbit. However, the force is perpendicular to the Moon's motion, so it does no work on the Moon. Similarly, a magnetic field changes the direction of a charged particle's motion but does no work on the particle. You can see that the magnetic force on a charge is a conservative force just like the electrostatic and gravitational forces.

When the charge's velocity vector is perpendicular to the magnetic field, Equation 21.2 reduces to

$$F_{mag} = |q|vB,$$

since sin 90° is 1. The direction of **F** for such a case is shown in Figure 21-14. The symbol ⊙ stands for a vector pointing up out of the page. Think of it as the view

The velocity in Equation 21.2 is the *relative* velocity between the charge and the field. This means that the charge alone, the field alone, or both the charge and field may be moving in an external reference frame.

If Equation 21.2 is solved for B, an analysis of the resulting units gives

$$\frac{N}{C \cdot m/s} = \frac{N \cdot s}{C \cdot m}.$$

This ratio in SI units is the tesla (T).

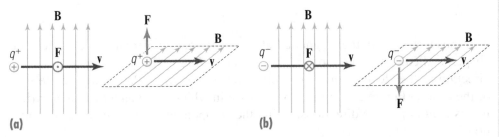

21-14 Magnetic force on a moving charge. The charge in (a) is positive; the charge in (b) is negative.

of the tip of an archery arrow pointing toward you. The symbol $\otimes$ denotes a vector pointing down into the page. It is the view you would have when looking at the feathers on an arrow pointing away from you. These symbols will be used throughout our discussion of magnetic fields to represent vectors in the third dimension.

21.8 Right-Hand Rule for Magnetic Force on a Positive Charge

The force vector ($\mathbf{F}_{mag}$) is always perpendicular to the plane defined by $\mathbf{B}$ and $\mathbf{v}$. You can find the direction of $\mathbf{F}_{mag}$ on a positive charge by another right-hand rule. Refer to Figure 21-15. Find the component of $\mathbf{v}$ that is perpendicular to $\mathbf{B}$, as in figure (a). Extend your right index finger in the direction of the magnetic field. Extend your thumb in the direction of the perpendicular velocity component. Then your middle finger extended at a right angle to the index finger and thumb points in the direction of the magnetic force vector ($\mathbf{F}_{mag}$). You may have to adjust the position of your hand in order to align your fingers in the required directions. Figure (c) shows how to position your hand to show the direction of the magnetic forces in figure (a). The direction of the force on a negative charge is opposite that on a positive charge.

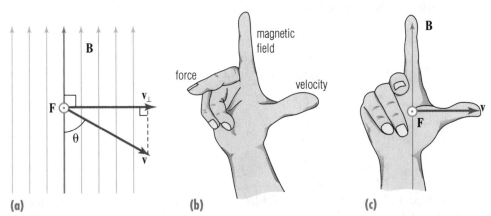

21-15 The right-hand rule for magnetic force on a moving positive charge

21.9 Velocity Selectors and Mass Spectrometers

An electrical field displaces a charged particle parallel to an electric field line. A magnetic field deflects a moving charged particle perpendicular to a magnetic field line. What happens when both an electric field and a magnetic field influence

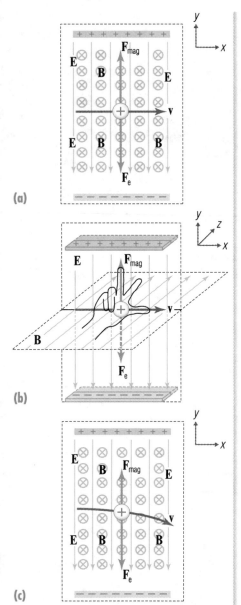

(a)

(b)

(c)

21-16 The motion of an electric charge through mutually perpendicular electric and magnetic fields; (a) magnetic forces are equal but opposite in direction; (b) three-dimensional view of balanced forces; (c) motion of charge when the electric force exceeds the magnetic force

A **mass spectrograph** is an instrument that analyzes and records the composition of mixtures or pure materials by determining the atomic or molecular masses of their constituent substances. Such devices are used in atmosphere quality monitors, for manufacturing quality assurance, and in forensic and archaeological research.

a charged particle? The net force on the particle is the vector sum of the electrical force and the magnetic force.

$$\Sigma \mathbf{F} = \mathbf{F_e} + \mathbf{F_{mag}} \qquad (21.3)$$

If the electric field, the magnetic field, and the velocity vector of the particle are all perpendicular to one another, the magnetic force will be either in the same direction as the electric force or in the opposite direction, depending on the sign of the charge. Figure 21-16 (a) shows a case in which the magnetic and electrical forces are in opposite directions. Using the components of the vectors, the Equation 21.3 becomes

$$\Sigma F_y = F_{e\,y} + F_{mag\,y}.$$

In Chapter 19 you learned that

$$\mathbf{F_e} = |q|\mathbf{E} \text{ and}$$
$$F_{e\,y} = -|q|E_y. \qquad (21.4)$$

Equation 21.4 is true if the electric field is oriented downward compared to the positive y-axis. If the charge q is positive, the force is downward. If the charge is negative, the force is upward.

Also, you know that for $\mathbf{v}$ (along the x-axis) perpendicular to $\mathbf{B}$ (along the z-axis),

$$F_{mag\,y} = |q|v_x B_z. \qquad (21.5)$$

Equation 21.5 is true for any possible combination of charge polarity (positive or negative), velocity component (right or left), and magnetic field direction (into or out of the page).

Therefore,

$$\Sigma F_y = -|q|E_y + |q|v_x B_z$$
$$\Sigma F_y = |q|(v_x B_z - E_y).$$

One analytical use of the arrangement of fields and forces in Figure 21-16 is as a **velocity selector.** In a velocity selector, the fields and the velocity are adjusted until the charged particles pass through the fields undeflected. In order for the particles to pass through undeflected (unaccelerated), the net force on them must be zero according to Newton's first law of motion:

$$\Sigma F_y = 0 \text{ N}$$
$$0 \text{ N} = |q|(v_x B_z - E_y)$$
$$0 \text{ N} = v_x B_z - E_y$$
$$E_y = v_x B_z$$
$$\frac{E_y}{B_z} = v_x, \text{ or}$$
$$\frac{E}{B} = v \qquad (21.6)$$

Only those particles with a speed of E/B will pass through the velocity selector undeflected. Charged particles moving at velocities other than this ratio will be scattered out of line with the x-axis. A plate with a small hole in it placed on the x-axis of the velocity selector can prevent scattered particles from emerging from the selector. Only undeflected particles with a specific speed pass through the hole.

The velocity selector is a component of a mass spectrograph. A charged particle emerges from the velocity selector into a region with only a magnetic field of precisely known strength oriented perpendicular to the particle's velocity. The magnetic field influences the particle to follow a semicircular path onto a

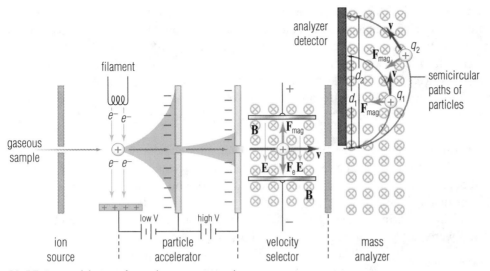

21-17 Functional diagram of a simple mass spectrograph

photographic film or into an electronic detector. The radius of this path is directly proportional to the particle's mass.

To find the exact relationship between path radius and the mass, remember the formula for centripetal acceleration:

$$a_c = \frac{v_t^2}{r}$$

The magnetic force must be

$$F_{mag} = ma_c$$

$$F_{mag} = m\frac{v_t^2}{r}.$$

However, when the particle exits the velocity selector and enters the spectrometer magnetic field, its velocity vector is perpendicular to the magnetic field. You know that when the particle's velocity and the magnetic field are perpendicular,

$$F_{mag} = |q|vB.$$

The particle's tangential speed and its velocity selector exit speed are the same. Therefore,

$$m\frac{v^2}{r} = |q|vB$$

$$mv^2 = |q|vBr$$

$$m = \frac{|q|Br}{v}. \tag{21.7}$$

The distance from the exit of the velocity selector to where the particle of interest enters the detector or is imaged on film is the diameter of its semicircular path, d. The radius (r) is half the diameter. If you know the charge on a particle, the magnetic and electric fields in the velocity selector, and the magnetic field in the spectrograph, you can find the mass of the particle. In order to get a detectable reading with this technique, many particles of the same mass must be present.

21.10 Discovery of the Electron

A velocity selector was used to discover the electron and find its charge-to-mass ratio. J. J. Thomson used a modified **cathode-ray tube (CRT)** to accomplish this. A typical cathode-ray tube consists of an evacuated tube with two electrodes in it.

21-18 A cathode-ray tube contains an inert gas to make the beam of electrons visible.

The cathode is a solid plate of metal; the anode is a larger plate with a small hole in it. They are connected to opposite terminals of a source of high voltage, but the circuit is not complete because a large vacuum gap exists between the electrodes within the tube. However, when a high potential difference (hundreds of volts) exists between the cathode and the anode, electrons will leave the cathode and stream toward the anode. Because the anode has a hole in it, many electrons shoot through the anode and on to the end of the tube before being attracted back to the anode.

Thomson added a velocity selector after the anode and a phosphorescent screen at the end of the tube. The electrons were attracted to the anode by a potential difference. They passed through the velocity selector and struck the screen. The collision of the beam of electrons caused a glow of light. Thomson did not know before his experiment that the rays that went through the anode were electrons. They were called *cathode rays* at the time.

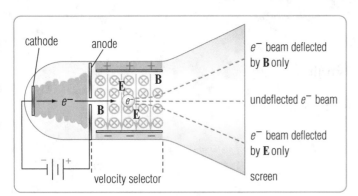

21-19 Description of J. J. Thomson's experiment

Thomson's experiment consisted of three steps. The first step was to connect a potential difference across the electrodes without activating the velocity selector. By doing this, he determined the position of the undeflected beam on the screen. In step two, he added either a magnetic field or an electric field, but not both. Figure 21-19 shows the resulting deflections for this step. Either would show that the cathode rays have a negative charge. Finally, Thomson put both an electric field and a magnetic field in the velocity selector. He adjusted the fields until the rays struck the screen at the undeflected position. Knowing the voltage (V), the electric field (E), and the magnetic field (B), Thomson could find the charge-to-mass ratio, q/m.

Thomson concluded that cathode rays were actually particles having mass because their deflection in electric and magnetic fields demonstrated that they had inertia and otherwise obeyed Newton's laws of motion.

21.11 Determining the Electron's Charge-to-Mass Ratio (q/m)

The work done by the potential difference in accelerating the electrons is

$$W = |q|V. \tag{21.8}$$

The work is also equal to the difference in the kinetic energy of the electrons. Since they accelerated almost from rest, the work is equal to their final kinetic energy,

$$W = \tfrac{1}{2}mv^2. \tag{21.9}$$

Set Equations 21.6 and 21.7 equal to each other:

$$|q|V = \tfrac{1}{2}mv^2$$

Multiply both sides by 2/m:

$$2V\frac{|q|}{m} = v^2 \tag{21.10}$$

The speed of the undeflected particle exiting the velocity selector is

$$v = \frac{E}{B}, \text{ so}$$

$$v^2 = \frac{E^2}{B^2}.$$

By substituting for v^2 in Equation 21.10, you find that

$$2V\frac{|q|}{m} = \frac{E^2}{B^2}.$$

Dividing both sides by $2V$, you get

$$\frac{|q|}{m} = \frac{E^2}{2VB^2}.$$

Thomson discovered that the q/m ratio was constant no matter what material the cathode was made of. He concluded that every material contains small, negatively charged particles with a charge-to-mass ratio of

$$\frac{|q|}{m} = 1.76 \times 10^{11} \text{ C/kg.}$$

In this way, J. J. Thomson discovered the electron.

21B Section Review Questions

1. What factors determine the magnitude and direction of the magnetic force on an electrical charge?

2. Why is it necessary to use the absolute value of the charge in Equation 21.2 when finding the magnitude of the magnetic force?

3. How do you determine the direction of the magnetic force on a charge moving through a magnetic field?

4. When a charged particle is propelled along a path that is perpendicular to mutually perpendicular magnetic and electric fields, what condition must exist for the particle to maintain a straight path?

5. Briefly discuss the principle of operation of a mass spectrograph.

6. a. What type of device was used by J. J. Thomson to determine the charge-to-mass ratio of an electron?

   b. Briefly describe its construction.

7. After completing his experiments, what fact led Thomson to conclude that all matter contains small, negatively charged particles he called electrons?

●●8. What atomic theory of matter had to be discarded with the discovery of the electron?

21C ELECTROMAGNETISM AND CONDUCTORS

21.12 Magnetic Force on Currents

A current is simply a collection of moving charges. What, then, is the magnitude of the magnetic force on a current-carrying wire segment placed in a magnetic field? In Section 21B you learned that the magnitude of the magnetic force on a moving charge is

$$F_{\text{mag}} = |q|vB \sin \theta.$$

The product qv for a single charge has the units $\text{C} \cdot \text{m/s}$. This is analogous to the product Il for a current of I in a wire of length l (units: $\text{C/s} \cdot \text{m}$). The latter product includes the speed and polarity of all the charges in the wire. The units are the same:

$$|q|v \Rightarrow \text{C} \cdot \text{m/s} \Rightarrow Il$$

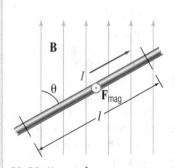

21-20 Magnetic force on a current-carrying conductor

In fact, experiments have shown that the force a magnetic field **B** exerts on a wire segment of length *l* carrying a current *I* is

$$F_{\text{mag}} = IlB \sin \theta, \qquad (21.11)$$

where θ is the smallest angle ($\leq$ 180°) between the current direction and **B**.

21.13 Right-Hand Rule for Magnetic Force on a Conductor

The direction of the force on a conductor is found in the same way as the direction of the force on an individual moving positive charge. Point your right thumb in the direction of conventional current flow (or its component) that is perpendicular to the magnetic field. Point your index finger at a right angle to your thumb in the direction of the magnetic field. Hold the middle finger perpendicular to the other two fingers to indicate the direction of the magnetic force on the conductor.

21.14 Rotating Loops

One problem with Equation 21.11 is that it applies only to a segment of a conductor. In practice, continuous currents exist only in complete circuits. Therefore, in order to make use of the magnetic force, conductors in the shape of a loop must somehow move within a magnetic field. In order to find the net force exerted on the loop conductor, you must sum the forces on each of its segments.

Consider a square conductor loop supplied with a continuous current (*I*), such as in Figure 21-21. The loop is positioned in a uniform magnetic field (**B**) as shown. The currents in segments *ab* and *cd* are in the same direction as **B** and −**B**, respectively. Therefore, sides *ab* and *cd* experience no magnetic force (sin 0° = sin 180° = 0).

The force on side *bc* is

$$F_{bc} = IlB \sin 90°$$
$$F_{bc} = IlB.$$

The direction of the force is into the page, as shown in the figure.

The force on side *da* is

$$F_{da} = IlB \sin 90°$$
$$F_{da} = IlB$$
$$F_{da} = F_{bc} = F.$$

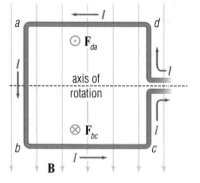

21-21 Magnetic force on a current loop

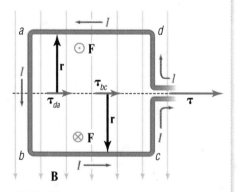

21-22 Torque on a current loop

However, the direction of **F**$_{da}$ is out of the page and opposite to the direction of **F**$_{bc}$. The sum of these forces is zero, so the square loop will not be translated from its position.

Although there is no net translational force on the loop, it will rotate if free to do so. Therefore, there must be a net torque on the circuit. Recall from Chapter 7 that the torque (τ) on an object results when a force is applied to a moment arm at a distance *r* from the axis of rotation. The formula for the magnitude of torque is

$$\tau = rF \sin \theta,$$

where θ is the smallest angle between **F** and **r** when the two vectors are placed tail to tail. In Figure 21-22 the torque on side *bc* is

$$\tau_{bc} = rF \sin 90°$$
$$\tau_{bc} = rF.$$

The direction of the torque is found by using the *right-hand rule for torques.* The torque on side *da* is

$$\tau_{da} = rF \sin 90°$$
$$\tau_{da} = rF.$$

The magnitudes of the torques on segments *BC* and *DA* are equal and in the same direction. Therefore, the total torque is

$$\tau = \tau_{bc} + \tau_{da}$$
$$\tau = rF + rF$$
$$\tau = 2rF.$$

In Figure 21-24, each segment of the square current loop is perpendicular to **B.** Therefore, all the sides experience forces of the same magnitude,

$$F = IlB.$$

The net torque is now zero, since all the forces are parallel to their respective moment arms (**r**) and the current loop does not rotate. The current loop in Figure 21-21 will tend to rotate until its plane is perpendicular to **B.** At that point it has no net torque, and so it continues to rotate at a constant speed. As the loop moves through the perpendicular position and is no longer perpendicular to **B,** the torque generated by the magnetic forces on the current loop tends to return it to a perpendicular position. Because this is a restoring force, the current loop will oscillate about its equilibrium position, perpendicular to **B.**

Circular current loops have no straight segments. You can get an idea of the direction of the force or torque on a circular circuit by approximating it with a polygon. The currents in segments *ab* and *ef* in Figure 21-25 are parallel to **B;** so there is no force on these segments. The direction of the forces on the other segments can be found from the right-hand rule and Equation 21.11. They are indicated in the figure. As you can see, the circular loop will also tend to rotate.

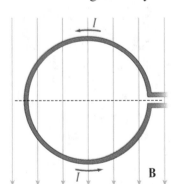
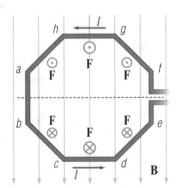

21-25 Analyzing the magnetic force on a circular current loop

21.15 Rotating Coils

Coils of wire behave like stacks of current loops. The wire may be wound around a solid core, or it may be wound without a core. Consider a coil of *n* wraps or windings. The force of the magnetic field on this coil is the same as the force of the field on *n* current loops with the same width and height. That is, for a coil, the force on a side is

$$F_{\text{mag}} = nIlB \sin \theta . \qquad (21.12)$$

A *galvanometer* uses the rotation of a coil in a magnetic field to measure current. Torque is proportional to force, which is proportional to current. Therefore,

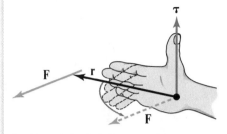

21-23 Right-hand rule for torques: For counterclockwise rotation, the torque vector points up; for clockwise rotation, the torque vector points down.

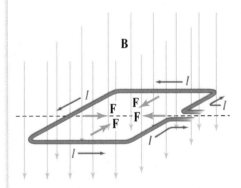

21-24 Magnetic forces on a current loop oriented perpendicular to the magnetic field

21-26 A coil of wire acts like a stack of individual current loops

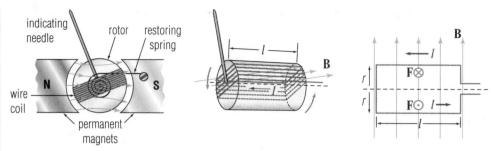

21-27 Functional diagram of a galvanometer

rotational torque can be used to measure current. The galvanometer consists of a coil wound on a cylindrical core, a spring mounted so that it opposes the turning of the coil, a needle to show the amount of turning, a scale, and a magnet with concave poles to produce a uniform magnetic field. When a current flows in the wire, a torque on the coil rotates it in the magnetic field. The rotation of the coil stretches the attached spring and turns the attached needle. The needle stops moving when the torque of the magnetic force is cancelled by the opposing torque applied by the stretched spring. Greater currents cause greater deflection of the needle. Opposite currents cause deflections in opposite directions. The zero-current position is at the middle of the scale. More sensitive galvanometers contain more turns of wire in their coils in order to increase the magnetic force for a given amount of current.

21.16 DC Motors

The DC motor is, in some ways, similar to the galvanometer. The purpose of the DC motor is to convert electrical energy to mechanical work. A motor, like a galvanometer, has a coil wound on a ferromagnetic core. The core is mounted on an axle supported by bearings to reduce friction. This assembly is called a **rotor.** The coil is surrounded by a magnetic field produced by permanent magnets or electromagnets mounted in the fixed frame of the motor, called the **stator.** The DC current source is connected to the rotating coils through a split ring mounted on the axle of the rotor. (A split ring is two conducting half-rings separated from each other by an insulated gap). Two electrical contacts called **brushes** ride on the split ring and carry the current supplied to the motor. At any given instant, each of the brushes contacts only one side of the split ring. The subassembly of the rotor containing the split ring is called a **split-ring commutator.**

When current flows through the motor circuit, the coil rotates. If the DC current were to remain constantly in one direction, the coiled winding would tend to rotate until it was perpendicular to the magnetic field, then oscillate around that position as discussed above. However, as the rotor turns, the split ring segments move with it. When the rotor has turned far enough, the split ring gap slides under each brush just as the coil becomes perpendicular to the field. Inertia carries the rotor a little farther, and the other commutator segments then contact the brushes. This action reverses the current flow through the rotor coils, maintaining the magnetic torque in the same direction as it was initially. The rotor continues another half turn, and the commutator again reverses the current, sustaining the torque in the correct direction. The DC motor continues to rotate, doing useful work.

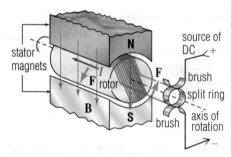

21-28 A DC motor uses a split-ring commutator to reverse the electric current in the rotating coil.

> The **commutator** of a DC motor permits continuous rotation of the motor in one direction without having to continually change the polarity of the voltage source connected to the motor.

21C Section Review Questions

1. What factors determine the magnitude of the magnetic force on a current moving through a straight conductor in a magnetic field?

2. Copy the following conductor loops onto your paper. Indicate, using proper symbols, the direction of the magnetic force on each segment of the loop.

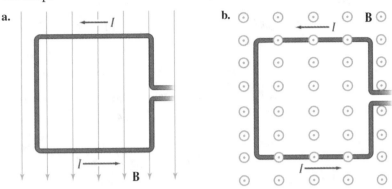

3. Which of the current loops in Question 2 will experience a torque? Explain your answer.

4. What factors determine the magnitude of the magnetic force on a coil of wire containing a current in a magnetic field?

5. The figure shows a single current loop mounted so that it can rotate. Explain what will happen when the switch is shut.

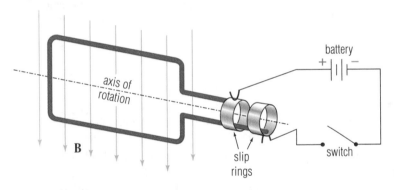

6. The figure shows a different current loop mounted so that it can rotate. Explain what will happen when the switch is shut.

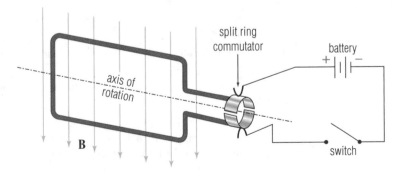

In Terms of Physics

lodestone	21.1		magnetic declination	21.5
pole	21.1		annual variation	21.5
law of magnetic poles	21.1		magnetosphere	21.6
gauss	21.2		solar wind	21.6
magnetic field vector (**B**)	21.2		Van Allen belts	21.6
tesla (T)	21.2		aurora	21.6
magnetic induction	21.3		velocity selector	21.9
relative permeability (κ_m)	21.3		mass spectrograph	21.9
ferromagnetic	21.3		cathode-ray tube (CRT)	21.10
paramagnetic	21.3		rotor	21.16
diamagnetic	21.3		stator	21.16
magnetic dipole moment (μ)	21.4		brush	21.16
domain	21.4		split-ring commutator	21.16
Curie temperature (T_C)	21.4			

Problem-Solving Strategies

21.1 (page 459) The direction of magnetic declination at your location can be determined by imagining the meridian line connecting your position and the geographic north pole. If the magnetic north pole is to the left (west) of this line, then the declination is west; if to the right (east) of the line, then declination is east.

21.2 (page 464) In order to find the angle between the velocity and the magnetic field vectors, place their vectors tail to tail. The angle θ is the smallest angle between the vectors.

21.3 (page 470) **The right-hand rule for torques:** Hold your right hand flat with the thumb extended at a right angle to the fingers. Point your fingers away from the axis of rotation parallel to the position vector **r**. Curl your fingers toward the direction of the force vector where it is applied to the moment arm. The torque vector points in the direction of your thumb.

Review Questions

1. Can you obtain isolated poles by breaking a long magnet in half? Explain.

2. Name one way to find the shape of a magnetic field.

3. **a.** Where is this magnet's north pole?
   b. Where is its field strongest?

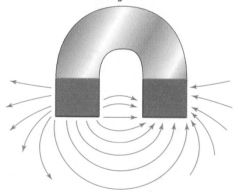

4. What vector represents a magnetic field?

5. How does a magnet attract an ordinary piece of metal?

6. How can you distinguish ferromagnetic steel from paramagnetic aluminum?

7. Bismuth has a relative permeability of 0.99983. Is this ferromagnetic, diamagnetic, or paramagnetic?

8. If you want a strong magnetic field, in what kind of material should you generate the field?

9. What is the best model for explaining magnetism?

10. Why does a magnetic compass point north?

11. Are the earth's magnetic poles at its geographic poles? Explain.

12. What is magnetic declination?

13. In what direction does the north-seeking end of a compass at the north geographic pole point?

14. In what direction does the north-seeking end of a compass at the north magnetic pole point?

15. What keeps harmful charged particles from the Sun from reaching the earth's surface?

16. What causes the auroras?

17. If the earth's magnetic field has always been decreasing as it has been observed to do since men began observing it, how old may it be?

18. For a charge of a given magnitude moving at a constant velocity through a uniform magnetic field, what is the difference in the magnetic force on the charge when it is positive compared to when it is negative?

19. Study the figure. In what direction will the charge be deflected?

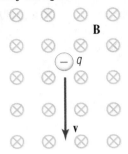

20. Describe a mass spectrometer.

21. Two kinds of particles are detected in a mass spectrometer sample. One kind of particle is detected at 9.5 cm from the entrance point. The other is detected at 13.8 cm from the entrance. Which kind of particle has greater mass?

22. A current-carrying conductor is aligned parallel to the magnetic field. What direction is the magnetic force on the conductor?

23. A current loop lies in a magnetic field so that all of its sides are perpendicular to the field lines. Will the loop rotate? Explain.

24. A circular current loop lies in a magnetic field so that it is in a plane parallel to the magnetic field lines. Will the loop rotate? Explain.

25. Why does a DC electric motor rotate continuously rather than oscillate?

True or False (26–40)

26. All magnets have both a north and south pole.

27. One tesla is approximately the strength of the terrestrial magnetic field at the earth's surface.

28. A magnetic field is not affected in any way by the material in which it is located.

29. The strength of a magnet depends not only on the dipole moments of its atoms but also on their orientation.

30. Magnetic declination over the surface of the earth is relatively constant from one decade to the next.

31. A charge at rest in a magnetic field will be accelerated perpendicular to the field lines.

32. A moving electric charge is deflected in the direction of a magnetic field line.

33. The closer the crossing angle between a charge's velocity and a magnetic field line is to 90°, the greater the magnetic force on the charge.

34. Magnetic force is a conservative force.

35. The right-hand rule for magnetic force is valid only for positive charges.

36. The torque vector on a current loop in a magnetic field points in the direction of rotation.

37. The formula for magnetic force in a coil of wire (Equation 21.12) assumes that all the coils have the same dimensions.

38. A galvanometer works by balancing an electromagnetic torque against a spring torque.

39. In a simple DC motor, the stator contains the electric current that drives the motor, and the rotor contains the magnetic field.

40. The purpose of a commutator is to reverse the current in the motor coil so that it can continue to turn in one direction.

◉41. If the elementary charge of an electron is -1.602×10^{-19} C, what is its mass?

◉42. A $+1.00 \times 10^{-6}$ C charge moves through a magnetic field, $B = 3.00 \times 10^{-5}$ T, initially perpendicular to the field, with a constant speed of 1.33×10^{-3} m/s. What is the magnitude of the magnetic force exerted on the charge?

◉43. A magnetic field exerts a force of 5.50×10^{-10} N on a -1.10 mC charge initially moving perpendicularly to the field at a constant speed of 2.50×10^{-2} m/s. What is B?

◉44. A point charge moving at a constant speed of 1.50×10^{-2} m/s perpendicular to a magnetic field of $B = 3.00 \times 10^{-4}$ T experiences a force of 9.00×10^{-12} N. What is the magnitude of the point charge?

◉45. A velocity selector has a magnetic field of 3.00×10^{-4} T and an electric field of 4.50 N/C. What velocity does it select?

◉46. You want a velocity selector to allow charged particles with a speed of 2.25×10^3 m/s to pass undeflected. The magnet is capable of producing a field with $B = 3.333 \times 10^{-4}$ T. What must the electric field strength be?

◉◉47. Prove that the unit ratio $\frac{\text{N} \cdot \text{s}}{\text{C} \cdot \text{m}}$ equals the tesla ($\text{V} \cdot \text{s/m}^2$).

◉◉48. If there are two kinds of particles detected in a mass spectrometer, such that one kind is twice as far from the beam entrance as the other, how do their masses compare? (Assume that both kinds of particles have the same velocity and the same charge when they enter the spectrograph.)

22A	Currents and Magnetic Fields	477
22B	Alternating Current	484
22C	AC Circuit Characteristics	490
Facet:	Diesel Locomotives	498
Biography:	James Clerk Maxwell	499

New Symbols and Abbreviations

permeability of free space (μ_0)	22.1	inductance (L)	22.11
magnetic flux (ϕ)	22.5	henry (H)	22.11
total magnetic flux (Φ)	22.5	mutual inductance (M)	22.11
weber (Wb)	22.5	inductive reactance (X_L)	22.17
alternating current (AC)	22.9	capacitive reactance (X_C)	22.17
root mean square (rms)	22.9	impedance (Z)	22.17
back emf (ε)	22.10		

Alternating current transformers are
essential for the transmission of
electrical power over great distances.

Electromagnetism

> *But he [Rehoboam] forsook the counsel of the old men, which they had given him, and consulted with the young men that were grown up with him, and which stood before him.* *1 Kings 12:8*

This is the key verse in the sad story of the division of the kingdom of Israel. The new king, Solomon's son Rehoboam, had been petitioned by the people to be released from the crushing tax burden imposed by his father. But he was *induced* by his wicked childhood friends to go against the reasonable request of the people, and he purposed to place an even greater tax burden on them.

Inducements to do evil come from many directions in your life. Unexpected consequences can occur when you succumb to these temptations. Similarly, scientists have found that magnetic fields can induce electrical currents in unexpected ways. Electronic circuits sometimes must be shielded from adverse magnetic fields in order to guarantee reliable performance. Therefore, be careful of the friends that you choose and be wary of the inducements they offer. Test everything against Scripture to see if it is true and right.

> Recall that conventional current is opposite to the flow of electrons in a wire.

22A CURRENTS AND MAGNETIC FIELDS

22.1 Magnetic Field Around a Conductor

As Oersted discovered in 1820, moving charges produce magnetic fields. The magnetic field generated by charges moving in a long, straight wire can be graphically represented by concentric circles around the wire. (Of course, each end of the long wire must be connected to a source of potential difference. If the connections are far from the middle of the wire, then the magnetic field near the wire at its middle is like that of an isolated wire.) You can find the direction of the magnetic field vector (**B**) near a wire by a new right-hand rule. Point your right thumb in the direction of the (conventional) current flowing through the wire and wrap your fingers around the wire (mentally, of course, unless the wire is insulated). Your curled fingers point in the direction that the field circles the wire. Figure 22-2 on the next page shows the direction of **B** as it relates to your fingers. This illustrates the *magnetic field vector **right-hand rule for conductors.***

22-1 The magnetic field surrounding a current-carrying conductor

477

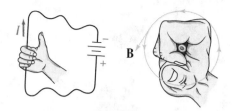

22-2 The right-hand rule when used to determine the magnetic field vector **B**

Jean-Baptiste Biot (1774–1862) was a French mathematician and scientist who investigated extensively in astronomy, cartography, geometry, thermodynamics, and magnetism. He held a number of prominent chairs in academic institutions and was even involved in a short-lived political rebellion.

Felix Savart (1791–1841) was a French professor who experimented with magnetism and acoustics. With Biot, he discovered that a magnetic field generated by a current-carrying conductor was proportional to the inverse square of the distance from the conductor.

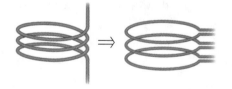

22-4 A solenoid coil is like a stack of individual current loops.

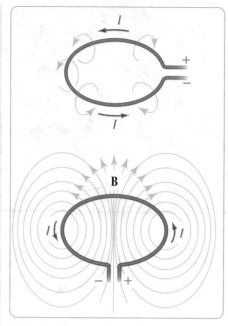

22-3 A circular current loop and its magnetic field

A circular current loop, such as the one in Figure 22-3, can be mentally divided into many small, nearly straight segments. Each segment contributes to the net magnetic field, and the field is simply the sum of the contributions from all segments. The magnetic field of each segment points straight up inside the loop and straight down on the outside. The entire magnetic field of a circular loop looks like the magnetic field of a bar magnet. Therefore, the magnetic field of a current loop is said to be a *magnetic dipole*.

The magnetic induction vector at a point in space produced by current flowing through a very short segment (Δl) of a long wire is given by the **Biot-Savart law:**

$$\Delta B = \frac{\mu_0}{4\pi} I \frac{\Delta l \sin \theta}{r^2} \qquad (22.1)$$

The **permeability of free space** (μ_0) is a measure of how effectively the magnetic field penetrates a vacuum. The angle θ is the angle between the wire segment and a line connecting the center of the segment with the point in space. The term r is the distance from the center of the segment to the point. The magnetic field vector (**B**) at the point representing the total field of the current loop is the sum of the magnetic field vectors (Δ**B**) at that point contributed by each tiny segment (Δl).

The Biot-Savart law is important because it demonstrates that the magnetic field generated by a small segment of a conductor is similar to a gravitational field generated by a small mass and an electric field generated by a point charge.

- Each is proportional to a constant related to the kind of field.
- Each is an inverse-square law.

22.2 Electromagnets

A single turn of a current-carrying wire can act as a type of magnet called an **electromagnet.** A **solenoid** is a series of wire coils used for this purpose. If the solenoid is wound tightly, it resembles a stack of independent current loops. The magnetic field of a solenoid with n turns is n times as strong as the magnetic field of a single current loop. Like the magnetic field of a single current loop, a solenoid's magnetic field is shaped like a bar magnet's field. This fact supports the model that bar magnets are magnetized by atomic currents.

22.3 SI Definition of the Ampere

The magnetic field of one long, straight wire will interact with the magnetic field of another long, straight wire. If the current in the wires is in the same direction, the wires attract each other. If the currents are in opposite directions, the wires repel each other. This interaction occurs because each wire is positioned in the other's magnetic field. Use the magnetic field vector right-hand rule to determine the direction of the field due to the other wire at each wire's location. Then use the first right-hand rule from Chapter 21 to determine the force on the wire. Point your thumb in the current direction and your index finger at a right angle to your thumb, parallel to the field vector, **B**, *from the other wire.* The middle finger, held at a right angle to the other two fingers, points in the direction of the force on the wire. In Figure 22-5, $\mathbf{F}_{1\rightarrow2}$, which represents the force of wire 1's magnetic field on wire 2, points toward wire 1. Similarly, the force of wire 2's magnetic field on wire 1, $\mathbf{F}_{2\rightarrow1}$, points toward wire 2. The wires are drawn together.

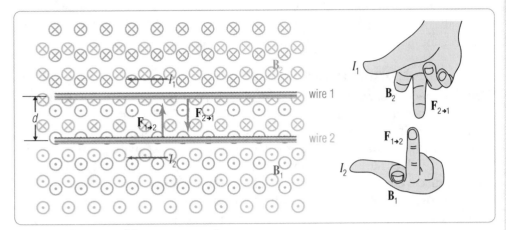

22-5 Forces between parallel current-carrying conductors

Each wire exerts on the other wire a force of

$$\frac{F_{\text{mag}}}{l} = \frac{kI_1I_2}{d},$$

where d is the distance between the wires and k is a constant related to the permeability of free space. This force gives a practical measurement of the ampere. The ampere was defined in Chapter 20 as the current in a conductor in which 1 C of charge flows past a given point in 1 s. However, a coulomb, defined as 6.24×10^{18} electron charges, is difficult to measure. To avoid this problem, the ampere has been redefined as follows: "1 A is the amount of current that, when it exists in each of two infinitely long wires separated by a distance of 1 m, causes each wire to exert on the other a force of 2×10^{-7} N for each meter of length." This is the present-day SI definition of the ampere. The forces can be measured accurately on a current balance (Figure 22-6). The coulomb has been redefined as the amount of charge that flows past each point of a conductor carrying a current of 1 A in 1 s.

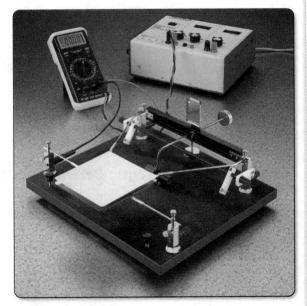

22-6 A current balance

22.4 Electromagnet Applications

Most magnets used today in industry are electromagnets, usually solenoids wound on ferromagnetic cores. Using a ferromagnetic core greatly concentrates a solenoid's magnetic field. Since electromagnets that have magnetic fields far stronger than those of permanent magnets can be constructed, they are used when strong magnetic fields are required. Electromagnets are also used when a continuous magnetic field is not desired. Doorbells, for instance, use electromagnets to ring their bells, and tiny electromagnets are used to read and write information on magnetic digital media.

22.5 Electromagnetic Induction

Since an electric current produces a magnetic field, scientists in the 1820s reasoned that perhaps a magnetic field produces an electric current. They experimented to see if they could prove this speculation. They discovered that no matter how strong the magnet is or how conductive the wire is, a stationary magnet will not produce electricity in a stationary current loop. Finally, in 1831 two experimenters discovered a way to generate electricity from magnetism. Michael Faraday and Joseph Henry independently discovered that a magnetic field must be changing with respect to the current loop to generate an electric current. This generation of electricity from magnetism is called **electromagnetic induction.**

Joseph Henry (1797–1878), an American physicist, was one of the founders of the National Academy of Sciences and the first Secretary of the Smithsonian Institution.

Faraday explained the results of his experiments using the concept of **magnetic flux (ϕ).** You may remember from Chapter 21 that the magnitude of the magnetic field vector **B** is the *magnetic flux density (B)*. Thus, **B** involves flux per unit area. In fact, the flux through an imaginary planar area is equal to the product of the area and the perpendicular component of **B** passing through the area.

$$\phi = \mathbf{B}_\perp A \tag{22.2}$$

If **B** is not uniform throughout the region, the region must be divided into smaller areas for which **B** is uniform. The total flux through the region is the sum of the fluxes through the smaller areas. For a coil, the **total magnetic flux (Φ)** is equal to the flux (ϕ) through each loop multiplied by the number of loops (n).

The unit of magnetic flux is the **weber (Wb).**

$$1 \text{ Wb} = 1 \text{ V·s}$$

Magnetic flux is measured in **webers (Wb).** From Equation 22.2 we see that

$$1 \text{ Wb} = 1 \text{ T·m}^2, \text{ but}$$

$$1 \text{ T} = 1\frac{\text{V·s}}{\text{m}^2}, \text{ so}$$

$$1 \text{ Wb} = 1\frac{\text{V·s}}{\cancel{\text{m}^2}} \cdot \cancel{\text{m}^2} = 1 \text{ V·s}.$$

Wilhelm Weber (1804–91) was a German physicist whose work helped Maxwell develop the electromagnetic theory of light.

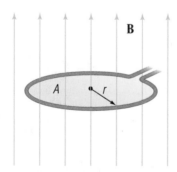

22-7 Magnetic flux depends on the strength of the field through a certain area of space.

EXAMPLE 22-1

Magnetic Field Strength: Determining Magnetic Flux

Find the magnetic flux through the current loop in Figure 22-7. The magnetic field (**B**) is perpendicular to the loop, and $B = 5.00 \times 10^{-2}$ T. The loop has a radius of $r = 0.100$ m.

Solution:
Use Equation 22.2:

$$\phi = B_\perp A$$

The vector **B** is perpendicular to the loop, so

$$B_\perp = B = 5.00 \times 10^{-2} \text{ T}.$$

The area of the loop is

$$A = \pi r^2 = \pi(0.100)^2$$
$$A \doteq 0.03141 \text{ m}^2.$$

Therefore, after substituting into Equation 22.2, ϕ is

$$\phi = (5.00 \times 10^{-2} \text{ T})(0.03141 \text{ m}^2)$$
$$\phi \doteq 1.570 \times 10^{-3} \text{ V·s}.$$

Since 1 V·s = 1 Wb, the flux is

$$\phi \approx 1.57 \times 10^{-3} \text{ Wb}.$$

22.6 Faraday's Law

Faraday's experiments showed that a difference in electric potential is induced in any current loop through which the flux changes. This result is called **Faraday's law.** Mathematically,

$$V_i = -\frac{\Delta\phi}{\Delta t}, \tag{22.3}$$

where V_i is the induced potential difference and $\Delta\phi$ is the change in flux during a time Δt. The term *induced potential difference* is sometimes replaced by *induced emf* in order to avoid confusion with the voltage drop across a circuit or circuit component.

What causes the induced current? A conductor has free electrons. When the conductor is moving through a nonuniform magnetic field, its free electrons are affected just as individual moving negative charges would be. For instance, the electron in Figure 22-8 is deflected as the figure shows. As it moves, it repels free electrons in the left side of the loop. The other electrons in the left side are deflected in the same direction as the electron shown. This deflection and repulsion establishes a conventional current (of positive charges) in the direction opposite to the electron flow.

One way the magnetic flux can be changed is by motion. A changing magnetic flux may be caused by moving the current loop through a nonuniform magnetic field, as in Figure 22-8, or by moving a magnetic field through a current loop. Both methods have the same deflecting effect. Only relative motion matters.

A second way the magnetic flux can be changed is by changing the current. A changing current produces a changing magnetic field, and this changing magnetic field has the same effect as motion. A current loop connected to a source of changing potential difference can induce a potential difference in another circuit loop. Figure 22-9 shows a possible arrangement for one current loop to induce current in another. At the instant the switch is closed, the magnetic field of the first loop expands and

In Figure 22-8, the conductor is moving to the right, which is the original motion of the charge in the magnetic field. Only after the magnetic force thus generated acts on the charge and deflects it according to the right-hand rule does a current become established in the conductor.

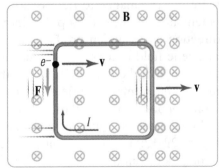

22-8 An electron in a conductor is deflected as it moves through a magnetic field.

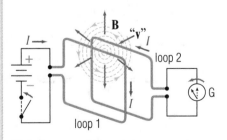

22-9 One current loop can induce a current in another loop.

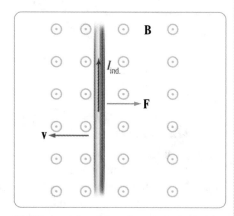

Heinrich F. E. Lenz (1804–65) was a Russian physicist who discovered the relationships between induced magnetic fields, voltage, and current when a conductor passes through a magnetic field.

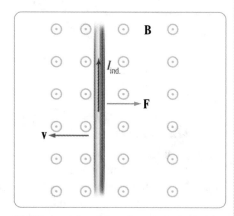

22-10 Lenz's law as it applies to a simple, straight conductor

Use the magnetic field vector right-hand rule with Faraday's law to determine the direction of the induced current in loop 2. Use the right-hand rule for conductors to determine the direction of the magnetic field around loop 2 and its inductive effect on the other loop.

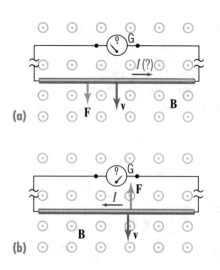

22-12 Lenz's law is a consequence of the conservation of energy.

"moves out" through the second loop. This change of the magnetic field generates a current in the second loop. When the current in the first loop reaches its steady-state value, the magnetic field stops changing. Then the current in the second loop decays to zero. At the instant the switch is opened, the magnetic field of the first loop collapses, moving past the second loop. This change generates another current in the second loop, in the opposite direction to the earlier current. A fairly continuous alternating current could be induced in the second loop by rapidly opening and closing the switch in the first loop, causing a continuously changing magnetic field.

22.7 Lenz's Law

The direction of an induced potential difference, or emf, is described by **Lenz's law:** The emf always opposes the change in the magnetic field that caused it. Consider a wire segment traveling through a nonuniform magnetic field. The magnetic flux around the wire changes with time. The change in magnetic flux induces a potential difference across the segment. The direction of the induced potential difference is such that the magnetic field exerts a force on the segment opposite to the direction of the segment's motion. The magnetic force on the induced current will halt the segment unless the segment is pushed or pulled along.

Another example using Lenz's law is the two-current-loop system we studied earlier. Figure 22-11 is a detail of the two current loops in Figure 22-9. In figure (a), the switch supplying loop 1 has just closed, so the current (I_1) and therefore the magnetic field in loop 1 rapidly increase to their steady-state value. The increasing magnetic field ($\Delta \mathbf{B}_1$) induces a potential difference in loop 2. The potential difference produces a current (I_2) opposite to the current in loop 1, which generates a changing magnetic field ($\Delta \mathbf{B}_2$). This changing magnetic field tends to induce a current in loop 1 in the direction opposite to I_1, which opposes the change in the magnetic field ($\Delta \mathbf{B}_1$) that caused I_2 in the first place. Therefore, the temporary induced magnetic field in loop 2 opposes the buildup of the magnetic field in loop 1.

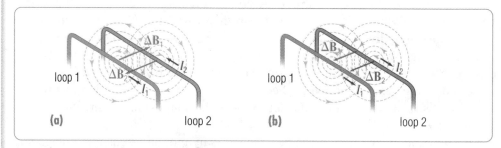

22-11 Lenz's law for induction in two loop circuits

On the other hand, in figure (b) the switch has just opened. The current (I_1), and therefore the magnetic field of loop 1, decrease. This change ($\Delta \mathbf{B}_1$) induces a potential difference in loop 2, which causes a changing current (I_2) in the *same* direction as I_1, which produces a changing magnetic field ($\Delta \mathbf{B}_2$). The induced magnetic field in loop 2 reinforces the dying current in loop 1 and thus opposes the change of the field from loop 1 that caused I_2; that is, it opposes the *decrease* in the original magnetic field. Therefore, the temporary induced magnetic field in loop 2 opposes the decay of the field around loop 1.

Lenz's law is a natural consequence of the conservation of energy. Study Figure 22-12: a wire segment connected by long wires to a galvanometer moves through a nonuniform magnetic field. If the induced current were in the direction shown in figure (a), the force of the magnetic field on the current would be in the direction of the segment's motion. The force would continue to move the segment

482   Chapter 22

even if it were not pulled along. The continuing motion would continue to generate electricity and would produce a perpetual motion generator. The conservation of energy requires that Lenz's law be obeyed. The actual direction of current and the resulting force on the current are shown in figure (b). Energy must be supplied to keep the segment moving and the induced current flowing.

22.8 Eddy Currents

When a conductor moves through a magnetic field, a potential difference is induced. If the conductor is a wire, the current flowing through it can travel in only one path, since the wire is very small. In a conductor other than a wire, such as a metal plate or bar, the current has many paths open to it. The current will divide into **eddy currents.** This name is derived from the current path's resemblance to eddies around a floating object. Figure 22-13 shows eddy currents in a conducting plate within a changing magnetic field.

The magnetic field exerts a force on eddy currents within the material to oppose the conductor's motion. Therefore, eddy currents are said to have a braking effect on a solid conductor's motion. A rotating conducting disk can be brought to a standstill by impressing on it a magnetic field, which induces eddy currents. The eddy currents, like friction, dissipate the kinetic energy of the plate as thermal energy.

Eddy currents are usually undesirable in electrical equipment. The strategy used to minimize eddy currents is to break up the area or volume of the conductor by interspersing conductors and insulators. In disks or plates, slits provide these breaks. Air is not a good conductor, and so the slits restrict the eddy currents. In solenoids and other coils, eddy currents in the core are a problem. Therefore, cores are laminated in a direction that is perpendicular to potential eddy currents. Thin strips of iron or other ferromagnetic (and conducting) materials are alternated with thin insulators to restrict the paths of eddy currents.

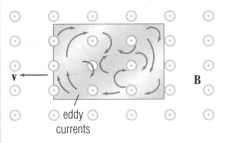

22-13 Eddy currents generated in a plate of metal

22-14 Transformers use laminations of thin strips of lacquered metal to minimize eddy current formation.

22A Section Review Questions

1. Name and describe the rule for determining the orientation of the magnetic field around a straight conductor carrying a continuous current.

2. What is the significance of the Biot-Savart law?

3. a. To what could you compare the shape of the magnetic field produced by a current-carrying loop or coil of wire?

   b. What is the name given to such a magnetic field?

4. a. What happens to the charges within a conductor placed in a magnetic field that changes strength with time?

   b. What law describes this observation?

5. What happens, according to Lenz's law, when a magnetic field changes around a conductor?

6. How can you change the magnetic field surrounding a conductor (i.e., produce a $\Delta\phi/\Delta t$)?

7. a. What occurs in large, thick conducting materials when a magnetic field passing through them changes?

   b. How can this effect be minimized?

•8. What diameter of a circular current loop would be required in order to produce an average magnetic flux of 2.50×10^{-5} Wb if the average field generated perpendicular to the loop is 6.30×10^{-2} T?

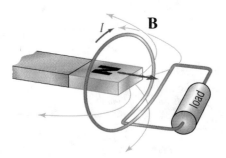

22-15 Generating a current with a magnet

Alternating current (AC) is electricity that is characterized by the periodic change of voltage polarity and current direction within the circuit.

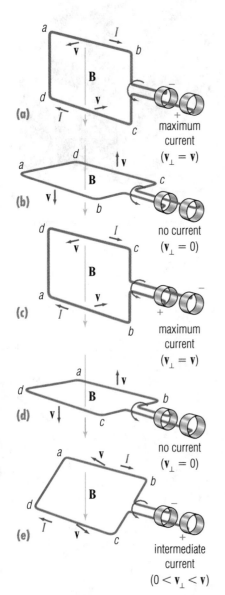

(a) maximum current ($\mathbf{v}_\perp = \mathbf{v}$)

(b)

(c) maximum current ($\mathbf{v}_\perp = \mathbf{v}$)

(d) no current ($\mathbf{v}_\perp = 0$)

(e) intermediate current ($0 < \mathbf{v}_\perp < \mathbf{v}$)

22-17 Steps in the production of alternating current

22.9 Generating Alternating Current

There are many ways to generate electricity using electromagnetic induction. A simple demonstration of a generator is a magnet and a loop of wire. The magnet is pushed in and out of the loop. This changing magnetic field induces a current in the loop. The current in this generator changes directions (alternates) during each motion. It tends to oppose the magnet's field when the magnet is approaching the loop and reinforce the magnet's field when the magnet recedes.

A basic representation of the most common type of generator works by rotating a wire loop (really one or more coils) in a magnetic field (Figure 22-16). A rotating coil also produces **alternating current (AC).** Figure 22-17 shows why the current alternates. When the wire loop rotates through the magnetic field in the direction of **v,** a clockwise current is induced in the loop. As the loop rotates, its orientation with respect to the magnetic field changes. In figure (a) the induced current flows through the loop from *a* to *b* to *c* to *d*. When the loop has rotated 180°, as figure (c) shows, the current (which is still clockwise) moves from *d* to *c* to *b* to *a*. Since each "edge" of the current loop is always connected to the same side of the load circuit through the slip rings, the current in the load circuit alternates.

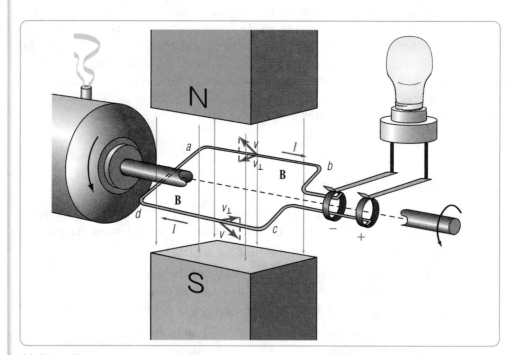

22-16 An AC generator

The rotating coil-core assembly of the generator is called the **armature.** The rotation of the armature coil in a generator is the result of an input of mechanical energy, which may come from one of several sources. Commercial generators often use steam turbines to turn the coils. Coal, natural gas, and nuclear reactions are used to produce steam to turn the turbines. Hydroelectric generators use falling water instead of steam, and wind turbines connected to generators use the wind to turn the coil. Highly efficient (but expensive to run) gas turbines are currently used to augment electric power requirements during peak load periods. Emergency electrical generators are often powered by diesel and gasoline engines.

The potential difference and current produced by any generator change in magnitude as well as in direction. The current is caused by the magnetic force on a moving charge. From Chapter 21, you learned that the magnitude of this force is

$$F_{mag} = |q|vB \sin \theta.$$

As the loop rotates, θ changes and F_{mag} has the form of a sine wave. Similarly, the current and potential difference produced by F_{mag} have sine-shaped graphs. Figure 22-18 is a graph of alternating current. The labeled points correspond to the steps in Figure 22-17.

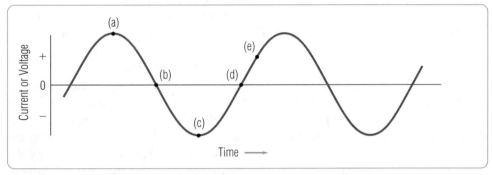

22-18 Generated alternating voltage and current versus time

Since the potential difference varies, it is helpful to know the *effective* or *rms* potential difference. The rms potential difference is the equivalent DC (nonvarying) potential difference that would produce the same power that the AC potential difference does. For a potential difference that varies as a sine function of time, the rms potential difference is

$$V_{rms} = V_{max}/\sqrt{2}, \tag{22.4}$$

where V_{rms} is the rms potential difference and V_{max} is the maximum AC potential difference. Similarly, the effective or rms current is

$$I_{rms} = I_{max}/\sqrt{2}. \tag{22.5}$$

A single, strictly sine waveform is not always the best for generating electricity, especially for large electrical motors and other loads. Three-phase generators, which use three pairs of electromagnets with varying magnetic fields, deliver a given amount of electrical power in a more compact way. Three wires connect the generator to the loads, each of which carries an alternating current that is 120° "out of phase" with the currents in the other wires. Peak positive values of current are produced whenever the rotor coil is aligned in the same direction in one of the three magnetic fields. In a three-phase generator, this alignment occurs three times in every rotation (every 120°). The induced emf for the entire generator is three out-of-phase sine waves that have been superimposed.

The circuit symbol for a source of DC potential difference consists of one or more voltaic cells. Because alternating current is usually produced by a rotating generator, AC voltage sources are given the symbol ⊸⊚⊸ in electrical schematics.

22.10 Direct Current Generation

The current generated in a rotating coil is always alternating current. In spite of this fact, it is possible to make a DC generator. For a DC motor, DC was turned into AC within the rotating coil by the use of a split-ring commutator. The same

It is somewhat difficult to compute energy, work, and power associated with any physical quantity that varies with time according to a sine or cosine function. A more useful quantity is the average (or mean) value of the varying quantity. Scientists and engineers therefore often determine the **root mean square (rms)** values of alternating current or voltage by squaring the maximum value (a square is always positive), dividing the squared quantity by 2 (finding the square's mean), then taking the square root to find the root mean square.

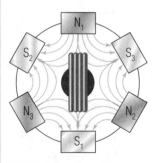

22-19 A three-phase AC generator

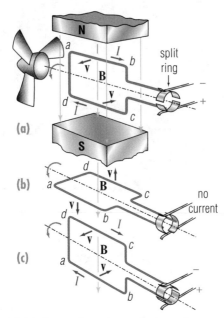

22-20 Steps in the production of pulsating direct current

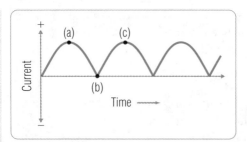

22-21 Pulsating direct current versus time

device can convert a generator's AC in the coil to DC at its output terminals. The current coming from the commutator will no longer change its direction, but it will still change its magnitude. The split ring is connected to the coil, and the output terminals are connected to the commutator by brushes. When the coil has turned far enough so that its current is reversed in the magnetic field, the brushes are in contact with the opposite sides of the split ring. The current therefore continues in the same direction. The current, which is plotted against time in Figure 22-21, is called **pulsating DC.**

A motor, a device that converts electrical energy to mechanical work, is fundamentally a generator run in reverse. Instead of the armature being rotated to generate electricity, it is supplied electricity that causes it to rotate. Since the armature of a motor rotates in a magnetic field, a potential difference, or emf, is induced across the armature. This induced emf opposes the potential difference supplied to the motor (Lenz's law); therefore, it is called **back emf (ε).** The current drawn by the back emf reduces the total current in the armature by opposing the supplied current. This "coincidence" provides several safety features. Back emf reduces the danger of the motor's overheating due to excessive current, since the wire in the armature winding presents little resistance to current flow. It also prevents the motor from increasing in angular speed unchecked until it flings itself apart.

22.11 Inductance

An armature is not unique in experiencing a back emf. Any coil or loop with a changing current produces a changing magnetic field. The changing field causes a change in magnetic flux through the coil; the result is an induced potential difference across the coil. This induced emf opposes the change in current in the coil, and so it is a back emf.

Consider a coil in a DC circuit. There is a switch in the circuit. At the instant the switch is closed, the current through the coil begins increasing. The increasing current causes an increasing magnetic field in the coil. The resulting back emf creates an induced current whose direction is opposite that of the main current in the circuit. This opposition causes the current in the circuit to take a measurable time—usually much less than one second—to reach its full value. Similarly, when the switch is opened, the decreasing magnetic field induces a current in the same direction as the main current. The induction of a back emf in a circuit element by a current flowing through that element is called **self-induction.**

The magnitude of a back emf depends on the geometry of the coil as well as the change in the current in the coil. A factor that takes into account the geometry is called the self-inductance, or simply the **inductance (L).**

$$L = \frac{n\phi}{I} \qquad (22.6)$$

The symbol I is the current, and n represents the number of turns in the coil. Just as for capacitance, the values of $n\phi$ and I are independent of each other—they depend only on the geometry of the magnetic coil. Such a coiled component in a circuit is called an **inductor.** The dimensions of inductance are units of magnetic flux (T·m² or Wb) divided by current (A). Inductance is measured in **henrys (H).**

There are special electrical circuits that can "condition" the pulsating DC output to establish a nearly constant DC voltage. The easiest way to generate DC is to use an AC generator and electronically convert the AC output to a nearly steady DC voltage. The alternator in your car is probably such a generator.

There are very few practical uses for pulsating direct current. Most of the time it must be smoothed into a nearly steady flow. There are also no practical uses for a pulsating Christian testimony. We must be sure our lives constantly demonstrate what we preach. "Take heed unto thyself, and unto the doctrine; continue in them: for in doing this thou shalt both save thyself, and them that hear thee" (I Tim. 4:16).

$$H \equiv \frac{T \cdot m^2}{A} = \frac{Wb}{A}$$

The magnetic flux in a given inductor is found by rewriting Equation 22.6:

$$LI = n\phi \qquad (22.7)$$

If the current through an inductor is varying (ΔI), then the magnetic flux will change with time as well ($n\Delta\phi$). Faraday's law indicates that an emf will be induced in the inductor:

$$\text{induced emf} = -\frac{n\Delta\phi}{\Delta t} \qquad (22.8)$$

If we substitute Equation 22.7 into Equation 22.8, we have

$$\text{induced emf} = -\frac{L\Delta I}{\Delta t}. \qquad (22.9)$$

The induced emf of an inductor can be found by either Equation 22.8 or 22.9.

22-22 Inductors are manufactured in a variety of shapes and sizes.

Problem-Solving Strategy 22.2

The formulas for induced emf have a negative sign because the emf opposes the applied potential difference in the circuit.

EXAMPLE 22-2

Determining Inductance

An inductor with ten coils produces a magnetic field with a flux of $\phi = 2.50$ Wb per coil when a current of $I = 1.25$ A flows through its coils. What is its inductance?

Solution:

This relationship of current and magnetic flux is determined by the structure of the inductor. Recall that ϕ is the flux per current loop, so the total magnetic flux (Φ) for the inductor is $n\phi$.

$$L = \frac{\Phi}{I} = \frac{n\phi}{I}$$

$$L = \frac{(10 \ \text{loops})(2.50 \ \text{Wb/loop})}{1.25 \ \text{A}}$$

$$L = 20.0 \ \text{Wb/A}$$

$$L = 20.0 \ \text{H}$$

When a varying magnetic field of one current loop links (passes through) another current loop, it induces a potential difference in the other loop. This process is called **mutual induction.** The resulting potential difference depends on the size, shape, and relative orientation of the current loops as well as the current change in the active inducing loop or coil. The geometric factors are accounted for in the **mutual inductance (M).**

$$M = \frac{n_A \phi_{B \to A}}{I_B} \quad \text{or} \quad M = \frac{n_B \phi_{A \to B}}{I_A},$$

where n_A is the number of turns in inductor A, n_B is the number of turns in inductor B, $\phi_{B \to A}$ is the flux that inductor B causes in inductor A, $\phi_{A \to B}$ is the flux that inductor A causes in inductor B, I_A is the current in inductor A, and I_B is the current in inductor B. Notice that the mutual inductance is the same no matter which inductor is the inducing component. The potential difference induced in inductor B by inductor A is given by either of the following equations:

$$\text{induced emf}_B = -\frac{n_B \Delta \phi_{A \to B}}{\Delta t}, \text{ or}$$

$$\text{induced emf}_B = -\frac{M \Delta I_A}{\Delta t}$$

The mutual inductance, like the self-inductance, is measured in henrys.

22.12 Inductors

An inductor is a circuit element whose purpose is to resist changes in current. It is represented in a circuit diagram by the symbol ⌒⌒⌒⌒. Inductors are usually coils of wire wound with either no core or a ferromagnetic core. Those with ferromagnetic cores have higher inductance than air-core inductors of the same size.

Recall that, in order for an inductor to function, the current must change with time through the inductor. In a DC circuit, current does not normally change with time except when the circuit is turned on or off; therefore, an inductor appears as a straight length of wire to direct current. Inductors are therefore important in circuits where current is expected to change with time, such as in alternating current circuits.

Inductors may be connected in series or in parallel. However, since two inductors in parallel have less inductance than either inductor alone, most circuits do not use inductors in parallel. Two inductors in series produce a total inductance equal to the sum of the individual inductances if the inductors are not close enough to cause mutual induction.

$$L_{\text{total}} = L_1 + L_2$$

If the inductors are close enough for mutual induction or wound on the same core, they may enhance or detract from each other's inductance. If the inductors are placed so that their windings are in the same direction, the total inductance is more than the sum of the individual inductances.

$$L_{\text{total}} = L_1 + L_2 + 2M$$

If they are placed so that their windings are in opposite directions, the total inductance is less than the sum of the individual inductances.

$$L_{\text{total}} = L_1 + L_2 - 2M$$

Inductors store energy in magnetic fields, just as capacitors store energy in electric fields. A capacitor and an inductor in series can theoretically keep a current alternating between themselves indefinitely if a current is begun in the circuit. In real circuits, the wires in the circuit and in the inductor have resistance that will dissipate the electrical energy as heat.

22.13 Transformers

Transformers are important induction devices. A **transformer** consists of two coils of insulated wire wound on a common ferromagnetic core. One coil, the *primary winding,* is attached to an AC source. The other coil, the *secondary winding,* is attached to an electrical load. The primary induces a potential difference across the secondary. The potential difference across each turn in the winding is the same.

The total inductance of a series of inductors is the simple sum of the inductors.

$$L_{\text{total}} = L_1 + L_2 + \ldots + L_n$$

Parallel inductors follow the same rules as parallel resistors. For two parallel-connected inductors,

$$L_{\text{total}} = \frac{L_1 L_2}{L_1 + L_2}.$$

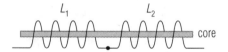

L_1 L_2

core

22-23 Inductors in series that reinforce each other

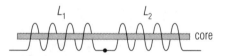

L_1 L_2

core

22-24 Inductors in series that oppose each other

How do you maximize inductance? Put two or three inductors (with ferromagnetic cores) in series. Put two or three praying Christians together, and what happens? "For where two or three are gathered together in my name, there am I in the midst of them" (Matt. 18:20).

If the voltage across each turn is V_{turn} and the primary has n_p turns, the potential difference across the primary is

$$V_p = n_p V_{turn}.$$

The secondary has n_s turns. Therefore,

$$V_s = n_s V_{turn}.$$

Dividing the secondary voltage by the primary voltage, you obtain the following:

$$\frac{V_s}{V_p} = \frac{n_s \cancel{V_{turn}}}{n_p \cancel{V_{turn}}}$$

$$V_s = V_p \left(\frac{n_s}{n_p} \right) \tag{22.10}$$

If the secondary has more turns than the primary, the potential difference across the secondary is larger than across the primary. The transformer therefore is a **step-up transformer.** If the secondary has fewer turns than the primary, as in Figure 22-25, the secondary potential difference is less than the primary potential difference. This transformer is a **step-down transformer.**

The power in the secondary *must* be the same as that in the primary in order to conserve energy. Recall that one of the electrical power equations is

$$P = VI.$$

Therefore, using the effective values of current and voltage,

$$V_s I_s = V_p I_p$$

$$\frac{V_s}{V_p} = \frac{I_p}{I_s}. \tag{22.11}$$

The current in the secondary winding is *smaller* than the primary current in a step-up transformer but *greater* than the primary current in a step-down transformer. These facts allow efficient transfer and safe use of electricity. High-current transmission of power would cause too much heating to be practical. Therefore, electricity suppliers use a step-up transformer to raise the voltage and lower the current for transmission. High voltage would be unsafe for home use, so electricity suppliers use substations with transformers to step down the voltage. The transformers make electricity both safe and economical to use. Transformers within electronic devices are used to step down the supplied 120 V to an appropriate level for the operation of the device.

Transformer cores are ferromagnetic, and most ferromagnetic materials are conductors. The core is therefore subject to induction of eddy currents. As stated before, unrestricted eddy currents create magnetic fields that interfere with the operation of the transformer. Therefore, transformer cores are laminated to restrict eddy currents without destroying the core's ferromagnetic properties (see Figure 22-26).

22B Section Review Questions

1. What kind of electricity is produced by a current loop, rotating in a magnetic field, that is connected by slip rings to an external circuit?

2. What is the rms value of an alternating quantity?

3. What prevents the coiled wire in the armature of a rotating DC motor from acting as a short circuit?

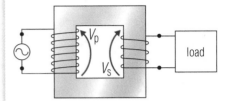

22-25 Schematic of a simple transformer and attached circuits

22-26 A transformer used for supplying current at low voltage to a household doorbell

Recall that power used in an electrical component is

$$P = VI.$$

High-voltage overland power transmission lines operate at voltages up to half a million volts AC. Power delivered to large industrial complexes is stepped down to around 12 800 volts. Transformers that supply residences step the voltage down to 220 volts, divided between two conductors and a ground wire. Two lines at 110 volts each are connected to the house.

4. What factors determine the induced emf generated across a circuit component containing coils of a conductor?

5. What does an inductor in a DC circuit act like?

⊛6. The inductance of an inductor is 15.0 H. If it normally has 0.90 A flowing through it, what is the total magnetic flux in the inductor?

⊛7. Calculate the total inductance for two inductors of 5.0 H each connected

　a. in series.

　b. in parallel.

⊛8. A transformer has 500 primary winding turns and 750 secondary winding turns.

　a. What function does this transformer perform?

　b. If the rms primary voltage is 120 VAC, what is the secondary voltage?

　c. If the load connected to the secondary side draws 1.5 A (rms), how many amps flow through the primary windings?

22C AC CIRCUIT CHARACTERISTICS

22.14 Resistor-Capacitor (RC) Circuits

Not all circuit elements react the same to AC and DC. Purely resistive circuits, which have already been discussed, do not distinguish between AC and DC. Other circuits do. A resistor-capacitor (RC) circuit is said to block DC but pass AC. The circuit in Figure 22-27 is a DC circuit. When the switch is positioned to *A*, current begins to flow from the DC voltage source through the resistor. Positive charge builds up on one side of the capacitor, and negative charge is attracted to the other side. When the capacitor is fully charged based on its capacitance and the potential difference across the capacitor (*C* = *Q*/*V*), current no longer flows anywhere in the circuit. The direct current flow is blocked. When the switch is positioned to *B*, the charge on the capacitor flows through the resistor, discharging the capacitor. The resistor transforms the electrical energy to thermal energy.

In an AC-RC circuit, the direction of the current changes regularly. The capacitor is charged in one direction until the AC source potential difference changes polarity. The positive side of the capacitor now faces the negative side of the AC source. The positive charges flow toward the lower potential, and the negative charges flow toward the higher potential. This process is repeated every time the current reverses. Although no current passes through the capacitor (ideally), there is a current throughout the circuit. Since a fully charged capacitor blocks the current, the ideal AC-RC circuit is more nearly achieved when a large capacitor (which is harder to charge completely) is used. Frequent changes in current direction also allow the current to flow more continuously by giving the capacitor less time to charge. At high current frequencies, a capacitor appears as a short (that is, a segment of wire).

22.15 Resistor-Inductor (RL) Circuits

A resistor-inductor (RL) circuit impedes AC more than DC. At the instant the switch in Figure 22-29 is positioned to *A*, the current through the inductor is increasing. The induced back emf opposes the circuit current so that it takes time to reach its full value. When the current has reached its full value, it is no longer

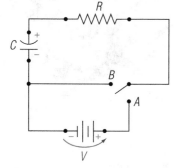

22-27 A DC-RC circuit

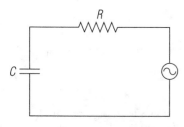

22-28 An AC-RC circuit

changing. Therefore, there is no more back emf, and circuit current is determined by the voltage drop across the resistor. At the instant the switch is positioned to *B*, the current in the inductor begins decreasing. The induced back emf sets up a current in the same direction as the original current. The induced current from the inductor produces thermal energy in the resistor as the electrical energy is dissipated.

In an AC-RL circuit, the current is continually changing directions and magnitude. The induced back emf opposes the changes in current. This constant opposition lowers the peak current below that which would exist without the inductor. Higher inductance produces a higher back emf. Faster current changes also cause a higher back emf and, therefore, a lower total current in the circuit. At high current frequencies, an inductor appears as an open point in the circuit.

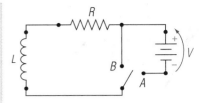

22-29 A DC-RL circuit

22.16 Inductor-Capacitor (LC) Circuits

An inductor-capacitor (LC) circuit is usually part of a much more complex circuit. If the capacitor is charged from another part of the circuit, its charges move around the LC portion of the circuit. This motion of charge produces an increasing current through the inductor. The induced back emf opposes the current and slows the capacitor's discharging. Then, when the capacitor's potential difference drops below the corresponding potential difference across the inductor, the inductor sets up a current to oppose the decrease in current. This induced current charges the capacitor with the opposite polarity to that which it was originally. The capacitor then begins to discharge again, and the process repeats itself. In real LC circuits, the electrical energy in the current is converted to thermal energy by the resistance of the wires in the circuit. The oscillating current rapidly dies out unless the voltage across the capacitor is periodically restored by the outside circuit.

Electrical frequency is measured in Hz. The United States power grid that supplies homes, schools, and businesses operates at a frequency of 60 Hz, which is extremely stable.

22.17 Resistor-Inductor-Capacitor (RLC) Circuits

A resistor-inductor-capacitor (RLC) circuit contains a capacitor, which blocks DC. It also contains an inductor, which carries small amounts of AC. The characteristics of a circuit containing resistors, inductors, and capacitors depend on the frequency (*f*) of the current changes in the circuit.

The frequency of a current, like the frequency of an oscillating spring, is the number of cycles it undergoes per second. One cycle for the current is the change from a maximum in one direction through a maximum in the other direction and back to another maximum in the first direction. Current frequency is measured in cycles per second, or hertz (Hz).

The frequency of a current flowing through a resistor is not affected by the presence of the resistor. Ohm's law is followed exactly. The voltage drop across the resistor produces a corresponding current that is exactly in phase. This result does not occur with the other two kinds of components.

As the alternating voltage changes across an inductor, its inductance tends to retard the buildup of current in response to the increasing voltage and delay the reduction of current as voltage decreases. The obstacle that an inductor's back emf presents to AC is called **inductive reactance (*X_L*).** The magnitude of inductive reactance is found from the formula

$$X_L = 2\pi f L. \tag{22.12}$$

Inductive reactance is higher for higher inductance and for higher current frequencies. For very high frequencies, an inductor appears almost like an open circuit. Since voltage must change before current can change in an inductor, we say that voltage *leads* current in an inductor.

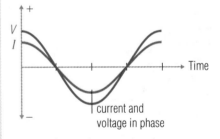

22-30 Voltage and current are in phase across a resistor.

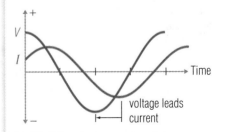

22-31 Voltage leads current in an inductor.

Electromagnetism **491**

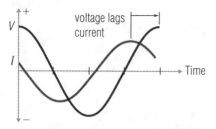

22-32 Voltage lags current in a capacitor.

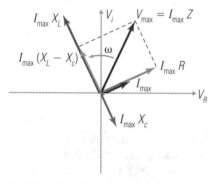

22-33 An RLC voltage phasor diagram is plotted on a coordinate system consisting of real and imaginary voltage axes. The whole plot rotates counterclockwise with time. The inductive reactance leads current by 90°, and the capacitive reactance lags current by 90°. The maximum circuit voltage is the vector sum of the total reactance and the circuit resistance, or the product of peak current and total impedance.

The peak voltage drop across each kind of circuit element follows the form

$$V = IZ,$$

where for a

resistor $Z_R = R,$

inductor $Z_L = iX_L = i2\pi fL,$ and

capacitor $Z_C = -iX_C = \dfrac{1}{i2\pi fC}.$

Do not concern yourself with the complex factors. This list is intended to show that impedance is related to AC resistance and reactance.

Capacitors also affect AC. As an alternating voltage is applied across a capacitor, current flows into and out of a capacitor in an attempt to neutralize the applied voltage. The potential difference across the capacitor plates doesn't equal circuit voltage until the current stops flowing. Therefore, capacitor voltage is said to *lag* current. A capacitor's tendency to impede voltage changes is called **capacitive reactance (X_C)**. The magnitude of capacitive reactance is calculated with the formula

$$X_C = \frac{1}{2\pi fC}. \tag{22.13}$$

Note that capacitive reactance decreases with greater capacitance and AC frequency. For very high frequencies, X_C approaches zero, and a capacitor appears to be a short in a circuit.

The effect of both resistance and reactance in a circuit is called **impedance (Z)**. The reactance and resistance add as though they were vectors at right angles to each other. Their "vectors" are called *phasors* and are shown in Figure 22-33. The impedance is therefore

$$Z = \sqrt{R^2 + X^2},$$

where X is the vector sum of the X_C and X_L phasors.

AC circuit impedance is analogous to the simple resistance in a DC circuit. Impedance opposes the flow of current in an AC circuit. For an AC circuit, Ohm's law becomes

$$V = IZ.$$

The magnitude of the total impedance of a series RLC circuit is

$$Z_{\text{total}} = \sqrt{R^2 + (X_L - X_C)^2}. \tag{22.14}$$

This equation shows that impedance is smallest when the resistance and the differences of the reactances are smallest. Although there is always some resistance in a real circuit, the total reactance can be zero. Therefore, the impedance of an RLC circuit is at a minimum when the difference of the reactances are zero.

In an AC-RLC circuit, the current is largest when the impedance is the smallest. The maximum current in such a circuit occurs when $X_L - X_C = 0$. This state in which there is zero reactance is called **resonance.** Resonance happens at one AC frequency for a circuit. The *resonance frequency (f_{res})* is found by setting the reactance term in Equation 22.14 equal to zero:

$$(X_L - X_C)^2 = 0 \ \Omega$$

Taking the square root of both sides gives

$$X_L - X_C = 0 \ \Omega, \text{ or}$$
$$X_L = X_C.$$

Since X_L is $2\pi f_{\text{res}}L$ and X_C is $1/(2\pi f_{\text{res}}C)$,

$$2\pi f_{\text{res}}L = \frac{1}{2\pi f_{\text{res}}C}. \tag{22.15}$$

Multiplying both sides by $f_{\text{res}}/(2\pi L)$ to isolate f_{res} on the left side gives

$$f_{\text{res}}^2 = \frac{1}{(2\pi)^2 LC}.$$

Now take the square root of both sides:

$$f_{res} = \frac{1}{2\pi\sqrt{LC}}$$

(22.16)

A circuit called a **tuning circuit** can be constructed so that the resistance is small. The resonance effect then makes the resonance current large compared to the current at other frequencies. If the circuit is connected to a voltage source at or near the resonance frequency, the current is able to accomplish whatever the circuit is designed to do. If the frequency of the voltage source is much different from the resonance frequency, the current is too small to be noticed. An obvious application of tuning circuits is in radios and televisions.

22.18 Electromagnetic Waves

Faraday, like many others who have proposed a theory, did not fully develop his field model of electricity. Much of our current understanding of electric fields comes from Faraday's younger contemporary, James Clerk Maxwell, a theorist. One of Maxwell's most successful predictions from the field theory was that electromagnetic waves exist. He reasoned that a changing electric field, like that of a moving charge, induces a magnetic field. A changing magnetic field induces a potential difference, which defines an electric field. If the electric field is changing nonuniformly—for instance, if an electric charge is accelerating—it induces a changing magnetic field. The changing magnetic field induces an electric field in the same direction as the original electric field, which in turn induces a magnetic field, and so on. If there is no circuit with resistance to convert the energy to thermal energy, these fields will continue to reinforce one another, forming an **electromagnetic (EM) wave.**

The mutually reinforcing fields are perpendicular to each other, and both are perpendicular to the direction in which the wave moves. The latter fact gives the wave the name *transverse electromagnetic wave.* Figure 22-34 shows a sinusoidal EM wave. This wave can be begun by oscillating charges in a rod called an **antenna.** The electric field oscillates in the same way as the charges. The charges must be accelerating, since they change direction. Therefore, the nonuniformly changing electric field induces a changing magnetic field. The resulting EM wave may look like Figure 22-35. The charges on the antenna are said to *radiate* because they produce an EM wave. In fact, all accelerating charges produce **electromagnetic radiation.** The waves have the same frequency as the oscillating charges.

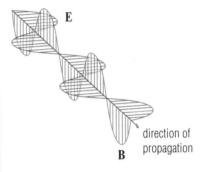

22-34 A transverse electromagnetic wave

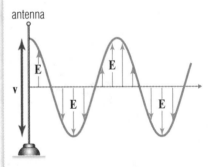

22-35 An antenna emits an EM wave.

22.19 Speed of Light

All electromagnetic waves propagate with the same speed, *c,* in a vacuum.

$$c \doteq 3.00 \times 10^8 \text{ m/s}$$

Light is an electromagnetic wave, and it travels at the same speed as the other electromagnetic waves. The speed *c* is therefore called the **speed of light.** Light travels fastest in a vacuum, so in any other medium its speed is less than *c.* The speed of electromagnetic waves in a medium is related to the electric permittivity (ε) and the magnetic permeability (μ) of the medium. A medium's permittivity and permeability interact to determine the speed with which any electromagnetic wave travels through the medium. This speed is

$$v_{EM} = \frac{1}{\sqrt{\mu\varepsilon}}.$$

(22.17)

The **speed of light (c)** is approximately 3.00×10^8 m/s in a perfect vacuum.

The speed of light may be the fastest speed our finite minds can comprehend, yet God's answers to the right kind of prayer are even faster. "And it shall come to pass, that before they call, I will answer; while they are yet speaking, I will hear" (Isa. 65:24).

EXAMPLE 22-3

Slowing Light: The Speed of Light in Water

Light travels through water, which has $\varepsilon = 1.77\varepsilon_0$ and $\mu = 1.00\mu_0$. (The permittivity and permeability values for a vacuum are ε_0 and μ_0, respectively.) What is its speed through water?

Solution:

$$v_{light} = \frac{1}{\sqrt{\mu\varepsilon}}$$

$$v_{light} = \frac{1}{\sqrt{(1.00\mu_0)(1.77\varepsilon_0)}}$$

Rearranging,

$$v_{light} = \frac{1}{\sqrt{\mu_0\varepsilon_0} \cdot \sqrt{1.77}}$$

But since $1/\sqrt{\mu_0\varepsilon_0}$ is the speed of light in a vacuum,

$$v_{light} = c \cdot \frac{1}{\sqrt{1.77}}$$

$$v_{light} = \frac{3.00 \times 10^8 \text{ m/s}}{\sqrt{1.77}}$$

$$v_{light} \doteq 2.254 \times 10^8 \text{ m/s}$$

$$v_{light} \approx 2.25 \times 10^8 \text{ m/s.}$$

The energy contained in an EM wave is reflected in the *Poynting vector*. The Poynting vector is a measure of the power passing through a unit area perpendicular to the wave's direction of motion. The Poynting vector has units of watts per square meter.

The electric field vector **E** is in the direction of motion of the oscillating particles in an antenna. In Figure 22-35, **E** is always vertical. The EM wave is said to be **polarized** in the direction of **E.** Since **E** is always on the same straight line lying in a plane along the path of propagation, it is *linearly* polarized.

22C Section Review Questions

1. If you wanted to increase the current flowing through the resistor of an AC-RC circuit segment, would you increase or decrease the size of the capacitor?

2. If you wanted to decrease the peak current in an AC-RL circuit segment, would you increase or decrease the size of the inductor?

3. Discuss the relationship of the factors that contribute to the inductive reactance of a circuit component.

4. Discuss the relationship of the factors that contribute to the capacitive reactance of a circuit component.

5. Discuss the characteristics of an electromagnetic wave. What kind of medium is required for the propagation of EM waves?

●6. What is the total impedance of a series-connected, 60 Hz RLC circuit where $R = 100.\ \Omega$, $L = 15.0$ H, and $C = 500.$ pF?

●7. In a series-connected 400 Hz RLC tuning circuit, what must be the size of the capacitor that will be in resonance with a 5.0 H inductor? (*Hint:* See Equation 22.15.)

●8. What is the resonance frequency of a tuning circuit that has a 8.80 H inductor and a 300. μF capacitor?

Chapter Review

In Terms of Physics

right-hand rule for conductors	22.1	inductance (L)	22.11
Biot-Savart law	22.1	inductor	22.11
permeability of free space (μ₀)	22.1	henry (H)	22.11
electromagnet	22.2	mutual induction	22.11
solenoid	22.2	mutual inductance (M)	22.11
electromagnetic induction	22.5	transformer	22.13
magnetic flux (φ)	22.5	step-up transformer	22.13
total magnetic flux (Φ)	22.5	step-down transformer	22.13
weber (Wb)	22.5	inductive reactance (X_L)	22.17
Faraday's law	22.6	capacitive reactance (X_C)	22.17
Lenz's law	22.7	impedance (Z)	22.17
eddy current	22.8	resonance	22.17
alternating current (AC)	22.9	tuning circuit	22.17
armature	22.9	electromagnetic (EM) wave	22.18
root mean square (rms)	22.9	antenna	22.18
pulsating DC	22.10	electromagnetic radiation	22.18
back emf (ε)	22.10	speed of light (c)	22.19
self-induction	22.11	polarized	22.19

Problem-Solving Strategies

22.1 (page 481) Recall that the term *emf* is not a true Newtonian force. Emf is used for describing potential differences that are induced in an electrical circuit by nonelectrical sources, such as magnetic fields.

22.2 (page 487) The formulas for induced emf have a negative sign because the emf opposes the applied potential difference in the circuit.

Review Questions

1. The figure shows a wire surrounded by its magnetic field. Is the current in the wire coming out of the page or going into the page?

2. How can you reverse the magnetic field of a current-carrying loop?

3. A moving magnetic field is needed to generate a current. Earth, which has a magnetic field, is moving rapidly through space. Why can we not generate electricity simply by constructing large coils on the ground, since we are so close to a moving magnetic field?

4. Some laboratory balances are equipped with magnets that establish a magnetic field through a plate of paramagnetic metal attached to the end of the balance arm. What is the magnet's function?

5. How can you minimize eddy currents within a conductor?

6. What is alternating current?

7. What is the principal difference between an AC generator and a DC generator?

⊚⊚8. What is the difference between an AC motor and a DC motor?

9. What is the rms or effective potential difference of an AC electrical supply?

10. Describe self-induction.

11. What is the difference between self-induction and mutual induction? Can they ever occur simultaneously?

12. Why does it make a difference whether two inductors in a circuit segment are wound in the same direction on a common core or not? (A diagram may help.)

13. By what principle do transformers work?

14. Why do power companies transmit electricity at high voltages?

15. Suppose you wish to increase the reactance of an RC circuit. Would you increase or decrease its capacitance?

16. If you want to increase the reactance of an RL circuit, should you increase or decrease its inductance?

17. Identify each circuit described below as RC, RL, LC, or RLC.
    a. It permits a large alternating current in a narrow band of current frequencies near its resonance frequency; it allows only a small current at other frequencies.
    b. It blocks DC but passes AC.
    c. It will, in theory, sustain an oscillating current for an unlimited time.
    d. It presents a greater obstacle to AC than to DC.

18. An RLC circuit has a resonance frequency (f_{res}) under certain conditions. Will the resonance frequency increase, decrease, or remain constant in the following situations?
    a. Another resistor is added in series.
    b. Both L and C are increased.
    c. Both L and C are decreased.
    d. L doubles while C halves.
    e. L is increased, but C stays the same.
    f. C decreases while L stays the same.

19. What is an electromagnetic wave?

True or False (20–33)

20. If a straight length of wire that is not connected to an external circuit moves through a magnetic field perpendicular to the magnetic field lines, a current will *not* flow in the conductor.

21. The magnitude of a magnetic flux is dependent on the size of the area perpendicular to the field lines.

22. Faraday's law concerns the potential produced by a voltaic cell.

23. According to Lenz's law, the current generated by a collapsing magnetic field around a conductor produces a field that opposes the original magnetic field.

24. Alternating current is produced by a loop rotating in a magnetic field when the segments cutting the field lines are connected to the same ends of the external circuit by slip rings.

25. The effective or rms voltage of an AC electricity supply is the average of the minimum and maximum voltages.

26. A split-ring DC generator produces constant direct current and voltage.

27. Back emf is the potential induced in the winding coils by a changing magnetic field that opposes the potential difference (and its current) that caused the magnetic field in the first place.

28. Inductance is dependent on the dimensions of the coiled conductor and the material in the core on which the wire is coiled.

29. A transformer converts AC to DC.

30. A circuit segment that acts more and more like a straight piece of wire with increasing frequency is an RC circuit.

31. To increase the alternating current in a circuit segment at a given supply frequency, you could add a large inductor to the segment.

32. The resonance frequency of an RLC circuit segment is reduced by reducing the capacitance of the circuit.

33. The speed of light (c) is constant.

⊚34. Using the equation of force for each of two parallel wires and the definition of the ampere, find the constant k.

⊚35. A 1200 W toaster is connected to 120 V. The wires in the 1.00 m toaster cord are 5.00×10^{-3} m apart. Since one wire carries current to the toaster and the other carries current away from the toaster, the currents are always in opposite directions.
    a. What is the current in the toaster circuit?
    b. What is the force between the wires in the cord?
    c. Will the wires attract or repel each other?

⊚36. In the figure, a 0.10 m square conduction loop is entirely within a uniform magnetic field with $B = 1.0 \times 10^{-5}$ T. The loop is traveling at a speed of 1.0 m/s.

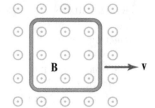

    a. What is ɸ at the instant shown in the drawing?
    b. One second later, the loop is still entirely within the magnetic field. What is ɸ now?
    c. Calculate Δɸ and V_i.
    d. Explain your answer for the magnitude and sign of V_i by showing what happens to the free electrons.

⊚37. In (a) a current loop with width $w = 0.100$ m and length $l = 0.300$ m has just begun to enter a uniform magnetic field $B = 1.5 \times 10^{-3}$ T. The rectangular loop has a length $l' = 0.100$ m inside the field. In (b), 1.0 s later, the entire loop has just entered the field.
    a. Find $\phi_{(a)}$.
    b. Find $\phi_{(b)}$.
    c. Find Δɸ and the potential difference induced in the loop in the second between (a) and (b).

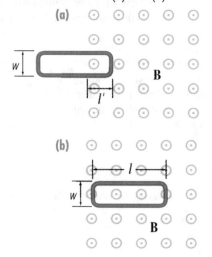

38. In residential circuits the rms potential difference is about 120 V. In the circuit pictured,

a. find the rms current through each branch. (Assume all loads are resistances.)

b. calculate the maximum voltage of the AC supplied to the house.

c. find the total rms current when switches 1, 2, and 3 are closed.

d. find the total rms current when all the switches are closed.

e. Will a 5 A fuse guard this circuit without blowing?

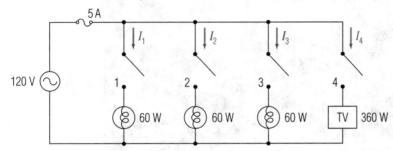

39. What is the inductance of a coil with fifteen turns if a current of 3.0 mA produces a flux of 2.0×10^{-10} Wb per coil?

40. A coil with an inductance of 6.00 μH carries a current of 10.0 mA. Over a period of 3.00 ms, the current increases to 12.0 mA. What potential difference is induced in that time?

41. Two coils of insulated wire are wound on opposite sides of a hollow square iron core. Coil 1 has six turns and an inductance $L_1 = 5.00$ μH. Coil 2 has three turns. The current is 0.500 A at $t_0 = 0.00$ s, and 0.00 A at $t = 2.00 \times 10^{-3}$ s. The mutual inductance $M = 1.00$ μH.

a. What potential difference is induced in coil 2 during that time?

b. What is the current induced in coil 2 at each time?

c. What is the potential difference in coil 1?

d. What is the back emf in coil 1?

42. An AC-RLC circuit has an rms potential difference $V_{rms} = 120$ V at a frequency of $f = 60.0$ Hz. $R = 10.0\ \Omega$; $L = 14.0/\pi$ mH; $C = 1.0/\pi$ mF.

a. Find the capacitive reactance, X_C.

b. Find the inductive reactance, X_L.

c. Assuming that the capacitor and inductor have no resistance, find the impedance Z for the circuit.

d. Find the rms current in the circuit.

e. Find the circuit's resonance frequency (f_{res}).

43. What is the ratio of the speed of light in Lucite (clear plastic) to the speed of light in air? Lucite's permeability is μ_0, and its permittivity is $2.5\ \varepsilon_0$. Air's permeability is μ_0, and its permittivity is ε_0.

44. Prove that the units of impedance and capacitance yield frequency units in Equation 22.16.

Diesel Locomotives

When you think of an electric generator, you probably imagine a huge rotating machine in a power plant. Such generators stay in the same place for most of their existence. However, there are some common portable generators. In fact, most vehicles carry their own generators. The alternator that charges your car battery is an AC generator. In rail mass transportation, locomotives carry diesel engines that generate electricity to turn their wheels.

Why do the diesel engines not turn the wheels directly? You may remember from the discussion of engines in Chapter 16 that most engines produce back-and-forth (reciprocating) linear motion. A *transmission* converts the linear motion to circular motion to turn the wheels. Steam engines used a rod connected directly from the piston to the wheels, and the reciprocating motion of the rod turned the wheels. This transmission led to uneven acceleration that was difficult to control. Automobiles use gears to change the engine's linear motion to circular motion, and some light locomotives also use gear transmissions.

But engines pulling heavy trains that consist of tens or even hundreds of cars cannot use gear transmissions because the range of gearing would have to be too great, and the mechanism would be too bulky to fit within a typical locomotive. Instead, most diesel engines use electric transmissions. That is, the diesel engine turns an electric generator. The electricity goes to motors mounted to the locomotive's axles that turn the wheels. This arrangement has many advantages. The electromagnets that create the fields through which the motor coils turn can be continuously varied from zero to the large values required to generate the torque necessary to turn the wheels of a heavily loaded train.

Diesel locomotives are capable of accelerating smoothly and providing the full range of power necessary for long-haul rail service. Diesel engines are inherently more economical to operate, do not need to make frequent stops for fuel or water, contaminate the air with fewer pollutants, and need less service. For these reasons and others, diesel locomotives have replaced rail steam engines in the United States and throughout most of the world.

James Clerk Maxwell (1831–79)

Many advances in science and technology stem from our understanding of electromagnetic waves. Radios, televisions, cellular phones, pagers, and the GPS, for example, send or receive messages through space using electromagnetic waves. Yet little more than a century ago few people even knew these waves existed. The first man to predict their presence was a Scottish physicist named James Clerk Maxwell.

Maxwell was reared by his father on a farm. His classmates at Edinburgh Academy made fun of his unsophisticated clothes and manners, nicknaming him "Dafty." But they were soon surprised on two counts. First, he was a swimmer, gymnast, rower, and horseman. But more than that, his brains matched his physique. He had a knack for mathematics and science: when he was only fifteen, he presented a paper on drawing ellipses to the Royal Society of Edinburgh. After studying mathematics at Cambridge, he became a professor at Marischal College and later at King's College in London. There Maxwell met Michael Faraday, whose lines of force fascinated him.

Maxwell's knowledge of mathematics enabled him to make major contributions in many fields, including the kinetic theory of heat, the velocity of gas molecules, human perception of color, and even the composition of Saturn's rings. In 1871 he was invited to become the first professor of experimental physics at Cambridge. There he used his genius to design and staff the famous Cavendish laboratory.

Maxwell's greatest accomplishment was to convert the vague field theory of Faraday (who was not a mathematician) into four valuable equations. For the first time in history, a physicist showed the basic similarity between light, electricity, and magnetism. Maxwell realized that light is simply a small band of electromagnetic radiation to which the human eye is sensitive.

His equations showed that the velocity of an electric signal is related to electric permittivity and magnetic permeability. When experimenters showed that electric signals travel at nearly the speed of light, he proved that the two velocities are theoretically equal because light results from variations within the electromagnetic field. He further predicted the existence of wave frequencies outside the range of visible light. His prediction was confirmed a few years later when Heinrich Hertz discovered radio waves.

Maxwell shared with Faraday not only a theory of field lines but also a faith in Jesus Christ as Savior. Like Faraday, Maxwell saw science as an attempt to find out more about God's creation. He strongly opposed the theory of evolution, which was becoming popular among scientists of his time. Like Faraday, Maxwell was both a good Christian and a good scientist. Maxwell was the best theoretical physicist in his century and is ranked with Newton and Einstein as one of the three greatest physicists of all time.

UNIT 5

Geometric Optics and Light

Although light is one of the most ubiquitous phenomena in nature and provides the majority of our sensory information about our surroundings, it was one of the least-understood phenomena until the late nineteenth and early twentieth centuries. Newton thought that light consisted of particles, but later experimentation confirmed that it acted as a wave. Light is now known to be just a small portion of the entire electromagnetic energy spectrum, a portion that our Creator has designated as the stimulus for sight.

In Unit 5, you will first investigate the principles of geometric reflection and refraction. You will also study the interesting phenomena resulting from the wave nature of light, such as diffraction and iridescence. The properties of color and color mixing will be reviewed. Finally, the characteristics of various optical instruments, such as telescopes and microscopes, will be discussed.

23A	Light and the Electromagnetic Spectrum	503
23B	Sources and Propagation of Light	507
23C	Reflection and Mirrors	513
Biography: Christiaan Huygens		511

New Symbols and Abbreviations

light-emitting diode (LED)	23.12	focal length (f)	23.21
angle of incidence (θ_i)	23.18	object distance (d_0)	23.21
angle of reflection (θ_r)	23.18	image distance (d_I)	23.21
principal focus (F)	23.21		

The Hobby-Eberly telescope at the McDonald Observatory in the Davis Mountains of Texas. Its mirror consists of 91 hexagonal, spherically curved, 1 m segments that yield a mirror diameter of 11 m and an effective aperture of 9.2 m. Each segment is supported by three computer-controlled actuators that adaptively adjust the mirror's surface shape to compensate for atmospheric motion. The telescope is primarily used for spectroscopy.

Light and Reflection

The Lord is my light and my salvation; whom shall I fear? (Ps. 27:1) Then spake Jesus again unto them, saying, I am the light of the world: he that followeth me shall not walk in darkness, but shall have the light of life. (John 8:12) Thy word is a lamp unto my feet, and a light unto my path. (Ps. 119:105) Ye are all the children of light, and the children of the day: we are not of the night, nor of darkness. (1 Thess. 5:5) That ye may be blameless and harmless, the sons of God, without rebuke, in the midst of a crooked and perverse nation, among whom ye shine as lights in the world. (Phil. 2:15)

It is clear that darkness is the absence of light—not the other way around. You cannot shine darkness into a lighted room with a "flashdark." Of the various forms of energy, light is most frequently mentioned in Scripture as an attribute of God's truth and presence. Scripture refers to the Lord Himself as light. Christians are the bearers of God's light because they have the Holy Spirit living within them. The world would be a spiritually dark place without Bible-believing Christians who radiate the love of Christ. Does your light shine into the world? Are you a beacon drawing people who are in darkness to Christ?

23A LIGHT AND THE ELECTROMAGNETIC SPECTRUM

23.1 Introduction

In this unit you will study a familiar form of energy—visible light. Perhaps you regard light more as an illuminator than as a form of energy. But consider these evidences that light is energy:

1. Light powers the chemical reactions that expose photographic film.
2. Light produces an electrical current in a photoelectric cell.
3. Light rotates a radiometer.
4. Sunlight exerts pressure on the tails of comets, causing them to point away from the Sun.
5. Light from a laser produces thermal energy in a material that it strikes.

23-1 A radiometer demonstrates that light energy can be converted into mechanical energy.

In each instance, light proves itself to be energy by doing work. On a microscopic level, light moves electrons, atoms, and dust particles. On a macroscopic level, the energy of a laser can be concentrated enough to destroy large objects.

23.2 The Electromagnetic Spectrum

Visible light is a member of a large family of waves called *electromagnetic waves.* Also included in the family are radio waves, microwaves, infrared waves, ultraviolet rays, x-rays, and gamma rays. All these waves travel at the same speed, 3.00×10^8 m/s. Looking at Figure 23-7 on page 506, you can see that radio waves have the longest wavelengths and gamma rays have the shortest. The boundaries between the various kinds of electromagnetic waves are arbitrary. A collection or distribution of a property such as wavelength that can change continuously without breaks is called a **spectrum.** The electromagnetic spectrum can refer to either wavelengths or the wave frequencies.

Waves with shorter wavelengths have higher frequencies. Because energy is proportional to frequency, waves with shorter wavelengths also have greater energies. Thus, radio waves are the least energetic, and gamma rays are the most energetic.

23.3 Radio Waves

The electromagnetic waves called *radio waves* include the waves that transmit television signals and long-range radio communications. These waves are generated by electrons accelerating in an *antenna.* Radio frequencies extend from around 300 GHz (10^9 Hz) down to undetectable (>0 Hz), and radio wavelengths may be from a few millimeters to many thousands of kilometers long. The longest radio waves created by man are in the ELF band (30 to 300 Hz). These radio waves are used by the military to communicate with submerged submarines. Television signals are transmitted in the VHF and UHF bands (30 MHz to 3 GHz). The highest frequency waves in the radio frequency band are called microwaves. Microwaves, which have frequencies from 1 to 300 GHz, are used in radar, line-of-sight communications, and microwave ovens. All radio waves are generated by accelerating charges in man-made or natural electrical circuits.

23.4 Infrared Waves

The next region of the electromagnetic spectrum is occupied by infrared (IR), or "below red," waves, with frequencies from 10^{11} to 10^{14} Hz (750 nm–1 mm). Infrared waves, often (incorrectly) called heat rays, are produced by the thermal motion of atoms. Since all matter has thermal motion, all matter emits infrared waves. Infrared rays, having frequencies lower than those of red light, are invisible to humans. Infrared is used for medical and industrial heat treatments, remote sensors in resource satellites and fire detection systems, and night imaging sensors.

23.5 Visible Light Waves

Visible light, the most familiar form of electromagnetic radiation and the subject of this chapter, includes only a narrow band of the electromagnetic spectrum, from 3.9×10^{14} to 7.7×10^{14} Hz (770 nm–390 nm, corresponding to the color spectrum from deep red to deep violet). Visible light is generated when energetic electrons in atoms fall to lower energy levels, emitting light in order to conserve energy.

As with the rest of the electromagnetic spectrum, the various colors of visible light form a continuous spectrum as well. Contrary to what you may see in typical drawings of rainbows, the visible spectrum contains more than seven colors.

23-2 Radio telescopes investigate the radio-frequency portion of the electromagnetic spectrum in space.

Radio-frequency waves are produced by accelerating charges in a conductor or conductive medium.

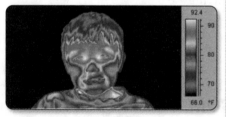

23-3 In infrared frequencies, warmer sources are brighter, while less warm spots are darker.

Infrared radiation is produced by the thermal motion of atoms and molecules.

Computer graphics circuits and monitors are capable of displaying millions of distinct colors. Theoretically, there is an infinite number of colors. We will discuss color and related properties of visible light in Chapter 26.

23.6 Ultraviolet Waves

Ultraviolet (UV), or "beyond violet," light is also produced by electrons returning to their lowest-possible energy levels, but it involves greater energy than visible light. Its frequencies are higher than those of visible violet light, ranging from around 10^{15} to 10^{17} Hz (300 nm–3 nm).

UV light is arbitrarily subdivided into three bands. Near UV (or UV-A) is the lowest frequency band and has the lowest energy UV. It forms about 90% of the solar UV that reaches the earth's surface. It cannot cause sunburns, but sufficient exposure can injure unprotected eyes. Near UV is used medically for treating certain skin disorders and for setting some dental polymer restorations. Also called "black light," it causes phosphorescent paints to glow and can identify compounds and minerals that fluoresce. Far UV (UV-B) is the middle band of UV energies. It is the predominant form of solar UV at midday and can cause sunburns. It is used industrially for quality assurance in manufacturing and for chemical conditioning. Extreme UV (UV-C) is the highest UV energy band. UV-C radiation from the Sun is completely absorbed in the atmosphere and does not reach the ground. It is used medically for germicidal treatments and for food, water, and air purification. Chronic exposure to any form of UV has been implicated as a cause of skin cancers.

23-4 Ultraviolet radiation is used to kill bacteria in many municipal water supplies.

Ultraviolet radiation is emitted when energetic electrons within an atom fall to their lowest-possible energy levels.

23.7 X-rays

X-rays, with frequencies from 10^{17} to 10^{21} Hz (3 nm–0.1 pm), are produced when high-energy electrons strike atoms and suddenly decelerate. Such radiation is technically called *bremsstrahlung,* or "braking radiation." The x-rays' most impressive characteristic is their ability to penetrate solid matter. X-rays are used for medical and industrial diagnostics, though precautions are necessary because x-rays are very damaging to biological systems. Fortunately, all but the most energetic solar and cosmic x-rays are absorbed by the atmosphere before they reach the earth's surface. As a result, astronomical x-ray observatories must be placed in orbit in order for scientists to study x-rays. From above Earth's atmosphere, they allow astronomers to examine the Sun's most energetic processes and to map the high-energy objects in the universe.

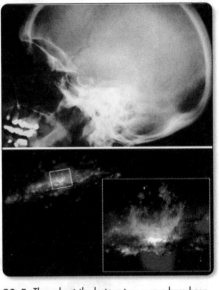

23-5 Throughout the last century, x-rays have been ranked among the most important medical diagnostic equipment. Today, space probes are extending the imaging of celestial objects into the x-ray spectrum.

X-rays are generated by high-energy free electrons as they smash into the electron cloud of an atom and slow down, releasing their excess energy.

God doesn't need an x-ray machine to see what we are. "Examine me, O Lord, and prove me; try my reins and my heart" (Ps. 26:2). An earlier meaning of "try" is to melt for the purpose of removing impurities. Therefore we can deduce that "try my reins" is saying, "God, test my obedience. Find my impurities and cleanse my innermost being." Only God can do this.

23.8 Gamma Rays

Although gamma rays, with frequencies of more than 10^{19} Hz (3 pm and shorter), are much like x-rays and overlap the x-ray spectrum band, their source is entirely different. Gamma rays are produced when an unusually high-energy proton or neutron returns to a lower energy level. This event usually occurs during a nuclear reaction or decay. Gamma rays are stopped only by very thick and dense material.

23-6 High-energy radioactive gamma sources are used to take photos of the internals of large metal structures to find flaws in welds and materials.

Gamma rays are emitted by energetic nuclear particles (protons and neutrons) as they reduce their energy state.

The extremely high-energy extraterrestrial gamma rays that almost reach the earth's surface (also called "cosmic rays") can cause significant damage to tissues that they pass through. Fortunately, we are exposed to very few of them. Decaying radionuclides in the earth's crust are the major source of gamma rays at the earth's surface. They make up a significant portion of what is called the "background" radiation that we all are exposed to throughout life. Gamma rays are difficult to produce on demand. A few highly radioactive isotopes (byproducts of nuclear reactions) emit sufficient gamma rays to be useful for inspecting the internals of thick metal structures and for performing biological experimentation.

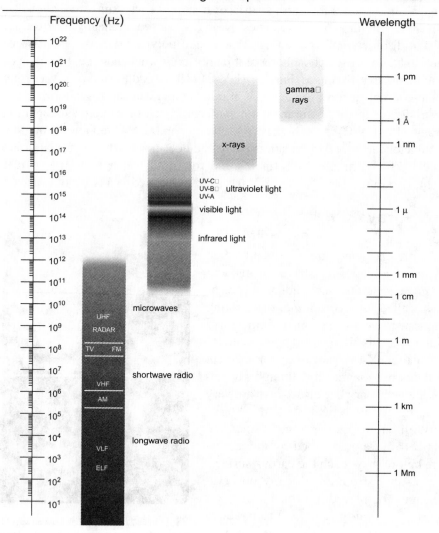

23-7 The main regions of the electromagnetic spectrum

23A Section Review Questions

1. Discuss at least three ways in which light energy can be converted into other forms of energy.

2. What do you call the continuous distribution of a physical property that varies, such as frequency or wavelength?

3. **a.** How are radio waves produced?

   b. What man-made radio waves have the longest and shortest wavelengths?

4. **a.** How are infrared waves produced?

   b. What is one technology that depends on IR radiation?

5. **a.** How are visible light waves produced?

   b. How many colors are there?

6. **a.** How are ultraviolet waves produced?

   b. What is one benefit and one hazard of UV radiation?

7. **a.** Discuss the origin of x-rays.

   b. What is their most useful application in technology?

8. **a.** What is the source of gamma rays?

   b. Why are they so difficult to produce in a controllable fashion?

23B SOURCES AND PROPAGATION OF LIGHT

Sources of Light

You have seen that light is produced when an energetic electron loses some energy. There exist many natural and artificial sources of light.

23.9 Incandescent Sources

Incandescent sources are objects that are heated until they glow. All bodies emit electromagnetic energy in the form of infrared due to the thermal motion of their atoms; however, hotter bodies emit EM radiation of higher frequencies than cooler bodies. At room temperature, a blackbody (a perfect radiator and absorber—see Chapter 15) emits radiant energy with the majority of its frequencies near 10^{13} Hz (mid-infrared). At 700 °C it glows red (visible light). As the blackbody's temperature continues to rise, it becomes orange, yellow, and so on, until at a temperature of about 1800 °C it glows yellowish-white. Incandescent lamps consist of a tungsten filament surrounded by gases such as nitrogen and argon. The filament heats rapidly, and the gas prevents chemical reactions that could destroy the filament. Incandescent sources emit radiation in all visible frequencies.

23-8 When a material is heated sufficiently, it will begin to glow in the visible region of the electromagnetic spectrum.

23.10 Gas-Discharge Tubes

Gas-discharge tubes also produce light. In these, a glass tube fitted with electrodes contains a gas sealed within it. When a large potential difference is established across the electrodes, a current flows through the tube using the valence (highest energy) electrons of the gas as the current carrier. These electrons jump to higher energy levels, then drop back, emitting visible light in the process. Each kind of gas emits light of a unique color because each gas has an atomic structure with a unique arrangement of electrons. For example, mercury vapor emits a bluish light, and sodium vapor emits a yellowish light. Unlike incandescent sources, gas-discharge tubes generally emit light at only a few discrete frequencies or in narrow bands of frequencies. So-called "neon lights" and sodium-vapor street lamps fall into this category.

The relationship that exists between the Kelvin temperature of a blackbody radiator and the wavelength of the radiation having maximum intensity is summarized in **Wien's law:**

$$\lambda_{max}T = 2.90 \times 10^{-3} \text{ m·K}$$

A broad spectrum of radiation may be emitted by a blackbody radiator, but λ_{max} is the wavelength having the greatest intensity.

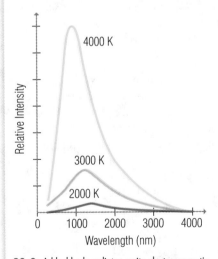

23-9 A blackbody radiator emits electromagnetic radiation in wavelengths that depend on its Kelvin temperature. One band of wavelengths has the greatest intensity for a specific temperature.

23-10 Automobile xenon-vapor headlamps are gas-discharge tubes.

23-11 Lasers are used to measure the motion of the atmosphere in order to compensate for optical distortion in modern telescopes.

23-12 LEDs are used for illumination and as indicators on many kinds of electronic equipment.

The firefly is a good example of the concept of *irreducible complexity* in nature. Certain arrangements in living systems appear to be designed in such a way that proper functioning can not occur unless all components of the arrangement are simultaneously present. These arrangements are evidence for design as opposed to gradual or even punctuated evolution.

A variation on the gas-discharge tube is the fluorescent light. The mercury vapor or other gas in the fluorescent tube emits ultraviolet light, rather than visible light. The glass tube is coated with *phosphors*—chemicals that emit visible light when struck by high-energy EM radiation. This indirect method of producing light is more efficient than heating a filament. A 22 W fluorescent fixture emits more light than a 75 W incandescent bulb.

23.11 Lasers

Lasers are devices that produce coherent light—light at a single frequency (i.e., monochromatic) where all the light waves from atoms in the laser source combine to form a single energetic electromagnetic wave. All other light sources produce noncoherent light (i.e., light waves that are not combined in phase), and most are polychromatic. Laser light is extremely intense because of its coherent nature. Lasers are used in many industrial, medical, and military applications, but not for area lighting.

23.12 Light-Emitting Diodes (LEDs)

A **light-emitting diode (LED)** is a solid-state electronic component manufactured from semiconductor materials that emits monochromatic light when a small potential difference is established across it. LEDs emit light only when the voltage polarity is in the right direction. They are important components in digital technology, where they function as light sources for a wide variety of applications, from simple on-off indicators to full-color digital displays. Their principal advantages are very low power requirements and efficiency. They can also be manufactured in extremely small sizes using microprocessor production techniques, making them suitable for miniature digital devices.

23.13 Cold Light

Cold light sources generate light with a minimum of heat, using chemical reactions. People have known for centuries that phosphorus can glow in the dark because of slow oxidation. In the past century, scientists developed several methods for producing "cold light" from chemical reactions using silicon-based chemicals such as siloxene and other proprietary chemicals. For example, one *chemiluminescent* product produces light when a complex organic chemical is combined with an oxidizer, such as hydrogen peroxide. The light can be used directly or it can activate a fluorescent dye to produce a wide range of colors. The ever-popular light sticks and glow-in-the-dark safety equipment use this cold light technology.

Bioluminescence is the production of cold light by living things. Some insects and deep-sea fish, squids, jellyfish, bacteria, protozoans, and fungi produce light ranging in color from blue to yellow-green. In these living creatures, an enzyme combines with another chemical in a biochemical reaction to generate light. These reactions are among the most efficient in nature, converting as much as 80% of the available chemical energy into light energy.

23-13 A firefly utilizes a bioluminescent chemical reaction to produce its flashing light.

The Speed of Light

23.14 Historical Setting

When emphasizing that something happens very quickly, people commonly say that it happens with the speed of light. How fast does light travel? Galileo, intrigued with this question, tried to measure the speed of light. One night he and his assistant held shaded lanterns on hilltops about a mile apart. Galileo uncovered his lantern. As soon as his assistant saw its light, he uncovered his own lantern, and Galileo recorded the time that passed between uncovering his own lantern and seeing the light from his assistant's lantern. He came to the conclusion that he had measured only his assistant's reaction time and that the speed of light is too great to measure, if not infinite.

Another scientist, Ole Roemer, found when investigating eclipses of the moons of Jupiter that the speed of light is not infinite. After much study, he made a schedule of eclipses. His schedule worked perfectly when the earth was on the same side of the Sun as Jupiter, but when it was on the far side of the Sun, the eclipses fell behind schedule. Roemer concluded that light takes twenty-two minutes to cross the earth's orbit and therefore does not move with infinite speed.

In 1849 Armand Fizeau, a French physicist, improved on Galileo's experiment. Instead of having an assistant a mile (1.61 km) away, Fizeau placed a mirror 8.63 km away. He placed a light behind a wheel with evenly spaced notches. He theorized that light would go out through one notch, strike the mirror, and return to the wheel. If the wheel rotated at the right speed, each burst of returning light would be blocked by the next tooth on the wheel. From the rotational speed of the wheel, Fizeau calculated the speed of light with only a 4% error.

Ole Roemer (1644–1710) was a Danish astronomer who made the first measurement of the speed of light through astronomical observations.

Armand Fizeau (1819–96) was a French physicist who made the first experimental measurement of the speed of light on the earth. He also discovered that radiant heat from the Sun behaves as wave motion.

23.15 Current Value for the Speed of Light

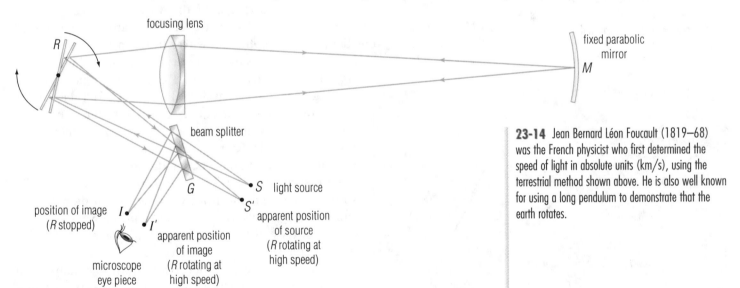

23-14 Jean Bernard Léon Foucault (1819–68) was the French physicist who first determined the speed of light in absolute units (km/s), using the terrestrial method shown above. He is also well known for using a long pendulum to demonstrate that the earth rotates.

Similar techniques involving rotating mirrors allowed experimenters to measure the speed of light with increasing accuracy. In 1862, Léon Foucault was the first to determine the speed of light in metric terms using a terrestrial method. Finally, however, directly measuring the speed of light gave way to an indirect method developed by Albert Michelson using interferometers (see Chapter 25). As you learned in Chapter 12, the speed of a wave of length λ and frequency f is found by

$$v = \lambda f.$$

Albert Michelson (1852–1931) was the first American to receive the Nobel Prize in Physics (in 1907) for his measurement of the speed of light to within 0.02 percent of its current value.

Light and Reflection **509**

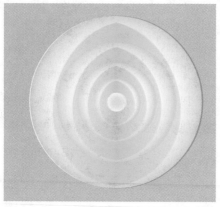

(a)   3-dimensional waves

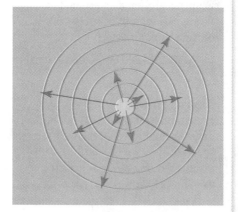

(b)   2-dimensional representation

23-15 A point light source emits spherical light waves. These are represented on a flat diagram as two-dimensional circles concentric with the source.

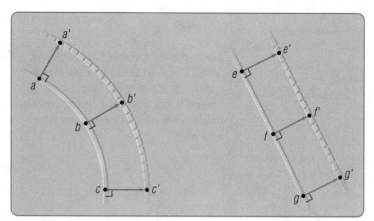

23-17 Points on a wave travel perpendicularly to the orientation of the wave front.

An *envelope* is a geometrical curve that is tangent to an associated set of curves or surfaces. Its purpose is to show graphically the net effect of the collection of curves.

The currently accepted value for the speed of light (*c*), 299 792 458 m/s (exactly), was approximated in the 1970s by measuring a laser light's wavelength and frequency with great precision and then multiplying them together. Eventually, the speed of light was known to a much higher precision than the meter itself was known. So, in 1983, the speed of light was defined to the nearest whole meter per second, and the meter was redefined as the distance that light travels in a vacuum in 1/299 792 458 s. The letter *c* is the abbreviation for the Latin *celeritus,* speed, and is usually rounded to 3.00×10^8 m/s.

As was just mentioned, the value of *c* is the speed of light in a vacuum. When traveling through matter, light travels more slowly than it travels through a vacuum. In most gases, light travels at nearly *c;* in liquids and solids, however, speed reduction is significant. Light travels through water with a speed of about (3/4)*c*, through glass with a speed of around (2/3)*c*, and through diamond with a speed of roughly (2/5)*c*.

Wave Characteristics of Light

23.16 Physical Description of Light Waves

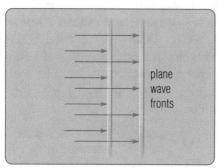

23-16 When a spherical wave front travels far enough from its source, it appears to be a nearly flat, or plane wave, front.

You learned in Chapter 22 that electromagnetic waves consist of self-supporting, changing electric and magnetic fields. The electric and magnetic field vectors are always at right angles to each other, and both are at right angles to the direction the wave travels. The changing electric field reinforces the magnetic field, and the changing magnetic field reinforces the electric field, perpetuating the wave.

Light travels outward in concentric **spherical waves** from a point source. Since this movement is hard to draw, light is usually shown in a two-dimensional diagram moving in a circular wave (see Figure 23-15). Like a water wave, light travels at equal speeds through a uniform medium in every direction from a source. The sphere (representing a single wave cycle) expands as the light travels from its source. At great enough distances, the sphere is so large that a small portion of its surface is nearly flat (a straight line segment in a two-dimensional diagram). A straight-line wave surface is called a **plane wave.**

The circles in figure (b) represent **wave fronts.** The wave displacement and velocity vectors are always perpendicular to the wave front in a uniform medium. For a point source, the wave front is spherical; the waves of light travel radially outward from the source and are concentric. For plane waves, the wave fronts are planar, and the light waves are parallel to each other (Figure 23-16).

If you know where the wave started and how fast it is going, you can easily find where it will be at a given time. But what if you do not know where the wave started? The Dutch physicist Christiaan Huygens expressed a principle to answer this question. Huygens postulated that every point on a wave front acts like a new source of small spherical waves, called wavelets. As the wavelets expand with time, their wave fronts coalesce into a new wave front, defined by the *envelope* of

Christiaan Huygens (1629–95)

The behavior of light puzzled scientists worldwide until a Dutchman named Christiaan Huygens proposed the wave theory of light. He suggested, for example, that an advancing wave front creates secondary spherical waves at each point. Huygens's principle, as it is now called, has remained an outstanding element of our modern wave theory. His ideas were the culmination of a lifetime of work in mathematics and physics.

Like many fortunate scientists of his time, Huygens had wealthy parents who wanted him to spend his life studying rather than making money. At age sixteen, Christiaan left home to study mathematics and law at the University of Leiden and the College of Breda. Four years later he returned to the family estate, where for the next twenty years he made discoveries in mechanics, astronomy, optics, and chronometry. His knowledge of mathematics enabled him to consider the technical complexities of his findings.

Huygens began his studies with astronomy. He and his brother Constantijn ground their own lenses for homemade telescopes. To correct the problem of spherical aberration in the standard single-lens system, they invented the Huygens eyepiece. With this improved telescope, Huygens studied Saturn more closely than anyone had before. After many years of observation, he discovered Saturn's largest moon, Titan, and realized that Saturn's rings are not connected to the planet.

During the same period, the later 1650s, Huygens's interest in the motion of pendulums led to the invention of a pendulum clock, which was accurate enough to measure time in minutes and seconds. He published his studies of the pendulum nearly twenty years later in the treatise *Horologium Oscillatorium* (Oscillating Timepiece). In this work he gave a thorough mathematical description of the working of his clock, of the motions of pendulums, and of the laws of "centrifugal force."

By 1666, Huygens had enough prestige to be elected a founding member of the French Académie des Sciences in Paris, and the Académie gave him a pension and an apartment in its building. He enjoyed Paris life and made friends with many intellectual leaders, including Pascal and Leibnitz (whom he tutored for three years). His stay in Paris was cut short in 1681, when illness forced him to return home to Holland. His hopes of returning to France were dashed in 1685 when Louis XIV revoked the Edict of Nantes, which had guaranteed Protestants freedom from persecution.

While he lived in Paris, Huygens (wrongly) came to the conclusion that light, like sound, is a wave moving through a medium. This medium could not be air, because light travels through a vacuum. So Huygens assumed that some substance permeates all space. He called this medium *ether*. He defined candlelight as the sum of the spherical waves from each tiny particle in the flame. Huygens's principle, that each point on a wave front generates secondary spherical waves, was a logical extension of his view of candle flame. Huygens's theory of light enabled him to explain satisfactorily the reflection and refraction of light.

The publication of Newton's *Principia* in 1687 caused a stir in the international scientific community. Two years later Huygens broke the monotony of his stay in Holland with a lengthy trip to England. Although he admired Newton's mathematics, Huygens disagreed with his action-at-a-distance theory of gravity. Nevertheless, Huygens failed to gain a following for his own mechanical theory of gravity.

Besides studying light and pendulums, Huygens spent much time studying collisions. He discovered the conservation of mv (momentum) and mv^2 (kinetic energy) in elastic collisions. He used the concept of center of mass, which he had studied extensively in pendulums, to explain the behavior of colliding bodies.

Huygens's contributions to the study of physics were many and varied. By questioning his surroundings and studying until he found satisfactory answers, he was able to build on the foundations laid by Galileo. Later scientists further built on what he had learned. Although some of his ideas were mistaken, many helped to form the basis of modern science.

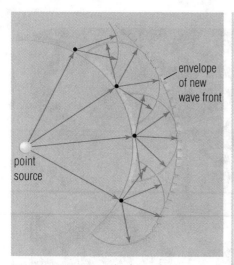

23-18 Huygens's principle of wave front propagation

the wavelets (see Figure 23-18). A spherical wave front propagates as a spherical front, and a plane front propagates as a plane front. This principle is called *Huygens's principle*.

23.17 Mathematical Description of Light Waves

An electromagnetic wave is mathematically described by time-dependent sine functions. The magnitude of the electric field strength (E) and the magnitude of the magnetic field vector (B) both act as sine waves. That is, they can be written as

$$E = E_{max} \sin \omega \text{ and}$$
$$B = B_{max} \sin \omega,$$

where ω is a time-dependent variable.

The equations are similar because the electric and magnetic fields are in phase. That is, both reach their maximum strength at the same time, and both reach their minimum strength at the same time.

Most significant in the mathematical description of light waves was the work of James Clerk Maxwell, who successfully related electricity, magnetism, and light. By putting together four known equations describing the behavior and interaction of electrical and magnetic fields, and by modifying one equation, he showed that light is related to electromagnetism.

The first equation, Gauss's law for electricity, states that the net electric field through a closed surface is proportional to the charge enclosed by the surface. The second equation, Gauss's law for magnetism, states that the net magnetic flux through a closed surface is zero. The third equation is Faraday's law, which states that a changing magnetic field induces an electric field. The fourth equation, the one Maxwell modified, shows how both changing electric fields and electric currents induce magnetic fields.

Maxwell's equations are as important to the study of electromagnetism as Newton's laws are to the study of mechanics, or as the laws of thermodynamics are to the study of heat. Using these equations, Maxwell predicted the existence of electromagnetic waves, deduced that light is an electromagnetic wave, and calculated the speed of light. These equations affect such diverse things as motors, generators, transformers, radios, televisions, radar units, and microwave ovens. Even the equations for E and B can be derived from Maxwell's equations.

23B Section Review Questions

1. **a.** Discuss how incandescent light emission occurs.

   b. List three sources of incandescent light not mentioned in the text.

2. What two measurable dimensions does Wien's law link for a blackbody radiator?

3. What is probably the most common gas discharge tube used for illumination?

4. What are the two key properties of laser light?

5. What makes bioluminescence such an elegant testimony of design in creation?

6. The speed of light and the length of a meter ultimately depend on what measurement?

7. Where do plane electromagnetic wave fronts come from?

8. What was the significance of Maxwell's contributions to understanding electromagnetic theory?

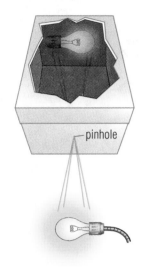

23C REFLECTION AND MIRRORS

23.18 Ray Optics

Light can be regarded as a group of **rays,** which are represented as lines drawn in the direction the wave is traveling, perpendicular to its wave front. Light rays travel in straight lines through any single material, but they may bend when passing from one material to another. We know that light travels in reasonably straight lines because objects cast sharp shadows. For example, a spherical Jovan moon casts a circular shadow on the planet it orbits. A pinhole viewer also demonstrates that light rays travel in straight lines. A shoe box with a pinhole in one end will display an inverted image of a light source placed outside it. Figure 23-19 shows why the image is inverted.

When light strikes an object, some of the light rays "bounce off" its surface. This change of direction is known as **reflection.** The way light rays reflect from a surface depends on the material the surface is made of. Irregular surfaces, such as paper, cloth, and brick, reflect rays in random directions. The sizes of the irregularities are much larger than one wavelength of the incident light. Reflection in random directions is called *diffuse reflection.* Polished opaque surfaces, such as mirrors or smooth metals, reflect rays in a predictable pattern. The dimensions of the irregularities on such surfaces are much smaller than one wavelength of light. Predictable reflection is called *regular* or *specular* reflection.

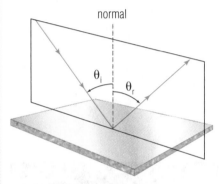

23-19 A pinhole projector produces an inverted and reversed image because rays travel in straight lines.

Consider a ray of light striking a mirror, as Figure 23-20 shows. The imaginary line perpendicular to the mirror's surface at the point of reflection is called the normal line, or just the *normal.* You first saw normal lines in Chapter 8 when studying friction and the normal force. The angle that the incoming ray makes with the normal is called the **angle of incidence (θ_i).** The angle that the reflected ray makes with the normal is called the **angle of reflection (θ_r).** How is the reflected ray related to the incoming ray? The answer is the two-part **law of reflection:**

1. The incoming (incident) ray, the normal, and the reflected ray all lie in the same plane.
2. The angle of incidence equals the angle of reflection.

23-20 The law of reflection

The word "specular" is derived from the Latin word *speculum,* a mirror.

23.19 Plane Mirrors

An image in a flat, or plane, mirror is formed by many rays of light, each obeying the law of reflection. Figure 23-21 shows two rays from the top of the object being reflected. The eye, detecting these rays, assumes that they come from the same point in straight lines; so the object appears to be located *behind* the mirror. Since the rays do not really come from behind the mirror, the image of the object in the mirror is said to be a **virtual image.** A virtual image in optics is one that exists only in the mind of the observer. The mind interprets the reflected rays as originating at a point that doesn't exist in reality. A virtual image cannot be projected onto a screen.

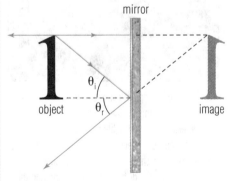

23-21 Reflection in a plane mirror and a virtual image

When you think of a mirror image, you probably think of an image with its left and right sides reversed. This idea is a semantic misnomer. Actually, the sides are not reversed. A mirror reflects your right side on your right and your left side on your left. Since your reflection is facing you, your mind considers your reflection another person and automatically orients the reflection accordingly. You naturally consider the image's right side (from the "image's perspective") to be a reflection of your left side and vice versa. So you assume that the image is reversed. In fact, the left side of the image (as you face it) is a reflection of your left side and the image's right side is a reflection of your right side. Your reflection is a "mirror image."

What happens when two plane mirrors are set at right angles? When an object is approximately equal distances from each mirror, it has *three* images. Images I_1 and I_2 are formed as they would be if the mirrors were separate. But where does the third image come from? Some light rays are reflected from both mirrors before meeting an observer's eye. These twice-reflected rays form the third image, I_3, which is reversed so that the object's left side is reflected to the object's right.

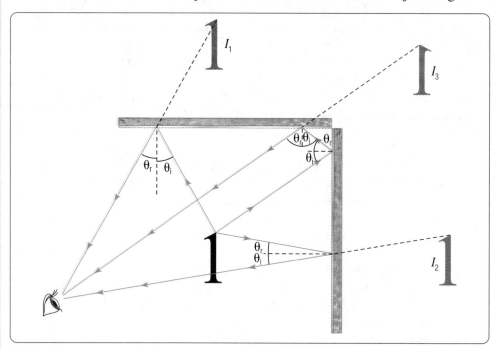

23-22 Reflection in two perpendicular mirrors

If the angle between the mirrors is reduced to 60°, five images appear. If the angle is decreased to 45°, an object will produce seven images. The number of images at a given angle is determined by the formula

$$n = \frac{360°}{\theta} - 1,$$

where θ is the angle between the mirrors in degrees. What if the mirrors face each other? Then $\theta = 0°$, and

$$n = \frac{360°}{0°} - 1,$$

$$n = \text{d.n.e.} \ (\rightarrow \infty).$$

There will be an unlimited number of images if the mirrors are properly adjusted. (You may have seen this effect at amusement parks or in furniture stores where wall mirrors are sold.)

23.20 Curved Mirrors

Mirrors with curved surfaces are useful in many different ways. You have probably seen such mirrors used as cosmetics mirrors, rearview mirrors in cars, safety mirrors in the aisles of stores, and as flashlight reflectors. Such mirrors may be **concave,** formed from part of the inside of a curved surface, or **convex,** formed from part of the outside of a curved surface. Mirrors that are exactly spherical are the easiest to manufacture and are very common. However, they produce slightly blurry images, an effect called **spherical aberration.** Parabolic mirrors avoid spherical aberration. These mirrors, like plane mirrors, follow the law of reflection. We will discuss both kinds of mirrors in the following sections.

23-23 Apollo program astronauts left several box-mirror laser reflectors on the surface of the Moon. Each reflector cell is the right-angle corner of a box. Such reflectors can reflect an incident ray in the direction from which it came. The Moon reflectors are used to measure the distance between Earth and the Moon to a precision within a few centimeters.

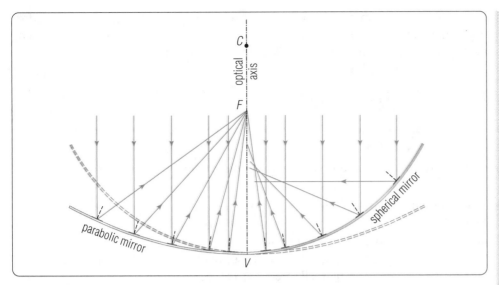

23-24 Spherical mirrors do not focus parallel light rays at a point. Parabolic mirrors do.

23.21 Concave Mirror Reflections

When parallel rays approach a spherical concave mirror, the reflected rays pass through a point called the **principal focus** (**F**) or *focal point*. The distance from the mirror to the principal focus is the **focal length** (*f*) of the mirror. The distance *R* is the *radius* of the mirror—the radius of a sphere that the mirror's surface is a part of. The point *C* is the *center* of the spherical surface. Every point on the surface of a spherical mirror is *R* units from *C*. The line through *F* and *C* intersects the mirror at its *vertex* (*V*) and is called the *principal* or **optical axis** of the mirror. The focal point is midway between *V* and *C*:

$$f = \frac{R}{2}$$

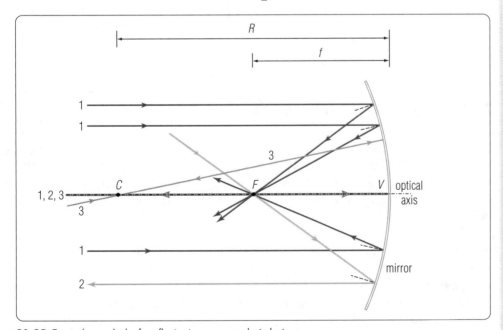

23-25 Terminology and rules for reflection in a concave spherical mirror

A visual image of a reflection is formed when certain rays from an object are reflected so that they come together at a point on the retina of your eye. In order to determine how the rays from an object are reflected, three specific rays from a point on the object are projected to the mirror's surface and then reflected according to the law of reflection. Where they intersect is the position of the image. The special rays obey the following rules (the corresponding rays are numbered in Figures 23-25 and 23-27 through 23-32):

1. Incident rays that are *parallel to the optical axis* are reflected through the principal focus (F).

2. Incident rays that *pass through the principal focus* are reflected parallel to the optical axis.

3. Incident rays that *pass through the center of curvature* (C) are reflected back through C.

Keeping these rules in mind, let us see where an image of an object will appear. The **object distance (d_O)** is the distance of the object from the mirror. The **image distance (d_I)** is the distance of the image from the mirror. For simplicity, we will assume that the object is located on the optical axis, although this discussion applies to any location where all three rays can be reflected from the mirror. There are six optically different object locations to consider.

Case 1: at "infinity" ($d_O \Rightarrow \infty$) or far enough from the mirror so that all rays from the object are nearly parallel

Case 2: beyond C ($d_O > R$)

Case 3: at C ($d_O = R$)

Case 4: between C and F ($f < d_O < R$)

Case 5: at F ($d_O = f$)

Case 6: between F and the mirror ($0 < d_O < f$)

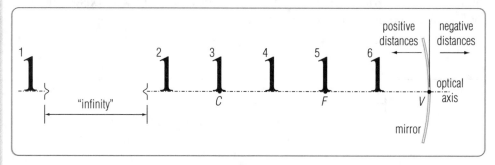

23-26 Six possible object locations that affect image location and orientation

Let us first consider Case 2, an object beyond the mirror's center of curvature, C. The object in Figure 23-27 is represented as a solid black numeral "1" at a distance d_O from the mirror, and its image is a gray "1" at a distance d_I from the mirror. By convention, all distances in front of the mirror are positive, and distances behind the mirror are negative.

To locate the object's image, trace the paths of two rays from the top of the numeral, one parallel to the optical axis and one through the focus. Where they meet, the image of the corresponding point on the object will form. As the figure shows, they intersect between C and F on the opposite side of the optical axis. As a result, the image is inverted compared to the object's orientation. Notice that the light rays converge at the point where your eye would have to be in order to see the image. Since the image location is in front of the mirror, it is called a **real image.** Real images can be focused on a screen.

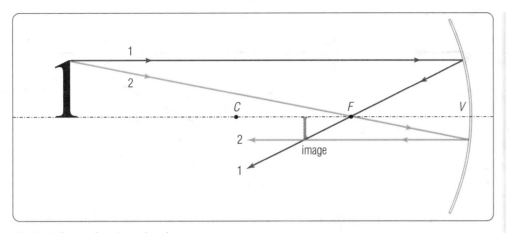

23-27 Reflection of an object when $d_0 > R$

The key difference between a real image and a virtual image is where they are (or appear to be) formed in relation to the mirror's surface. Real images are formed by the convergence of light rays from a real point in front of the mirror's surface, while virtual images appear to be formed by the divergence of light rays from a virtual point behind the mirror's surface.

Now consider Case 4, where the object is placed between C and F and the image appears beyond C ($d_I > R$), as Figure 23-28 shows. Notice that the image here is larger than the object. Whenever $d_O < d_I$, the image is larger than the object. However, when $d_O > d_I$, the image is smaller than the object. In both Case 2 and Case 4, the image is upside-down, or inverted.

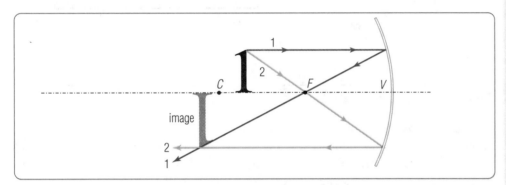

23-28 Reflection of an object when $f < d_0 < R$

For Case 3, when the object is placed at C (Figure 23-29), its image will also appear at C but will be inverted so that the object and its image are tail to tail. The image in this arrangement is also real and is the same size as the object.

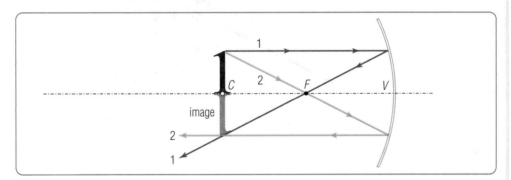

23-29 Reflection of an object when $d_0 = R$

Real images are formed for Cases 2 through 4, but not for the other three. Suppose the object is at infinity (Case 1). At that distance *all* the rays that reach the mirror from the object are essentially parallel to the optical axis. Therefore, all the rays are reflected through the focus, F. The image is formed where the rays meet—at the focus. So, when an object is at infinity, its image is a miniscule point at the focus.

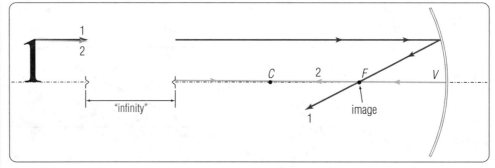

23-30 Reflection of an object located at an infinite distance from the mirror

Suppose, instead, that the object is at the focus (Case 5). Rays from every point of the object are reflected so that they never intersect. The image is said to be formed at infinity and is infinitely large. It cannot be seen, so it does not exist.

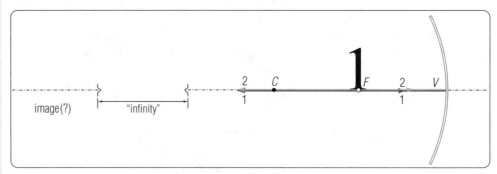

23-31 Reflection of an object located at F $(d_O = f)$

Finally, consider an object between F and the mirror (Case 6). As Figure 23-32 shows, the rays do not converge. However, as with the plane mirror, the eye sees the rays as diverging from the same point, apparently behind the mirror. All points of the image are in the same vertical relationship as the object, so the image is upright, or *erect*. This image, like that in a plane mirror, is virtual—not real.

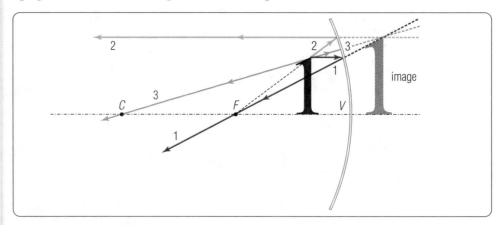

23-32 Reflection of an object when $d_O < f$

23.22 Finding Image Position

Can the exact location of an image be predicted? Yes, if the mirror's focal length and the position of the object are known. The **mirror equation** is

$$\frac{1}{d_O} + \frac{1}{d_I} = \frac{1}{f},$$

(23.1)

where d_O is the distance between the object and the mirror, d_I is the distance between the image and the mirror, and f is the mirror's focal length. For laboratory mirrors, these distances are usually measured in centimeters.

Problem-Solving Strategy 23.1
Distances in front of the mirror are assumed to be positive; distances behind the mirror are negative. Orientations upright are considered positive; inverted orientations are considered negative.

EXAMPLE 23-1

Good Reflection: Finding the Image Distance

Locate the image formed by a concave mirror whose focal length is 10.0 cm if the object is 35.0 cm from the mirror.

Solution:
You know that $d_0 = 35.0$ cm and $f = 10.0$ cm, and you want to find d_I. Use Equation 23.1:

$$\frac{1}{d_0} + \frac{1}{d_I} = \frac{1}{f}$$

$$\frac{1}{35.0 \text{ cm}} + \frac{1}{d_I} = \frac{1}{10.0 \text{ cm}}$$

$$\frac{1}{d_I} = \frac{1}{10.0 \text{ cm}} - \frac{1}{35.0 \text{ cm}}$$

Multiply by the lowest common denominator of $(70.0 \text{ cm})d_I$.

$$\frac{(70.0 \text{ cm})d_I}{d_I} = \frac{(70.0 \text{ cm})d_I}{10.0 \text{ cm}} - \frac{(70.0 \text{ cm})d_I}{35.0 \text{ cm}}$$

$$70.0 \text{ cm} = 7.00d_I - 2.00d_I$$

$$70.0 \text{ cm} = 5.00d_I$$

$$d_I = 14.0 \text{ cm}$$

The image is 14 cm from the mirror. Since $f = 10.0$ cm and $R = 2f = 20.0$ cm, the image is between F and C.

EXAMPLE 23-2

Virtual Image Distance

Locate the image of an object placed 4.0 cm from a concave mirror with focal length $f = 8.0$ cm.

Solution:
This is a case in which $d_0 < f$. Use Equation 23.1 again.

$$\frac{1}{4.0 \text{ cm}} + \frac{1}{d_I} = \frac{1}{8.0 \text{ cm}}$$

$$\frac{1}{d_I} = \frac{1}{8.0 \text{ cm}} - \frac{1}{4.0 \text{ cm}}$$

Multiply by the lowest common denominator, $(8.0 \text{ cm})d_I$.

$$\frac{(8.0 \text{ cm})d_I}{d_I} = \frac{(8.0 \text{ cm})d_I}{8.0 \text{ cm}} - \frac{(8.0 \text{ cm})d_I}{4.0 \text{ cm}}$$

$$8.0 \text{ cm} = d_I - 2.0 \, d_I$$

$$8.0 \text{ cm} = -d_I$$

$$-8.0 \text{ cm} = d_I$$

A negative d_I means that the image is *behind* the mirror; it is a virtual image.

EXAMPLE 23-3

Loss of Focus: Object at Infinity

Locate the image of an object at infinity if the concave mirror's focal length is 18 cm.

Solution:

You can do this one of two ways. First, if you know that the image of an object at infinity is at the focal point, you can simply say that

$$d_I = f = 18 \text{ cm.}$$

Or you can use Equation 23.1 to work it out.

$$\frac{1}{d_0} + \frac{1}{d_I} = \frac{1}{f}$$

But $d_0 = \infty$, so

$$\frac{1}{\infty} + \frac{1}{d_I} = \frac{1}{f}.$$

As d_I approaches ∞, $1/\infty$ approaches 0:

$$0 + \frac{1}{d_I} = \frac{1}{f}$$

$$d_I = f = 18 \text{ cm}$$

Problem-Solving Strategy 23.2

No object is truly at an infinite distance from a mirror. However, you can assume that an object is at infinity if its incident light rays are nearly parallel.

23.23 Magnification

For all spherical mirrors, the height of the image (H_I) relates to the height of the object (H_O) by this equation:

$$\frac{H_I}{H_O} = -\frac{d_I}{d_O} \tag{23.2}$$

The negative sign maintains the correct sign relationship. The object distance and object height are always positive for mirrors. For real images, image distance is always positive because it is in front of the mirror, and image height is always negative because it is inverted. Virtual images have negative image distances and positive heights because they are erect.

Notice in Equation 23.2 that if the image is farther from the mirror than the object is, the image must be larger than the object to maintain the proportionality. If the image is closer to the mirror than the object is, the image is smaller than the object. The **magnification** of an image is the absolute value of the image height to the object height:

$$m = \left| \frac{H_I}{H_O} \right| \tag{23.3}$$

Problem-Solving Strategy 23.3

The magnification of an object may be greater or less than 1 but never negative as long as an image is formed. Magnification may be zero only when an object at an infinite distance forms a point image at the principal focus.

EXAMPLE 23-4

Supersizing: Magnification

Find the size of the image of a 15.0 cm object that is 30.0 cm from a concave mirror whose focal length is 20.0 cm. What is the magnification?

Solution:

You must first find d_I using Equation 23.1, then H_I using Equation 23.2.

$$\frac{1}{30.0 \text{ cm}} + \frac{1}{d_I} = \frac{1}{20.0 \text{ cm}}$$

$$\frac{1}{d_I} = \frac{1}{20.0 \text{ cm}} - \frac{1}{30.0 \text{ cm}}$$

Multiply by $(60.0 \text{ cm})d_I$.

$$\frac{(60.0 \text{ cm})d_I}{d_I} = \frac{(60.0 \text{ cm})d_I}{20.0 \text{ cm}} - \frac{(60.0 \text{ cm})d_I}{30.0 \text{ cm}}$$

$$60.0 \text{ cm} = 3.00 \, d_I - 2.00 \, d_I$$

$$60.0 \text{ cm} = d_I$$

Solve for image height.

$$\frac{H_I}{H_0} = -\frac{d_I}{d_0}$$

$$H_I = -\frac{H_0 d_I}{d_0} = -\frac{(15.0 \text{ cm})(60.0 \text{ cm})}{30.0 \text{ cm}}$$

$$H_I = -30.0 \text{ cm}$$

The real image ($d_I > 0$) is 30.0 cm tall and inverted ($H_I < 0$).

Find the magnification using Equation 23.3.

$$m = \left|\frac{H_I}{H_0}\right| = \left|\frac{-30.0 \text{ cm}}{15.0 \text{ cm}}\right| = 2.00\times$$

The image is magnified to twice the object height.

23.24 Spherical Mirror Aberration and Parabolic Mirrors

For a concave spherical mirror, not all incoming rays parallel to the optical axis are reflected so that they actually pass through the focal point. Although light rays aligned with the optical axis are reflected directly back to the focal point, light rays that are off-axis (but still parallel to the axis) are reflected to points slightly in front of the focal point, toward the mirror. The farther away from the vertex of the mirror the incident ray is, the farther in front of the focal point the reflected ray lies. For this reason, images formed by a highly curved spherical mirror are blurry. As we mentioned earlier, this problem is called spherical aberration, and it exists in every spherical mirror. One method of compensating for spherical aberration is to use a relatively flat spherical mirror (R is large compared to the mirror's diameter). Because the curvature is small, the reflected light rays focus close to the focal point. This method can minimize spherical aberration, but it cannot eliminate it because it is a built-in characteristic of spherical mirrors.

A *paraboloidal* reflecting surface is not subject to spherical aberration. A paraboloid is formed by rotating a parabola around its axis of symmetry. Geometrically, every incident light ray parallel to the optical axis of the mirror is reflected through the principle focus, F. The location of the effective center of curvature (C) for a parabolic mirror is defined in the same way as for spherical mirrors: $R = 2f$. Parabolic mirrors are the most common light-gathering element of Newtonian reflector telescopes.

The symbol "×" is often used as a unit of magnification "power." A more formal way is to refer to the number of diameters of magnification.

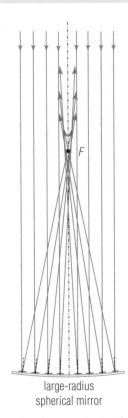

large-radius
spherical mirror

23-33 A spherical mirror with a long focal length is relatively flat. The reflected rays tend to focus very close to the geometrical focal point ($R/2$).

Instead of focusing rays to form an image, parabolic mirrors sometimes spread rays. When a light source is at the principal focus, the mirror reflects parallel rays. This arrangement is used for flashlights, spotlights, searchlights, and automobile headlights.

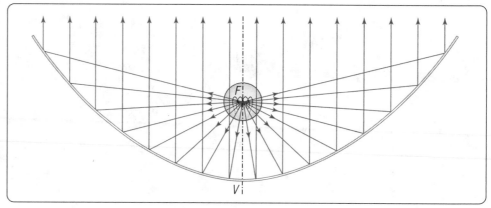

23-34 Reflection of a light source at the focus of a parabolic reflector

23.25 Convex Mirrors

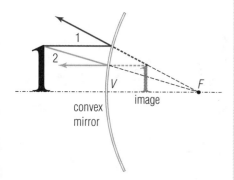

23-35 Reflection in a spherical convex mirror

Convex mirrors (mirrors that bulge out) can produce only one type of image. The image is always virtual (behind the mirror), erect, and smaller than the object. The principal focus is *behind* the mirror, again at $f = R/2$. The rules that you used for concave mirrors must be modified slightly. First, an incoming ray that is parallel to the optical axis has a reflection that, if extended through the mirror, would pass through the focus. Second, an incoming ray that would, if extended through the mirror, pass through the focus has a reflection parallel to the optical axis.

Because convex mirrors reflect over a wide range of angles and their images are smaller than the objects, they can give a panoramic view of their surroundings. They are therefore used as rearview mirrors, especially for trucks, as shoplifting deterrents in stores, and to prevent forklift collisions at busy intersections in factories and warehouses. The earth's oceans also act as convex mirrors. Astronauts see an image of the sun reflected in the oceans.

Equation 23.1 applies to convex as well as concave mirrors. Of course, for convex mirrors, f and d_I are negative numbers. Computing image size using Equation 23.2 is also valid for convex mirrors.

EXAMPLE 23-5

Through a Looking Glass: Virtual Images

Locate the image of a 2.0 cm object that is 10.0 cm from a convex mirror whose focal length is 10.0 cm. How high is the image?

Solution:

$d_0 = 10.0$ cm and $f = -10.0$ cm

Use Equation 23.1 to find d_I.

$$\frac{1}{d_0} + \frac{1}{d_I} = \frac{1}{f}$$

$$\frac{1}{10.0 \text{ cm}} + \frac{1}{d_I} = \frac{1}{-10.0 \text{ cm}}$$

$$\frac{1}{d_I} = -\frac{1}{10.0 \text{ cm}} - \frac{1}{10.0 \text{ cm}}$$

$$\frac{1}{d_I} = -\frac{2}{10.0 \text{ cm}} = -\frac{1}{5.0 \text{ cm}}$$

$$d_I = -5.0 \text{ cm}$$

Solve for image height using Equation 23.2:

$$\frac{H_I}{H_0} = -\frac{d_I}{d_0}$$

$$H_I = -\frac{H_0 d_I}{d_0} = -\frac{(2.0 \text{ cm})(-5.0 \text{ cm})}{10.0 \text{ cm}}$$

$$H_I = +1.0 \text{ cm}$$

The image is 5.0 cm behind the mirror and 1.0 cm high. The image is erect, which is another clue that it is a virtual image.

Problem-Solving Strategy 23.4

Since magnification is always positive, you must use the absolute value of image distance when finding the magnification of virtual images.

23C Section Review Questions

1. What aspect of the way light propagates makes it relatively simple to study reflection?

2. **a.** Discuss the law of reflection.

   b. What are the two kinds of reflection?

3. **a.** Is your image reflected from a plane mirror really reversed? Explain your answer.

   b. What is meant by a mirror image?

4. Describe how to determine the location of an object's image after reflection by a concave spherical mirror.

5. Why is a virtual image always erect and a real image always inverted?

⊛6. In a high school laboratory, a spherical concave mirror 10.0 cm in diameter has a focal length of 35.0 cm. If a paper clip 4.70 cm long is mounted vertically in a stand on the mirror's optical axis 40.0 cm from the center of the mirror, where will the paper clip's image be? What will be its orientation?

⊛7. Find the height of the paper clip's image if the paper clip in Question 6 is positioned 15.0 cm from the concave mirror. What is the magnification of the image at this point?

⊛8. Solve Question 6 assuming that the mirror is a convex mirror.

In Terms of Physics

spectrum	23.2	concave	23.20
Wien's law	23.9	convex	23.20
light-emitting diode (LED)	23.12	spherical aberration	23.20
spherical wave	23.16	principal focus (F)	23.21
plane wave	23.16	focal length (f)	23.21
wave front	23.16	optical axis	23.21
ray	23.18	object distance (d_0)	23.21
reflection	23.18	image distance (d_I)	23.21
angle of incidence (θ_i)	23.18	real image	23.21
angle of reflection (θ_r)	23.18	mirror equation	23.22
law of reflection	23.18	magnification	23.23
virtual image	23.19		

Problem-Solving Strategies

23.1 (page 519) Distances in front of the mirror are assumed to be positive; distances behind the mirror are negative. Orientations upright are considered positive; inverted orientations are considered negative.

23.2 (page 520) No object is truly at an infinite distance from a mirror. However, you can assume that an object is at infinity if its incident light rays are nearly parallel.

23.3 (page 520) The magnification of an object may be greater or less than 1 but never negative as long as an image is formed. Magnification may be zero only when an object at an infinite distance forms a point image at the principal focus.

23.4 (page 523) Since magnification is always positive, you must use the absolute value of image distance when finding the magnification of virtual images.

Review Questions

1. What is visible light?

2. Name three types of waves in the electromagnetic spectrum.

3. **a.** Which electromagnetic waves are the most energetic?
   b. Which are the least energetic?

4. Which kinds of rays are produced by objects radiating thermal energy?

5. Which of the following is not a direct source of visible light?
   a. incandescence
   b. electrical current in a wire
   c. gas-discharge tubes
   d. chemical reactions

6. What is the shape of a light wave front from a point light source when it is initially formed?

7. Whose work provided a mathematical description of light waves?

8. Did the scientist who described light waves mathematically devise the equations that he used?

9. What is a *normal,* in a discussion of reflection?

10. State the law of reflection.

11. **a.** Where is the image in a plane mirror reflection?
    b. Is it real or virtual?

12. What is a virtual image?

13. How is your mind tricked into thinking that the left and right sides of an image in a plane mirror are reversed?

14. Suppose you cut out a letter *B* and hold it up so that you read the letter correctly. Will its reflection in a mirror behind it be backwards?

15. Where should you place an object in front of a concave mirror if you want its image to be larger than it is?

16. Where should you place an object if you want its image in a concave mirror to be virtual?

17. Name two uses for convex mirrors.

True or False (18–29)

18. Radio waves are always less energetic than visible light waves.

19. Red light has a higher frequency than blue light.

20. Extreme ultraviolet light from the Sun is not a danger to life at the surface of the earth.

21. High-energy gamma rays measurable at the earth's surface probably originate from outside the earth.

22. Single-frequency light is obtained from gas discharge and laser light.

23. Coherent light is obtained from cold light sources.

24. The speed of light is a fundamental physical constant. It is approximately 3.00×10^8 m/s wherever it is measured.

25. Spherical wave fronts eventually become plane waves if allowed to propagate long enough and far enough.

26. Light rays travel in straight lines through uniform media.

27. The incident ray, the reflected ray, and the normal at the point of reflection all lie in the same plane.

28. Ideally, all of the incoming light rays that are parallel to the optical axis of a concave mirror are reflected through the same point.

29. A convex mirror cannot produce a real image.

⊚**30.** Calculate the number of images produced by two plane mirrors at a right angle.

⊚**31.** How many images are produced by two plane mirrors at a 30° angle?

⊚**32.** An object is located 30.0 cm from a concave mirror with a focal length of 10.0 cm.
   a. Where will its image be?
   b. Will the image be larger or smaller than the object?

⊚**33.** An object placed 20.0 cm from a concave mirror has an image of the same size at 20.0 cm from the mirror. What is the mirror's focal length?

⊚**34.** A 6.00 cm long pencil is placed 30.0 cm from a concave mirror with a focal length of 10.0 cm. Where is the image and how tall is it?

⊚**35.** A 2.00 cm high eraser is 20.0 cm from a convex mirror with a focal length of 5.00 cm. Where and how tall is its image?

24A	Theory of Refraction	527
24B	Application of Refraction—Lenses	534
Facet:	Fresnel Lenses	545

New Symbols and Abbreviations

index of refraction (n)	24.1
angle of refraction (θ_r)	24.2
angle of incidence (θ_i)	24.2
critical angle (θ_c)	24.3

Digital cameras still require an optical lens to form the image from which a picture is made.

Refraction 24

In this chapter, you will learn about the bending of light and about magnification with lenses. In every case, magnification results from some agent changing our perception of the object by the use of an optical device, such as a lens, or just by your attention being focused on something. When you are taught to draw by an experienced art teacher, you are first instructed to intently observe the subject of your drawing, noting its overall shape, its proportions, the flow and intersections of its lines, its color values, and so on. Your subject fills your consciousness to the exclusion of everything else as you consider how you are to render it in your drawing. You mentally magnify the object. This kind of magnification is what the psalmist was referring to when he wrote that he would magnify the Lord with a song of thanksgiving. As he prayerfully gave thanks, God's presence in his heart and mind grew—was magnified—because he was concentrating only on Him. A Christian needs to frequently seek opportunities to magnify the Lord through song, through the reading of His Word, and through prayer.

24A THEORY OF REFRACTION

24.1 Cause of Refraction

When light travels through matter, it moves more slowly than it does in a vacuum. Some media (kinds of matter through which light traverses) slow light down more than others. When light travels from one medium to another in which its speed is different, the light rays bend. This bending is called **refraction.**

An illustration may help you to understand why light refracts. Imagine a company of soldiers marching in formation. Each column of soldiers represents a light ray, and each row of soldiers represents a wave front. The soldiers are marching on a level concrete surface at an angle to a boundary with an expanse of tall grass and mucky ground. As each soldier reaches the grass, he is forced to slow down. He cannot keep up with the adjacent soldier who is still on the paved surface. Therefore, the row bends. At the same time, the soldiers in the grass are trying to keep aligned with the soldiers in their row, so the column also bends.

concrete

marsh and tall grass

24-1 Refraction can be illustrated by a group of soldiers marching into tall grass from a smooth concrete surface while attempting to keep formation.

The word **interface** is not strictly an optical term. It is useful in any study of something occurring at the boundary between two different substances, or even between systems.

Problem-Solving Strategy 24.1

When the two media at an interface are named, the first contains the incident ray, and the second contains the refracted ray. In an air-glass interface, the ray is moving from the air into the glass.

The **index of refraction (n)** (or the *refractive index*) of a medium is the ratio of the speed of light in a vacuum to the speed of light in the medium. Notice that indices of refraction are dimensionless because the units cancel.

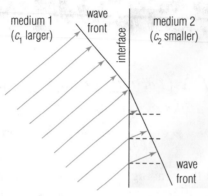

24-2 Light refracting as it passes from one medium into an optically denser one

Similarly, when light passes from one medium to another at an angle, its wave fronts bend. The light rays, which are always perpendicular to the wave front, bend too. Figure 24-2 shows the direction that a light ray bends. When light passes from a medium in which its speed is greater to a medium in which its speed is less, it bends toward a *normal* line (a line drawn perpendicular to the boundary at the point where the ray crosses the boundary). When light goes from a medium in which it moves slower to a medium in which it is faster, it bends away from the normal. The boundary between two media is called the **interface** between the media. Light travels slower in a medium that is optically denser. The optical density of a medium is related to, but not directly proportional to, its mass density.

How much does a medium slow light? The optical density of a medium is indicated by its **index of refraction (n)**, which is the ratio of the speed of light in a vacuum to the speed of light in the medium.

$$n = \frac{c_{\text{vac}}}{c_{\text{med}}} \qquad (24.1)$$

From the medium's refractive index, you can calculate the speed of light in the medium.

TABLE 24-1	
Selected Indices of Refraction	
Medium	*n*
air	1.0003
hydrogen	1.0001
carbon dioxide	1.0004
ice	1.309
methyl alcohol	1.33
water	1.333
ethyl alcohol	1.36
fluorite	1.43
Lucite	1.51
glass, crown	1.52
sodium chloride	1.54
glass, flint	1.64
diamond	2.42

EXAMPLE 24-1

Speed Zone: Speed of Light in Water

Water's refractive index is 1.33 at room temperature. What is the speed of light in water?

Solution:

$$n_{\text{water}} = 1.33$$

$$n_{\text{water}} = \frac{c_{\text{vac}}}{c_{\text{water}}}$$

$$c_{\text{water}} = \frac{c_{\text{vac}}}{n_{\text{water}}} = \frac{3.00 \times 10^8 \text{ m/s}}{1.33}$$

$$c_{\text{water}} \doteq 2.2\underline{5}5 \times 10^8 \text{ m/s}$$

$$c_{\text{water}} \approx 2.26 \times 10^8 \text{ m/s}$$

24.2 Description of Refraction

In 1621 Willebrord Snell discovered a mathematical relationship between the **angle of refraction** (θ_r) (the angle between the normal and the refracted ray) and the **angle of incidence** (θ_i) (the angle between the normal and the incident ray in the first medium). This relationship is known as **Snell's law:**

$$n_i \sin \theta_i = n_r \sin \theta_r \qquad (24.2)$$

In this equation n_i is the index of refraction of the medium containing the incident ray, n_r is the index of refraction of the medium containing the refracted ray, θ_i is the angle of incidence, and θ_r is the angle of refraction.

Figure 24-3 shows why Snell's law is true. A light ray is being refracted as it travels from the first medium to the second medium, and a circle is drawn with its center at the point of refraction. Both $\overline{AB}$ and $\overline{BC}$ are radii of the circle; therefore, $AB = BC$. The dotted lines $\overline{AE}$ and $\overline{CD}$ are perpendicular to the normal and are called *semichords,* since they each form half a chord of the circle. Results of experiments that measure various angles of incidence and refraction at a glass-air interface and the corresponding ratios of *AE/CD* are tabulated in Table 24-2.

TABLE 24-2				
θ_i (°)	θ_r (°)	*AE* (cm)	*CD* (cm)	*AE/CD*
10	6.7	3.43	2.30	1.51
20	13.3	6.84	4.60	1.49
30	19.6	10.00	6.71	1.49
40	25.2	12.83	8.52	1.51
50	30.7	15.32	10.2	1.50
60	35.1	17.32	11.5	1.51
70	36.6	18.79	12.5	1.50
80	40.6	19.69	13.02	1.51

The ratio *AE/CD* is a constant, within experimental error. It is also close to the ratio of the index of refraction of glass to the index of refraction of air,

$$\frac{AE}{CD} = \frac{n_{glass}}{n_{air}}.$$

How is *AE/CD* related to the angles of incidence and refraction? Notice that $\triangle BDC$ and $\triangle BEA$ are right triangles. Recall that the angles and sides of right triangles can be related by the use of sine ratios. The sine of an acute angle in a right triangle is defined as

$$\sin \theta = \frac{\text{opposite}}{\text{hypotenuse}}.$$

In Figure 24-3,

$$\sin \theta_i = \frac{AE}{AB} \text{ and } \sin \theta_r = \frac{CD}{BC}.$$

Since $AB = BC$, $\sin \theta_i$ can be written as

$$\sin \theta_i = \frac{AE}{BC}.$$

The ratio $\sin \theta_i / \sin \theta_r$ is

$$\frac{\sin \theta_i}{\sin \theta_r} = \frac{AE/\cancel{BC}}{CD/\cancel{BC}} = \frac{AE}{CD}.$$

Willebrord Snell (1580–1626) was a Dutch mathematician best known for showing that different materials have different indices of refraction.

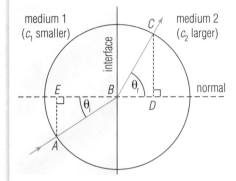

24-3 Derivation of Snell's law

The symbol $\overline{AE}$ represents the line segment between endpoints A and E. The magnitude of the line segment is indicated simply by the endpoints (AE).

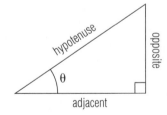

24-4 The relationship of the sides to an angle in a right triangle

We know that

$$\frac{AE}{CD} = \frac{n_{glass}}{n_{air}}, \text{ so}$$

$$\frac{\sin \theta_i}{\sin \theta_r} = \frac{n_{glass}}{n_{air}}. \tag{24.3}$$

Equation 24.3 is true for any combination of transparent materials. If a light ray passes from a medium containing the incident ray with an index of refraction n_i to a second medium with an index of refraction n_r, then

$$\frac{\sin \theta_i}{\sin \theta_r} = \frac{n_r}{n_i}.$$

By cross-multiplying, you get the equation

$$n_i \sin \theta_i = n_r \sin \theta_r, \tag{24.4}$$

which is the mathematical statement of Snell's law.

EXAMPLE 24-2

Finding the Angle of Refraction, θ_r

Find the angle of refraction when an incident ray of $\theta_i = 30.0°$ passes from air into ethyl alcohol (subscript ETOH).

Solution:
From Table 24-1 you find that $n_{ETOH} = 1.36$.

$$n_i \sin \theta_i = n_r \sin \theta_r$$

Dividing by n_r, you get the following:

$$\frac{n_i \sin \theta_i}{n_r} = \sin \theta_r$$

$n_{air} = n_i = 1.00$; $n_{ETOH} = n_r = 1.36$; $\theta_i = 30.0°$

$$\sin \theta_r = \frac{1.00 \sin 30.0°}{1.36}$$

$$\sin \theta_r \doteq 0.3676$$

$$\theta_r = \sin^{-1}(\sin \theta_r) \doteq 21.\underline{5}6°$$

$$\theta_r \approx 21.6°$$

The angle of refraction in the ethyl alcohol is about 21.6° from the normal.

EXAMPLE 24-3

Finding an Unknown Index of Refraction

A ray of light passes through the wall of a glass container ($n_{glass} = 1.50$) to the liquid in the container at the glass-liquid interface. The angle of incidence is 35.0°, and the angle of refraction is 40.0°. What is the liquid's index of refraction?

Solution:

$$n_i \sin \theta_i = n_r \sin \theta_r$$

Divide by $\sin \theta_r$ to isolate n_r:

$$\frac{n_i \sin \theta_i}{\sin \theta_r} = n_r$$

$$n_{glass} = n_i = 1.50; \; \theta_i = 35.0°; \; \theta_r = 40.0°$$

$$n_r = \frac{1.50 \sin 35.0°}{\sin 40.0°}$$

$$n_r \doteq 1.3\underline{38}$$

$$n_r \approx 1.34$$

The liquid's index of refraction is about 1.34.

24.3 Critical Angle of Incidence

Let's look at Snell's law a little closer:

$$n_i \sin \theta_i = n_r \sin \theta_r$$

If n_i is greater than n_r, then $\sin \theta_i$ must be less than $\sin \theta_r$ to maintain the equality. Therefore, light moving from one medium to another with a lower index of refraction has an angle of refraction larger than its angle of incidence. Such a boundary has interesting effects on light with large angles of incidence.

When a light ray passes from a medium of low *n* to one of higher *n*, the ray bends toward the normal. When the ray passes into a medium of lower *n*, the ray bends away from the normal.

EXAMPLE 24-4

Losing the Refracted Ray: Critical Angle

A light ray passes from water to air with an angle of incidence of 48.61°. What is its angle of refraction?

Solution:

$$n_i \sin \theta_i = n_r \sin \theta_r$$

Divide by n_r to isolate $\sin \theta_r$.

$$\frac{n_i \sin \theta_i}{n_r} = \sin \theta_r$$

$$\theta_r = \sin^{-1}\left(\frac{n_i \sin \theta_i}{n_r}\right)$$

$$n_{water} = n_i = 1.333; \; n_{air} = n_r = 1.000; \; \theta_i = 48.61°$$

$$\theta_r = \sin^{-1}\left(\frac{1.333 \sin 48.61°}{1.000}\right)$$

$$\theta_r = 90.0°$$

The refracted ray will travel along the interface boundary and be lost.

Problem-Solving Strategy 24.3

Critical angles exist only when the refraction index of the medium containing the incident ray is larger than that of the medium containing the refracted ray ($n_i > n_r$).

What if the angle of incidence in the example were greater than 48.61°? Then $\sin \theta_r > 1.000$. But no angle has a sine greater than 1. Therefore, the ray is not refracted at all but is reflected back into the water. This phenomenon, called **total internal reflection,** occurs at a water-air interface whenever light rays have an angle of incidence greater than 48.61°. This angle of incidence is called the **critical angle (θ_c).** Every interface where light passes from a medium with a higher

Problem-Solving Strategy 24.4

Snell's law is valid as long as the critical angle at the interface of the two media is not exceeded.

index of refraction into a medium with a lower index of refraction has a critical angle of incidence at which light is refracted 90° along the interface surface. At angles of incidence greater than the critical angle, the light experiences total internal reflection.

EXAMPLE 24-5

Finding the Critical Angle

Find the critical angle for a glass-air boundary.

Solution:
You want to find θ_i for $\theta_r = 90.0°$.

$$n_i \sin \theta_i = n_r \sin \theta_r$$

Divide by n_i to isolate $\sin \theta_i$:

$$\sin \theta_i = \frac{n_r \sin \theta_r}{n_i}$$

$$\theta_i = \sin^{-1}\left(\frac{n_r \sin \theta_r}{n_i}\right)$$

$n_{glass} = n_i = 1.50$; $n_{air} = n_r = 1.00$; $\theta_r = 90.0°$

$$\theta_i = \sin^{-1}\left(\frac{1.00 \sin 90.0°}{1.50}\right) \doteq 41.\underline{8}1°$$

$$\theta_i \approx 41.8°$$

The critical angle for the glass-air interface is about 41.8°.

Total internal reflection has many uses. One application is in **fiber optics,** a technology in which a light beam is directed into a glass fiber. The light is internally reflected totally within the fiber, even if the fiber bends, as long as it is not bent sharply. A fiber used in this way is called a *light pipe.* Bundles of fibers in the form of cables transmit high-speed digital signals across great distances. The advantages of fiber optics in digital telecommunications are the higher signal-to-noise ratio possible with the optical signal and the fact that fiber optics are impervious to *electromagnetic interference (EMI)* along the cable, which can degrade a standard electrical or radio-frequency signal. Fiber optics are also used for optical inspection of inaccessible locations within machinery and the human body. Doctors regularly use fiber-optic devices for observing surgical procedures within small cavities in a patient's body as well as for carrying precision laser beams required for delicate surgery.

24-5 Laser light delivered by a fiber-optic cable is used for extremely precise brain surgery for which a scalpel could not be used.

24.4 Phenomena Involving Refraction

We have been discussing the refraction of light as though all light refracts in the same way. Actually, different frequencies of light have different indices of refraction in the same materials. These differences occur because the wavelengths of light toward the blue end of the spectrum are closer to the dimensions of the particles in matter, so those colors tend to interact more strongly with matter and are slowed more than colors toward the red end of the spectrum. In our examples and tables, we have been using an average index of refraction, based on a yellowish-colored light. Since each frequency of light has a different index of refraction, different frequencies with the same angles of incidence have different angles of refraction. If white light—light of all frequencies—is incident on a refractive interface, the different frequencies will not emerge from the interface at the same angle. Instead, they will be spread out, or *dispersed,* displaying a spectrum of colors.

A prism is a triangular solid, usually made of glass, that causes **dispersion.** When white light enters the glass, the different frequencies are refracted at different angles. Emerging from the glass, the rays of different frequencies separate even more. (For glass with parallel edges, the

24-6 A prism disperses light

second edge would reverse the dispersion, not continue it.) Since different frequencies have different colors, it is easy to see the dispersion of white light through a prism. The dispersion can be reversed temporarily with a second wider prism placed upside down in the dispersed beam. However, reconstruction of the original light beam is not easily done, because it requires a series of prisms and lenses.

An atmospheric rainbow results from dispersion only when specific conditions exist. Numerous water droplets must be illuminated directly by sunlight, and the observer must be between the Sun and the water droplets. The colors of the sunlight are dispersed by being refracted twice and reflected once inside each raindrop (see Figure 24-7). The arching shape of the rainbow is due to the geometry of the effect. The observer sees the light rays returning from the water droplets within a certain angle of an imaginary line that passes from the Sun through the observer's head, called the *antisolar line.* Since blue light is refracted more than red, it appears about 40° from the antisolar line. Red light, being least refracted, appears about 42° away. The other colors fall between. If you were in an airplane observing a rainbow, you could see a complete circle of colors. Because the ground intersects the circle under most circumstances, you see only the upper arch. Secondary rainbows can sometimes be observed about 51° from the antisolar line. These rays are reflected twice inside the water drop, reversing the order of the colors in the process.

Refraction is often the cause of optical illusions. If you dip a stick into a glass of water, it looks bent and distorted in size. As Figure 24-8 (b) shows, the ruler looks bent because the eye

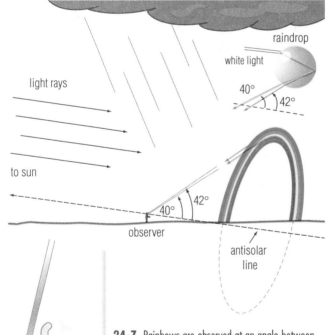

24-7 Rainbows are observed at an angle between 40° and 42° from the antisolar line.

24-8 A partially immersed ruler appears bent due to refraction.

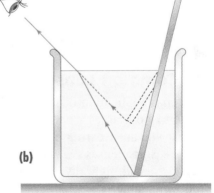

(b)

receives refracted rays of light from the stick. The mind is designed so that it interprets light rays as straight lines. Therefore it extrapolates the rays back to where they would have met if they had not been refracted, and it constructs the image at that location.

Refraction can occur in the atmosphere when layers of hot air are close to layers of cool air. Air's index of refraction varies with temperature and humidity—hot air is less dense, and cold air is more dense. Index of refraction generally increases with air density, so light moving from the cool air to the hot, or vice versa, is refracted. In extreme cases this atmospheric refraction makes light rays from the sky appear to come from the ground, or rays from an object near the ground appear to come from the sky. The former is called a mirage, which is the refraction of the sky and horizon that gives the appearance of a reflection from the surface of water in the desert. The latter, called an *inferior mirage* or *looming*, causes ships and structures to appear to float above the horizon and makes distant lighthouses and ships visible that should be hidden by the earth's curvature.

Minor refraction in the atmosphere due to atmospheric movement makes the stars "twinkle." Beyond our atmosphere, the stars are seen to shine steadily. Similarly, on hot days distant objects viewed through the unstable air over a black road shimmer because of refraction.

Snell's law assumes that a distinct interface exists between two uniform media. The word "uniform" is key. If the optical characteristics of a medium gradually change with position or time, then a ray passing through the medium is going to be refracted in curves rather than in angles between straight segments. This kind of refraction is responsible for many atmospheric visual phenomena and the propagation of sound through the deep ocean basins.

24A Section Review Questions

1. How does refraction occur?
2. What measurable quantities are considered in Snell's law?
3. What is the angle of refraction when the incident ray is at the critical angle for the two materials forming the interface?
4. Explain why red light is refracted the least and blue light is refracted the most.
⊛5. Calculate the speed of light in diamond.
⊛6. A new thermoplastic is being tested to determine its properties. A laser beam enters a sample with an incident angle of 35.0°. The beam is refracted at an angle of 21.3° to the normal. What is the plastic's refractive index?
⊛7. What is the critical angle for a diamond-air interface?
⊛8. Imagine that you are sitting on the bottom of a swimming pool looking upward toward the sky. What is the maximum angle from the vertical at which you can see the sky?

24B APPLICATION OF REFRACTION—LENSES

24.5 Common Lenses

One familiar application of refraction is in lenses—disks of transparent material with one or both sides machined into a three-dimensional curved surface. It is likely that you are wearing corrective lenses for your eyes, either glasses or contact lenses. Some lenses *converge,* or bring together, light rays, while other lenses *diverge,* or spread, light rays. Common lenses and their effects are summarized in the following table.

TABLE 24-3

Lens Type		Effect	Lens Type		Effect
Plane convex		converging	Plane concave		diverging
Double convex		converging	Double concave		diverging

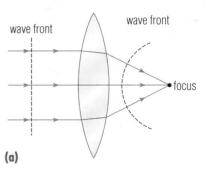

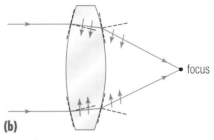

24-9 Converging lens function considering (a) a wave front and (b) ray paths.

24.6 Converging Lenses

Notice that a lens which is thicker in the middle than at the edges is a **converging lens.** Figure 24-9 shows why. Assume that the wave front is planar at first. The rays near the edges of the lens have to go through less of the lens; thus they are slowed less than the rays near the middle, which must go through more lens. Therefore, the outer wavelets "get ahead" of the inner wavelets, and the wave-front envelope curves. The wave fronts are refracted so that after passing through the lens they meet at a point called the *focus* or *focal point* of the lens.

One ray's refraction can also show how the rays converge. The ray strikes the lens, and, since it is entering a material with a higher index of refraction than air, it bends toward the normal. When the ray leaves the lens, it enters the air, which has a lower index of refraction than the lens; so the ray bends away from the normal. Both refraction paths tend to bend the ray toward the focus.

A thin converging lens is much like a concave mirror. The same terminology is used for both. Like concave mirrors, converging lenses have a focus at a distance f from their centers on the optical axis. For symmetrical spherical lenses, f is one-half the radius (R) of the lens's curvature. One difference between mirrors and lenses is that the lenses' real image is located on the opposite side of the lens from its object. We will always assign the object distance (d_O) a negative value. The image may be real or virtual at a distance d_I from the center of the lens. If the image is real, d_I will have a positive value; if the image is virtual, it will appear to be on the same side of the lens as the object and will have a negative d_I. The mirror equation must be modified to relate these quantities correctly:

Problem-Solving Strategy 24.5

All lens distances should be measured from the center of the lens, not from the surface.

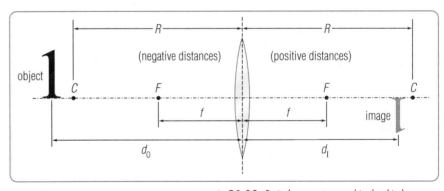

24-10 Optical parameters used in the thin lens equation

$$-\frac{1}{d_O} + \frac{1}{d_I} = \frac{1}{f} \quad (24.5)$$

Since only thin lenses obey this equation, it is called the **thin-lens equation.**

The height of a thin-lens image can be calculated using an equation similar to Equation 23.2 but without the negative sign on the right. This equation is valid if the medium on both sides of the lens is the same, as in the case of a hand magnifier.

$$\frac{H_I}{H_O} = \frac{d_I}{d_O} \quad (24.6)$$

Magnification of a thin lens uses the same formula as Equation 23.3:

$$m = \left| \frac{H_I}{H_O} \right| \quad (24.7)$$

A *thin lens* consists of two refracting surfaces having small distance between them compared to (a) image and object distances and (b) the curvature radii of the surfaces. The diagrams of lenses in this chapter are shown relatively thick for illustrative purposes.

As with mirrors, lens magnification is denoted by the word "power" or the symbol "×." A lens that achieves a ten-fold magnification is a ten-power (10×) lens.

EXAMPLE 24-6

Maintaining Focus: Using the Thin-Lens Equation

A double convex thin lens has a focal length of 16.0 cm. An object 6.00 cm high is placed 40.0 cm from the lens. Find (a) the image distance and (b) the height of the image.

Solution:

a. Find the image distance using Equation 24.5:

$$-\frac{1}{d_0} + \frac{1}{d_I} = \frac{1}{f}$$

$$\frac{1}{d_I} = \frac{1}{f} + \frac{1}{d_0} = \frac{1}{16.0 \text{ cm}} + \frac{1}{-40.0 \text{ cm}}$$

Multiply by the common denominator $(80.0 \text{ cm})d_I$.

$$\frac{(80.0 \text{ cm})d_I}{d_I} = \frac{(80.0 \text{ cm})d_I}{16.0 \text{ cm}} - \frac{(80.0 \text{ cm})d_I}{40.0 \text{ cm}}$$

$$80.0 \text{ cm} = 5.00d_I - 2.00d_I = 3.00d_I$$

$$d_I = \frac{80.0 \text{ cm}}{3.00}$$

$$d_I = 26.6\overline{6} \text{ cm}$$

$$d_I \approx 26.7 \text{ cm}$$

b. Calculate the height of the image:

$$\frac{H_I}{H_0} = \frac{d_I}{d_0}$$

$$H_I = H_0\frac{d_I}{d_0}$$

$H_0 = 6.00 \text{ cm}$; $d_0 = -40.0 \text{ cm}$; $d_I = 26.7 \text{ cm}$

$$H_I = 6.00 \text{ cm}\frac{26.6\overline{6} \text{ cm}}{-40.0 \text{ cm}}$$

$$H_I \doteq -4.00\underline{0} \text{ cm}$$

$$H_I \approx -4.00 \text{ cm.}$$

The real image is about 26.7 cm from the lens and 4.00 cm high (inverted).

Problem-Solving Strategy 24.6

Remember to include the negative sign in front of the object term in the thin-lens equation. This sign is required because the object distance is always negative for lenses.

Christians are admonished to magnify God's work. "Remember that thou magnify his work, which men behold" (Job 36:24). How can Christians magnify God's work? By living their lives so men cannot ignore or deny the miraculous change He has wrought in their lives.

Chapter 23 explained six cases for the concave mirror. The same six cases may be applied to the converging lens, with the same results. Table 24-4 lists the characteristics of the image for each location of the object, and Figure 24-11 shows how each case produces an image.

TABLE 24-4

Object Location	Image Location	Image Size	Orientation	Type
$d_0 = \infty$	$d_I = f$	minuscule	inverted	real
$d_0 > R$	$f < d_I < R$	$H_I < H_0$	inverted	real
$d_0 = R$	$d_I = R$	$H_I = H_0$	inverted	real
$f < d_0 < R$	$d_I > R$	$H_I > H_0$	inverted	real
$d_0 = f$	d.n.e. (∞)	d.n.e.	—	—
$d_0 < f$	$d_I < 0$	$H_I > H_0$	erect	virtual

Note: "d.n.e." means "does not exist" (value is infinitely large).

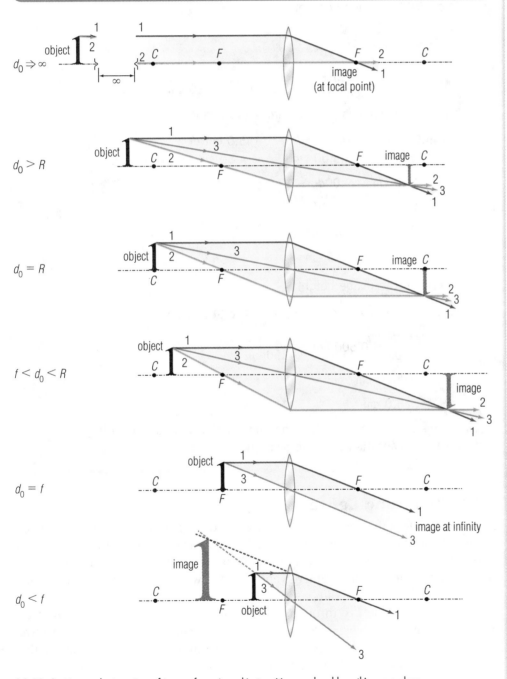

24-11 Positions and orientations of images for various object positions produced by a thin convex lens

The last case, where $d_O < f$, is most effective for magnifying. The image appears to be on the same side of the lens as the object; therefore, the image is virtual. A magnifying glass uses a converging lens to magnify an object that is less than one focal length away from the lens.

EXAMPLE 24-7

Virtually Larger: Magnification in a Thin Lens

A magnifying glass has a focal length of 5.00 cm. If an object 0.500 cm high is placed 4.50 cm from the lens, (a) where is the image and (b) how large is it?

Solution:

a. Calculate the image position:

$$-\frac{1}{d_O} + \frac{1}{d_I} = \frac{1}{f}$$

$$\frac{1}{d_I} = \frac{1}{f} + \frac{1}{d_O} = \frac{1}{5.00 \text{ cm}} + \frac{1}{-4.50 \text{ cm}}$$

Multiply by the common denominator $(45.0 \text{ cm})d_I$.

$$\frac{(45.0 \text{ cm})d_I}{d_I} = \frac{(45.0 \text{ cm})d_I}{5.00 \text{ cm}} - \frac{(45.0 \text{ cm})d_I}{4.50 \text{ cm}}$$

$$45.0 \text{ cm} = 9.00d_I - 10.0d_I$$

$$45.0 \text{ cm} = -1.0d_I$$

$$d_I \doteq -45.0 \text{ cm}$$

$$d_I \approx -45 \text{ cm}$$

b. Determine the image height:

$H_0 = 0.500$ cm; $d_I = -45.0$ cm; $d_I = -4.50$ cm.

$$H_I = H_0\frac{d_I}{d_O} = 0.500 \text{ cm} \cdot \frac{-4.50 \text{ cm}}{-4.50 \text{ cm}}$$

$$H_I \doteq +5.00 \text{ cm}$$

$$H_I \approx +5.0 \text{ cm}$$

The image appears to be 45 cm behind the lens and is 5.0 cm high (erect). It has been magnified by a factor of 10.

24.7 Diverging Lenses

A lens with edges thicker than the middle is a **diverging lens.** Consider a planar wave front approaching a diverging lens. The light near the edges of the lens must pass through more of the lens than the light near the middle of the lens. Therefore, the wavelets near the center emerge ahead of the wavelets near the ends, and the wave-front envelope is curved.

Following a single ray through the lens also shows why the light diverges. When a ray enters the lens, it bends toward the normal, as Figure 24-12 (b) shows. When it emerges from the lens, it bends away from the normal. You can see that both of these refractions direct the ray away from the optical axis; therefore the rays diverge.

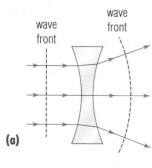

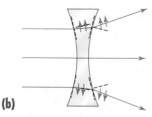

24-12 (a) Wave front refracted by diverging lens; (b) Light rays diffracted by diverging lens

The diverging lens behaves much like a convex mirror. Also like a convex mirror, a diverging lens produces only one kind of image: virtual, erect, and smaller than the object. The thin-lens equation describes the location of the image. Because the virtual image appears to be on the same side of the lens as the object, diverging lenses are given negative focal lengths.

Diverging lenses produce only virtual images. For this reason, the virtual image is always on the same side of the lens as the object, which means the image distance has a negative value.

EXAMPLE 24-8

Divergent Point of View: Image in a Diverging Lens

An object 8.00 cm high is 30.0 cm from a lens with a focal length of -10.0 cm. Find (a) the position and (b) the height of the image.

Solution:

a. Calculate the image position:

$d_0 = 30.0$ cm and $f = -10.0$ cm.

$$-\frac{1}{d_0} + \frac{1}{d_I} = \frac{1}{f}$$

$$\frac{1}{d_I} = \frac{1}{f} + \frac{1}{d_0} = \frac{1}{-10.0 \text{ cm}} + \frac{1}{-30.0 \text{ cm}}$$

Multiply by the lowest common denominator, $(30.0 \text{ cm})d_I$.

$$\frac{(30.0 \text{ cm})d_I}{d_I} = -\frac{(30.0 \text{ cm})d_I}{10.0 \text{ cm}} - \frac{(30.0 \text{ cm})d_I}{30.0 \text{ cm}}$$

$$30.0 \text{ cm} = -3.00d_I - 1.00d_I = -4.00d_I$$

$$d_I = -7.50 \text{ cm}$$

b. Determine image height:

$H_0 = 8.00$ cm; $d_I = -7.50$ cm; $d_0 = -30.0$ cm.

$$H_I = H_0\frac{d_I}{d_0} = 8.00 \text{ cm} \cdot \frac{-7.50 \text{ cm}}{-30.0 \text{ cm}}$$

$$H_I = +2.00 \text{ cm}$$

The image appears to be 7.50 cm behind the lens and is 2.00 cm high and erect.

Problem-Solving Strategy 24.7
Images formed by diverging lenses are always virtual and erect.

Remembering the following rules for signs will help you to use the thin-lens equation. These rules apply strictly to images formed by single lenses.

1. All object distances (d_O) are negative (behind the lens).
2. All real images are in front of the lens. Their image positions (d_I) are positive.
3. All virtual images are behind the lens, hence their positions are negative.
4. Focal lengths of converging lenses are positive; focal lengths of diverging lenses are negative.
5. Objects are assumed to be erect (positive heights). Real images are inverted (negative heights) while virtual images are erect (positive heights).

From his point of view, an observer is in front of the lens and the object is behind the lens. Thus, all positions in front of the lens have positive distances, and all positions behind the lens have negative distances.

Refraction **539**

24.8 Combining Lenses

So far, we have discussed only forming an image with a single lens. How does a combination of lenses affect light? Telescopes, microscopes, binoculars, and cameras all use combinations of lenses to form an image. If one lens receives light that has passed through another lens, it reacts just as it would to any other light. A real image formed by the first lens is treated as an object by the next lens. Figure 24-13 shows a lens producing an image I_1 of the object. The next lens takes I_1 as its object and produces a second image, I_2. The location and size of I_1 can be found by applying the thin-lens equation to the object's distance from the first lens. The location and size of I_2 can be found by applying the thin-lens equation to I_1 and the second lens.

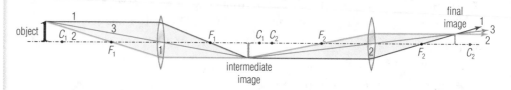

24-13 An image formed by a combination of lenses

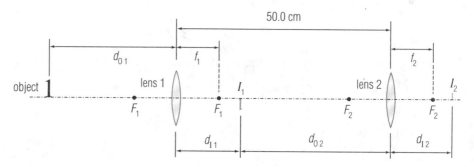

24-14 Schematic of the lenses and images for Example 24-9

EXAMPLE 24-9

Scoping It Out: Combining Lenses

A 1.00 cm object is 30.0 cm from a lens with a focal length of 10.0 cm. The first lens is 50.0 cm from a second lens with a focal length of 10.0 cm. What are the sizes and locations of (a) the image from the first lens and (b) the image from the second lens?

Solution:

a. For I_1, $d_{0\,1} = -30.0$ cm, $f_1 = 10.0$ cm:

$$\frac{1}{d_{I\,1}} = \frac{1}{f_1} + \frac{1}{d_{0\,1}} = \frac{1}{10.0\ \text{cm}} + \frac{1}{-30.0\ \text{cm}}$$

Multiply by $(30.0\ \text{cm})d_{I\,1}$:

$$\frac{(30.0\ \text{cm})d_{I\,1}}{d_{I\,1}} = \frac{(30.0\ \text{cm})d_{I\,1}}{10.0\ \text{cm}} - \frac{(30.0\ \text{cm})d_{I\,1}}{30.0\ \text{cm}}$$

$$30.0\ \text{cm} = 3.00d_{I\,1} - 1.00d_{I\,1} = 2.00d_I$$

$$d_{I\,1} = +15.0\ \text{cm}$$

Find the height of I_1:

$$H_{I\,1} = H_{0\,1}\frac{d_{I\,1}}{d_{0\,1}} = 1.00 \text{ cm} \cdot \frac{+15.0 \text{ cm}}{-30.0 \text{ cm}}$$

$$H_{I\,1} = -0.500 \text{ cm}$$

The real image, I_1, is 15.0 cm from lens 1 between lenses 1 and 2. Since there is 50.0 cm between the lenses, I_1 is 35.0 cm from lens 2.

b. For the second lens, $d_{0\,2} = -35.0$ cm, $f_2 = 10.0$ cm:

$$\frac{1}{d_{I\,2}} = \frac{1}{f_2} + \frac{1}{d_{0\,2}} = \frac{1}{10.0 \text{ cm}} + \frac{1}{-35.0 \text{ cm}}$$

Multiply by $(70.0 \text{ cm})d_{I\,2}$:

$$\frac{(70.0 \text{ cm})d_{I\,2}}{d_{I\,2}} = \frac{(70.0 \text{ cm})d_{I\,2}}{10.0 \text{ cm}} - \frac{(70.0 \text{ cm})d_{I\,2}}{35.0 \text{ cm}}$$

$$70.0 \text{ cm} = 7.00d_{I\,2} - 2.00d_{I\,2} = 5.00d_{I\,2}$$

$$d_{I\,2} = +14.0 \text{ cm}$$

Find the height of the second image:

$$H_{I\,2} = H_{0\,2}\frac{d_{I\,1}}{d_{0\,1}} = -0.500 \text{ cm} \cdot \frac{+14.0 \text{ cm}}{-35.0 \text{ cm}}$$

$$H_{I\,2} = +0.200 \text{ cm}$$

The second image, I_2, is 14.0 cm beyond the second lens and is 0.200 cm high and erect. This *real* image is erect because the object (I_1) was inverted.

Problem-Solving Strategy 24.8

In a compound lens system, it is *not* necessary for the refracted rays from the first lens to form a focused image before they enter the second lens.

24.9 Correcting Lens Aberrations

Spherical mirrors cannot reflect parallel light rays to a single point because of spherical aberration. Similarly, spherical lenses cannot refract rays near their edges according to the rules you have studied. The *spherical aberration* of a lens produces blurry images. Several solutions to help prevent this blurriness are possible. First, the edges of the lenses may be covered by an iris diaphragm, an opaque ring which reduces the effective diameter of the lens, called its **aperture.** A smaller diameter reduces the total lens curvature in the light path, thus reducing the effect of this kind of aberration. However, iris diaphragms restrict the amount of light through the lens, so they cannot be used where much light is required, as in telescopes. Another solution is to make the two halves of a double convex lens asymmetrical. The difference in curvature can cause peripheral rays to focus properly. The most satisfying approach to the problem, however, is to use two lenses so that the second lens corrects the problem of the first. Lenses of different indices of refraction are often able to compensate for each other's aberrations. The finest optical instruments use this last approach to correct spherical aberrations.

24-15 Refractor telescopes depend entirely on combinations of lenses to collect light and form a magnified image.

Our sin nature is an aberration in mankind introduced at the Fall. Christians must rely on the "corrective lens" of the Holy Spirit to permit the light of Christ in them to shine forth undistorted. "But ye shall receive power, after that the Holy Ghost is come upon you: and ye shall be witnesses unto me both in Jerusalem, and in all Judea, and in Samaria, and unto the uttermost part of the earth" (Acts 1:8).

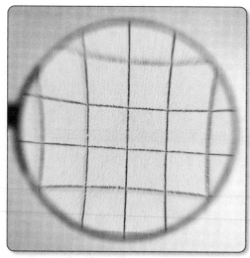

24-16 Simple spherical lenses have inherent optical aberrations that degrade the images they produce. Note the blurring and color dispersion near the edge of the lens.

24-17 Achromatic lenses come in many different combinations of elements in order to provide the correct combination of aperture and focal length.

The term *achromatic* means "without color."

Another kind of problem that lenses have is called **chromatic aberration**—the unwanted dispersion of colors. The solution for this problem is the same as the solution for spherical aberration. Simply combine two lenses of different materials so that the second lens can recombine the colors dispersed by the first lens. For example, a converging crown-glass ($n \approx 1.5$) lens bonded to a diverging flint-glass ($n \approx 1.6$) will exhibit little or no spherical or chromatic aberrations. Such a combination is called an *achromatic doublet.*

24.10 Special Lenses

Lenses are not restricted to the shapes discussed above. One special kind of lens is the meniscus lens, so-called because it has a concave and a convex surface. Meniscus lenses may be either converging or diverging, depending on whether their center or their edges are thicker. If the lens's center is thicker, it is a converging lens; if its edges are thicker, it is a diverging lens. Meniscus lenses are used in eyeglasses to allow for movement of the eyelashes and to reduce distortions near the edges due to movement of the eyeball. Converging lenses are used for farsightedness, and diverging lenses correct nearsightedness.

24-18 A meniscus lens, like the kind used in eyeglasses

24B Section Review Questions

1. Discuss the differences between the shape of a converging lens and that of a diverging lens.
2. **a.** When you are considering the positions of the object and the image, what sign is conventionally given to the object distance (d_O)?
   b. Is this position in front of or behind the lens in relation to the observer?
3. What restriction exists on the use of the image-height and magnification equations (24.6 and 24.7)?
4. Where must the object be in relation to the lens for the lens to be used as a magnifier?
5. What can a manufacturer do to correct a lens for spherical and chromatic aberrations?

6. What is the size of the image of the Sun produced by a lens with a 20.0 cm focal length? (*Note:* Use the Earth's and Sun's astronomical data found in the appendices.)

7. In a school laboratory, a physics student observes that a 12.0 cm candle 36.0 cm from a thin lens produces on a screen a sharp inverted image that measures 3.0 cm high. What is the focal length (*f*) and the radius of curvature (*R*) of the lens?

8. What is the magnification of the lens in Question 7?

Chapter Review

In Terms of Physics

refraction	24.1	fiber optics	24.3
interface	24.1	dispersion	24.4
index of refraction (n)	24.1	converging lens	24.6
angle of refraction (θ_r)	24.2	thin-lens equation	24.6
Snell's law	24.2	diverging lens	24.7
total internal reflection	24.3	aperture	24.9
critical angle	24.3	chromatic aberration	24.9

Problem-Solving Strategies

24.1 (page 528) When the two media at an interface are named, the first contains the incident ray, and the second contains the refracted ray. In an air-glass interface, the ray is moving from the air into the glass.

24.2 (page 530) You can avoid rounding error when using Snell's law to compute angles of incidence or refraction by taking the inverse sine (arcsine) of the right-hand side of the equation. For example, the angle of refraction is found by

$$\theta_r = \sin^{-1}\left(\frac{n_i \sin \theta_i}{n_r}\right).$$

24.3 (page 531) Critical angles exist only when the refraction index of the medium containing the incident ray is larger than that of the medium containing the refracted ray ($n_i > n_r$).

24.4 (page 532) Snell's law is valid as long as the critical angle at the interface of the two media is not exceeded.

24.5 (page 535) All lens distances should be measured from the center of the lens, not from the surface.

24.6 (page 536) Remember to include the negative sign in front of the object term in the thin-lens equation. This sign is required because the object distance is always negative for lenses.

24.7 (page 539) Images formed by diverging lenses are always virtual and erect.

24.8 (page 541) In a compound lens system, it is *not* necessary for the refracted rays from the first lens to form a focused image before they enter the second lens.

Review Questions

1. What is refraction?

2. What is the surface where water and glass meet called?

3. State Snell's law.

4. a. What is a critical angle?

   b. Does it apply to the interface between any two materials? Explain.

5. What happens to a light ray at an interface when its angle of incidence is greater than the critical angle?

6. How is internal reflection applied in long-distance communication technology?

7. Why does a prism transform a beam of white light to a band of colors?

8. If a person needs a large correction in his glasses for near or far vision, what could be one undesirable consequence of using a material with a low index of refraction for the lenses?

9. Which color is on the outside or on top of a primary rainbow?

10. Based on what you know about refraction in air, does the ocean horizon normally appear closer or farther away than it would appear if air had an index of refraction of $n = 1$? Assume that the air is coolest and densest near the ocean's surface. Explain your answer.

11. You are looking through a pair of binoculars at a distant lighthouse that appears to be floating above the water's surface at the horizon. What kind of optical phenomenon is this?

12. Classify the following lenses as converging or diverging.

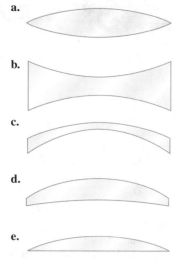

   a.

   b.

   c.

   d.

   e.

13. Where are real images formed by lenses in relation to their objects?

14. Why does the thin-lens equation have a negative sign in front of the object-related term?

15. If you want to magnify an object, where should you put it with respect to the focal point of the lens?

16. Can a diverging lens be used as a magnifying glass? Explain your answer.

17. Some scuba divers have large lenses manufactured to their eyeglass prescription glued to their mask faceplates to help them see more clearly. What must be considered when making these lenses that is not true for the divers' glasses or contact lenses?

18. The Johnson-Sea-Link is a deep-diving research mini-submersible. The pilot sits inside a 1.5 m diameter clear acrylic sphere whose wall is 15 cm thick. With the minisub sitting on its support ship's deck, will the submarine's pilot appear smaller or larger to an outside observer than he really is?

19. An object is viewed through two converging lenses so that neither the object nor its image is between a lens and its focus. Is the resulting image from the second lens real or virtual? erect or inverted?

20. Can a virtual image be an object for a lens? If not, show why not with a ray diagram. If it can, give an instance from everyday life.

True or False (21–30)

21. A light ray moving through a glass-air interface will bend toward the normal.

22. A light ray from a star that passes straight down the optical axis of a refractor telescope will be refracted.

23. The path taken by a ray of light across the interface between two media is the same regardless of the direction the light is moving.

24. Dense materials have an index of refraction less than 1.

25. Reflection of sunlight by water droplets is the only process responsible for the formation of atmospheric rainbows.

26. When using the thin-lens equation, the object distance is always positive.

27. An image that is smaller than its object is indicated by a negative magnification.

28. A virtual image cannot be focused on a screen.

29. A negative image height implies that the image is real.

30. When combining lenses, the refracted rays from the first lens do not *have* to form a focused image before they enter the second lens.

⊚31. Find the speed of light in each of the following materials.
    a. crown glass
    b. ice
    c. hydrogen
    d. flint glass

⊚32. Given the speed of light in each of the following materials, find its index of refraction.
    a. ethyl acetate ($c_{ETAC} = 2.19 \times 10^8$ m/s)
    b. corn syrup ($c_{syrup} = 2.01 \times 10^8$ m/s)
    c. iodine crystal ($c_I = 8.98 \times 10^7$ m/s)
    d. ruby ($c_{ruby} = 1.69 \times 10^8$ m/s)

⊚33. A ray of light moves from water into salt. Will the angle of refraction be greater or less than the angle of incidence? Why?

⊚34. The angle of incidence of a ray of light at an air-fluorite interface is 45.0°. At what angle will it be refracted?

⊚35. A ray of light with an angle of incidence of 0.0° in flint glass enters crown glass. At what angle is it refracted?

⊚36. Light from a laboratory lamp falls on the surface of a beaker of methyl alcohol at an angle of 30.0° and is refracted at an angle of 22.1° with the normal. What is the index of refraction of methyl alcohol?

37. A waterproof lamp is submerged at the bottom of a beaker containing water. One light ray approaches the water-glass interface at an angle of 45.0° to the normal and has an angle of refraction of 34.6°. What is the glass's index of refraction?

38. At what angle would a light ray have to approach a flint glass-diamond interface to have a refraction angle of 30.0°?

39. Find the critical angle for the following interfaces if the light ray passes
  a. from flint glass into crown glass.
  b. from diamond into water.
  c. from fluorite into air.

40. A 0.500 cm ant is observed through a hand magnifier lens ($f = 10.0$ cm). Find the location, size, orientation (erect or inverted), and type (real or virtual) of the ant's image when the ant's distance from the lens is
  a. 10 000.0 cm.
  b. 25.0 cm.
  c. 20.0 cm.
  d. 15.0 cm.
  e. 10.0 cm.
  f. 5.00 cm.

41. Which of the following will magnify an object more: placing it 9.5 cm from a lens with a focal length of 10.0 cm or placing it 5.0 cm from the lens? (*Hint:* Find d_i for each case first.)

42. A 10.0 m high tree is 100.0 m from a diverging lens with a focal length of −10.0 cm.
  a. Where is the tree's image?
  b. How high is it?
  c. Is it real or virtual?
  d. Is it erect or inverted?

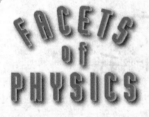

Fresnel Lenses

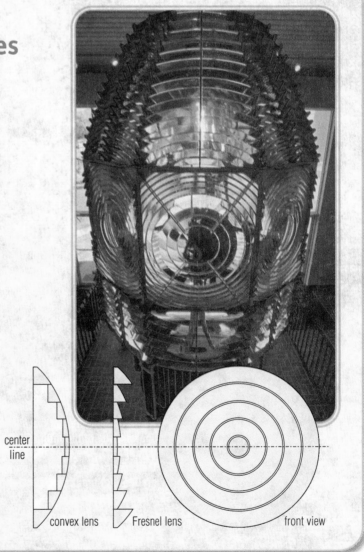

Lighthouses and navigational beacons must produce strong beams of light, but they need not be the same strength in all directions simultaneously. A single beam, sweeping across the sea, is enough to warn passing ships of any danger. Beacons use less energy if their light is directed in a single beam.

How can a beacon's light be concentrated in one direction? The easiest method might seem to be a lens. However, the lens required to focus a strong beam of light is so large and heavy that it tends to be impractical for most navigational aides such as buoys and unmanned light towers.

While studying this problem in the early 1800s, Augustin Jean Fresnel (1788–1827), a French physicist, decided to modify the normal lens. Instead of making the lens a solid portion of a sphere, Fresnel cut its curved surface into rings of uniform width and thickness. Then he laid all the rings together on a flat surface and fastened them together. This lens focuses light as well as the original solid, but it is much lighter weight and easier to make.

Today Fresnel lenses are popular wherever strong, lightweight lenses are needed. For example, Fresnel lenses work well in spotlights and in overhead projectors. Modern Fresnel lenses are usually stamped or molded from glass or plastic.

center line

convex lens   Fresnel lens   front view

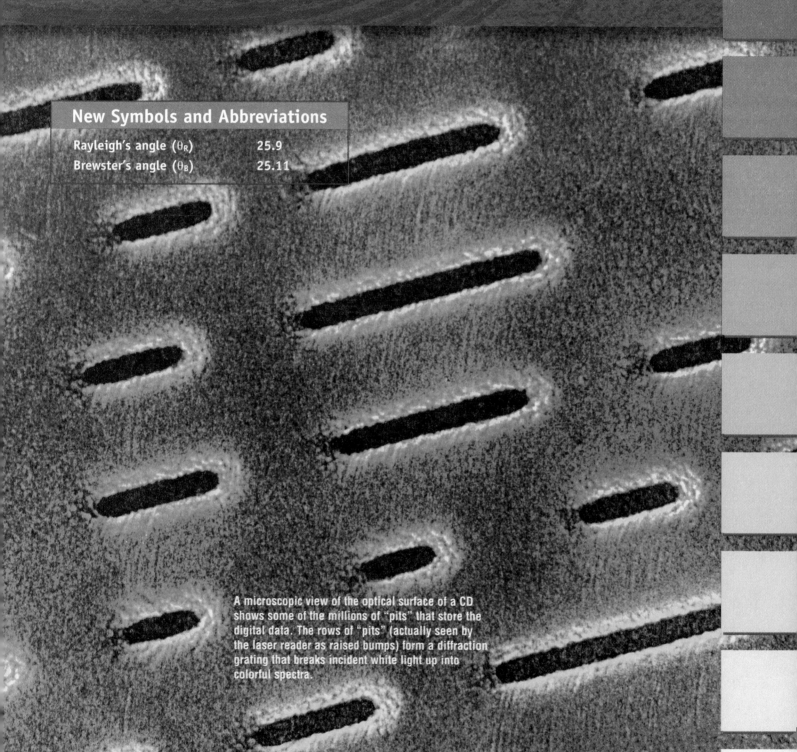

25A	Wave Interference	547
25B	Diffraction	558
25C	Polarization of Light	562
Biography:	Thomas Young	550
Facet:	Lasers	556
Facet:	Optical Testing	565

New Symbols and Abbreviations

| Rayleigh's angle (θ_R) | 25.9 |
| Brewster's angle (θ_B) | 25.11 |

A microscopic view of the optical surface of a CD shows some of the millions of "pits" that store the digital data. The rows of "pits" (actually seen by the laser reader as raised bumps) form a diffraction grating that breaks incident white light up into colorful spectra.

Wave Optics

> *For other foundation can no man lay than that is laid, which is Jesus Christ. Now if any man build upon this foundation gold, silver, precious stones, wood, hay, stubble; Every man's work shall be made manifest: for the day shall declare it, because it shall be revealed by fire; and the fire shall try every man's work of what sort it is. If any man's work abide which he hath built thereupon, he shall receive a reward. If any man's work shall be burned, he shall suffer loss: but he himself shall be saved; yet so as by fire.* I Corinthians 3:11–15

In becoming more Christlike, we should be motivated to please Him more and more by our actions. Works that we accomplish in selfless devotion to Him are described as gold, silver, and precious stones. What we do in selfish devotion to ourselves is wood, hay, and stubble. The collection of our works is very similar to a beam of unpolarized light—light that contains waves oriented in all different directions. A polarizing filter allows only those light waves that are oriented parallel to its transmission axis to pass through. The verses above remind us that, in a similar way, our works must pass through the fire of God's holy judgment before we receive any reward. Are your works precious in God's sight? Will they survive His "polarizing filter?"

25A WAVE INTERFERENCE

25.1 Describing Interference

Before the nineteenth century, scientists debated whether light is made up of particles or of waves. Isaac Newton believed that, in spite of some wavelike characteristics, light could be best explained as a stream of particles. Later scientists hesitated to question Newton's ideas. In 1801, however, Thomas Young showed conclusively that light has wavelike properties that a particle model cannot explain.

One unique characteristic of waves is called **interference.** Suppose two students hold a rope stretched between them. First one student, then the other snaps

25-1 Constructive interference of pulses in a rope

A limited burst of wave activity is called a *pulse*. The duration of a pulse is short compared to the travel time of the wave.

The **superposition principle** states that the net displacement of a point in a medium at a given instant is the vector sum of the displacements of all waves passing through the point at that instant.

the rope upward, starting a *pulse* toward the rope's center. What happens when the pulses meet? For an instant, they become one displacement equal to the sum of the displacements from the pulses. Then they separate, and each pulse continues along its original path. The increased amplitude in this case indicates **constructive interference.**

Suppose, instead, that one student snaps the rope upward and the other student snaps the rope downward. The pulses again travel toward the rope's center. Now what happens when they meet? For an instant, if the pulses have the same amplitude, the rope will have zero displacement. The downward pulse's amplitude is *subtracted* from the upward pulse's amplitude. Although the rope has no displacement, the wave is still moving, and after an instant each pulse continues on its original path. This case demonstrates **destructive interference.** The fact that the displacements of two impulses at their meeting place add to give the rope's displacement is called the **superposition principle.**

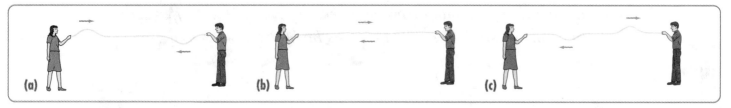

25-2 Destructive interference

In Acts 27 the ship Paul was on ventured into an area where two wave patterns met with constructive interference. The ship got stuck on a sand bar, and the huge waves destroyed it. "And falling into a place where two seas met, they ran the ship aground; and the forepart stuck fast, and remained unmoveable, but the hinder part was broken with the violence of the waves" (41).

The superposition principle applies to waves in any kind of media and to electromagnetic waves—large and small—and to any number of waves. If the waves are in phase—that is, if their troughs and crests (lowest and highest points, respectively) coincide—they will interfere constructively. If they are completely out of phase—the crests of one wave coincide with the troughs of the other—they will interfere destructively. If the waves are neither in phase nor completely out of phase, they will experience some constructive interference and some destructive interference. Many different effects can result from wave interference, but the most distinctive is called an *interference pattern.*

If light is a wave, it should show interference, obeying the superposition principle. When two light waves interact, they may show constructive interference, destructive interference, or, more probably, both. Areas of constructive interference appear brighter; areas of destructive interference appear dimmer than light from either original wave. Does light really show interference? Does light plus light ever equal darkness?

25.2 Young's Double-Slit Experiment

In 1801 Thomas Young demonstrated light interference in his "double-slit" experiment. Light interference is more striking when waves come from two sources that are in phase. Since light sources are hard to synchronize, Young split a single source into two parts. He also found that the effect he was observing was much clearer if he used colored filters to restrict the wavelengths of light.

Young's experiment involved three opaque screens. One screen was pierced by a single pinhole, which generated a point-source of light for the distant second screen. The second screen had two closely spaced narrow slits. The third screen was a white screen he used to observe the interference patterns. Sunlight falling on the filter and first screen passed through the pinhole. The wave front arriving at the second screen was essentially a plane wave. As the waves spread out from the slits, a pattern of alternating light and dark bands (or *fringes*) appeared on the third screen.

Why does the interference pattern appear? According to Huygens's principle, each slit in the second screen acts as a source of light, radiating in all directions. Since the light comes from the pinhole in the first screen, the light leaving each slit in the second screen is in phase with light leaving the other slit and of the same amplitude. Notice in Figure 25-4 that the bright spot in the center, called the *central maximum,* is the same distance from each slit. Thus, light reaching the central maximum from the slits is in phase, creating constructive interference. At the maximum labeled $n = 1$ (the first order maximum), the light from the farther slit travels exactly one wavelength farther than the light from the nearer slit, so they are still in phase. The maxima mark areas where the *path difference*—the difference in the distances that the light from the two sources travels—is an integer multiple of the light's wavelength.

The interference *maxima* are separated by dark areas *(minima)*. The light reaching the minimum next to the central maximum from the farther slit travels exactly one-half wavelength farther than light from the nearer slit. The light is completely out of phase, and it destructively interferes; therefore, the area is dark. Young reasoned that physical particles cannot interact and destroy each other as these light rays seemed to be doing, so he concluded that light acts like a wave.

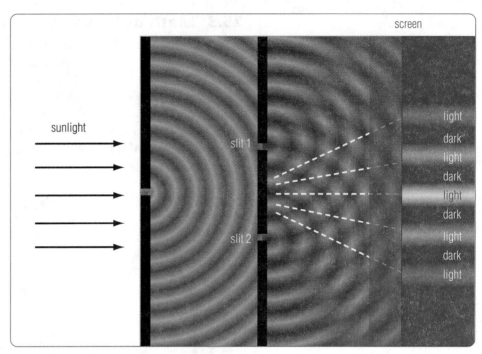

25-3 Schematic of Young's double-slit experiment

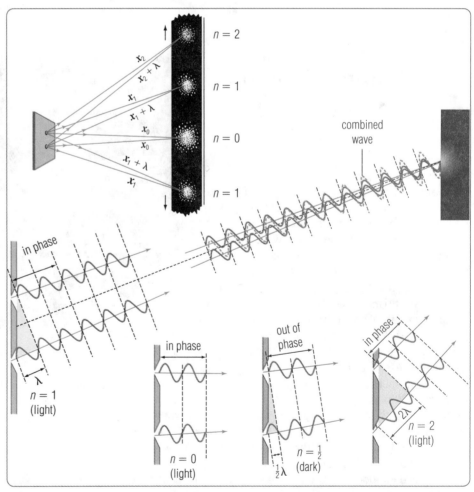

25-4 Constructive and destructive interference in Young's experiment

In an interference pattern, the bright fringes are called *maxima* and the dark fringes are called *minima*.

25.3 Math of Wave Interference

Is it possible to calculate where maxima and minima will appear if you know the light's wavelength? Yes, it is possible if you ensure that the distance between the slits is far smaller than the distance between the second and third screens. Figure 25-5 shows the conditions that produce an interference maximum. The rays of light from slits 1 and 2 are labeled rays 1 and 2, respectively. We have arranged the rays so that there is a maximum on the screen at their intersection (point E). Therefore, ray 2 must be longer than ray 1 by $n\lambda$, where n is an integer. If we drop

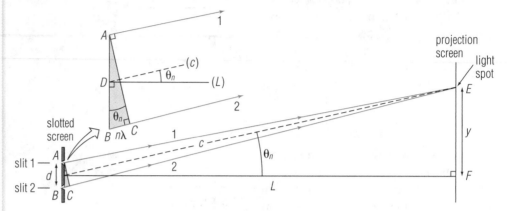

25-5 Conditions that produce a bright fringe in Young's experiment

Thomas Young
1773–1829

English scientists after Newton believed that he was almost infallible. Because they inferred that Newton advocated the particle theory of light, so did they. When the Dutch physicist Huygens proposed the wave theory as an alternative, no Englishman would listen to this foreigner's unproven ideas. However, in the early nineteenth century, experiments forced these scientists to see that only a wave theory could explain many of light's properties.

Thomas Young, although trained as a doctor, had abilities in many fields. A child prodigy, he learned to read when he was only two, read through the Bible twice by the time he was four, and spoke several languages before he began his medical studies. While still a student, Young formulated the theory that the lens of the eye focuses by changing thickness. The twenty-one-year-old student was elected a fellow of the Royal Society. After studying in three countries, Young received his medical degree from Göttingen.

The young doctor was interested in the way human senses work. He was one of the first scientists to propose that receptors in the human eye receive only three colors—red, green, and blue. He also studied the ear's reception of sound waves.

From his studies, Young became convinced that light is a transverse wave. His double-slit experiment in 1801 proved that light undergoes interference, something that only waves can do. At first, other English scientists refused to listen to Young, who they thought was contradicting Newton. Eventually, however, experiments like Young's double-slit experiment proved Young's position.

Young's interests were not limited to the human body or light. For example, he analyzed the elastic properties of solids. With his expertise in languages, Young became a leading Egyptologist and was among the first to decipher the Rosetta Stone. In fact, one of his last publications was a work about Egypt, not physics.

a perpendicular line from slit 1 to ray 2, ray 2 is divided into two segments—one with the same length as ray 1 and one of length $n\lambda$ (see the inset to Figure 25-5). This segment (BC) forms a right triangle ABC with an angle θ_n at vertex A. The sine of θ_n is

$$\sin\theta_n = \frac{\text{opposite}}{\text{hypotenuse}} = \frac{n\lambda}{d}. \tag{25.1}$$

The symbol n represents the *order number* of the maximum as well as the number of wavelengths in the path difference between the two rays. The central maximum is the zeroth order ($n = 0$) maximum. The first order maxima have $n = 1$, and so on.

Since the separation distance of the screens (L) is much greater than the distance between the slits (d), rays 1 and 2 and line $\overline{DE}$ (the line segment connecting the center of the slits to the maximum) are essentially parallel. Therefore, $\overline{AC}$ is perpendicular to $\overline{DE}$, and $\overline{DF}$ is perpendicular to $\overline{AB}$. Angle EDF is equal to θ_n, and triangle DFE is similar to triangle ACB. The distance y is the distance between the maximum being studied and the central maximum. The tangent of θ_n for the larger triangle is

$$\tan\theta_n = \frac{\text{opposite}}{\text{adjacent}} = \frac{y}{L}. \tag{25.2}$$

For small angles,

$$\tan\theta_n \approx \sin\theta_n.$$

Therefore,

$$\sin\theta_n \approx \frac{y}{L}.$$

By substituting y/L for $\sin\theta_n$ in Equation 25.1, we get

$$\frac{y}{L} = \frac{n\lambda}{d} \text{ (for maxima)}. \tag{25.3}$$

This equation is especially useful for determining the distance between slits or the wavelength of light, assuming the other parameters are known.

For analyzing the *minima* between maxima, the ray path length difference is in half-wavelengths. Therefore, you substitute $n - \frac{1}{2}$ for n, and Equation 25.3 changes to

$$\frac{y}{L} = \frac{(n - \frac{1}{2})\lambda}{d} \text{ (for minima)}. \tag{25.4}$$

In this equation, y is still the distance from the minimum to the central maximum, L is the screen separation, and d is the slit separation.

Problem-Solving Strategy 25.1

The symbol n in the interference equation is equal to two quantities: (1) the difference between path lengths in whole wavelengths of the light that produce the maximum and (2) the order number of the same maximum.

Problem-Solving Strategy 25.2

When analyzing the *minima*, be sure to use the order number (n) of the next higher maximum in the interference formula.

EXAMPLE 25-1

Determining Wavelength Using Interference Patterns

Red light illuminates a screen containing two slits separated by 0.500 mm. The distance between the screens is 3.00 m, and the distance between the central maximum and the first-order maximum is 0.400 cm. Find the light's wavelength.

Solution:

Since you are given the distance to a *maximum,* use Equation 25.3:

$$\frac{y}{L} = \frac{n\lambda}{d}$$

$$\lambda = \frac{y}{L} \cdot \frac{d}{n}$$

Wavelengths of visible light are conveniently measured in nanometers (1 nm = 10^{-9} m).

Since the problem involves a *first-order maximum*, $n = 1$. Convert all length dimensions to meters.

$$\lambda = \frac{(4.00 \times 10^{-3}\ \text{m})(5.00 \times 10^{-4}\ \text{m})}{(3.00\ \text{m})(1)}$$

$$\lambda = 6.66\overline{6} \times 10^{-7}\ \text{m}$$

$$\lambda \approx 6.67 \times 10^{-7}\ \text{m} \ \text{(or 667 nm)}$$

You can check your answer by noting that light with a wavelength of 667 nm is red (see Table 26-3, p. 574).

EXAMPLE 25-2

Determining Slit Spacing

Monochromatic yellow light with a wavelength of 580. nm illuminates a double-slit screen. The centers of the minima adjacent to the central maximum are 0.435 m from the center of the pattern. The projection screen is 120. cm from the slits. Calculate the distance between the slits.

Solution:

You are working with the distance of a *minimum;* therefore, use Equation 25.4. The next greater maximum is 1. Convert all distances to meters.

$$\frac{y}{L} = \frac{(n - \frac{1}{2})\lambda}{d}$$

$$d = \frac{L(n - \frac{1}{2})\lambda}{y}$$

$$d = \frac{(1.20\ \text{m})(1 - \frac{1}{2})(580. \times 10^{-9}\ \text{m})}{4.35 \times 10^{-4}\ \text{m}}$$

$$d = 8.00 \times 10^{-4}\ \text{m}$$

The slits are 0.800 mm apart.

25.4 Thin Film Interference

Did you ever wonder why a soap bubble or a film of spilled gasoline displays a shimmering color? The answer is interference. In a thin film, light reflects from both the top and the bottom film surfaces. Depending on the path difference and the light's wavelength, the reflected waves may interfere constructively or destructively. If white light illuminates the film, some wavelengths (colors) at each point on the film will probably be the right wavelength to interfere constructively. Other wavelengths will interfere destructively; therefore, only the constructively interfering colors will be seen. Since most films vary in thickness, different colors will be reflected by different portions of the film. Therefore, swirling bands of color can be seen in the film as air currents move the film around.

A thin lens resting on a flat glass plate produces interference patterns. These patterns are caused by interference between light reflected from the flat plate below the lens and light reflected from the curved lower surface of the lens. The thickness of the air below the lens increases from the center of the lens out to the edges. Thus there are concentric "rings" of air under the lens, each ring having a

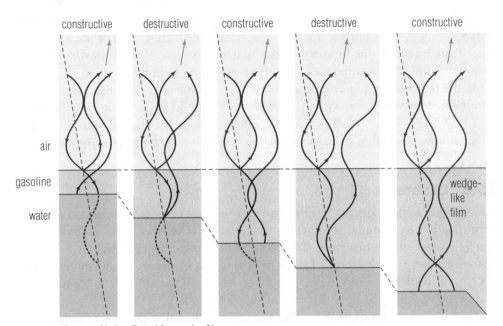

25-6 Interference of light reflected from a thin film

different width. Where constructive interference occurs, it produces a bright ring of light. Where destructive interference occurs, a dark ring appears. These rings are called *Newton's rings* after Isaac Newton, who first observed them. Interference also occurs between two flat glass plates separated by a tiny wedge of air. The light is reflected above and below the air wedge, and the two reflected rays interfere. White light separates into colors when reflecting from this device.

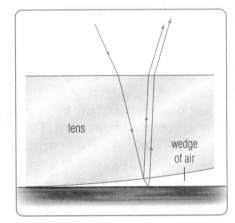

25-7 Light rays producing Newton's rings

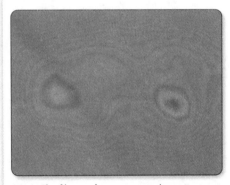

25-8 Thin-film interference in water between two sheets of glass

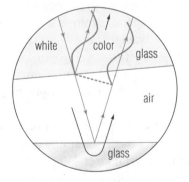

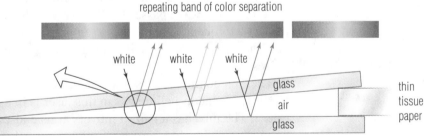

25-9 Interference of rays reflected from two glass plates separated by a thin wedge of air

25.5 Interferometers

An **interferometer** uses the interference of light to make precise measurements of distance. A monochromatic (one-color) light source directs its light toward a partially silvered mirror (mirror A), which reflects half of the light to mirror 1 and lets the rest pass to mirror 2. (The plane glass plate simply ensures that light from both paths passes through the same total thickness of glass.) Mirrors 1 and 2 reflect the light back through the mirror to a screen. If mirrors 1 and 2 are the same distance

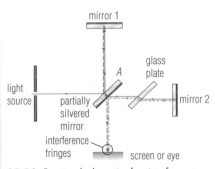

25-10 Functional schematic of an interferometer

from mirror A, then light rays from both paths travel the same distance and should arrive in phase, with only constructive interference. If they are not the same distance from mirror A, the light will make an interference pattern from which an experimenter could determine the path difference of the light rays. Albert Michelson made the first working interferometer and used it to measure the meter to a greater precision than was possible before. He determined that a meter was equal to 1 553 165.5 wavelengths of red light emitted from a plasma of cadmium. (The meter is now defined as the distance light travels in a vacuum in $\frac{1}{299\,792\,458}$ s.)

25.6 Lasers

Interference also makes the laser possible. The word *laser* is an acronym for *l*ight *a*mplification by *s*timulated *e*mission of *r*adiation. The laser emits *coherent* light that is almost monochromatic. The method of producing laser light generates a near-linear beam. Therefore, a laser's light is intense and can remain concentrated over great distances. Incandescent light sources, on the other hand, emit many wavelengths, and their light disperses in all directions.

There are many kinds of lasers, but they all use the same principle. Some atoms can retain excess energy in their electron orbitals for extended periods of time. If they absorb additional energy from a wave of light, they can emit an identical light ray in the same direction as the original and in phase with it. This result is amplification—getting two rays of light for one. A laser has a source of energy (a light "pump"), a mass of material to be energized, and mirrors at each end to reflect the spontaneously emitted light back and forth through the material to stimulate the emission of more light.

The original laser, built by Theodore Maiman in 1960, was a *crystal laser* with ruby as the emitting material. The ruby rod, about the size of a short pencil, has exactly parallel ends. The front of the rod is partially silvered to reflect some light but to allow the laser to emit some light. The back of the rod is fully silvered. Surrounding the rod is a spiral flash tube like those used in photography. The flash tube "pumps" energy into the ruby to raise the average energy level of the ruby atoms. Some atoms emit light spontaneously, stimulating others to emit light. The light rays that are aligned axially with the rod reflect back and forth between the mirrors. The distance between the mirrors determines the specific frequency that will be reflected and constructively reinforced—all other wavelengths eventually die out. As the wave passes back and forth, it is reinforced by further emissions of the same frequency and phase. Eventually, enough energy is gained to emit a burst of light through the partially mirrored end of the rod. Such a laser emits light in pulses.

The first continuous laser was a *gas laser* built by Ali Javan and his co-workers in 1961. The neon-helium gas in this laser is in a long quartz tube containing electrodes. The tube has a pair of mirrors at its ends, as in the crystal laser. When high voltage is connected across the electrodes, the helium is ionized. Its electrons energize the neon atoms that emit light, and thus stimulate the emission of more in-phase light rays. Since the voltage is continuous, the laser light is also continuous. Gas lasers also emit more nearly monochromatic light than crystal lasers; for this reason they are used more for communications and measurements, where frequency stability is important. Crystal lasers emit more energy, so they are used for drilling holes and welding metals.

Semiconductor or *diode lasers* also produce laser light directly from electrical energy. Since these lasers are tiny—often less than 1 mm in each dimension—they are frequently used in such electronics applications as the "reader" for digital information on a compact disc (CD).

Theodore Maiman (1927–) is an American physicist who invented the ruby laser in 1960.

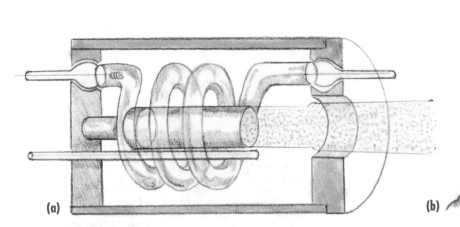

25-11 (a) A schematic of the first ruby laser. (b) Ali Javan (1928–), the Iranian-American physicist who invented the gas laser in 1961

The potential uses for lasers are nearly unlimited. Lasers have found a place in precision macro- and micro-scale measurement. Lasers are used in medicine for delicate bloodless surgery. High speed photography, data recording and reading, metallurgy, photocopying, sensors, barcode readers, and even energy weapons all use laser technology. See the facet on the next page for more information on this subject.

25A Section Review Questions

1. Patterns of ocean waves sometimes cross each other such that there are extraordinarily tall waves formed where individual waves intersect. What specific wave phenomenon causes these waves to form?

2. Discuss what produces the alternating bright and dark fringes in Young's experiment.

3. While conducting an experiment similar to Young's, an experimenter noted third-order maxima. If the wavelength of light used is λ, what is the length difference of the ray paths that form the maxima?

4. Why do lasers emit monochromatic light?

⊙5. Young's experiment is performed in a school laboratory using 660. nm light and slits 5.00×10^{-6} m apart. What is the angular separation between the central maximum and the second-order maximum?

⊙6. Referring to Question 5, if the slits are 1.00 m from the projection screen, what is the distance of the second-order maximum from the central maximum?

⊙7. You are performing Young's experiment using 540. nm light and slits that are separated by 7.50×10^{-6} m. What is the distance between the central maximum and the first-order minimum if the projection screen is 1.25 m from the slits?

⊙8. a. Will the spacing of interference maxima be greater for longer or shorter wavelengths of light for a given experimental setup?

   b. Work out an example to justify your answer.

Lasers

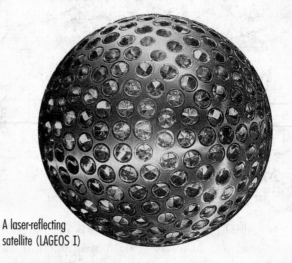

A laser-reflecting satellite (LAGEOS I)

Men have long dreamed of producing a pure beam of light powerful enough to vaporize any matter. When Theodore Maiman introduced the first successful laser in 1960, the world realized that his invention had even more potential than they had dreamed. (Lasers were at first called a solution waiting for a problem.) Not only did the laser concentrate tremendous energy in one direction, but it also generated a nearly monochromatic light (one frequency of "pure color") that could be pulsed with incredible speed. Within four decades many different types of lasers were being produced for a wide variety of jobs.

Because laser light is virtually monochromatic, lasers can measure fine details from almost any distance. Scientists have measured the speed of light one hundred times more accurately than ever before. Now high-flying planes can distinguish something as small as the steps in a football stadium. Improved interferometers can measure wavelengths and minute linear displacements. For example, three interferometers, located 15 miles from California's San Andreas fault, detect changes in strain across the fault as small as ten-millionths of an inch. Any detectable strain indicates a build-up of stress, which alerts seismologists to a possible earthquake.

Because laser light is coherent (a single frequency and phase) and is generated by axial reflections within a tube or crystal, the beam forms a nearly perfectly straight line. It is ideal for surveys and large construction projects. Land engineers find the laser indispensable when drilling tunnels and laying pipelines, and jet engineers align their 200-foot jigs (aligning tools) at accuracies up to a few hundredths of a centimeter. Lasers also monitor the production of precision tools, and they gauge super-fine wire.

The beam from a laser can travel long distances without losing its identity. Can you imagine having a "flashlight" that could illuminate something thousands of miles away? If you knew the time the light took to reach an object and bounce back to you, you could figure out the object's distance. Scientists do just that. Erbium-doped glass lasers are used to measure the height of clouds, and lasers on airplanes measure the altitude of the plane. Some pulsed ruby lasers on earth track distant satellites to an accuracy of a few centimeters! Something even more unbelievable happened after the first men who landed on the Moon, "Buzz" Aldrin and Neil Armstrong, placed multi-prism corner reflectors on the Moon during the *Apollo 11* mission. Laser-equipped Earth stations were able to pick up beams bounced off the reflectors and precisely measure the distance from Earth to the Moon. Instruments taking readings three times each day determined that the Moon is gradually receding from Earth at the rate of a few centimeters per century.

Industries routinely employ laser energy in drilling, cutting, and welding. The energy in one pulse of a "mode-locked" laser can produce tens of trillions of watts—more power in one instant than is being generated by all the earth's power plants combined. Pulsating ruby lasers blaze small holes through the three hardest substances on Earth: sapphires and rubies for bearings in watches as well as diamonds for wire-drawing dies.

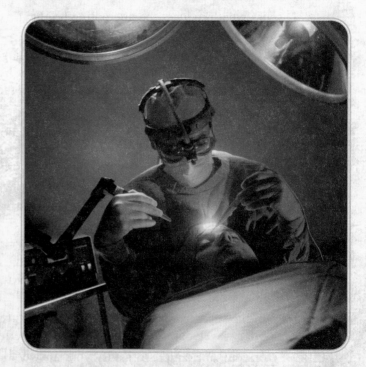

A hologram

Other kinds of lasers drill holes in tiny circuit boards and cut grooves so that the boards can be broken into individual circuits. The ultimate in ink erasers, lasers remove ink without touching the paper. Clothing manufacturers cut materials quickly and precisely. Construction crews can cut 5-centimeter-thick stainless steel at 40 centimeters per minute and weld 2 centimeter sheets at 30 centimeters per minute.

Microsurgery is making advances almost every day. Without touching the skin, lasers remove both tattoos and skin blemishes, which otherwise would remain for life. "Bloodless surgery" is becoming more practical—as lasers remove tumors and other tissue, the heat cauterizes the outlying blood vessels and keeps the incision sterile. Microsurgeons "spot weld" detached retinas or narrow the laser beam to operate on chromosomes within single cells.

The military has taken special interest in the potential power of the laser beam. All modern tanks are equipped with laser range finders to determine the target's exact distance. Laser-guided shells (such as "smart bombs") home in on targets that have been illuminated by ground YAG (yttrium aluminum garnet) lasers: YAG lasers are ideal because they produce an invisible infrared beam. Gas-injection lasers shine a light on the enemy to make him an easy target. Recently, air-, ground-, and sea-borne high energy lasers have been developed for use in strategic and tactical ballistic missile defense. Such weapons were the stuff of science fiction stories for nearly a hundred years.

High-speed photography takes advantage of the rapid pulses achieved by crystal lasers. You probably cannot detect the time your own camera takes to expose each flash because it is very fast. But these high-tech cameras can expose a frame in one ten-trillionth of a second! Unlike hand-held video cameras, which record images at a rate of 30 frames per second (fps), the best laser cameras take as many as one billion fps. This breakthrough helps in studying ultra-fast, momentary phenomena such as the formation of vapor during laser welding.

When three-dimensional pictures (called *holograms*) were first developed in 1947, Dennis Gabor lacked a coherent light source. Now that pure colors are available from lasers, interest in holography has been renewed. Unlike conventional photography, 3-D photography does not use lenses. Instead, two beams of monochromatic light are directed to a photographic plate. One beam reflects off the subject before it joins the other beam on the plate. The resulting interference pattern is unrecognizable until the developed film is illuminated by another beam. An image then forms that looks just like the original. Looking into the plate is like looking through a window: as you move from side to side you can see different angles of the subject.

Even if the hologram is broken into tiny pieces, you can still see the whole picture in one fragment, in the same way that you could still see through one window pane after all the other panes had been covered with dirt. The picture will be less clear, however. The immense advantage of holograms is the depth of information that can be stored on only one photograph. Moving particles like pollen and fog droplets cannot be photographed more than once before they move, so only a hologram could provide the detailed information needed for motion studies.

Lasers can perform tasks either difficult or impossible by other means. Research continues in all fields to find new uses for lasers. Communications researchers are investigating how to send even more information by laser light waves via fiber optic networks. The band of light waves is 100 000 times greater than the present band in the microwave frequency range. In other words, one laser beam could carry as much information as all the existing radio channels on Earth carry!

Laser applications soon may involve the initiation of controlled fusion reactions, the production of metallurgical plasmas, photochemical reactions, and phase transitions of matter. In the distant future even 3-D television might be possible.

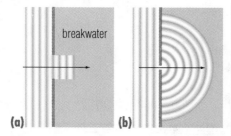

breakwater

(a)   (b)

25-12 Water wave crests showing (a) no diffraction and (b) diffraction around the edges of the gap

DIFFRACTION

25.7 Describing Diffraction

To early experimenters, light seemed to travel in straight lines and cast sharp shadows. Other waves, such as ocean and sound waves, were known to **diffract;** that is, they spread into a region behind an obstruction where, if they traveled in straight lines, a "shadow zone" should exist. For example, consider water waves approaching a gap in a harbor breakwater. If water waves had to travel in straight lines, waves inside the breakwater, such as those in Figure 25-12 (a), would result. However, since water waves diffract, the resulting waves are like those shown in figure (b).

Light's apparent lack of diffraction convinced Newton that light cannot be a wave. But Young's experiment proved that light does act like a wave. How can these statements be reconciled? The solution lies in light's short wavelengths. Water waves diffract only around obstacles whose dimensions in the direction of wave propagation are small compared to the wavelength. With the long wavelengths of most water waves (measured in meters), most obstacles are small compared to the wavelengths. In like manner, light diffracts only around objects that are small compared to its wavelengths. However, obstacles on the order of several hundred nanometers in size are too small to be seen without a microscope; therefore, you cannot see light diffracting around them. Light does not noticeably diffract around most visible objects.

Diffraction is a form of interference from parts of a wave front encountering an obstacle. The interference produces light and dark bands parallel to the obstacle. How is this interference different from the interference in Young's experiment? Young's experiment involved the interference of two point sources of light. In the study of diffraction, the interference of an entire wave front will be considered.

Diffraction can be separated into four categories: edge, obstacle, opening, and grating diffraction. The shadow of a knife edge under monochromatic light illustrates *edge diffraction.* Instead of a clear-cut shadow, bands of light and darkness appear inside the area expected to be in shadow. The light bends around the edge into the shadowed region.

Obstacle diffraction results when a small object is in the path of a beam of monochromatic light. Again the object has no clear-cut shadow, but it has alternating light and dark bands inside the shadowed area. For a small spherical object, the shadow has a bright spot in the center.

Opening diffraction occurs when light passes through a small opening. When you look through a single slit, alternating bands of dark and light appear in the opening. For a circular opening, alternating rings of dark and light appear in the opening. You can observe opening diffraction for yourself with no special equipment. Place the tips of your thumb and index finger about 0.5 cm apart, and look through the opening at a light source with your hand a few centimeters from your eye. Now gradually narrow the space between your fingers. As the opening closes, dark lines parallel to the surface of your finger and thumb will appear. This is opening diffraction. You can also see this diffraction by shining light through an opening onto a screen. Figure 25-13 shows the resulting pattern.

Grating diffraction occurs when light passes through a piece of glass scratched with evenly spaced fine parallel lines, called *rulings.* Between the rulings the glass is still transparent; so light can easily pass between them. The rulings obstruct the path of light. The light passing through the grating forms a diffraction pattern with minima and maxima.

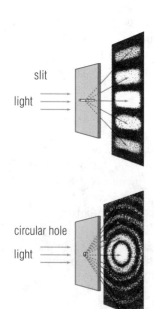

slit

light

circular hole

light

25-13 Opening diffraction

You can calculate the locations of bright regions for gratings the same way you calculated bright regions for interference patterns. The fact that there are now many sources of light (unruled areas) instead of two makes little difference. Figure 25-15 shows the light waves that constructively interfere to form a maximum. The path difference to a maximum between two waves from *adjacent* rulings is an integer multiple of wavelength ($n\lambda$); the path difference for *any* two rulings is an integer multiple of $n\lambda$. The geometry for any two lines is the same as Figure 25-15, from which the following equation was obtained:

$$\sin\theta_n = \frac{n\lambda}{d}$$

This equation is identical to Equation 25.1. The angle θ_n is the same as it was for Young's experiment with interference patterns. It is the angle between the imaginary line connecting the center of the grating to the central maximum on the projection screen and the line connecting the grating to the nth-order maximum. Equation 25.1 is especially useful for grating diffraction.

Diffraction gratings are so-named from the collection of closely spaced linear scratches, or rulings, whose appearance is similar to a fireplace grating.

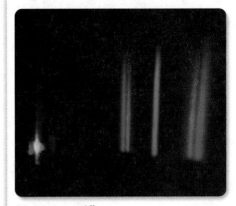

25-14 Grating diffraction

EXAMPLE 25-3

Diffraction Rules: Wavelength from Diffraction

A diffraction grating has 6000 ruled lines per centimeter. Find the wavelength of light that places the first-order maxima at $\theta_1 = 23.0°$ from the central maximum (see Figure 25-16).

Solution:
You know that $n = 1$ (the *first-order maximum*), $\theta_1 = 23.0°$, $d = 1.00$ cm/6000 lines $= 1.\underline{67} \times 10^{-4}$ cm $= 1.\underline{67} \times 10^{-6}$ m (assuming that the width of the ruling lines is negligible, the spacing between rulings is just the reciprocal of the line density). You need to find λ. Use Equation 25.1.

$$\sin\theta_n = \frac{n\lambda}{d}$$

$$\lambda = \frac{d\sin\theta_n}{n} = \frac{(1.\underline{67} \times 10^{-6}\ \text{m})\sin 23.0°}{1}$$

$$\lambda \doteq 6.5\underline{13} \times 10^{-7}\ \text{m}$$

$$\lambda \approx 651\ \text{nm}$$

The light's wavelength is about 651 nm, so its color is in the red region of the visible spectrum.

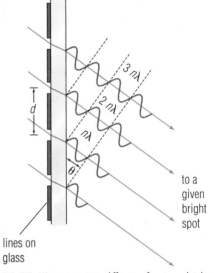

25-15 Waves in grating diffraction forming a bright fringe

EXAMPLE 25-4

Determining the Location of Diffraction Maxima

Light with a wavelength of 480. nm illuminates a grating with 6000 lines per centimeter. At what angle are the second-order diffraction maxima?

Solution:
You need to find θ_2.

$$\sin\theta_n = \frac{n\lambda}{d}$$

$$\theta_n = \sin^{-1}\left(\frac{n\lambda}{d}\right)$$

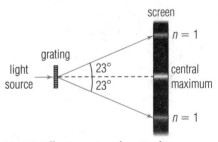

25-16 Diffraction grating information for Example 25-3

Problem-Solving Strategy 25.3

When you determine SDs associated with diffraction grating ruling densities, the number of lines is *counted* and therefore does not affect SDs in calculations. Its reciprocal is a distance, however, so assume that grating spacing has three SDs.

$$\theta_2 = \sin^{-1}\left[\frac{2(480. \times 10^{-9} \text{ m})}{1.\underline{6}7 \times 10^{-6} \text{ m}}\right]$$

$$\theta_2 \doteq 35.\underline{1}6°$$

$$\theta_2 \approx 35.2°$$

The second-order fringes will be at about 35.2° from the central maximum.

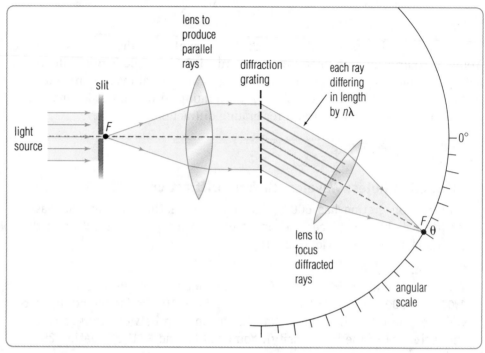

25-17 A diffraction grating spectrometer

25.8 Spectrometers

One important application of diffraction is the *grating spectroscope,* which separates light into its component wavelengths. Since the diffraction angle is dependent on wavelength, each color has a maximum at a different angle. The diffraction grating, like a prism, generates a continuous color spectrum when illuminated by white light. The light entering the spectroscope first passes through a **collimator.** A collimator adjusts the shape of the incident light beam and makes the rays parallel. Light emerging from the collimator lens directs parallel light rays toward the diffraction grating. From the grating, the light is dispersed into a diffraction pattern consisting of a bright first-order maximum and dim higher-order fringes. A rotatable eyepiece detects light parallel to its optical axis. At each angle of a first-order maximum, the eyepiece detects light of a different wavelength, so the angles on a fixed scale can be labeled with the wavelength that will be detected there.

Spectrometers are most useful when one is measuring the line spectra of sources that emit just a few wavelengths of light, such as sodium vapor lamps.

25.9 Optical Resolving Power

Lenses also show diffraction effects. In the ordinary use of lenses, you can assume that the image of a point is a point, but actually it is a group of concentric rings. This phenomenon makes it difficult for a lens to distinguish two points of light that are close together. Such a problem is evident in astronomical telescopes that attempt to separate—or *resolve*—two stars optically close together. The ability to separate close objects is called the **resolving power** of the lens. Contrary to what you may think, the resolving power of a lens or system of lenses is *not* a function

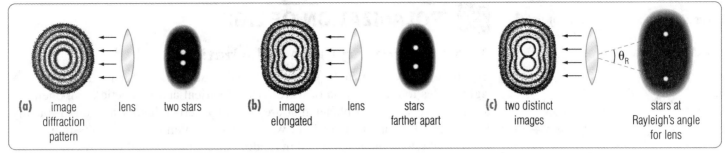

(a) image diffraction pattern   lens   two stars   **(b)** image elongated   lens   stars farther apart   **(c)** two distinct images   stars at Rayleigh's angle for lens

of the total magnification available, but rather of the diffraction fringes produced in the system.

There is a limit to how close point-like objects can be if they are to be resolved as separate images. Lord Rayleigh, an English physicist, showed that two points must be separated by at least an angle of θ_R for a given lens to show them as distinct images. The angle θ_R is called **Rayleigh's angle** and is computed by

$$\theta_R = 1.22\frac{\lambda}{d}, \tag{25.5}$$

where λ is the wavelength of light coming from the point sources, d is the diameter of the lens, and θ_R is given in radians. Any points with greater angular separation can be clearly resolved. For a given wavelength of light, it is clear that resolving power is inversely proportional to the diameter of the lens. Therefore, the larger the lens (or mirror, since this relationship is applicable to both kinds of optical components), the sharper and more detailed the image.

25B Section Review Questions

1. How does diffraction differ from refraction?

2. **a.** If the number of rulings per centimeter in a diffraction grating increases, will the diffraction maxima get closer together or farther apart?

   b. Explain your answer.

3. What is the purpose of a grating spectrometer?

4. What would happen to the image of two stars that could barely be resolved if you placed a circular mask over the end of the telescope that reduced the aperture to half its original diameter?

5. In a diffraction-grating experiment, what is the spacing of the grating rulings if the angular displacement of the first order maximum is 31.33° from the central maximum and you are using light with a wavelength of 650. nm?

6. What is the highest possible order number (n) for interference maxima if the diffraction grating is ruled at 12 000 lines/cm, and you are using red light ($\lambda = 700.$ nm)? (Remember that n must be a whole number.)

7. What is the theoretical Rayleigh angle for the 60.0 mm objective lens of a refractor telescope when viewing yellow stars (effective wavelength $\lambda = 580.$ nm)? (Remember to compute your answer in radians.)

8. How many seconds of an arc is the Rayleigh angle in Question 7? (*Hint:* Degrees are further subdivided into minutes and seconds.)

25-18 Two stars must have a minimum angular separation in order to be resolved by a given lens. The angle shown is exaggerated for clarity. A 10 cm lens can theoretically resolve two point-like objects about 1 second of an arc apart.

John William Strutt, Lord Rayleigh (1842–1919) was an English physicist who won the Nobel Prize in Physics in 1904 for his investigations of the densities of gases.

Problem-Solving Strategy 25.4
Remember that Rayleigh's angle (θ_R) is given in radians.

The ability to separate details in an optical image is a function of the aperture of the lens or mirror. The larger the aperture, the smaller the angular separation objects may have and still be resolved by the optical system. This is the reason astronomers seek to build larger telescopes—to separate extremely distant, close-together celestial objects in photographic images.

Transverse waves in a beam of light are **polarized** if they oscillate in parallel planes.

25-19 Unpolarized light consists of waves with different directions of **E.**

Étienne Louis Malus (1775–1812) was a French physicist who discovered that reflected light is polarized. He coined the word "polarize" because he accepted the particle theory of light and believed that the particles had to have poles in order to show polarization.

The double-minded man is like dichroic (polarizing) material. Sometimes you can see God's pure "polarized" light shining through him, and sometimes you can't. "A double minded man is unstable in all his ways" (James 1:8). To be the most effective channel of God's influence in the world, you must ensure that your "transmission axis" is aligned with God's light.

Problem-Solving Strategy 25.5

Brewster's angle (θ_B) may be given in degrees or radians.

Sir David Brewster (1781–1868) was a Scottish physician who invented the kaleidoscope and discovered the phenomenon of light reflected at specific angles.

25C POLARIZATION OF LIGHT

25.10 Describing Wave Polarization

You have learned that light waves are *transverse waves;* that is, they vibrate at right angles to the direction of travel. The electrical and magnetic components of a light wave vibrate in planes that are mutually perpendicular as well. Typically, the electrical components of light waves (*E*-waves) emitted by atoms in a source of light vibrate in planes oriented in random directions, even if the rays themselves all travel in the same direction (see Figure 25-19). Such randomly oriented light waves are called **unpolarized.** If the *E*-waves in a beam of light are *not* randomly oriented, the light is **polarized.**

Light can be polarized in a number of ways. If all the electric field vectors (**E**) and the wave velocity vector (**v**) lie in the same plane, we say that the wave is *plane polarized.* Looking at a fixed point in space where a light wave passes through it, we understand that the *E*-wave oscillates in a line superimposed on the point.

Some crystalline materials polarize light in other ways. When two plane-polarized waves are superimposed on one another, the resultant **E** vector rotates around the point in space in a plane perpendicular to the direction of propagation. In these cases, the composite wave is called a *circularly* or *elliptically* polarized wave.

Since light leaving a source is usually unpolarized, something must act on it to polarize it. There are four ways to polarize light: *reflection, double refraction, selective absorption,* and *scattering.*

25.11 Reflected Polarization—Malus's and Brewster's Laws

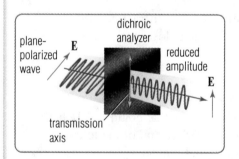

25-20 Malus's law describes the effect a dichroic analyzer has on the intensity of a beam of plane-polarized light.

In 1808 the French physicist Étienne Malus discovered that when a light beam strikes a pane of glass or the surface of water, the reflected light is polarized. He determined this fact after he looked through a special crystal of the mineral calcite, called *Iceland spar,* at light reflecting off building windows. As he rotated the crystal, reflections were extinguished at certain angles. Iceland spar and other materials transmit polarized incoming light in a preferred direction, called the **transmission axis.** Only light with waves parallel to the transmission axis passes through. Such polarizing materials are called **dichroic. Malus's law** states that the intensity of a polarized light beam transmitted by a dichroic material (the *analyzer*) is a function of the angle between the wave's plane of polarization and the analyzer's transmission axis.

When unpolarized light reflects off a surface, some of the light is reflected and some is refracted into the substance. The light that is reflected tends to contain more waves that are horizontally polarized, but the refracted rays tend to contain more vertically polarized waves. The proportion of each kind depends on the angle of incidence and the materials in contact at the reflection interface.

David Brewster discovered that for any pair of materials in contact at an interface, the reflected light is completely polarized when the reflected light rays in the incident medium are 90° from the refracted light rays in the other medium. At this special angle of incidence, called **Brewster's angle (θ_B),** the reflected beam consists entirely of horizontally polarized waves, but the refracted beam contains mostly (but not all) vertically polarized waves. This discovery is called **Brewster's law.** The mathematical form of Brewster's law is

$$\tan \theta_B = \frac{n_r}{n_i}, \tag{25.6}$$

where θ_B is the Brewster angle of incidence that produces the greatest polarization, and n_r and n_i are the refraction indexes of the media containing the refracted and the incident rays, respectively. Thus, you can easily calculate the Brewster angle of incidence that produces the greatest polarization in the reflected light beam.

EXAMPLE 25-5

Reflecting on Polarization: Brewster's Angle

What is Brewster's angle for an air-water interface? Use the indices of refraction from Table 24-1 on page 528.

Solution:

You find that $n_{air} = 1.0003$ and $n_{water} = 1.333$. Use Equation 25.6:

$$\tan \theta_B = \frac{n_r}{n_i} = \frac{n_{water}}{n_{air}}$$

$$\theta_B = \tan^{-1}\left(\frac{n_{water}}{n_{air}}\right) = \tan^{-1}\left(\frac{1.333}{1.0003}\right)$$

$$\theta_B \doteq 53.\underline{11}°$$

$$\theta_B \approx 53.1°$$

When the Sun is about 37° above the horizon (90° − 53.1°), its rays reflecting off the surface of water are horizontally polarized.

Problem-Solving Strategy 25.6

Remember that Brewster's angle is the incident angle of the incoming ray of light. The 90° angle is measured between the reflected ray and the refracted ray.

Erasmus Bartholinus (1625–98) was a Danish mathematician who wrote the first scientific description of the polarization of light in a crystal of Iceland spar calcite.

25-22 Double refraction in Iceland spar calcite

25.12 Polarization by Double Refraction

A second method of polarizing light uses a property of many mineral crystals called *double refraction* (or *birefringence*), which was first observed by Erasmus Bartholinus in 1669. It was so named because the indices of refraction are different through the different crystal planes. Some common examples of crystals with this property are aragonite, calcite, ice, mica, quartz, selenite, sugar, and topaz. In such a crystal, a refracted ray is split into two rays—the ordinary (*O*) ray and the *extraordinary* (*E*) ray. The ordinary ray follows Snell's law of refraction exactly. The extraordinary ray does not refract according to Snell's law and may be refracted at any angle, depending on the microscopic structure of the crystal. It does emerge parallel to the ordinary ray, however. Both rays are polarized so that their planes of polarization are at right angles to each other.

A *Nicol prism,* invented in 1828 by William Nicol, is made from a calcite crystal that has been cut and cemented together again with Canada balsam, an adhesive often used in optical applications. The calcite produces an *O*-ray and an *E*-ray,

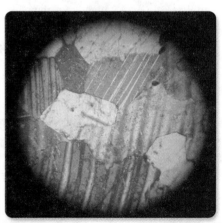

25-23 Minerals can be identified in a thin section of rock sample, using crossed Nicol prisms, when viewed through a polarizing microscope.

William Nicol (1768–1851) was a Scottish physicist.

25-24 When two sheets of selectively absorptive material are placed at right angles to each other in the optical path, they will absorb all of the originally unpolarized light.

25-25 Try this: Look through two Polaroid sunglasses held at right angles to each other and note the significant reduction of light intensity.

Dr. Edwin H. Land (1909–91) was an American researcher and industrialist who made many contributions to the science, technology, and art of optics. He held some 20 honorary doctorate degrees but never completed a formal college course of study.

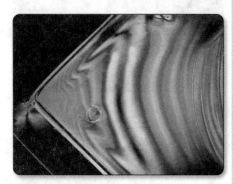

25-26 Crossed polarizers were used to reveal a flaw in this material.

but the *O*-ray is reflected by the Canada balsam so that the *E*-ray continues alone. This device produces polarized light. In use, two Nicol prisms are placed in the path of an unpolarized beam of light with a thin section of the material of interest between them. The polarized light from the first prism enters the sample and is rotated by its crystal lattice. The amount of rotation is detected by rotating the second prism until it transmits the maximum amount of light, showing that its transmission axis is aligned with the plane of the light's polarization. The amount of rotation is characteristic for each crystalline material and is useful for identifying unknown substances.

25.13 Polarization by Selective Absorption

A few materials not only doubly refract but also absorb one of the refracted rays. These materials naturally act like a Nicol prism, although they may emit the *O*-ray rather than the *E*-ray. This behavior is called selective absorption. Two crystals of tourmaline, for example, turned at 90° to each other, will together absorb all the light that is directed through them. The first will emit a polarized *O*-ray, and the second is turned so that it will absorb that ray.

Polaroid filters, invented by Edwin Land, selectively absorb a wide selection of optical wavelengths. This means that incident, unpolarized white light will be partially absorbed and partially transmitted with a preferred polarization. Certain Polaroid sunglasses are designed to reduce the overall intensity of sunlight by screening about half of the unpolarized sunlight rays. Others are designed to absorb horizontally polarized light reflected off water and glass. Such materials are found in windshields and in sunglasses used by fishermen, boaters, and seamen.

25.14 Polarization by Scattering

Scattering is the final way to polarize light. The colors you see in the sky are a result of the scattering of sunlight by air molecules and dust particles in the air. Short wavelengths are scattered more than long wavelengths; when you look at the sky, you see the greatly scattered blue color. At sunrise and sunset, sunlight must travel a much longer path through the atmosphere to get to your eye. Much more scattering of the shorter wavelengths takes place, so only the longer orange and red wavelengths remain visible.

When you look in directions that are about 90° from the Sun, you see mostly scattered light. You can show that this light is polarized by using a selectively absorptive material such as a Polaroid filter or lens. Polarized light illuminating a Polaroid filter must be aligned with the filter's transmission axis in order to pass through. If the sunlight were unpolarized, then about the same amount of light would pass through the Polaroid filter no matter at what angle you held it. However, the intensity of light observed through the filter changes as the filter is rotated, and so the scattered light must be partly polarized.

Some materials become doubly refracting when they have been stressed. Polarized light can be used to detect stresses. If two polarizing filters are aligned 90° to each other (they are *crossed*), no light passes through them. However, if a material placed between them changes the polarity of the light, as a doubly refracting material can, light will be transmitted through the second filter. Thus, if a material that has been stressed is placed between two crossed polarizers, the assembly will transmit light in a pattern showing the stressed areas. If the material has not been stressed, the assembly will not transmit light, and the object will appear dark. Polarized light is used to detect areas of high stress in models of complicated structures where mathematical calculations would be difficult.

Optical Testing

When designing buildings, bridges, aircraft, ships, and so forth, engineers employ a multitude of mathematical formulas to calculate the loads and the resulting stresses and strains on each member of their design. Analyzing loads is a difficult procedure. How can you estimate the maximum weight of trucks, buses, and cars that will ever be on your bridge at any one time? To make sure their design will do what it is supposed to do, engineers design in a "safety factor." A safety factor of three, for example, means your design will handle loads three times as heavy as your analysis assumes for maximum anticipated loads in use.

With complex designs there are serial loads; that is, the load on one member may depend on the loads on other connecting members. This relationship often makes precise calculations of stresses and strains impossible. In the past it was customary to increase the safety factor to compensate for possible impreciseness, but this also increased the construction cost. Recent advances in photoelastic plastics and holography are helping engineers to accurately analyze complex stress patterns and increase the precision of their calculations.

Photoelastic plastics are created with one special property. Under normal conditions they reflect or transmit light evenly in all directions, but under stress they doubly refract light. Polarizers can be positioned to block out all light except the second refracted beam that follows a slightly different path. The engineer places a small-scale plastic model of his design in the light path between two crossed polarizers. At first the unstressed model will appear black. But as the engineer applies different pressures to the model, the stressed areas will appear as light regions. The intensity or color of the light is proportional to the strain in the material. This information can help the engineer verify or improve his design.

Another new aid in analyzing stress is the hologram. First, an engineer makes a hologram of the machine part or model. After the hologram is completed, he returns it to the laboratory where it was exposed. He then illuminates both the hologram and the object. When the engineer superimposes the hologram on an image of light reflected directly from the object, the image is uniform. However, if the object is stressed, an interference fringe pattern forms in the area of strain. The advantage of this testing method is that the full-sized object made out of the actual materials can be tested. The engineer can experiment with many complex and shifting stresses.

As industrial materials and designs continue to be improved, photoelasticity and holography will be great helps in testing designs and keeping construction costs at a minimum.

25C Section Review Questions

1. Describe the difference between polarized and unpolarized light.
2. What happens to the different components of a light wave that reflect off the surface of glass?
3. How can you tell that a crystalline substance causes double refraction?
4. What happens to the energy in the part of the light wave that is *not* transmitted through a Polaroid filter?
5. At what angle should you align the transmission axis of a Polaroid filter in order to screen out the majority of direct sunlight reflected from a lake's surface?
6. What is the principle cause for the red color of sunsets?
⊙7. What is the Brewster's angle for an air–flint glass interface? The incident light ray is in air. See Table 24-1 for indices of refraction.

Chapter Review

In Terms of Physics

interference	25.1	Rayleigh's angle (θ_R)	25.9
constructive interference	25.1	unpolarized	25.10
destructive interference	25.1	polarized	25.10
superposition principle	25.1	transmission axis	25.11
interferometer	25.5	dichroic	25.11
diffract	25.7	Malus's law	25.11
collimator	25.8	Brewster's angle (θ_B)	25.11
resolving power	25.9	Brewster's law	25.11

Problem-Solving Strategies

25.1 (page 551) The symbol n in the interference equation is equal to two quantities: (1) the difference between path lengths in whole wavelengths of light that produce a maximum, and (2) the order number of the same maximum.

25.2 (page 551) When analyzing the *minima*, be sure to use the order number (n) of the next higher maximum in the interference formula.

25.3 (page 560) When determining SDs associated with diffraction grating ruling densities, the number of lines is *counted* and therefore does not affect SDs in calculations. Its reciprocal is a distance, however, so assume that grating spacing has three SDs.

25.4 (page 561) Remember that Rayleigh's angle (θ_R) is given in radians.

25.5 (page 562) Brewster's angle (θ_B) may be given in degrees or radians.

25.6 (page 563) Remember that Brewster's angle is the incident angle of the incoming ray of light. The 90° angle is measured between the reflected ray and the internally refracted ray.

Review Questions

1. What is wave interference?
2. Two pulses of equal amplitude that are completely out of phase move toward each other. Will they annihilate each other? Explain.
3. The two pulses in each figure travel toward each other. Will they interfere constructively or destructively?

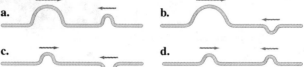

a.   b.

c.   d.

4. Why did Young's experiment demonstrate that light acts as a wave?

5. What does an interference pattern from two close slits look like?

6. The figure shows a double-slit experiment. Is point P dark or bright? Why?

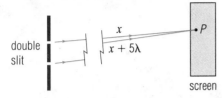

7. **a.** What factors can affect the distance between adjacent interference maxima in a double-slit experiment?
b. Discuss the effect of each factor.

8. Describe briefly why gasoline spilled on water looks multi-colored.

9. Referring to Question 8, why does a thick or opaque film not produce the same multicolored effect?

10. What are some characteristics of laser light?

11. An interferometer was instrumental in refining the precision of which physical constant?

12. What is one common use for diode lasers?

13. What is diffraction?

14. Why did Newton not notice light's diffraction?

15. Name three kinds of diffraction.

16. What does diffraction produce at the center of the shadow of a small, spherical object?

17. Which kind of diffraction produces a color spectrum?

18. What limits the ability of an optical system to resolve two visual objects that are separated by a small angle?

19. What is polarized light?

20. Name two ways to polarize light.

21. Discuss how you could determine the theoretical index of refraction of an opaque material using reflected polarization. (For example, the index of refraction for gold is 0.47.)

True or False (22–34)

22. When two wave pulses of opposite phases are superposed, the waves disappear.

23. Monochromatic light produces clearer maxima in Young's experiment.

24. The interference maxima in Young's experiment can be brought closer together by decreasing the wavelength of light used in the experiment.

25. The thinner a film is, the closer together the observed interference bands are.

26. The length of the crystal rod in a laser determines the wavelength of light emitted.

27. The reason you can hear someone speaking to you from behind you is mostly due to sound diffraction.

28. A grating spectrograph will produce bright, distinct lines of different colors when analyzing pure white light.

29. A telescope is said to have a high resolving power if two stars that appear to be very close to each other can be optically separated.

30. A larger Rayleigh's angle would be required to resolve two blue stars than to resolve two red stars.

31. Direct, unscattered light from the Sun is unpolarized.

32. Brewster's angle is the angle of incidence that produces the strongest reflection.

33. The direction of the extraordinary ray in calcite is dependent on the orientation of the mineral's crystal lattice.

34. Acting a little crazy, you wear your Polaroid sunglasses upside down. The filtering effect of the sunglasses no longer works in this position.

⊛35. An opaque screen is pierced by two slits separated by a distance of 0.250 mm. A second screen, 4.00 m away, reflects the interference pattern. The light has a wavelength of 600. nm.
a. How far is the center of the first dark minimum from the central maximum?
b. How far is the fifth bright maximum from the central maximum?
c. Is the edge of the screen, 25.0 cm from the central maximum, bright or dark?

⊛36. Monochromatic light falls on two small holes that are 3.00×10^{-4} m apart in a window shade. Their interference pattern encounters a wall 1.50 m away. If the first bright fringe is 2.40 mm from the central maximum, what is the light's wavelength?

⊛37. A laser light illuminates a screen with two circular holes 5.00×10^{-4} m apart and is projected onto a screen, 10.0 cm in diameter, that is 5.00 m away. The central maximum is at the circular screen's center. Two bright spots are at the perimeter of the screen on opposite sides. The wavelength of the light is 500. nm.
a. What is the highest-ordered maximum on the screen? ($n = ?$)
b. What is the highest-ordered minimum on the screen?
c. Are the spots at half the screen's radius from the central maximum the $n^{th}/2$ maxima? (n = highest order number of maxima on the screen)

⊛38. How closely should you space the lines on a diffraction grating if you want the first-order maximum from green light ($\lambda = 550.$ nm) to be at 29.7°?

⊛39. A light beam directed into a spectroscope with a diffraction grating ruled in 10 000 lines per centimeter has a wavelength of 500. nm. At what angle is its first-order maximum?

⊛40. A spectroscope contains a diffraction grating with 7500 rulings per centimeter. What wavelengths of light and infrared have a first-order maximum at the following angles?
a. 20.0°
b. 25.0°
c. 30.0°
d. 35.0°
e. 40.0°

⊛41. Judging from your answers to question 40, will a spectroscope have a linear scale? That is, will the same change in angle always indicate the same change in wavelength?

⊛42. What is the index of refraction of a new plastic material if its Brewster's angle is 18.5°? The incident ray is in air.

26A	Intensity and Color	569
26B	Optical Instruments	577
Facet:	Microscopy	578

New Symbols and Abbreviations

luminous intensity (I_L)	26.1	lumen (lm)	26.3
candela (cd)	26.1	illuminance (E)	26.4
steradian (sr)	26.1	lux (lx)	26.4
luminous flux (Φ)	26.3		

The 10 m Keck telescope and its sister Keck II on top of Mauna Kea in Hawaii are the largest optical and infared telescopes in the world.

Using Light

For God, who commanded the light to shine out of darkness, hath shined in our hearts, to give the light of the knowledge of the glory of God in the face of Jesus Christ. II Cor. 4:6

One of the objectives of this chapter is to describe how the intensity of visible light is measured. The intensity of a source of light is an intrinsic property of the source. The apostle Paul reminds us that the source of spiritual light in a dark world is the person of the Lord Jesus, who fully manifests the glory of the Father. The spiritual light that God shined in the hearts of the apostles gave them the knowledge of God and enabled them to show that light to others by preaching the gospel of salvation. This Light is not just a dim illumination. Even to the most righteous, God's glory is an all-consuming fire of billions of suns. God would not even permit Moses to behold His full glory because the intensity would have destroyed him (Exod. 33). Have you beheld the glory of God in the person of Jesus Christ? Are you, in your words and your manner of life, radiating that light to others?

26A INTENSITY AND COLOR

26.1 Light-Source Intensity—The Candela

How much light does a bulb produce? Light bulb manufacturers rate their bulbs in watts, a unit of power. Does a 100 W bulb produce four times as much light as a 25 W bulb? No. Although the amount of light a bulb emits is related to the power the bulb consumes, the relationship is not direct. The brightness of a light source is called the **luminous intensity (I_L).** How do scientists and engineers measure luminous intensity?

Through the nineteenth century, a light source was compared to a standard candle to determine its luminous intensity. The luminous standards used by each country varied, and no candle flame has a constant brightness. The first real attempt to standardize luminous intensity was based on carbon-arc incandescent lamps. However, even these were not stable enough to provide the standard required by scientists. A theoretical standard based on a blackbody radiator at the freezing point of platinum (2045 K) was proposed in 1933. This standard was too difficult to achieve in the laboratory, so a new standard was developed based on

The unit **steradian (sr)** in this definition is a conical solid angle (an angle measured in three dimensions), with its vertex at the center of a sphere, that cuts off a circular area on the surface of the sphere equal to the square of the sphere's radius.

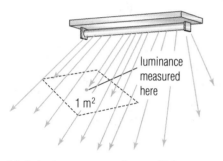

$A = R^2$

R

1 steradian

sphere of radius R

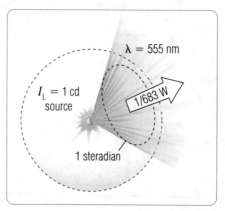

$\lambda = 555$ nm

$I_L = 1$ cd source

1/683 W

1 steradian

26-1 Luminous intensity is the rate of light energy emitted by a light source (its power) in a specific direction.

radiometry in 1979. The **candela (cd)** is now defined as *the luminous intensity, in a given direction, of a source that emits monochromatic radiation of frequency 540×10^{12} hertz and that has a radiant intensity in that direction of $\frac{1}{683}$ watt per steradian.* Let's look at this definition in detail.

The human eye is not equally sensitive to all frequencies of visible light. Therefore, the frequency that our eyes are most sensitive to was chosen for the definition. Frequency was used rather than wavelength because wavelength varies depending on the material light passes through. A frequency of 540×10^{12} Hz corresponds to a wavelength of 555 nm (in a vacuum), which is greenish-yellow light.

The fraction $\frac{1}{683}$ makes the current definition compatible with the previous definition based on the freezing point of platinum.

The requirement to measure the radiated power within a solid angle called a **steradian (sr)** makes the candela unit compatible with other SI power units.

An incandescent light bulb gives off roughly one candela for each watt it consumes. A 25 W bulb produces 21 cd; a 50 W bulb, 55 cd; and a 100 W bulb, 125 cd. Thus, a 100 W bulb produces more than five times as much light as a 25 W bulb produces. That is, a high-wattage bulb uses its power more efficiently than a low-wattage bulb.

26.2 Extended Lighting Sources—Luminance

The definition of luminous intensity assumes a point light source. How do you specify the intensity of a real light source such as a light bulb, overhead fluorescent lighting unit, or even the Sun? The rate of flow of light energy reaching a surface in a given direction from the source is called the **luminance** of the light source. Luminance is measured at a point and has units of power per unit of area, or cd/m².

26.3 Luminous Flux—The Lumen

We mentioned that the luminous intensity of a light source is a measure of the illumination strength in a particular direction, but how do you measure the total amount of light that a source gives off? This quantity is called the **luminous flux (Φ)** around the light source. Luminous flux is independent of the direction of measurement. It is measured in units called **lumens (lm),** which is the product of candelas and steradians (cd·sr). In other words, 1 lumen is the luminous flux of a 1 cd light source in 1 sr. Since there are 4π sr in a sphere, a 1 cd point light source has a total luminous flux of 12.57 lm ($4\pi = 12.57$). Luminous flux is a measure of power, as is luminous intensity. Both are derived from the SI unit for power, the watt.

26.4 Illumination at a Point—Illuminance

Lighting engineers want to know how bright the light falling on a surface appears to the human eye at a particular point. The lighting required for a desk in an office differs from that required for a sidewalk at night. You know from experience that the effectiveness of a source of light varies with the distance from the source. For example, headlights of an approaching car seem to grow brighter as the car nears.

26-2 Luminance measures the rate of light energy falling on a unit surface.

luminance measured here

1 m²

TABLE 26-1

Luminance (cd/m²)	Source
0.03	white paper in moonlight
2900	surface of the Moon
3200	clear sky at midday
5000	candle flame
6000	fluorescent lamp
25 000	white paper in sunlight
10^7	tungsten filament
2×10^9	surface of the Sun

1 lm = 1 cd $\times$ 1 sr

The distance of an object from a light source also affects the amount of light the object receives, with the light intensity greater at shorter distances from the source than farther away. The lighting value at a particular point is measured by the **illuminance (E)**. The illuminance at a distance from a light source is directly proportional to the luminous flux of the source and inversely proportional to the square of the distance from the source. In equation form,

$$E = \frac{\Phi}{r^2}, \tag{26.1}$$

where r is the distance from the light. The units of illuminance are lumens per square meter (lm/m^2), which are called **lux (lx)**. A 1 cd source produces an illuminance of 1 lux at every point that is 1 m from the source. Illuminance is directly proportional to luminous intensity. For example, a 2 cd light gives twice as much light at a given distance as a 1 cd light gives at the same distance. Notice that Equation 26.1 is an inverse square law. That is, an object that is $2x$ meters from a light source receives only one-fourth the light received by an object that is x meters from the source.

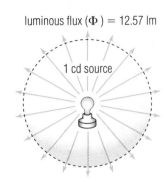

luminous flux (Φ) = 12.57 lm

1 cd source

26-3 Luminous flux measures the total rate of light energy emitted by a source in all directions.

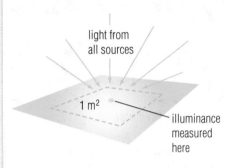

light from all sources

1 m²

illuminance measured here

26-4 Illuminance is the total illumination at a point on a surface

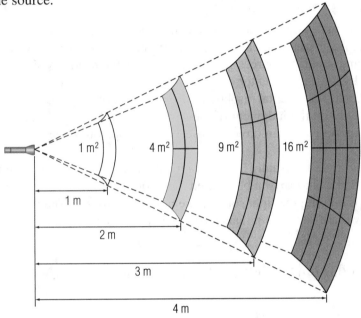

1 m² 4 m² 9 m² 16 m²

1 m

2 m

3 m

4 m

26-5 As the distance between a point light source and an illuminated surface increases, illuminance decreases because light spreads out over a greater area.

TABLE 26-2

Illuminance (lm/m^2)	Conditions
0.003	starlight
0.2	full moonlight
100	precision work, artificial light
1000	office near window
10 000	bright overcast day
100 000	clear sky at midday

$$1 \text{ lx} = \frac{1 \text{ lm}}{1 \text{ m}^2}$$

EXAMPLE 26-1

Calculating Illuminance

What is the illuminance at a point 5.0 m from a light source emitting 600. lm?

Solution:
Use Equation 26.1:

$$E = \frac{\Phi}{r^2}$$

$$E = \frac{600. \text{ lm}}{(5.0 \text{ m})^2}$$

$$E = 24 \text{ lm/m}^2 \text{ or } 24 \text{ lux}$$

How Bright a Bulb? Determining Intensity

How strong a light source, in lumens, is needed to give an illuminance of 80. lx at a distance of 1.5 m?

Solution:
Use Equation 26.1:

$$E = \frac{\Phi}{r^2}$$

$$\Phi = Er^2$$

$$\Phi = (80.\ \text{lm/m}^2)(1.5\ \text{m})^2$$

$$\Phi = 120\ \text{lm}$$

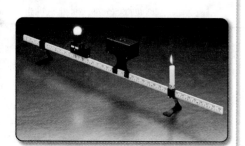

26-6 A laboratory photometer

The photometer described here is a *comparator photometer*. Other kinds of photometers, including those called light meters by photographers, measure incident light either from all directions at a point or from a specific direction.

26.5 Comparing Light Sources

The intensity of an unknown light source can be determined by comparing it to a known source. An instrument called a **photometer** receives beams of light from the unknown and standard sources and either displays them side by side in an eyepiece viewed by the experimenter or measures their intensities electronically and provides the results via a digital display. The former method is less accurate because it depends heavily on the judgment of the experimenter, but it is more instructive because the experimenter can directly observe how apparent intensity varies with distance.

The photometer setup usually includes a meter stick, the standard and unknown light sources, and a comparison box called the photometer head or a digital detector. The clamps holding the light sources are adjusted until the comparison box or detector receives the same illuminance from each light source. That is,

$$E_{\text{unk}} = E_{\text{std}}.$$

Since luminous flux (Φ) is proportional to luminous intensity (I_L), we can say that

$$E \propto I_L/d^2,$$

$$E_{\text{unk}} \propto \frac{I_{L\,\text{unk}}}{d_{\text{unk}}^2} \text{ and } E_{\text{std}} \propto \frac{I_{L\,\text{std}}}{d_{\text{std}}^2}, \text{ or}$$

$$\frac{I_{L\,\text{unk}}}{d_{\text{unk}}^2} = \frac{I_{L\,\text{std}}}{d_{\text{std}}^2}.$$

Rearranging,

$$I_{L\,\text{unk}} = I_{L\,\text{std}}\frac{d_{\text{unk}}^2}{d_{\text{std}}^2}. \tag{26.2}$$

Since you can read d_{unk} and d_{std} from the meter stick, and you know I_{std}, you can calculate I_{unk}. Notice that you do not have to know E_{unk} to find I_{unk}.

Problem-Solving Strategy 26.1

Use the source's luminous intensity (in candelas) rather than its luminous flux (in lumens) when using the photometer equation.

Finding an Unknown Intensity

A standard 100. cd bulb that is 32.0 cm from the photometer detector has the same luminous intensity as an unknown bulb that is 25.0 cm from the photometer head. Find the unknown's luminous intensity.

Solution:
Use Equation 26.2:

$$I_{L\ unk} = I_{L\ std}\frac{d_{unk}^2}{d_{std}^2}$$

$$I_{L\ unk} = (100.\ cd)\frac{(25.0\ cm)^2}{(32.0\ cm)^2}$$

$$I_{L\ unk} \doteq 61.\underline{0}3\ cd$$

$$I_{L\ unk} \approx 61.0\ cd$$

26.6 Transmittance

Materials differ in their ability to transmit light. *Transparent* materials, such as glass, water, air, Lucite, and cellophane, transmit light so that you can see clearly through them. Since a transparent material does reflect and absorb some light, a thick piece of transparent material is harder to see through than a thin piece of transparent material. The **transmittance** of an object is the ratio of the transmitted luminous flux to the incident flux.

A *translucent* material, like frosted glass, tissue paper, or cloth, transmits light but distorts it so that you cannot see clearly through it. An *opaque* material, such as wood, metal, or concrete, does not transmit light at all unless the material is extremely thin. The divisions between transparent and translucent materials and between translucent and opaque are not distinct, so the thickness or smoothness of an object may determine which category it belongs in.

26.7 Color

If you doubt the importance of color, look around you and try to imagine the world as if it were a black-and-white photograph. Scientists long ago determined that the majority of an average adult human's sensory information is received through his eyes. It is therefore understandable that people have developed color photography, paints, dyes, colored inks, color television, and color digital displays in order to fully employ this sense.

26-7 Without color, much visual information becomes uninteresting or may even be lost, as shown in the monochrome half of the picture.

TABLE 26-3

Color	Wavelength, λ (nm)
violet	400–450
blue	450–500
green	500–570
yellow	570–590
orange	590–610
red	610–700

What you see as color is actually different frequencies of light. You can receive this light in three ways: directly from its source, by reflection from a surface, or by transmission through a material. When light reaches you directly from its source, its color depends on the energy that each source atom emits. The light's frequency, which determines its color, is related to the energy change of the electrons by the equation

$$E = hf, \tag{26.3}$$

where the coefficient of the frequency (f) is *Planck's constant (h)*. As you can see, larger changes in electron energy produce higher energy emissions that are observed as higher frequencies of light.

Light that reflects from a surface or passes through a medium may have some of its frequencies attenuated (lessened) or removed. The energy from the absorbed light heats the material that absorbs it.

The wavelength (or frequency) of light is not the only factor that determines how the color appears to you. Colors are specified by hue, saturation, and brightness. *Hue* refers to the mixture of wavelengths in the light. The basic color type is determined by hue. *Saturation* is a measure of the amount of neutral colors present, such as white or gray, that tend to dull the color. The difference between a greenish-tinged gray and brilliant kelly green is saturation. *Value, brightness,* or *lightness* is how dark or light the color appears. A nearly black color has no value, while a nearly white color has a very high value. The term can also refer to the amplitude of the light, where high-amplitude light appears lighter than low-amplitude light. Figure 26-8 provides two different ways of graphing the colors seen by the human eye. Note that the eye is not equally sensitive to all colors or color ranges.

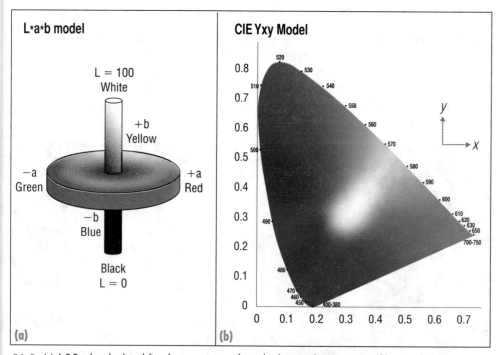

26-8 (a) A 3-D color wheel modeling the approximate relationship between hue (perimeter of horizontal circle), saturation (radial distance from center), and value (vertical position). (b) Two-dimensional representation showing range of color saturation and hue perceived by the human eye.

26-9 A color is perceived in relation to its surroundings.

An object's color also depends on the colors of surrounding objects. That is, you see color in contrast to its surroundings.

26-10 Colored lights combine in an additive way to produce colors on a white surface.

26.8 Additive Color Mixing

White light is usually a mixture of all wavelengths, but some combinations of two or three wavelength bands produce a light that looks white to us. For example, yellow and blue lights, green and magenta (bluish red) lights, or red and cyan (greenish blue) lights combine to produce white light. Since, when combined together, these pairs appear to produce the complete spectrum (i.e., white light), they are called **complementary colors.**

If beams of red, blue, and green light shine on a white screen so that they overlap, the region where all overlap is white. Combinations of these three colored lights can produce any color in the spectrum, so they are called **additive primary colors.** Notice that combining red and green light gives yellow, combining red and blue light gives magenta, and combining green and blue light gives cyan. No combination of additive primary colors can produce black. Only the total absence of light is perceived as black.

The **additive primary colors** are red, blue, and green.

26.9 Colored Objects

Why does a red object appear red? The answer is that the light coming from it to your eye is red. There are three reasons the light may be red: the object may be illuminated by only red light; it may reflect only red light and absorb all other colors; or, if it is not opaque, it may transmit only red and absorb all other colors. If the object is illuminated with white light, it must absorb all colors but red. The substance in an object that absorbs certain colors and reflects others is called **pigment.** The pigment is named by the color it reflects.

Suppose you illuminate an object containing red pigment with red light. It will continue to reflect or transmit only red light. However, since the object is illuminated with only red light, all light reaching it will be reflected or transmitted; therefore, the object will still be red. What happens if you illuminate the object with green light? Since the object receives no red light, and it absorbs all the rest, it will not reflect or transmit light. The object will appear black.

26.10 Subtractive Color Mixing

Figure 26-11 on the next page shows white light from behind passing through a yellow filter, then a cyan filter to your eyes. When the light passes through the yellow filter, the filter absorbs all but the yellow light. Yellow light can be thought of as a combination of red light and green light. The cyan filter absorbs all but cyan light, that is, all but blue and green light. However, in this example only yellow (red and green) light reaches it. The cyan filter absorbs the red light and transmits the green light. Therefore, the light coming through the overlapping filters will be green. This process is called *subtractive* color mixing because the remaining color is the light that pigments have not absorbed (subtracted) from the original light.

Subtractive color mixing also governs the color of reflected light. For example, if an object contains both yellow and cyan pigments, it will absorb all light reaching it except green, which it will reflect. For subtractive mixing, you can produce all colors from the **subtractive primary colors:** cyan, magenta, and yellow (also called the CMY color model). A mixture of yellow and magenta pigments reflects orange light; a mixture of cyan and magenta pigments reflects purple; and, as you have seen, a mixture of cyan and yellow pigments reflects green. The correct mixture of the three subtractive primary colors produces black. Note that only white pigments (those that do not absorb *any* colors) can reflect white light. No mixture of subtractive primary pigments can produce white.

The **subtractive primary colors** are cyan, magenta, and yellow.

26-11 Subtractive color mixing. Notice that the light that has passed through both the yellow and cyan filters appears green.

26A Section Review Questions

1. **a.** Compare the terms *luminous intensity, luminous flux, luminance,* and *illuminance*.

    b. Which term is important to an architect who wants to know how much light is needed in a hallway?

2. Both luminous intensity and luminous flux involve a dimension associated with energy. What is the unit of this dimension, and how does it relate to the candela and the lumen?

3. Using a photometer to compare two light sources, what can you conclude about the intensity of a lamp if it produces the same illuminance at twice the distance as the standard lamp?

4. **a.** Which has the greater amount of energy for a given amplitude, red light or blue light?

    b. Explain your answer.

5. What color term classifies the colors grass green or apple red?

6. **a.** If possible, how would you produce the color *black* on a white surface by mixing primary additive colors or their complements?

    b. Explain.

7. Which subtractive pigment colors are required to produce a white-colored surface in white light?

⊛**8.** What is the illuminance of an omnidirectional lamp that produces 251 lm on a flat surface 3.0 m from and facing the lamp?

OPTICAL INSTRUMENTS

Optical instruments use one or more lenses, mirrors, or prisms to form an image that is enlarged for direct observation or is to be recorded in some way. For example, microscopes, telescopes, cameras, binoculars, projectors, and periscopes are optical instruments. Looking at two of these should demonstrate the principles that all optical instruments use.

26.11 The Microscope

A **microscope** magnifies minute objects in order to allow observers to see fine details that are closer together than the resolving power of our eyes. Magnifying lenses have been used since the time of the ancient Greeks, but not until the seventeenth century were small, accurately shaped lenses like those in a modern microscope made. The first microscopes were simple, with only one converging lens. Anton van Leeuwenhoek (pronounced *lay-u-wen-hook*) constructed a simple microscope through which he could see blood capillaries, red corpuscles, muscle fibers, hairs, skin cells, nerve cells, and reproductive cells. In his microscope a double convex lens with a short focal length was mounted in a frame with a handle. A specimen was mounted on one side of the lens and focused with a screw adjustment that moved the specimen nearer to or farther from the lens.

26-12 Dutch biologist Anton van Leeuwenhoek (1632–1723) and one of his microscopes

Although Leeuwenhoek was the first to observe many microscopic structures, his use of a single lens limited the magnification power of his microscope. Using two lenses "in series" produces an overall magnification equal to the product of the individual magnifications of the lenses. This strategy underlies the **compound microscope,** which is essentially two short–focal length converging lenses. The compound microscope acts as two simple microscopes. The first lens, the *objective,* receives light from the object, producing a magnified inverted real image (located beyond *F* of the objective lens). The second lens, the *ocular* (eyepiece), uses the real image of the objective as *its* object and magnifies it, producing a much enlarged virtual image. The image in a standard compound microscope is inverted top to bottom and right to left.

The objective is usually a combination of lenses. The first lens has a short focal length so that it magnifies an object placed near it, and the other lenses correct the aberrations of the first. Because the ocular magnifies the image produced by the objective, the objective is the most important component in determining the overall quality of the image. Most compound microscopes you will see in a school laboratory have three separate objectives mounted in a rotatable turret nosepiece so that the user can choose low, medium, or high magnification.

Leeuwenhoek has been mistakenly called the "father of the microscope." Simple and compound microscopes had been in use for several decades before his birth. In fact, he was inspired by Robert Hooke's discovery, using an improved compound microscope, of cells and microscopic creatures. Leeuwenhoek's contributions to science were his methodical and careful observations, which he wrote about in correspondence with other scientists, and his ability to generate interest in microscopy.

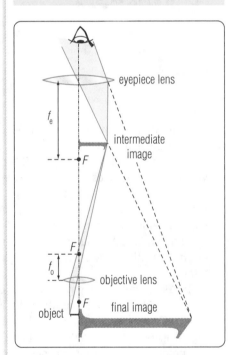

26-13 Rays of light passing through a microscope

The total magnification of a microscope is the product of the magnification powers of its objective and ocular. For example, a microscope with a 10× objective and a 5× ocular produces a total magnification of 50×.

The ocular is also usually a combination of lenses, so there is as little distortion as possible. Better-quality microscopes are provided with a selection of oculars to increase the choices of magnification. Some microscopes are used with two oculars *(binocular)* to reduce the strain of observing for long periods with a single eye. If it has two objectives as well as two oculars, the instrument is a *stereo binocular microscope,* or simply a **stereo microscope.** A stereo microscope permits depth perception. Such microscopes are used for dissection, geology field sample identification, and manufacturing quality inspections when low magnifications, spatial perception, or opaque objects are involved. A stereo microscope has an additional prism system in the light path that restores proper up-down, right-left orientation to make manual work using the microscope natural.

Research microscopes are often built with the ability to record or transmit images observed by the user. Such systems include built-in or add-on film and video cameras, or projection screens. Typically, there is a splitting prism in the optical path between the objective and the ocular. Part of the light is transmitted to the ocular and part to the camera. Some instruments have selectors that permit moving the prism completely into or out of the optical path to provide as much light as possible to the ocular or the camera.

26-14 A stereo microscope

FACETS of PHYSICS

Microscopy

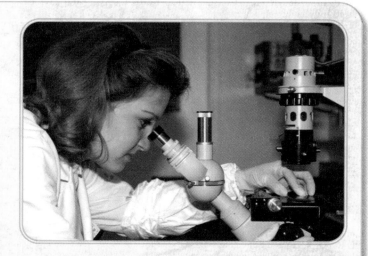

Following the invention of the electron microscope, many people expected the light microscope to become an oddity of the past—like a Ford Model T. Yet both electron and conventional light microscopes had severe limitations when used for studying live organisms. Because the inherent contrast in the cells of organisms is very low, they must be killed and stained before they can be observed. So a biologist cannot be certain how much organisms' structures have changed as a result of death; many structures disappear or remain invisible after death.

In 1934 the Dutch physicist Frits Zernike (Nobel Prize in Physics, 1953) solved this dilemma when he invented the *phase-contrast microscope.* By diffracting light through an organism rather than reflecting light off it, this microscope can produce high-contrast images directly from living, moving cells. Both it and the interference microscope have become standard tools in all serious biological studies.

The phase-contrast microscope takes advantage of the different indexes of refraction in specimens. Light passing through a specimen arrives at the objective lens a little behind the light that passes around it. A transparent plate between the specimen and the objective lens increases the phase difference. As the different light rays interfere with each other, an image is formed. Biologists have been able to study and film dynamic events such as cell division and cell response to foreign agents such as chemicals. Interesting discoveries have also been made about hormones; for example, the discovery of phytochrome pigments in plants has helped us to understand how plants detect and follow the Sun.

The *interference microscope* also uses the slowing of light to form clear images of transparent objects. However, the light is divided into two beams, one that passes through or reflects off the object and another that does not. The beams later recombine to form an interference pattern depicting the object. Though the interference microscope is more complex and more expensive than the phase-contrast microscope, its advantages far outweigh its disadvantages. It can vary the phase to produce images that are sharper, have more depth, and reveal striking "relief" effects. Contour maps of information about the mass of material per unit volume have greatly aided the study of cell theory (protozoans, slime molds, cytoplasm, etc.) and of metallurgy (the development of strong alloys).

26.12 The Telescope—Refractors

A **telescope** makes distant objects seem nearer. The inventor of the first telescope is not known. Several men in the Netherlands had claimed to be the inventor, but Hans Lipperhey, a Dutch lensmaker, was the first to apply for a patent from his government in 1608 and make the instrument well known. Having received a description of this instrument, Galileo built his own telescope and observed the night sky with it in 1609. He thus became the first person to see the Moon's mountains, Saturn's rings, Venus's phases, and the four largest moons of Jupiter. When he studied the Milky Way, Galileo saw for himself that "the host of heaven cannot be numbered" (Jer. 33:22).

Galileo's first telescope had little magnifying power. Like a microscope, it had an objective lens and an eyepiece. The objective was a double convex converging lens. Its ocular was a double concave diverging lens, which produced a virtual, erect image of the objective's image. Since the ocular was placed *between* the objective's focal point and the lens, the image remained erect throughout the entire optical path. However, the telescope had an extremely narrow *field of view*. Galileo continued to improve his telescope until he reached a magnification of about thirty power.

Eventually, scientists realized that they could get greater magnification if both lenses were double convex. Today's "all-lens" telescopes, or **refracting telescopes,** use two double-convex converging lenses. The objective lens is a long–focal length lens that gathers light from the object, so it works better if its diameter is large. A telescope is described by the diameter of its objective—a 75 cm telescope has an objective whose diameter is 75 cm. The standard ocular forms a virtual, erect image of the objective's inverted image; therefore, the final image is inverted compared to the object.

The magnifying power of a telescope is the ratio of the angle the image subtends, or takes up (θ_I, Figure 26-16 (a)), to the angle the object subtends (θ_O, Figure 26-16 (b)). A planet's image in a telescope is much easier to see than the planet itself because it occupies a greater proportion of your field of vision. The ratio of θ_I/θ_O, the magnifying power, is equal to the ratio of the focal lengths of the objective (f_o) and the ocular (f_e), as expressed by Equation 26.4 on the next page.

Hans Lipperhey (1570?–1619) was a Dutch spectacle maker.

26-15 One of Galileo's telescopes

The *field of view* of a telescope, measured in degrees, is the maximum angular size of an object that can be observed.

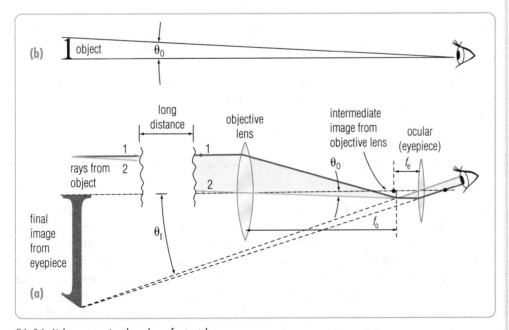

26-16 Light rays passing through a refractor telescope

The largest refractor telescope in the world is the 102 cm telescope in the Yerkes Observatory in Williams Bay, Wisconsin.

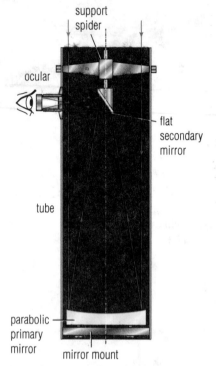

26-17 The Newtonian form of the reflector telescope uses a small, flat secondary mirror at the main mirror focus that directs the image through a hole in the side of the tube.

James Gregory (1638–75) was a Scottish mathematician.

At the time of this textbook's printing, the world's largest single-mirror reflector telescope is the Subaru reflector operated by the National Astronomical Observatory of Japan (NAOJ) atop Mauna Kea volcano in Hawaii. The largest general-purpose segmented-mirror reflectors are the two Keck telescopes operated jointly by the California Institute of Technology, the University of California, and NASA (see the photograph on page 568).

$$\text{magnifying power} = \frac{\theta_I}{\theta_O} = \frac{f_o}{f_e} \qquad (26.4)$$

For example, if the objective's focal length is 100. cm and the ocular's focal length is 1.25 cm, the telescope's magnifying power is

$$\text{magnifying power} = \frac{f_o}{f_e} = \frac{100. \text{ cm}}{1.25 \text{ cm}} = 80.0\times.$$

As you learned in Chapter 25, the larger the lens, the greater the resolving power. This is true because the larger aperture has greater light-gathering ability. You would expect that the best refracting telescopes would have very large objective lenses. However, there is a practical limit to the size of a lens. If the lens is much more than 1 m in diameter, gravity causes the center of the lens to sag, distorting its shape and the images it produces. Thus, the largest telescopes have objective lenses that are around 1 m in diameter.

26.13 Reflector Telescopes

Because a refractor's objective can be supported only at its edges, large lenses have a sagging problem. To circumvent the problem of distortion and other optical aberrations associated with large lenses, the **reflecting telescope** was developed. Reflector telescope mirrors can be supported across the entire back surface; thus much larger and more massive light-gathering elements are possible (exceeding diameters of 8.3 m for single mirrors and 10 m for segmented mirrors).

In 1663 James Gregory first suggested building a reflecting telescope, and in 1668 Isaac Newton made the first working model. One drawback to the reflecting telescope is that the mirror forms a real image at its focal point on the same side as the object. In order to observe the image, a person has to somehow get between the object and the mirror. How, then, does the image reach a human eye without obstructing light from the object? For large astronomical telescopes, there is an observation "cage" at the focal point of the mirror. The focal point is where an astronomer or a camera may receive the image. Since each part of the mirror receives light from every point of the object, this cage reduces the total amount of light reaching the mirror, but it does not create a hole in the middle of the image. For a large telescope, this reduction of light is tolerable.

For small telescopes, on the other hand, placing an observer at the focal point would block too much light. Newton's means of observing the image was to place a small plane mirror at an angle to the principal axis near the focal point of the mirror. The image was reflected perpendicularly to the optical axis of the mirror through a hole in the side of the telescope tube. An ocular mounted in the hole focused the light reflected by the plane mirror so that an observer could use the telescope. This arrangement is called the **Newtonian reflector** form of the reflecting telescope.

26.14 Folding the Light Path—Cassegrainian Telescopes

Another method of observing the image in a small reflecting telescope was devised by the French inventor Cassegrain. In this telescope, a convex mirror at the mirror's focal point reflects the image back through a hole in the main concave mirror to an eyepiece behind the mirror. The **Cassegrainian** form of the reflecting telescope is especially convenient in larger amateur telescopes since its eyepiece is behind the concave mirror. With a larger Newtonian telescope, the observer must sometimes use a platform or ladder to reach the eyepiece, which is located near the open end of the telescope tube.

26.15 The Schmidt-Cassegrainian Telescope

One problem introduced by the basic Cassegrainian form is spherical aberration from the convex mirror. This problem was solved by Bernhard Schmidt. He used a spherically-ground main concave mirror and installed a "correcting plate" at the open end of the telescope tube to cause the incoming rays to focus at a point when reflected by the spherical mirror, thus canceling the spherical aberration problem. The design was originally applied to the Newtonian form but was quickly adapted to the Cassegrainian form, producing what is called the **Schmidt-Cassegrainian** form of reflector telescope. This is a very popular telescope among amateurs because the focal length of the main mirror can be "folded up" within a shorter tube, larger apertures are possible in a more compact form, and the interior of the tube is sealed from dust and thermal currents that affect the optical path of an open-tube telescope.

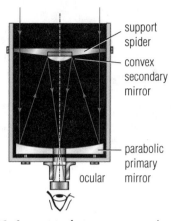

26-18 Cassegrainian focus; a convex secondary mirror at the main mirror focus directs the image through a hole in the main mirror.

Cassegrain (1652?–1712?) Almost nothing is known of this man—even his first name is in doubt, for he is variously known as Jacques, Guillaume, or N. Cassegrain. His only documented accomplishment was in France, where he designed the telescope that bears his name. He didn't even build a telescope of his design—that was first accomplished by James Short (1710–68).

26-19 The interior of a Keck observatory dome. The position of each of the 36 segments of the 10.0 m primary mirror can be controlled independently by computer to compensate for the restless motion of the atmosphere.

26.16 Advantage of Size

Why do astronomers prefer large telescopes to small ones? At least three reasons are apparent. First, a large telescope gathers more light than a small telescope, and, therefore, produces a brighter image. The light-gathering power of a telescope depends on the area of its objective, or the square of its objective's radius, since $A_{\text{circle}} = \pi r^2$. Therefore, a 200 cm telescope gathers four times as much light as a 100 cm telescope and can detect stars that are only one-fourth as bright as the faintest star a 100 cm telescope can detect.

Second, a large telescope has better resolution than a small telescope. Remember from Chapter 25 that the smallest angular distance between two objects that will allow the objects to be resolved is

$$\theta_R = 1.22\frac{\lambda}{d}. \tag{26.5}$$

Objects with a smaller angular separation can be seen as separate objects with an objective or mirror of a larger diameter (d). Smaller instruments tend to blur two close objects, making them look like a single object.

Third, a large telescope can magnify finer details better than a small telescope. Magnification is dependent on focal length, and large telescopes have longer practical focal lengths than smaller ones. Because of the lower resolving power in a smaller telescope, magnifying an indistinct image beyond a certain power produces no more detail.

Bernhard Schmidt (1879–1935) was born in Estonia, but he was educated in and lived in Germany. He was an optical technician for a prominent German observatory.

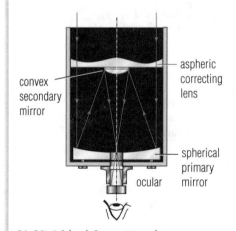

26-20 A Schmidt-Cassegrainian telescope

Using Light **581**

26B Section Review Questions

1. Discuss the purpose (the optical function) of the following instruments:
   a. camera
   b. periscope
   c. telescope
   d. microscope

2. What was the main limitation of Leeuwenhoek's microscope?

3. a. What is the orientation of the image produced by a standard compound microscope?
   b. If you move a specimen slide to the left while observing it through the microscope, which way does the image move?

4. What kind of microscope would you want to use if you were a U.S. mint worker verifying that the anti-counterfeiting microprinting on twenty-dollar bills was printed correctly?

5. Refractor telescopes are limited in size by the mass and diameter of their objective lenses. Explain why the need to correct spherical aberration limits the size of the objective in a refracting telescope.

6. Does the secondary mirror in a Newtonian reflector make the image in the ocular erect? Explain your answer.

⊙7. What is the magnification of a refractor telescope having a 1.25 m focal length objective and a 25 mm focal length ocular?

⊙⊙8. Describe any image-producing instruments that are not optical instruments within the definition given in this chapter.

Chapter Review

In Terms of Physics

luminous intensity (I_L)	26.1	pigment	26.9
candela (cd)	26.1	subtractive primary color	26.10
steradian (sr)	26.1	microscope	26.11
luminance	26.2	compound microscope	26.11
luminous flux (Φ)	26.3	stereo microscope	26.11
lumen (lm)	26.3	telescope	26.12
illuminance (E)	26.4	refracting telescope	26.12
lux (lx)	26.4	reflecting telescope	26.13
photometer	26.5	Newtonian reflector	26.13
transmittance	26.6	Cassegrainian reflector telescope	26.14
complementary color	26.8	Schmidt-Cassegrainian telescope	26.15
additive primary color	26.8		

Problem-Solving Strategies

26.1 (page 572) Use the source's luminous intensity (in candelas) rather than its luminous flux (in lumens) when using the photometer equation.

26.2 (page 574) The equation $E = hf$ uses frequency rather than wavelength because wavelength changes as light passes through various forms of matter.

1. What is the difference between luminous intensity and illuminance?

2. **a.** Consider a standard incandescent light bulb. Is all the illumination it provides useful if the bare bulb is used primarily as a reading light?
   b. What is used to improve the bulb's efficiency by directing more of its light onto the page of the book?

3. **a.** What does a photometer measure?
   b. If one side of a comparator photometer screen is brighter than the other, does that necessarily mean that the lamp on that side is brighter than the lamp on the other? Explain.

4. Which kind of optical material transmits light but distorts the path of light rays so that an image cannot be formed?

5. Why is it not possible for a material to have a transmittance greater than 1?

6. **a.** What is color?
   b. Does color exist in nature in the absence of a mind to perceive it?

7. The energy in a light wave is proportional to what property of light?

8. What property of color describes how dull or brilliant a given color may be?

9. Under what circumstances may colors be additively combined?

10. Under what circumstances may colors be subtractively combined?

11. What is the minimum number of primary and complementary colors that may be added in order to produce white?

12. List the subtractive primary colors. Are they the same as the additive primary colors?

13. What is the minimum number of primary and complementary colors you must combine in order to produce black?

14. **a.** List three optical instruments.
    b. Why is a photometer not an optical instrument according to the definition given in this chapter?

15. What is the difference between a simple microscope and a compound microscope?

16. Is a telescope's image larger or smaller than the object?

17. What is a refracting telescope?

18. Why are the world's largest telescopes, such as the 10 m Keck telescopes, not refracting telescopes?

19. Where in relation to the object and the primary light-gathering element does a Newtonian reflecting telescope's image form?

20. Describe two ways to bring the image in a small reflecting telescope to an observer's eye.

21. Give one advantage that a large telescope has over a small one.

True or False (22–33)

22. The total amount of light a bulb produces is measured simply in watts.

23. Luminous intensity is the measure of the rate of light energy emitted in a particular direction.

24. The apparent or observed intensity of a light source is inversely proportional to the distance from the source.

25. Opaque materials do not transmit light.

26. Shorter wavelengths of light have less energy than longer wavelengths.

27. A dark burgundy color has less hue than a light burgundy color.

28. Adding two complementary additive colors produces a primary color.

29. Subtractive color mixing only works when light interacts with pigments—not with light by itself.

30. The image of a microscope is smaller than the object.

31. The shorter the focal length of a telescope's objective, the higher its resolution.

32. A Schmidt-Cassegrainian telescope is generally easier to use than a Newtonian telescope of equal aperture.

33. A telescope with a mirror three times the diameter of a smaller telescope can view stars one-ninth the intensity of those viewable by the smaller.

◉34. What is the illuminance on a surface that is 2.0 m from a 1600 lm source?

◉35. What luminous intensity is needed to produce the following illuminances? (*Hint:* First determine the illuminance at 1 m.)
    a. 100. lx at 1.50 m
    b. 333 lx at 3.00 m
    c. 1200. lx at 5.00 m
    d. 1500. lx at 0.400 m

◉36. Find the unknown luminous intensity ($I_{L\ unk}$) for the following cases:
    a. An unknown bulb at $d_{unk} = 0.070$ m has the same illuminance as the standard source of $I_{L\ std} = 1.00$ cd at $d_{std} = 0.350$ m.
    b. $I_{L\ std} = 1.00$ cd; $d_{std} = 0.250$ m; $d_{unk} = 0.500$ m
    c. $I_{L\ std} = 1.00$ cd; $d_{std} = 0.0100$ m; $d_{unk} = 0.200$ m
    d. $I_{L\ std} = 1.00$ cd; $d_{std} = 0.100$ m; $d_{unk} = 0.700$ m

◉◉37. What influence does the ratio of a telescope's focal length to the diameter of its objective (or mirror) have on its performance?

◉◉38. Consider Question 6b again. Discuss the philosophical position that some scientists take when they state that if one cannot perceive something (either directly with the senses or with the aid of instruments), then it has no reality. Is this a biblical philosophy?

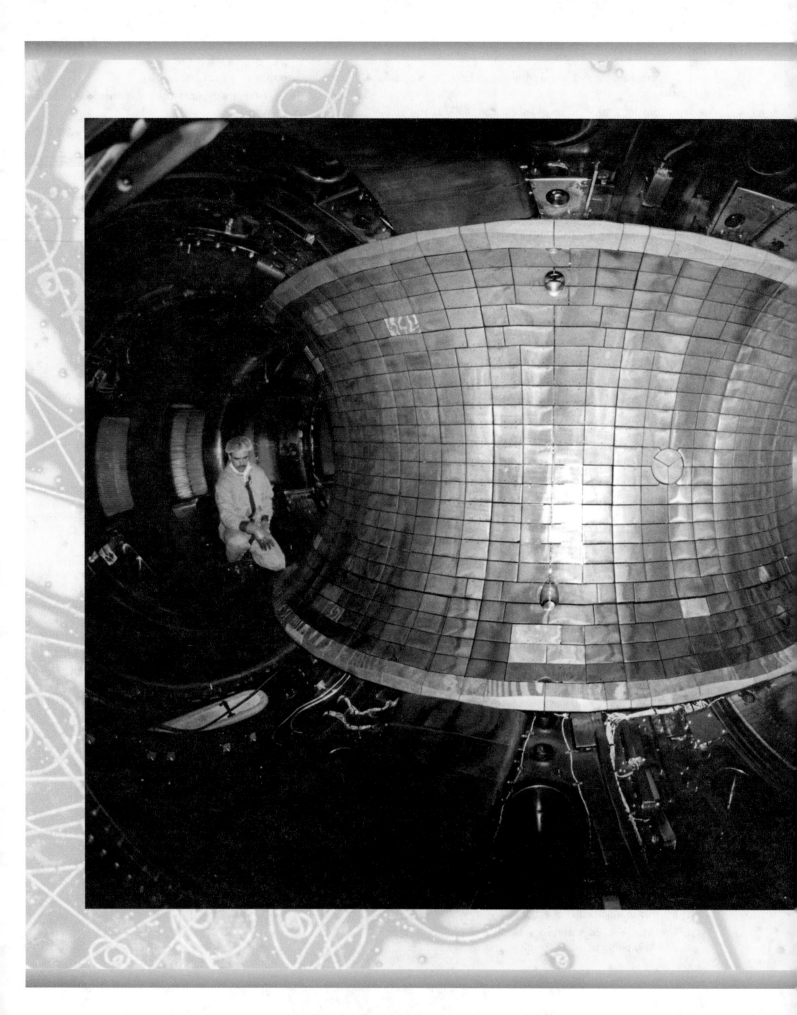

UNIT 6

Modern Physics

The dawning of the twentieth century revealed to scientists that the universe was much more complicated than indicated by the laws of classical physics. For one thing, energy didn't come in a continuous range of values—it was quantized. The atom wasn't the ultimate particle of matter, and matter itself was mostly space. Further research uncovered sources of energy that have become the most efficient commercial fuels as well as the most devastating weapons of war. Today, scientists are still struggling with Einstein's theory of relativity and seeking the theory that will unify all the forces of nature in one grand theory.

In Unit 6, you will just scratch the surface of the various theories of relativity and coordinate system transformations. You will learn how matter and energy are quantized and are interchangeable. After reviewing the structure of the atom, you will study nuclear reactions in some detail and learn about the truly fundamental particles that make up all matter.

27A	Galilean Relativity	587
27B	Special Relativity	593
27C	General Relativity	602
Biography:	Albert Einstein	592
Facet:	Gravitational Red Shift	604

New Symbols and Abbreviations

relativistic factor (γ)	27.10
proper length (l_0)	27.11
rest mass (m_0)	27.12
rest-mass energy (E_0)	27.16

Some of the first spectra of distant galaxies ever obtained with the European Southern Observatory's (ESO) visible multi-object spectrograph (VIMOS). More than 220 distant galaxies were observed simultaneously in this first-ever image. These spectra allow astronomers to obtain the redshift, a measure of distance, as well as to assess the physical status of the gas and stars in each of these galaxies. According to the theory of relativity, the laws of physics are the same in these distant galaxies as they are here on Earth.

> *Where wast thou when I laid the foundations of the earth?*
> *declare, if thou hast understanding. Who hath laid the measures*
> *thereof, if thou knowest? or who hath stretched the line upon it?*
> *Whereupon are the foundations thereof fastened?*
> *or who laid the corner stone thereof?* Job 38:4–6

Thus God begins a series of questions to His servant Job that immediately establishes the infinitely subordinate and ignorant position of a man relative to Himself. As you read through the final chapters of the Book of Job, God shows that there is not one area of His creation where man even begins to approach being His equal in knowledge.

In the latter half of the nineteenth century, scientists believed that all the important information about the universe had been discovered. Only the minor details needed to be filled in. The emergence of Einstein's theory of relativity in the first decades of the twentieth century was one of several key developments that shook the very fabric of science. Newton's laws of motion were not the final explanation of a mechanistic universe after all. Although some point to this failure of science to explain the physical world as justification to reject science, the Christian should take comfort in the fact that we serve an almighty Creator, and studying His creation through science is a humbling and wonderful privilege.

27A GALILEAN RELATIVITY

27.1 Phenomenological Relativity

An avid runner runs along, thinking of nothing but the speed he can achieve. His objective is a six-minute mile, sustained over ten miles. Breathing heavily, he keeps close track of his pedometer and his stopwatch. Finally, he is reaching his objective! He has maintained his six-minute-mile speed for nine miles—nine and one-half miles—almost there! Then, his wife walks past his treadmill to call him to supper.

The man was running fast relative to the belt of his treadmill but was motionless compared to his exercise room. Which standard—the treadmill or the exercise room—is correct? Neither. Only relative motion has meaning, since there is no

We used the word "phenomenological" in Chapter 1 to describe any event or entity that could be observed or measured.

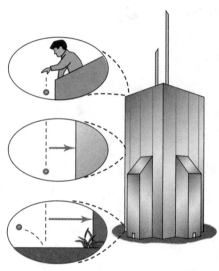

27-1 According to reactionary scientists of the sixteenth century, a ball dropped on a moving Earth could not fall straight down.

valid standard of "motionless-ness" to which motion can be compared. This truth is emphasized in the study of objects in one reference frame (set of points used to locate objects) as seen from another reference frame that is moving with respect to the first, called **relativity.** The principle of no absolute rest applies to all realms of physics, including mechanics, optics, thermodynamics, and electromagnetics. The idea originated with Galileo Galilei, who applied it to mechanics.

27.2 Galileo's Observations

In Galileo's time, Copernicus's heliocentric model of the solar system was the subject of much debate. If Copernicus were right, Earth must move and move rapidly. Many scholars who disagreed with Copernicus declared that this conclusion was obviously wrong. As an example, they said, suppose Earth were moving rapidly, as Copernicus said, and a man dropped a ball from the top of a tall building. After being released, the ball would fall straight down, but as it fell, Earth would move from under it and the ball would eventually land a noticeable distance from the building. However, a ball dropped from a building actually lands close to the building. Therefore, they reasoned, Earth cannot be moving.

Galileo's answer to this argument was that the laws of mechanics are the same in steadily moving reference frames as in resting reference frames. This idea is called **Galilean relativity.** He pointed out that before the ball is released, it travels at the same velocity that Earth and the building travel. There are no horizontal forces on the falling ball, so its horizontal velocity does not change. The ball continues to travel along with the building as it falls, and it lands close to the building.

Galileo's explanation was corroborated by an experiment using a reference frame that everyone agreed was moving—a ship. As the ship was sailing with constant velocity, a man dropped a weight from the mast. The anti-Copernicans predicted that the weight would fall toward the back of the ship, but Galileo predicted that it would fall beside the mast. The weight landed next to the mast, so Galileo's argument was proved right.

A proper analysis of the weight's fall must take into account all of its velocities. To predict the path of the weight as seen from the shore, you must add the velocity

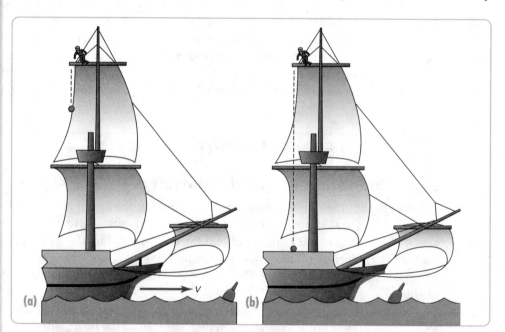

27-2 An experiment proved that on a ship moving at a constant velocity with respect to the earth a dropped weight appears to fall straight down when observed from the deck of the ship.

of the weight as seen by a man on the ship to the velocity of the ship as seen by the man on the shore. This becomes more obvious when the velocities are in the same direction. Suppose a ship is moving away from a shore with a speed

$$v_s = 10 \text{ km/h}.$$

A man on the ship begins walking toward the front of the ship at the speed of

$$v_m = 3 \text{ km/h}.$$

How fast is the man on the ship moving with respect to the observer on the shore? After a little thought, the answer should be obvious.

$$v = v_s + v_m$$
$$v = 10 \text{ km/h} + 3 \text{ km/h}$$
$$v = 13 \text{ km/h}$$

This is a prediction of Galilean relativity.

27-3 The crewman's total velocity is $\mathbf{v}_s + \mathbf{v}_m$.

27.3 Galilean Coordinate Transformations

How is Galilean relativity useful? Let's use a contrived scenario to illustrate its usefulness. A physics teacher, exhausted after a long but fruitful year of teaching, takes a vacation at the beach. He relaxes in his floating lounge chair just off shore in the gentle swell, reading Newton's *Principia*. Completely absorbed in his book, he is unaware that the current has carried him well out to sea. His absence is noticed, a rescue is initiated, and the Coast Guard officer standing where the teacher was last seen on the beach directs a helicopter to fly directly away from that point, parallel to the current, at 50 m altitude.

The officer's coordinate system consists of three dimensions and time—*x, y, z,* and *t.* He chooses to align his *x*-axis horizontal, parallel to the direction of the current. Then his *y*-axis is horizontal and perpendicular to the *x*-axis, and his *z*-axis is vertical. The origin for his system is located at his position.

The helicopter's coordinate system (using primed symbols) is parallel in all respects to the officer's, but the helicopter's origin is located at its (changing) position.

Eventually, the helicopter sights the physics teacher in his lounge chair 200 m ahead ($x' = +200$ m), 50 m to the left ($y' = +50$ m), and 50 m below ($z' = -50$ m). The pilot reports the teacher's location as (200 m, 50 m, -50 m). The officer looks at the point $x = 200$ m, $y = 50$ m, and is perplexed, wondering how the teacher can be at $z = -50$ m (below sea level) from his position on the shore. Obviously, he has assumed that the helicopter's coordinate system is identical to his own—that

$$x = x', y = y', \text{ and } z = z'.$$

But the coordinates are actually different.

How can the officer convert the helicopter's coordinate system into his own? After studying Figure 27-6, you can see that both the officer and helicopter pilot will agree that the teacher's position is 50 m to the left. Therefore, $y = y'$. The helicopter is flying at a constant altitude of $z = 50$ m, so

$$z = z' + 50 \text{ m}.$$

If the helicopter flew over the officer at time $t = t' = 0$, then at that time x and x' were equal. The distance between the origins of the two systems, Δx, is just the distance the helicopter travels in time Δt.

$$\Delta x = v\Delta t$$

27-4 The officer's coordinate system

27-5 The helicopter's coordinate system

Relativity   **589**

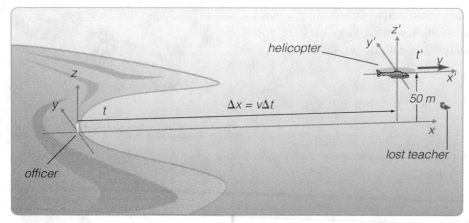

Therefore, according to the officer on the shore,

$$x = x' + v\Delta t.$$

The officer can translate the coordinates from the helicopter to his own system using this **Galilean transformation:**

$$x = x' + v\Delta t,$$
$$y = y',$$
$$z = z' + 50 \text{ m, and}$$
$$t = t'$$

Similarly, if the officer on the shore gives coordinates to the helicopter, the pilot must translate them. His transformation will be exactly like the officer's transformation, except that he sees the shore 50 m *below* him *receding* with a velocity of −*v*. His translation is

$$x' = x + (-v)\Delta t,$$
$$y' = y,$$
$$z' = z - 50 \text{ m, and}$$
$$t' = t.$$

27-6 Locating a position in two different reference frames

Problem-Solving Strategy 27.1

Time in Galilean transformations is always assumed to be the same in every reference frame.

$$t = t'$$

Problem-Solving Strategy 27.2

Note that accelerations may occur within inertial reference frames, but the reference frames themselves cannot be accelerating.

27.4 Inertial Reference Frames

The laws of mechanics—those known to Galileo and those discovered since his time—have the same mathematical form in every reference frame that does not accelerate. Such a reference frame is called an **inertial reference frame.** As long as the helicopter in the previous illustration moves with a constant velocity, mechanical experiments carried out in the aircraft will give the same result as they would have if carried out on the "stationary" shore. Without looking out the window, the teacher (still avidly reading his book) would be unable to tell whether the helicopter is moving at a constant speed or hovering. If he dropped the book, it would appear to him to fall straight down in either case. Any other mechanical experiments will also give the same results whether the helicopter is moving in a straight line at constant speed or not moving at all. There is, in fact, no mechanical experiment that can determine absolute motion. Only relative motion has meaning in mechanics.

27.5 Problems Arising from Galilean Relativity

Scientists of the nineteenth century understood that mechanics cannot determine absolute motion, but they thought that optics could. All the waves they had studied had been disturbances in a medium, as sound is a disturbance in the air. They assumed that light was the same. However, light travels in an apparent vacuum in space. To explain this phenomenon, scientists theorized that a colorless, odorless, weightless substance called the **luminiferous** (light-bearing) **ether** filled all space and all transparent objects. The ether was assumed to be at rest, with the earth and other bodies moving through it; so it would provide a reference frame that was absolutely at rest. For example, if a light were shined from the back of a moving ship to the front, it would have to travel through the stationary ether at 3.0×10^8 m/s. The light's speed relative to the ship would be less—it would equal the speed of light minus the ship's speed. Therefore, it would take longer for the light to reach the front on a moving ship than on a stationary ship.

Like every other theory, the ether theory needed to be tested. Every theory makes predictions, but the test of the theory is whether its predictions are borne out by experiment. If they are not, the theory is modified or discarded. From the ether theory, Albert A. Michelson predicted that it would take light longer to travel a path parallel to the earth's motion through the ether than to travel a path of the same length perpendicular to that motion.

Michelson used an interferometer, shown in Figure 27-7, to test his prediction. A light source, *L,* at the left directs light toward a partially silvered mirror, which reflects some light to *R* and allows the rest of the light to continue to *T.* Mirrors *R* and *T* reflect the light back to *C,* which directs the light to the detector telescope, *D.* Michelson reasoned that if the instrument were oriented so that the light rays from *L* to *T* were traveling in the same direction as the motion of the earth through the ether, the round-trip journey for the light would take a maximum time. The light going by way of *R* would take less time. When the light reached the detector, the light from the two paths would be out of phase, and the detector would detect an interference pattern. He also reasoned that the relative phase of the light from the two paths would change if the device were rotated, since the path toward *T* would no longer be parallel to the earth's motion through the ether. The diffraction pattern would therefore change. Michelson's interferometer was sensitive enough to detect changes much smaller than the calculated change. Detecting this change would corroborate the ether theory.

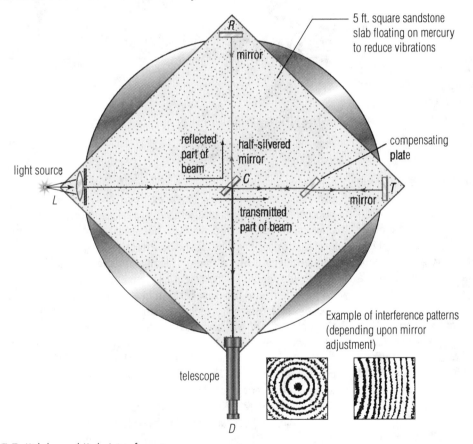

27-7 Michelson and Morley's interferometer

Michelson and his partner, Edward W. Morley, tried repeatedly to find the predicted change. According to the ether theory, if either path were aligned along the direction of the earth's motion, the interference pattern would change. To achieve this alignment, the experimenters rotated the instrument gradually through a full circle, and they performed the experiment at different times of year, when the

Edward W. Morley (1838–1923) was an American chemist.

The failure of the Michelson-Morley experiment is an excellent demonstration that an experiment does not have to prove a hypothesis true to be considered successful. This experiment was a success in that it provided the corroborating data supporting some of the predictions of Einstein's theories of relativity.

earth's motion through hypothetical ether was in different directions. In every case they obtained the same result: the interference pattern remained constant. The prediction of the ether theory was not borne out by this or any further experiment. The theory obviously had to be modified or discarded, so it was discarded.

27A Section Review Questions

1. Discuss what is meant by *relativity*.
2. What is the main premise of Galilean relativity?
3. What did the anti-Copernicans fail to take into account when describing the path of a falling object?

ALBERT EINSTEIN (1879–1955)

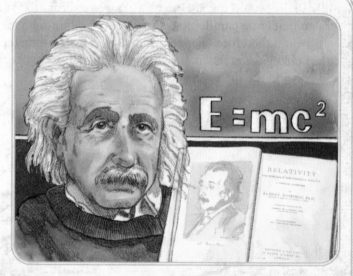

The best-known modern physicist is Albert Einstein. When you hear his name, you probably picture a genius scribbling incomprehensible equations on a chalkboard. At one time it was commonly reported that only a dozen people could understand Einstein's higher mathematics, but today his theories have become a fundamental part of every college course in theoretical physics.

Einstein's Jewish father and uncle ran a small engineering plant in Munich, Germany. The young boy did not perform well at school because he disliked the rigid discipline of his German teachers. It was his uncle who stimulated his passion for mathematics and science. When his family moved to Switzerland, Albert managed to pass the Polytechnic Academy entrance exam on his second try. After graduation he found a quiet job at the patent office in Berne, where he spent his spare time puzzling over the nature of the universe.

In 1905, the 26-year-old clerk published four papers that forever changed man's understanding of space, time, mass, and the universe. The first paper earned him a doctorate from the University of Zurich. The next three papers won him world renown. One gave a theoretical explanation for Brownian motion, a mystery since it had been discovered in 1827. The next paper suggested that light consists of both particles (quanta) and waves; his quantum theory of light explained why light striking a metal stimulates it to emit electrons (called the photoelectric effect). The last paper introduced his special theory of relativity with the famous equation $E = mc^2$.

Soon this unknown clerk was in demand as an instructor at several universities. He changed professorships almost once a year until he finally settled in Berlin. The position offered by the Prussian Academy of Sciences freed him to research and publish papers on general relativity, heat, and other topics in physics. In 1921 he received the Nobel Prize for Physics for his explanation of the photoelectric effect. The Royal Society of London elected him a foreign member the same year. Einstein was now a leader in the scientific community.

But the rise of the Nazi Party in Germany coincided with the ebb of Dr. Einstein's scientific output. He spoke out against the prospects of war and Nazi persecution of the Jews. When Hitler became chancellor of Germany in 1933, Einstein left for the United States and later became a U.S. citizen. Though he continued to receive honors throughout his life, he no longer was at the forefront of discoveries in physics. During the last thirty years of his life he parted company with the younger generation of physicists, who believed that a unified field theory was impossible. Einstein remained convinced that the electromagnetic field and gravity are two forms of the same phenomenon. He never found the secret formula, which has eluded scientists to this day. Nevertheless, Einstein did not believe that he had wasted his time searching for this unification. He had, he declared, showed future scientists where not to look.

4. **a.** What is the mathematical process that must be performed in order to convert positions in one coordinate system to another coordinate system?

   b. Do the coordinate systems need to be stationary to accomplish this conversion?

5. **a.** What are unaccelerated coordinate systems called?

   b. Give an example of an accelerated coordinate system as well as an unaccelerated one.

6. **a.** During the late nineteenth century, what was considered the absolute reference frame for all motion?

   b. What experiment disproved this theory?

⊙**7.** At time t_0, a car is traveling west at a constant 95 km/h on flat terrain. Its coordinate system is defined by $x = x_0 - (26.4 \text{ m/s})\Delta t$, y, z, and t. Another car starting 5.0 km to the east of the first car at time t_0 heads directly north at 60. km/h. Determine the transformation equations to describe the position of the first car within the reference frame of the second car. In other words, $x = ?$, $y = ?$, $z = ?$, and $t = ?$.

27B SPECIAL RELATIVITY

27.6 Assumptions of Special Relativity

The ether theory presupposed that a fixed universal reference frame existed. The theory's failure was readily explained by another model, the **special theory of relativity.** Albert Einstein, the theory's originator, stated that there could be no preferred reference frame and therefore no ether. His theory was corroborated by the Michelson-Morley experiment's negative results. The theory consists of two assumptions, or postulates:

1. All the laws of physics have the same mathematical form in all *inertial* reference frames.

2. The speed of light in a *vacuum* is the same in all reference frames.

The first postulate is simply an extension of Galilean relativity, which said that the laws of mechanics are the same in every inertial reference frame. Einstein extended the rule to include thermodynamics, electromagnetism, and optics. He assumed that *no* physics experiment can distinguish between a moving reference frame and a stationary one.

In contrast, the second postulate is contrary to all our experience. If a car is approaching you, and a person in the car throws a ball at you, the ball will reach you sooner than it would if the car were stationary. However, Einstein said that if the person in the car turned on the headlights instead of throwing a ball, the light would reach you at the same time that it would if the car had been stationary. Relative motion of the source of the light and the observer has no effect on the observer's measurement of the speed of light.

27.7 The Lorentz Transformation

The special theory of relativity does not agree with the Galilean transformation, so a new transformation is needed. However, since normal occurrences follow the Galilean transformation, the new transformation must reduce to the Galilean transformation at normal speeds. The new transformation, called the **Lorentz transformation,** is as follows:

Hendrik Lorentz (1853–1928) was a Dutch physicist who won the Nobel Prize in 1902.

$$x = \frac{x' + v\Delta t'}{\sqrt{1 - (v/c)^2}}$$

$$y = y'$$

$$z = z'$$

$$t = \frac{t' + vx'/c^2}{\sqrt{1 - (v/c)^2}} \qquad (27.1)$$

In these equations, v is the speed of the object measured within the observer's frame of reference.

At normal speeds, v is much smaller than the speed of light in a vacuum (c), so the terms $(v/c)^2$ and v/c^2 become very small compared to the other quantities. For everyday situations, the transformation reduces to

$$x = x' + v\Delta t,$$

$$y = y',$$

$$z = z', \text{ and}$$

$$t = t',$$

which is the Galilean transformation. "Much smaller than c" includes most speeds under $0.1c$. No observable terrestrial object has ever come close to $0.1c$, so everyday experience does not include Lorentz transformations. Subatomic particles that approach the speed of light follow the Lorentz transformation. A particle whose speed approaches the speed of light is said to have a **relativistic** speed.

27.8 Predictions of Special Relativity

To test the special theory of relativity, simply use it to make predictions. Some of the predictions are contrary to experience, or "common sense," but the circumstances described by the theory are outside normal experience. Experiments have so far confirmed the predictions. It should not be surprising that not everything in the universe conforms to man's common sense. The universe was designed by God, not by man. Einstein's theory is presently man's best explanation for what God has made. Only God knows how nearly correct it is. However, you should try to understand as well as you can the world God made, even if the current theories do not appeal to you. The special theory of relativity leads to four interesting predictions.

1. Simultaneity—If two observers are moving at a relativistic speed with respect to each other, two events that appear to be simultaneous to one observer will not appear to be simultaneous to the other.

2. Time dilation—A clock moving at a relativistic speed with respect to an observer will appear to be running slow.

3. Length contraction—When a stationary observer sees an object moving at a relativistic speed, the object's length parallel to the direction of motion appears to be shortened.

4. Mass increase—The mass of an object appears to increase with increasing speed.

27.9 Simultaneity

One prediction of the special theory of relativity is that if one observer says that two events in two places are simultaneous, another observer moving with respect to the first will disagree. Consider the following example. A rocket boat is traveling parallel to the shore at a speed near that of light. A man on the shore, S, stands halfway between the seaside corners (A and B) of a pier. A woman in the boat, S', sits at the midpoint of the boat. There are strobe lights at A and B. When A is opposite A', the strobe at A goes off. When B is opposite B', the strobe at B goes off. The observers judge when the strobes went off by when they see the light.

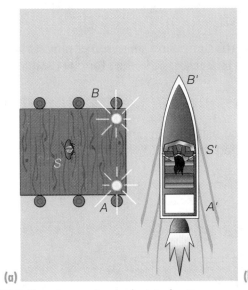
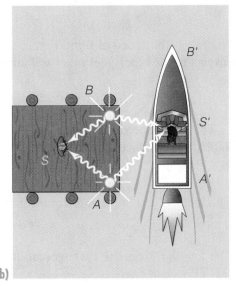

(a)   (b)

27-8 These observers disagree about simultaneity.

Suppose S receives the light from both strobes at the same time. He decides that the strobes went off simultaneously and that, therefore, the boat is as long as the width of the pier. The boat, which is moving with a speed near that of light with respect to S, moves between the time the strobes go off and the time the light reaches S, Figure 27-8 (b). From the perspective of S, the light from B has less distance to travel to S' than to S, so B's light reaches S' first. The light from A has farther to travel to S' than to S, so A's light reaches S first. Therefore from the perspective of S, the light from B reaches S' before the light from A.

The observer S' agrees that she received light from B before she received light from A. Since she knows that the light had the same distance to travel from both points, she assumes that the strobe at B went off before A. She realizes that B' must have reached B before A' reached A, so she concludes that her boat is longer than the width of the pier. She disagrees with S in two observations: S says the flashes were simultaneous; S' says B flashed first. Also, S says that the boat and the pier are the same length; S' says that the boat is longer.

The essence of this discussion on simultaneity is this: If two events are simultaneous in a given frame of reference, they will not be simultaneous in a frame of reference moving with respect to the first.

27.10 Time Dilation

The problem of simultaneity brings up two more predictions of special relativity: time dilation and length contraction. Time **dilation** means that an observer perceives a clock moving with respect to him as running slow. Consider a moving "light clock."

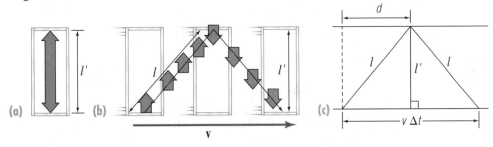

27-9 A moving light clock seen (a) from its own reference frame and (b) from a reference frame in which it is moving. Figure (c) gives the geometry for determining l.

Time is measured by the number of oscillations a beam of light makes between two mirrors. Figure 27-9 (a) shows how an observer at rest with respect to the clock sees it. He measures the time for one oscillation as

$$\Delta t' = \frac{2l'}{c}, \tag{27.2}$$

where c is the speed of light. Figure 27-9 (b) shows how an observer moving with respect to the clock sees one oscillation. He measures the time for one oscillation as

$$\Delta t = \frac{2l}{c}. \tag{27.3}$$

The distance l relative to l' can be found using trigonometry. The distance d is half of the distance traveled by the clock in time Δt.

$$d = \tfrac{1}{2}v\Delta t \tag{27.4}$$

The length l' can be found from Equation 27.2:

$$l' = \tfrac{1}{2}\Delta t' c \tag{27.5}$$

The length l, from the Pythagorean theorem, is

$$l = \sqrt{d^2 + l'^2}. \tag{27.6}$$

Substituting the values d and l' from Equations 27.4 and 27.5 results in

$$l = \sqrt{\left(\tfrac{1}{2}v\Delta t\right)^2 + \left(\tfrac{1}{2}\Delta t' c\right)^2}.$$

Since both terms contain $\left(\tfrac{1}{2}\right)^2$, it can be factored out.

$$l = \tfrac{1}{2}\sqrt{(v\Delta t)^2 + (c\Delta t')^2}$$

The time elapsed is equal to the distance the light traveled, $2l$, divided by its speed:

$$\Delta t = \frac{2l}{c} = \frac{2\left[\tfrac{1}{2}\sqrt{(v\Delta t)^2 + (c\Delta t')^2}\right]}{c}$$

Squaring both sides yields

$$(\Delta t)^2 = \frac{(v\Delta t)^2 + (c\Delta t')^2}{c^2} = \frac{(v\Delta t)^2 + c^2(\Delta t')^2}{c^2}$$

$$(\Delta t)^2 = \frac{(v\Delta t)^2}{c^2} + (\Delta t')^2.$$

Rearranging,

$$(\Delta t')^2 = (\Delta t)^2 - \frac{(v\Delta t)^2}{c^2} = (\Delta t)^2 - (\Delta t)^2\frac{v^2}{c^2}.$$

Factor out $(\Delta t)^2$:

$$(\Delta t')^2 = (\Delta t)^2\left(1 - \frac{v^2}{c^2}\right) = (\Delta t)^2\left[1 - \left(\frac{v}{c}\right)^2\right]$$

Solving for $(\Delta t)^2$ gives

$$(\Delta t)^2 = \frac{(\Delta t')^2}{1 - (v/c)^2}.$$

Taking the square root of each side results in

$$\Delta t = \frac{\Delta t'}{\sqrt{1 - (v/c)^2}}. \tag{27.7}$$

The factor

$$\frac{1}{\sqrt{1-(v/c)^2}} \qquad (27.8)$$

occurs so often in the study of relativity that it is called the **relativistic factor (γ)**. Therefore, Equation 27.7 becomes

$$\Delta t = \gamma \Delta t'. \qquad (27.9)$$

In Equation 27.9 the factor γ is always greater than 1, which means that the relativistic clock's time interval $\Delta t'$ is observed to be the longer time interval Δt when measured by the observer's own clock. A clock moving with respect to an observer always seems to him to show less elapsed time than, say, his wrist watch. That is, a moving clock always appears to run slow to an observer in the "stationary" reference frame (the one he is observing from). A person in the reference frame of the "moving" clock will note the clock of the "stationary" observer, which appears to him to be moving, as running slow also.

27.11 Length Contraction

Special relativity also predicts that the observed dimensions of a moving object are affected by its relativistic velocity. *Length contraction* is related to time dilation. If an object is moving with respect to an observer, its length along the direction of motion appears shorter. The length of an object as measured by an observer moving with the object (i.e., in the object's reference frame) is its **proper length (l_0)**. Its length as measured by an observer in a "stationary" reference frame is l, which is less than l_0. The exact relationship is

$$l = l_0\sqrt{1-(v/c)^2}, \text{ or}$$

$$l = \frac{l_0}{\gamma}. \qquad (27.10)$$

Notice that only the dimension parallel to the direction of motion is affected, and the effect is noticeable only to an observer outside the moving system.

27.12 Mass

Another prediction of special relativity is that mass increases with speed. This requirement is necessary for the conservation of momentum. The mass of an object as measured by one who is at rest with respect to the object is its **rest mass (m_0)**. The mass of the same object measured by one who is moving with respect to the object *appears* to be

$$m = \frac{m_0}{\sqrt{1-(v/c)^2}} = \gamma m_0. \qquad (27.11)$$

27.13 The Practical Effect of γ

For relativistic time, length, and mass, the velocity-dependent factor that describes the changes observed by the hypothetical "stationary" observer is γ. To describe your observations of systems moving at "normal" speeds, this factor must be nearly 1. Consider two reference frames with a relative velocity of 10^4 m/s (10 km/s, or nearly the escape speed from Earth). The denominator of γ becomes

$$\sqrt{1-(v/c)^2} = \sqrt{1-(10^4 \text{ m/s}/3 \times 10^8 \text{ m/s})^2}$$

$$= \sqrt{1-(10^8 \text{ m/s}/9 \times 10^{16} \text{ m/s})}$$

The factor

$$\gamma = \frac{1}{\sqrt{1-(v/c)^2}}$$

appears in every Lorentz transformation in special relativity.

(a)

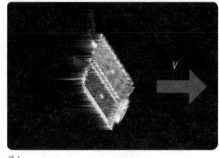

(b)

27-10 Length contraction at relativistic speeds as seen by a "stationary" observer. Figure (a) is observed moving at speeds <<0.1c; figure (b) is observed moving at speeds >>0.1c.

Note that mass is an intrinsic property of an object and cannot change. Therefore, the only mass that needs to be considered is the rest mass (m_0). The apparent increase of mass is related to an object's momentum at relativistic speeds in an observer's frame of reference.

$$= \sqrt{1 - 1.\overline{1} \times 10^{-9}}$$

$$\doteq 0.999\,999\,999\,4.$$

The factor γ is then

$$\gamma = \frac{1}{\sqrt{1 - (v/c)^2}} = \frac{1}{0.999\,999\,999\,4}$$

$$\doteq 1.000\,000\,001.$$

Most measurements are not made to a precision of nine decimal places, so an observer of a rocket moving at 10^4 m/s, for example, will not notice any relativistic effects at all.

27.14 The Theoretical Consequences of γ

The predictions of the special theory of relativity about objects traveling at the speed of light are interesting. Consider two observers, one in a spaceship moving at the speed of light and the other in a "stationary" space station. Even though only one observer is being propelled, each views the other as moving at the speed of light in his own reference frame. Each observer has a large clock that can be seen by the other observer. The moving observer notes that 1 s elapses on his clock ($\Delta t = 1$ s). How much time does he see elapse on the other's clock?

$$\Delta t = \frac{\Delta t'}{\sqrt{1 - (c/c)^2}}$$

$$\Delta t = \frac{\Delta t'}{\sqrt{1 - (1)^2}}$$

$$\Delta t = \frac{\Delta t'}{0} \Rightarrow \text{d.n.e. (infinitely large time interval)}.$$

According to the observer in the spaceship, the other observer's time in the space station is standing still. The other observer on the space station also says that while 1 s passed on his clock the "moving" observer's time seems to be standing still.

Each observer holds a meter stick parallel to the direction of relative motion. The observer on the spaceship measures the other observer's stick as

$$l = (1 \text{ m})\sqrt{1 - (c/c)^2}$$

$$l = 0 \text{ m}.$$

Not only does the "stationary" observer's meter stick have zero length, the entire space station appears as a vanishingly thin line perpendicular to the direction of relative motion! (Don't ask how you can read the meter stick—this is all hypothetical anyway.) The "stationary" observer thinks that the moving observer's stick (and the observer himself) has no length, either.

Finally, both observers bounce a 1 kg ball on their respective floors and measure the mass of the ball based on the conservation of momentum of the ball's collision with the floor. Each finds that

Recall that momentum is the product of mass and velocity:

$$p = mv$$

$$m = \frac{m_0}{\sqrt{1 - (c/c)^2}} = \frac{m_0}{0}$$

$$m \Rightarrow \text{d.n.e. (mass becomes infinitely large)}.$$

Each observer says that the other observer's ball has infinite mass. This observation prohibits any object from accelerating at, or even approaching, the speed of light.

Since an apparent infinite mass requires an infinite outside force to accelerate it and there are no natural infinite forces, nothing can exceed the speed of light *in a vacuum*. (However, some particles can go faster than the speed of light in a medium; for example, some subatomic particles can go through water faster than light can.)

27.15 Velocity Addition in Special Relativity

Another prediction of special relativity concerns velocity addition. Consider two reference frames with observer O' moving at a relative velocity of $\mathbf{v}$ to observer O. Observer O' launches a rocket with a speed of $\mathbf{v'_r}$ with respect to O', parallel to the relative velocity $\mathbf{v}$. What is the rocket's velocity with respect to the observer O in the other reference frame? The speed of anything is simply the change in distance divided by the change in time. For the rocket,

$$v_r = \frac{\Delta x}{\Delta t}. \tag{27.12}$$

From the Lorentz transformations (Equations 27.1):

$$\Delta x = \frac{\Delta x' + v\Delta t'}{\sqrt{1 - (v/c)^2}}$$

$$\Delta t = \frac{\Delta t' + v\Delta x'/c^2}{\sqrt{1 - (v/c)^2}} \tag{27.13}$$

Therefore, dividing Δx by Δt gives the speed of the rocket (v_r) in observer O's frame of reference:

$$\frac{\Delta x}{\Delta t} = \frac{\Delta x' + v\Delta t'}{\sqrt{1 - (v/c)^2}} \cdot \frac{\sqrt{1 - (v/c)^2}}{\Delta t' + v\Delta x'/c^2}$$

$$v_r = \frac{\Delta x' + v\Delta t'}{\Delta t' + v\Delta x'/c^2} \tag{27.14}$$

This can be simplified somewhat by dividing numerator and denominator by $\Delta t'$, giving

$$v_r = \frac{\Delta x'/\Delta t' + v}{1 + v\Delta x'/c^2\Delta t'}. \tag{27.15}$$

But $\Delta x'/\Delta t'$ is simply v'_r (in observer O''s reference frame).

$$v_r = \frac{v'_r + v}{1 + (vv'_r/c^2)} \tag{27.16}$$

As long as v'_r and v are less than c (which must be the case, since nothing can exceed the speed of light in a vacuum), v_r (the speed of the rocket as observed by O) is less than c. At low speeds, the denominator is so close to 1 that the equation reduces to

$$v_r = v'_r + v,$$

which is the familiar Galilean transformation (assuming all speeds are in the same direction).

27.16 Mass-Energy Equivalence

The most familiar prediction of special relativity is expressed by the equation

$$E = mc^2, \tag{27.17}$$

where E is a particle's total energy and m is its relativistic mass (the mass it appears to have when moving at a high speed). The kinetic energy of the moving particle is

$$E_k = mc^2 - m_0c^2. \tag{27.18}$$

Problem-Solving Strategy 27.4

Relativistic velocity additions always yield velocities that are less than or equal to the speed of light. If you obtain a total summed speed greater than c, you can be sure that you made a mistake.

The **rest-mass energy** is sometimes called just *rest energy*.

When the particle is at rest, therefore, it has a **rest-mass energy (E_0)** of

$$E_0 = m_0c^2.$$ (27.19)

Equation 27.19 shows a most useful prediction of special relativity: mass is a form of energy. Mass can be converted to other kinds of energy, and other kinds of energy can be converted to mass. Any addition of energy to a particle results in an increase in its mass.

27.17 Experimental Evidence for Special Relativity

Experiments have supported the predictions of special relativity. One of these experiments involves *muons,* particles somewhat smaller than protons. Muons produced in laboratories (with low relativistic speeds) have a half-life of 2.2×10^{-6} s. That is, after 2.2×10^{-6} s, half of the muons have decayed into even smaller subatomic particles. Muons, which are produced in the atmosphere from cosmic ray interactions, may fall toward the earth with speeds of $0.9c$ and more.

At $0.9c$, a muon travels almost 600 m in one half-life. Therefore, you would expect to find half as many muons reaching the earth's surface as there are 600 m above the earth's surface. Actually, much more than half reach the earth. The explanation is that to an observer on Earth the muons' time is running slow. When we say that 2.2×10^{-6} s have passed, only 9.6×10^{-7} s have passed in the reference frame of the fast muons (their clocks run slower). Since the muons' time determines their half-life, they last longer than we would expect.

The muons "see" the earth rushing toward them at $0.9c$. Why does the earth travel 600 m in only 9.6×10^{-7} s? Actually, it does not, since the muons also "see" a length contraction. The 600 m to them is around 260 m. The measured proportion of muons reaching the earth agrees well with these calculations.

Another specific prediction of the theory is that the mass of a particle emerging from a particle accelerator is more than its rest mass. This prediction can be tested by a mass spectrometer (Chapter 21). A larger mass requires a stronger magnetic field to deflect it. By using a spectrometer scientists have determined that particles going near the speed of light may have masses that are hundreds or thousands of times larger than their rest masses.

27-12 An atomic clock is the most accurate time measure available. Four of these clocks were used to test the clock paradox.

Special relativity also predicts what is known as the **clock paradox.** Two observers, each with a clock, begin at rest with respect to each other. At first, the clocks are synchronized. One observer, O', decides to travel. He goes at a constant speed to his destination, then returns to the other observer, O, at the same speed. The observer O has not moved. When O' returns, they compare their clocks. They discover that the traveler's clock registered less time for the journey than the other observer's clock.

The paradox is that it seems we have defined a preferred frame of reference. We can tell which observer traveled. Why can we not view the trip as O' staying still

27-11 Particle accelerators can generate huge relativistic velocities and masses for subatomic particles.

and O traveling in the opposite direction? The solution to the paradox is that O stayed in one inertial frame, but O' was in three different reference frames. He began in O's reference frame. On his trip to his destination, O' was in a reference frame moving in one direction with respect to O. On his return trip, O' was in a reference frame moving in the opposite direction with respect to O. Finally, O' returned to O's reference frame. This changing of reference frames makes O''s clock slower than O's clock.

The clock paradox was corroborated in an experiment using atomic clocks on jets. Four clocks were compared to atomic clocks that would not travel. Then the four went around the world on jets. They began in a reference frame at rest with respect to the earth, changed to a reference frame moving with respect to the earth, and finally returned to rest with respect to earth. When the traveling clocks were again compared to the clocks that stayed in place, the traveling clocks were a fraction of a second slower than the resting clocks. The results of the experiment agreed with the prediction of special relativity.

27.18 Applications of Special Relativity

You can apply the predictions of special relativity to better understand your surroundings and how they can be used. The statement of mass-energy equivalence is probably the most useful result of the theory. Since mass is a form of energy, it is not surprising that sometimes mass is replaced by another form of energy in certain processes.

For instance, the mass of an atom's nucleus is less than the sum of the rest masses of the particles that compose the nucleus. The "lost" mass is called the **mass defect** due to nuclear **binding energy.** When the particles come together to form the nucleus, they give up some of their mass as another form of energy. If two smaller nuclei merge, each gives up some of its mass. This kind of reaction, called *fusion*, produces large quantities of energy that have potential use. On the other hand, before a particle can leave the nucleus, it must gain enough energy to restore its full mass. If the binding energy is large, it is difficult for a particle to gain enough energy to leave the nucleus. Therefore, nuclei with large binding energies are more stable than nuclei with small binding energies. We will study nuclear stability further in Chapter 29.

A technological application of nuclear binding energy is nuclear power. Nuclear power encompasses both its peaceful use in electrical power generation and propulsion and its use in weapons of war. Mass is a concentrated form of energy. One gram of matter potentially contains 9×10^{13} J of energy. If mass could be totally transformed into electrical energy and sold at one cent per kilowatt-hour, a 1 g paper clip would produce energy worth $2.5 million. Actually, only a small fraction of mass is converted to energy in nuclear reactions. Even that fraction contains great potential for energy release, as is shown by the power of nuclear weapons. Nuclear bombs involve an uncontrolled conversion of mass to other forms of energy, while nuclear power plants provide a controlled reaction.

You may have gathered from the discussion of the Lorentz transformations earlier in this section that the differences in time, length, velocity, and mass between reference frames were virtual—illusions—due to differences in observer reference frames. This conclusion is not accurate. If it were possible to actually measure each of those parameters in the other reference frame, they would show the change predicted by the transformation equations. The measured dimension would be as "real" as any other measurement in the observer's frame of reference.

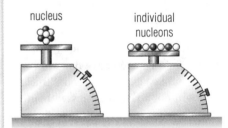

27-13 The mass of the entire nucleus is less than the sum of the individual particle's rest masses.

Some people believe they are spiritually special because of moral relativity. They say, "I'm better than some of the members in my church." Or, "I'm a good person. I've never robbed a bank or killed anybody." There probably is no clearer instruction against moral relativity than this verse: "But they measuring themselves by themselves, and comparing themselves among themselves, are not wise" (II Cor. 10:12).

27B Section Review Questions

1. What are the two postulates of the special theory of relativity?

2. If two spaceships are approaching each other at 0.5c, and one ship shines a laser beam at the other, how fast is the light in the laser beam moving when measured by the second ship?

3. Describe four predictions made by the special theory of relativity.

4. Explain why it is impossible for physical objects to attain the speed of light in a vacuum.

⊛5. The *Voyager 1* space probe is the fastest-moving man-made object, traveling through space at about 17.1 km/s.

    a. What fraction of c is this spacecraft moving?

    b. Do relativistic effects apply to *Voyager 1?*

⊛6. A meteoroid with a rest mass of 1.00 g is moving at $0.8c$ in an observer's reference frame. What is the mass of the meteoroid according to the observer?

⊛7. In a test of Einstein's special theory of relativity, a space probe traveling at $0.65c$ relative to the earth fires a projectile at a speed of $0.70c$ relative to the probe in the same direction as the probe's movement. What is the projectile's speed as measured from Earth?

⊛8. a. If the 5.5 g mass of a U.S. quarter were converted completely into energy, how many joules would be produced?

    b. How many meters could this energy move a railway car against a 10 000 N rolling friction force if all the energy were converted into mechanical work?

27C GENERAL RELATIVITY

27.19 The Limitation of Special Relativity

Special relativity is special because it deals only with inertial reference frames. Actually, no known reference frame is completely free from acceleration. The velocity of the earth continually changes as it revolves around the Sun. Every point on the earth's surface is accelerated as it rotates around the earth's axis. Inertial frames are usually defined as those that do not accelerate with respect to the earth. It is difficult, if not impossible, to define a truly inertial reference frame. Fortunately, Einstein was able to extend his special theory to all reference frames, accelerated as well as inertial.

27.20 Gravity = Acceleration

The **general theory of relativity** states mainly that acceleration and gravity are indistinguishable. For example, neither a voyager in a spaceship far from any significant mass nor a parachutist in free fall feels any gravitational force. The space traveler is not in a gravitational field. The other man is experiencing unrestrained acceleration (no normal force). It is not possible for a person to tell which is his condition if he cannot observe his surroundings.

The statement of the last sentence in the first paragraph is conditional, since the Foucault pendulum experiment *could* indicate whether one was in an accelerating spaceship or at rest on a rotating planet.

For another example, consider the distant spaceship accelerating at a rate of 9.81 m/s². The astronaut standing on the deck experiences the same weight that he would if he were standing on the earth's surface. If he were to release a ball inside the ship, it would appear to accelerate toward the deck. No mechanical experiment that he could perform inside the spaceship

27-14 Acceleration makes a ball appear to fall.

could reveal whether he was accelerating in free space or at rest on the earth (assuming the earth did not rotate).

An interesting outgrowth of this example is gravity's effect on light. If the space traveler shines a flashlight on the far wall of his accelerating spaceship, the light beam will seem to be deflected. The light beam initially has an upward speed equal to the spaceship's speed as the light leaves the flashlight. The ship's speed increases before the light reaches the opposite wall, so the ship travels farther upward than the light does in that time. Therefore, the light strikes the wall at a lower level than it started at.

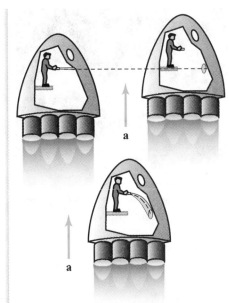

27-15 Acceleration makes a light beam bend.

27.21 Gravitational Lensing

General relativity predicts that gravity will attract light. The earth's gravity is too small and too restricted in space to have a noticeable effect on the fast-moving light beams. However, the effect of the Sun's gravity is observable on starlight. Starlight coming past the Sun is deflected; therefore, stars near the limb of the Sun appear to be in the wrong places. Most of the time you cannot see light from stars near the Sun because the Sun's light is too intense, but during a solar eclipse these stars become visible. Observations of stars during eclipses have confirmed this prediction.

Further proof of this prediction of the general theory was the discovery, through deep space photography using the Hubble Space Telescope, of the phenomenon called **gravitational lensing.** Massive, distant galaxies contain enough matter to deflect the light of even more distant objects beyond them that would not be directly visible from Earth. Distorted images of the farther object appear at small angular separations from the intervening galaxy, as light rays on opposite sides of the galaxy are bent toward Earth. Figure 27-16 shows an example of gravitational lensing.

27-16 Hubble photograph of gravitational lensing by a galaxy of an even more distant galaxy

27.22 Black Holes

A strong gravitational field may prevent even light from escaping it. A body with such a strong gravitational field is called a **black hole.** Black holes are thought to be the remnants of super-massive stars that collapse due to exhaustion of hydrogen and the cessation of the normal fusion processes that power a star. It is theorized that, as the star violently collapses, its matter is crushed into a mass of subatomic particles. Astronomers believe that the shock wave of rebounding gases from this hyper-dense core blasts away the outer layers of the star in a supernova. The remaining matter is so dense and massive that not even light can escape. Since no light ever leaves a black hole, black holes cannot be observed directly. However, their intense gravitational fields accelerate ionized dust and gases for great distances, generating tell-tale electromagnetic signatures that are observable. Detection of such electromagnetic evidence, as well as the wobble of companion stars, have all but confirmed the existence of these strange objects. Some astronomers believe that super-massive black holes are the central gravitational anchor that holds galaxies together.

27-17 X-ray photograph of the torus dust ring surrounding a supermassive black hole at the center of the spiral galaxy M77 (NGC 1068) in the constellation Cetus

27.23 Other Evidence for the General Theory

Other observations have confirmed the validity of the general theory of relativity. The planet Mercury is the planet closest to the Sun in our solar system. The alignment of the major axis of its elliptical orbit changes (precesses) about $\frac{1}{6}$ of a degree per century. The majority of this precession is due to the gravitational

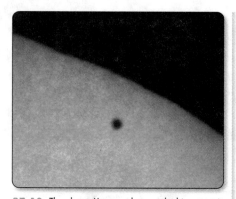

27-18 The planet Mercury photographed in a transit of the Sun. Precession of Mercury's orbit is due in part to the Sun's immense gravitational field, according to the general theory of relativity.

effects of the other inner planets, but a small amount of the angular change could not be explained without resorting to the general theory of relativity.

The general theory predicts that clocks in stronger gravitational fields run slower than in weaker fields. Therefore, your watch will be fast after completing a transcontinental trip in an airplane compared to what it would have been if you had taken a car. For the same reason, clocks in the basement of a skyscraper run slower than clocks on the roof. Atomic clocks have verified this prediction.

Gravity causes wavelengths of light to shift depending on whether the light is leaving a strong gravitational body or falling toward it. Outgoing light experiences a **gravitational red shift** as its wavelengths are stretched by gravity, but inbound light shows a *gravitational blue shift* as its wavelengths are shortened. Gravitational red shift was confirmed by experiment in 1954, and gravitational blue shift was observed as early as 1920 using the companion star to Sirius.

27C Section Review Questions

1. What is the main restriction on the special theory of relativity?
2. Discuss the main point of Einstein's general theory of relativity.
3. If you were completely enclosed in a windowless cubicle, what experiment could you perform to prove that you are at rest on the earth rather than accelerating at *g* through space?
4. Discuss three predictions of the general theory of relativity.
5. a. What is the primary observational property of a black hole?
   b. How do scientists believe black holes can be detected?
6. a. What is a gravitational red shift?
   b. How was this prediction of the general theory of relativity confirmed?

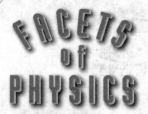

Gravitational Red Shift

One conclusion reached from Einstein's general theory of relativity is that gravity affects light. As light passes through a gravitational field, its path changes slightly, and some of its energy is expended in escaping the field. Therefore, the light emerges from the field with less energy.

As you learned in the last chapter, light's energy is related to its frequency by the equation

$$E = hf,$$

where *h* is Planck's constant. Light with high energy has high frequency, and light with low energy has low frequency. Therefore, light's frequency decreases after it passes through a gravitational field. Because the lowest visible frequency that light can become is red, this change, predicted by Einstein, is called Einsteinian red shift, to distinguish it from Doppler shift.

Only a very strong gravitational field removes enough energy to cause a noticeable red shift. Observers first detected this shift in light traveling past Sirius B, an extremely dense star called a white dwarf.

Gravitational red shift is presently used to detect features of the universe with unusually strong gravitational fields. Astronomers believe they have identified black holes and other exotic objects in space by the way they affect light.

In Terms of Physics

relativity	27.1	proper length (l_0)	27.11
Galilean relativity	27.2	rest mass (m_0)	27.12
Galilean transformation	27.3	rest-mass energy (E_0)	27.16
inertial reference frame	27.4	clock paradox	27.17
luminiferous ether	27.5	mass defect	27.18
special theory of relativity	27.6	binding energy	27.18
Lorentz transformation	27.7	general theory of relativity	27.20
relativistic	27.7	gravitational lensing	27.21
dilation	27.10	black hole	27.22
relativistic factor (γ)	27.10	gravitational red shift	27.23

Problem-Solving Strategies

27.1 (page 590) Time in Galilean transformations is always assumed to be the same in every reference frame.
$$t = t'$$

27.2 (page 590) Note that accelerations may occur within inertial reference frames, but the reference frames themselves cannot be accelerating.

27.3 (page 597) Remember that γ is always greater than 1.

27.4 (page 599) Relativistic velocity additions always yield velocities that are less than or equal to the speed of light. If you obtain a total summed speed greater than c, you can be sure that you made a mistake.

Review Questions

1. What is relative when considering relativity?

2. What does a Galilean transformation do?

3. Describe an inertial reference frame.

4. **a.** What theory did the Michelson-Morley experiment test?
   b. Did it confirm the theory?

5. What are the assumptions of special relativity?

6. Suppose you and a friend are in identical spacecraft traveling at $0.9c$ with respect to each other.
   a. If you can see your friend's clock, will you say that it runs faster, slower, or the same as your clock?
   b. Will your friend say that your clock is faster, slower, or the same as his clock?
   c. You measure your friend's spacecraft and find it about half as long as yours. Will your friend agree with this measurement?

7. What mathematical factor accounts for the differences in mass, length, velocity, and elapsed time of an object traveling at a relativistic speed when observed from a "stationary" frame of reference?

8. Give two examples of an experiment that supports a prediction of the special theory of relativity.

9. What properties of a system are observed to be affected by relativistic speeds?

10. What is "special" about special relativity?

11. What is the main postulate of general relativity?

12. Discuss the effects of gravitational red shift on light.

13. Give two examples where light is affected by extreme gravity.

True or False (14–25)

14. The concept of relativity, that physical laws can be the same in different reference frames moving with respect to each other, was invented by Einstein.

15. If you throw a baseball while riding a bicycle, the ball's speed relative to the ground is the sum of the bicycle's speed and the speed that you threw it.

16. Relativity means that all motion can be described relative to one absolute frame of reference.

17. Mass is a form of energy.

18. Two events that occur simultaneously in one inertial reference frame will be observed to be simultaneous in all inertial reference frames.

19. You could get a few extra minutes of sleep each night if you could charter a very fast vehicle with a sleeping compartment that would return you to your house in time to go to school.

20. You and a friend are traveling in the same very fast spacecraft. You face each other, so that one of you faces the front and the other faces the back. Each of you would appear to be like flattened photographs to the other.

21. If all the subatomic particles in the universe were separated, the universe would have more mass than it does now.

22. There is no experimental difference between occupying a position in a gravitational field and experiencing an equal acceleration in a place with negligible gravity.

23. Since light bends under a high acceleration or gravity, it must experience inertia. Objects that exhibit inertia have mass. Therefore, light must have mass.

24. Black holes cannot be directly observed, but their presence is believed to produce observable effects.

25. The light we see coming from a very massive star is more blue than the light emitted at the star's surface.

●26. You are riding in a minivan that is crossing a bridge over a railroad track. A freight train engine is passing underneath in a perpendicular direction from right to left at the same moment.

(You wave at the engineer.) The van is moving at 10.0 m/s and the train is moving at 7.0 m/s. If the bridge is 12 m above the track, write the Galilean translation equations of the minivan's position coordinates with respect to the train's frame of reference (let the train's coordinates be primed).

27. A rocket traveling at 0.75c with respect to the earth launches a satellite at a speed of 0.75c in the same direction with respect to the rocket. What is the satellite's speed with respect to the earth?

28. You are in a spacecraft moving at 0.90c. The deck you are standing on is perpendicular to the direction of motion. How tall would you appear to an observer in a "stationary" space station? Be sure to include your rest length (height) as well as your relativistic length in your answer.

29. Refer to Question 19. You normally go to bed at 10 P.M. and awake at 6 A.M.—a total of 8 hours of sleep. How fast would your relativistic sleeping car have to go in order to give you an extra hour of sleep by 6 A.M.? (*Hint:* You need to get 9 hours of sleep while the clocks back home indicate only 8 hours have passed by the time you get back from your trip.)

30. A spacecraft with a rest length of 20.0 m shoots by you at 0.250c.
   a. How long does the vessel appear to you?
   b. How long does it take to go by, according to your stop-watch?

31. A linear particle accelerator is a long, tube-like device that contributes to the study of nuclear matter by accelerating subatomic particles to high velocities and smashing them into target nuclei. One such linear accelerator is 3.0 km long. How long does it appear to an electron moving at 0.999 999c?

28A	Quantum Theory	609
28B	Quantum Mechanics and the Atom	614
28C	Modern Atomic Models	619
Biography:	Max Planck	611
Facet:	Electron Microscope	620
Biography:	Niels Bohr	627

New Symbols and Abbreviations

mass number (A)	28.12	principal quantum number (n)	28.21
proton (p)	28.12	azimuthal quantum number (l)	28.21
atomic number (Z)	28.12	magnetic quantum number (m)	28.21
neutron (n)	28.12	spin quantum number (m_s)	28.21
electron volt (eV)	28.14		

A "stadium" of iron atoms on a substrate of copper atoms. The structure was created by moving individual atoms into position with a scanning-tunneling electron microscope (STM). The image was obtained using the same instrument.

Quantum Physics 28

Of all the topics you will study in physics, the most strange and wonderful will be quantum physics. It would not be exaggerating to say that the discovery of quantized energy, the photon, and wave-particle duality severely disturbed man's understanding of the universe. Scientists are continually surprised by the unexpected complexity and unanticipated properties of nature. At the same time, they presuppose that there is a cause-and-effect relationship for everything. Otherwise, meaningful science could not be undertaken. Every discovery leads mankind closer to understanding the underlying principles of creation. At the same time, it is clear that the product of an infinite Creator will never be completely understood by a finite mind. There will always be surprises on the horizon.

28A QUANTUM THEORY

28.1 Blackbody Radiation

As we mentioned in the previous chapter, most physicists near the end of the nineteenth century believed that few major discoveries remained to be made in physics. The accepted theories at that time in mechanics, thermodynamics, and electromagnetics, now called *classical physics,* appeared to be able to explain all experimental facts. Physics, it seemed, was complete. Then, all at once, some facts came to light that challenged many basic assumptions of physics. Classical physics, which had seemed nearly perfect, failed. One area that classical physics failed to explain was blackbody radiation.

Every object whose temperature is above absolute zero emits thermal radiation. Some bodies are more efficient radiators than others. You will remember that a perfect emitter of thermal radiation is called a *blackbody.* Although perfect emitters do not exist, *cavities*—voids in opaque materials that have small openings to the outside—are a close approximation. The energy emitted by a cavity is called **blackbody radiation.** The rate energy is emitted follows Stefan's law,

$$S = \sigma T^4, \tag{28.1}$$

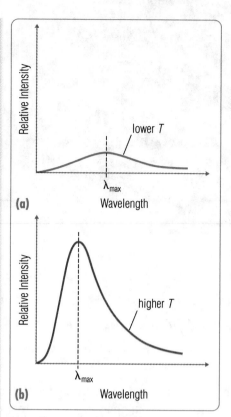

28-1 Energy emitted by a blackbody vs. wavelengths emitted at (a) low temperature; (b) high temperature

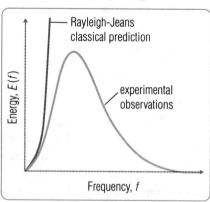

28-2 At higher frequencies, the Rayleigh-Jeans function does not agree with the observed energies and frequencies emitted from a blackbody at a given temperature.

where S is the rate of radiated energy emitted per square meter, σ is the Stefan-Boltzmann constant, and T is the cavity's absolute temperature.

28.2 Wien's Displacement Law

A cavity emits a continuous spectrum of radiation that includes radiation of all frequencies. At higher temperatures the cavity emits more of its radiation at higher frequencies with shorter wavelengths. Figure 28-1 shows the relative intensity of energy emitted at each wavelength. Figure (a) is for a lower absolute temperature than figure (b). The symbol λ_{max} represents the wavelengths at which the most energy is emitted.

In some ways the predictions of classical physics were correct. For instance, *Wien's displacement law* says that the most intense radiant energy emitted by a cavity progresses to shorter wavelengths as the cavity's temperature rises:

$$\lambda_{max}T = 2.8978 \times 10^{-3} \text{ m·K} \qquad (28.2)$$

Since frequency increases as wavelength decreases, Wien's displacement law predicts the experimental finding that Figure 28-1 shows.

28.3 The Rayleigh-Jeans Function

Another formula based on classical physics, developed by Lord Rayleigh and James Jeans, was much less successful. The function below is their attempt to find the relationship between frequency and the energy emitted at that frequency from a blackbody radiator at a given temperature. If $E(f)$ is the emitted energy as a function of frequency, the Rayleigh-Jeans function is

$$E(f) = \frac{8\pi f^2 kT}{c^3}. \qquad (28.3)$$

The symbol k is *Boltzmann's constant,* and c is the speed of light. All other quantities in Equation 28.3 are constant at a given Kelvin temperature. Figure 28-2 shows that the function was only partly successful in its attempt to describe the energy profile for a blackbody radiator's spectrum. The function provided a close fit to the observed curve for low frequencies. However, as the figure shows, Equation 28.3 predicts that an infinite amount of energy will be emitted at high frequencies. This prediction is not only untrue but also preposterous. The failure of this function, based on classical electromagnetics and thermodynamics, is called the **ultraviolet catastrophe.**

Max Planck (1858–1947)

Modern theories about the atom's structure stem from the quantization of energy. Though just a little more than one hundred years old, this idea has come to dominate all branches of modern physics. Many investigators have added to our knowledge of quanta, but most credit must go to the man who first introduced it, Max Planck.

Planck's specialty was thermodynamics. After studying physics and mathematics in Munich and Berlin, he presented his doctoral thesis on thermodynamics. He later became a professor of theoretical physics in Berlin, where he served for thirty-seven years.

Planck's interest was drawn to the problem of blackbody radiation, the thermal radiation that a completely black cavity emits. He tried to solve it with classic principles of thermodynamics. However, the only equation that agreed with his experiments required a radical new theory of physics. On October 19, 1900, he announced that the energy of radiation is proportional to its frequency ($E = hf$, where h is a constant, named after Planck). To explain his theory, he proposed that blackbodies emit radiation in "packets" of energy, called quanta.

He wondered whether the equation was an accurate reflection of nature or merely a product of artificial manipulations. However, physicists began to solve many other problems with the quantum theory. Einstein explained the photoelectric effect, and Bohr improved Rutherford's atomic model. Planck was awarded the Nobel Prize in Physics in 1918 for his discovery of quantum theory.

Although his discovery revolutionized physics, Planck himself remained a conservative physicist. Throughout his life he was reluctant to accept his own concept of quanta in blackbodies, and he spoke out against its application in other fields. He belonged to an older generation of physicists who supported classical theories and feared that their young followers were chasing unsound new ideas.

28.4 Planck and Quantized Energy

Because classical physics could not explain the results of experiments with blackbody radiation, Max Planck proposed "out of desperation" (in his words) a non-classical equation that agreed with experimentation:

$$E(f) = \frac{8\pi f^2}{c^3} \cdot \frac{1}{e^{hf/kT} - 1} \tag{28.4}$$

Equation 28.4 is called *Planck's blackbody spectrum.* It produces a curve extremely close to experimental observations. As noted in previous chapters, the symbol h is *Planck's constant,*

$$h = 6.626 \times 10^{-34} \text{ J·s},$$

and k is Boltzmann's constant. To derive Equation 28.4, Planck assumed that energy is emitted as **quanta,** or certain discrete values proportional to the frequency of the radiation. These allowed values are given by the equation

$$E = nhf, \text{ where } n = 1, 2, 3, \ldots \tag{28.5}$$

The idea that energy is not continuous but can take on only certain values seemed unreasonable to classical physicists. Waves, in classical theory, could have any energy. To assume that radiant energy is **quantized** (emitted as quanta) implies that it originates not as waves but as particles. Planck himself had trouble accepting the quantum hypothesis. However, Planck's equation agreed well with experiments. It was therefore eventually accepted that energy is quantized.

28.5 Photons and the Photoelectric Effect

Planck intended his idea of quanta to apply only to the *emission* of energy, but Albert Einstein expanded the idea to energy in transit. Light, Einstein said, travels in packets of definite amounts of energy called **photons,** each with an energy of *hf.* This extension of Planck's quantum theory explained another of the failures of classical physics—the photoelectric effect.

Light, especially ultraviolet light, falling on a polished metal surface can dislodge electrons from the metal. This is called the **photoelectric effect,** and the electrons so dislodged are called *photoelectrons.* According to classical theory, a light wave falls on the metal and distributes its energy over the surface of the metal. When an electron has absorbed enough energy to overcome the metal's bonds, it leaves the metal. If it has absorbed more energy than it needs to break the metal's bonds, the electron acquires kinetic energy. From the classical theory you would expect that

1. any frequency of light can liberate electrons if it is intense enough;
2. there is a measurable time between the instant the light strikes the metal and the first electrons' ejections;
3. a more intense light gives electrons more energy.

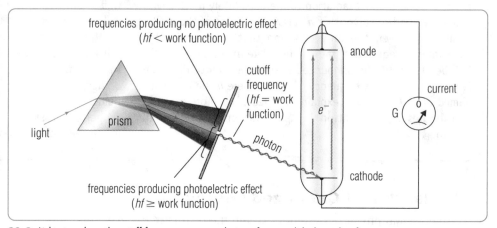

28-3 Light at or above the cutoff frequency removes electrons from a polished metal surface.

28.6 Photoelectric Work Function

The quantum theory of Planck and Einstein, on the other hand, explains the photoelectric effect in terms of photons. As light strikes a metal target, a few photons will probably strike electrons. However, it is unlikely that more than one photon will strike the same electron at the same time. Therefore, each electron receives an energy of *hf.* If this energy is greater than the **work function** that an electron must gain to escape the metal, the electron leaves the metal and acquires some kinetic energy. If the energy is less than the electron's work function, the electron remains in the metal and soon dissipates its extra energy. From this theory, you would expect that

1. the frequency of light must be such that *hf* is greater than the work function for any electrons to be liberated;
2. there may be no time between the instant the light strikes the metal and the first electrons' ejections;
3. a more intense light ejects more electrons at the same maximum energy.

The photoelectric work function principle provides an interesting spiritual parallel. Jesus said, "I am the light of the world" (John 8:12), but we know that Satan himself is transformed into an angel of light (II Cor. 11:14). The question is, What is your spiritual "work function" that motivates you to service? Does the false light of the enemy induce grudging, low-energy effort that is more selfserving than Christ-serving, or do you instantly respond wholeheartedly to only the perfect light of your Savior?

28.7 Photoelectric Cutoff Frequency

To investigate the photoelectric effect, experimenters used cathode-ray tubes. Electrons are emitted when light falls on a polished cathode. The number of photoelectrons is measured by the current through the tube. Experiments showed that the current is zero for light of frequencies below a **cutoff frequency,** which depends on only the cathode material. Light below this frequency, regardless of intensity, will not liberate any electrons. Light above the cutoff frequency will liberate at least a few electrons immediately ($< 10^{-10}$ s) after striking the cathode, no matter how weak the light. These findings supported the quantum theory.

28.8 Experimental Support for the Quantum Theory

A cathode-ray tube can also be used to measure the kinetic energy of the electrons. When the tube is connected as Figure 28-4 shows, the external voltage opposes the motion of the photoelectrons. The voltage is increased until the current is zero. Then even the most energetic electrons cannot reach the anode. A voltage V_0 will obstruct the motion of electrons with energies up to

$$E_0 = eV_0, \qquad (28.6)$$

where e is the magnitude of the charge on an electron (the fundamental charge). If V_0 stops the current, then E_0 is the kinetic energy of the most energetic electrons. Experiments showed that E_0 depends on only the light's frequency, not its intensity. In other words, classical predictions were wrong, but the quantum theory agreed with experiments.

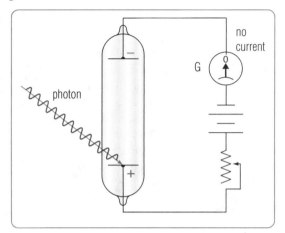

28-4 Because of the opposing voltage, none of the dislodged electrons reaches the anode.

28.9 The Compton Effect

Another experiment corroborated the quantum theory. Arthur Compton discovered some unexpected lines in an x-ray diffraction spectrum. He found that he could completely explain the lines by assuming that x-ray photons, with a momentum of E/c, collided with electrons in the diffracting crystal. Such a collision would transfer some of the x-ray's energy to an electron and change the x-ray's direction. With less energy, the x-ray would have a lower frequency. The changes of the x-ray's frequency and direction are called the **Compton effect.** Compton's interpretation of his discovery strengthened the quantum theory.

Arthur Holly Compton (1892–1962) was an American physicist. He is considered a pioneer of high-energy physics and won a Nobel Prize in 1927.

28A Section Review Questions

1. In a blackbody radiator, how is the wavelength of the most intense radiant energy related to the blackbody's Kelvin temperature?

2. What was the *ultraviolet catastrophe?*

3. Why did Max Planck doubt the validity of his spectrum equation?

4. What was one result of Planck's quantum theory that was inconsistent with the classical wave theory of electromagnetic radiation?

5. Discuss how electromagnetic radiation incident on a metal surface produces the photoelectric effect.

6. What is the critical variable in producing the photoelectric effect?

⊛7. According to Wien's displacement law, what is the most intense wavelength of light radiated by a blackbody at 1.80×10^3 K?

⊛8. Assuming the Sun is a blackbody radiator, what is the temperature of its photosphere if $\lambda_{max} = 550.$ nm (yellow light)?

28B QUANTUM MECHANICS AND THE ATOM

28.10 The Electrical Atom

Before the 1890s most physicists believed that atoms were hard spheres that could not be divided. However, in 1897 J. J. Thomson discovered that the charge carriers in different conductors have the same charge-to-mass ratio. This evidence suggested that all atoms contain the same charge carriers, now called *electrons*. Since atoms are electrically neutral, Thomson suggested that an atom is a ball of positive charge with electrons scattered evenly throughout, like raisins in the English plum pudding. Thomson's model came to be known as the **plum-pudding model** of the atom.

28.11 The Nuclear Atom

Thomson's atomic model was soon challenged. Ernest Rutherford studied atoms by directing alpha particles, particles emitted naturally by certain atoms, at metal foils that were a few atoms thick. Alpha particles are about eight thousand times as massive as electrons and have a positive charge twice as large as the electron's negative charge. According to Thomson's model, an alpha particle should be slightly deflected by the positive charge throughout the atom and perhaps by a collision with an electron. Rutherford and his research team found, however, that most of the particles were not deflected at all. Of those that were deflected, some were deflected slightly, some at large angles, and a few even bounced directly back the way they had come. According to Thomson's model, this back deflection was impossible. Therefore, Thomson's model had to be changed.

Rutherford proposed a new atomic model based on the results of his experiment. Since most alpha particles were undeflected, he suggested that an atom is mostly empty space. The positive charge, he said, is concentrated in a small but massive **nucleus.** The electrons orbit the nucleus as planets orbit the Sun. The masses of most nuclei are at least as large as the mass of an alpha particle. In fact, Rutherford later showed that an alpha particle is a helium nucleus. It is not surprising that the few alpha particles striking a massive nucleus bounced off like BBs hitting a bowling ball.

28.12 Isotopes

The nucleus is made up of two major kinds of particles, collectively called **nucleons.** The total number of nucleons in a given nucleus is its **mass number (A).** One kind of nucleon is the **proton (p),** a positively charged particle with nearly two thousand times the mass of an electron but with the same magnitude of charge. The number of protons in a nucleus, called the **atomic number (Z),** determines what element it is. For instance, all oxygen atoms contain eight protons in their nuclei, while gold atoms have seventy-nine protons. The other kind of nucleon is the **neutron (n),** a neutral particle that is slightly more massive than a proton. In general, the higher the atomic number, the larger the number of neutrons that may be present in the nucleus.

Atoms of the same element with different mass numbers (different numbers of neutrons) are called **isotopes** of the element. The *isotopic symbol* for an element includes the mass number at the top of the element symbol and may include its atomic number at the bottom as well (see Figure 28-6). For instance, carbon (C) has six protons ($Z = 6$). One isotopic form of carbon has six neutrons; so its mass number is twelve ($A = 12$), and the isotope is called carbon-12. Another carbon isotope has eight neutrons and $A = 14$, so it is called carbon-14.

Another term for describing different combinations of protons and neutrons in a nucleus is **nuclide.** While "isotope" and "nuclide" are often used interchangeably,

Ernest Rutherford (1871–1937) was a physicist from New Zealand and is considered to be the Father of Nuclear Physics. He discovered that some elements become different elements through radioactive decay. He won a Nobel Prize in 1908 for his study of radioactive decay and the chemistry of radioactive substances.

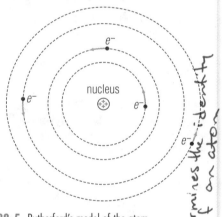

28-5 Rutherford's model of the atom

Rutherford's experiments indicated that the nucleus is on the order of 10^{-5} the diameter of the atom itself.

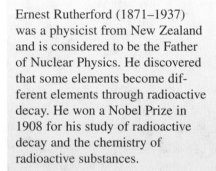

28-6 Isotopic notation used for distinguishing different nuclear forms of an element

we will use "isotope" when discussing different atomic species of the same element and "nuclide" when discussing isotopes of different elements. For example, oxygen-18, carbon-14, and nitrogen-16 are all radioactive nuclides.

28.13 The Bohr Atomic Model

Although Rutherford's atomic model was useful, it did have a serious fault. Electrons (imagined to be tiny BBs) orbiting a nucleus are accelerating because their velocities are continually changing direction. In classical electromagnetic theory, Maxwell demonstrated that accelerating charges radiate electromagnetic energy. Then, as electrons radiate electromagnetic energy, they lose total energy, which is a combination of their kinetic energy of motion and potential energy relative to the nucleus. If they continuously lose energy as they orbit, they will slow down and "lose altitude," eventually spiraling into the nucleus. The atom collapses. However, atoms do not collapse. Therefore, Rutherford's theory needed to be modified.

Another problem with Rutherford's theory was its failure to account for the energy observed to be emitted by a low-pressure ionized gas at high temperature. Unlike other glowing materials, the gas does not emit a continuous spectrum of radiated energy. Instead, each gas emits certain characteristic frequencies of radiation. Since these frequencies can be shown in a spectrograph as lines, a gas is said to emit a **line spectrum.** According to Rutherford's model, an electron may emit radiation of any frequency. His model does not explain why low-pressure incandescent gases emit only a few characteristic frequencies.

28.14 Electron Shells

Both of these problems—that electrons do not continuously radiate and the existence of line spectra (implying that atoms *do* radiate specific energies under certain circumstances)—led Niels Bohr to propose a new atomic model. Bohr postulated that electrons are restricted to certain circular orbits characteristic of an atom of a given element. The radii of these orbits were functions of an integer number. In other words, the orbit radii were quantized. He further postulated that while electrons remain in these allowed orbits, they do not radiate. The energy in each of these orbits was also quantized by the same integer numbers.

The allowed electron orbits in Bohr's model are called **electron shells** and represent levels of energy. Electrons in the shell nearest the nucleus, the **ground state,** have the least energy (E_1). (Similar to gravitational potential, E_1 is a relatively large negative number.) Electrons in shells farther away from the nucleus have more energy (smaller negative numbers) and are called **excited states.** The shells are designated by integer numbers from the ground state, $n = 1$, to a theoretical orbit so far from the nucleus that electrons have zero potential energy with respect to the nucleus ($n = \infty$). The shells become closer together as n increases. If an electron's energy is positive with respect to a nucleus, it is not bound to that nucleus.

Each shell is associated with an energy E_n. When an electron moves from a shell with greater energy (E_n) to a shell of less energy (E_{n-x}), it emits a single photon with an energy of

$$E_{\text{phot}} = hf = -(E_{n-x} - E_n). \tag{28.7}$$

The frequency of the emitted photon can be found by solving Equation 28.7 for f:

$$f = -\frac{E_{n-x} - E_n}{h} \tag{28.8}$$

Now consider an electron moving from an energy of E_n to a higher energy of E_{n+x}. To accomplish this move, the electron must gain an energy of

$$\Delta E = E_{n+x} - E_n. \tag{28.9}$$

The lifetime of an atom with a continuously spiraling electron would be about 10^{-11} s.

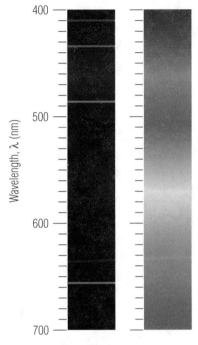

28-7 Line spectra of hydrogen gas

The energy of an electron is measured in **electron volts (eV)**—the energy required to move a fundamental charge through a potential difference of 1 volt. The energy of the ground state in hydrogen is −13.6 eV.

Recall that the energies of electron shells are negative and $E_{n-x} < E_n$. Equation 28.8 will always yield a positive number for energy levels associated with the same nucleus.

when an electron moves to a higher level of energy it must absorb a specific amount of energy

Equation 28.9 implies that the electron must absorb a photon of $hf = \Delta E$ in order to move to the higher energy shell.

28.15 Evidence that Supports the Bohr Model

Bohr's atomic model explained the line spectra of hydrogen-like gases (gases containing atoms with only one electron). By extension, he proposed that each atom in a gas has shells in which electrons have different energies. To radiate, an electron must move from one shell to another. The difference in energy between these shells determines the frequency of the radiation absorbed or emitted. Since all atoms of the same gas have the same energy levels, each gas can radiate only certain frequencies that are related to the differences between those levels. At low gas pressures, the photons from different atoms of a given gas do not interfere to form other frequencies; so only the frequencies that would be emitted by any single atom are detected. Different elements have electron shells with different energies and thus give off light of different frequencies.

There are several lines in a spectrum because several transitions are possible. For instance, an electron in the fourth shell may fall to the third shell, the second shell, or the first shell. Photons produced by any transition *to* a given shell have frequencies of the same order of magnitude. A group of frequencies caused by a change to the same shell is called a **spectral series.** For instance, in hydrogen-like atoms, transitions to the ground state produce frequencies in the ultraviolet portion of the spectrum and are part of the **Lyman series.** Transitions from $n > 2$ to the $n = 2$ shell produce visible light and are part of the **Balmer series.**

For an electron to move to a higher energy level, it must absorb a specific amount of energy. The highest energy associated with a given nucleus is $n = \infty$. This energy level has been assigned a value of zero. With greater energy, the electron is free from any atom. The bound electrons have lower, and therefore negative, energy. The energy required to remove an electron from the n shell of an atom is

$$\Delta E = E_\infty - E_n, \text{ so}$$
$$\Delta E = -E_n. \quad (28.10)$$

If this electron is the one in the highest energy level of a neutral atom (a *valence* electron), then $-E_n$ is called the **first ionization energy** of the atom—the energy needed to remove the most loosely held electron in the atom.

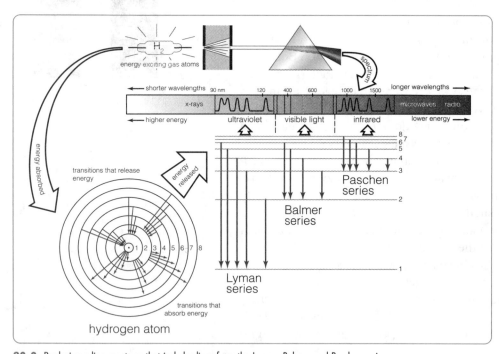

28-8 Producing a line spectrum that includes lines from the Lyman, Balmer, and Paschen series

Bohr's model says that atomic electron energy is quantized, just as Planck and Einstein said that electromagnetic energy is quantized. However, some predictions of the classical theory (assuming that energy is continuous) were also correct. How could both theories make correct predictions? The *correspondence principle* states that new theories must give the same results as old theories where the old theories were supported by experiments. Bohr's model and classical theories give the same results for the outermost energy levels of large atoms. In the Bohr atomic model, the energy levels get closer together as *n* increases. For large *n* the levels are so close together that the energy appears to be continuous. Thus, when *n* is large, Bohr's theory and the classical theory predict the same results.

28.16 X-ray Spectra

When atoms, especially ones with large atomic numbers, are placed in the path of high-speed electrons, high-energy, high-frequency radiation in the x-ray portion of the spectrum is emitted. Figure 28-9 shows two types of x-rays. The lower, smooth continuous curve represents photons emitted as the colliding electrons slow down within the cloud of electrons surrounding the atom as they pass through the atom. This radiation is called *bremsstrahlung*. Occasionally, the colliding electrons remove ground-state electrons from the atom. Since electrons in an atom tend to seek the lowest possible energy level, an electron in a higher shell ($n \geq 2$) will promptly fill the vacancy in the ground state. In doing so, the electron emits an extra–high-energy x-ray. The frequencies of the x-ray line spectra are characteristic of the bombarded, or target, atom. These characteristic lines are superimposed on the normal continuous spectrum (see Figure 28-9).

Bremsstrahlung radiation is the opposite process to the photoelectric effect. In the former, energy is added to an atom by an energetic electron, and it emits a photon. In the latter, a photon is absorbed by an atom, which then emits an electron.

28.17 Particles as Waves

If waves (like light) can be considered to be particles (like photons) under some circumstances, why cannot particles act like waves at times? This question caused Louis de Broglie to calculate what a "matter wave" would be like. He predicted that the wavelength of a matter wave would be

$$\lambda = \frac{h}{mv}. \tag{28.11}$$

This wavelength is so small for ordinary-sized objects that it cannot be detected by diffraction effects. Diffraction appears only when a wave passes an obstacle whose width is nearly the wavelength. The smallest obstacles that can be used are atoms, which have radii on the order of 10^{-10} m. Matter wavelengths are much smaller than this.

The *correspondence principle* states that new theories must give the same results as old theories where the old theories were supported by experiments.

Bremsstrahlung is a compound German word meaning "braking" and "radiation." The "braking" refers to the slowing down of the electrons as they fly through an atom.

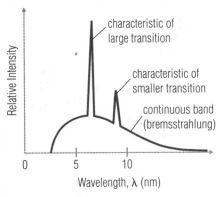

28-9 An x-ray spectrum resulting from a specific energy of colliding electrons

Louis de Broglie (1892–1987) was a French physicist who won the Nobel Prize in 1929 for his discovery of the wave nature of electrons.

> **EXAMPLE 28-1**
>
> ### Wave Runner: Calculating a Matter Wavelength
> Calculate the wavelength of a 75 kg athlete running at a speed of 5.0 m/s.
>
> **Solution:**
> You are given the runner's speed and his mass. Use Equation 28.11:
>
> $$\lambda = \frac{h}{mv} = \frac{6.626 \times 10^{-34} \text{ J} \cdot \text{s}}{(75 \text{ kg})(5.0 \text{ m/s})}$$
>
> $$\lambda \doteq 1.\underline{7}6 \times 10^{-36} \frac{(\text{kg} \cdot \text{m}^2/\text{s}^2) \cdot \text{s}}{\text{kg} \cdot \text{m}/\text{s}}$$
>
> $$\lambda \approx 1.8 \times 10^{-36} \text{ m}$$
>
> A wavelength on the order of 10^{-36} m is undetectable by any known instrument.

28-10 Electron standing waves correspond to Bohr's orbits

An undetectable wave is not very useful. However, matter waves are undetectable only for ordinary-sized objects. The waves associated with electrons, for instance, are detectable. An electron traveling at 3 000 000 m/s ($10^{-2}c$) has a wavelength of about 2×10^{-10} m. This is near x-ray wavelength, so, like x-rays, electron waves should be diffracted by a crystal. Davisson and Germer directed high-speed electrons at a nickel crystal. They found a pattern identical to that caused by a wavelength in the x-ray range.

28.18 Schrödinger Wave Equation

Because waves are sets of time-dependent positions, rather than a single point, they are described by functions. Erwin Schrödinger developed an equation that uses the *wave function, ψ,* of an electron's wave. For one dimension, this equation is

$$-\frac{\hbar^2}{2m} \cdot \frac{\partial^2 \psi}{\partial x^2} + U\psi = i\hbar \frac{\partial \psi}{\partial t}, \qquad (28.12)$$

where $\hbar = h/2\pi$, m is the electron's mass, U is the electron's potential energy function, i is the complex root ($\sqrt{-1}$), and the entire right side is the total energy function. The term $\partial^2 x/\partial x^2$ is used in calculus to describe change in the wave's motion with respect to the x-axis. Similarly, the term $\partial \psi/\partial t$ describes the rate of change of some aspect of the wave with time. This equation is a *differential equation.* That means its solution is itself a function rather than a point or two. Equations as this one require such advanced mathematics to solve for even simple cases that most scientists use computers to solve them. However, *Schrödinger's wave equation* gives some useful information about an electron whose total and potential energies are known.

The function ψ for an electron's wave gives the possible location of the particle. The electron can be thought of as being somewhere along the waveform. The electron is more likely to be where the wave has a high amplitude than where it has a low amplitude. It will not be where the wave's amplitude is zero. The waveform is therefore called a *probability wave.*

The relationship between the wave and the electron's location is used to explain the existence of allowed orbits in atoms. Using the Bohr model, if an orbit's circumference is a whole number of wavelengths, the wave will reinforce itself around the orbit. The electron is likely to be in that orbit. If, however, the orbit's circumference is not a whole number of wavelengths, the wave will destructively interfere and cancel itself. The electron is unlikely to be there. Thus, Bohr orbits are those orbits that can contain a whole-number multiple of the electron's wavelength.

According to modern theory, the orbits around a nucleus are not as clear-cut as in Bohr's model. There are three-dimensional regions of greater probability of finding an electron at the distance of the Bohr orbits. However, it is possible to find the electrons outside of Bohr orbits—even in the nucleus or completely free from the atom. Electron locations are thought to be better described as **orbitals** than as clear-cut orbits. The sum total of all possible electron orbitals in an atom produce an indistinct region around the atom called the **electron cloud.**

28.19 Heisenberg Uncertainty Principle

Physicists are forced to deal with probability waves because they cannot locate an electron and know its momentum at the same time. Werner Heisenberg showed that this limitation is a result of making an observation of subnuclear objects rather than due to imperfections of instruments. To observe an object, at least one photon must be reflected or diffracted by it. Many photons may be reflected by ordinary objects without noticeably changing the object or its motion. However, an electron's motion is changed unpredictably by an interaction with a single photon. If you set up an experiment to find an electron's location, the interaction with a photon will change its momentum so that you cannot predict where the electron will be later. Heisenberg showed that even under ideal conditions, the product of the uncertainties in measuring the electron's position and momentum is greater than $\hbar/2$. As the uncertainty in position decreases (the measurement becomes more precise), the uncertainty in momentum increases. This concept is called the **Heisenberg uncertainty principle.**

Werner Karl Heisenberg (1901–76) was a German physicist who won the Nobel Prize in 1932 for the development of quantum physics and the principle of *indeterminacy.*

28.20 Wave-Particle Duality

The uncertainty principle illustrates the principle of **complementarity:** the concepts of waves and particles are both needed to describe reality. In the measurement of an electron's position, the electron acts as a wave until a photon strikes it. Then, like a particle, it is moved by the collision. Both concepts are required to explain observations.

For a Christian, the concept of wave-particle duality should not be surprising any more than the fact that Jesus was simultaneously both human and God. "And the Word was made flesh, and dwelt among us, (and we beheld his glory, the glory as of the only begotten of the Father,) full of grace and truth" (John 1:14). Remember that God's creation reflects the nature and character of its Creator.

28B Section Review Questions

1. What experiment conclusively proved that atoms are *not* a solid matrix of positive charge with embedded negative charges?

2. Approximately how large is an atom's nucleus compared to the atom itself?

3. What are isotopes of an element?

4. What key prediction of classical electromagnetism was violated in Rutherford's planetary nuclear model of the atom?

5. Why do atoms radiate specific wavelengths of light under certain situations?

6. a. What energy transitions produce the Lyman series of line spectra in hydrogen-like atoms?

   b. In what portion of the electromagnetic spectrum do these emissions fall?

7. What causes bremsstrahlung radiation?

8. What did the combination of de Broglie's and Schrödinger's theories do to the classical particle theory of matter?

●9. What is the matter wavelength of the *Voyager 1* spacecraft ($m = 815$ kg) traveling at 17.1 km/s?

28C MODERN ATOMIC MODELS

28.21 The Quantum Numbers

Just as the equation $y = 3x + 2$ has ordered pairs of numbers such as (2, 8) as unique solutions, the Schrödinger wave equation can be solved to give four **quantum numbers** that uniquely specify each electron's energy in an atom. The **principal quantum number (n)** (also called the *energy level number*) stands for the probable distance of the electron from the nucleus. As you learned before,

$$n = 1, 2, 3, 4, \ldots$$

and it identifies the energy level that the electron occupies.

The **azimuthal** (or sublevel) **quantum number (l)** refers to the angular momentum of the electron. Its possible values depend on n:

$$l = 0, 1, 2, \ldots n - 1$$

In other words, for every energy level n, there are n azimuthal numbers.

The **magnetic** (or orbital) **quantum number (m)** relates to the electron's orientation as detected by a magnetic field. The possible values are

$$m = -l, -(l - 1), \ldots 0, \ldots l - 1, l$$

The **spin quantum number (m_s)** stands for the direction of the electron's spin. It has two possible values: *no integer value*

$$m_s = +\tfrac{1}{2} \text{ or } -\tfrac{1}{2}$$

An analogy may help you to understand these numbers. The n is like the distance a planet orbits a sun. The number l is analogous to a planet's angular momentum, mvr, as it orbits (which is related to its distance from the sun and its energy). The m-number shows the orientation of the plane of the planet's orbit. For $m = 0$, the orbital plane contains the sun. For other values of m, the planet may be

> Each quantum number has a maximum number of allowable values:
>
> n: no limit
>
> l: n (for each n)
>
> m: $2l + 1$ (for each sublevel)
>
> m_s: 2 (for each m)

> In nuclear physics parlance, positive spin is "up," and negative spin is "down." Try to remember that these words do not have their strictly physical meanings in this context.

FACETS of PHYSICS

Electron Microscope

An electron microscope

When Leeuwenhoek first used refracted light to magnify unseen "animalcules," the scientific world was astounded. For the next two centuries glass lenses were continually refined in an effort to increase the resolving power of the light microscope. The improved lenses focused more light on the specimen and spread the refracted light farther, thus increasing the magnification. Smaller and smaller structures were discovered. But by the turn of the twentieth century, one weakness in the light microscope prevented scientists from making any more discoveries: a wave of light cannot resolve anything smaller than its own wavelength of about 10^{-7} m.

Theoretically, electrons, which have wavelengths as low as 10^{-10} m, could replace light because electrons can also be refracted. But no one was able to discover a practical lens until 1933, when Ernst Ruska developed the first successful electron microscope. Ruska used a magnetic "lens" rather than a glass lens to focus and spread the stream of electrons. The refracted electrons could not be viewed directly because, unlike light, they are invisible to the human eye. The image must be photographed, projected onto a fluorescent viewing screen, or reproduced electronically.

Electron microscopes provide a wealth of information about ultrasmall structures. The *transmission microscope*, which passes electrons through the specimen, can view large molecules such as DNA and even large atoms. Metallurgists gain valuable knowledge about crystals and metal surfaces, and medical laboratories examine viruses, bacteria, and pollen grains. But live specimens must be killed, treated, and sliced to a maximum of 1 micron (10^{-6} m) in thickness before the transmission microscope will work.

A helpful supplement to this type of electron microscope is the *scanning electron microscope (SEM)*, which reflects an

$\displaystyle \text{\Large$\star$} \; {}^{11}_{6}C \rightarrow {}^{0}_{-1}e + {}^{11}_{7}N$ — completes the disintegration of carbon 11 as it gives off a Beta Particle.

symmetrical about a position axis or in some other configuration (this picture does not fit the mechanical analogy very well). Finally, m_s gives the direction in which the planet rotates about its own axis—its spin. For the closest orbit, the planet must orbit at one speed and it has only one possible orbital plane. Each farther orbit has more freedom. This analogy is not perfect, but it may help you to understand quantum numbers.

EXAMPLE 28-2

Using Quantum Numbers
What values of l are allowed if $n = 5$?

Solution:
The allowable numbers are $l = 0, 1, 2, \ldots n - 1$. Therefore,

$$l = 0, 1, 2, 3, 4 \text{ (five numbers)}.$$

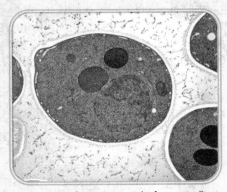

A transmission electron micrograph of a yeast cell.

Even newer kinds of microscopes use the ability of electrons to "tunnel" across a vacuum gap between two charged surfaces. No image is formed directly by the electrons. Instead, a charged needle is slowly moved over the prepared surface of a specimen at extremely small distances. As the distance varies between the specimen and the needle, the current of electrons that "tunnel" across the gap varies in

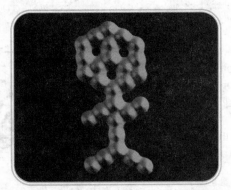

"Carbon monoxide man." An STM image of carbon monoxide molecules on a platinum substrate.

electronic beam off the surface of the specimen. Specimens are much easier to prepare, and they do not need to be sliced up. Minuscule structures do not have to be removed from large specimens such as flies, mosquitoes, and butterflies. Some studies simply show the beauty and economy of God's creation, and others reveal practical details about the human body that could not be observed before. The inner ear, taste buds, and the color rods inside the eye are only three examples.

A scanning electron micrograph of the head of a cigar beetle.

proportion to the distance. The current signal is stored by the digital memory of the instrument along with the position of the needle. An image of the surface is reconstructed by plotting the tunneling distance versus position. With such a *scanning-tunneling (STM)* electron microscope, the surfaces of small atoms have been indirectly imaged.

Otto Stern (1888–1969) was a German physicist whose work demonstrated the wave nature of atoms and molecules. He won the Nobel Prize in 1943.

Walther Gerlach (1889–1979) was a German physicist.

EXAMPLE 28-3

By the Numbers: Sets of Quantum Numbers

List all possible sets of quantum numbers for $n = 2$, $l = 1$.

Solution:

$n = 2$

$l = 1$

For $l = 1$, $m = -1$, 0, or $+1$

$m_s = +\frac{1}{2}$ or $-\frac{1}{2}$

Sets:

2, 1, -1, $+\frac{1}{2}$	2, 1, -1, $-\frac{1}{2}$
2, 1, 0, $+\frac{1}{2}$	2, 1, 0, $-\frac{1}{2}$
2, 1, $+1$, $+\frac{1}{2}$	2, 1, $+1$, $-\frac{1}{2}$

This means that there are six possible electron energies in the second sublevel of the second energy level.

28.22 Electron Ordering Using Quantum Numbers

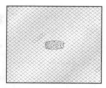 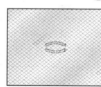

(a) Classically predicted   **(b)** Observed

28-11 Target deposition pattern in the Stern-Gerlach experiment that demonstrated the existence of electron spin. (a) Pattern predicted by classical theory; (b) Observed pattern

The idea that an electron responds to magnetic fields as though it were spinning about its own axis was demonstrated by the Stern-Gerlach experiment in 1922. In this experiment, neutral silver atoms were propelled through a narrow rectangular collimating window into an asymmetrical magnetic field and then deposited on a metal screen. Classical electromagnetic theory predicted that a single spot would form. Instead, the experimenters observed the pattern shown in Figure 28-11 (b), indicating that the atoms had two different magnetic moments. After considering the evidence and performing additional experiments, scientists realized that the *electrons* in the atoms had two different magnetic moments, which were analogous to spin in opposite directions.

Electrons seek the lowest possible energy level. However, even nonenergetic atoms have electrons in energy levels above the ground state. This phenomenon is explained by the **Pauli exclusion principle,** which states that no two electrons in the same atom can have identical sets of quantum numbers. For an atom with a single energy level in its ground state, $n = 1$; so the remaining electron quantum numbers are either $l = 0$, $m = 0$, and $m_s = +\frac{1}{2}$ or $l = 0$, $m = 0$, and $m_s = -\frac{1}{2}$. There are only two possible sets of quantum numbers for the ground state electrons. Therefore, the lowest energy level in the ground state can hold only two electrons.

Electrons occupy the sublevel with the lowest energy that is open to them. In the planet analogy, a planet in a smaller orbit with a greater angular momentum may have more energy than a planet in a larger orbit with less angular momentum. Similarly, an electron in the fifth energy level with a high l-number has more energy than an electron in the sixth energy level with a low l-number. The azimuthal numbers are given the name *sublevels*. For convenience, the sublevels are labeled *s, p, d, f, g, h,* and so on for $l = 0, 1, 2, 3, 4, 5, \ldots$, respectively. Each principle energy level has "room" for the number of sublevels allowed by the number of *l*-values for that energy level number (*n*). For example, the fifth energy level (*n* = 5) has 5 allowable azimuthal (*l*) numbers. Therefore, sublevels *s, p, d, f,* and *g* may exist in the fifth energy level if the element's atomic number is large enough.

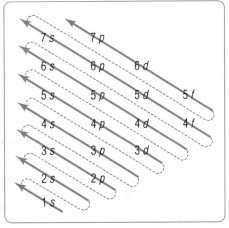

28-12 The arrows go from lower-energy to higher-energy sublevels.

Sublevels are identified by their principle quantum number and the sublevel letter. The fourth sublevel (*f*) of the fifth energy level (5) is denoted 5*f*. Similarly, the first sublevel of the second energy level is denoted 2*s*. The number of electrons actually present in the sublevel is denoted by a superscript on the sublevel letter. For instance, the presence of two electrons in 5*f* is denoted $5f^2$ (spoken, "five-eff-two"). Figure 28-12 orders the energies of the sublevels from least energy to greatest. The sublevels of atoms in their lowest energy state are filled with electrons in this order.

Each type of sublevel can hold a definite number of electrons. The *s* sublevel (*l* = 0), for instance, can hold two; $l = 0, m = 0, m_s = \frac{1}{2}$ and $l = 0, m = 0,$ $m_s = -\frac{1}{2}$. The total number of electrons each subshell can contain is

$$2(2l + 1)\, e^-.$$

The locations of electrons for a specific element are given by a string of shorthand notations, in order of energy. For instance, the eight electrons in minimum-energy oxygen atoms are represented by $1s^2 2s^2 2p^4$.

28.23 The Periodic Table

The arrangement of electrons in an atom determines its chemical and many of its physical properties. Particularly important are the electrons in the highest occupied energy level, the *valence electrons*. These electrons are the most easily removed and added. Therefore, they are the ones involved in chemical bonding. More than 100 years ago, scientists organized the elements into a useful table that relates repeating (periodic) similarities in both their electron structures and chemical properties. This table is called the **periodic table** (Appendix H). It represents one of the most impressive pieces of evidence that the basic building blocks of matter are designed.

The periodic table is arranged so that element atomic numbers increase in rows, called *periods* or *series,* from left to right. Interestingly, the elements in each vertical column, or chemical *group,* usually have the same number of valence electrons. For this reason, the elements in a group typically have similar chemical properties. The periods of the table are numbered according to the highest occupied energy level (when the atom is in its lowest energy state).

Figure 28-13 on the next page is a reproduction of the periodic table omitting the element symbols. Instead, the blocks indicate the sublevel where each electron is added with increasing atomic number. Referring back to Figure 28-12, you can see that sublevels are not filled from lowest to highest in order. Rather, certain sublevels are filled in higher energy levels before other sublevels in lower energy

Wolfgang Ernst Pauli (1900–58) was an Austrian physicist who made significant contributions to quantum theory. He was first to predict the existence of the neutrino. He won the Nobel Prize in Physics in 1925 for his exclusion principle.

The term "filling" refers to the way electrons are added to the electronic structure of atoms of elements arranged in a tabular fashion, such as in the periodic table (see Appendix H). Such an order assumes that the atoms are in their lowest energy state. In an actual atom, individual electrons will occupy allowed sublevels according to their excited energies. For a highly energetic atom, lower sublevels may be briefly vacated as electrons move from one energy level to another.

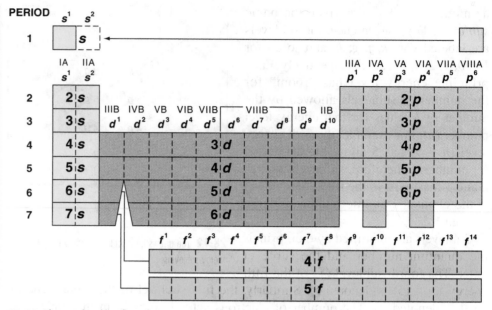

28-13 The periodic table reflects the arrangements of electrons within atoms by their position on the table

levels. The rules that govern the filling order and determine valence electron structure are normally introduced in high school chemistry courses. You should note that in each period, the valence electrons normally occupy the *s* and *p* sublevels.

28.24 Applications of Quantum Physics

Masers and Lasers

The quantization of energy provides a way to get *coherent*—single phase and frequency—radiation. Most radiation sources emit radiation of different frequencies. If one or more electrons in each atom in a material could all be raised to the same energy level and then stimulated to emit radiation at the same time, the result would be coherent radiation. This condition does not readily occur.

In the middle of the twentieth century, Charles Townes discovered a practical way to produce nearly coherent radiation. A photon passing an excited (unusually energetic) electron stimulates the electron to emit radiation in phase with the photon. However, keeping the electrons excited until a photon passes is difficult because excited electrons tend to radiate almost immediately. It is necessary to find a relatively stable excited energy level, called a *metastable state.* Not every material is capable of establishing metastable states. Those that do tend to be crystalline or molecular materials that have bonding orbitals that hold electrons for relatively long times. The electrons are "pumped" to an energy slightly greater than the metastable energy. They immediately fall to the metastable state, where they remain briefly until a photon passes. Then they emit a photon in phase with the first. If the process of pumping more electrons continues, and the photon beam is allowed to bounce back and forth through the material, two things happen: more electrons radiate photons, and the photons constructively interfere to produce a single, high-amplitude wave of a single (monochromatic) wavelength. The result is strong, coherent radiant energy. Townes called his device, which emitted microwaves, a **maser** (*m*icrowave *a*mplification by *s*timulated *e*mission of *r*adiation). An optical maser, emitting light, is called a *laser* (see the facet on lasers in Chapter 25).

Masers have many uses. Masers are often used as amplifiers for radio astronomy—the study of astronomy by analyzing celestial low-frequency radiation, such as radio waves. Hydrogen masers are used as extremely stable short-term time standards for scientific research and calibration of national time standards. Masers occur naturally in the gas nebulae surrounding certain stars and galaxies.

Semiconductors

Quantum theory provides another explanation of conduction. This theory takes into account the way atoms in a solid interact with each other. The inner electrons are strongly bound to the individual atoms, but the valence electrons interact to form *energy bands*. The wave-like valence electrons in their lowest energy states reinforce each other in what is called the *valence band*. For some elements, the electron waves at slightly higher energies interfere destructively, prohibiting the existence of electrons at those energies. This range of energies forms a *forbidden gap* for the element's valence electrons. At even higher energies, the electrons are free to escape from the atom in what is called the *conduction energy band*.

In conductors, the valence energy band overlaps the conduction band. Electrons can easily move from one orbital to another and from one atom to another if they are given a little energy. There is no forbidden band between the valence energy band and the conduction band. In an insulator, the valence energy level is nearly full of electrons. The combined interaction of the electron waves reduces the allowable energy states significantly. Therefore, a large forbidden gap exists between the valence band and the conduction band. For an electron to escape, it must gain a very large amount of energy. The amount of energy necessary to cause an insulator to readily conduct electricity is usually sufficient to melt or destroy the material first.

In a semiconductor, as in an insulator, the electrons nearly fill the valence energy level, reducing the magnitude of the valence energy band. Unlike an insulator, however, the forbidden gap is small. It is not too difficult to give electrons enough energy to reach the next conduction energy band. Therefore, semiconductors conduct after a moderate addition of energy.

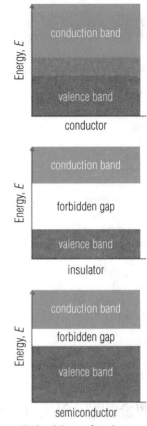

28-14 The band theory of conduction

28C Section Review Questions

1. **a.** Describe the four quantum numbers that are solutions to the Schrödinger wave equation.

   b. What do these numbers uniquely identify?

2. Why can't all the electrons in an atom occupy the ground state energy level?

3. Discuss why the second energy level of an atom in the second period might not contain all 8 electrons allowed by the quantum number rules for $n = 2$.

4. What is the outermost energy level being filled by atoms in the fourth period?

5. Which energy levels are incompletely filled in the element iron?

6. **a.** What is an optical maser called?

   b. What are the principle uses of some nonoptical masers?

7. How does quantum mechanics explain the mechanism of conduction and semiconductors?

In Terms of Physics

blackbody radiation	28.1	ground state	28.14
ultraviolet catastrophe	28.3	excited state	28.14
quanta	28.4	electron volt (eV)	28.14
quantized	28.4	spectral series	28.15
photons	28.5	Lyman series	28.15
photoelectric effect	28.5	Balmer series	28.15
work function	28.6	first ionization energy	28.15
cutoff frequency	28.7	orbital	28.18
Compton effect	28.9	electron cloud	28.18
plum-pudding model	28.10	Heisenberg uncertainty principle	28.19
nucleus	28.11	complementarity	28.20
nucleon	28.12	quantum number	28.21
mass number (A)	28.12	principal quantum number (n)	28.21
proton (p)	28.12	azimuthal quantum number (l)	28.21
atomic number (Z)	28.12	magnetic quantum number (m)	28.21
neutron (n)	28.12	spin quantum number (m_s)	28.21
isotope	28.12	Pauli exclusion principle	28.22
nuclide	28.12	periodic table	28.23
line spectrum	28.13	maser	28.24
electron shell	28.14		

Problem-Solving Strategies

28.1 (page 622) Follow these steps to determine all of the possible quantum numbers for a given energy level:
1. List all of the azimuthal numbers for the associated principal quantum number. The number of possible sublevels equals the numerical value of the energy level.

2. For each azimuthal number, determine the possible orbital numbers. There are $2l + 1$ orbitals for each l number.
3. For every combination of n, l, and m, there are two possible spin numbers.

Review Questions

1. What was the ultraviolet catastrophe?
2. What does the term *quantized* mean?
3. **a.** What is a photon?
   b. Is it a part of classical or quantum mechanics?
4. Describe the photoelectric effect.
5. Describe Thomson's "plum-pudding" model of the atom.
6. Describe Rutherford's model of the atom.
7. Name the major subatomic particles—or nucleons—in a nucleus.
8. A nitrogen ($Z = 7$) atom has seven neutrons. What would you call this isotope?
9. What were Bohr's postulates about electrons?
10. Why do we not notice the wave nature of ordinary objects?
11. Why can scientists not locate an electron precisely and know its momentum precisely at the same time?
12. What are valence electrons?
13. Must all sublevels of an energy level be filled before any in the next level are? Explain.

True or False (14–26)

14. The power of radiant energy from a blackbody emitter is proportional to the fourth power of the emitter's Kelvin temperature.
15. According to Wien's displacement law, as the temperature of a blackbody rises, the peak frequency of the radiant spectrum is lower.
16. Planck's blackbody spectrum equation was predicted by classical physics.
17. Photoelectrons from a specific metal come only in certain energies.
18. Rutherford's experiment demonstrated that the atom was mostly empty space.
19. Elements ^{16}N and ^{16}O are both nuclides of the same isotope.
20. Line spectra are characteristic of low-pressure glowing gases.
21. A line spectrum series consists of all possible wavelengths produced by electrons dropping *to* a specified energy level.
22. It is impossible to see the wave characteristic of normal forms of matter such as baseballs because their wavelengths are too long to be noticed.

23. An electron with a principal quantum number of 3 could have any of 14 combinations of quantum numbers.

24. Valence electrons are the electrons that occupy the highest energy sublevel in the atom, regardless of the energy level in which they are found.

25. Groups of elements on the periodic table have the same number of energy levels in the lowest energy state of their atoms.

26. Coherent waves have the same amplitude and the same frequency.

27. What is the peak wavelength of radiation emitted from a snowball ($t = 0.0$ °C) after it is placed in a dark, black box (i.e., no reflected light)?

28. Find the de Broglie wavelength of a 0.149 kg baseball traveling at 40.0 m/s.

29. How many electrons in one atom can have the quantum numbers $n = 1$, $l = 0$, $m = 0$, $m_s = \frac{1}{2}$?

30. How many electrons in one atom can have the quantum numbers $n = 2$, $l = 1$?

31. After conducting some research, write a paper or give an oral report on one of the following topics:
    a. the significance of Planck's constant to the order of nature
    b. the properties of a wave function
    c. how doping affects the conduction characteristics of a semiconductor
    d. a biography of one of the scientists mentioned in this chapter

Niels Bohr (1885–1962)

At the turn of the twentieth century, the world's leading researchers concentrated on the atom and its structure. Models were constantly being proposed, accepted, proven inadequate, and reformulated. By 1913, a twenty-eight-year-old Dane named Niels Bohr proposed the model on which our present-day concept of the atom is based.

Bohr was born in Copenhagen, Denmark. His father, a professor at the University of Copenhagen, encouraged him to pursue his scientific interests and abilities. Bohr's genius became apparent when he attended the university. His doctoral thesis has become a classic on the electron theory of metals. Upon graduation, Bohr journeyed to England to work under the famous researcher J. J. Thomson, but he was soon attracted by the experiments of Ernest Rutherford.

While working in Rutherford's laboratory, Bohr became interested in the radiation emitted by hydrogen atoms. To explain what he observed, Bohr applied the quantum theory to Rutherford's nuclear model of the atom. His radical new theory argued that atoms exist in stable energy states and that they emit radiation only during the transition between two of these states. Every scientist before that time had assumed that atoms emit energy continuously rather than in quanta. But only Bohr's new theory could explain both the spectra of light emitted by hydrogen atoms and the activity of an atom's outer electron shell.

He returned to his old university to direct the newly created Institute for Theoretical Physics. In 1922 he received the Nobel Prize in Physics for his atomic model. He was already a member of the Danish Academy of Sciences and was soon elected a foreign member of the Royal Society of London. However, Bohr did not rest on his past achievements. He continued to study the atom and proposed modifications to his model, concluding that no model could ever predict the exact location of an electron at any one time. This belief sparked a lifelong dialogue between Bohr and other leading scientists, such as Albert Einstein, who strongly disagreed.

When Germany overran Denmark during World War II, Bohr was forced to leave Copenhagen and seek refuge in England. He then journeyed to the United States to assist the scientists at Los Alamos. After the war he returned to his homeland, where, like many of the Los Alamos scientists, he was a lifelong opponent of nuclear war. Bohr is held to be one of the greatest thinkers in the history of science.

29A	Radiation and Radioactivity	629
29B	Radioactive Decay	636
29C	Nuclear Reactions	642
29D	Elementary Particles	647
Facet:	Radioactive Halos	639
Facet:	Cosmic Rays	653

New Symbols and Abbreviations

alpha radiation (α)	29.3	dose (D)	29.9
beta radiation (β)	29.3	roentgen (R)	29.9
gamma radiation (γ)	29.3	quality factor (QF)	29.9
positron (β^+)	29.5	dose-equivalent (DE)	29.9
activity (A)	29.9	sievert (Sv)	29.9
curie (Ci)	29.9	decay constant (λ)	29.10
becquerel (Bq)	29.9	half-life ($t_{1/2}$)	29.10
gray (Gy)	29.9		

The interior of a nuclear reactor of a commercial power plant during refueling. The reactor core is visible at the bottom of the photo. Note the faint blue glow of Cherenkov radiation surrounding a spent fuel module as it is removed from the reactor.

Nuclear Physics 29

In this chapter, you will study nuclear decays and fissions, in which susceptible atomic nuclei spontaneously eject particles or break up into smaller parts. The remaining fragments have higher entropy (show less order) than the original intact nucleus.

The above passage from I Kings describes the "fissioning" of the kingdom of Israel into two parts. The nation had started out as a strong, unified collection of twelve families. However, with time they became subject to God's judgment because of unconfessed and unrepented iniquity in their leadership that affected the entire nation. The splitting of Israel into the southern and northern kingdoms resulted in two smaller and weaker nations. Their spiritual power and influence continued to decline, and eventually Israel was carried away into the Babylonian captivity.

A Christian needs to confess his iniquities daily and repent from those things contrary to God's commandments. In this way he will avoid the "fissioning" of fellowship with his Savior.

29A RADIATION AND RADIOACTIVITY

29.1 Discovery of Radiation

Many important scientific discoveries were made by accident. For instance, Wilhelm Roentgen was studying electricity with a cathode-ray tube when he noticed that a nearby sample of barium platinocyanide was **fluorescing,** or emitting visible light. Materials fluoresce only after they have absorbed energy from light or other sources. The barium platinocyanide fluoresced even in a darkened room, as long as the cathode-ray tube was operating. Roentgen concluded that the cathode-ray tube emitted some previously unknown rays that excited the materials to fluoresce. He called the rays *x-rays.*

Henri Becquerel reasoned that if x-rays *cause* fluorescence, they may also be *emitted* in fluorescence. X-rays, like visible light, can expose film. Becquerel decided to wrap photographic film in black paper so that no light could enter. He

Wilhelm Roentgen (1845–1923) was a German physicist who in 1901 won the first Nobel Prize ever awarded in physics.

Antoine Henri Becquerel (1852–1908) was a French physicist who won the Nobel Prize in 1903 for his discovery of spontaneous radiation in uranium.

The terms associated with nuclear decays are often incorrectly used. *Radiation* consists of the particles and rays emitted by an unstable nucleus. *Radioactivity* is the act of emitting such particles and rays and is measured as the time rate of radiation emission. *Radioactive* is the property describing a material that emits radiation.

The change of one nucleus into another, which would effectively change the element, was the age-old aspiration of alchemists. Their goal was to *transmute* a "base" metal, such as lead, into precious metal, such as gold.

29-1 Marie Sklodowska Curie (1867–1934) was a French physicist. She was the first female professor of physics at the Sorbonne. She won two Nobel Prizes: in 1903 she shared the prize with two others (one was her husband) for her work on the spontaneous radiation discovered by Becquerel; in 1911 she won the Nobel Prize in Chemistry for her discovery of and investigations into the chemistry of radium and polonium.

intended to place a fluorescent material on the wrapped film and set the experiment in the sunshine. The Sun's ultraviolet light would cause the material to fluoresce, and any x-rays produced would expose the film through its protective wrapper.

Becquerel chose uranium potassium sulfate as a fluorescent material. The day he was to set the equipment in the sunshine was cloudy, so he put the uranium and film into a drawer to store it until a sunny day. The next several days were also cloudy. Finally, Becquerel decided to develop the film without putting it in the sunshine. He expected at the most a weak exposure, but he found a surprisingly strong image. Investigating further, Becquerel found that uranium will emit film-exposing radiation even if it has not been exposed to an external energy source for months or years.

Other scientists soon began studying this strange form of energy. Marie Curie discovered two new elements in uranium ore that produced even stronger photographic exposures. She named these polonium and radium—the former after her home country of Poland. She also coined the term **radioactivity** for the emission of these mysterious rays.

The short-term rates of emission of radiation have been found to be stable, remaining the same regardless of changes in the temperature, pressure, or chemical form of the radiator. Emission rates are found to depend only on the kind of atoms and the number present. Therefore, scientists concluded that radioactivity comes from the nucleus of an atom. A radioactive nucleus emits a particle or electromagnetic energy or both and may become another element. Radioactive nuclei are therefore unstable—they are likely to change into other kinds of nuclei.

29.2 Cause of Radioactivity

What makes a nucleus unstable? A better question is, "What makes a nucleus stable?" You learned in earlier chapters that nuclei are composed of positively charged protons and neutral neutrons. One positive charge repels another, and the electrostatic force is one of the strongest forces in nature. So it is astounding that up to ninety-two protons can be found in naturally occurring nuclei. What holds protons together in a nucleus?

Since protons do manage to remain together in a nucleus, physicists postulated a strong nuclear force that binds protons and neutrons together. This force must apply to all nucleons, regardless of charge. It must be stronger than the electrostatic repulsion of adjacent protons. Experiments have shown that this force has a range of 10^{-15} m. This distance is smaller than some nuclei.

In large nuclei, the electrical repulsion between protons on opposite sides of the nucleus is stronger than the strong nuclear force between the same protons. Therefore, those protons tend to repel each other. Neutrons surrounding each proton help to stabilize the nucleus by contributing to the strong nuclear force but not to electrical repulsion. Notice in Figure 29-2 the number of neutrons versus the number of protons for stable nuclei. For lighter nuclei, there are about the same number of neutrons as protons. For heavier nuclei, more neutrons are required per proton for stability. For nuclei with more than eighty-three protons, no number of neutrons has been found to provide stability. These unstable nuclei are radioactive.

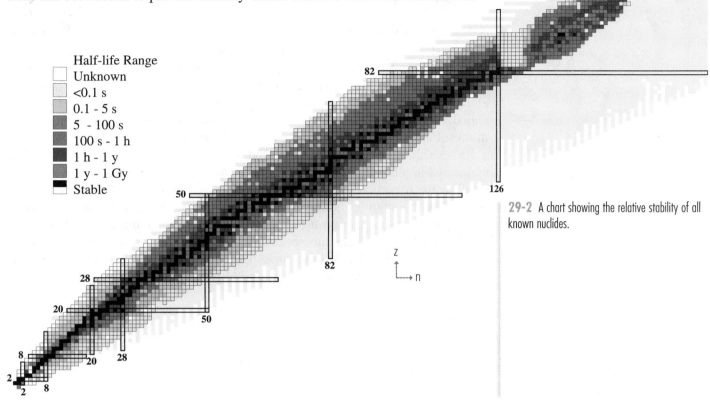

29-2 A chart showing the relative stability of all known nuclides.

29.3 Kinds of Radioactivity

Not all emissions from naturally radioactive materials are the same, as a simple experiment can demonstrate. Figure 29-3 shows a schematic of such an experiment set up in a vacuum. A lead block has a hole drilled in it, and a mixture of various radioactive materials is placed in the hole. Not far from the hole is a series of three sheets of photographic film. A lead plate with a small collimating hole is placed between the radioactive source and the first sheet of film. Between the block and the film is an electric field (a magnetic field could work as well). The sources are allowed to radiate, after which the film (sheet A) is developed. The film is exposed in four areas, apparently by four types of radiation deflected by the electric field. One kind of radiation is greatly deflected toward the positive plate. A second kind of radiation is deflected toward the negative plate an equal distance as the first. A third is slightly deflected toward the negative plate. The fourth kind of radiation is undeflected.

The four types of radiation also differ in penetrating power. If a piece of heavy paper is placed between the collimator and film B, the radiation that formed the slightly deflected spot on sheet A in Figure 29-3 does not appear on film B, while

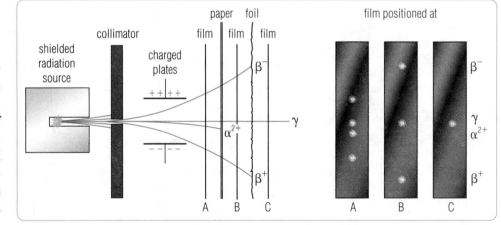

29-3 Alpha, beta, and gamma radiation, demonstrating charge and penetrating ability

the others do. This "missing" radiation is called **alpha radiation (α)**. If film C, which follows a thin sheet of metal foil, is developed, the highly deflected spots also disappear. These forms of radiation are called **beta radiation (β)**. The radiation that penetrated both coverings, forming the undeflected spot, is called **gamma radiation (γ)**.

the most penetrating

29.4 Alpha Decay

The least penetrating radiation, α-radiation, is deflected slightly toward the negative plate of the electric field. It was found to be made up of positively charged particles with a mass-to-charge ratio of about four thousand times the electron's mass-to-charge ratio. These massive positively charged particles are called *α-particles*. Rutherford studied them by placing a source of α-particles, uranium sulfate, in a thin-walled tube, or *ampule*. He sealed this alpha source within a glass vacuum tube containing electrodes (similar to Figure 29-4) that could be connected to a high voltage source. He left the tube unenergized for several days to allow the α-particles to accumulate as a low-pressure gas. Then, he energized the tube with a high DC voltage between the plates. The electric current through the α-particle vapor caused it to glow incandescently, emitting a line spectrum. Rutherford discovered that this spectrum exactly matched the line spectrum for the element helium. He decided that an α-particle was a helium atom stripped of its two electrons, or just a particle consisting of two protons and two neutrons.

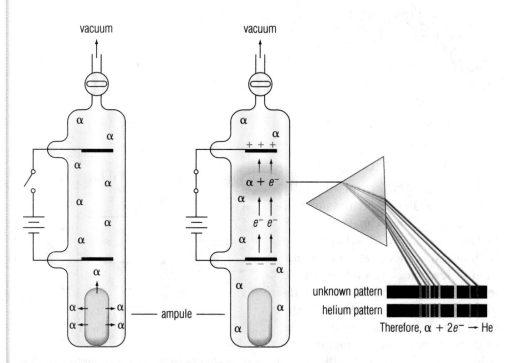

29-4 Incandescent alpha particles emit helium's line spectrum.

It should not be surprising that a nucleus is changed when it emits an α-particle. For instance, when a uranium nucleus loses the two protons in the α-particle, it is no longer the element uranium. This α-decay is expressed by the equation

$$^{238}_{92}\text{U} \rightarrow\, ^{234}_{90}\text{Th} + \,^{4}_{2}\text{He},$$

in which $^{4}_{2}\text{He}$ is the isotopic symbol for an α-particle. By this decay, uranium-238 changes to thorium-234. Notice that both mass and nuclear charge must be conserved during a nuclear decay.

In some textbooks, this reaction is written simply as

$$^{238}_{92}\text{U} \rightarrow\, ^{234}_{90}\text{Th} + \alpha.$$

In the above form, the conservation of mass and charge is not as obvious but is implied by what the α symbol represents.

29.5 Beta Decay

The next radiation, β-radiation, was found to be made up of particles with the same charge-to-mass ratio as electrons. However, the particles seemed to be split between those with the normal negative charge of an electron and a variety with a positive charge. All the evidence accumulated about β-particles indicated that they were negative and positive electrons. However, these electrons come from the nucleus. The negative nuclear electrons were named β⁻-particles and the positive electrons were named **positrons (β⁺).** Collectively, electrons and positrons are called beta particles. The evidence about β-decay shows that when a β⁻-particle is emitted, a neutron disappears and a proton is left in the nucleus. If a positron is emitted, a proton disappears and a neutron is left in its place. Beta decay, like alpha decay, changes the atomic number of the nuclide, and thus the element. For instance, carbon-14 decays to nitrogen-14 by the equation

$$^{14}_{6}C \rightarrow {}^{14}_{7}N + \beta^{-}, \text{ or}$$
$$^{14}_{6}C \rightarrow {}^{14}_{7}N + {}^{0}_{-1}e,$$

where $_{-1}^{0}e$ is a convenient symbol for an electron to demonstrate conservation of mass and charge in a nuclear reaction.

The symbol for a *positron* in a nuclear equation would be $_{+1}^{0}e$.

29.6 Gamma Decay

The most penetrating radiation, γ-radiation, resembles x-rays. For instance, our hypothetical experiment shows that γ-rays, like x-rays, are not affected by electric fields. Also, both x-rays and γ-rays can be diffracted by optically opaque crystalline materials. Therefore, γ-rays are assumed to be high-frequency electromagnetic radiation. The dividing line between x-rays and γ-rays is arbitrary in terms of energy and frequency. They are differentiated primarily by their source. X-rays originate in the electron cloud of large atoms when energetic free electrons collide with them. Gamma rays are generated by nucleons releasing excess energy as they transition from very high energy levels to very low energy levels. Gamma rays usually accompany α- or β-emissions, but they may occur as isolated events as well.

Since no nucleons are changed during a gamma emission, the element's isotope remains the same. Therefore, an atom may undergo gamma decays repeatedly without a change in its identity.

29.7 Natural Fission

A fourth possible, although naturally rare, form of radioactivity is **fission,** the splitting of a nucleus into two similarly sized nuclei, called *fission fragments,* and a few individual particles. Spontaneous fission occurs only in nuclei with very large mass numbers at statistically random intervals. Fission may also occur when energy is added to a large nucleus by the absorption of a gamma ray or speeding particles. Because susceptible nuclei are so far apart in natural ores, and the density of free neutrons and gamma rays (which themselves are produced from other fissions) is so low, fission rarely occurs in nature.

Fission can be produced artificially by bombarding certain nuclei with neutrons. We will discuss artificial fission used for power generation later in this chapter.

In 1972 mineralogists discovered the evidence of extensive natural fission within a uranium ore deposit in the small country of Gabon, located on the west coast of Africa on the equator.

29.8 Detection of Radiation

In order to study radioactivity, you must have some way of detecting radiation. Photographic film can detect large numbers of particles, but this method is cumbersome and time-consuming. More modern imaging systems use fluorescing screens, but the equipment is hardly portable. Several instruments developed early in the twentieth century can detect even single α- or β-particles and individual γ-rays. All these instruments depend on the principle that energetic radiation particles and rays can remove electrons from atoms as they collide with them. In fact, nuclear radiation is classified as **ionizing radiation** to differentiate it from nonionizing radiation, such as microwaves.

The basic alpha-beta-gamma detector is often called the *Geiger counter,* first invented in 1911 by Hans Geiger. The radiation detection component of the instrument consists of a specially designed electronic tube called a *Geiger-Mueller (G-M) tube.* It is a sealed, thin-walled glass chamber containing a metallic cylinder and a central wire conductor insulated from the cylinder. If it is designed to detect alpha particles, the end of the chamber will be sealed with a very thin mylar film instead of glass. The tube is filled with an inert gas. The detector is connected to a source of several hundred volts DC and to measuring circuits. In operation, when a particle or ray passes through the gas, some of the atoms are ionized. The ions and electrons are accelerated to the electrodes by the electrical potential, colliding with other gas atoms in the process and producing additional charges in a cascade effect. The large momentary current produces a pulse in the circuit. The pulses are counted over time and displayed as a rate on a meter. Usually an audio speaker installed in the instrument produces the characteristic clicking sound associated with Geiger counters. More rapid clicks indicate more ionizing events in the G-M tube. Figure 29-5 shows several modern beta-gamma portable detectors based on the Geiger principle.

Other portable and fixed radiation detectors have been designed for specific purposes. Some instruments have been designed to detect fast and slow neutrons. Such instruments are used to detect power levels in nuclear reactors. Portable neutron detectors must be shielded from other forms of radiation, so their detectors are usually contained in heavy cylindrical "pigs" of lead and are not easily carried.

Workers in areas containing radioactive sources must be able to conveniently monitor their exposure in real time. They carry instruments called pocket

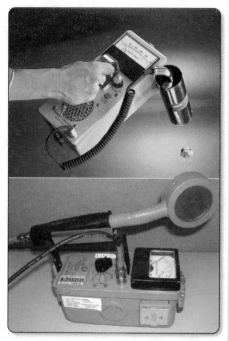

29-5 Portable radiation detection instruments. (a) This instrument can detect only beta and gamma radiation. (b) This instrument has a thin mylar window for detecting alpha particles.

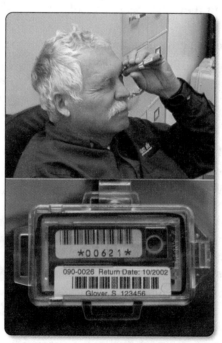

29-6 Two types of personal radiation detection instruments. (a) A worker checking a pocket dosimeter; (b) A thermoluminescent dosimeter (TLD), which must be read by a special machine.

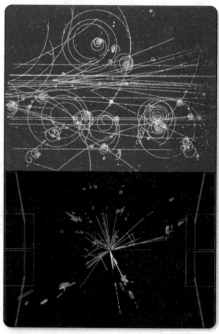

29-7 (a) Bubble chamber particle tracks; (b) Modern high-energy particle collider display

dosimeters that they can view as they work to see how much radiation they have been exposed to. For longer-term monitoring, workers also carry personal dosimeters that must be read by a machine.

In nuclear research, several types of instruments, in addition to ionizing detectors, have been developed to detect nuclear particles. The purpose of research detectors is to display the nuclear particle in such a way that its energy and momentum can be calculated. These detectors often are surrounded by magnetic and electric fields that produce spiral tracks (in the case of charged particles) for this purpose. A *cloud chamber* contains a cold gas near its vaporization point. Gases tend to condense around ions, so the gas will condense first along a particle's ion trail. A *bubble chamber* contains a liquid near its vaporization point. The liquid will vaporize first around ions, causing the particle to leave a trail of bubbles in the liquid. A *spark chamber* contains closely spaced conducting plates maintained at a high potential difference with respect to one another. A spark easily jumps along a trail of ions such as a particle leaves. The trails of particles in cloud, bubble, and spark chambers can be photographed for future study. These trails are much larger than the particles, just as the contrail—the cloudy trail that follows a jet—is often visible when the jet is too small to be visible. The newest instruments include digital detectors and processors that can reconstruct the track of a particle in three dimensions and display the track on computer terminals.

29.9 Units of Radiation and Radioactivity

Radioactivity and radiation can be measured in a seemingly bewildering array of units, depending on the property of concern. The most elementary measurement is the number of particles or rays produced per unit time. This *decay rate* is called the **activity (A)** of the sample. The older unit for activity is the **curie (Ci),** where one curie is the activity of about 1 g of radium, or 3.70×10^{10} disintegrations per second. The current SI unit for activity is the **becquerel (Bq),** where

$$1 \text{ Bq} = 1 \text{ disintegration per second.}$$

Decay rates say nothing about how much energy is deposited in material exposed to radiation. In order to quantify this, in 1975 a new SI unit was established called the **gray (Gy).** One gray is equal to one joule of energy absorbed by one kilogram of material. The number of grays quantifies the **dose (D)** of radiation received by a material.

The ionizing effect of electromagnetic radiation on a sample of air is measured with a unit called the **roentgen (R):**

$$1 \text{ R} = 25\,800 \text{ coulombs of charge per kilogram of air}$$

While the roentgen is of academic interest, scientists and health physicists are more interested in the effects of ionizing radiation on biological tissues. The ionization of biological molecules in cells damages or destroys cellular function. Some kinds of radiation cause more damage than others in living tissue because they travel short distances and deposit more energy per unit path length. For example, a highly penetrating γ-ray of a certain energy will deposit less energy per centimeter of path length through living tissue than a similar energy short-range α-particle. Health physicists have established a method to account for the differences in damage caused by radiation called the **quality factor (QF).** The magnitude of the quality factor is dependent on the kind of radiation, its energy, and the type of biological tissue receiving the dose of radiation. β-radiation and γ-radiation are assigned a QF around 1 since they deposit relatively little energy per unit path length, while α-radiation has a QF of about 20.

In nuclear physics, a single ionizing emission is considered a disintegration or decay of the nucleus, even if the emission is just a gamma ray.

The older unit of absorbed energy was the *rad (rad)*.

$$1 \text{ Gy} = 100 \text{ rad}$$

The Gy is equivalent to J/kg, or m^2/s^2 in base SI units.

Louis Harold Gray (1905–65) was a British physicist whose later career focused on medical physics and radiobiology.

TABLE 29-1

Quality Factors for Various Forms of Radiation

Radiation	QF
x-rays, β, γ	1
slow p, n (10^3 eV)	2–5
fast p, n (10^6 eV)	5–10
α-particles	20

The older dose-equivalent unit was called the rem (*r*oentgen *e*quivalent *man*). 1 Sv = 100 rem

The Sv is equivalent to J/kg, or m^2/s^2 in base SI units.

A measure of the comparative amount of damage is the **dose-equivalent (DE)** of the radiation. The DE is the product of the radiation dose and the quality factor of the radiation:

$$DE = D \times QF$$

Dose-equivalents are measured in SI units called **sieverts (Sv),** named after a Swedish radiologist who extensively studied the effects of radiation on living organisms.

The effects of radiation on organisms, especially people, are difficult to assess for a variety of reasons. One measurable effect that has been identified is the slight increase of cancer rates in large populations of workers exposed over periods of time to low levels of ionizing radiation. For this reason, workers in nuclear power plants, medical radiation personnel, crew members in nuclear-powered ships, nuclear weapons personnel, and nuclear materials processing personnel are closely monitored, and their exposure records are retained indefinitely in order to assess the long-term effects of radiation.

TABLE 29-2

Typical Dose-Equivalents for Various Radiation Exposures

Exposure	DE
Natural background radiation (cosmic, soil)	100–200 mrem/y (1–2×10^{-3} Sv/y)
chest x-ray (bone)	50 mrem (5×10^{-4} Sv)
dental x-ray (bone)	2 mrem (2×10^{-5} Sv)

The International Commission on Radiation Protection recommends a whole body dose-equivalent of no more than 0.500 rem (5.00×10^{-3} Sv) per year for the general public and 5.00 rem (5.00×10^{-2} Sv) per year for radiation workers.

Bone is considered the tissue most sensitive to radiation damage because, compared to other kinds of tissues, it has the least ability to replace damaged cells.

29A Section Review Questions

1. Why would it be incorrect to say, "This mineral is emitting strong radioactivity"?

2. Discuss why radioactive decay is a natural form of transmuting one element into another.

3. What seems to be the main factor in determining the stability of an atom's nucleus?

4. Why could breathing in a tiny particle of an alpha-emitting isotope that lodges in your lung cause more biological damage than receiving the same dose of gamma radiation over your whole body?

5. Discuss how beta decay changes the nucleus.

6. What is the main difference between gamma rays and x-rays?

7. a. Why is nuclear radiation classified as ionizing radiation?

   b. Describe the standard instrument used for detecting all types of ionizing radiation.

8. a. What is the SI unit for radioactivity?

   b. What is the SI unit for dose-equivalents?

⊙9. How many chest x-rays would it take to exceed the recommended annual dose of ionizing radiation for the general public?

29B RADIOACTIVE DECAY

29.10 Radioactive Decay Law

Naturally radioactive nuclei emit radiation at apparently unchangeable rates. For instance, 1 g of recent organic carbon typically has an activity of 0.231 Bq: about every four seconds, on the average, it emits a β-particle. If the carbon-14 in the sample continues to emit radiation for a long time, its atoms will eventually change to mostly nitrogen-14 mixed in with the stable carbon-12 atoms of the sample.

As it decays over time, the sample's activity decreases because there are fewer carbon-14 atoms to radiate. This means that the decay rate of the sample is proportional to the amount of original material present, or

$$\frac{\Delta N}{\Delta t} \propto N, \tag{29.1}$$

where N is the number of a certain kind of decaying nuclei, and $\Delta N / \Delta t$ is the rate of decay of the nuclei in disintegrations per unit time. A proportionality constant can be added to give us an equation:

$$\frac{\Delta N}{\Delta t} = -\lambda N \tag{29.2}$$

The factor λ is called the **decay constant.** The units of the decay constant are inverse time, such as s^{-1} or y^{-1}. The negative sign before the right-hand side indicates that the decay rate results in a decreasing number of decaying atoms with time. Note that the activity of the radioactive nucleus (A) is $|\Delta N / \Delta t|$ because activity is reported simply as disintegrations per unit time.

Radioactive decay is a *statistical* process. While it is fairly straightforward to predict how much of a sample of billions of nuclide atoms will change to another nuclide with time, there is no way to determine exactly when a particular nucleus will experience a decay. This aspect of nuclear decay seems to be completely random.

Through calculus, it can be shown that the number of nuclei remaining at a certain time after an initial number of nuclei begin to decay can be found from the equation

$$N = N_0 e^{-\lambda \Delta t}, \tag{29.3}$$

where N_0 is the original number of decaying atoms, and N is the number of atoms after a time interval Δt, which is the total elapsed time since the original sample N_0 began to decay. As Equation 29.3 shows, the number of undecayed nuclei remaining with time decreases exponentially. Figure 29-8 is a typical decay curve of a sample of radioactive nuclei.

The time it takes to reduce the original number of radioactive atoms by half is called the isotope's **half-life ($t_{1/2}$).** Clearly, this value is more useful than the decay constant for indicating how rapidly a particular isotope will decay away. An isotope's half-life can be determined from Equation 29.3 by calculating the time it takes for N to equal $N_0/2$:

$$N = N_0 e^{-\lambda \Delta t}$$

$$\frac{N_0}{2} = N_0 e^{-\lambda t_{1/2}}, \text{ or}$$

$$\frac{1}{2} = e^{-\lambda t_{1/2}} \tag{29.4}$$

Taking the natural logarithm of both sides of Equation 29.4 gives

$$\ln 1 - \ln 2 = -\lambda t_{1/2}.$$

Since $\ln 1 = 0$, we can rearrange the remaining terms to solve for $t_{1/2}$:

$$t_{1/2} = \frac{\ln 2}{\lambda}$$

$$t_{1/2} = \frac{0.693}{\lambda} \tag{29.5}$$

Equation 29.5 relates the half-life of a radioactive nuclide to the decay constant. The half-life is constant for a given nuclide.

The **decay constant (λ)** is characteristic of a specific isotope and *decay mode.* An isotope may decay via alpha-, beta-, or gamma-emissions, each of which has a different decay constant.

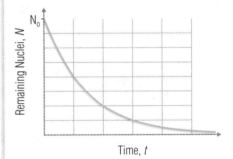

29-8 A radioactive decay curve showing the number of undecayed nuclei remaining with time.

Problem-Solving Strategy 29.1

Here is a brief review of the laws of logarithms:

If $n^y = x$, then $\ln_n x = y$
$\ln e = 1$
$\ln 1 = 0$
$\ln x \cdot y = \ln x + \ln y$
$\ln x/y = \ln x - \ln y$
$\ln x^y = y \ln x$

It can be demonstrated that Equation 29.3 is valid not only for the numbers of nuclides left over time but also for the decay rate or activity of the nuclides. Since activity is easier to measure than the actual number of nuclide atoms present in a sample, the equation's more useful form is

$$\frac{\Delta N}{\Delta t} = \left(\frac{\Delta N}{\Delta t}\right)_0 e^{-\lambda \Delta t}, \text{ or}$$

$$A = A_0 e^{-\lambda \Delta t}, \tag{29.6}$$

where A is activity in becquerels.

29.11 Radioactive Dating

Knowing the number of half-lives that have passed for a given sample of radioactive nuclide atoms can quickly indicate how much radioactivity is left. After one half-life there is half, $(\frac{1}{2})^1$, the original activity. After two half-lives there is $(\frac{1}{2})^2$, or $\frac{1}{4}$ the original activity. After three half-lives, $(\frac{1}{2})^3$, or $\frac{1}{8}$, and so on. Conversely, scientists claim that if the original activity of a sample is known or can be reliably estimated, the number of half-lives that elapsed to produce the activity observed in the present can be determined. This number of half-lives then determines the age of the sample since it was formed. This concept is the basis for **radioactive dating.**

For instance, it takes carbon-14's activity 5730 years to decrease by half ($t_{1/2} = 5730$ y). Therefore, if a sample begins with an activity of 0.5 Bq, after 5730 years its activity will be only 0.25 Bq. After another 5730 years the activity will halve again, to 0.125 Bq.

Because no physical or chemical change is known to affect a nuclide's half-life, radioactive nuclides (or *radionuclides*) are used to date objects. If a person knows the activity of the radionuclide at the "beginning" (A_0), how much activity there is now (A), and the radionuclide's half-life ($t_{1/2}$), he can calculate the time that has passed since the "beginning" (Δt) using Equation 29.6. Solving the equation is straightforward. Measuring the present activity of a radionuclide can be done with a variety of instruments, and the decay constants are known for nearly all possible radionuclides. The problem comes in determining the original activity of the radionuclide in a sample.

For most samples, the amount of the original radionuclide, the **parent nuclide,** is calculated from the amount of the radionuclide remaining in the sample today plus the amount of the **daughter nuclide,** the element that the radionuclide becomes after decaying. For instance, if you have a sample with ten grams of carbon-14 and ten grams of nitrogen-14, you might conclude that the sample originally contained twenty grams of carbon-14. The following discussion reveals why this conclusion could be wrong.

The problem of determining how much of the daughter nuclide was in the original sample is a serious one for those who use radioactive dating techniques. The normal way of resolving the problem is to assume that the elements have always had nuclides in the same proportion that they do today. For example, the element strontium (Sr) has nuclides 84, 87, 86, and 88 in a 1:13:18:147 ratio today. Suppose a geologist finds a sample that includes 1 g of rubidium-87 (Rb-87), which decays into strontium-87. In the same sample are 0.01 g of Sr-84, 0.63 g of Sr-87, 0.18 g of Sr-86, and 1.47 g or Sr-88. The other nuclides of strontium are in their normal ratio, but there should be only 0.13 g of Sr-87. Therefore, the geologist (or his colleague, a radioactivity expert) concludes that the "extra" 0.5 g of Sr-87 is **radiogenic** (was produced by radioactive decay). The radioactivity expert knows how much Rb-87 decays into 0.50 g of Sr-87; therefore, he can calculate the original amount of Rb-87 and the time it has taken to decay.

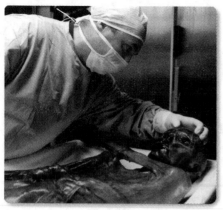

29-9 The age of the mummified remains of the "Ice Man" was determined using radiocarbon dating.

RADIOACTIVE HALOS

Radioactive halos in mineral crystals are spherical regions of discoloration surrounding a minute radioactive source of material called an *inclusion*. Alpha-decays of the radioactive nuclei in the inclusion send out alpha particles with several MeV (mega-electron-volts) of kinetic energy. As the alpha particles plow through their surroundings, they knock hundreds of thousands of electrons loose and cause damage to the crystalline structure. Since the alpha particles will eventually lose all their kinetic energy, their range depends on their starting energy. Different decays produce rings of different radii. The photo shows a Po-218 halo.

The decay chain of Po-218 includes three isotopes that emit alpha particles—Po-218, Po-214, and Po-210. Other nuclides in the chain undergo beta decay, but beta particles are not usually energetic enough to cause halos. This decay chain ends with

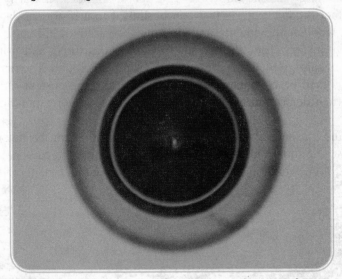

A Po-218 halo magnified 1450×. A radioactive inclusion in biotite mica (from Sweden) produces spherical regions of discoloration, which appear as circles in this photograph.

Pb-206, which is a stable nucleus with a long half-life. Thus there will be an alpha particle from the Po-218, one from the Po-214, and one from the Po-210. Experiments have determined the energies of alpha particles, which travel known distances before losing all their energy. Hence, the largest of the three rings shown in the picture goes with a 7.69 MeV alpha, the middle ring with a 6.00 MeV alpha, and the small ring with a 5.30 MeV alpha.

Robert V. Gentry, who is the scientist with the most experience in studying these halos, worked at Oak Ridge National Laboratory for over a decade. Gentry found that, because of the short half-lives of Po-218 (3.10 minutes) and Po-214 (163.7 microseconds), the existence of these halos was very difficult to explain using evolutionary timeframes. Halos form only in crystallized, nonmolten rock. In order for all three halos to form, the rock must have already been crystallized when Po-218 was placed there. If there had been any significant cooling period, because of its short half-life, virtually no Po-218 or Po-214 would have been left to generate any halo. The only reasonable explanation for this evidence is that the crystallized rock, along with the polonium inclusions, was created instantaneously rather than by uniformitarian processes.

Some scientists believe the polonium halos are really generated by uranium inclusions, since polonium is also in the decay chain of uranium-238. According to this theory, the daughter Po-218 migrated and formed halos at its new position. However, Gentry found no alpha particle pits and fission tracks in the crystals surrounding the polonium halos, which would have been expected if migration were true. Thus, Gentry showed that the polonium could not have originated from the migration of uranium atoms into the mineral crystal. Instead, he favors a Creation model in which the polonium was created in place in the rocks.

Reference: Robert V. Gentry, *Creation's Tiny Mystery*. (Knoxville, TN: Earth Science Associates, 1992).

In calculating the original amount of a radionuclide in a sample, it is necessary to make several assumptions:

1. The proportions of nuclides for each element have remained constant throughout the life of the sample.

2. The sample has not been contaminated with either the parent or the daughter nuclide originating from outside the sample (e.g., by diffusion or through transport by ground water).

The nuclear reaction equation for the production of carbon-14:

$$^{14}_{7}N + ^{0}_{0}\gamma \rightarrow ^{14}_{6}C + ^{0}_{+1}e$$

3. None of the parent or daughter nuclides have escaped from the sample (e.g., diffusion or leaching by ground water).

The second and third assumptions are often checked using the first assumption; that is, a person assumes that if any element entered or left the sample, it did so in the normal proportions.

29.12 Problems with Radioactive Dating

Keeping these assumptions in mind, let us examine carbon-14 dating. Carbon-14 is continually being produced in the atmosphere by cosmic rays interacting with atmospheric nitrogen-14, causing positron emissions. Radioactive carbon, like carbon-12, forms carbon dioxide and is incorporated into all living tissues. While the tissue is alive, it exchanges carbon with the atmosphere and so has the same proportion of carbon-14 as the atmosphere has. When an organism dies, it stops exchanging carbon with the atmosphere. Its own carbon-14 continues to decay. Theoretically, it is possible to determine when an organism dies by its present C-14 activity.

To use carbon-14 dating, you must know the atmosphere's C-14/C-12 ratio when the organism died. The present-day ratio is approximately 1.20×10^{-12}. However, the ratio has not been constant throughout the earth's history. It is possible to calibrate the dating technique for periods up to five thousand years ago. To do this, you must determine the age of an organic object by linking it to a specific date in history. For instance, you determine the age of a tree at the time it was felled or died in a historical context (building construction, battle, fire, etc.) by counting its rings (assuming that one ring is formed per year—not necessarily a good assumption). Other organic objects that can be linked to a historical date, such as textile fabrics, wooden implements, and mummies are used to calibrate the carbon ratio throughout history. Then the activity of C-14 is measured, and from this value the number of carbon-14 atoms present in the sample is found. The ratio of C-14/C-12 can thus then be calculated. Any deviation from the current ratio must be considered when determining ages in materials showing corresponding activities per gram.

This procedure is the reason why meaningful C-14/C-12 ratios can be determined for only the last five thousand-or-so years. Within that time the carbon-14 dating method can give somewhat reliable ages. However, before that time there is

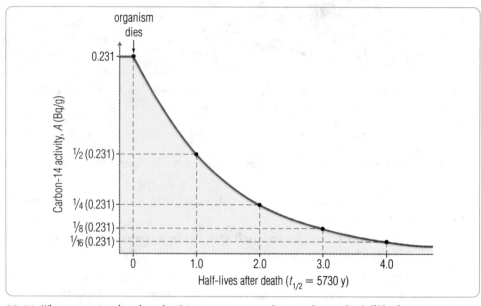

29-10 When an organism dies, the carbon-14 present in its tissues begins to decay with a half-life of 5730 y.

no way to determine age from the C-14/C-12 ratio of the atmosphere without making unsupportable assumptions, so carbon-14 dating does not give reliable dates prior to historical times. Of course, for all carbon-14 dates, an investigator must be careful to avoid contaminated samples and samples that do not represent equilibrium conditions with atmospheric levels of carbon-14 and carbon-12. For example, seawater contains dissolved calcium carbonate with almost no carbon-14. Thus, animals living in this water are surrounded by an environment with a C-14/C-12 ratio very different from that found in the atmosphere. Standard radiocarbon dating techniques have assigned living mollusks ages of thousands of years!

Many factors can affect the C-14/C-12 ratios in organic materials. Plants prefer C-12 to C-14, so plant material seems older than it actually is. Assumptions must be made about the constancy of the cosmic ray flux over time that produces C-14 from N-14 high in the atmosphere. The concentration of N-14 in the atmosphere may vary with time. Other factors, such as volcanic activity that introduces C-12 into the air and the burial of organic material that removes C-12 from the organic life cycle, must be considered as well. Both of these latter effects were significant at the time of the Genesis Flood. The Industrial Revolution was responsible for the introduction of additional carbon dioxide into the atmosphere. Atomic weapons testing in the twentieth century increased the concentration of C-14 above natural levels. As you can see, simple assumptions that would make C-14 dating reliable are not possible.

29.13 Geochronology: Determining the Age of the Earth

Carbon dating is valid only for formerly living things. Scientists resort to other dating methods when dating inorganic rocks and, ultimately, the age of the earth. Dating the earth's rocks by any method is called **geochronology.** Geologists may use index fossil, radiometric, or stratigraphic methods for obtaining relative or absolute ages for rock. One radiometric method uses the decay of potassium-40 to argon-40 (K-Ar method). This decay has a half-life of 1.3 billion years.

One difficult aspect about the potassium-argon method of dating is that the daughter nuclide, Ar-40, is a noble gas that does not chemically react with other substances and is, therefore, free to move around. Scientists assume that molten magma ejects gas atoms and molecules, or *degassifies,* as it reaches shallow depths or the surface, so there is no argon in the rock at the time it solidifies. They assume, then, that any argon found in the rock is radiogenic because they also assume that the argon atoms become trapped in the mineral crystal lattices after solidification. Neither of these assumptions is supported in laboratory experiments or field observations of recent eruptions. For example, the potassium-argon method has been used to date volcanic rocks that were observed to have formed within the past two hundred years. Due to the high concentrations of Ar-40, the age obtained was several million years—much older than the known age. These observations show conclusively how unreliable the potassium-argon dating method is.

Perhaps the best-known method for dating rocks is the uranium-lead method. Uranium-235 decays to lead-207, and uranium-238 decays to lead-206. Since both nuclides of uranium and lead usually occur in the same sample, they are used to check each other. Again, geologists assume that nonradiogenic lead nuclides have always occurred in the same proportions and that any lead-206 or -207 above the expected proportions is radiogenic. The age of the rock is calculated from the decay of both uranium nuclides, and the ages obtained from the two methods are compared. They are often significantly different.

A serious problem with the uranium methods is that uranium dissolves in acid solutions, but lead does not. This fact is used to purify uranium in the mining

By mentioning the various methods of *geochronology,* we are not giving credence to any of them. These are methods utilized by the scientific community based on the prevailing old-Earth evolutionary model. *Index fossil* dating assumes that certain fossils are representative of specific biological epochs in the geologic column. *Stratigraphic* dating depends on the uniformitarian assumption that identifiable rock layers (strata) were laid down in a specific order keyed to the geologic column. *Radiometric* dating uses ratios of parent and daughter radionuclides to obtain absolute ages for rocks.

29-11 Lava rocks from eruptions of Mt. Ngauruhoe in New Zealand within the past 60 years have been dated as old as 3.5 million years using the K-Ar method.

process. Ground water is often slightly acidic, and uranium samples are not isolated from ground water. There is a possibility that significant amounts of uranium are carried away from the sample in ground water.

As you can see, there are significant technical difficulties with just obtaining valid ages from radionuclide dating techniques. Other issues surrounding the process include the lack of scientific honesty when interpreting the dates. Many cases have been identified by both Creationist and non-Creationist scientists in which valid age data was discarded by researchers because it did not agree with their preconceived notions of what the date *should* be. If all the data were considered, the computed age of a rock sample might be a lot younger, a lot older, or indeterminate. The only scientific conclusion one can make is that radioactive dating of the earth's rocks will not produce meaningful ages unless the underlying assumptions can be assured in some way and the data is honestly interpreted.

Nowhere else in naturalistic science do presuppositions intentionally displace objectivity more than in *geochronology*—dating the age of the earth. Biological evolution presumes billions of years for the first glimmer of cellular activity to be recorded in the geologic column. Accordingly, geologists (supported by cosmologists) provide the data that supports a 4.5 billion-year age of the earth.

29B Section Review Questions

1. True or False: The activity of a specific isotope is independent of the amount of the isotope.

2. How does the amount of a radioactive isotope in a sample vary with time?

3. How does isotope half-life vary with the magnitude of the decay constant?

4. Discuss the assumptions that are made when using a radioactive dating technique.

5. Why is radiocarbon dating useful only for objects less than about 5000 years old?

⊚6. Write the nuclear decay equations for the alpha decay of U-238 followed by beta decays to U-234.

⊚7. How many years would it take a sample of highly radioactive cobalt-60 to decay to 1/1000 of its initial activity? The half-life of Co-60 is 5.3 y.

⊚8. Assuming that the C-14/C-12 ratio has not varied over recorded history, about how old is an axe handle that emits 1.6×10^{-1} Bq/g? Refer to Figure 29-10.

29C NUCLEAR REACTIONS

29.14 Nuclear Energy

The mass of a nucleus is less than the sum of the masses of the individual nucleons that compose the nucleus. The difference in mass, called the **mass defect,** is converted to other forms of energy when the nucleus is assembled. To separate the nucleons, energy at least equal to the mass defect must be supplied to it. This energy is also called the *binding energy* of the nucleus. The binding energies of different nuclei may be compared by dividing the total binding energy by the number of nucleons. The binding energy per nucleon is plotted versus the number of nucleons in Figure 29-12. In a nuclear reaction, binding energy in the form of mass may be converted to other forms of energy or vice versa, but total energy is conserved.

Many nuclear reactions can be forced to occur. Often the purpose of the reaction is to produce useful energy converted *from* mass. Such a reaction is called

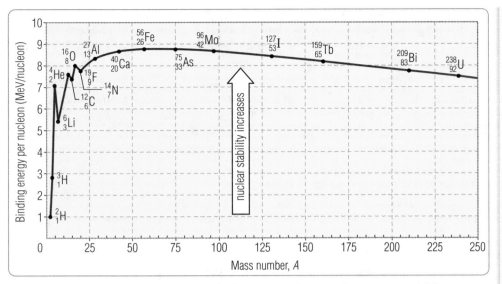

29-12 Graph of binding energy per nucleon. Note that the maximum energy per nucleon (greatest mass defect) occurs when $A \approx 60$. This means that atoms in this range have the greatest nuclear stability.

exoergic. **Endoergic** reactions, on the other hand, *produce* mass at the expense of other forms of energy.

How can you predict whether a reaction is exoergic or endoergic? Look at Figure 29-13. Find the reactants and the products on the graph. If the products are *uphill* from the reactants, the reaction is *exoergic*. This means that the products have less total mass than the original nucleus due to the higher binding energy per nucleon. The mass difference was released as energy in the reaction. If the products are *downhill* from the reactants, the reaction is *endoergic*.

There are two types of exoergic reactions, fission and fusion. In fission, a large nucleus splits into smaller nuclei. This reaction is exoergic as long as both the reactant and the product nuclei have mass numbers greater than about 60. Fission reactions start with reactants far to the right side of Figure 29-13, and products are generally to the right of the peak of the graph. In **fusion,** two or more small nuclei merge to form a larger nucleus. The reactants are at the far left end of the curve in the figure, and products are to the left of the curve's peak. This reaction is exoergic as long as the reactants and the product have mass numbers less than about 60.

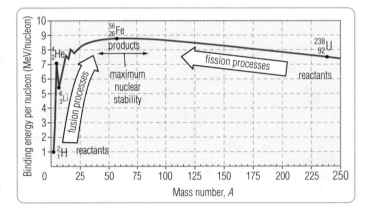

29-13 Fission and fusion products are uphill from reactants on a binding energy per nucleon graph.

29.15 Fission Reactions

At our present state of technology it is easier to force a fission reaction than a fusion reaction. Some materials—notably U-235 and Pu-239—undergo fission when they are bombarded with neutrons. The nucleus absorbs a neutron, then splits. The fission produces more neutrons. If there are enough fissionable nuclei, it is likely that at least one of the new neutrons will strike another nucleus and cause it to split. A series of sequential fissions caused by fission neutrons is called a **chain reaction.** Since each U-235 fission produces an average of 2.47 neutrons, the potential exists for the number of fissions to increase with each step (see Figure 29-14). The minimum amount of pure material containing fissionable nuclei necessary to sustain a chain reaction is called a **critical mass.** If a chain reaction produces a constant number of fissions per unit time, the reaction is said to be *critical,* or *at criticality.* If the reaction rate increases with time, it is said to be *supercritical;* if it decreases with time, the reaction is *subcritical.*

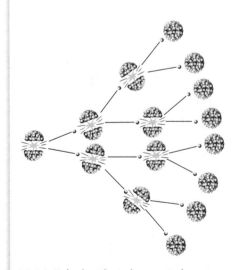

29-14 Under the right conditions, a single uranium fission can produce many fissions in a supercritical chain reaction during a brief interval of time.

The time delay between the production of a neutron from fission until the next fission occurs is on the order of micro- to nanoseconds. Therefore, reaction rates can increase very rapidly (or conversely, die out rapidly) if appropriate conditions exist in the fissionable material.

When a fast neutron elastically collides with a hydrogen nucleus in a water molecule, classical mechanics shows us that the neutron's rebound velocity is reduced to as little as half of what it was before the collision. This occurs because the neutron's original momentum is divided between two particles with similar masses (a lone proton has nearly the same mass as a neutron). Collisions with heavier nuclei leave the neutron with more of its original velocity, which means many more collisions are required to slow it to thermal energies. In this case, the neutron has a greater potential to "leak out" of the fuel without causing a fission.

Most U.S. nuclear power plants operate by the fissioning of uranium-235. This reaction begins when a U-235 nucleus captures a neutron and undergoes fission, producing more neutrons. Natural uranium ore contains about 0.7% U-235, and the remaining 99.3% is U-238. U-238 does not fission under neutron bombardment. The uranium used as fuel for commercial power plants is a mixture of about 3% U-235 and about 97% U-238. The percentage of U-235 is artificially increased through a very expensive chemical and mechanical process called *enrichment.* The U-238 does not undergo fission, but it does absorb neutrons. U-238 is more likely to absorb fast neutrons than U-235, while U-235 is more likely to absorb slow, or *thermal,* neutrons. Most of the neutrons produced by fission are fast. In order to slow the neutrons down so that they are more likely to cause a fission in U-235, a **moderator** material is interspersed among the uranium to slow the neutrons. In most power reactors, the moderator is simply water, although some older designs use carbon or some other material.

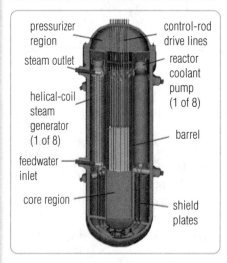

pressurizer region
steam outlet
helical-coil steam generator (1 of 8)
feedwater inlet
core region

control-rod drive lines
reactor coolant pump (1 of 8)
barrel
shield plates

29-15 A cutaway of a nuclear reactor showing the location of the fuel and control rods, reactor vessel parts, and cooling water flow

With water as a moderator, a critical chain reaction can be established and easily sustained. The goal is to get exactly the same number of fissions from each preceding "generation" of fissions. A few more, and the reaction rapidly runs out of control. A few less, and the reaction quickly dies out. The mass of uranium fuel exposed to the neutron flux in a reactor is the critical mass, just enough to sustain the critical reaction. The water moderator contains hydrogen atoms that quickly slow down the fast fission neutrons to thermal energies. Some neutrons leak out of the reactor, but enough are reflected back into the fuel to sustain the reaction. If too many fissions occur, producing higher temperatures in the water, the water density decreases, allowing more neutrons to leak out and causing the reaction to go subcritical. Then, as the water cools, its density increases slightly, reflecting more neutrons and increasing neutron level back to criticality. This "feedback" process is very rapid and makes reactors stable.

A mechanical device that contains an appropriate arrangement of nuclear fuel and moderator is called a **nuclear reactor.** In order to establish criticality in a reactor and to compensate for fuel-atom burnout as the atoms are depleted by fission, **control rods** are used. They are made of an alloy of hafnium, cobalt, or boron that has a high affinity for thermal neutrons but will not fission itself. Neutrons removed by these nuclear "poisons" cannot cause another fission. The rods are remotely operated and can be manually raised, lowered, or automatically and rapidly inserted to shut down the reactor in an abnormal situation.

Another kind of fission reactor uses plutonium-239. Pu-239 is mixed with U-238. Both Pu-239 and U-238 absorb fast neutrons, so there is less need of a moderator. When Pu-239 absorbs a neutron, it splits, producing energy and more neutrons. Some neutrons strike other Pu-239 atoms, while other neutrons are absorbed by U-238. The resulting U-239 soon undergoes β-decay to form Pu-239. If the reactor is designed to produce more Pu-239 than it consumes, it is called a **breeder reactor.** Breeder reactors were used in the United States to produce nuclear weapon–grade plutonium. Since nuclear weapons programs have been curtailed in the U.S., breeder reactors have been deactivated in this country. In other countries, such as France, breeder reactors are used to produce plutonium for power reactors because fissionable uranium is scarce.

An *uncontrolled* fission reaction is the basis of the nuclear bomb. Where a nuclear reactor uses just the critical mass of a U-235/U-238 mixture and sustains a critical reaction, a bomb requires a *supercritical* mass of nearly pure (>99% enriched) U-235 or Pu-239. To avoid a spontaneous nuclear reaction, the fissionable material is separated into two or more subcritical masses. When the bomb is to be detonated, high explosives force the masses together. The rapidly expanding chain reaction instantaneously produces far more neutrons than are needed to sustain just a critical reaction. This *prompt-critical* reaction releases immense amounts of energy in a brief moment.

29.16 Fusion Reactions

Fusion reactions usually involve hydrogen or helium as the reactant. In one fusion process, four hydrogen nuclei go through several steps to emerge as a helium nucleus. Other reactions produce lithium, another light nucleus. This reaction releases huge amounts of energy. Because of the energy involved and the abundance of hydrogen and helium in the Sun, most scientists believe that this fusion reaction produces the Sun's energy. The hydrogen-helium fusion converts much more mass into energy than fission reactions do. However, scientists have not yet found a way to produce usable power from the reaction.

The "energy of the future" is thought to be fusion reactions. Researchers are trying to produce a self-sustaining fusion reaction that lasts for more than a few milliseconds and that produces usable energy. The main obstacle is that to induce fusion, the light nuclei must be heated to millions of kelvins. No known material has a melting point of more than a few thousand kelvins; therefore, the superheated plasmas cannot be held in a material container. Researchers are using two methods to initiate a fusion reaction. One is creating a high-temperature plasma of hydrogen or helium and containing the ions in a magnetic "bottle." A **tokamak** is a fusion reactor built on this principle. The other method involves forcing tiny pellets together with such force that fusion takes place. These *inertial fusion* reactors use lasers to drive the nuclei together so rapidly that they cannot escape before the reaction occurs. No reactor has succeeded in generating more energy than it

29-16 The first successful test of a fission bomb at Alamogordo, New Mexico in 1945 was code-named "Trinity."

Is there such a thing as a chain reaction of witnesses? This may be what Paul had in mind when he wrote to Timothy, "And the things that thou hast heard of me among many witnesses, the same commit thou to faithful men, who shall be able to teach others also" (II Tim. 2:2). If one Christian could start this (i.e., win one person to the Lord each month, and those people did the same), in 33 months the entire population of the world would be converted in a miraculous chain reaction of regeneration.

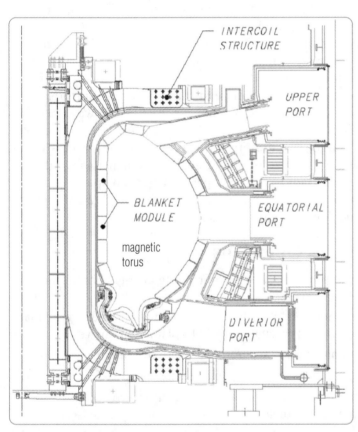

29-17 A cross-section schematic of one-half of the proposed ITER fusion reactor showing the chamber containing the magnetic "bottle"

INTERCOIL STRUCTURE

UPPER PORT

BLANKET MODULE

magnetic torus

EQUATORIAL PORT

DIVERTOR PORT

29-18 The first thermonuclear bomb tested by the United States was 500 times more powerful than the "Trinity" test shown in Figure 29-16.

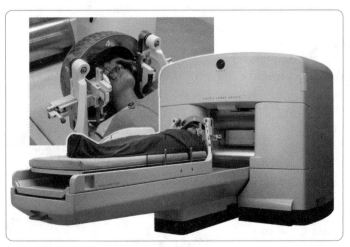

29-19 A Gamma Knife machine uses multiple beams of gamma rays aimed at a precisely located point in a tumor within the patient's head.

consumes to initiate reaction, although scientists believe they are close to attaining this important milestone.

Like fusion reactors, fusion bombs work on the principle of creating heavier nuclei out of lighter. These *thermonuclear* devices employ a complicated process of first creating deuterium and tritium nuclei (^2H and ^3H, respectively) powered by the detonation of a miniature fission bomb within the warhead, then driving these nuclei together under extremely high temperatures and pressures generated by x-rays emanating from other materials in the warhead. A fusion detonation then occurs. The whole reaction takes place in less than a few billionths of a second, producing hundreds of times more energy than a fission bomb of equivalent mass.

29.17 Nuclear Medical Tools

Medicine, as well as energy production, has benefited from the discovery of radioactivity. Gamma therapy for cancer uses γ-rays emitted by cobalt-60 sources to kill cancer cells. A machine directs the γ-rays at the cancer. Nearby healthy tissue is shielded with lead or the beam is collimated to deliver radiation precisely to the tumor. The radiation doses are usually spread out over time to kill the cancer without causing radiation sickness.

Since radionuclides have the same chemical properties as stable nuclides, they are absorbed by the body in the same way. For instance, iodine is absorbed mostly by the thyroid. If a doctor suspects that the thyroid is working improperly, he may give the patient a dose of radioactive iodine. The doctor then traces the iodine with a radiation detector. If the proper proportion does not arrive at the thyroid, the doctor's suspicion is confirmed. Radionuclides used in this way are called **tracers.** In a similar way, radioactive "bullets" can be created that target specific organs or tissues where cancers occur. These bullets are specially formulated compounds containing intensely radioactive isotopes that bind to specific sites on cancer cells. The radiation kills the sensitive cancer cells while doing little damage to healthy cells.

29C Section Review Questions

1. Explain why isotopes with the greatest binding energy per nucleon have the most stable nuclei.

2. Do exoergic reactions yield product nuclei with a total mass more than or less than the original reactant nuclei? Explain your answer.

3. Why would an unrestrained supercritical chain reaction be hazardous?

4. What is the purpose of a moderator in a reactor that uses U-235 for fuel?

5. If U-238 doesn't fission when it absorbs a neutron, why is the majority of the fuel in a uranium-fueled reactor U-238?

6. a. Discuss the differences between fission and fusion reactions.
   b. Which produces more energy per reaction?

7. What is the greatest technological hurdle in developing fusion power?

●8. If a radionuclide having a half-life of 18.0 h were injected into a patient for medical therapy, how long would the doctor have to retain the patient under observation in order to reduce the radionuclide activity remaining in the patient's body to 1/100th of its initial level? (This is done to minimize exposure to family members and others who might come into close proximity with the patient.)

29.18 Early Evidence of More Subatomic Particles

energy is conserved in nuclear reactions

In some nuclear reactions, it appeared at first that energy was not conserved. For instance, a β-particle does not have enough kinetic energy from a beta decay to account for the difference in mass between a neutron and the resulting proton and electron β-particle. (The proton remains fixed in the nucleus, so only the β-particle has kinetic energy.) Because the conservation of energy had not been violated before, Pauli suggested in 1931 that there may be another particle emitted in a β-decay. He proposed that this neutral particle should have no rest mass and would account for the "missing" energy. The following year, Enrico Fermi named this mysterious particle the *neutrino* ("little neutral one"). How could experimenters confirm or disprove Pauli's suggestion? Unknown to researchers, neutrinos could travel through several light-years of matter before being absorbed. They left no tracks in the standard cloud, bubble, and spark chambers available at the time. Eventually, researchers designed an experiment that detected neutrinos produced by a nuclear reactor. The first neutrino interactions were detected in 1956. In this way, twenty-five years after it was first predicted, the existence of the neutrino was confirmed.

29.19 Nuclear Building Blocks

Nuclear reactions and particle accelerators have shown that more than 300 **elementary particles,** particles smaller than atoms, may exist. These particles include the familiar proton, neutron, and electron, but some of these particles are themselves made of even more basic particles. Just as with whole nuclei, elementary particles enter into reactions with each other and decay as well.

Physicists use two basic rules when studying these particles. First, every reaction will happen at some time unless it is prohibited by a conservation law. Even if a particular reaction has never been observed, physicists presume that if no known conservation law prohibits it, a reaction can and does occur. Thus, physicists have found that the classical conservation laws governing energy, momentum, angular momentum, and electric charge apply to the elementary particles too.

The second rule of elementary particles is that every particle has an **antiparticle;** that is, for every particle there is another particle of the same mass and spin, but its charge and magnetic moment are opposite. For example, physicists studying nuclear decays discovered the positron, a particle with the same mass as an electron but with opposite charge. The positron could also be called the *antielectron*. The symbol for an antiparticle is the particle's symbol under a bar. If the particle is charged, the antiparticle symbol bears the opposite sign, as well. For example, the symbol for an antiproton is $\bar{p}$.

When a particle collides with its antiparticle, both completely disappear in a burst of energy equivalent to their rest masses plus whatever kinetic energy they possessed. This process is called *pair annihilation.*

It is theoretically possible for antimatter—atoms consisting of antiprotons, antineutrons, and positrons—to exist. Anti-hydrogen was even synthesized at the CERN particle physics laboratory in 1995. However, if antimatter meets ordinary matter, both will be annihilated. Naturally occurring antimatter has not been observed anywhere in the universe—only "normal" matter. This poses a serious problem for supporters of the big bang origin of the universe. Presuming that all matter in existence originated with the condensation of particles from the intense energy "soup" shortly after the big bang occurred, there should have been an equal number of particles and antiparticles formed. Since these particles were uniformly

It is interesting to note that there are far more elementary particles forming the nucleons of the atom than there are elements forming all of matter!

The opposite process to pair annihilation is *pair production.* This occurs when a high-energy x-ray photon passes close to a nucleus and disappears, producing an electron and a positron pair.

mixed together, many (one would think, most) particles should have been quickly eliminated in pair-annihilation events, leaving nothing but high-energy photons in a dark universe. This obviously is not the case. Therefore the only reasonable explanation is that the universe was created with only "normal" matter.

29.20 Elementary Particle Zoo

Since there are so many elementary particles, scientists needed a way to classify them. Particles have been classified according to rest mass and spin. The category with the least rest mass contains only one particle, the photon, which has zero rest mass and a spin of 1. The next category contains the **leptons,** or lightweight particles, which all have a spin of $1/2$. The electron, muon, tauon, and all neutrinos are leptons. Including their antiparticles, there are twelve leptons.

The heaviest particles are contained within the **hadron** group. This is subdivided into two subgroups. The **mesons,** or medium-weight particles, have a spin of 0 or 1 and rest masses that are intermediate between electrons and nucleons. They include pions, kaons, and etas. The **baryons,** the heaviest particles, have a spin of $1/2$ or $3/2$. Protons, neutrons, lambdas, sigmas, xis, and omegas are baryons. Each of these particles has an antiparticle. There is also a host of extremely short-lived particles that have not been mentioned. Table 29-3 lists some properties of the major leptons and hadrons.

TABLE 29-3

Characteristics of Selected Elementary Nuclear Particles[†]

Family Name	Particle Name	Particle Symbol	Antiparticle Symbol	Rest Energy (MeV)	Particle Lifetime
Leptons	Electron	e^-	e^+	0.511	$>4.6 \times 10^{26}$ y
	Muon	μ^-	μ^+	105.7	2.20×10^{-6} s
	Tauon	τ^-	τ^+	1777	2.91×10^{-13} s
	Electron neutrino	ν_e	$\bar{\nu}_e$	$<3 \times 10^{-6}$	
	Muon neutrino	ν_μ	$\bar{\nu}_\mu$	<0.19	
	Tauon neutrino	ν_τ	$\bar{\nu}_\tau$	<18.2	
Hadrons					
Meson	Pion (Light, unflavored)	π^-	π^+	139.6	2.6×10^{-8} s
		π°	π°	135.0	8.4×10^{-17} s
	Kaon (Strange)	K^+	K^-	493.7	1.2×10^{-8} s
		K°	$\bar{K}^\circ$	497.6	
Baryon	Proton	p	$\bar{p}$	938.3	$>1.6 \times 10^{25}$ y
	Neutron	n	$\bar{n}$	939.6	885.7 s
	Lambda	Λ	$\bar{\Lambda}$	1116	2.6×10^{-10} s
	Sigma	Σ^+	$\bar{\Sigma}^-$	1189	0.80×10^{-10} s
		Σ°	$\bar{\Sigma}^\circ$	1193	7.4×10^{-20} s
		Σ^-	$\bar{\Sigma}^+$	1197	1.5×10^{-10} s
	Xi	Ξ°	$\bar{\Xi}^\circ$	1315	2.9×10^{-10} s
		Ξ^-	$\bar{\Xi}^+$	1321	1.6×10^{-10} s
	Omega	Ω^-	Ω^+	1672	0.82×10^{-10} s

[†]K. Hagiwara *et al.*, Phys. Rev. D **66**, 010001 (2002) and 2003 off-year partial update for the 2004 edition available on the PDG WWW pages (URL: http://pdg.lbl.gov/)

Only a few of the elementary particles are stable. Even the neutron, a major part of the atom, has a half-life of only 14.8 minutes when it is not combined with protons. Protons and electrons have such extremely long half-lives that they are considered stable. That is, their half-lives are many orders of magnitude longer than the known age of the universe by anyone's reckoning. Why are protons and electrons stable? No one really knows, although scientists have postulated conservation principles to account for this puzzling fact. However, if protons and electrons were not stable, atoms would not be stable, and the universe would soon disintegrate. God has arranged atoms so that the universe will continue as long as He wants it to.

29.21 Quarks and Strange Flavors

After scientists discovered the existence of elementary particles, they believed that they had discovered the basic building blocks of the universe. (Does this sound familiar?) They noted that every elementary particle was a combination of two or more other elementary particles. In other words, there was no difference between "fundamental" elementary particles and "composite" elementary particles because all were components of each other.

Then in 1963, scientists Murray Gell-Mann and George Zweig suggested that the hadron elementary particles could be subdivided into even more fundamental entities that are now called **quarks.** There are six kinds, or *flavors,* of quarks having definite properties. In order to distinguish the flavors, scientists chose some whimsical names for their distinctiveness rather than descriptiveness of any given property. The six flavors in order of increasing mass are *up, down, strange, charm, top* (or *truth*), and *bottom* (or *beauty*). The leptons do not appear to have any internal structure, which qualifies them as truly fundamental particles along with quarks. The names of leptons are now called flavors as well. Table 29-4 lists the quark flavors and their charges. Note that every quark has its associated antiquark with an opposite fractional charge of the same magnitude as the quark.

The up, down, and strange quarks were discovered first. When combined they form the hadron particles. Scientists recognized that since some of the baryons were charged, the quarks must have fractional charges in order to combine in such a way that they produce unit elementary charges in the larger particles. Figure 29-20 shows how several of the hadrons are formed by various quarks.

The individual masses of quarks are determined by complicated calculations that take into account their kinetic energies in particle-photon interactions. Even though they cannot be observed directly, their existence has been experimentally verified repeatedly through the application of conservation laws in high-energy particle experiments.

Quarks can exist only inside of hadrons. They cannot exist independently like electrons and the other leptons can, no matter how much energy is expended to tear the hadron particle apart. This theory is called **quark confinement.** The forces that hold quarks tightly within the hadron particles are described by a property called *color charge,* or just *color.* The color charges are analogous to electric charges between the quarks and there are three of them: red, blue, and green. The color charges between quarks may be the source of the strong nuclear force between hadrons (protons and neutrons) in the nuclei of atoms.

Are there even more basic particles than quarks? Scientists today will not make any predictions one way or the other. Christians can be sure that, ultimately, it is God that sustains the universe through His providence manifested in these subatomic particles.

Not All Hadrons are stable elementary Particles. for example Neutrons

Murray Gell-Mann (1929–) is an American physicist who won a Nobel Prize in 1969 for his work on elementary particles.

George Zweig (1937–) is an American physicist and neurobiologist.

TABLE 29-4		
Quarks		
Flavor	**Symbol**	**Charge**
up	u	$+\frac{2}{3}e$
down	d	$-\frac{1}{3}e$
strange	s	$-\frac{1}{3}e$
charm	c	$+\frac{2}{3}e$
top	t	$+\frac{2}{3}e$
bottom	b	$-\frac{1}{3}e$

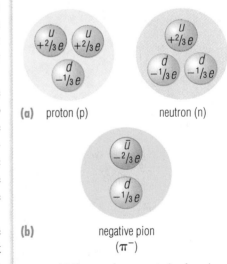

29-20 (a) Three quarks are required to form the baryons, such as the proton and neutron. (b) A quark and antiquark are required to form the mesons, such as the negative pion.

29.22 Exchange Particles

All the forces in nature have been placed in four categories: gravitational, weak nuclear, electromagnetic, and strong nuclear forces, in order of increasing strength. Each kind of force can be understood in terms of an **exchange particle**. If one object transfers an exchange particle to another object, there is a force between the two objects.

The weakest force is the *gravitational force*. If the strong force had a strength of 1, gravity would have a strength of 10^{-45}. However, gravity's range is unlimited. The predicted exchange particle for gravity is the massless, uncharged *graviton,* which has not been observed. The force of gravity is the only force that affects all particles. In spite of its relative weakness, however, gravity is most important in long-distance interactions. For example, gravity holds planets, solar systems, and galaxies together.

The *weak nuclear force* is much stronger than gravity but still much weaker than the strong nuclear force. If the strong force were 1, the weak force would be 10^{-8}. The range of the weak force is the shortest range—only 10^{-17} m, or less than 1/1000 the diameter of the nucleus. The exchange particle for the weak force was predicted to be the heavy *W-particle*. The W-particle was first detected in the 1980s. This particle is more than 100 times more massive than the proton. It comes in three varieties: W^+, W^-, and Z^0, with charges of $+e$, $-e$, and 0 respectively. The weak force is involved only in the transmuting of particles that exist within the nucleus. For instance, β-decay in neutrons and protons is mediated by the weak force among protons, neutrons, and electrons/positrons.

The *electromagnetic force* is nearly as strong as the strong force. If the strong force were 1, the electromagnetic force would be 10^{-3}. Its range is unlimited. The exchange particle for the electromagnetic force according to the theory is the photon, which has been detected. Electromagnetic forces affect all charged particles. For example, they hold electrons within orbitals around an atom's nucleus.

The strongest force is the *strong nuclear force.* Its range is short—just 10^{-15} m. The predicted exchange particle, the *pion* (π meson), has been detected. This particle is about 270 times more massive than an electron. Like the W-particle, a pion may have a charge of $+e$, $-e$, or 0. The strong force affects baryons and mesons. For example, it holds the nucleus of an atom together.

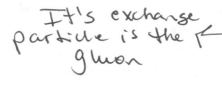
It's exchange particle is the gluon

The strengths of the four fundamental forces that have been identified are related in the exact ratio that our present universe requires. If the ratios of the forces were even one order of magnitude different—for example, if the relative strength of gravity were 10^{-46} or 10^{-44} instead of 10^{-45}—the universe would be radically different from the one we know, and life could not exist. Atheistic scientists are amazed that the universe is exactly right for us. Bible believers, however, realize that God has created the universe harmoniously in a way that preserves His creation and brings honor to Himself.

29D Section Review Questions

1. a. What was the first elementary particle discovered?
   b. What class of elementary particles is it a member of?
2. a. Discuss the characteristics of an antiparticle.
   b. Are particles and their antiparticles compatible?
3. a. What seems to be the relative proportion of antimatter to "normal" matter in the universe?
   b. What does this proportion say about the origin of the universe?

4. What group of particles are the building blocks of the hadron elementary particles?

5. What three quarks would have to be combined in order to produce an antiproton?

6. a. What is an exchange particle?

  b. Suggest a reason why scientists believe that the graviton exists.

Chapter Review

In Terms of Physics

fluoresce	29.1	geochronology	29.13
radioactivity	29.1	mass defect	29.14
alpha radiation (α)	29.3	exoergic	29.14
beta radiation (β)	29.3	endoergic	29.14
gamma radiation (γ)	29.3	fusion	29.14
positron (β^+)	29.5	chain reaction	29.15
fission	29.7	critical mass	29.15
ionizing radiation	29.8	moderator	29.15
activity (A)	29.9	nuclear reactor	29.15
curie (Ci)	29.9	control rod	29.15
becquerel (Bq)	29.9	breeder reactor	29.15
gray (Gy)	29.9	tokamak	29.16
dose (D)	29.9	tracer	29.17
roentgen (R)	29.9	elementary particle	29.19
quality factor (QF)	29.9	antiparticle	29.19
dose-equivalent (DE)	29.9	lepton	29.20
sievert (Sv)	29.9	hadron	29.20
decay constant (λ)	29.10	meson	29.20
half-life ($t_{1/2}$)	29.10	baryon	29.20
radioactive dating	29.11	quark	29.21
parent nuclide	29.11	quark confinement	29.21
daughter nuclide	29.11	exchange particle	29.22
radiogenic	29.11		

Problem-Solving Strategies

29.1 (page 637) Here is a brief review of the laws of logarithms:

If $n^y = x$, then $\ln_n x = y$

$\ln e = 1$

$\ln 1 = 0$

$\ln x \cdot y = \ln x + \ln y$

$\ln x/y = \ln x - \ln y$

$\ln x^y = y \ln x$

Review Questions

1. Describe the three kinds of nuclear radiation.

2. a. What happens to a nucleus's atomic number when it emits a β-particle?
   b. What happens to its mass number?

3. Name the unidentified product in the reaction: $^{22}_{11}\text{Na} \rightarrow ^{22}_{10}\text{Ne} + ?$

4. How does the energy of γ-rays generally compare to that of x-rays?

5. Name two detectors of radiation.

6. What kind of instrument provides real-time indication of a radiation worker's dose in millirem or sieverts?

7. What kind of mathematical curve does radioactive decay follow when you plot activity versus time?

8. What is the half-life of a radionuclide?

9. What can carbon dating, when used properly, reveal about the life of an organism?

10. Why is carbon dating more reliable than, for instance, potassium-argon dating?

11. Give one weakness of
    a. the K-Ar dating method.
    b. the U-Pb dating method.

12. Thermodynamically speaking, what kind of reactions are fission and fusion?

13. What is the difference between fission and fusion?

14. If *A* produces *B* and *C,* will the reaction be endoergic or exoergic?

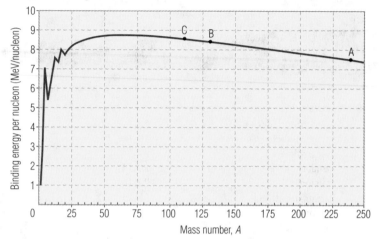

15. What is a chain reaction?

16. Why does a U-235 fission reactor need a moderator?

17. What do control rods do?

18. What undergoes fission in a breeder reactor?

19. Describe two methods of initiating fusion in a controlled fashion.

20. What is the main obstacle to practical fusion power?

21. Which early-discovered elementary particle hardly interacts with matter at all?

22. Explain why pair annihilation argues against the plausibility of the big bang origin of the universe.

23. What is the significance of the names used for the six flavors of quarks?

24. Make a table of the four natural forces, listing them from weakest to strongest. In the table, provide columns for relative strength (strong force = 1), effective distance, associated exchange particle, and whether the exchange particle has been detected.

True or False (25–40)

25. Both x-rays and natural radiation were discovered by accident.

26. The total number of neutrons determines the stability of the nucleus.

27. A more penetrating form of radiation will have a higher quality factor than a form that is stopped at the surface of the skin.

28. A person can receive essentially zero sieverts of radiation per year by living as far away from a nuclear reactor as possible.

29. Nothing seems to change radioactivity of a given amount of a nuclide.

30. The larger the decay constant, the shorter the half-life of an isotope.

31. As a young-earth Creationist, you would expect the oldest possible radiocarbon date to be around one C-14 half-life.

32. The steadiness of cosmic ray intensity over thousands of years has little to do with carbon-14 dating techniques.

33. Fission reactions are exoergic but fusion reactions are endoergic.

34. A chain reaction does not *have* to be a supercritical reaction.

35. Control rods function by removing neutrons from the reactor so that they cannot cause more fissions.

36. One function of a breeder reactor is to produce more U-235.

37. There are only a few known elementary particles.

38. The neutrino was the first lepton elementary particle discovered.

39. The antiproton has a charge of –*e*.

40. Force exchange particles are included in the collection of elementary particles.

⊙41. A bismuth-212 nucleus ($^{212}_{83}$Bi) emits an α-particle.
    a. What is the daughter product's mass number, *A*?
    b. What is the daughter product's atomic number, *Z*?
    c. Use the periodic table in Appendix H to find the name of the daughter product. Write it as ^A_ZX where X is the product's proper abbreviation.

⊙42. If a certain nuclide has a half-life of 5.4 y, what is its decay constant?

⊙43. Berillium-8 has a decay constant of 6.93×10^{15} s^{-1}. What is its half-life?

⊙44. Construct a positive pion using any two of up, down, or strange quarks.

⊙⊙45. After doing some research in other references, write the reactions of one possible decay chain leading from U-238 to Pb-206. (There are at least six.)

⊙⊙46. Where did the term "quark" originate from?

⊙⊙47. Write a paper discussing the experiment and describe the data that confirmed the existence of the neutrino.

⊙⊙48. Give a report on the latest research in refining the quark model, neutrino research, or the grand unification theory.

Cosmic Rays

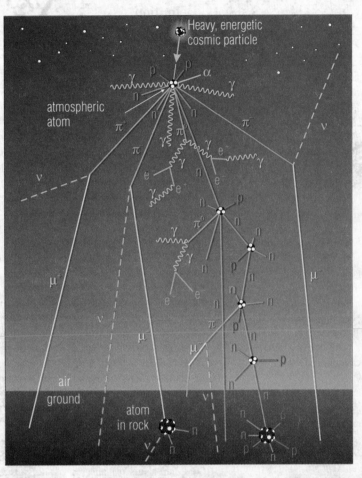

What is the main source of the high-energy particles found on the earth? It is not a particle accelerator, as you might think. Long before artificial accelerators had been developed, astronomers discovered that a steady stream of high-energy radiation from outer space bombards the earth day and night. At first they thought it was gamma radiation; so they called the radiation cosmic *rays*.

But they were wrong. Instead of being rays, the radiation was a mixture of about 10% electrons and 90% atomic nuclei (83% of these were protons, 16% helium, and 1% from all other elements on the periodic table). What first intrigued astrophysicists about this super-energetic radiation was the subnuclear particles that it produced after colliding with atmospheric nuclei. In a sense they were witnessing the opposite of a nuclear explosion—massive energy was being replaced by tiny particles of matter! These particles (positrons, muons, pions, and neutrinos) had never been seen before because they require such concentrated energy and are unstable (neutral pions last about 10^{-16} s).

Most *primary cosmic rays* (radiation from outer space) never make it to the earth's surface because they run into atmospheric nuclei first. Even at high altitudes, only one in a thousand survive the trip. Therefore, most observatories study the rays indirectly by examining the particles produced during atmospheric collisions. When a ray strikes a nucleus, its kinetic energy is dispersed in one of two ways: either the energy turns into pions with kinetic energy, or the ray splits the nucleus (a fission) and imparts kinetic energy to the fragments. The pions and fission fragments are called *secondary cosmic rays*. They in turn decay or sometimes collide with other nuclei to form more secondary cosmic rays, some of which will eventually reach the earth.

The energy in some of the primary rays is incredible. A typical particle carries billions of electron volts (10^9 eV or GeV). By contrast, the particles at room temperature have a kinetic energy of 1/20 eV, and particles from a nuclear explosion carry 10^7 eV (or 10 MeV). A few electrons and positrons exceed 10^{15} eV, which cannot be measured directly on Earth. The highest-energy particle measured so far had 10^{20} eV. That is enough energy, in one nuclear particle, to raise a 1 kg mass 1 m! Such high-energy particles begin a chain reaction when they enter our atmosphere, producing an electron shower, or *cascade shower*, similar to meteor showers. In some instances, one particle has produced 10 billion new particles that spread over a 1000 m range before reaching the earth.

After the development of artificial accelerators, astronomers stopped looking at subnuclear physics in space and began inquiring about the source of cosmic rays and how the rays are accelerated. The issue remains a mystery of astronomy, and many hypotheses have been proposed to account for the rays.

Only one thing is certain: the radiation is deadly to life on Earth. Except for our atmosphere's cushion, we would all die from radiation poisoning. Because astronauts lack this protection, the cabin and space suits must be made of heavy, thick material to impede radiation. Currently the protection is enough to block most rays, and the flights are so short that the limited exposure is not threatening. But new designs and materials will have to be developed for lengthy trips to places such as Mars.

APPENDIX A

Base and Derived* Units of the SI†

Dimension	Name	Symbol	Definition
Base Units			
length	meter	m	The meter is the length of the path traveled by light in a vacuum during a time interval of 1/299 792 458 of a second.
mass	kilogram	kg	The kilogram is the unit of mass; it is equal to the mass of the international prototype of the kilogram.
time	second	s	The second is the duration of 9 192 631 770 periods of the radiation corresponding to the transition between the two hyperfine levels of the ground state of the cesium-133 atom.
electric current	ampere	A	The ampere is that constant current which, if maintained in two straight parallel conductors of infinite length, of negligible circular cross section, and placed 1 meter apart in vacuum, would produce between these conductors a force equal to 2×10^{-7} newton per meter of length.
thermodynamic temperature	kelvin	K	The kelvin, the unit of thermodynamic temperature, is the fraction 1/273.16 of the thermodynamic temperature of the triple point of water.
amount of substance	mole	mol	The mole is the amount of substance of a system which contains as many elementary entities as there are atoms in 0.012 kilogram of carbon-12. When the mole is used, the elementary entities must be specified and may be atoms, molecules, ions, electrons, other particles, or specified groups of such particles.
luminous intensity	candela	cd	The candela is the luminous intensity, in a given direction, of a source that emits monochromatic radiation of frequency 540×10^{12} hertz and that has a radiant intensity in that direction of (1/683) watt per steradian.
The Radian and Steradian			
plane angle	radian	rad	The radian is the plane angle between two radii of a circle that cuts off on the circumference an arc equal in length to the radius.
solid angle	steradian	sr	The steradian is the solid angle that, having its vertex in the center of a sphere, cuts off an area of the surface of the sphere equal to the radius of the sphere squared.
Derived Units			
absorbed dose	gray	Gy	$J/kg = m^2 \cdot s^{-2}$
activity (radionuclide)	becquerel	Bq	s^{-1}
capacitance	farad	F	$C/V = A^2 \cdot s^4 \cdot kg^{-1} \cdot m^{-2}$
Celsius temperature	degree	°C	K
dose equivalent	sievert	sv	$J/kg = m^2 \cdot s^{-2}$
electric charge	coulomb	C	$A \cdot s$
electric potential difference, emf	volt	V	$J/C = kg \cdot m^2 \cdot A^{-1} \cdot s^{-3}$
electric resistance	ohm	Ω	$V/A = kg \cdot m^2 \cdot A^{-2} \cdot s^{-3}$
energy, work, quantity of heat	joule	J	$N \cdot m = kg \cdot m^2 \cdot s^{-2}$
force	newton	N	$kg \cdot m \cdot s^{-2}$
frequency	hertz	Hz	s^{-1}
illuminance	lux	lx	$lm/m^2 = cd \cdot m^{-2}$
inductance	henry	H	$Wb/A = kg \cdot m^2 \cdot A^{-2} \cdot s^{-2}$
luminous flux	lumen	lm	$cd \cdot sr$
magnetic flux	weber	Wb	$V \cdot s = kg \cdot m^2 \cdot A^{-1} \cdot s^{-2}$
magnetic flux density	tesla	T	$Wb/m^2 = kg \cdot A^{-1} \cdot s^{-2}$
plane angle	radian	rad	$m \cdot m^{-1} = 1$
power, radiant flux	watt	W	$J/s = kg \cdot m^2 \cdot s^{-3}$

Base and Derived* Units of the SI (continued)

Dimension	Name	Symbol	Definition
Derived Units (continued)			
pressure, stress	pascal	Pa	$N/m^2 = kg \cdot m^{-1} \cdot s^{-2}$
solid angle	steradian	sr	$m^2 \cdot m^{-2} = 1$
Experimentally Derived Units			
particle energy	electron-volt	eV	The electron-volt is the kinetic energy acquired by an electron in passing through a potential difference of 1 V in vacuum; $1 \text{ eV} = 1.602\,177\,33 \times 10^{-19}$ J
particle mass	atomic mass unit	u	The atomic mass unit is equal to 1/12 of the mass of an atom of the nuclide ^{12}C; $1 \text{ u} = 1.660\,540\,2 \times 10^{-27}$ kg.
Other Units Used with the SI			
pressure	bar	bar	$1 \text{ bar} = 0.1 \text{ MPa} = 100 \text{ kPa} = 10^5$ Pa
level of power	bel	B	1 B is the log level of power when $P/P_0 = 10$, where P is measured power and P_0 is the reference power.
	decibel	dB	$1 \text{ dB} = 0.1 \log (P/P_0)$

*Derived and Other units listed are those used in this textbook

†NIST (URL: http://physics.nist.gov/Pubs/SP811/cover.html)

APPENDIX B

Commonly Used Unit Abbreviations

Unit	Abbreviation	Unit	Abbreviation
ampere	A	joule	J
atomic mass unit	u	kelvin	K
atmosphere	atm	kilogram	kg
coulomb	C	meter	m
day	d	minute	min
decibel	dB	newton	N
degree (angle)	°	ohm	Ω
degree Celsius	°C	pascal	Pa
degree Fahrenheit	°F	radian	rad
electron-volt	eV	revolution	rev
farad	F	second	s
gram	g	tesla	T
henry	H	volt	V
hertz	Hz	watt	W
hour	h	year	y

APPENDIX C

Unit Conversions

The conversion factor required to convert a dimension in the left vertical column to the dimension in the top row is found in the table where the two dimensions cross. The factor may be used in dimensional analysis.

Length	cm	m	km	in.	ft	mi
1 cm =	1	10^{-2}	10^{-5}	0.3937	3.281×10^{-2}	6.214×10^{-6}
1 m =	100	1	10^{-3}	39.37	3.281	6.214×10^{-4}
1 km =	10^5	1000	1	3.937×10^4	3281	0.6214
1 in. =	2.540	2.54×10^{-2}	2.540×10^{-5}	1	8.333×10^{-2}	1.578×10^{-5}
1 ft =	30.48	0.3048	3.048×10^{-4}	12	1	1.894×10^{-4}
1 mi =	1.609×10^5	1609	1.609	6.336×10^4	5280	1

1 angstrom (Å) $= 1.000 \times 10^{-10}$ m	1 nautical mile (nm) $= 6080$ ft $= 1852$ m
1 light year (ly) $= 9.461 \times 10^{15}$ m	
1 astronomical unit (ua) $= 1.496 \times 10^{11}$ m	

Mass	g	kg	u	oz	lbm	t
1 g =	1	0.001	6.022×10^{23}	3.527×10^{-2}	2.205×10^{-3}	1.102×10^{-6}
1 kg =	1000	1	6.022×10^{26}	35.27	2.205	1.102×10^{-3}
1 u =	1.660×10^{-24}	1.660×10^{-27}	1	5.855×10^{-26}	3.660×10^{-27}	1.829×10^{-30}
1 oz =	28.35	2.835×10^{-2}	1.708×10^{25}	1	6.250×10^{-2}	3.125×10^{-5}
1 lbm =	453.6	0.4536	2.732×10^{26}	16	1	5.000×10^{-4}
1 t =	9.072×10^5	907.2	5.465×10^{29}	3.2×10^4	2000	1

Time	s	min	h	d	y
1 s =	1	1.667×10^{-2}	2.778×10^{-4}	1.157×10^{-5}	3.169×10^{-8}
1 min =	60	1	1.667×10^{-2}	6.944×10^{-4}	1.901×10^{-6}
1 h =	3600	60	1	4.167×10^{-2}	1.141×10^{-4}
1 d =	8.640×10^4	1440	24	1	2.738×10^{-3}
1 y =	3.156×10^7	5.260×10^5	8.766×10^3	365.25	1

Pressure	atm	in.$_{Hg}$	torr	Pa	mb	lbf/in.^{2}
1 atm =	1	29.92	760	1.013×10^5	1013	14.70
1 in.$_{Hg}$ =	3.342×10^{-2}	1	25.4	3386	33.86	0.4912
1 torr =	1.316×10^{-3}	3.937×10^{-2}	1	133.3	1.333	0.01934
1 Pa =	9.869×10^{-6}	2.953×10^{-4}	7.501×10^{-3}	1	0.01	1.450×10^{-4}
1 mb =	9.869×10^{-4}	2.953×10^{-2}	0.7501	100	1	1.450×10^{-2}
1 lbf/in.^{2} =	6.805×10^{-2}	2.036	51.71	6.895×10^3	68.95	1

Volume	cm^3	m^3	L	in.^{3}	ft^3
1 cm^3 =	1	1.0×10^{-6}	1.000×10^{-3}	6.102×10^{-2}	3.532×10^{-5}
1 m^3 =	10^6	1	1000	6.102×10^4	35.32
1 L =	1000	1.000×10^{-3}	1	61.02	3.532×10^{-2}
1 in.^{3} =	16.39	1.639×10^{-5}	1.639×10^{-2}	1	5.787×10^{-4}
1 ft^3 =	2.832×10^4	2.832×10^{-2}	28.32	1728	1

1 U.S. fl. gal. = 4 U.S. fl. qt. = 8 U.S. pt. = 128 U.S. fl. oz. = 231 in.^{3}
1 British imperial gal. = 277.4 in.^{3}

APPENDIX D

Physical Constants

Quantity	Symbol	Value
Fundamental		
Atomic mass unit	u	1.661×10^{-27} kg
Avogadro's number	N_A	6.022×10^{23} mol^{-1}
Boltzmann's constant	k	1.381×10^{-23} J/K
Electron mass	m_e	9.109×10^{-31} kg
Fundamental charge	e	1.602×10^{-19} C
Gas constant (SI units)	R	8.315 J/(mol·K)
Gravitational constant	G	6.673×10^{-11} N·m^2/kg^2
Neutron mass	m_n	1.675×10^{-27} kg
Permeability of free space	μ_0	$4\pi \times 10^{-7}$ T·m/A
Permittivity of free space	ε_0	8.854×10^{-12} C^2/(N·m^2)
Planck's constant	h	6.626×10^{-34} J·s
	$\hbar \equiv {}^h\!/_{2\pi}$	1.055×10^{-34} J·s
Proton mass	m_p	1.673×10^{-27} kg
Speed of light in vacuum	c	2.998×10^8 m/s
Stefan-Boltzmann constant	σ	5.671×10^{-8} W/(m^2·K^4)
Wien displacement law constant	$\lambda_{max}T$	2.898×10^{-3} mK
Other Useful Constants		
Absolute zero	0 K	-273.15 °C
Atmospheric pressure, standard	1 atm	1.013×10^5 Pa
Coulomb constant	k	8.988×10^9 N·m^2/C^2
Density of air	—	1.20 kg/m^3 (20 °C and 1 atm)
Density of water	—	1.000×10^3 kg/m^3 (20 °C and 1 atm)
Gravitational acceleration, standard	$|g|$	9.807 m/s^2 (average, near Earth's surface)
Latent heat of vaporization of water	L_v	22.57×10^5 J/kg
Latent heat of fusion of water	L_f	3.335×10^5 J/kg
Rest energy of 1 u	—	931.5 MeV
Speed of sound in air	—	343 m/s (20 °C and 1 atm)
Triple point of water	273.16 K	0.01 °C (at 0.006 atm)

APPENDIX E

Planetary and Astronomical Data

Object	Radius (km)	Mass (kg)	Density (g/cm³)	Orbital Radius* (ua)	Rotational Period (d)	Orbital Period† (y)
Sun	696 000	1.989×10^{30}	1.409	—	25.6	—
Mercury	2440	3.302×10^{23}	5.43	0.3871	58.65	0.2408
Venus	6052	4.869×10^{24}	5.24	0.7233	243.01	0.6152
Earth	6378	5.974×10^{24}	5.515	1.0000	0.9973	1
(Moon)	1738	7.348×10^{22}	3.34	0.002 570	27.32	0.074 80
Mars	3397	6.419×10^{23}	3.94	1.5237	1.026	1.8809
Jupiter	71 492	1.899×10^{27}	1.33	5.2028	0.4135	11.862
Saturn	60 268	5.685×10^{26}	0.70	9.5388	0.4375	29.458
Uranus	25 559	8.663×10^{25}	1.30	19.1914	0.65	84.01
Neptune	24 764	1.028×10^{26}	1.76	30.0611	0.768	164.79
Pluto	1151	1.5×10^{22}	1.1	39.5294	6.387	248.54

*Mean orbital radius

†University Corporation for Atmospheric Research (UCAR). ©1995–99, 2000 The Regents of the University of Michigan; ©2000–02 University Corporation for Atmospheric Research. (URL: http://www.windows.ucar.edu/tour/link=/the_universe/uts/orbits_data.html&edu=high)

Other Data

Astronomical unit: 1 ua = 1.496×10^{11} m
Light year: 1 ly = 9.461×10^{15} m
Parsec: 1 pc = 3.086×10^{16} m   Lunar gravitational acceleration: $|g_{Moon}| = 1.62$ m/s²
(A parsec is the astronomical distance unit defined as the distance from Earth where stellar parallax caused by the earth's revolution about the Sun is 1 second of an arc.)

APPENDIX F

Moments of Inertia for Selected Regular Objects

Object	Description	Formula	Notes
	Thin ring Thin cylindrical shell	$I = mr^2$	Thickness of ring $\ll r$
	Thick ring Thick cylindrical shell	$I = \frac{1}{2}m(r_1^2 + r_2^2)$	r_1 and r_2 are radii of inner and outer respectively
	Solid disk Solid cylinder $\rangle$ center axis Cylinder	$I = \frac{1}{2}mr^2$	
	Solid disk; transverse axis	$I = \frac{1}{4}mr^2$	
	Solid sphere	$I = \frac{2}{5}mr^2$	
	Thin, spherical shell	$I = \frac{2}{3}mr^2$	Thickness of wall $\ll r$
	Thin, solid rod—axis at middle	$I = \frac{1}{12}mL^2$	L is length of rod
	Thin, solid rod—axis at end	$I = \frac{1}{3}mL^2$	L is length of rod
	Rectangular plate	$I = \frac{1}{12}m(a^2 + b^2)$	a and b are the length and width; Axis is perpendicular through center of face ab

APPENDIX G ♦ Mathematical Reference

This appendix is provided as a ready reference for most of the mathematical principles required for this textbook. It is not intended to be exhaustive.

A. Mathematical Symbols

1. *Relational Symbols*

Symbol	Definition
$\equiv$	is defined as
$=$	equals exactly; assigns a value to
$\neq$	does not equal
$\doteq$	a rounded or cut off calculator result (not rounded to correct significant digits)
$\approx$	answer rounded to correct significant digits
$\sim$	approximately equal to; result is representative of actual value
$<>$	less than; greater than
$\leq \geq$	less than or equal to; greater than or equal to
$<<>>$	much less than; much greater than
$\Leftrightarrow$	if and only if;
$\Rightarrow$	implies; "if a, then b" is $a \Rightarrow b$
$\propto$	proportional to

2. *Other Symbols*

Symbol	Definition
$\pm$	plus or minus—value following may be positive or negative
$\lvert x \rvert$	absolute value of x; magnitude of x
Δx	change in x; final value minus initial value, or $\Delta x = x_2 - x_1$
Σx	summation of x; the sum of all values represented by x
$x \to 0$	x approaches zero; the magnitude of x becomes as small as desired
∞	infinity; the value is as large as desired

B. Rounding Rules

1. Determine the position of the digit to which the number must be rounded. This digit will be referred to as the last significant digit (the "last SD"). The rules for determining significant digits are discussed in Chapter 2.

2. Note the digit to the right of the last SD and round according to the following rules:

   a. If the number is 0 through 4, drop all digits to the right of the last SD. If necessary, substitute the dropped numbers with zeros as placeholders to locate the decimal point but do not write in the decimal point.

   b. If the number to the right of the last SD is 5 through 9, add 1 to the last SD and drop the remaining digits, replacing them with zero placeholders if required.

   c. Do not consider any digits other than the one immediately to the right of the last SD. Specifically, do *not* begin rounding from the right-most digit. This often results in a final digit that has been rounded up in error.

3. Examples:

   Round 37 547 m to the nearest 1000 m. *Answer:* 38 000 m

   Round 1.448 cm to the nearest 0.1 cm. *Answer:* 1.4 cm

C. Scientific and Engineering Notation

1. Scientists and engineers must often deal with extremely large or small numbers that involve many zeros for locating the decimal point. To make working with such numbers easier, a shorthand method of writing them is used called *scientific* or *engineering notation*.

2. In *scientific notation,* the number is written in the form $M \times 10^n$. The value of M can be positive or negative but its magnitude must be greater than or equal to 1 and less than 10 ($1 \leq \lvert M \rvert < 10$). The exponent, n, is an integer (not a fractional or decimal number) and can be positive or negative as well.

3. In order to convert a number to scientific notation, move the decimal point left or right as necessary until only one nonzero digit appears to the left of the decimal point. The number

of places that the decimal point has moved is the exponent (n). If you moved the decimal to the left, the exponent is positive. If you moved the decimal point to the right, the exponent is negative. Retain in M all significant digits present in the original number. All nonsignificant trailing zeros should be dropped. For example:

Decimal Form	Improper	Proper
0.000 057	0.57×10^{-4}	5.7×10^{-5}
34 500 000	345×10^5	3.45×10^7
176.00	1.76×10^2	1.7600×10^2

As you can see in the third example, scientific notation can be used to indicate that the trailing zeros in a number are significant.

4. If all the trailing zeros in a small number are significant, indicate this fact by a decimal point following the final zero.

5400 cm	2 SDs	50 m	1 SD
5.40×10^3 cm	3 SDs	50. m	2 SDs

5. *Engineering notation* is similar to scientific notation in that it has both a decimal and an exponential part. However, engineering notation is used when the orders of magnitude of a group of numbers are to be compared. The decimal part, M, may be any value as long as the power of 10 is the same for each number being compared.

Example: Compare the masses of the planets Mercury, Earth and Jupiter in scientific notation: Mercury, 3.302×10^{23} kg; Earth, 5.974×10^{24} kg; Jupiter, 1.899×10^{27} kg

In engineering notation:
Mercury, 0.3302×10^{24} kg; Earth, 5.974×10^{24} kg; Jupiter, 1899×10^{24} kg

You can quickly see the relative masses between the planets using engineering notation.

6. *Operations on numbers in scientific notation*

Addition and Subtraction

Numbers written in scientific notation must have the same powers of 10 before you can add or subtract them. If the exponents are not the same, convert one of the numbers before you perform the operation. Converting the exponent term of the smaller number to the larger exponent is usually a good practice. Report the result in proper scientific notation.

Example: Add 3.5×10^4 kg and 7.0×10^3 kg.

7.0×10^3 kg = 0.7×10^4 kg   1. Convert one of the numbers so that they have the same exponent.

$$
\begin{array}{r}
3.5 \times 10^4 \text{ kg} \\
+ \ 0.7 \times 10^4 \text{ kg} \\
\hline
4.2 \times 10^4 \text{ kg}
\end{array}
$$

2. Add, following the rules for significant digits, if applicable.

3. If necessary, convert answer to proper scientific notation.

Multiplication and Division

Multiplication and division can be done on numbers in scientific notation without changing their form. Perform the operation on the decimal (M) factors first. Then multiply or divide the exponential terms following the rules for exponents. The rules for exponents are more fully explained later in this appendix. Convert the resulting number to standard scientific notation.

Examples:

a. $(2.5 \times 10^7)(1.4 \times 10^{-3}) = (2.5 \times 1.4)(10^7 \times 10^{-3})$
$$= 3.5 \times 10^{7+(-3)}$$
$$= 3.5 \times 10^4$$

b. $\dfrac{2.7 \times 10^5}{3.0 \times 10^{-2}} = \dfrac{2.7}{3.0} \times \dfrac{10^5}{10^{-2}} = 0.90 \times 10^{5-(-2)} = 0.90 \times 10^7$
$$= 9.0 \times 10^6$$

Square roots

To find the square root of a number in scientific notation, rewrite the number if necessary so that the exponent is an even number, then find the square root of M and divide the exponent by 2.

$$\sqrt{6.4 \times 10^7} = \sqrt{64 \times 10^6} \qquad \text{Change the exponent to an even power.}$$
$$= \sqrt{64} \times \sqrt{10^6} \qquad \text{Separate the radicand into its two parts.}$$
$$= 8.0 \times 10^3 \qquad \text{Take the square root of both parts and change the result to standard scientific notation, if required.}$$

D. Logarithms and Exponential Operations

1. Any positive real number, n, can be expressed as a base number, a, raised to some exponent, b. In an equation, this can be written as

$$n = a^b.$$

2. The number b is called the *logarithm* of n to the base a, so a logarithm is nothing more than an exponent. This expression is written as:

$$\log_a n = b$$

3. Any positive number may be the base of a logarithm, but you are likely to come across only certain numbers used as bases in physics. Logarithms of base 10 are called *common logarithms,* while those of the base e (2.71828 . . .) are called *natural logarithms.* The common log of n is expressed as

$$\log n = b,$$

where the base (a) is understood to be 10. The natural log of n is expressed as

$$\ln n = b,$$

where the base of ln is understood to be e.

4. The logarithm of numbers greater than 1 are positive decimal values. The logarithms of numbers between 0 and 1 are negative decimals. The logarithm of 1 is zero (0). You cannot obtain a logarithm for zero, and for a negative number, you may obtain only the logarithm of its magnitude.

5. Examples using a calculator:

   a. $\log 25 \doteq 1.3979 \ldots$ (equivalent to $10^{1.3979} \sim 25$)

   b. $\log 0.4 \doteq -0.3979 \ldots$ (equivalent to $10^{-0.3979} \sim 0.4$)

   c. $\ln 25 \doteq 3.2188 \ldots$ (equivalent to $e^{3.2188} \sim 25$)

   d. $\ln 0.4 \doteq -0.9162 \ldots$ (equivalent to $e^{-0.9162} \sim 0.4$)

6. *Operations with Logarithms.* The following list of operations is valid for both common and natural logarithms.

   a. $\log 10 = 1$; $\ln e = 1$

   b. $\log 1 = 0$

   c. $\log ab = \log a + \log b$

   d. $\log a/b = \log a - \log b$

   e. $\log a^b = b \log a$

   f. $\log \sqrt[b]{a} = (\log a) \div b$

7. *Rules of Exponents.* These rules look similar to the laws of logarithms but are more general because they can apply to any base, including variable symbols.

   a. $x^1 = x$

   b. $x^0 = 1$

   c. $1/x^a = x^{-a}$

   d. $x^a \cdot x^b = x^{a+b}$ (bases must be the same)

   e. $x^a/x^b = x^{a-b}$ (bases must be the same)

   f. $\sqrt[a]{x} = x^{1/a}$

E. Angular Measure and Converting Angles

1. Angular measure provides the means to quantify direction and circular motion.

2. You are most familiar with the unit of angular degree, symbol °. There are 360° in a full circle. Because of the way a degree is defined, the kinds of mathematical operations that can be performed on angular measurements using this unit are limited.

3. The derived SI unit for angular measure is the *radian (rad)*. The radian is preferred in physics because the unit may be used in any legitimate mathematical operation involving angular measure, for it is actually *dimensionless* (has no SI units associated with it). A "dimensionless unit" may seem to be contradictory, so let's see how it is derived.

4. A radian is equal to the central angle of a circle formed by two radii that subtend (cut off) an arc of the circle equal in length to its radius. See Figure G-1. Since 1 rad = r meters/r meters = 1, radians have no SI base units and are dimensionless.

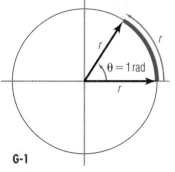

G-1

5. You know that the circumference of a circle is equal to $2\pi r$. Accordingly, there are 2π radians in a full circle, or $360° = 2\pi$ rad. This relationship can be used to develop a formula for converting degrees to radians and radians to degrees:

$$360° = 2\pi$$
$$180° = \pi$$
$$1 = \frac{\pi}{180°} \qquad\qquad \frac{180°}{\pi} = 1$$
$$\theta_{rad} = \frac{\pi}{180°}\,\theta_{deg} \qquad\qquad \theta_{rad}\frac{180°}{\pi} = \theta_{deg}$$

6. Examples:
Convert 38° to radians:

$$38°\cdot\frac{\pi}{180°} \doteq 0.6632 \text{ (rad)}$$

Convert 1.5π to degrees:

$$1.5\pi\cdot\frac{180°}{\pi} = 270°$$

F. Geometric Figures

This section summarizes some basic geometric figures that are referred to in this textbook and formulas that are associated with them.

Circle of radius r	area = πr^2	perimeter = $2\pi r$
Sphere of radius r	surface area = $4\pi r^2$	volume = $\frac{4}{3}\pi r^3$
Triangle	area = $\frac{1}{2}bh$	
Rectangle	area = wl	perimeter = $2(w + l)$
Rectangular prism ($B = w\cdot l$)	lateral area = $2h(w + l)$	volume = $wlh = Bh$
Right circular cylinder of radius r and height h	lateral area = $2\pi rh$	volume = $\pi r^2 h$
Right circular cone of radius r and height h	lateral area = $\pi r\sqrt{r^2 + h^2}$	volume = $\frac{1}{3}\pi r^2 h$

G. Algebra and Trigonometry

The rules and operations of algebra and trigonometry are extensive, and discussing them in any detail is beyond the scope of this appendix. If you need to brush up on some aspect of solving algebraic equations or trigonometric identities, you should review the applicable sections of a mathematics textbook such as ALGEBRA 2 *for Christian Schools* or PRECALCULUS *for Christian Schools*.

H. Functions

1. One characteristic of God's creation that man discovered was that many phenomena can be described by simple equations that relate a physical quantity to one or more others. For example, if you know the speed of a car and how long it has been traveling at that speed, you can tell how far the car traveled during the time interval.

2. Such relationships are called *functions* because a set of input numbers (called the *independent* variable set or *domain*) produces a unique set of output numbers (called the *dependent* variable set or *range*).

3. You will study a variety of functions in physics. In this course they will be limited to a few simple types. We will discuss their characteristics and graphs in this section.

4. *Linear functions* are the simplest type. They represent a direct proportional relationship between the independent and dependent variables and produce a straight-line graph. Linear functions may be written in the form

$$y = mx + b,$$

where m is the *slope* of the line and b is the coordinate where the line crosses the y axis. If the relationship between the independent and dependent variables is *direct,* then the line passes through the origin of the graph ($b = 0$). Figures G-2 (a) and G-2 (b) show a linear function and a direct function, respectively. One example of a linear function is the expansion of a thin metal rod with temperature as it is heated.

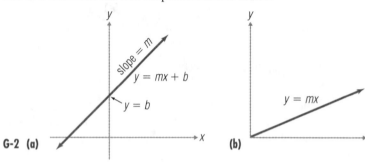

G-2 (a)   **(b)**

5. *Quadratic functions* are relationships in which the dependent variable is a function of the square of the independent variable. Graphs of quadratic functions are parabolas. Quadratic functions are written in the form

$$y = ax^2 + bx + c,$$

where a, b, and c are constants. An example of a quadratic function is the position of a ballistic projectile with time. Figure G-3 shows the graph of a quadratic function.

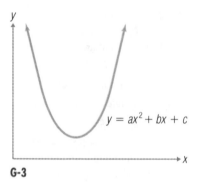

G-3

6. *Inverse functions* are relationships in which, as the independent variable increases, the dependent variable decreases in a reciprocal fashion. Such relationships are called inverse proportions. Graphs of inverse functions are hyperbolas. These functions may be written in the form of either

$$y = \frac{k}{x} \text{ or } yx = k,$$

where k is a proportionality constant. An example of an inverse function is Boyle's law. Figure G-4 is a graph of an inverse function. Note that neither variable ever reaches zero nor crosses its axis in the inverse function of a physical phenomenon.

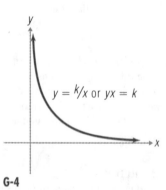

G-4

7. *Exponential functions* are very common in natural processes. In a basic exponential function, the dependent variable is the result of some base number raised to the power equal to the independent variable. If the values of the independent variable are not restricted, then the dependent variable may take any positive real number value. These functions can be written in the form of either

$$y = n^x \text{ or } y = n^{-x},$$

where n is a positive real number. The magnitude of n determines how steep or how shallow the curves are. For natural processes, n is usually the base of the natural logarithms, e. The graphs of the exponential functions form sweeping rising or falling curves (see Figure G-5). Note that when $x = 0$, both curves pass through $y = 1$.

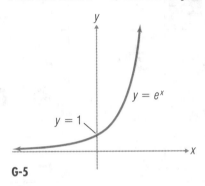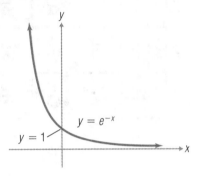

G-5

8. *Periodic functions* are functions that are repetitive or show an oscillation of the dependent variable over the domain of the independent variable. Such functions include the sine, cosine, and tangent functions. The equation form of these functions varies depending on the operator and argument but generally follows the form

$$y = A \text{ } op \text{ } (\theta + k) + c,$$

where A, k, and c are all constants and the operator $op \text{ } (\text{ })$ may be any periodic function operator, such as a trigonometric ratio. The quantity within the parentheses may be an angle or an angular frequency that is dependent on time. An example of a periodic physical phenomenon is the displacement of a pendulum from its vertical hanging position. Figure G-6 shows two graphs of typical periodic functions.

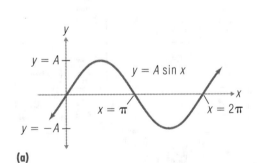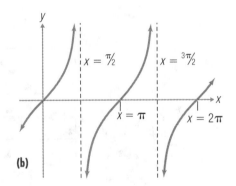

(a)   **(b)**

G-6 (a) A sine function, (b) A tangent function

APPENDIX H ◆ Periodic Table of the Elements

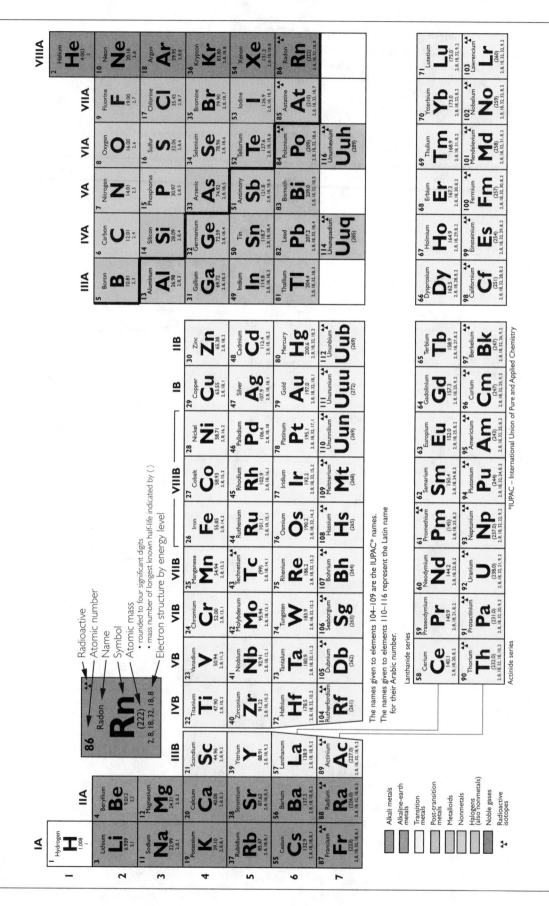

APPENDIX I

Prefixes for Powers of 10

Prefix	Symbol	Factor	Example
exa-	E	$\times 10^{18}$	—
peta-	P	$\times 10^{15}$	—
tera-	T	$\times 10^{12}$	terawatt (TW)
giga-	G	$\times 10^{9}$	gigabyte (GB)
mega-	M	$\times 10^{6}$	megajoule (MJ)
kilo-	k	$\times 10^{3}$	kilometer (km)
hecto-	h	$\times 10^{2}$	(rarely used in physics)
deka-	da	$\times 10^{1}$	(rarely used in physics)
(base)	—	$\times 10^{0}$	—
deci-	d	$\times 10^{-1}$	decibel (dB)
centi-	c	$\times 10^{-2}$	centimeter (cm)
milli-	m	$\times 10^{-3}$	millivolt (mV)
micro-	μ	$\times 10^{-6}$	micropascal (μPa)
nano-	n	$\times 10^{-9}$	nanometer (nm)
pico-	p	$\times 10^{-12}$	picofarad (pF)
femto-	f	$\times 10^{-15}$	—
atto-	a	$\times 10^{-18}$	—

APPENDIX J

Greek Alphabet

Letter	Lower-case	Upper-case	Letter	Lower-case	Upper-case
alpha	α	A	nu	ν	N
beta	β	B	xi	ξ	Ξ
gamma	γ	Γ	omicron	o	O
delta	δ	Δ	pi	π	Π
epsilon	ε or ϵ	E	rho	ρ	P
zeta	ζ	Z	sigma	σ	Σ
eta	η	H	tau	τ	T
theta	θ	Θ	upsilon	υ	Υ
iota	ι	I	phi	ϕ	Φ
kappa	κ	K	chi	χ	X
lambda	λ	Λ	psi	ψ	Ψ
mu	μ	M	omega	ω	Ω

A

absolute pressure (P) Pressure measured with respect to a perfect vacuum; gauge pressure plus 1 atm

absolute zero (0 K) The minimum theoretical temperature; the temperature at which all thermal motion ceases

acceleration (a) The time-rate of change of velocity; $\Delta v/\Delta t$

accuracy An assessment of how close a measurement is to the accepted or known value

action-reaction force pair The two equal but opposite forces that exist when two systems act on each other

activity (A) The rate of emission of particles or energy from a radioactive nucleus

actual mechanical advantage (AMA) The ratio of a machine's output force to its input force

additive primary color Red, green, or blue light; various combinations can produce any other color, including white

adhesion Attraction between molecules of different substances

adiabatic A property that allows no energy exchange across the boundary of a system

airfoil The cross-sectional shape of a part, such as a wing or rudder, that produces lift by causing different rates of fluid flow and employing Bernoulli's principle

alpha radiation (α) Alpha particles emitted by radioactive nuclei

alternating current (AC) An electric current that regularly changes direction

ammeter A device that measures and displays the current flowing through a circuit

amorphous solid A solid without any definite crystalline structure

ampere (A) The SI fundamental unit of electrical current; a current of 1 A in each of two parallel wires of infinite length separated by 1 m will produce a force of 2×10^{-7} N on each wire

amplitude (A) The maximum displacement from the equilibrium position of a system experiencing simple harmonic motion; applies to the medium or fields that carry periodic waves

angle of incidence (θ_i) The angle between the incident ray of light and the normal line at the point of incidence

angle of reflection (θ_r) The angle between the reflected ray of light and the normal line at the point of incidence

angle of refraction (θ_r) The angle between the refracted ray of light and the normal line at the interface between two media

angular acceleration (α) The time-rate of change of angular velocity; $\Delta\omega/\Delta t$

angular momentum (L) The momentum of a rotating system that is dependent on the angular velocity and mass distribution of the system

angular speed (ω) The time-rate of change of angular position; $\Delta\theta/\Delta t$

angular velocity (ω) The vector representing the time-rate of change of angular position in a specified direction

annual variation The annual change of direction of the earth's magnetic flux vector at a specified location

antenna A device in which charged particles accelerate either to form an electromagnetic wave or in response to an electromagnetic wave

antiparticle An elementary particle whose properties are identical to those of the named particle in every other respect but are opposite in sign

aperture The opening through which light is admitted into an optical instrument

apparent weight The weight measured by an accelerated scale

Archimedes' principle The buoyant force of a fluid on an object is equal to the weight of the fluid the object actually displaces when it is immersed therein.

armature The component that generates the primary rotating magnetic field in an electric motor

astronomical unit (ua) The unit of astronomical measure equal to the mean distance between Earth and the Sun

atmosphere (atm) The average pressure of the atmosphere at sea level; 1 atm = 1.013×10^5 Pa

atom A neutral particle with a centrally located nucleus consisting of protons and neutrons, surrounded by electrons; the smallest representative entity of an element

atomic mass unit A convenient unit of mass equal to $^1/_{12}$ the mass of a carbon-12 atom

atomic number (Z) The number of protons in the nucleus of an atom

aurora A polar light display resulting from the interaction between high-energy charged particles trapped in Earth's magnetic field and upper atmospheric gas molecules

average acceleration ($\bar{a}$) The rate of change of a system's velocity during an interval of time

Avogadro's constant (N_A) The number of atoms in exactly 12 g of carbon-12; the number of atoms whose mass in grams is the same number as the weight of one atom in atomic mass units; $N_A \doteq 6.022 \times 10^{23}$ particles

azimuthal quantum number (l) The quantum number that tells which orbital or subshell an electron occupies

B

back emf (ε) A potential difference induced across an inductor that opposes the change in current through the inductor

balanced forces Concurrent forces that have a zero vector sum

ballistic The portion of a projectile's trajectory where only external forces act on it

ballistic pendulum A device used to measure the energy of ballistic projectiles through conservation of momentum

Balmer series A series of wavelengths of light produced when an electron falls to the first energy level in an atom

bar A non-SI unit for measuring atmospheric pressure; 1 atm = 1.013 bar

barometer An instrument that measures and displays atmospheric pressure

baryon A class of elementary particles including nucleons and heavier particles

battery A source of electrical potential difference consisting of two or more voltaic cells connected in series

becquerel (Bq) The SI unit of radioactivity; 1 Bq = 1 emission per second

Bernoulli's equation The equation that shows the inter-relationship between pressure, speed, and elevation for a flowing noncompressible fluid

beta radiation (β) Electrons and positrons emitted by radioactive nuclei

bias (Perspective) A lack of objectivity, a prejudice, possessing opinions not based upon facts; (Electrical) Applying a slight positive or negative charge to an electronic component

bimetallic strip Two metals welded together to form a long, narrow strip, each metal having a different coefficient of expansion that causes the strip to bend in response to temperature changes

binding energy (Atomic) The amount of energy that must be added to remove a particular electron from an atom; (Nuclear) The amount of energy that a particle must give up in order to join a nucleus; energy equivalent to the nuclear mass defect

Biot-Savart law Defines the relationship between the current in an infinitesimal portion of a conductor and the magnetic field induced by it

blackbody A perfect emitter and absorber of electromagnetic radiation

blackbody radiation The electromagnetic radiation emitted by a blackbody

black hole An astronomical body so dense and massive that even light cannot escape its gravitational field

block and tackle system An arrangement of fixed and movable pulleys connected by ropes that provides a mechanical advantage for exerting a force

Boltzmann's constant (k) A proportionality constant for expressing the kinetic energy of a particle of an ideal gas per degree kelvin; $k = 1.381 \times 10^{-23}$ J/K

Boyle's law The pressure of a certain mass of an ideal gas is inversely proportional to its volume if the temperature is held constant; $PV = k$

breeder reactor A fission reactor using Pu-239 as a fuel that stimulates the transmutation of U-238 to Pu-239 and thus produces more fuel than it consumes

Brewster's angle (θ_B) The special angle of incidence where, when two materials are in contact, the reflected light rays from the incident medium are exactly 90° from the refracted rays of the other medium; The reflected beam will be entirely horizontally polarized, and the refracted beam will be mostly vertically polarized.

Brewster's law A law that states that the angle of incidence that best polarizes reflected light is given by $\tan \theta_B = n_i$, the index of refraction of the reflecting medium

bridge circuit An instrument used to determine the unknown resistance of a component placed in a series-parallel relationship with three known resistances

Brownian motion The random, chaotic movements of microscopic particles dispersed in a fluid medium

brush One of a set of electrical contacts in motors that directs the current to the rotor through a split ring commutator or slip rings

buoyant force An upward force exerted by a fluid on an object immersed in the fluid

C

caloric theory The theory, now obsolete, that thermal energy is a material fluid that flows from hot objects to cold ones

calorie (cal) The amount of thermal energy required to raise the temperature of 1 g of water 1 °C

calorimeter An insulated apparatus within which thermal processes can be quantitatively evaluated with minimal influence from the surroundings

candela (cd) The SI fundamental unit of light intensity; 1 candela is the intensity of light in a specified direction that radiates in monochromatic light with a frequency of 540×10^{12} hertz and that has a radiant intensity in that direction of 1/683 watt per steradian

capacitance (C) The amount of charge per volt of potential difference that a capacitor can store

capacitive reactance (X_C) The effective resistance of a capacitor to current

capacitor A device, containing two conducting plates separated by an insulator, that is used to store charge

capillarity The movement of a liquid through a narrow passage or tube caused by the adhesion of the molecules in the liquid to the surface with which it is in contact

Carnot cycle A theoretical heat cycle that operates at the greatest possible efficiency between two given temperatures

Cassegrainian reflector telescope A reflecting telescope with a folded light path that places the eyepiece behind the center of the primary mirror

cathode-ray tube (CRT) An evacuated glass envelope containing at least two oppositely charged plates at high electrical potential; Electrons are boiled off the negative plate (the cathode) and propelled by the voltage difference toward the anode. In modern CRTs, external magnetic fields shape and direct the resulting electron beam (cathode rays) toward a phosphor-coated surface for displaying information.

center of buoyancy The center of mass of the fluid displaced by an object immersed in the fluid

central force A force acting between two objects along the line connecting their centers; a conservative force

centripetal acceleration (a_c) The center-seeking acceleration that results when a central force holds a system in a circular path

centripetal force (F_c) A central force responsible for accelerating a system in a circular path

chain reaction A nuclear fission process in which neutrons produced by fissions trigger more fissions

Charles's law The volume and absolute temperature of a certain mass of an ideal gas are directly proportional to one another when the pressure is held constant; $V = kT$

chemical compound A combination of two or more different elements in definite proportions that cannot be separated by ordinary physical means

chromatic aberration Dispersion of colors in the image by a lens because different wavelengths of light have different indexes of refraction

circuit A complete conducting path for a current

circuit breaker A reusable safety device that automatically acts to break the circuit if the current becomes too high

circuit component Any part of an electrical circuit that performs a specific electrical function other than just conducting current

circular motion Two-dimensional motion in a circular path; includes both revolving and spinning motions

classical physics The division of physics that encompasses the three major areas of mechanics, electrodynamics, and thermodynamics, all of which were well developed prior to the establishment of quantum physics in the twentieth century

clock paradox The relativistic principle that when one clock moves away from and then back toward another clock, it runs slow compared to the second clock

closed system A system that can exchange energy but not matter with its surroundings

coefficient of friction (μ) A measure of the resistance to relative motion of two materials in contact with each other; $\mu = f/N$ (dimensionless)

coefficient of linear expansion (α) A measure of the amount a material expands in one direction per degree change in temperature (dimensionless)

coefficient of viscosity (η) A measure of a fluid's resistance to flow under a constant force and laminar conditions

coefficient of volume expansion (β) A measure of the amount a material expands in three dimensions per degree change in temperature; $\beta = 3\alpha$ (dimensionless)

cohesion Mutual attraction between atoms or molecules of a substance

collimator An optical device that adjusts the shape of an incident light beam to make the rays parallel

combined gas law A law that combines Boyle's, Charles's, and Gay-Lussac's laws; $PV/T = k$

complementarity The principle that the concepts of waves and particles are both needed to describe the properties of matter under all conditions

complementary colors Pairs of additive colors that, when combined, produce white light—yellow and blue, green and magenta, red and cyan

component The abbreviated name for a component vector

component vector One of two (or three) vectors that are parallel to the designated coordinate axes that are summed to produce the resultant vector

compound microscope A microscope that uses two or more lenses or groups of lenses

compression zone In a sound wave, the region where the medium particles are closer together than average; the "peak" of the sound wave

Compton effect The changes in frequency and direction of photons by collisions with electrons

concave Shaped as a portion of the inner surface of a sphere

conduction The flow of thermal energy or electric current from one object to another through contact

conductor A material consisting of particles with loosely held electrons that allows the easy transmission of thermal energy or electricity

conservation law A principle that states a certain dimensional quantity remains constant even though different forms of it may vary through a process

conservation of mechanical energy A principle of an ideal system that has no energy losses to drag or friction; total energy is conserved as it is translated from kinetic to potential or vice versa

conservation of momentum The law that states that unless an unbalanced external force acts on a system, the total momentum of the system is conserved

conservative force A force that can convert kinetic energy to potential energy and vice versa without loss

conserved A dimensional quantity remains constant

constructive interference The superposition of wave forms reinforces, producing larger oscillations

contact (charging) One of two ways to charge an object; when a charged object touches an uncharged object and electrons are transferred from the one to the other

contact force A force caused by the repulsion of the atoms of one object for the atoms of another object as they touch; a force that can be applied only through the physical contact of two systems

control rod A bladed rod made of a material that can absorb neutrons produced by a nuclear chain reaction; the nuclear reaction rate depends on its position in the reactor core

convection The transfer of thermal energy from one place to another by the movement of particles

converging lens A lens that focuses light rays at a point

conversion factor A multiplication factor that converts the units of a quantity to other units of the same dimension while preserving the size of the quantity

convex Shaped as a portion of the exterior surface of a sphere

cosine θ In a right triangle, the ratio of the leg adjacent to the acute angle θ and the hypotenuse

coulomb (C) The SI unit of electrical charge; $1\,C = 1\,A\cdot s$

Coulomb's law The law relating the electrical force between two point charges to the magnitude of their charges and the distance between them

critical angle (θ_c) For light traveling from a medium with a higher index of refraction to one with a lower, it is the angle of incidence that produces an angle of refraction of 90° (the light ray is lost parallel to the interface surface)

critically damped oscillation A truncated oscillation that barely overshoots the equilibrium point

critical mass The smallest mass of a fissionable material that can sustain a nuclear chain reaction

crystalline solid A solid in which the atoms or molecules occur in a regular, repeating pattern

curie (Ci) A non-SI unit for the decay rate or radioactivity; $1\,Ci = 3.70 \times 10^{10}\,Bq$

Curie temperature (T_C) The temperature at which a permanent magnet loses its magnetic field

current The flow of electrical charges

cutoff frequency The lowest frequency of radiation that will eject photoelectrons in a material

cycle One complete path of a system which is repeated in periodic motion

D

damped harmonic oscillator A system that oscillates periodically with decreasing amplitude due to an opposing force acting on the system

damping Reducing the amplitude of periodic motion through friction or other resistance

daughter nuclide The nuclide resulting from a nuclear particle decay

decay constant (λ) A proportionality constant relating the rate of decay of a particular radioactive nuclide to the number of that nuclei present

decibel (dB) A non-SI unit of relative intensity; The smallest difference in loudness that the average human ear can detect is 1 dB.

deduction A conclusion arrived at using deductive reasoning

deductive reasoning A process of reasoning from the general to the specific; forming a conclusion that is necessarily true if the supporting statements of fact (premises) are true

density (ρ) A measure of the concentration of matter; expressed as the ratio of an object's mass to its volume

derived unit In the SI, any unit that may be expressed in terms of other derived or fundamental SI units

destructive interference The superposition of wave forms results in a decrease in overall amplitude or cancellation of the wave forms at a point

diamagnetic Not attracted by magnets; having a magnetic permeability (μ) less than 1

diathermic The property of being an ideal conductor of thermal energy

dichroic A property of some materials that polarizes light passing through them into one or two separate rays

dielectric The insulator in a capacitor

dielectric constant (κ) The ratio of a given electric field strength in a vacuum to the same field passing through a particular material

difference vector (ΔV) The resultant of a vector subtraction operation; $\Delta V = V_2 - V_1$

diffraction The bending of waves around the edges of openings, objects, or gratings that have dimensions similar to the wavelength of the wave

dilation The observed lengthening of a system's time as it moves at a relativistic speed compared to an observer at rest; a prediction of the special theory of relativity

dimension One of three coordinate directions needed to define how an object is perceived (i.e., length, width, or height); any measurable property (mass, force, time, etc.)

direct current (DC) An electric current that flows in only one direction

dispersion The refraction of different frequencies of light at different angles by transparent materials

displacement (d) The distance and direction between an object's position at one time and its position at a later time; in oscillatory motion, the distance and direction of the system from its equilibrium position; $d = r_2 - r_1$

disturbance A pulse or oscillation in a medium

diverging lens A lens that spreads light rays apart

domain A microscopic magnetic region of a metal that contributes to the overall magnetic field of the material

Doppler effect The alteration of wave frequency emitted by a source as perceived by an observer moving relative to the source

dose (D) The amount of energy absorbed per unit of mass by a substance exposed to nuclear radiation; measured in grays (Gy)

dose-equivalent (DE) A measure of the damage done by exposure to nuclear radiation; The type of radiation and the kind of material being radiated are used to qualify the dose.

driven oscillations Oscillations resulting from a periodic addition of energy at the natural frequency of the system

dry cell A voltaic cell with electrodes immersed in an electrolytic paste instead of a liquid electrolyte

dynamics The branch of mechanics that mathematically describes the causes of motion

E

eddy current A circulating current induced by an oscillating magnetic field in a two- or three-dimensional conductor

efficiency (η) The ratio of useful work output to work input for a machine

elastic collision A collision in which kinetic energy is conserved

elastic (Young's) modulus (E) The ratio of stress (σ) to strain (ε)

elastic potential energy Energy possessed by a system with respect to an elastic force

electrical component Any device in an electrical or electronic circuit, except those whose only function is to conduct current

electrical ground A source or sink of charge large enough that adding or subtracting a finite amount of charge will not change its charge; The earth is the ultimate electrical ground.

electrical potential (V) The amount of work per unit charge that is needed to move a charge a certain distance r from a point charge; $V = U_e/q$

electric field The area around a charged object in which it can exert an electrical force

electric field strength (E) The direction and magnitude of the electric field at a point in space

electrode A conductor that allows an electrical current to enter or leave an electrolytic cell (or other apparatus)

electrodynamics The study of the causes and effects of current electricity

electromagnet A solenoid with a core of ferromagnetic material

electromagnetic (EM) wave A self-propagating wave consisting of mutually perpendicular electric and magnetic component fields

electromagnetic force One of the four fundamental forces in nature; manifested in the forces exerted between atoms

electromagnetic induction The generation of electricity from magnetism or, conversely, of magnetism from electricity

electromagnetic radiation Energy transferred by electromagnetic waves

electromagnetism The field of study involving any aspect of electricity and magnetism

electron A negatively charged particle-wave entity that usually occupies the space immediately surrounding the nucleus of an atom

electron cloud The space region around an atom where the atom's electrons are likely to exist

electron shell A model describing any one of a set of atomic energy levels occupied by electrons having the same principal quantum number

electron volt (eV) The kinetic energy acquired by a fundamental charge accelerated through a potential difference of one volt

electroscope An instrument that can detect and measure charge

electrostatic force The attraction or repulsion force between two electrically charged objects

electrostatic unit (esu) A non-SI unit of charge; permits the constant k in the Coulomb's law equation to be 1

element A pure substance that cannot be broken down into more elementary parts by ordinary chemical means; Its atoms all have the same atomic number.

elementary particle A particle smaller than an atom

endoergic A nuclear reaction that converts other forms of energy to mass

energy A property of a system that gives it the ability to do work

entropy (S) The measure of randomness or disorder in a specified portion of the universe; a measure of the likelihood that a system will be in a particular state

equation of flow continuity In fluid dynamics, the equation that states the product of a fluid's speed and the cross-sectional area of its flow tube is constant

equilibrant force A force that yields a zero force sum when added to one or more concurrent forces

equilibrium position The rest position of a system

equipotential surface An imaginary surface in an electrical field on which the electrical potential is the same at all points

error The difference between a measured value and the actual or accepted value

escape speed The minimum initial radial speed that an object must possess in order to permanently escape a gravitational influence

evaporation The process by which energetic liquid-phase molecules spontaneously enter the gaseous phase

exchange particle In the particle exchange theory of forces, a particle that, when passed from one entity to another, results in a force between the entities

excited state The condition of an electron that is in an energy level higher than its lowest possible energy state

exoergic A nuclear reaction that converts mass to other forms of energy

experiment The process of testing a hypothesis under carefully controlled conditions, meticulously recording the method and results, and drawing conclusions by inductive or deductive reasoning or both

F

farad (F) The SI unit of capacitance; 1 F = 1 C/V

Faraday's law The law that states that a change in magnetic flux around a conductor induces a potential difference in the conductor

ferromagnetic The property of certain metals strongly attracted to magnets; having a magnetic permeability (μ) much greater than 1

fiber optics The use of glass fibers to carry light by total internal reflection

field model A diagrammatic (rather than mathematical) approach to the study of certain phenomena, such as magnetic or electrostatic fields

field theory That the gravitational attraction exerted by matter and the attraction or repulsion of magnetic or charged particles are properties of the space surrounding the matter (the field)

first-class lever A lever in which the fulcrum is between the resistance load and the effort force

first ionization energy The energy required to remove the highest-energy electron in a neutral atom

first law of thermodynamics The physical law that states that heat transferred to or from a system is equal to the sum of the change of the system's internal energy and the work the system does on its surroundings; One consequence is that energy and matter are conserved in all processes.

fission The process of splitting a massive nucleus, usually with the release of great amounts of energy and two large, similar-mass fragments

flow tube An imaginary volume of moving fluid bounded by streamlines

fluid mechanics The application of Newtonian principles to fluids; subdivided into hydrostatics and hydrodynamics

fluoresce To release energy, usually visible light, after exposure to ultraviolet rays or higher energy radiation

focal length (f) The distance from the center of a lens or mirror to its principal focus

force board A circular laboratory apparatus designed to demonstrate the summation of concurrent forces

force A push or a pull on a system

formula unit A single molecule of a molecular chemical compound; the smallest ratio of elements in an ionic compound

frame of reference A coordinate system that an observer chooses to consider stationary

free-body diagram A diagram that isolates systems to analyze the forces on each system

free electron An electron not associated with a particular nucleus; a bonding electron in a metal

free fall Falling under the influence of only gravity

frequency (f) Number of cycles completed per unit time

friction (f) The contact force between two surfaces that opposes motion

fundamental The lowest dominant frequency in a complex sound, such as a single tone produced by a musical instrument

fundamental charge (e) The magnitude of electrostatic charge on a single electron or proton

fundamental force One of four constituent forces in nature—gravitational, electromagnetic, strong nuclear, and weak nuclear forces

fuse A one-time use electrical safety device that melts to break a circuit if the current becomes too high

fusion The nuclear process of combining two or more smaller nuclei into one larger nucleus, releasing great amounts of energy

G

Galilean relativity The idea that the laws of mechanics are the same in all nonaccelerating reference frames

Galilean transformation A change in coordinates from one inertial frame to another at nonrelativistic speeds between the reference frames

galvanic cell A source of electrical potential difference created by dissimilar metals immersed in an electrolyte

galvanometer A device that measures very small electric currents

gamma radiation (γ) Gamma rays emitted by radioactive nuclei

gauge pressure (P_g) The pressure of a fluid system with reference to ambient atmospheric pressure

gauss A non-SI unit of magnetism; used for weak magnetic fields; $10\,000$ gauss $= 1$ tesla

Gay-Lussac's law The pressure of a confined gas is proportional to its absolute temperature, provided its volume is held constant; $P = kT$

general theory of relativity The idea that acceleration and gravity are indistinguishable and that the laws of mechanics are valid in all reference frames with appropriate transformation rules

geocentric theory The theory that the earth is the unmoving center of the universe

geochronology Any of several methods used to estimate the age of individual rocks and the earth itself

gravitational force The force between two systems exerted across a distance due to their masses

gravitational lensing The bending of light from more distant celestial objects around closer massive galaxies

gravitational potential Gravitational potential energy per unit mass at a specific height or distance from the source of gravity; $U_g/m = |g|h$

gravitational red shift The apparent stretching of light wavelengths due to the proximity of an extremely strong source of gravity

gray (**Gy**) The SI unit of radiation dose to organic tissues; 1 Gy $= 1$ J absorbed/kg tissue

ground fault interrupter (**GFI**) A circuit breaker used in some electrical circuits to protect people by shutting off electricity if even a minute current to ground is detected

ground state The lowest possible electron energy level in an atom

H

hadron The family of elementary atomic particles that includes the mesons and the baryons

half-life ($t_{1/2}$) The amount of time required for half of the number of a particular radioactive nuclei in a sample to undergo a certain kind of decay

harmonic A sound wave whose frequency is an integer multiple of the principal or fundamental frequency

heat (Q) Thermal energy transferred from one system to another

heat capacity (C) The amount of thermal energy required to raise the temperature of a system 1 degree

heat death The uniformitarian theory that, given enough time, all of the energy in the universe will become unavailable for work, and the temperature will be uniform at a few kelvins

heat engine A device that converts thermal energy to mechanical work

heat pump A device that moves thermal energy from a cold reservoir to a hot reservoir

heat reservoir A theoretical body that is assumed to remain at a constant temperature no matter how much energy is added to it or taken from it

Heisenberg uncertainty principle The principle that it is impossible to know the momentum and exact position of a subatomic particle at the same time

heliocentric theory The theory that the Sun is at the center of the solar system

henry (H) The SI unit of inductance; $1\ H = 1\ Wb/A$

hertz (Hz) The SI unit of frequency; $1\ Hz = 1\ s^{-1}$

Hooke's law Within the elastic limit of a material, the restoring force is directly proportional to the displacement of the system from its equilibrium position

Hooke's law force Any nonconstant force exerted by or on a system that obeys Hooke's law

horizontal projection The motion of a projectile having constant horizontal velocity and zero initial vertical velocity

hydraulic device A simple machine that transmits a force via an enclosed liquid

hydrodynamics The study of fluids in motion

hydrometer A weighted apparatus that indicates the density of a liquid on a calibrated scale by the depth that it floats

hydrostatics The study of fluids at rest

hypothesis An educated guess at the solution to a scientific problem

I

ideal fluid A hypothetical incompressible fluid that is assumed to flow smoothly (without turbulence), without friction, and with a velocity that is constant at every point in the fluid

ideal gas A hypothetical gas that behaves exactly according to the ideal gas law

ideal gas law Relates the pressure, volume, temperature, and amount of an ideal gas at moderate conditions; $PV = nRT$

ideal gas process A process that allows the use of ideal gas relationships or formulas

ideal pulley A grooved, frictionless wheel and axle used to change the direction of a tension force on a string, rope, cable, or chain

ideal spring A hypothetical spring that has no internal resistance, has no mass, and is perfectly elastic

ideal string A hypothetical string that has no mass and does not stretch

illuminance (E) The amount of light an object receives from a light source; proportional to the intensity of the source and inversely proportional to the square of its distance from the source

image distance (d_I) The distance of the image from the center of the mirror or lens

impedance (Z) The resistance to current flow in an AC component or circuit; depends on the resistance and/or reactance of the component

impulse (I) The change in momentum due to the application of a force; $\mathbf{I} = \Delta\mathbf{p} = \Sigma\mathbf{F}\Delta t$

inclined plane A category of simple machines that includes wedges, screws, and ramps; a sloped surface

index of refraction (n) The ratio of the speed of light in a vacuum to the speed of light in a given medium; a measure of a medium's optical density

inductance (L) A measure of a circuit component's induced potential difference in the presence of a changing current

induction (Reasoning) Arriving at a conclusion using inductive reasoning; (Electrostatics) Causing a segregation of charges in an object without touching it with another charged object

inductive reactance (X_L) The opposition an inductor presents to a changing flow of current

inductive reasoning A form of logic that derives general principles from specific facts or instances; concludes what is most probably true until new evidence proves otherwise

inductor A circuit component that produces an induced potential difference due to a change in current

inelastic collision A collision in which kinetic energy is not conserved

inertial reference frame An unaccelerated coordinate system in which Newton's laws hold exactly

instrument (Scientific) Any device built for the purpose of observing physical phenomena; for scientific research, usually includes some means of making measurements; (Other) In general, any device required to perform a complex, delicate, or artistic task

insulator A material having no mobile electrons, so that it does not easily conduct thermal energy or electricity

intensity (I) The strength of a wave; generally, proportional to the square of the wave's amplitude

interface The boundary between two media in contact

interference The superposition of two or more waves that exist in the same place at the same time

interferometer An instrument that uses wave interference to determine the characteristics of the waves or to measure distance

internal energy (U) The sum total of energy consisting of the potential and kinetic energies associated with the particles of a substance

ion An atom or group of atoms that has lost or gained electrons, resulting in a net charge on the particle

ionizing radiation Any form of radiation energetic enough to remove electrons from atoms, creating ions; All forms of nuclear radiation and x-rays are considered ionizing radiation.

isobaric A process or condition at constant pressure

isochoric A process or condition at constant volume

isolated system A system that allows no exchange of energy or matter with its surroundings

isothermal A process or condition at constant temperature

isotopes Atoms of the same element that have different numbers of neutrons

joule (J) The SI unit of energy and work; $1\ J = 1\ N{\cdot}m$

Joule's law The power dissipated in a resistor is equal to the product of the resistance of the resistor and the square of the current through the resistor; $P = I^2 R$

kelvin (K) The SI fundamental unit of temperature; the unit of temperature on the absolute scale equivalent to 1 degree Celsius

Keplerian constant (K) For a given planetary system, the ratio of the cube of planet's orbital semi-major axis and the square of its orbital period is a constant; $K = R^3/T^2$

Kepler's laws Mathematical rules that geometrically describe the motions of the planets

kilogram (kg) The SI fundamental unit for mass; One kilogram is defined as the mass of a manufactured Pt-Ir alloy standard maintained at the International Bureau of Weights and Measures in Sèvres, France.

kilowatt-hour (kWh) The unit of electrical energy used by utilities to sell electricity

kinematics The branch of mechanics that describes how systems move without addressing what causes motion

kinetic energy (K) The mechanical energy possessed by a moving system; $K = \frac{1}{2}mv^2$

kinetic friction (f_k) The friction force that opposes the motion of a moving object; $f_k = \mu_k N$

kinetic theory of matter A theory that states that matter consists of tiny particles in constant motion; The properties of matter are consequences of that motion. Also called the kinetic-molecular theory

Kirchhoff's rules Rules describing the relationships among voltage drops in a circuit and relationships among currents in different branches of a circuit

laminar Smooth, layer-like, with continuous streamlines; used to describe fluid flow

latent heat of fusion (L_f) The amount of thermal energy required to change a certain mass of a solid at its melting point to a liquid at the same temperature

latent heat of vaporization (L_v) The amount of thermal energy required to change a certain mass of a liquid at its boiling point to a gas at the same temperature

law A statement (usually mathematical) describing a consistent relationship among physical quantities in nature

law of charges Like charges repel; unlike charges attract

law of magnetic poles Like poles repel; opposite poles attract

law of moments One method for establishing rotational equilibrium; $F_1 L_1 = F_2 L_2$

law of reflection In ray theory, the angle of incidence to the normal equals the angle of reflection

Lenz's law An induced potential difference in a conductor opposes the change in the magnetic field that caused it

lepton The class of lightweight elementary particles that includes the electron, neutrino, and muon

lever A device that can pivot about a fulcrum, transmitting a force at one end to a load at the other

Leyden jar An early device consisting of a lead-lined jar used to store electrical charge

light-emitting diode (LED) A semiconductor device that emits light when a minimum specified potential difference is established across its terminals

line of force One of an infinite number of imaginary lines defining the direction and density of the field at all points surrounding an electric charge or magnet

line spectrum A spectrographic display consisting of lines representing electromagnetic emissions in discrete frequencies or wavelengths

lodestone A natural magnet of an iron oxide ore

longitudinal wave A wave in which the medium oscillates parallel to the direction of the wave's propagation

Lorentz transformation A change of system coordinates from one inertial frame to another that is applicable at any relative speed between the reference frames

lumen (lm) The SI unit of luminous flux, equal to a 1 cd light source in 1 sr; $1\ lm = 1\ cd{\cdot}sr$

luminance The rate of flow of light energy reaching a surface in a given direction; units of cd/m^2

luminiferous ether A hypothetical fluid believed by nineteenth-century scientists to be the medium for propagation of light waves

luminous flux (Φ) The measure of the total rate of light energy produced by a source; units of lumens (lm)

luminous intensity (I_L) The intrinsic brightness of a light source; the power of light energy emitted in a specific direction; units of candelas (cd)

lux (lx) The SI unit for illuminance; $1\ lx = 1\ lm/m^2$, or the illuminance of a 1 cd light source at 1 m distance

Lyman series A series of wavelengths emitted as excited electrons fall to the second energy level in an atom

machine A device that changes the magnitude or direction of a force or both.

magnetic declination For a given location on the earth's surface, the angle between the directions to true north and magnetic north

magnetic dipole moment (μ) The vector representing the magnetic effect of the electron current in an atom or molecule

magnetic field vector (B) Defines the strength and direction of the magnetic field at a specified point in space; units of teslas (T)

magnetic flux (φ) The strength of a magnetic field through a given area; units of webers (Wb)

magnetic induction The alignment of atomic magnetic moments in a nonmagnet along an external magnetic field

magnetic quantum number (m) The third quantum number, which defines the orientation of an electron's orbital

magnetosphere The earth's effective magnetic field

magnification The ratio of the height of an image to the height of its object

Malus's law The intensity of a polarized light beam transmitted by a dichroic material (the analyzer) is a function of the angle between the wave's plane of polarization and the analyzer's transmission axis.

manometer An apparatus that measures the pressure of a fluid by allowing it to support a liquid column against atmospheric pressure

maser A device that produces a coherent beam of microwave radiation

mass defect The difference between the mass of a nucleus and the sum of the masses of the particles from which it was formed; equivalent to the binding energy of the nucleus

mass number (A) The sum of the number of protons and neutrons in the nucleus of an atom

mass spectrograph A device that uses electric and magnetic fields to determine the masses of particles in a sample

matter Any created substance that occupies spatial volume and has mass; one of the three fundamental dimensions of creation

mechanical equilibrium A condition that exists when the vector sum of all forces and torques on a system is zero

mechanics The study of the motion of macroscopic objects; the description of (kinematics) and causes of (dynamics) motion

medium The substance that oscillates as a wave travels through it

meson A "medium-weight" class of elementary particles, including pions and etas

meter (m) the distance traversed by light in a vacuum in *exactly* $1/299\,792\,458$ second.

microscope An optical instrument designed to magnify extremely small objects

millibar (mb) A non-SI unit meteorologists use for measuring and reporting atmospheric pressure; one thousandth of a bar; 1 mb = 100 Pa

mirror equation The reciprocal of the distance between the object and the mirror plus the reciprocal of the distance between the image and the mirror equals the reciprocal of the mirror's focal length

moderator A material containing light particles (such as water) with which neutrons can collide to slow down, interspersed with fuel in a U-235 fission reactor

modern physics As distinguished from classical physics, all physics discoveries made since approximately 1900

mole (mol) The SI fundamental unit when a specific number of atoms or molecules of a substance is to be measured

molecule Two or more covalently bonded atoms that exist as a separate, distinct, independent unit

moment arm (L) The perpendicular distance between the pivot point and the line of the applied force when computing torque; $L = r\cos\theta$, where θ is the smallest angle between the force and the position vectors placed tail-to-tail.

moment of inertia (I) A characteristic of any rotating system that is defined by the distribution of mass from the rotational center

momentum (p) Newton's "quantity of motion"; The vector consisting of mass multiplied by velocity; $\mathbf{p} = m\mathbf{v}$

mutual inductance (M) A measure of a circuit element's contribution to the induced potential difference in another circuit element

mutual induction Induction of a potential difference across one circuit element as a result of a change in another circuit element

N

natural oscillation frequency (f₀) The frequency at which a system oscillates if it is given a single pulse of energy

neutron (n) A neutral particle in the nucleus of an atom, with slightly more mass than a proton

Newtonian reflector telescope A reflecting telescope using a flat secondary mirror to project the image to the eyepiece mounted on the side of the telescope tube

Newton's law of universal gravitation The law that describes the attractive force between all masses

Newton's laws of motion Principles defining the motion of "normal-sized" objects at nonrelativistic speeds

nonconservative force A force is classified as nonconservative if the total work accomplished on a system over a closed course is not zero or if the magnitude of the work done depends on the path taken.

noncontact force Any force that acts on a system without the necessity of physical contact

nonuniform circular motion Circular motion that involves a change in angular speed during a time interval

normal force (N) When a system is supported by a surface, it is the force exerted by the surface on the system that is perpendicular to the surface.

nuclear reactor A device containing an arrangement of nuclear fuel, a moderator, and control mechanisms that is used as a heat source for a connected steam plant

nucleon A proton or neutron bound in the nucleus of an atom

nucleus The dense central part of an atom made up of protons and neutrons; The nucleus contains virtually all of the atom's mass but occupies only a small portion of the atomic volume.

nuclide A specific isotope of an element; the term often used when comparing isotopes of different elements

O

object distance (d₀) The distance of the object from the center of the mirror or lens

objectivity An ideal state or quality of thinking that is not influenced by emotions, prejudices, or preconceived ideas

observation A deliberate application of the senses to a phenomenon, with or without the aid of instruments

ohm (Ω) The SI unit of electrical resistance; $1\ \Omega = 1$ V/A

ohmmeter An instrument that measures electrical resistance

Ohm's law The law that states that the potential difference across a circuit or circuit component is equal to the product of the current through it and its resistance; $V = IR$

open system A system that allows the exchange of both energy and matter with its surroundings

optical axis The imaginary line that passes through the optical center of a lens or mirror and its center of curvature

orbital The probable location and energy of an electron defined by its first three quantum numbers

orbital motion Circular motion about an axis that does not pass through the moving object

order of magnitude The value of a numerical quantity expressed to the nearest power of 10; a comparison of two values expressed to the nearest integer power of 10

overcurrent A condition of an electrical circuit where the current flowing in it exceeds the current it is designed to safely carry

overdamped oscillation A partial oscillation where damping is so strong that the system returns directly to its equilibrium position with no overshoot

P

paradigm A set of assumptions, concepts, values, beliefs, and practices that controls the way a person or a group of persons perceives reality

parallel The type of electrical circuit that has more than one simultaneous path for the current to and from the voltage source

paramagnetic A material that is slightly attracted to magnets; having a magnetic permeability (μ) slightly greater than 1

parent nuclide A radioactive nuclide that emits alpha or beta particles and becomes a different nuclide

partially elastic collision A collision in which the colliders change some of their kinetic energy to other forms of energy but do not stick together

pascal (Pa) The SI unit of pressure; $1\ \text{Pa} = 1$ N/m^2

Pascal's principle The principle that an enclosed fluid exerts the same pressure in all directions at the same depth

Pauli exclusion principle No two electrons in the same atom can have identical sets of quantum numbers

pendulum A mass suspended at some distance from a pivot point that is free to swing back and forth

pendulum arm (l) The distance of the center of mass of a pendulum from the system pivot point

percent error The measurement error given as a percentage of the known or accepted value

performance coefficient (K) A numerical rating of the efficiency of a refrigeration engine or heat pump

periodic motion Motion that continuously repeats itself

periodic table A table of elements, arranged in order of increasing atomic number, relating their electron structures and chemical similarities

period (T) The time required to complete one cycle; a horizontal row in the periodic table of the elements

permeability of free space (μ_0) A measure of how effectively a magnetic field penetrates a vacuum

permittivity of free space (ε_0) A measure of how effectively an electric field penetrates a vacuum

permittivity of material (ε) A measure of a material's ability to resist the formation of an electric field within it

phase diagram A graph showing the relationship between the phases of a substance and controlling factors such as temperature, pressure, or composition

phenomenon An observable event or process

photoelectric effect Light of an appropriate frequency striking polished metal ejects electrons from the metal.

photometer An instrument that compares the luminous intensity of an unknown light source to a standard light source

photon An electromagnetic "particle" with energy hf and zero rest mass

physical pendulum A real pendulum, where some mass is distributed along the pendulum arm

pigment A substance that selectively absorbs certain wavelengths of light; Its color comes from the wavelengths it reflects or transmits.

pitch The distance between two adjacent threads on a screw; the quality of highness or lowness of sound that depends upon the fundamental frequency

Planck's constant (h) One of the four fundamental constants; $h = 6.626 \times 10^{-34}$ J·s; the step-size of quantization; All wave energies are multiples of h.

plane wave A wave with a flat, planelike wave front

plate One of the conductors in a capacitor; the anode in a vacuum tube

plum-pudding model The atomic model that views an atom as a sphere of positive charge in which negative electrons are embedded

polarity marking The + sign indicates a point of higher potential. The − sign indicates a point of lower potential.

polarization A condition of a light beam where all the E-waves oscillate in other than random directions

polarized (Electrical) Exhibiting two or more opposing electrical poles; applies also to magnetic materials; (Electromagnetic radiation) Exhibiting a preferred alignment of the E-wave after interacting with matter

pole A region or location where magnetic flux or electric charge is concentrated

position-time graph A graph of an object's position versus time

position vector (r) A vector with its tail at the origin of a coordinate system and its head at the position of an object at a given time

positron (β^+) A particle with the mass of an electron but a positive charge; an antielectron

potential difference (V) The difference of the electrical potential between two locations

potential energy (U) Mechanical energy that an object possesses because of its position or condition with respect to a conservative force

pound-force (lbf) The unit of force in the English Engineering System; the gravitational force on 1 lbm

pound-mass (lbm) The unit of mass in the English Engineering System

power (P) The time-rate at which work is performed; $P = W/\Delta t$

precision How close several measurements of the same quantity are to each other; how many decimal places a measuring instrument can yield

prejudice A preconceived and uninformed judgment or conviction; the condition or existence of such

primary cell A voltaic cell that uses an irreversible chemical reaction to supply electricity

principal focus (F) For a mirror or lens, the point at which incoming light rays parallel to the principal axis converge

principle of inertia An object in motion will continue in its original state of motion unless or until an outside agent acts on it; attributed to Galileo Galilei

principal quantum number (n) The first quantum number, denoting the energy level, or shell, of an electron

projectile Any flying object that is given an initial velocity and then allowed to fall under the influence of external forces, such as gravity

proper length (l_0) The length of an object as measured by an observer who has the same velocity as the object

proton (p) A positively charged particle in the nucleus of an atom with 1836 times more mass than an electron

pulsating DC A current that periodically oscillates between zero and a maximum but does not change direction

P-V diagram A graph on which the pressure of a working gas is plotted as a function of the gas's volume

Q

quality The ear's interpretation of the mixture of fundamentals and harmonics in a sound wave

quality factor (QF) A radiation dose correction factor based on the kind of radiation (α, β, or γ) and kind of tissue exposed in calculating the dose-equivalent

quanta Plural of "quantum"

quantized Produced or transmitted as quanta

quantum mechanics The study of matter and radiation on an atomic level in view of their quantized nature

quantum number One of four numbers that describe the location and energy of an electron in an atom; arguments of the Schrödinger wave equations

quark The most fundamental or elementary atomic particle; There are six kinds, or "flavors," all of which, with their antiquarks, compose the hadrons.

quark confinement The theory that quarks can exist only inside hadrons and are held there by a property called color charge

R

radian (rad) The SI unit of angular measure; in a circle, the central angle that subtends an arc on the circle equal in length to the radius of the circle

radiation The transportation of energy or matter without the use of an intervening medium

radioactive dating Estimating the age of a material by comparing the concentration of a radioactive nuclide and its daughter products, and calculating the amount of time necessary for the measured proportion to occur; requires making a variety of assumptions, many of which are usually not scientifically supportable

radioactivity The act of emitting rays and particles from an unstable nucleus; the time rate of such emissions

radiogenic A substance derived from radioactive decay

range (R) The horizontal distance covered by a projectile

rarefaction zone In a sound wave, the region where the medium particles are farther apart than average; the "trough" of a sound wave

ray A line representing the direction of travel of a light wave; It is perpendicular to the wave front.

Rayleigh's angle (θ_R) For a given lens, the smallest angle between two objects, such that the lens is able to resolve them into two separate images; depends on lens diameter

real image An image formed by light rays that actually meet at the location of the image

recoil The reaction force of a projectile on the device that launches it

reference angle (α) The acute angle between a vector and its horizontal component vector

reference direction The direction from which the vector angle is measured; usually the direction of the positive x-axis or to the north pole, depending on the situation

reflecting telescope A telescope that uses a parabolic or spherical mirror as the principal light-gathering element

reflection The bouncing of a wave off a surface

refraction The bending of waves as they pass from a medium of one transmission density to another medium with a different density

refracting telescope A telescope that uses a lens as the principal light-gathering element

refrigeration engine A heat engine that moves thermal energy from the cold reservoir to the hot reservoir

regelation The melting of ice by the application of extreme pressure

relative permeability (κ_m) The ratio of the magnetic permeability of the material and the permeability of free space (μ_0)

relative uncertainty A comparison of the magnitude of the uncertainty to the size of the measurement

relativistic Occurring at speeds of about $0.1c$ or greater

relativistic factor (γ) A mathematical factor that takes into account the speed of an object relative to a stationary observer in order to apply the rules of special relativity

relativity The idea that it is impossible to distinguish between moving systems and those at rest using the laws of classical mechanics if the appropriate transformations are made

resinous Electrical behavior like that of amber

resistance (R) An electrical property that quantifies an object's opposition to current flow

resistivity (ρ) A measure of the degree to which a given material impedes current flow at a given temperature

resistor Any electrical component that depends primarily on its material to impede the flow of current through it, rather than its shape or construction

resolving power The ability to separate two objects close in angle into two separate images

resonance The state of an AC oscillator circuit with zero impedance; the state of an oscillating system driven at its natural frequency; the transfer of energy from one object to another with the same natural frequency

rest mass (m_0) The mass of an object at rest as measured by an observer who is also at rest with respect to the object

rest-mass energy (E_0) The total energy of an object at rest; $E_0 = m_0 c^2$

restoring force ($\mathbf{F}_r$) A force that acts parallel to a system's motion to return it to its equilibrium position

resultant The vector sum of two or more vectors

reversible A controlled, quasi-static process that operates in either direction between two distinct states; a cyclic process that can be operated in reverse (e.g., the Carnot cycle)

rheostat A manually controlled variable resistor

right-hand rule for circular motion The right thumb points in the positive vector direction when the fingers wrap around the rotational axis in the direction of rotation

right-hand rule for magnetic force For a positive charge moving through a magnetic field, when the right index finger is extended in the direction of the magnetic field vector $\mathbf{B}$ and the right thumb is extended at a right angle to the index finger in the direction of the charge's velocity component perpendicular to the magnetic field ($\mathbf{v}_\perp$), then the right middle finger extended perpendicular to the other two indicates the direction of the force on the charge. The force on a negative charge is in the opposite direction.

right-hand rule for conductors When the thumb points in the direction of conventional current flow in a conductor, the fingers wrap around the conductor in the direction of the magnetic field vector

right-hand rule for torques The right extended thumb points in the direction of the torque vector along the axis of rotation when the fingers wrap from the direction of the position vector to the direction of the applied force

right triangle A triangle containing a 90° angle

roentgen (R) A non-SI unit of the ionizing effect of electromagnetic radiation on air; 1 R = 2.58×10^4 C/kg of air

rolling friction (f_r) An internal force that tends to oppose the rolling motion of a system

root mean square (rms) The effective (rather than the peak) value of an alternating quantity

rotational equilibrium The condition of a system when the sum of all torques is zero; The system will rotate either at a constant speed or not at all.

rotor That part of an electric motor, and many electrical instruments, that rotates under the influence of electromagnetic forces

S

scalar A quantity that can be completely described with only one piece of numerical information; a number that can be positive, negative, or zero

scalar component The scalar value of a component vector

Schmidt-Cassegrainian telescope A reflecting telescope having a spherical primary mirror, a convex secondary mirror, a rear-mounted ocular lens, and a front correcting plate to eliminate spherical aberration

scientific methodology A systematic way of finding the answer to a question about the observable world

scientific problem An observation that contradicts an accepted law or theory

screw A metal shaft surrounded by a helically wrapped inclined plane

second-class lever A simple machine where the load is between the fulcrum and the effort force

second law of thermodynamics The physical law that states that during any energy transformation, some energy becomes unusable

second (s) The SI fundamental unit of time; defined as exactly 9 192 631 770 cycles of oscillating gaseous cesium-133 atoms

self-induction The induction of a back emf in a circuit component by current changes in that component

semiconductor A substance that has electrical properties intermediate between conductors and insulators

semi-major axis (R) Half of the distance from a planet's perihelion to its aphelion

series A circuit with a single path of current flow

shear modulus (G) The ratio of the shear stress to the shear strain

SI Abbreviation for *Système International d'Unités (International System of Units)*; the system of standardized units used by scientists worldwide

sievert (Sv) The SI unit for radiation dose-equivalent calculations

significant digit (SD) In a measurement, any of the digits that is known for certain based on the instrument's scale markings, plus one estimated digit (a decimal fraction of the smallest scale increment)

similar triangles Triangles with congruent corresponding angles

simple harmonic motion (SHM) Periodic motion perpetuated in an ideal system by a restoring force

sine θ In a right triangle, the ratio of the length of the side opposite to the acute angle θ to the length of the hypotenuse

Snell's law A law governing the angles of incidence and refraction of light across the interface between two media

solar wind The stream of charged particles emitted by the Sun

solenoid An electromagnet consisting of a coiled conductor surrounding a ferromagnetic metal core

space A volume in which matter may be found; one of the three fundamental aspects of creation

special theory of relativity The idea that (1) all of the laws of physics have the same form in all nonaccelerated frames of reference, and (2) the speed of light in a vacuum is the same in all reference frames

specific gravity (s.g.) The ratio of a sample's density to the density of water

specific heat (c_{sp}) The amount of thermal energy required to raise the temperature of 1 g of a substance 1 °C

spectral series A series of wavelengths of radiation emitted by electrons of any energy falling to the same energy level

spectrum The ordered distribution of some aspect of a physical system; an arrangement of electromagnetic emissions by wavelength or frequency

speed of light (c) Refers to the speed of any electromagnetic radiation; generally means the speed of light in a vacuum, which is approximately 3.00×10^8 m/s—the maximum speed attainable in the physical universe

spherical aberration Distortion of an image due to the geometric impossibility of focusing parallel rays at a single point by a spherical mirror or lens

spherical waves Waves emitted by a point source

spin Rotation of an object about an axis that passes through the object

spin quantum number (m_s) The fourth quantum number, denoting the direction in which an electron spins

split-ring commutator A device mounted on the shaft of a rotor consisting of two conducting half-rings separated by an insulator and connected to the coil winding conductor; Brushes resting on the commutator carry current to or from the coils rotating in a magnetic field. When the brushes switch half-rings, they reverse the current through the commutator.

spring constant (k) The constant in Hooke's law for a spring, showing how much force is required to cause a given amount of compression or extension in the spring

standard temperature and pressure (STP) A set of agreed-upon conditions for using the ideal gas laws; 273.15 K and 1 atm

state variable A physical variable that defines the state of a system; a path-independent variable

static electricity Related to or producing stationary electrostatic charges

static friction (f_s) Friction acting on an object at rest, opposing any forces that act to initiate motion

stator The stationary part of an electric motor, and many electrical instruments, that contains the magnets or coils that envelop the rotor

Stefan-Boltzmann constant (σ) A proportionality constant relating radiant energy emitted by a body to its absolute temperature

Stefan's law The rate of thermal emissions radiated by an object is proportional to the fourth power of its absolute temperature.

step-down transformer A transformer designed such that the input coil has more turns than the output coil, therefore output AC voltage is lower than the input voltage

step-up transformer A transformer designed such that the input coil has fewer turns than the output coil, therefore output AC voltage is higher than the input voltage

steradian (sr) The conical solid angle, with its vertex at the center of a sphere, that cuts off a circular area on the surface of the sphere equal to the radius of the sphere squared

stereo binocular microscope A microscope that has two objectives and two eyepieces, producing a stereoscopic image

storage cell A voltaic cell that uses a reversible chemical reaction to produce electricity

strain (ε) In an object subject to mechanical stress, the ratio of the amount of change in a dimension to the original dimension

streamline In fluid flow, an imaginary line that shows the path of fluid particles

stress (σ) Also called normal stress; the ratio of the magnitude of the force applied perpendicular to a specified surface area and the area itself

strong nuclear force The short-range force between nucleons that holds the nucleus together

sublimation A change of phase directly between the solid and gaseous states, in either direction, bypassing the liquid phase

subtractive primary color One of three pigments that, when added in various combinations, produce all colors by the absorption of incident or transmitted light; yellow, magenta, and cyan

superconductor A material that, at temperatures near 0 K, offers no detectable resistance to current

superposition principle Waves in a given medium will generate constructive or destructive interference in proportion to how much they are in or out of phase, their amplitude, and how closely they match in frequency; applies to any kind of wave or medium

surface tension The tendency for the surface of a liquid to present a barrier to penetration due to unbalanced forces on the surface particles

surroundings In thermodynamics, everything outside the designated boundaries of a system

system A part of the universe isolated, at least mentally, for study

T

tangent θ In a right triangle, the ratio of the length of the leg opposite the acute angle θ and the leg adjacent to the angle

tangential acceleration (a_t) The time-rate of change of the tangential velocity; $\Delta v_t / \Delta t$

tangential velocity (v_t) The velocity vector whose magnitude is equal to the tangential speed and whose orientation is perpendicular to the rotating position vector

teleological The use of ultimate purpose to explain phenomena

telescope An optical instrument that makes distant objects appear to be nearer using lenses or mirrors to collect light

tension (T) A force that tends to stretch or pull an object apart

terminal velocity For a falling object, the velocity at which air resistance and its weight balance; the limit of the velocity at which an object can fall through a fluid

tesla (T) The SI unit of magnetic field intensity; $1 \text{ T} = 1 \text{ Wb/m}^2$

test charge A positive charge small enough that it does not distort an electric field that is used to determine the direction of the electric field vector

theoretical mechanical advantage (TMA) The ratio of the output force of a machine to its input when friction is not present

theory A partially verified explanation to a scientific problem that relates a number of different observations

thermal efficiency (ε) The ratio of the absolute value of the net work done by a heat engine to the thermal energy absorbed by the engine from the hot reservoir; $\varepsilon = |W_{cycle}|/|Q_H|$

thermal equilibrium Two systems are in thermal equilibrium when they are separated by a diathermic boundary yet they exchange no thermal energy—they must be at the same temperature.

thermocouple A thermoelectric device that employs two dissimilar metals joined so that a change in temperature causes a change in electrical potential across the contact surface

thermodynamics The study of the sources and utilization of heat and its conversion to other forms of energy

thermodynamic system A system studied for its thermodynamic properties and where its total mechanical energy is assumed to be constant

thermometer A device that uses a thermometric property of a material to measure and display temperature

thin-lens equation A formula relating object distance, image distance, and lens focal length that is valid only for thin lenses

third-class lever A simple lever machine in which the effort force is applied between the fulcrum and the load

third law of thermodynamics The principle that absolute zero is unattainable

thrust The impulsive force accompanying the expulsion of propellant exhaust that accelerates a rocket or aircraft

time A nonphysical continuum that orders the sequence of events and phenomena; one of the three fundamental dimensions of the universe

tokamak A doughnut-shaped fusion reactor using electromagnetic fields to contain high temperature hydrogen plasma

torque (τ) The effect of a force applied at a distance from an axis of rotation; the product of a force and its moment arm

torr A non-SI unit of pressure; 1 torr of atmospheric pressure will support a column of mercury 1 mm high in an evacuated tube.

torsion balance A sensitive instrument for measuring an unknown force by balancing it against a known torsional force

total acceleration (a_{total}) In circular motion, the vector sum of tangential and centripetal accelerations

total internal reflection Reflection of the light ray at the interface when the incident angle in the optically denser medium exceeds the critical angle for the two media

total magnetic flux (Φ) The sum of the flux produced by each coil in a solenoid or inductor; net magnetic flux in the core of an inductor

total mechanical energy (E) The sum of the potential energy and kinetic energy of a system

tracer A small amount of a radioactive isotope used to detect the movement of an element through a biological system

traction (f_t) The frictional force that opposes the propulsive force of a system acting on the ground; the force that permits motion over ground

trajectory The path of a projectile through air or space

transformer An electrical device consisting of two coils of wire wound on a ferromagnetic core, such that the output AC voltage is different than the input voltage; a similar device that isolates the current in one AC circuit from another while allowing energy to flow between them

transmission axis The direction through the crystal lattice of a dichroic material that transmits the E-field of light undiminished; defines the polarization of plane-polarized light

transmittance The measure of transparency equal to the ratio of transmitted luminous flux to the incident flux

transverse wave A wave in which the oscillating component vibrates at right angles to the direction of the wave's travel

triple point The single condition of pressure and temperature for many substances where all three phases exist simultaneously

tuning circuit A resonating circuit that responds significantly only to a narrow band of input frequencies

turbulent Fluid flow characterized by discontinuous or irregular streamlines

U

ultraviolet catastrophe The discovery that classical physical formulas incorrectly predict that a blackbody will radiate an infinite amount of energy at high frequencies

unbalanced forces Concurrent forces that sum to a nonzero resultant force; the unequal forces on an object that cause the object to accelerate

uncertainty The estimated maximum value by which a measurement might differ from the true measured value; in this textbook, assumed to be $\pm \frac{1}{2}$ the place value of the estimated significant digit

uniformitarianism The belief that physical processes have always followed the same laws—that is, the present is the key to the past; allows for no miracles or other supernaturally caused physical events in the earth's history

uniformly accelerated motion Motion where the average acceleration equals the instantaneous acceleration; Acceleration during the time interval of interest is constant.

uniform circular motion Circular motion where the angular speed is constant

unit analysis As a problem-solving method, the cancellation of units in factors on one side of an equation so that the remaining units can be compared to the expected units in the solution

universal gas constant (R) The product of Boltzmann's and Avogadro's constants; in SI units, $R = 8.135 \text{ m}^3 \cdot \text{Pa}/(\text{K} \cdot \text{mol})$

universal gravitational constant (G) The proportionality constant in Newton's law of universal gravitation; $G = 6.673 \times 10^{-11} \text{ N} \cdot \text{m}^2/\text{kg}^2$

unpolarized A condition where the E-waves in a beam of light oscillate in randomly oriented planes

V

Van Allen belt Any of several toroidal regions of charged particles encircling the earth high above the atmosphere, trapped by the earth's magnetic field

Van der Waals gas A gas that obeys the more complex Van der Waals equation that predicts the relationship between pressure, volume, and temperature at extremes of temperature and/or pressure where the ideal gas law fails

vaporization point The temperature at which all the liquid molecules have sufficient kinetic energy to escape from the liquid phase and enter the gaseous phase

vapor pressure The equilibrium pressure exerted by the gaseous phase of a substance in contact with its liquid or solid phase in an enclosed container under specified conditions of pressure and temperature

vector A physical quantity that requires two pieces of information in order to be completely described—a scalar value and a direction relative to a reference direction

vector angle (θ) The direction of a vector from a reference direction, expressed in degrees or radians

velocity (v) The time-rate of change in position in a specified direction; $\Delta r/\Delta t$

velocity selector An analytical device that uses superimposed electric and magnetic fields to deflect all particles in a collimated beam except those with a speed of E/B

velocity-time graph A graphical plot of velocity (usually one component) versus time; may be used to evaluate displacement and accelerations during selected time intervals

vernier scale An auxiliary graduated scale on a measuring instrument used to increase the precision of measurements

virtual image An image formed by light rays that appear to diverge from points behind the mirror or lens; an optical illusion formed by the tendency of the human mind to assume light rays travel in straight lines

virtual particle In the exchange particle theory of forces, a nonmaterial particle that is believed to carry energy and create forces; Known virtual exchange particles are photons, gluons (pions), and the W/Z-particles.

viscosity The "thickness" of a fluid or the tendency of a fluid to resist flowing

vitreous Electrical behavior like that of glass

voltage drop The change in electrical potential across a circuit component that causes current to flow through it

voltaic cell An electrochemical cell in which a spontaneous reaction produces an electrical potential and a source of electrons

voltmeter A device that measures the potential difference between two points in a circuit

volt (V) The SI unit used to measure electrical potential and potential difference; $1 \text{ V} = 1 \text{ J/C}$

W

watt (W) The SI unit of power; $1 \text{ W} = 1 \text{ J/s}$

wave For physical waves, an oscillatory disturbance traveling through a medium; for electromagnetic waves, a self-propagating interaction between oscillating electric and magnetic fields

wave front A model of wave propagation; a surface perpendicular at all points to the direction of a wave's travel

wavelength (λ) The distance between corresponding points on adjacent waveforms

wave speed (v) The speed with which a wave travels; a property of the medium, not of the origin or shape of the wave

weak nuclear force The force that enables all lepton and quark particles and their antiparticles to interchange energy, mass, and charge, effectively allowing them to change into each other

weber (Wb) The SI unit of magnetic flux; $1 \text{ Wb} = 1 \text{ V} \cdot \text{s}$

weight (F_w) A measure of the force of gravity acting on an object's mass at the earth's surface

Wien's law In a blackbody radiator, the wavelength of maximum radiant intensity is a function of the blackbody's kelvin temperature.

work (W) (Mechanical) The energy transferred to or from a system as a force acts on it through a displacement; $W = Fd \cos \theta$

work-energy theorem The work done on a system by all external forces is equal to the change in the system's kinetic energy.

work function The minimum electromagnetic energy (hf) needed to eject the most loosely bound electrons from a metal surface

X, Y, Z

zeroth law of thermodynamics Two systems in thermal equilibrium with the same third system are in thermal equilibrium with each other.

zero vector (0) Any vector with zero magnitude; useful for determining difference vectors originating at the origin of the coordinate system

INDEX

A

absolute pressure (P), 371
absolute zero, 311
acceleration (a), 59–60
accelerometer, 123
accuracy, 31–32
achromatic doublet, 542
action-reaction force pair, 129
activity (A), 635
actual mechanical advantage (AMA), 224
additive primary color, 575
adhesion, 289
adiabatic, 332, 353
aeolipile, 355–56
Aetna, Mount, 98
airfoil, 387
Aldrin, "Buzz," 556
alpha (α) particle, 632
alpha radiation (α), 631–32
alternating current (AC), 484
ammeter, 442
amorphous solid, 291
ampere (A), 430
Ampère, André Marie, *457*
amplitude (A), 259
ampule, 632
analyzer (polarizer), 562
angle of
    incidence (θ_i), 513
    reflection (θ_r), 513
    refraction (θ_r), 529
angular acceleration (α), 142–43
angular momentum (L), 250–51
angular speed (ω), 139
angular velocity (ω), 141
annual variation, 459
anode, 431
anoxia, 375
antenna, 493
antielectron, 647
antiparticle, 647
antiprotons ($\bar{p}$), 647
antisolar line, 533
aperture, 541
Apollo 11, 556
Apollo 15, 114
apparent weight, 171–73
a priori, 3
Aquinas, Thomas, *7*
Archimedes, 378
Archimedes' principle, 378
argument (of an operator), 80
Aristarchus of Samos, *151*
Aristotle, *7*, 113, **135**, 151
armature, 484
Armstrong, Neil, 556
ASAT (missile system), 62

astronomical unit (ua), 153
atmosphere (atm), 371
atom, 287
atomic clock, 25, 600
atomic number (Z), 614
aurora (borealis/australis), 461
average acceleration ($\bar{a}$), 96
Avogadro, Lorenzo R. A.C., *27*, 288
Avogadro's constant (N_A), 288, 319
azimuthal quantum number (l), 620

B

back emf (ε), 486
Bacon, Francis, *7*
balanced forces, 117
ballistic, 98
ballistic pendulum, 244
Balmer, Johann Jakob, 616
Balmer series, 616
bar, 371
barometer, 377
Baron Rayleigh (John William Strutt), *561*, 610
Bartholinus, Erasmus, *563*
baryon, 648
battery, 432
baumé, 381
Becquerel, Antoine Henri, *629*
becquerel (Bq), 635
Bernoulli, Daniel, *386*
Bernoulli's equation, 386
beta radiation (β), 631–32
bias, 3
bimetallic strip, 306
binding energy, 601, 642
binocular, 578
bioluminescence, 508
biomechanics, 107
Biot, Jean-Baptiste, *478*
Biot-Savart law, 478
birefringence, 563
Black, Joseph, *329*
Blackbird (SR-71), 56
blackbody, 341
blackbody radiation, 609
black hole, 603
block and tackle system, 227
Bohr, Niels, 615, **627**
boiling point, 297
Boltzmann, Ludwig, *319*
Boltzmann's constant (k), 319, 610
Boyle, Robert, 316, *329*
Boyle's law, 316–17
Brackett series, 616
Brahe, Tycho, 116, *152*
breeder reactor, 644
bremsstrahlung, 505
Brewster, Sir David, *562*

Brewster's angle (θ_B), 562–63
Brewster's law, 563
bridge circuit, 443
brightness, 574
brix, 381
Brown, Robert, *328*
Brownian motion, 328
brush (commutator), 472
bubble chamber, 634–35
buoyant force, 378

C

caloric theory, 329
calorie (cal), 330
calorimeter, 333
Canadarm, 129
canal ray, 400
candela (cd), 570
capacitance (C), 421
capacitive reactance (X_C), 492
capacitor, 419–20
    parallel connected, 422
    parallel-plate, 420
    polarized, 424
    series connected, 423
capillarity, 296
cardinal direction, 75
Carnot, Nicolas Sadi, *18*, **344**
Carnot cycle, 358
Cartesian coordinate, 82
cascade shower, 653
Cassegrain, Jacques, *581*
Cassegrainian reflector telescope, 580
Cassini, 248
cathode, 431
cathode ray, 468
cathode ray tube (CRT), 467
Cavendish, Henry, *156*
Celsius, Anders, *311*
center of buoyancy, 380
central force, 217
central maximum, 549
centripetal acceleration (a_c), 140
centripetal force (F_c), 146
chain reaction, 643
Charles, Jacques, *314*
Charles's law, 314–15
chemical compound, 287
chromatic aberration, 542
chronometer, 27
Church, Roman Catholic, 67
circuit, 432
circuit breaker, 446
circuit component, 433
circular motion, 138–41
circularly polarized, 562
classical physics, 15, 17–18

clock paradox, 600
closed system, 353
clothoid loop, 160
cloud chamber, 635
coefficient
    of friction (μ), 175
    of linear expansion (α), 305
    of viscosity (η), 388
    of volume expansion (β), 306
coherent, 554, 624
coherent light, 554
cohesion, 289
cold reservoir, 355
collimator, 560
color
    charge, 649
    visual, 573–74
Columbia, 216, 233
combined gas law, 318
comparator photometer, 572
complementarity, 619
complementary color, 575
components, 82
component vector, 82
compound microscope, 577
compression zone, 273
Compton, Arthur Holly, *613*
Compton effect, 613
concave, 514
concurrent, 119
conduction, 625
conduction energy band, 339
conductor, 401, 423
conservation law, 215
conservation of mechanical energy, 208–9
conservation of momentum, 236
conservative force, 201
constructive interference, 548
contact (charging), 404
contact force, 122
control rod, 644
convection, 339–40
convention, 75
conventional current, 430
converge, 534
converging lens, 535–38
conversion factor, 29
convex, 514
Copernicus, Nicolaus, *152*
Coriolis effect, 267
correspondence principle, 616–17
cosine θ, 80–81
coulomb (C), 406, 431
Coulomb, Charles, *17*
Coulomb's law, 405–6
critical (nuclear reaction), 643
critical angle (θ_c), 531
critically damped oscillation, 269

critical mass, 643
crossed filter, 564
crystal laser, 554
crystalline solid, 290
curie (Ci), 635
Curie, Marie Sklodowska, *630*
Curie, Pierre, *458*
Curie temperature (T_C), 458
current (I), 429
cutoff frequency, 613
cycle, 259

D

Dalton, John, *287*
damped harmonic oscillator, 269
damping, 258
Darwin, Charles, *9*
daughter nuclide, 638
Davisson, David J., *618*
Davy, Humphrey, *329*
de Broglie, Louis, *617*
de Buffon, Compte, *9*
De Magnete (On Magnets), 399
de Maricourt, Petrus P., *455*
De motu (On Motion), 7
decay constant (λ), 637
decay mode, 637
decay rate, 635
decibel (dB), 277
deconditioning, 188
deduction, 7
deductive reasoning, 7
degassify, 641
degree
    Celsius (°C), 311
    Fahrenheit (°F), 311
density (ρ), 370
derived unit, 28
destructive interference, 548
*Dialogue Concerning the Two Chief
    World Systems—Ptolemaic and
    Copernican*, 8
diamagnetic, 457
diathermic, 348
dichroic, 562
dielectric, 419
dielectric constant (κ), 420–21
difference vector, 59, 78–79
diffraction,
    edge, 558
    grating, 558
    obstacle, 558
    opening, 558
diffuse reflection, 513
digital thermometer, 310
dilation, 595
dimension, 24
dimensional analysis, 29–30
dipole, 456
Dirac, Paul, *618*

direct current (DC), 430
dispersion, 533
displacement (d), 50
disturbance (wave), 272
diverge, 534
diverging lens, 538–39
domain, 458
dominion mandate, 6
Doppler, Christian Johann, *278*
Doppler effect, 278
dose (*D*), 635
dose-equivalent (DE), 636
dosimeter, 634–35
double refraction polarization, 562,
    563–64
drift velocity, 430
driven oscillation, 270–71
dry cell, 433
Du Fay, Charles-François, *399*
dynamic equilibrium, 297
dynamics, 50

E

ear squeeze, 373
earthquake, 276
earthquake wave (P, S, & L), 276
eddy currents, 483
edge diffraction, 558
efficiency (η), 228
effort arm, 225
effort force, 225
Einstein, Albert, *18*, **592**, 593
elastic collision, 240
elastic limit, 293
elastic (Young's) modulus (*E*), 291
elastic potential energy, 200, 206
electrical component, 421
electrical ground, 405
electrical potential, 416
electric field, 413
electric field strength (E), 414
electrick, 398
electrification, 398
electrode, 431
electrodynamics, 429
electrolytic solution, 430
electromagnet, 478
electromagnetic (EM)
    force, 121
    induction, 480
    interference (EMI), 532
    radiation, 493
    wave, 493
electromagnetism, 17, 477
electromotive force (emf), 431
electron, 400
electron cloud, 618
electron shell, 615
electron volt (eV), 615
electroscope, 403

electrostatic force, 405
electrostatic unit (esu), 406
element, 287
elementary particle, 647
Ellef Ringnes Island, 459
elliptically polarized, 562
endoergic, 643
energy, 191
energy band, 625
English Engineering System (EES),
    131
enrichment, 644
entropy (*S*), 361–62
equation of flow continuity, 384
equations of motion
    first, 61–62
    second, 63–64
    third, 64–65
equilibrant force, 119
equilibrium position, 195
equipotential surface, 418
error, 31
escape speed, 219
evaporation, 296
Everest, Mount, 374, 375
exchange particle, 650
excited state, 615
exoergic, 643
experiment, 8
Explorer I, 460
extraordinary (*E-*) ray, 563

F

Fahrenheit, Daniel Gabriel, *311*
farad (F), 421
Faraday, Michael, *17*, 294, **410**, 413,
    480
Faraday's law, 481–82
ferromagnetic, 457
fiber optics, 532
fiducial point, 312
field model, 413
field of view, 579
field theory, 122
first-class lever, 225
first ionization energy, 616
first law of thermodynamics, 348–49
fission, 633
fission fragment, 633
Fizeau, Armand, *509*
flavor (of quarks), 649
flow tube, 382
fluid, 295
    gas: 288–89, 297–98
        adhesion, 289
        cohesion, 289
        ideal gas model, 288
        sublimation, 298
        triple point, 299
        vaporization point, 297

liquid:
    liquid capillarity, 296
    surface tension, 296
    transition to/from, 294–95
    triple point, 299
    vapor pressure, 297
fluid mechanics, 370
fluoresce, 629
focal length (*f*), 515
focal point, 535
focus, 535
forbidden gap, 625
force, 117
    balanced, 117
    central, 217
    centripetal (F_c), 146
    conservative, 201
    contact, 122
    effort, 225
    electromotive (emf), 431
    electrostatic (F_e), 405
    equilibrant, 119
    friction (f), 174
    fundamental, 121
    gravitational (F_g), 121, 137, 154,
        650
    Hooke's law, 195
    lines of, 413
    magnetic (F_{mag}), 464–65
    nonconservative, 201
    noncontact, 122
    normal (N), 169
    nuclear, 121, 650
    pound- (lbf), 131
    restoring (F_r), 258, 262–64
    strong nuclear, 121, 650
    table, 123
    unbalanced, 118, 146
    weak nuclear, 121, 650
formula unit, 287
Foucault, Jean Bernard Léon, 267,
    509
Foucault pendulum, 266
frame of reference, 94
Franklin, Benjamin, *399*
free-body diagram, 163
free electron, 339
free fall, 66
frequency (*f*), 259
Fresnel, Augustin Jean, *545*
Fresnel lens, 545
friction (f), 174
fringe, 549
fuel cell, 452–53
    alkaline, 452
    PEM, 452
fundamental (acoustic), 277
fundamental charge (*e*), 400
fundamental force, 121
fuse, 445
fusion, 601, 643, 645–46

G

Galilean moons, 137
Galilean relativity, 588
Galilean transformation, 590
Galilee, 459
Galileo Galilei, 7–8, **67**, 114, 262, 309, 509, 588
Gall, Franz Joseph, *1*
Galvani, Luigi, *431*
galvanic cell, 431
galvanometer, 442
gamma radiation (γ), 631–32
gas laser, 554
gauge pressure (P_g), 371
gauss, 456
Gauss, Carl Friedrich, 462
Gay-Lussac, Joseph Louis, *314*
Gay-Lussac's law, 317–18
Geiger, Hans, *634*
Geiger counter, 634
Geiger-Mueller tube, 634
Gell-Mann, Murray, *649*
general theory of relativity, 602
Gentry, Robert V., 639
geocentric theory, 151
geochronology, 641–42
Gerlach, Walther, *622*
Germer, Lester Halbert, *618*
Gilbert, William, **399**, 459
Goldstein, Eugen, 400
Gould, Stephen Jay, *10*
gram (g), 26, 28
Grand Coulee Dam, 213
grating
    diffraction, 558
    ruling, 558–59
    spectroscope, 560
gravitational
    blue shift, 604
    force (F_g), 121, 137, 154, 650
    lensing, 603
    potential, 204–5
    red shift, 604
graviton, 650
gray (Gy), 635
Gregory, James, *580*
ground fault interrupter (GFI), 447
ground state, 615
group (chemical), 623

H

hadron, 648
half-life ($t_{1/2}$), 637
halo, 639
harmonics, 277
Harrison, John, 27
heat (Q), 330

heat
    capacity (C), 330
    death, 363
    engine, 349–50
    pump, 359
    reservoir, 355
    specific (c_{sp}), 331
Heisenberg, Werner Karl, *618–19*
Heisenberg uncertainty principle, 618–19
heliocentric theory, 151
henry (H), 486
Henry, Joseph, *480*
Hero of Alexandria, *355*
hertz (Hz), 259
Hertz, Heinrich Rudolf, *259*
Hooke, Robert, *195*
Hooke's law, 195
Hooke's law force, 195
Hoover Dam, 368
horizontal projection, 99
Horologium Oscillatorium (Oscillating Timepiece), 511
hue, 574
humidity
    absolute, 298
    relative, 298
Hutton, James, *9*
Huygens, Christiaan, *274*, 510, **511**
hydraulic device, 375
hydrodynamics, 370
hydrometer, 380–81
hydrostatics, 370
hypothesis, 12–13, 14

I

Iceland spar (calcite), 562
ideal
    fluid, 382
    gas, 314
    gas law, 319–20
    gas process, 354
    pulley, 166
    spring, 195
    string, 164
illuminance (E), 570–71
image distance (d_i), 516, 535
impedance (Z), 492
impulse (I), 235
inclined plane, 223, 224
index fossil dating, 641
index of refraction (n), 528
inductance (L), 486
induction
    electrical, 404
    logic, 8
inductive reactance (X_L), 491
inductive reasoning, 8
inductor, 486
inelastic collision, 240, 243–244

inertia (principle of), 115
inertial fusion, 645
inertial reference frame, 590
inferior mirage, 534
insulators, 401
intensity (I), 277
intercardinal direction, 75
interface, 528
interference pattern, 548
interference microscope, 578
interferometer, 553–54
internal energy (U), 348
ion, 287
ionizing radiation, 634
irrational number, 32
irreducible complexity, 508
isobaric, 354
isochoric, 354
isolated system, 353
isotherm, 358
isothermal, 353
isotope, 614
isotopic symbol, 614

J

Javan, Ali, *555*
Jeans, Sir James H., *610*
Joule, James Prescott, *13*, **325**, 329
joule (J), 192
Joule's law, 435

K

Keck telescope, 568, 580
Kelvin, Lord (William Thomson), *312*
kelvin (K), 312
Kepler, Johannes, 116, *152*
Keplerian constant (K), 153
Kepler's laws, 152–53
kilogram (kg), 26
kilowatt-hour (kWh), 435
kinematics, 50
kinetic energy (K), 198
kinetic friction (f_k), 175–76
kinetic theory of matter, 290
Kirchhoff, Gustav, *441*
Kirchhoff's rules, 441–42
Krakatoa (volcano), 273

L

laminar, 382
Land, Edwin H., *564*
laser, 554–55, 624
    crystal, 554
    diode, 554
    gas, 554
    semiconductor, 554
    YAG, 557
latent heat of fusion (L_f), 336

latent heat of vaporization (L_v), 336
Lavoisier, Antoine, *329*
law, 14
    Biot-Savart, 478
    Boyle's, 316–17
    Brewster's, 563
    Charles's, 314–15
    combined gas, 318
    conservation, 215
    Coulomb's, 406
    Faraday's, 481–81
    first; of thermodynamics, 348–49
    Gay-Lussac's, 317–18
    gravitational, 121
    Hooke's, 195
    ideal gas, 319–20
    Joule's, 435
    Kepler's, 152–53
    Lenz's, 482–83
    Malus's, 562
    Newton's
        first; of motion, 124
        of gravitation, 121, 155
        of motion, 124
        second; of motion, 125
        third; of motion, 128
    of charges, 399
    of magnetic poles, 455
    of moments, 149, 225
    of reflection, 513
    Ohm's, 434
    second; of thermodynamics, 355
    Snell's, 529
    Stefan's, 341
    third; of thermodynamics, 359–60
    Wien's, 507, 610
    zeroth; of thermodynamics, 348
length contraction, 597
Lenz, Heinrich F. E., *482*
Lenz's law, 482–83
lepton, 648
lever, 225–26
Leyden jar, 419
light pipe, 508
light emitting diode (LED), 532
lightness (of color), 574
lines of force, 413
line spectrum, 615
Lipperhey, Hans, *579*
load, 225
load cell, 123
lodestone, 455
longitudinal wave, 273
looming, 534
Lord Rayleigh (John William Strutt), 510, *561*
Lorentz, Hendrik, *593*
Lorentz transformation, 593–94
lumen (lm), 570
luminance, 570
luminiferous ether, 590

luminous flux (Φ), 570
luminous intensity (I_L), 569
lux (lx), 571
Lyell, Charles, *9*
Lyman, Theodore, *616*
Lyman series, 616

M

machine, 222
magnetic
  declination, 459
  dipole, 478
  dipole moment (μ), 458
  field vector (**B**), 456
  flux (φ), 480
  flux density |**B**|, 456
  force ($\mathbf{F}_{mag}$), 464–65
  induction, 456
  permeability (μ), 457
  quantum number (*m*), 620
magnetosphere, 460
magnification, 520
Maiman, Theodore, *554, 556*
Malus, Étienne Louis, *562*
Malus's law, 562
manometer, 377
map coordinates, 83
maser, 624–25
mass, 26
  atomic, 288
  defect, 601, 642
  number (*A*), 614
  propellant, 237
  spectrograph, 466
matter, 24, 26
  kinetic theory of, 290
  nature of, 286–87
  particles of, 287
maximum, 549
Maxwell, James Clerk, *17,* 121, **499,** 512
mechanical equilibrium, 125
mechanical equivalent of heat, 330
mechanics, 17, 49
medium, 272
mega-electron volt (MeV), 639
memes, 2
meson, 648
metalloid, 401
metastable state, 624
meter (m), 24
Michelson, Albert A., *509, 591*
microscope, 577–78
millibar (mb), 371
Millikan, Robert, *400*
minimum, 549
mirror equation, 518
moderator, 644
modern physics, 15, 18
mole (mol), 288

molecule, 287
moment arm (*L*), 148
moment of inertia (*I*), 267
momentum (p), 233
monopole, 456
Morley, Edward W., *591*
motion
  celestial, 114
  natural, 113
  terrestrial, 114
muon, 600
mutual inductance (*M*), 487

N

National Astronomical Observatory of Japan (NAOJ), 580
naturalism, 9–10
natural oscillation frequency (f_0), 270
neutrino, 647
neutron (*n*), 614
Newcomen, Thomas, *356*
Newton, Isaac, *9,* **110,** 116, 154, 580
Newtonian reflector telescope, 580
newton (N), 126–27
Newton's law of universal gravitation, 155
Newton's laws of motion, 124–30
Newton's rings, 553
Nicol, William, *563*
Nicol prism, 563
nonconservative force, 201
noncontact force, 122
nonuniform circular motion, 142
normal, 170, 513, 528
normal force (N), 169
north magnetic pole, 458
Novum Organum (The New Organon), 7
nuclear force, 650
nuclear reactor, 644
nucleon, 614
nucleus, 614
nuclide, 614

O

object distance (d_O), 516, 535
objective, 577
objectivity, 3–4
observation, 11–12
obstacle diffraction, 558
ocular, 577
Oersted, Hans Christian, *457*
ohm (Ω), 433
Ohm, Georg Simon, *434*
ohmmeter, 442
Ohm's law, 434
opaque, 573
opening diffraction, 558
open system, 353

operator (mathematical), 618
optical axis, 515
orbital (electron), 618
orbital motion, 138
ordered pair, 50
order number, 551
order of magnitude, 42
ordinary (*O*-) ray, 563
origin (of the universe), 4–5
overcurrent, 445
overdamped oscillation, 269

P

pair annihilation, 647
Pantheon, 273
paradigm, 4
parallel, 436–37
paramagnetic, 457
parent nuclide, 638
partially elastic collision, 240
Pascal, Blaise, *375*
pascal (Pa), 291, 371
Pascal's principle, 375
Paschen series, 616
path difference, 549
Pauli, Wolfgang Ernst, *623*
Pauli exclusion principle, 622
pendulum, 262
pendulum arm (*L*), 262
penstock, 212
percent error, 32
performance coefficient (*K*), 359
period (*T*), 153
period (chemical), 623
periodic motion, 258
periodic table, 623
permeability of free space (μ_0), 457, 478
permittivity of free space (ε_0), 421
permittivity of material (ε), 421
Pfund series, 616
phase-contrast microscope, 578
phase diagram, 295
phasor, 492
phenomenon, 2
Philosophiae Naturalis Principia Mathematica (Principia), 116
phosphor, 508
photoelastic plastic, 565
photoelectric cutoff frequency, 613
photoelectric effect, 612
photoelectron, 612
photometer, 572
photon, 612
phrenology, 1–2
physical pendulum, 267
piezoelectric crystal, 431
pigment, 575
pion, 650

pitch
  screw thread, 225
  sound, 277
Planck, Max, **611**
Planck's blackbody spectrum, 611
Planck's constant (*h*), 574, 611
plane polarized, 562
plane wave, 510
plate (capacitor), 419
plum-pudding model, 614
poise, 388
polar coordinates, 138
polarity marking, 430
polarization
  circular, 562
  double refraction, 563–64
  elliptical, 562
  plane, 562
  reflected, 562–63
  scattering, 564
  selective absorption, 564
Polaroid filter, 564
pole (magnetic), 455
position-time graph, 52–54
positron (β^+), 633
potential difference (ΔV), 430
potential energy (*U*), 200
pound-force (lbf), 131
pound-mass (lbm), 131
power (magnification) (×), 521, 536
power (*P*), 196–197
Poynting vector, 494
precession, 267
precision, 32–33, 36
prejudice, 2
pressure gauge, 123
primary cell, 432
primary cosmic rays, 653
principal axis, 515
principal focus (F), 515
principal quantum number (*n*), 619
Principia, 589
probability wave, 618
projectile, 98
prompt-critical reaction, 645
proper length (l_0), 597
proportional limit, 293
prosthetic leg, 268
proton (p), 614
pseudoscience, 2
Ptolemaeus, Claudius (Ptolemy), *151*
pulley, 226
pulsating DC, 486
pulse, 548
P-V diagram, 350–51

Q

quality, 277
quality factor (QF), 635
quanta, 611

quantized, 611
quantum mechanics, 18
quantum number, 619–21
   azimuthal (l), 620
   energy level (n), 619
   magnetic (m), 620
   orbital (m), 620
   principal (n), 619
   spin (m_s), 620
   sublevel (l), 620
quark, 649
quark confinement, 649
quartz clock, 27
quasi-static process, 349

R

radian (rad), 39
radiation, 340, 341, 630
radioactive, 630
radioactive dating, 638
radioactivity, 630
radiogenic, 630
radiometric dating, 641
radionuclides, 638
ramjet, 56
ramp (inclined plane), 223, 224
range (R), 99
Rankine, William, 312
rarefaction zone, 273
ray, 513
Rayleigh, Lord (John William Strutt), *561, 610*
Rayleigh's angle (θ_R), 561
real image, 516
recoil, 236
red shift
   Einsteinian, 604
   gravitational, 604
reference angle (α), 82
reference direction, 74–75
reflected polarization, 562–63
reflecting telescope, 580
reflection, 513
Reflections on the Motive Power of Fire, 344
reformer (fuel cell), 453
refracting telescope, 579
refraction, 527
refrigeration engine, 359
regelation, 295
relative permeability (κ_m), 457
relative uncertainty, 33
relativistic (speed), 594
relativistic factor (γ), 597
relativity, 588
resinous, 398
resistance (R), 433
resistance temperature detector (RTD), 443
resistivity (ρ), 433

resistor, **438–40**
   parallel connected, 438
   series connected, 438
resolve, 560
resolving power, 560
resonance
   (f_{res}), 492
   oscillations, 270
   RLC circuits, 492
rest mass (m_0), 597
rest-mass energy (E_0), 600
restoring force (F_r), 258, 262–64
rest position, 258
resultant, 77
reversible, 357
rheostat, 443
right-hand rule
   circular motion, 141
   conductor field (solenoids), 477
   force on conductor, 470
   magnetic force on charge, 465
   torque, 470
right triangle, 80
Roemer, Ole, *509*
Roentgen, Wilhelm, *629*
roentgen (R), 635
rolling friction (f_r), 179–80
root mean square (rms), 485
rotational equilibrium, 149
rotor, 472
Rumford, Count (Benjamin Thomson), *11–12*, **324**
Ruska, Ernst, 620
Rutherford, Ernest, *614*

S

St. Elmo's fire, 399
Saint Helens, Mount, 191
saturation, 574
Savart, Felix, *478*
Savery, Thomas, *356*
scalar, 74
scalar components, 82
scanning electron microscope (SEM), 620–21
scanning-tunneling electron microscope (STM), 621
scattering polarization, 564
Schmidt, Bernhard, *581*
Schmidt-Cassegrain telescope, 581
Schrödinger, Erwin, *618*
Schrödinger's wave equation, 618
science, 3
scientific methodology, 11
scientific problem, 12
Scott, David, Col., USAF, 114
scramjet, 56
screw, 224
second (s), 25
secondary cosmic rays, 653

second-class lever, 225
second law of thermodynamics, 355
selective absorption polarization, 564
self-induction, 486
semichords, 529
semiconductor, 401
semi-major axis (R), 153
semimetal, 401
series (chemical), 623
series (connection), 436
shear modulus (G), 293
Short, James, *581*
SI, 23–24
sievert (Sv), 636
significant digit (SD), 35–36
similar triangle, 80
simple harmonic motion (SHM), 258
Simplicio, 67
simultaneity, 594–95
sine θ, 80
Snell, Willebrord, *529*
Snell's law, 529
solar wind, 460
solenoid, 478
solid
   amorphous, 291
   crystalline, 290
   transition to/from, 294–95
   triple point, 299
south magnetic pole, 458
space, 24
spark chamber, 635
special theory of relativity, 593
specific density, 331
specific gravity (s.g.), 370
specific heat (c_{sp}), 331
spectral series
   Balmer, 616
   Brackett, 616
   Lyman, 616
   Paschen, 616
   Pfund, 616
spectrum, 504
specular reflection, 513
speed of light, 493, 510
spherical aberration
   lens, 541
   mirror, 514
spherical wave, 510
spin, 138
spin quantum number (m_s), 620
split-ring commutator, 472
spring constant (k), 195
spring scale, 123
standard kilogram, 26–27
standard meter, 25
standard temp. & press. (STP), 320
state variable, 352
static electricity, 398
static friction (f_s), 176–77
stator, 472

Stefan, Josef, *341*
Stefan-Boltzmann constant (σ), 341
Stefan's law, 341
step-down transformer, 489
step-up transformer, 489
steradian (sr), 570
stereo microscope, 578
Stern, Otto, *622*
storage cell, 432
strain (ε), 291
stratigraphic dating, 641
streamline, 382
stress (σ), 291
strong nuclear force, 121, 650
Strutt, John William (Lord Rayleigh), *561, 610*
subcritical, 643
sublevels, 622–23
sublimation, 298
subtractive, 575–576
subtractive primary color, 576
superconductor, 433
supercritical, 643
superposition principle, 548
supersonic combustion ramjet, 56
surface tension, 296
surroundings, 349
sympathy (magnetic), 399
system, 124
Système International d'Unités (SI), 24

T

Tacoma Narrows Bridge, 270
Talos (interceptor missile), 56
talus, 101
tangent θ, 81
tangential acceleration (a_t), 143
tangential velocity (v_t), 139
teleological, 8
telescope, 579–81
tension (T), 163
terminal velocity, 216
Tesla, Nicola, *456*
tesla (T), 456
test charge, 414
theoretical mechanical advantage (TMA), 222
theory, 14
thermal efficiency (ε), 358
thermal equilibrium, 348
thermal neutrons, 644
thermocouple, 310
thermodynamic process, 353
thermodynamics, 18
thermodynamic system, 353
thermometer, 309
thermometric property, 309
thermonuclear device, 646
thermoscope, 309

thin lens, 535
thin-lens equation, 535
third-class lever, 226
third law of thermodynamics, 359–60
Thomson, Benjamin (Count Rumford), 11–12, **324**
Thomson, Sir Joseph John, *400*, 614
Thomson, William (Lord Kelvin), *312*
Three Gorges Dam, 286
thrust, 237
thrust reverser, 130
time, 25
time dilation, 595–97
tokamak, 645
torque (τ), 148
torr, 371
Torricelli, Evangelista, *371*
torsion balance, 156
total acceleration (a_{total}), 263
total internal reflection, 531
total magnetic flux (Φ), 480
total mechanical energy (E), 208
tracer, 646
traction (f_t), 174
trajectory, 98
transformer, 488–89
translucent, 573
transmission (mechanical), 498
transmission axis, 562
transmittance, 573
transmute, 630
transparent, 573
transverse wave
    light, 562
    vibration, 273
triple point, 299
tuning circuit, 493
turbulent, 382
twaddle, 381

U

ultraviolet catastrophe, 610
unbalanced forces, 118, 146
unbalanced torque, 149
uncertainty, 33
uncontrolled fission, 645
uniform circular motion, 139
uniformitarianism, 9, 362
uniformly accelerated motion, 60
unit analysis (bridge), 29–30
unit cell, 290
universal gas constant (R), 320
universal gravitation constant (G), 155
unpolarized, 562
Ussher, James, *8*

V

vacuum dielectric, 420
valence electrons, 623

value, 574
Van Allen, James Alfred, *460*
Van Allen belt, 460
van der Waals, Johannes D., *322*
van der Waals gas, 321–22
van Leeuwenhoek, Antoni, *577*
vaporization point, 297
vapor pressure, 297
vector, 73–88
vector angle (θ), 74–75
vector diagram, 74, 75
Vehicle Assembly Building (VAB), 286
velocity (v), 58–59
velocity selector, 466
velocity-time graph, 59–60
ventricular fibrillation, 446
vernier scale, 33, 34
vibration graph, 273
virtual image, 513
virtual particle, 122
viscosity (η), 388
vitreous, 398
volt (V), 417
Volta, Count Alessandro, *431*
voltage drop, 436
voltaic cell, 431
voltmeter, 442
Voyager, 220

W

Watt, James, 197, *356*
watt (W), 197, 435
wave, 272
    front, 510
    gamma rays, 505–6, 631–32
    infrared, 504
    plane, 510
    radio, 504
    sound, 275, 277
    ultraviolet, 505
    visible light, 504–5
    x-ray, 505
waveform graph, 273
wavelength (λ), 274
wave speed (v), 274
weak nuclear force, 121
weber (Wb), 480
Weber, Wilhelm, *480*
wedge, 224
weight (F_w), 119
Wien, Wilhelm, *400, 610*
Wien's law, 507
work (W), 192
work-energy theorem, 199
work function, 612
W-particle, 650
Wright brothers, 130

X, Y, Z

YAG laser, 557
Yerkes Observatory, 580
Young, Thomas, *292*, **548, 550**
Young's modulus, 291
Zernike, Frits, 578
zeroth law of thermodynamics, 348
zero vector (**0**), 94
Zweig, George, *649*

Photo Credits

The following agencies and individuals have furnished materials to meet the photographic needs of this textbook. We wish to express our gratitude to them for their important contribution.

Adam Hart-Davis/DHD Multimedia Gallery
Altizer, Suzanne
American Medical Association
Ames Research Center
Anglo-Australian Observatory/Royal Observatory Edinburgh
AP/WIDE WORLD PHOTOS
Arbor Scientific
Arrow-Technologies, Inc.
Artemis Images
Aquionics, Inc.
Behringer, R.P./Duke University
Bjerk, John
Britten, William/lighthousegetaway.com
Brooks/Cole, a division of Thomson Learning
Bureau International des Poids et Mesures
Caltech
Carrier Corporation
Carvalho, Tina/University of Hawaii
Celestron
CERGA in the French Riviera Observatory
CERN, Geneva, Switzerland
Cherry, Gregory W./Cherry Optical Holography
Corbis
Coss, Dr. Tom
Curtis, Jan
Dennis Kunkel Microscopy, Inc.
Department of Terrestrial Magnetism, Carnegie Institution, Washington
Dieman, Paul
Digital Vision
Dilyard, Matt/Courtesy of The College of Wooster
Dutch, Dr. Steve/University of Wisconsin
Eastman Chemical Division
Edgar Fahs Smith Collection, University of Pennsylvania Library
Edmund Industrial Optics
Edwards AFB History Office
Elekta, Inc.
Energy Northwest
Flammarion, Camille
Espenak, Fred www.MrEclipse.com
European Southern Observatory
Firestone, Richard/Lawrence Berkeley National Laboratory
Fitzgerald, Jim/Courtesy of Mansion House Maple Syrup
French, Peter
Glatzmaier, Gary A., University of California
General Dynamics Electric Boat
GE Medical Systems
General Motors
Gentry, Dr. Robert V.
Giraudon/Art Resource, NY
GOMACO Corporation
Goshen College
Grand American Road Racing Association
Guidice, Rick
Hansen, Brenda
Harris, Marty/McDonald Observatory
Heinselmann, Keith/Lawrence Berkeley National Laboratory
Hemera Technologies, Inc.

Hinton, Peter/Ely Diocesan Association of Church Bell Ringers
Howell, Daniel/Duke University
IBM Research, Almaden Research Center
Istituto e Museo di Storia della Scienza
ITER
Jagiellonian Library
Jenkins, John www.sparkmuseum.com
Koh, Tracey ButterflyUtopia.com
Krutein, Wernher/photovault.com
Landis, Joyce
Laser Technology Inc.
Lawrence Berkeley National Laboratory
Library of Congress
Lloyd, Jim, University of Florida, Institute of Food and Agriculture Sciences
Moss, Greg
Musee du Louvre, Paris
The Museum of Questionable Medical Devices
NASA
NASA, Goddard Space Flight Center
NASA/JPL/Caltech
National Anthropological Archives, Smithsonian Institution
National Archives
National Institute of Standards and Technology (NIST)
National Library of Medicine
National Maritime Museum, London
National Museum of American History, Smithsonian Institution
National Museum of Photography Film & Television/Science & Society Picture Library
National Oceanic and Atmospheric Administration (NOAA)
National Parks Department in Washington, DC
National Rifle Association (NRA)
NCAR/NSF
Neiconi, Emanuel
Netzsch Instruments, Inc.
Oak Ridge National Laboratory for the U.S. Department of Energy
Ocean Remote Sensing Group, John Hopkins University Applied Physics Laboratory
OHAUS
Orbital Sciences Corporation
Otto Bock Health Care
Parker, David, 1997/Science Photo Library
Perry, Susan
Pflug, Kathy
PhotoDisc/Getty Images
Rambach, Glenn/Third Orbit Power Systems, Inc.
Robert Morris Photography
Robl, Ernest H.
Rodon Products, Inc., Huntington Beach, CA
Samec, Dr. Ron
Science Kit and Boreal Labs
Six Flags over Georgia
STEM Labs, Inc.
Strainoptic Technologies, Inc.
UC Berkeley Analog Devices, Inc./Boris Murmann
University of Oklahoma Libraries
University of Utah Andrology Laboratory
Unusual Films

Chapter 15

© 2003 Hemera Technologies, Inc. All rights reserved 326; Unusual Films 328(all), 333(top); Courtesy of Netzsch Instruments, Inc. 333(bottom); Suzanne R. Altizer 335

Chapter 16

Image By: Wernher Krutein/photovault.com 346; NASA 355, 364; Carrier Corporation 359; Visuals Unlimited, Inc. 362; University of Utah Andrology Laboratory 365

Chapter 17

U.S. Bureau of Reclamation 368; © 2003 Hemera Technologies, Inc. All rights reserved 369; Paul Dieman 381(left); Jim Fitzgerald/Courtesy of Mansion House Maple Syrup 381(right); Unusual Films 383(both), 387(all), 388(bottom); U.S. Air Force photo by Staff Sgt. Andy Dunaway 388(top); Courtesy of General Motors and Wieck Media Services, Inc. 389(left); Michael Tweed/Courtesy of General Motors and Wieck Media Services, Inc. 389(right)

Unit IV

Jan Curtis 394–95

Chapter 18

PhotoDisc/Getty Images 396; © 2003 Hemera Technologies, Inc. All rights reserved 397; Ward's Natural Science Establishment, Inc. 398(left); Unusual Films 398(center, right), 402(both), 403

Chapter 19

PhotoDisc/Getty Images 412; Unusual Films 425

Chapter 20

UC Berkeley Analog Devices, Inc./Boris Murmann 428; Matt Dilyard/Courtesy of The College of Wooster 431; Edgar Fahs Smith Collection, University of Pennsylvania Library 434; Unusual Films 443, 445, 447(right); Joyce Landis 446(both), 447(left); NASA 452; Courtesy of Glenn Rambach/Third Orbit Power Systems, Inc. 453(top); Courtesy of General Motors 453(bottom)

Chapter 21

NASA 454; Unusual Films 455, 471; Joyce Landis 458; Dept. of Terrestrial Magnetism, Carnegie Institution, Washington 459; U.S. Geological Survey 460(map); NASA 460(bottom); Gary A. Glatzmaier, Professor of Earth Sciences, University of California, Santa Cruz 462; John Jenkins, www.sparkmuseum.com 468

Chapter 22

U.S. Department of Energy 476; Courtesy of VWR International 479; Unusual Films 483, 489; Courtesy Rodon Products, Inc., Huntington Beach, CA 487; Eastman Chemical Division 498

Unit V

© Tracey Koh/ButterflyUtopia.com 500–1(butterfly); PhotoDisc/Getty Images 500–1(background)

Chapter 23

Marty Harris/McDonald Observatory 502; Image Courtesy of Arbor Scientific 503; David Parker, 1997/Science Photo Library 504(top); Courtesy of NASA/JPL/Caltech 504(bottom); Compliments of Aquionics, Inc. 505(top); © 2003 Hemera Technologies, Inc. All rights reserved 505(center); NASA 505(bottom), 506, 514; PhotoDisc/Getty Images 507; Kathy Pflug 508(top); NASA, Goddard Space Flight Center 508(center); Susan Perry 508(bottom left); Photo by Jim Lloyd, University of Florida, Institute of Food and Agriculture Sciences 508(bottom right)

Chapter 24

Unusual Films 526, 533(bottom), 542(top left, bottom); NASA 532; Adam Hart-Davis/DHD Multimedia Gallery 533(top); Telescope photo provided by Celestron, a manufacturer of telescopes and related accessories, binoculars, spotting scopes and microscopes. Product information available at www.celestron.com 541; Photo Courtesy of Edmund Industrial Optics 542(top right); Photo by William Britten lighthousegetaway.com 545

Chapter 25

Copyright Dennis Kunkel Microscopy, Inc. 546; Unusual Films 553, 559, 563(bottom left), 564(top); NASA 556(top); PhotoDisc/Getty Images 556(bottom); Gregory W. Cherry/Cherry Optical Holography 557; Joyce Landis 563(top); Dr. Steve Dutch/University of Wisconsin—Green Bay 563(bottom right); Strainoptic Technologies, Inc. 564(bottom); Daniel Howell and R.P. Behringer, Duke University 565

Chapter 26

Peter French 568; Courtesy of VWR International 572; Digital Vision 573; PhotoDisc/Getty Images 578(both); Istituto e Museo di Storia della Scienza 579; W.M. Keck Observatory 581

Unit VI

U.S. Department of Energy 584–85

Chapter 27

European Southern Observatory 586; NASA 597(both), 603(both), 604(bottom); Courtesy DELPHI experiment, CERN, Geneva, Switzerland 600(left); National Institute of Standards and Technology 600(right); © 1973 Fred Espenak, www.MrEclipse.com 604(top)

Chapter 28

Courtesy: IBM Research, Almaden Research Center. Unauthorized use not permitted. 608, 621(top right); NASA 620; Lawrence Berkeley National Laboratory 621(top left); Tina Carvalho/University of Hawaii 621(bottom)

Chapter 29

Courtesy of Energy Northwest 628; Edgar Fahs Smith Collection, University of Pennsylvania Library 630; Richard Firestone/Lawrence Berkeley National Laboratory 631(top); Courtesy of VWR International 634(top left); Keith Heinselmann/ Lawrence Berkeley National Laboratory 634(bottom left); Arrow-Technologies, Inc. 634(center top); Lawrence Berkeley National Laboratory 634(center bottom); Courtesy CERN, Geneva, Switzerland 634(top right); Courtesy DELPHI experiment, CERN, Geneva, Switzerland 634(bottom right); AP/WIDE WORLD PHOTOS 638; Used by permission of Dr. Robert V. Gentry 639; NOAA 641; Courtesy of Oak Ridge National Laboratory, managed by UT-Battelle, LLC, for the U.S. Dept. of Energy 644; U.S. Department of Energy 645(top), 646(top); ITER 645(bottom); Elekta, Inc. 646(bottom)